NANOMATERIALS FOR SUPERCAPACITORS

NANOMATERIALS FOR SUPERCAPACITORS

Editor

Ling Bing Kong

School of Materials Science and Engineering
Nanyang Technological University
Singapore

CRC Press is an imprint of the
Taylor & Francis Group, an **informa** business

A SCIENCE PUBLISHERS BOOK

CRC Press
Taylor & Francis Group
6000 Broken Sound Parkway NW, Suite 300
Boca Raton, FL 33487-2742

© 2018 by Taylor & Francis Group, LLC
CRC Press is an imprint of Taylor & Francis Group, an Informa business

No claim to original U.S. Government works

Printed on acid-free paper
Version Date: 20170912

International Standard Book Number-13: 978-1-4987-5842-0 (Hardback)

This book contains information obtained from authentic and highly regarded sources. Reasonable efforts have been made to publish reliable data and information, but the author and publisher cannot assume responsibility for the validity of all materials or the consequences of their use. The authors and publishers have attempted to trace the copyright holders of all material reproduced in this publication and apologize to copyright holders if permission to publish in this form has not been obtained. If any copyright material has not been acknowledged please write and let us know so we may rectify in any future reprint.

Except as permitted under U.S. Copyright Law, no part of this book may be reprinted, reproduced, transmitted, or utilized in any form by any electronic, mechanical, or other means, now known or hereafter invented, including photocopying, microfilming, and recording, or in any information storage or retrieval system, without written permission from the publishers.

For permission to photocopy or use material electronically from this work, please access www.copyright.com (http://www.copyright.com/) or contact the Copyright Clearance Center, Inc. (CCC), 222 Rosewood Drive, Danvers, MA 01923, 978-750-8400. CCC is a not-for-profit organization that provides licenses and registration for a variety of users. For organizations that have been granted a photocopy license by the CCC, a separate system of payment has been arranged.

Trademark Notice: Product or corporate names may be trademarks or registered trademarks, and are used only for identification and explanation without intent to infringe.

Library of Congress Cataloging-in-Publication Data

Names: Kong, Ling Bing, editor.
Title: Nanomaterials for supercapacitors / editor Ling Bing Kong, School of
    Materials Science and Engineering, Nanyang Technological University,
    Singapore.
Description: Boca Raton, FL : CRC Press, Taylor & Francis Group, [2017] | "A
    Science Publishers book." | Includes bibliographical references and index.
Identifiers: LCCN 2017033996 | ISBN 9781498758420 (hardback : alk. paper)
Subjects: LCSH: Supercapacitors--Materials. | Nanostructured materials.
Classification: LCC TK7872.C65 N35 2017 | DDC 621.31/50284--dc23
LC record available at https://lccn.loc.gov/2017033996

Visit the Taylor & Francis Web site at
http://www.taylorandfrancis.com

and the CRC Press Web site at
http://www.crcpress.com

Preface

Global research on energy and energy related issues become more and more intensive, due to the significantly increased investment. This is because of the limited availability of conventional energy resources and the green house problem. While searching for new energy is an important research topic, energy storage is also critical simultaneously. As one of the main energy storage devices, supercapacitors have been studied extensively and intensively all around the world. It is not doubted that the performances of supercapacitors, such as specific capacities, cycling stabilities, energy densities, powder densities, cost-effectiveness, etc., are all closely associated with the properties and characteristics of materials. Besides intrinsic properties, extrinsic properties of materials can be tailored to meet the requirements through controlling material processing and synthesis. Until now, various materials, including carbon materials, conducting polymers and metal oxides, have been explored as electrodes for the applications of supercapacitors.

While there have been a large number of books on supercapacitors published in recent years, no one has focused on the processing and characterization of electrode materials of supercapacitors. More importantly, with the development of nanotechnology and nanomaterials, significant progress has been achieved in developing high performance and high stability supercapacitors, which could find applications in various fields. This book aims at providing a thorough overview on progress and achievements in the development of supercapacitors in recent years, with focusing on two aspects, i.e., (i) synthesis and performances of nanomaterials as electrodes of supercapacitors and (ii) newly emerged flexible supercapacitors and microsupercapacitors. The first one is covered in Chapters 3–5, while the second one is presented in Chapters 6 and 7, together with a brief introduction (Chapter 1) and fundamental issues of capacitors (Chapter 2). From materials point of view, discussion is conducted from synthesis to characterization and electrochemical performances, with an effort to build up the underlying interrelationships among them.

Potential readers of this book could be very diverse, including senior undergraduate students, postgraduate students, researchers, designers, engineers, professors, and program/project managers, from areas and fields of materials science and engineering, applied physics, chemical engineering, materials manufacturing and design, institutes, research founding agencies, etc. It can also be used as reference book or text book for postgraduate students and training materials for engineers and workers.

Contents

Preface   **v**

1. Introduction   **1**

Ling Bing Kong, Wenxiu Que, Lang Liu, Freddy Yin Chiang Boey, Zhichuan J. Xu, Kun Zhou, Sean Li, Tianshu Zhang and *Chuanhu Wang*

2. Basic Concepts of Supercapacitors   **4**

Ling Bing Kong, Wenxiu Que, Lang Liu, Freddy Yin Chiang Boey, Zhichuan J. Xu, Kun Zhou, Sean Li, Tianshu Zhang and *Chuanhu Wang*

3. Carbon Based Supercapacitors   **28**

Ling Bing Kong, Wenxiu Que, Lang Liu, Freddy Yin Chiang Boey, Zhichuan J. Xu, Kun Zhou, Sean Li, Tianshu Zhang and *Chuanhu Wang*

4. Oxide Based Supercapacitors: I-Manganese Oxides   **162**

Ling Bing Kong, Wenxiu Que, Lang Liu, Freddy Yin Chiang Boey, Zhichuan J. Xu, Kun Zhou, Sean Li, Tianshu Zhang and *Chuanhu Wang*

5. Oxide Based Supercapacitors: II-Other Oxides   **277**

Ling Bing Kong, Wenxiu Que, Lang Liu, Freddy Yin Chiang Boey, Zhichuan J. Xu, Kun Zhou, Sean Li, Tianshu Zhang and *Chuanhu Wang*

6. Flexible Supercapacitors   **422**

Ramaraju Bendi, Vipin Kumar and *Pooi See Lee*

7. Microsupercapacitors   **448**

Ling Bing Kong, Wenxiu Que, Lang Liu, Freddy Yin Chiang Boey, Zhichuan J. Xu, Kun Zhou, Sean Li, Tianshu Zhang and *Chuanhu Wang*

Index   **515**

Introduction

Ling Bing Kong,[1,7,] Wenxiu Que,[2] Lang Liu,[3] Freddy Yin Chiang Boey,[1,a] Zhichuan J. Xu,[1,b] Kun Zhou,[4] Sean Li,[5] Tianshu Zhang[6] and Chuanhu Wang[7]*

The whole world has been focusing on energy related research for a quite while, because it is widely acknowledged that the conventional energy source on the earth will be used up sooner or later. Another issue closely related to energy usage is so-called global warming, which is attributed to the emission of carbon dioxide when using the conventional energies (fossil oils and coals). To address these problems, one approach is to increase energy efficiency, so that the conventional energy could be used for a longer time, while the energy consumption can be largely reduced. However, this is not the fundamental solution to solve the problems. Therefore, searching for new energy becomes more important. New energy is generally known as renewable energy, sustainable energy, clean energy, alternative energy and so on, which is also called. Renewable energy includes solar energy, wind energy, geothermal energy and so on. One of the problems that we encounter with renewable energy is that they are not continuously available. For example, solar energy is only available during the day (sun-shining) time, while electricity is required in the evening and at night. As a result, it is desired to store the renewable energy when it is unavailable but is required.

Devices for electrical energy storage are batteries and capacitors, both of which are electrochemical storage devices. Generally, specific energy and energy densities of supercapacitors are only about 10% of

---

[1] School of Materials Science and Engineering, Nanyang Technological University, 50 Nanyang Avenue, Singapore 639798.
[a] Email: mycboey@ntu.edu.sg
[b] Email: xuzc@ntu.edu.sg
[2] Electronic Materials Research Laboratory and International Center for Dielectric Research, Xi'an Jiaotong University, Xi'an 710049, Shaanxi Province, People's Republic of China.
   Email: wxque@xjtu.edu.cn
[3] School of Chemistry and Chemical Engineering, Xinjiang University, Urumqi 830046, Xinjiang, People's Republic of China.
   Email: llyhs1973@sina.com
[4] School of Mechanical and Aerospace Engineering, Nanyang Technological University, 50 Nanyang Avenue, Singapore 639798.
   Email: kzhou@ntu.edu.sg
[5] School of Materials Science and Engineering, The University of New South Wales, Australia.
   Email: sean.li@unsw.edu.au
[6] Anhui Target Advanced Ceramics Technology Co. Ltd., Hefei, Anhui, People's Republic of China.
   Email: 13335516617@163.com
[7] Department of Material and Chemical Engineering, Bengbu Univresity, Bengbu, Peoples's Republic of China.
   Email: bbxywch@126.com
* Corresponding author: elbkong@ntu.edu.sg

2 *Nanomaterials for Supercapacitors*

those of batteries, whereas they have power densities that are larger than those of batteries by factors of 10–100. As a consequence, supercapacitors usually have much shorter charge/discharge cycles. Moreover, supercapacitors can withstand more charge and discharge cycles as compared with batteries. This book aims to provide an overview on the progress in the development of supercapacitors, with a focus on synthesizing and characterization of nanomaterials that have been used as electrodes of supercapacitors.

Electrochemical Supercapacitor (ES) was initially to mean the electric double-layer present at the interface between a conductor and a solution of electrolyte in contact with the conductor. Established by Hermann von Helmholtz and further developed by Gouy, Chapman, Grahame and Stern, the electric double-layer theory has been used to understand electrochemistry, describing the electrochemical processes that take place at the electrostatic interfaces between an electrode material and an electrolyte. With that, various electrochemical theories and technologies have been developed, such as supercapacitors, batteries and fuel cells, which have become hot research topics especially in recent years.

As the first and second generation of capacitors, electrostatic and electrolytic capacitors, were used to store charges of μF–pF in Direct Current (DC) circuits or filter the frequencies in Alternating Current (AC) circuits. Supercapacitors are third generation of such devices that initially acted as electrolytic capacitors for low voltage applications. In most cases, carbon materials were employed as the electrodes.

With the strong development of mobile electronics and alternative energy vehicles, electrochemical energy storage systems are required to have much higher power densities, which have significantly driven the research on supercapacitors. Enormous progress has been achieved in the area of supercapacitors, with researches and studies focusing on electrode materials, composites, hybridizations and electrolytes, in order to improve their performances and decrease their production costs.

Meanwhile, design and component optimization of the supercapacitors have also been well understood, which resulted in significant improvement in supercapacitor performances, especially energy density. Accordingly, as more advanced supercapacitors emerged, with electrodes made of new electroactive materials, such as metal oxides and conducting polymers, to be incorporated with various carbon materials, in the form of composites. In pseudocapacitors, the electroactive materials delivered electrochemical reactions at the interfaces between the electrode materials and electrolyte. The electrochemical reaction mechanisms include adsorption, intercalation and reduction–oxidations (redox), which is responsible for the high energy density of pseudocapacitors.

This book aims to offer an overview on the development and progress of supercapacitors in the last decades, with focusing on the aspects of materials (Chapters 3–5) and special types of devices (Chapters 6 and 7). Chapter 1 serves as a brief introduction on general information of capacitors and supercapacitors, while Chapter 2 provides a theoretical description with regard to the basic concepts and fundamental understanding of capacitors and supercapacitors.

In Chapter 3, carbon materials, including carbon blacks (CAs), carbon nanotubes (CNTs), graphene oxide and graphene, are comprehensively discussed as electrodes of supercapacitors. Processing and characterization of CAs will be first discussed, where various technologies and strategies, including physical, chemical and hybrid methods, are presented and compared. Properties, including specific surface area, types of pores, pore size and pore size distribution, which are related to the processing methods, have been evaluated. After that, electrochemical performances of CAs as electrode of supercapacitors are systematically examined. However, due to the electrochemical double layer capacitive effect, CAs based electrodes have limited level of specific capacities. As a result, pseudocapacitive components, such as conducting polymers and metal oxides, have been incorporated with CAs, in order to achieve sufficiently high capacitive performances for practical applications.

Similar to CAs, CNTs that are used as electrodes of supercapacitors have also been in different forms. Meanwhile, both MWCNTs and SWCNTs have been examined. For instance, pure CNTs can be made into fiber, sheets/papers or 3D architectures, while they are also integrated with conducting polymers and metal oxides to develop pseudocapacitors with higher electrochemical activities. Graphene and Graphene Oxide (GO), as the third group of nanocarbons, are then presented, which are also found in various forms. Due to the poor electrical conductivity, GO is usually reduced to graphene, which is also known as reduced GO or rGO. All these 2D nanosheets can be combined with other components.

Introduction 3

As the most promising candidate of electrodes for supercapacitors, manganese oxides, including MnO_x, Mn_3O_4, Mn_2O_3, MnO_2, are systematically discussed in Chapter 4. Synthesis and characterization of manganese oxides, in forms of thin films and nanoparticles, are reviewed. Attempts have been made to establish interrelationships among synthetic processing, properties (such as oxidation state, morphology, particle size and size distribution) and electrochemical behaviours of the materials. Because of their low conductive characteristics, it is natural to combine manganese oxides with various nanocarbons and conducting polymers to form hybrid nanostructures, which make full utilization of high electrical conductivity of these materials, so as to develop supercapacitors with higher energy storage capabilities. Other oxides, including V_2O_5, Fe_2O_3, Fe_3O_4, Co_3O_4, NiO, CuO, ZnO, WO_3, MoO_3, RuO_2, $NiCo_2O_4$ and ferrites, are discussed in Chapter 5. Similarly, they can be synthesized with various morphologies, like nanoparticles, nanowires, nanorods, nanotubes, nanosheets, nanoneedles, nanoflakes and nanoflowers. In most cases, they are also incorporated with other components, such as CNTs, graphenes and conductive polymers, to form 0D–3D nanostructured materials.

In Chapter 6, the latest developments in flexible supercapacitors, based on carbon materials, conducting polymers and cellulose, will be presented and discussed. Flexible and rechargeable energy storage devices are considered to be crucial components for fully flexible electronic devices. Flexible and stretchable electronic devices have emerged as new generation technologies for wearable and implantable applications, including wearable displays, high performance sportswear and bio-integrated devices. In order to meet the requirements of portable and wearable electronics, the development of supercapacitors is moving towards improving the performance in flexibility and stretchability.

Micropower sources and small-scale energy storage devices have been in large demand in recent years, with the rapid development in miniaturized and portable electronic devices. Micropower devices and systems will have applications in implantable biosensors, remote and mobile environmental sensors, nanorobotics, microelectromechanical systems (MEMS) and wearable electronics. As compared with the conventional supercapacitors, microsupercapacitors have much smaller dimensions and scales. The last chapter covers this important research topic. Microsupercapacitors, based on thin film electrodes with thicknesses of ≤ 10 μm and array microelectrodes with micrometer sizes, will be mainly discussed, in terms of fabrication, processing and characterization.

Keywords: Renewable energy, New energy, Sustainable energy, Clean energy, Alternative energy, Conventional energy, Supercapacitors, Batteries, Energy storage, Energy density, Power density, Energy efficiency, Reduction–oxidations (redox), Direct Current (DC) circuits, Alternating Current (AC) circuits

Basic Concepts of Supercapacitors

Ling Bing Kong,[1,*] *Wenxiu Que,*[2] *Lang Liu,*[3] *Freddy Yin Chiang Boey,*[1,a] *Zhichuan J. Xu,*[1,b] *Kun Zhou,*[4] *Sean Li,*[5] *Tianshu Zhang*[6] *and Chuanhu Wang*[7]

Brief Introduction

This chapter aims to provide a general description on the fundamental aspects regarding supercapacitors (capacitors), which are important when developing new nanomaterials for such applications.

General Descriptions of Capacitors

An electric capacitor usually consists of two conductive plates, known as electrode and made of metals, between which a piece of dielectric material is inserted as shown in Fig. 2.1 [1]. Dielectrics that can used to construct a capacitor include air (vacuum), oiled paper, mica, glass, porcelain and various titanates. As an external voltage is applied across the two electrodes, the charging process occurs, during which, positive charges are accumulated on the positive electrode, whereas negative charges are accumulated on the negative electrode. After the external voltage is removed, both the positive and negative charges are still at the electrodes. Therefore, the capacitor separates the electrical charges. Once the two electrodes are connected with a conductive wire, the discharging process takes place, during which the positive and

---

[1] School of Materials Science and Engineering, Nanyang Technological University, 50 Nanyang Avenue, Singapore 639798.
[a] Email: mycboey@ntu.edu.sg
[b] Email: xuzc@ntu.edu.sg
[2] Electronic Materials Research Laboratory and International Center for Dielectric Research, Xi'an Jiaotong University, Xi'an 710049, Shaanxi Province, People's Republic of China.
  Email: wxque@xjtu.edu.cn
[3] School of Chemistry and Chemical Engineering, Xinjiang University, Urumqi 830046, Xinjiang, People's Republic of China.
  Email: llyhs1973@sina.com
[4] School of Mechanical and Aerospace Engineering, Nanyang Technological University, 50 Nanyang Avenue, Singapore 639798.
  Email: kzhou@ntu.edu.sg
[5] School of Materials Science and Engineering, The University of New South Wales, Australia.
  Email: sean.li@unsw.edu.au
[6] Anhui Target Advanced Ceramics Technology Co. Ltd., Hefei, Anhui, People's Republic of China.
  Email: 13335516617@163.com
[7] Department of Material and Chemical Engineering, Bengbu Univresity, Bengbu, Peoples's Republic of China.
  Email: bbxywch@126.com
* Corresponding author: elbkong@ntu.edu.sg

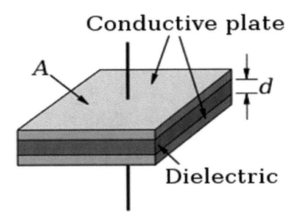

Fig. 2.1. Schematic diagram of a parallel-plate capacitor with dielectric materials to be inserted.

negative charges will be combined through the conductive wire. As a result, the capacitor can be used to store and deliver the charges.

Capacitor voltage

Electric charge

The functions of a capacitor include separation, storage and delivering of charges. It is well known that all physical objects contain both positive and negative charges. However, a physical object is generally at a neutral state, due to the same number of positive charges and negative charges. Once a net charge is present in some areas of an object because of the unbalanced charge equilibrium, some areas will have charges that have the same quantity but opposite signs.

As is well known, two charges repel each other with the same sign and mutually attract with opposite signs. The interaction between a positive and a negative charge is shown in Fig. 2.2. According to Coulomb's law, the electrostatic force between two charges can be described as:

$$F = \frac{q_+(-q_-)}{4\pi\varepsilon_0 r^2} = -\frac{q_+ q_-}{4\pi\varepsilon_0 r^2}, \tag{2.1}$$

where F is the electrostatic force, q_+ and q_- are the magnitudes of the positive and negative charges, r is the distance between these two charges, ε_0 is the dielectric constant of vacuum (air) and the negative sign means that the force is attractive instead of repelling. Because the charge magnitude in Eq. (2.1) can be represented by using the elementary charge of electron, the charge q can also be written as:

$$q = ne, \tag{2.2}$$

where $n = \pm 1, \pm 2, \pm 3, \ldots\ldots$, while e is the elementary charge constant that is equal to 1.602×10^{-19} C.

Electric field and potential

As seen in Fig. 2.2, a positive charge (q_+) can emit electric flux, while a negative charge (q_-) absorbs electric flux, so as to form electric field. In this case, the negative charge can feel the strength of the electric field generated by the positive charge, which is given by:

$$E = \frac{F}{q_-}, \tag{2.3}$$

6 *Nanomaterials for Supercapacitors*

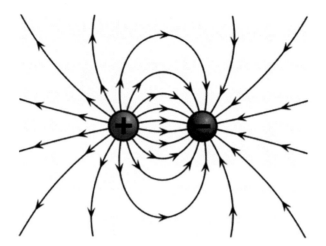

Fig. 2.2. Field lines of electric force that radiate outward from the positive charge and inward toward the negative charge.

where E is the electric field strength of the positive charge (q_+). Combination of Eqs. (2.1) and (2.3) yields the electric field strength, given by:

$$E = -\frac{q_+ q_-}{4\pi\varepsilon_0 r^2} \frac{1}{(-q_-)} = \frac{q_+}{4\pi\varepsilon_0 r^2}. \tag{2.4}$$

Correspondingly, an electric potential (V_{q_+}) at the position of the negative charge can be considered as the work done by moving the negative charge from its position through a distance (r) toward the positive charge, which is given by:

$$V_{q_+} = Er. \tag{2.5}$$

According to Eqs. (2.4) and (2.5), the electric potential related to the positive charge is expressed as:

$$V_{q_+} = \frac{q_+}{4\pi\varepsilon_0 r}. \tag{2.6}$$

Equation (2.6) suggests that the level of the electric potential is gradually decreased with increasing distance from the charge. Extremely, if the distance r becomes to infinite, the electric potential equals to be zero. Otherwise, if the distance becomes zero, the electric potential is infinite, regardless of the magnitude of the charge.

Electric potential of capacitor

In Fig. 2.2, as positive charge of Q_+ is uniformly distributed on the positive planar electrode, the negative charge of Q_- is on the negative electrode, where $|Q_+| = |Q_-|$. The two planar plate electrodes have an area of A and are separated by a distance of d, while the vacuum in between the two planar electrodes has a dielectric constant of ε_0 [2]. According to Gauss' law, the electric field outside of the planar plate is given by:

$$E = \frac{Q_+}{\varepsilon_0 A}. \tag{2.7}$$

The potential difference between the electrodes can be represented by the work that is required as the charge is moved from the positive electrode to the negative electrode. Similarly, the work or potential difference between the two electrodes (ΔV) can be derived with the following equation:

$$\Delta V = Ed. \tag{2.8}$$

Combination of Eqs. (2.7) and (2.8) yields the following equation:

$$\Delta V = \frac{Q_+ d}{\varepsilon_0 A} . \tag{2.9}$$

The potential difference between the two electrodes of a capacitor can be expressed as V instead of ΔV, i.e.:

$$V = \frac{Q_+ d}{\varepsilon_0 A} . \tag{2.10}$$

Capacitance

As shown in Fig. 2.1, a capacitor consists of two conducting parallel plates that are separated by vacuum or a dielectric material, with a distance of d. As a capacitor is charged, the two plates have equal magnitudes of charge but with opposite signs, so that a zero net charge is present on the capacitor. Because the plates are high electrically conductive, the charges are uniformly distributed on the surfaces, which resulted in a potential difference V. In this case, the total charge Q, $Q = |Q_+| + |Q_-|$, is proportional to the potential difference V, through a proportionality constant C, given by:

$$Q = CV. \tag{2.11}$$

The proportionality constant is called capacitance, which is actually a measure of the amount of charge that is required to create a potential difference between the two plates. Capacitance has a unit of farads (F), with $F = 1$ coulomb/volt (C/V). Combination of Eqs. (2.10) and (2.11) yields capacitance, given by:

$$C = \frac{Q}{V} = \frac{\varepsilon_0 EA}{Ed} = \frac{\varepsilon_0 A}{d} . \tag{2.12}$$

Equation (2.12) indicates that capacitance is proportional to the dielectric constant of vacuum and surface area of the plate electrode, but inversely proportional to the distance between the two planar electrodes. For other capacitors, such as cylindrical, spherical and isolate sphere capacitors, the capacitance will have different expressions. For example, if a long cylindrical coaxial capacitor has a length L, inner conducting cylinder radius of a and the outer cylinder radius of b, the capacitance is calculated as follows:

$$C = 2\pi\varepsilon_0 \frac{L}{\ln(b/a)} . \tag{2.13}$$

A spherical capacitor, with an inner sphere radius of a and outer sphere radius of b, has capacitance that is given by:

$$C = 4\pi\varepsilon_0 \frac{ab}{b-a} . \tag{2.14}$$

An isolated sphere capacitor has a single isolated spherical conductor with a radius of R, leading to capacitance that can be expressed as:

$$C = 4\pi\varepsilon_0 R . \tag{2.15}$$

Equations (2.12–2.15) imply that the capacitance is linearly proportional to the dielectric constant ε_0, if the capacitor has a fixed configuration and dimension.

8 *Nanomaterials for Supercapacitors*

If a piece of material is inserted into the capacitor, with a dielectric constant ε. The capacitance is thus changed according to the ratio of $\varepsilon/\varepsilon_0$, which is defined as the relative dielectric constant, i.e., $\varepsilon_r = \varepsilon/\varepsilon_0$. Therefore, Eq. (2.12) becomes to be:

$$C = \frac{Q}{V} = \frac{\varepsilon A}{d} = \frac{\varepsilon_r \varepsilon_0 A}{d}. \tag{2.16}$$

Accordingly, relative dielectric constant of vacuum is $\varepsilon_r = \varepsilon_0/\varepsilon_0 = 1$, so that materials always have a relative dielectric constant of larger than 1. Because ε is always larger than ε_0, the capacitance of a given capacitor can be simply increased by using materials with larger dielectric constant, without varying its dimension and configuration. In other words, to achieve the same level of capacitance, if a material with larger dielectric constant is used, the capacitor dimension can be proportionally reduced.

Dielectric constant

Usually, dielectric materials are characterized by using two parameters: (i) leakage conductivity σ_l and (ii) relative dielectric constant ε_r. The leakage conductivity is related to resistance of the material R (Ω), thickness d (cm) and dielectric surface area A (cm^2), through the following equation:

$$\sigma_l = \frac{d}{RA}. \tag{2.17}$$

Relative dielectric constant is also known as relative permittivity. According to Eq. (2.1), if two charges q_1 and q_2 are separated by a small distance r in vacuum, the electrostatic force F_0 is given by:

$$F_0 = \frac{|q_1||q_2|}{4\pi\varepsilon_0 r^2}. \tag{2.18}$$

If a piece of material is inserted, Eq. (2.18) becomes to be:

$$F_m = \frac{|q_1||q_2|}{4\pi\varepsilon r^2}. \tag{2.19}$$

With Eqs. (2.18) and (2.19), relative dielectric constant of the material (ε_r) is can be derived as:

$$\varepsilon_r = \frac{F_0}{F_m} = \frac{\varepsilon}{\varepsilon_0}. \tag{2.20}$$

As a vacuum capacitor is fully charged, the voltage difference across the two electrodes is V_0. Once a piece of material is inserted, the voltage difference will become to be V. Therefore, dielectric constant of the material can be given by:

$$\varepsilon_r = \frac{V_0}{V}. \tag{2.21}$$

Because there is always $V < V_0$, $\varepsilon_r > 1$.

Dielectric polarization

If a material can be polarized by external electric fields, it can be used to construct capacitors. As a material has at least one of the following properties [3], it is considered to be polarizable. These properties include (i) having covalently bonded molecules that exhibit a certain degree of natural electron polarities to withstand electrical stress and orbital deformation induced by the external electrical field, (ii) containing polarizable ions with ion centers to be displaced at electrical stress, (iii) demonstrating dipole polarity to cause dipole rotations and (iv) possessing domain polarity with entire domains that can be rotated when subject to an external electric field.

Charging–discharging

To charge a capacitor, the capacitor is integrated into an electrical circuit with an external power source, e.g., a battery. In a closed loop circuit, a flow of electrons is generated through the circuit, due to the electromotive force (emf) created by the battery. In this case, the positive electrode loses electrons to the positive terminal of the battery, so that the plate is positively charged. Meanwhile, electrons flow from the negative terminal of the battery to accumulate on the negative electrode, which is thus negatively charged. The number of electrons accumulated on the negative electrode is the same as that of positive charges accumulated on the positive electrode. The charging process continuously proceeds until the potential between the two electrodes from the initial value of zero to the potential difference between the battery terminals.

DC and AC currents

During the charging and discharging process, there is a charge flow within the plates through the external wires, which is known as electrical current and defined as the flow of electric charge across a point or area, in units of Coulombs per second, i.e., $C \cdot s^{-1}$, or amperes (A). In this case, besides the current generated due to the flow of electrons through the conducting wires, there is also a current created because of the transportation of ions through the electrolyte. This discussion will focus on the flow of electrons through solid metallic conductors.

In the presence of an electric potential gradient, electrons in metals acted as mobile charge carriers that flow in the direction from lower potential to higher potential. Therefore, the electric potential gradient is called electromotive or driving force for electrons to flow through a closed electric circuit or a connected electric load. According to definition, the direction of electrical current is the flow direction of positive charge, so that the electron flow direction is opposite to the current direction. There are two types of currents: (i) Direct Current (DC) and (ii) Alternating Current (AC).

DC is characterized by a unidirectional flow of electrons from a low potential area to high potential one. If there are two points, 1 and 2, within a circuit wire loop, the potential at point 1 is V_1 and that at point 2 is V_2, i.e., the potential difference or voltage difference will be $V = |V_2 - V_1|$. As the current (I) flows from point 1 to point 2, I and V can be linked with the following equation:

$$R = \frac{V}{I}.$$

(2.22)

This is known as Ohm's Law, where R is the electric resistance in ohm (Ω), if the unit of current is the ampere (A) and the potential difference is voltage (V).

In comparison, AC has directional change in the current flow, because the direction of the electromotive force is continually reversed [4]. In other words, AC is generated by the forced rotation of a loop conductor by using a magnetic field. During the rotation, alteration of the polarity occurs at a continuous oscillating frequency, which varies sinusoidally as a function of time. An induced potential $\tilde{V}$ through the loop is thus given by:

$$\tilde{V} = \tilde{V}_m \sin \varpi_d t ,$$

(2.23)

where $\tilde{V}_m$ is the amplitude of the oscillation, i.e., the maximum value, ω_d is the angular frequency of the rotating loop and t is time. Owing to the periodic change in potential, the alternating current $\tilde{I}$ will sinusoidally oscillate over time, at the same angular frequency, which can be expressed as:

$$\tilde{I} = \tilde{I}_m \sin(\varpi_d t - \varphi), \tag{2.24}$$

where $\tilde{I}_m$ is the maximum amplitude of current oscillation and φ is the phase constant to describe a situation in which the current is out of phase with the potential. The Ohm's law of Eq. (2.22) is also applicable to AC. The resistance (R) of an object can be expressed as $R = l/\sigma \cdot S$, where l is the length of the object, σ is the conductivity and S is the cross-sectional area.

Charging

To charge the capacitor, it should be connected to a potential or current source in series, so that the plates of the capacitor can be rapidly charged over time. If a resistor is inserted between the capacitor and the potential or current source, the charging time will be prolonged. In this case, an important relation, known as RC time constant, which can be used to measure the charge and relaxation times for capacitor charging, will be developed [4]. Figure 2.3 shows a circuit to derive the RC constant the Kirchoff's law, given by:

$$V^0 - IR - \frac{q}{C} = 0, \tag{2.25}$$

where V^0 is the voltage of the battery. By substituting $I = dq/dt$ into Eq. (2.25) and rearranging the equation into a first order differential, the function satisfying the equation, with the initial condition of $q = 0$ at $t = 0$, is thus given by:

$$q = CV^0 \left(1 - e^{-t/RC}\right). \tag{2.26}$$

Once the time approaches infinity, a charge $q = CV^0$ can be arrived. Using Eq. (2.11) to substitute the charge for potential, Eq. (2.26) can be written in terms of the potential V_p across the capacitor plates, due to the driving voltage, which is given by:

$$V_p = V^0 \left(1 - e^{-t/RC}\right). \tag{2.27}$$

Meanwhile, a relation between the charge time and the current flowing through the circuit can also be obtained, by deriving the derivative of q with respect to t in Eq. (2.26), i.e.:

$$\frac{dq}{dt} = I = \frac{V^0}{TR} e^{-t/RC}. \tag{2.28}$$

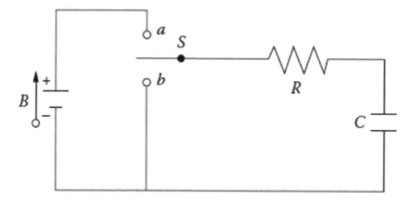

Fig. 2.3. Electric circuit consisting of a resistor and a capacitor connected in series, together with a Switch (S), an elemental Resistor (R), a Capacitor (C) and a Battery (B).

Here, the product RC is called capacitive time constant, which is denoted as $\tau_c = $ RC. For example, as $t = \tau_c = $ RC, Eq. (2.26) becomes to be $q = 0.63CV^0$, which implies that only 63% of the total charge can be achieved at this condition. Generally, a capacitor is considered to have full charge after five time constants.

Discharging

Once the capacitor is fully charged to the potential that is equivalent to that of the power source V^0, it can be charged to a loop circuit with a resistor. In the discharging loop, the fully charged capacitor will act as the power source [4], with the sum of the potentials to be given by:

$$\frac{q}{C} + \frac{dq}{dt}R = 0.$$ (2.29)

To first-order differential, the solution in terms of charge q is given by:

$$q = q_0 e^{-t/RC},$$ (2.30)

where the initial charge q_0 of the capacitor is equal to V^0C. Also, the discharge current can be obtained by differentiating Eq. (2.30) with respect to time t, i.e.:

$$I = \frac{q_0}{RC} e^{-t/RC}.$$ (2.31)

If the energy used to charge the capacitor and capacitive charge stored by the capacitor have no leakage or dissipation, they will be retained indefinitely until discharged [1]. However, this is not the case in real practice.

Energy storage

For an electrostatic charge to develop along the plates of a capacitor, work must be done by an external driving force. At the beginning of the charging, the net charge between the plates of the capacitor is zero. When a potential is applied, the charge will accumulate on the conducting plates. However, as an electric field develops between the plates it becomes more difficult to accumulate like charges and subsequently the process requires more work. This work, done by an external power source such as a battery, transfers energy into electric potential energy E and is stored in an electrical field within the dielectric material. Recovery of this stored energy is achieved by discharging the capacitor into a circuit. To calculate the work done by a potential difference V^0 for charge transfer on a capacitor, Eq. (2.32) is used where the incremental change (dq) in charge requires incremental work (dW):

$$dW = Vdq = \frac{q}{C}dq.$$ (2.32)

The total work performed, and thus the total potential energy stored in the capacitor, is:

$$E = \int dW = \frac{1}{C} \int_0^q qdq = \frac{q^2}{2C}.$$ (2.33)

Note that the capacitance C in Eq. (2.33) is independent of charge and can be taken out of the integral [2]. Combining Eq. (2.11) with (2.33), a more familiar form of the energy stored in a capacitor can be obtained:

$$E = \frac{q^2}{2C} = \frac{1}{2}C\left(V^0\right)^2.$$ (2.34)

Ideally, the energy stored in a capacitor and capacitive charge stored by the capacitor do not leak or dissipate and are retained indefinitely until discharged [1]. However, in practice, due to the leaking of dielectric material, the self-discharge rate of the capacitor is faster relative to batteries.

Electrochemical Double-Layer (ECDL) Supercapacitors

Electrode and electrolyte interfaces

As mentioned earlier, an electrochemical cell is constructed with two electrodes, while the space between them is filled with an electrolyte, which can be either solid or a solution. Usually, solid state electrolytes acted as both ionic conductors and separator to isolate the positive and negative electrodes. However, if solution electrolytes are employed, an inert porous separator should be used, which allows the electrolyte ions to pass through, but blocks the solvents. An electrochemical capacitor is very similar to an electrochemical cell in structure, but there is no electron transfer across the interface.

Figure 2.4 shows a schematic diagram of electric double-layer capacitor. Because of Coulomb's force, the positive charges accumulated on the positive electrode will attract the same number of negative charges near the electrode in the electrolyte. However, there is always heat fluctuation in the electrolyte, so that the charges carried by the ions have an uneven distribution, which resulted in the presence of net negative charges in the electrolyte near the electrode. As a result, an electric double-layer is formed, due to the charge balance between the electrode and the electrolyte.

At the same time, another double-layer is formed, because there is an equal number of negative charges accumulate at the negative electrode, near which an equal number of net positive charges is in the adjacent electrolyte, in order to maintain the electrical neutrality of the system. As a result, a double-layer capacitor has two electric double-layers, with each at the positive electrode–electrolyte interface and negative electrode–electrolyte interface.

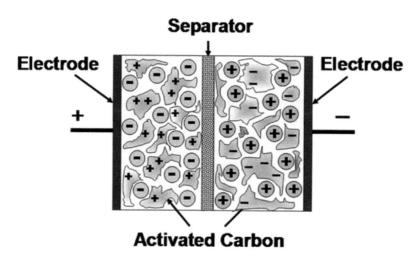

Fig. 2.4. Schematic diagram of electric double-layer supercapacitor.

Interface of electrode and electrolyte solution

In modern supercapacitors, the electrolytes could be aqueous solutions, non-aqueous electrolyte solutions and ionic liquids. Therefore, the electric double-layer discussed here applies to all three of these electrolyte solutions. Electric double-layer is formed, because the positive or negative charge developed along the interface between the electrode and electrolyte solution is balanced by oppositely charged solution ions near the electrode surface in the solution through the Coulomb's force. Due to thermal fluctuation in the solution, the net negative ions are distributed in such a way that the concentration is gradually decreased from near the electrode surface to the bulk of the solution. This scattered layer together with the electrode positive charge results in diffusing the double-layer or diffuse layer, which is usually known as the Gouy point charge layer or the Gouy–Chapman model. Thickness of the diffuse layer is dependent on several factors, including (i) temperature, (ii) concentration of the electrolyte, (iii) number of charges carried by the ions and (iv) dielectric constant of the electrolyte solution.

Normally, the thickness of the diffuse layer is increased with increasing temperature, but decreased with increasing electrolyte concentration, number of the charges carried by the ion and dielectric constant. Specifically, if the temperature is very low, while the other three parameters are sufficiently high, the diffuse layer will be very thin. In this case, an array of negative ions near the electrode surface is formed, leading to a compact electric double-layer, which is known as the Helmholtz layer. In practice, the two types of layers are present together, thus leading to the Stern–Grahame model. Moreover, the Helmholtz layer can be further considered to consist of two layers, i.e., (i) inner plane and (ii) outer plane.

It is well known that the potential drop across the double-layer can be expressed by using the potential difference ($\Delta\phi_{M/S}$) between the inner potential at a point on the metal electrode (ϕ_M) and the inner potential at the end point of the diffuse layer in electrolyte solution (ϕ_S), i.e., $\Delta\phi_{M/S} = \phi_M - \phi_S$, which is denoted as absolute potential difference. The potential drop can also be presented as outer potential difference ($\Delta\psi_{M/S}$), which is usually simplified as $\Delta\psi$, as shown in Fig. 2.5. The outer potential difference is also known as Volta potential. Other names include Volta potential difference, contact potential difference and outer potential difference. The outer potential can be linked with inner potential through the following expression:

$$\phi = \psi + \chi, \tag{2.35}$$

where ϕ is the inner potential while χ is the surface potential related to short range effects caused by the adsorbed ions and oriented water molecules. Because the outer potential can be measured experimentally, whereas the surface potential is not measurable, the inner potential drop is not a parameter that can be measured. The double-layer potential drop can be expressed as either $\Delta\phi_{M/S}$ or $\Delta\psi_{M/S}$, i.e., $\Delta\phi_{M/S} = \Delta\psi_{M/S}$. The overall double-layer potential drop is given by:

$$\Delta\phi_{M/S} = \Delta\chi_{O/M} + \Delta\chi_{M/O} + \Delta\psi_{M/H} \\ + \Delta\chi_{O/H} + \Delta\chi_{H/O} + \Delta\psi_{H/S}\Delta\chi_{O/S} + \Delta\chi_{S/O} \tag{2.36}$$

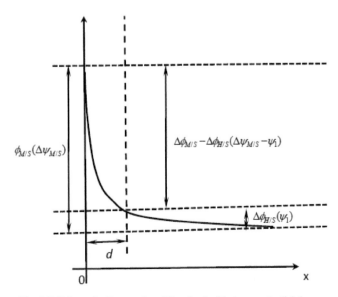

Fig. 2.5. Schematic diagram describing the double-layer potential drops.

14   *Nanomaterials for Supercapacitors*

Since there is $\Delta\chi_{O/M} = -\Delta\chi_{M/O}$, $\Delta\chi_{O/H} = -\Delta\chi_{H/O}$ and $\Delta\chi_{O/S} = -\Delta\chi_{S/O}$, it is true that $\Delta\phi_{M/S} = \Delta\psi_{M/H} + \Delta\psi_{H/S} = \Delta\psi_{M/S}$.

The potential drop across the diffuse layer is considered to be the outer potential drop, instead of inner potential drop, thus $\Delta\psi_{H/S} = \psi_1$. Therefore, $\Delta\psi_{M/S}$ and ψ_1 are the potential drops across the entire double-layer and diffuse layer, respectively. As a result, the potential drop across the Helmholtz layer becomes to be $\Delta\psi_{M/S} - \psi_1$, while the potential across the entire double-layer is given by:

$$\Delta\psi = \left(\Delta\psi_{M/S} - \psi_1\right) + \psi_1 . \tag{2.37}$$

If the unit charge quantity accumulated on the electrode side or in the electrolyte solution side can be expressed as q on a unit area

$$q = \frac{Q}{A}, \tag{2.38}$$

where A is the real electrode surface area in contact with the electrolyte solution. Generally, the reciprocal of the overall double-layer differential capacitance is expressed as:

$$\frac{1}{C_{dl}} = \frac{d\Delta\psi}{dq} = \frac{d\left(\Delta\psi - \psi_1\right)}{dq} + \frac{d\psi_1}{dq} = \frac{1}{C_H} + \frac{1}{C_{diff}}, \tag{2.39}$$

where C_H is the differential capacitance of the Helmholtz layer, while C_{diff} is the differential capacitance of the diffuse layer, i.e.:

$$C_H = \frac{dq}{d\left(\Delta\psi_{M/S} - \psi_1\right)}, \tag{2.40}$$

$$C_{diff} = \frac{dq}{d\psi_1} . \tag{2.41}$$

Gouy–Chapman–Stern (GCS) model

According to the Gouy–Chapman–Stern (GCS) model [5], the concentration of ion ($C_{i(x)}$) within the diffuse layer at a point x, known as the Boltzmann distribution, is given by:

$$C_{i(x)} = C^0 \exp\left(-\frac{z_i F \psi_x}{RT}\right), \tag{2.42}$$

where C^0 is bulk concentration of the ion i, F is Faraday's constant, ψ_x is the potential at x point, R is the gas constant and T is temperature. Therefore, the net charge per unit volume, ρ, with a unit of $\mu C \cdot cm^{-3}$, is derived as:

$$\rho = \sum z_i F C_{i(x)} = \sum z_i F C^0 \exp\left(-\frac{F\psi_x}{RT}\right). \tag{2.43}$$

The volume charge can be linked with the potential through the following equation:

$$\frac{d^2\psi_x}{dx^2} = -\frac{\rho}{\varepsilon_r \varepsilon_0}, \tag{2.44}$$

where ε_r is relative dielectric constant of the electrolyte solution, while ε_0 is dielectric constant of vacuum, so that there is following equation regarding to Eq. (2.43):

$$\frac{d^2\psi_x}{dx^2} = -\frac{1}{\varepsilon_r\varepsilon_0}\sum z_i FC^0 \exp\left(-\frac{z_i F\psi_x}{RT}\right). \tag{2.45}$$

Mathematically, there is:

$$\frac{d^2\psi_x}{dx^2} = \frac{1}{2}\frac{d}{d\psi_x}\left(\frac{d\psi_x}{dx}\right)^2, \tag{2.46}$$

Equation (2.45) can be expressed as:

$$d\left(\frac{d\psi_x}{dx}\right)^2 = -\frac{2}{\varepsilon_r\varepsilon_0}\sum z_i FC^0 \exp\left(-\frac{z_i F\psi_x}{RT}\right)d\psi_x. \tag{2.47}$$

With the conditions $x \to \infty$, $\psi_x = 0$ and $d\psi_x/dx = 0$, Eq. (2.47) can be further rewritten as:

$$\begin{aligned}\left(\frac{d\psi_x}{dx}\right)^2 &= -\frac{2RT}{\varepsilon_r\varepsilon_0}\sum C^0\left[\exp\left(-\frac{z_i F\psi_x}{RT}-1\right)\right]\\ &= \frac{2RTC^0}{\varepsilon_r\varepsilon_0}\left[\exp\left(\frac{z_i F\psi_x}{RT}\right)-\exp\left(-\frac{z_i F\psi_x}{RT}\right)\right].\end{aligned} \tag{2.48}$$

At the point of $x = d$, $\psi_x = \psi_1$, Eq. (2.48) becomes to be:

$$\left(\frac{d\psi_x}{dx}\right)^2_{x-d} = \frac{2RTC^0}{\varepsilon_r\varepsilon_0}\left[\exp\left(\frac{z_i F\psi_x}{RT}\right)-\exp\left(-\frac{z_i F\psi_x}{RT}\right)\right]. \tag{2.49}$$

Integrating Eq. (2.46) from $x = d$ to $x = \infty$ yields the total net charge within the diffuse layer:

$$\left(\frac{d\psi_x}{dx}\right)_{x=d} = \frac{1}{\varepsilon_r\varepsilon_0}\int_{x=d}^{x=\infty}\rho dx = -\frac{q}{\varepsilon_r\varepsilon_0}. \tag{2.50}$$

By combining Eqs. (2.49) and (2.50), the total net charge within the diffuse layer becomes to be the following expression:

$$q = \sqrt{2\varepsilon_r\varepsilon_0 RTC^0}\left[\exp\left(\frac{z_i F\psi_1}{RT}\right)-\exp\left(-\frac{z_i F\psi_1}{RT}\right)\right]. \tag{2.51}$$

Theoretical differential capacitance

By differentiating Eq. (2.51) with respect to the potential drop, differential capacitance of the diffuse layer can be derived as:

$$C_{diff} = \frac{dq}{d\psi_{H-S}} = \frac{z_i F}{2RT}\sqrt{2\varepsilon_r\varepsilon_0 RTC^0}\left[\exp\left(\frac{z_i F\psi_1}{RT}\right)-\exp\left(-\frac{z_i F\psi_1}{RT}\right)\right]. \tag{2.52}$$

It thus suggested that the differential capacitance can be modified by controlling either the concentration of the electrolyte solution or the potential drop at both positive and negative potential directions. As the potential drop across the diffuse double-layer is zero, the differential capacitance approaches the minima, which known as the electrode potential at zero charge (pzc).

According to Eq. (2.52), as the potential drop across the diffuse layer is close to zero, it can be simplified as:

16   *Nanomaterials for Supercapacitors*

$$C_{diff} \approx \frac{z_i F}{RT} \sqrt{2\varepsilon_r \varepsilon_0 RTC^0} \; . \tag{2.53}$$

This implies that, if $\psi_1 = 0$, the differential capacitance of the diffuse layer is proportional to the square root of concentration and dielectric constant of the electrolyte solution. Moreover, noting the definition of capacitance, if C_{diff} is described as:

$$C_{diff} \approx \frac{\varepsilon_r \varepsilon_0}{L_{diff}} , \tag{2.54}$$

where the thickness of the diffuse layer (L_{diff}) can be derived from Eq. (2.53), given by:

$$L_{diff} = \frac{1}{z_i F} \sqrt{\frac{\varepsilon_r \varepsilon_0 RT}{2C^0}} \; . \tag{2.55}$$

Since the thickness of the Helmholtz layer is constant, the differential capacitance is not changed with the potential drop. By combining Eqs. (2.39) and (2.52), differential capacitance of the entire double-layer can be derived as:

$$
\begin{aligned}
C_{dl} &= \frac{C_H C_{diff}}{C_H + C_{diff}} \\
&= \frac{C_H \dfrac{|z| F}{2RT} \sqrt{2\varepsilon_r \varepsilon_0 RTC^0} \left[\exp\left(\dfrac{|z| F \psi_1}{2RT} \right) + \exp\left(-\dfrac{|z| F \psi_1}{2RT} \right) \right]}{C_H + \dfrac{|z| F}{2RT} \sqrt{2\varepsilon_r \varepsilon_0 RTC^0} \left[\exp\left(\dfrac{|z| F \psi_1}{2RT} \right) + \exp\left(-\dfrac{|z| F \psi_1}{2RT} \right) \right]} \; .
\end{aligned}
\tag{2.56}
$$

According to the definition of capacitance, the differential capacitance of the Helmholtz layer is given by:

$$C_H = \frac{\varepsilon_r \varepsilon_0}{d} , \tag{2.57}$$

where the thickness of the Helmholtz layer (d) is close to the diameter of a water molecule, i.e., ~ 0.2 nm.

By combining Eqs. (2.40) and (2.51), the relationship between the Helmholtz layer and the diffuse layer potential drops can be expressed as:

$$\psi - \psi_1 = \frac{1}{C_H} \sqrt{2\varepsilon_r \varepsilon_0 RTC^0} \left[\exp\left(\frac{|z| F \psi_1}{2RT} \right) - \exp\left(-\frac{|z| F \psi_1}{2RT} \right) \right] . \tag{2.58}$$

Usually, the potential of the Helmholtz layer is lower than that of the diffuse layer when diluted electrolyte solutions are used. As the concentration of the electrolyte solution is increased, the potential of the diffuse layer becomes lower than that of the Helmholtz layer. Therefore, the potential drop of the entire double-layer is dominated by that of the diffuse layer at low concentrations, while at high concentrations that of the Helmholtz layer becomes dominating. In addition, the potential drop across the Helmholtz layer is a function of the square roots of both the electrolyte concentration and the dielectric constant ($\varepsilon_r \varepsilon_0$).

Supercapacitor parameters

Capacitance

The overall capacitance of a supercapacitor can be described as two differential capacitances that are connected in series. Assuming that the capacitance related to the positive electrode is given by:

$$C_{dl,p} = \frac{C_{H,p}C_{diff,p}}{C_{H,p} + C_{diff,p}}, \tag{2.59}$$

while the negative electrode is expressed as:

$$C_{dl,n} = \frac{C_{H,n}C_{diff,n}}{C_{H,n} + C_{diff,n}}, \tag{2.60}$$

the capacitance of the supercapacitor (C_{dl}^T) can be derived as:

$$C_{dl}^T = \frac{C_{dl,p}C_{dl,n}}{C_{dl,p} + C_{dl,n}}. \tag{2.61}$$

If two identical electrodes are used to construct the electrochemical double-layer supercapacitors, i.e., $C_{dl,p} = C_{dl,n}$, there is:

$$C_{dl}^T = \frac{1}{2}C_{dl,p} = \frac{1}{2}C_{dl,n}. \tag{2.62}$$

This means that the total capacitance of a double-layer supercapacitor is half of the capacitance of the individual electrode.

Equivalent Series Resistance (ESR)

For an ideal capacitor, as a sinusoidal alternative current is applied to it, the output voltage is out of phase by 90°, which is independent on the frequency. However, for supercapacitors, the output voltage is usually out of phase that is slightly less than 90°, which implies that it is coupled with an equivalent series ohmic resistor, known as the Equivalent Series Resistance (ESR). Generally, ESR consists of contact resistances between the current collector and the electrode layer, the electrode layer inter-particles due to their porous and particulate characteristics, external lead contact, the electrolyte and dielectric loss. Figure 2.5 shows an equivalent circuit of a supercapacitor with ESR. ESR is an important parameter used to characterize the performances of supercapacitors.

The complex impedance of this equivalent circuit (Z_{cell}^j), shown in Fig. 2.5, is given by: which is the complex AC impedance of the supercapacitor cell, can be expressed as Eq. (2.121) if the equivalent series resistance is expressed as:

$$Z_{cell}^j = R_{ESR} - j\frac{1}{2\pi f C_{dl}^T}, \tag{2.63}$$

where R_{ESR} is the equivalent series resistance and f is the AC frequency in hertz. The magnitude of Z_{cell}^j is thus obtained as:

$$\left|Z_{cell}^j\right| = \sqrt{\left(R_{ESR}\right)^2 + \left(\frac{1}{2\pi f C_{dl}^T}\right)^2}. \tag{2.64}$$

Charging and discharging

Charging and discharging of supercapacitors can be conducted (i) at a constant cell voltage to measure the cell current changes versus time, (ii) at a constant current to measure the cell voltage changes versus time or (iii) at constant power by varying the current with decreasing voltage [6].

18 *Nanomaterials for Supercapacitors*

Charging at constant voltage

It is assumed that the supercapacitor is at a zero charge state before charging is started, i.e., the voltage across the supercapacitor is zero. After charging for $t \geqslant 0$, the overall charging current of the supercapacitor cell (i_{cell}) can be derived to be:

$$i_{cell} = \frac{E}{R_{ESR} + R_p} + \frac{R_p E}{R_{ESR}\left(R_{ESR} + R_p\right)} \exp\left(-\frac{R_{ESR} + R_p}{R_{ESR} R_p C_{dl}^T} t\right), \tag{2.65}$$

where R_p is called leakage resistance. Therefore, the voltage drop across the supercapacitor is given by:

$$V_{sc} = \frac{R_p E}{R_{ESR} + R_p}\left[1 - \exp\left(-\frac{R_{ESR} + R_p}{R_{ESR} R_p C_{dl}^T} t\right)\right]. \tag{2.66}$$

At $t = 0$, Eq. (2.65) can be simplified as:

$$i_{cell} = \frac{E}{R_{ESR}}, \tag{2.67}$$

which is the maximum charging current. At $t = \infty$, i.e., the moment when the supercapacitor is fully charged, the charging current becomes to be:

$$i_{cell} = \frac{E}{R_{ESR} + R_p}. \tag{2.68}$$

This implies that there is always an extra current density necessary to overcome the self-discharging due to the presence of R_p, even in the case that the supercapacitor is fully charged. If the leakage current is close to zero, i.e., $R_p \to \infty$, the charging current i_{cell} will be given by:

$$i_{cell} = \frac{E}{R_{ESR}} \exp\left(-\frac{t}{R_{ESR} C_{dl}^T}\right). \tag{2.69}$$

As a result, the charging current will approach zero, at $t \to \infty$, so that $i_{cell} \to 0$. In this case, it is not necessary to have an extra current density to overcome the self-discharging. Generally, R_p is much higher than R_{ESR}, i.e., $R_p \gg R_{ESR}$, so that Eq. (2.65) can be simplified to Eq. (2.69).

For an ideal supercapacitor, i.e., $R_{ESR} = 0$, and $R_p = \infty$, the current will be infinite at a constant voltage. In practice, $R_{ESR} \to 0$ and $R_p \neq \infty$, the supercapacitor voltage can also be achieved very quickly, once the constant voltage is applied, as shown in Eq. (2.66).

Charging at constant current

As the charging process of a supercapacitor is carried out at a constant current density, I_{cell}, after charging for $t \geq 0$, the charging voltage, V_{cell}, is given by:

$$V_{cell} = I_{cell} R_{ESR} + I_{cell} R_p\left[1 - \exp\left(-\frac{t}{R_p C_{dl}^T}\right)\right]. \tag{2.70}$$

In this case, at $t = 0$, $V_{cell} = I_{cell} R_{ESR}$, at $t = \infty$, $V_{cell} = I_{cell} R_{ESR} + I_{cell} R_p$. Therefore, after fully charging, the voltage across the supercapacitor (V_{sc}) reaches the maximum value, denoted as V_{sc}^0, which is equal to $I_{cell} R_p$. If leakage current is zero, i.e., $R_p \to \infty$, Eq. (2.70) becomes to be:

$$V_{cell} = I_{cell}R_{ESR} + I_{cell}\frac{t}{C_{dl}^T}.$$ (2.71)

Accordingly, the time (t_c) that is needed to charge a supercapacitor to a desired voltage (V_T) can be described as:

$$V_{cell} = V_T / t_c = C_{dl}^T \left(\frac{V_T - I_{cell}R_{ESR}}{I_{cell}} \right).$$ (2.72)

This equation indicates that the cell voltage linearly increased with charging time. When there is no ESR, i.e., $R_{ESR} \rightarrow 0$, the cell voltage is given by:

$$V_{cell} = I_{cell}R_p \left[1 - \exp\left(-\frac{t}{R_p C_{dl}^T} \right) \right].$$ (2.73)

For an ideal supercapacitor, i.e., $R_{ESR} = 0$ and $R_p = \infty$, the cell voltage at a constant current is expressed as:

$$V_{cell} = I_{cell}\frac{t}{C_{dl}^T}.$$ (2.74)

Discharging at constant resistance

As a supercapacitor is discharged at a load resistance (R_L) for $t \geq 0$, voltage (V_{sc}) and current density (i_{sc}) are given by:

$$V_{sc} = V_{sc}^0 \exp\left[-\frac{R_p + R_{ESR} + R_L}{R_p(R_{ESR} + R_L)C_{dl}^T}t \right],$$ (2.75)

$$i_{sc} = \frac{V_{sc}^0 (R_p + R_{ESR} + R_L)}{R_p(R_{ESR} + R_L)} \exp\left[-\frac{R_p + R_{ESR} + R_L}{R_p(R_{ESR} + R_L)C_{dl}^T}t \right].$$ (2.76)

Accordingly, the cell current passing through R_L and the cell voltage across the load resistance are derived to be:

$$i_{cell} = \frac{V_{sc}}{R_{ESR} + R_L} = \frac{V_{sc}^0}{R_{ESR} + R_L} \exp\left[-\frac{R_p + R_{ESR} + R_L}{R_p(R_{ESR} + R_L)C_{dl}^T}t \right],$$ (2.77)

$$V_{cell} = V_{sc} - I_{cell}R_{ESR} = V_{sc}^0 \left(1 - \frac{R_{ESR}}{R_{ESR} + R_L} \right) \exp\left[-\frac{R_p + R_{ESR} + R_L}{R_p(R_{ESR} + R_L)C_{dl}^T}t \right].$$ (2.78)

Therefore, if $R_L = 0$, i.e., there is no load, i_{cell} is equal to the current flowing through R_{ESR} while V_{cell} is zero. If there is no leakage current, i.e., $R_p \rightarrow \infty$, Eqs. (2.77) and (2.78) are further simplified to:

$$i_{cell} = \frac{V_{sc}^0}{R_{ESR} + R_L} \exp\left[-\frac{t}{(R_{ESR} + R_L)C_{dl}^T} \right],$$ (2.79)

$$V_{cell} = V_{sc}^0 \left(1 - \frac{R_{ESR}}{R_{ESR} + R_L}\right) \exp\left[-\frac{t}{(R_{ESR} + R_L)C_{dl}^T}\right].$$

(2.80)

Discharging at constant voltage

As a supercapacitor is discharged at a constant voltage (E), which is slightly lower than the cell voltage (V_{sc}^0), the supercapacitor voltage (V_{sc}) and current density (i_{sc}), after discharging for $t \geq 0$, are given by:

$$V_{sc} = V_{sc}^0 - \frac{V_{sc}^0(R_{ESR} + R_p) - ER_p}{R_{ESR} + R_p}\left[1 - \exp\left(-\frac{R_{ESR} + R_p}{R_{ESR}R_pC_{dl}^T}t\right)\right],$$

(2.81)

$$i_{sc} = \frac{V_{sc}^0(R_{ESR} + R_p) - ER_p}{R_{ESR} + R_p}\exp\left(-\frac{R_{ESR} + R_p}{R_{ESR}R_pC_{dl}^T}t\right).$$

(2.82)

Correspondingly, the cell voltage (V_{cell}) and cell current density (I_{cell}) can be obtained as:

$$V_{cell} = E,$$

(2.83)

$$i_{cell} = -\frac{E}{R_{ESR} + R_p} + \left[\frac{V_{sc}^0}{R_{ESR}} - \frac{ER_p}{R_{ESR}(R_{ESR} + R_p)}\right]\exp\left(-\frac{R_{ESR} + R_p}{R_{ESR}R_pC_{dl}^T}t\right),$$

(2.84)

If $R_p \to \infty$, Eq. (2.84) is simplified to:

$$i_{cell} = \frac{V_{sc}^0 - E}{R_{ESR}}\exp\left(-\frac{1}{R_{ESR}C_{dl}^T}t\right).$$

(2.85)

Discharging at constant current

When discharging at a constant current (I_{cell}), the supercapacitor voltage (V_{sc}) and current density (i_{sc}), after discharging for $t \geq 0$, are derived as:

$$V_{sc} = V_{sc}^0 - (V_{sc}^0 + I_{cell}R_p)\left[1 - \exp\left(-\frac{t}{R_pC_{di}^T}\right)\right],$$

(2.86)

$$i_{sc} = \frac{V_{sc}^0 + I_{cell}R_p}{R_p}\exp\left(-\frac{t}{R_pC_{di}^T}\right).$$

(2.87)

Accordingly, the cell voltage (V_{cell}) and cell current density (I_{cell}) are given by:

$$V_{cell} = -I_{cell}R_{ESR} + V_{sc}^0 - (V_{sc}^0 + I_{cell}R_p)\left[1 - \exp\left(-\frac{t}{R_pC_{di}^T}\right)\right],$$

(2.88)

$$i_{cell} = -I_{cell}.$$

(2.89)

Basic Concepts of Supercapacitors   21

If $R_p \to \infty$, Eq. (2.88) becomes to be:

$$V_{cell} = -I_{cell} R_{ESR} + V_{sc}^0 - I_{cell} \frac{t}{C_{dl}^T} \cdot$$
(2.90)

The time (t_{fd}) that is necessary to fully discharge the supercapacitor, i.e., $V_{cell} = 0$, given by:

$$t_{fd} = C_{dl}^T \frac{V_{sc}^0 - I_{cell} R_{ESR}}{I_{cell}} \cdot$$
(2.91)

Therefore, as the discharging time is longer than t_{fd}, negative cell voltage will be present, which is practical. This means that there is always $t \, ? \, t_{fd}$ for the two equations to be valid.

Charging and discharging curves at constant current

According to Eq. (2.90) for constant charging and Eq. (2.88) for constant discharging, the charging and discharging curves can be obtained, with four variables, i.e., I_{cell}, R_{ESR}, R_p and C_{dl}^T. Each of them can be described as a function of time, by keeping the rest of the three to be fixed.

Energy and power densities

Energy density

For a double-layer supercapacitor, the energy density is given by:

$$E = \int_0^q V_{sc} dq = \int_0^q \frac{q}{C_{dl}^T} dq = \frac{1}{2} \frac{q^2}{C_{dl}^T} = \frac{1}{2} \frac{\left(C_{dl}^T V_{sc}\right)^2}{C_{dl}^T} = \frac{1}{2} C_{dl}^T V_{sc}^2,$$
(2.92)

where q is the total number of charge that is stored in the supercapacitor with a unit of $C \cdot cm^{-2}$, while C_{dl}^T is the double-layer capacitance of the supercapacitor with a unit of $F \cdot cm^{-2}$. Practically, specific energy density is more useful, which is a definition given by:

$$E_m = \frac{1}{2} \frac{C_m}{m} V_{sc}^2 = \frac{1}{2} C_{sp} V_{sc}^2,$$
(2.93)

$$E_m = \frac{1}{2} \frac{C_m}{M} V_{sc}^2 \cdot$$
(2.94)

E_m in Eq. (2.93) is based on the mass of the active material in the electrode layer, while that in Eq. (2.95) is calculated from the mass of the whole supercapacitor, both with a unit of $Wh \cdot kg^{-1}$.

As a supercapacitor is fully charged, the maximum voltage (V_{sc}^0) will be approached. Accordingly, maximum specific energy densities will be obtained as:

$$E_{m,\max} = \frac{1}{2} \frac{C_m}{m} \left(V_{sc}^0\right)^2 = \frac{1}{2} C_{sp} \left(V_{sc}^0\right)^2,$$
(2.95)

$$E_{m,\max} = \frac{1}{2} \frac{C_m}{M} \left(V_{sc}^0\right)^2 \cdot$$
(2.96)

It is well known that the usable voltage range in practice is limited by the linear voltage drop during the discharging process, which generates additional circuitry limitations. Due to the quadratic potential

22 Nanomaterials for Supercapacitors

drop, the stored energy is depleted by 75%, before the voltage reaches 50% of the usable range. Therefore, in order to use the last 25% energy that is stored in the supercapacitor, the voltage needs to be up-converted and regulated to a desired level, so that it is useful for efficient functioning of the circuit or load. As a result, the maximum usable energy of a supercapacitor is calculated within a voltage window from V_{sc}^0 to half of V_{sc}^0, thus leading to [7]:

$$E_{m,usable} = \frac{3}{8} \frac{C_m}{M} \left(V_{sc}^0 \right)^2.$$
(2.97)

For a given electrolyte solution, the effect of capacitance of the electrode material on specific energy density is described as:

$$\Delta E_{m,\max} = E_{m,\max} \frac{\Delta C_m}{C_m},$$
(2.98)

$$\Delta E_{M,\max} = E_{M,\max} \frac{\Delta C_m}{C_m}.$$
(2.99)

As an electrode material is selected, the effect of the electrolyte solution on specific energy density is given by:

$$\Delta E_{m,\max} = 2E_{m,\max} \frac{\Delta V_{sc}^0}{V_{sc}^0},$$
(2.100)

$$\Delta E_{M,\max} = 2E_{M,\max} \frac{\Delta V_{sc}^0}{V_{sc}^0}.$$
(2.101)

Power density

Power density is a measure to represent the rate to deliver the energy stored in the supercapacitor to an external load, which is defined as:

$$P_m = \frac{I_{cell} V_{cell}}{m}.$$
(2.102)

Since I_{cell} has a unit of $A \cdot cm^{-2}$, if m has a unit of $kg \cdot cm^{-2}$, the unit of P_m will be $W \cdot kg^{-1}$. As the supercapacitor is discharged at a constant current of I_{cell}, power density can be expressed as:

$$P_m = \frac{I_{cell}}{m} \left[-I_{cell} R_{ESR} + V_{sc}^0 - \left(V_{sc}^0 + I_{cell} R_p \right) \left(1 - \exp\left(-\frac{t}{R_p C_{dl}^T} \right) \right) \right].$$
(2.103)

This equation implies that the power density is decreased with increasing discharging time. In this case, the maximum specific power density is given by:

$$\frac{\partial P_m}{\partial I_{cell}} = \frac{1}{m} \left[-2I_{cell,\max} \left(R_{ESR} + R_p \left(1 - \exp\left(-\frac{t}{R_p C_{dl}^T} \right) \right) \right) + V_{sc}^0 \exp\left(-\frac{t}{R_p C_{dl}^T} \right) \right] = 0,$$
(2.104)

where $I_{cell,\max}$ is the maximum current density at the maximum power density, which is derived as:

$$I_{cell,max} = \frac{1}{2} \frac{V_{sc}^0 \exp\left(-\dfrac{t}{R_p C_{dl}^T}\right)}{R_{ESR} + R_p\left(1 - \exp\left(-\dfrac{t}{R_p C_{dl}^T}\right)\right)}. \tag{2.105}$$

Accordingly, the maximum voltage is given by:

$$V_{cell,max} = \frac{1}{2} V_{sc}^0 \exp\left(-\frac{t}{R_p C_{dl}^T}\right). \tag{2.106}$$

Therefore, the maximum specific power density can be expressed as:

$$P_{m,max} = \frac{1}{4m} \frac{\left(V_{sc}^0\right)^2 \left[\exp\left(-\dfrac{t}{R_p C_{dl}^T}\right)\right]^2}{R_{ESR} + R_p\left(1 - \exp\left(-\dfrac{t}{R_p C_{dl}^T}\right)\right)}. \tag{2.107}$$

In this case, as the discharge process is to be started, i.e., $t = 0$, the maximum power density is given by:

$$\left(P_{m,max}\right)^{t=0} = \frac{1}{4m} \frac{\left(V_{sc}^0\right)^2}{R_{ESR}}. \tag{2.108}$$

If $R_p \to \infty$, Eq. (2.108) is simplified to:

$$P_{m,max} = \frac{1}{4m} \frac{\left(V_{sc}^0\right)^2}{R_{ESR} + t/C_{dl}^T}. \tag{2.109}$$

Ragone plot

Obviously, the higher the energy and power densities, the higher the performances the supercapacitor would have. However, a high energy density does not always mean a high power density. The relationship between energy and power densities is usually described as the Ragone plot, with the specific power density to be presented versus the specific energy density [8]. To obtain Ragone plots, it is generally assumed that the external load of a supercapacitor cell is R_L, while there is no leakage current, i.e., $r \to \infty$.

Because of the presence of ESR, the specific energy density (E_L) available for the load is given by:

$$E_L = \frac{1}{2} C_{sp} \left(V_{sc}^0\right)^2 \frac{R_L}{R_L + R_{ESR}} = E_{m,max} \frac{R_L}{R_L + R_{ESR}}. \tag{2.110}$$

Accordingly, specific power density (P_L) available to the external load can be derived as:

$$P_L = \frac{V_L^2}{mR_L} = \frac{\left(V_{sc}^0\right)^2 R_L}{m\left(R_L + R_{ESR}\right)} = \frac{\left(V_{sc}^0\right)^2}{4mR_{ESR}} \frac{4R_L R_{ESR}}{\left(R_L + R_{ESR}\right)^2} = P_{m,max} \frac{4R_L R_{ESR}}{\left(R_L + R_{ESR}\right)^2}. \tag{2.111}$$

24 *Nanomaterials for Supercapacitors*

By combining Eqs. (2.110) and (2.111), the Ragone plot as the relationship between energy and power densities can be expressed as:

$$E_L = \frac{1}{2} E_{m,\max} \left(1 + \sqrt{1 - \frac{P_L}{P_{m,\max}}} \right). \tag{2.112}$$

Usually, the available specific energy is decreased with increasing load or decreasing load resistance (R_L), which results in an increase in specific power density, until the load resistance approaches of the level of ESR, i.e., $R_L = R_{ESR}$. Similarly, an increase in the load or decrease in load resistance also leads to a decrease in the power density. As a result, a trade-off is necessary when considering power and energy densities, in practical applications.

Electrochemical Pseudocapacitors

Overview

In double-layer supercapacitors, the capacitance is related to the accumulation and separation of net electrostatic charge at the electrode–electrolyte interface. In this case, the net negative charges (mainly electrons) are accumulated at the surface of the electrode, while positive charges with an equal number (mainly cations) are accumulated near surface of the electrode facing the electrolyte solution, forming electric double-layers such as the Helmholtz and diffuse layers [9]. Therefore, the level of the capacitance is thus determined by surface characteristics of the electrode materials. However, there is always a limitation in the magnitude of capacitance.

It has been found that there are some electrochemically active materials that could offer much high capacitance, which are known as pseudocapacitance. In pseudocapacitors, an electrochemical reduction–oxidation reaction, or redox reaction, takes place to facilitate faradic charge transfer in the electrode, which thermodynamically and kinetically are favorable [9]. In this case, the change in quantity of the charge due to the reaction (dq) is dependent on the change in potential (dV), while the ratio dq/dV is defined as the pseudocapacitance. In addition, the charging and discharging of a pseudocapacitor consist of two processes: (i) double-layer charging/discharging and (ii) electrochemical redox reaction.

Electrode–electrolyte interface

Electrochemical behaviors

If the redox materials and the reaction sites are uniformly distributed in the electrode layer and both the oxidant (O_X) and the reductant (R_d) are insoluble in the electrolyte solution, a redox process can be described as:

$$O_X + ne^- \Leftrightarrow R_d, \tag{2.113}$$

where n is the number of electrons that are transferred in the redox reaction. According to electrochemical thermodynamics, the reversible electrode potential induced by the reaction is represented by the Nernst equation [10].

$$E = E^0_{O_X/R_d} + \frac{RT}{nF} \ln \left(\frac{C_{O_s}}{C_{R_d}} \right), \tag{2.114}$$

where $E^0_{O_X/R_d}$ is the standard electrode potential at 25°C and 1.0 atm of the redox reaction, C_{O_X} and C_{R_d} are the concentrations of O_X and R_d in the electrode layer with a unit of mol·cm^{-3}, E is the electrode potential in V, R is the universal gas constant to 8.314 J·K^{-1}·mol^{-1} and T is the absolute temperature. It is assumed that the electrode layer initially contains only oxidant at a concentration of $C^0_{O_X}$, R_d will be produced, while the concentration of the oxidant is reduced to C_{O_X}, so that Eq. (2.114) becomes to be:

$$E = E^0_{O_X/R_d} + \frac{RT}{nF}\ln\left(\frac{C_{O_X}}{C^0_{O_X} - C_{O_X}}\right). \tag{2.115}$$

Accordingly, there is the following equation:

$$\frac{C_{O_X}}{C^0_{O_X} - C_{O_X}} = \exp\left[\frac{nF}{RT}\left(E - E^0_{O_X/R_d}\right)\right]. \tag{2.116}$$

Further rearranging yields:

$$C_{O_X} = \frac{C^0_{O_X}}{1 + \exp\left[\frac{nF}{RT}\left(E^0_{O_X/R_d} - E\right)\right]}. \tag{2.117}$$

This means that as the electrode potential E changes over time, the concentration of oxidant will be changed correspondingly, so that a current flow is generated through the electrode [11], with a current density to be given by:

$$i = \frac{nFAd}{A}\frac{dC_{O_X}}{dt} = nFd\frac{dC_{O_X}}{dt}, \tag{2.118}$$

where A is the geometric area of the electrode layer and d is thickness of the electrode layer. By substituting the differentiated Eq. (2.117) with respect to time to Eq. (2.118), the current density is obtained as:

$$i = \frac{n^2 F^2}{RT} dC^0_{O_X} \frac{\exp\left[\frac{nF}{RT}\left(E^0_{O_X/R_d} - E\right)\right]}{\left\{1 + \exp\left[\frac{nF}{RT}\left(E^0_{O_X/R_d} - E\right)\right]\right\}^2}\frac{dE}{dt}, \tag{2.119}$$

where dE/dt is known as the potential scan rate. According to the definition of capacitance, the electrode potential dependent pseudocapacitance induced by a redox reaction is thus given by:

$$C_{pc}(E) = \frac{idt}{dE} = \frac{i}{v} = \frac{n^2 F^2}{RT} dC^0_{O_X} \frac{\exp\left[\frac{nF}{RT}\left(E^0_{O_X/R_d} - E\right)\right]}{\left\{1 + \exp\left[\frac{nF}{RT}\left(E^0_{O_X/R_d} - E\right)\right]\right\}^2}. \tag{2.120}$$

In practice, the reaction redox sites are not uniform distributed, so that the redox reaction becomes quasi-reversible. In this case, it is necessary to introduce a factor, expressed as $-g(C_{O_X}/C^0_{O_X})$, standing for the lateral interaction energy [12]. Therefore, Eq. (2.116) becomes to be:

$$\frac{C_{O_X}}{C^0_{O_X} - C_{O_X}} = \exp\left[\frac{nF}{RT}\left(E - E^0_{O_X/R_d}\right) - g\frac{C_{O_X}}{C^0_{O_X}}\right]. \tag{2.121}$$

Similarly, Eq. (2.117) can be rewritten as:

$$C_{O_X} = \frac{C^0_{O_X}}{1 + \exp\left[\frac{nF}{RT}\left(E^0_{O_X/R_d} - E\right) + g\frac{C_{O_X}}{C^0_{O_X}}\right]}. \tag{2.122}$$

26 *Nanomaterials for Supercapacitors*

As a result, Eq. (2.120) becomes to be:

$$C_{pc}(E) = \frac{n^2 F^2}{RT} dC_{O_X}^0$$

$$\frac{\exp\left[\frac{nF}{RT}\left(E_{O_X/R_d}^0 - E\right) + g\frac{C_{O_X}}{C_{O_X}^0}\right]}{\left\{1 + \exp\left[\frac{nF}{RT}\left(E_{O_X/R_d}^0 - E\right) + g\frac{C_{O_X}}{C_{O_X}^0}\right]\right\}^2 + g\exp\left[\frac{nF}{RT}\left(E_{O_X/R_d}^0 - E\right) + g\frac{C_{O_X}}{C_{O_X}^0}\right]} \tag{2.123}$$

If the redox reaction is due to the intrinsic pseudocapacitance of the materials, an average value is generally used for it, which is given by:

$$C_{pc}(E) = \frac{AM_{mw}}{m} \sum_{j=1(E_{on})}^{j=n(E_{off})} C_{pc}(E_j)$$

$$= \frac{n^2 F^2}{RT} AM_{mw} dC_{O_X}^0$$

$$\sum_{j=1(E_{on})}^{j=n(E_{off})} \frac{\exp\left[\frac{nF}{RT}\left(E_{O_X/R_d}^0 - E_j\right) + g\frac{C_{O_X}}{C_{O_X}^0}\right]}{\left\{1 + \exp\left[\frac{nF}{RT}\left(E_{O_X/R_d}^0 - E_j\right) + g\frac{C_{O_X}}{C_{O_X}^0}\right]\right\}^2 + g\exp\left[\frac{nF}{RT}\left(E_{O_X/R_d}^0 - E_j\right) + g\frac{C_{O_X}}{C_{O_X}^0}\right]}, \tag{2.124}$$

where A is the electrode geometric area, d is thickness of the electrode layer, M_{mw} is molecule weight of the redox material and m is mass of the redox material in the whole electrode layer. E_{on} and E_{off} are the onset and offset potentials of the wave.

Usually, the difference between the onset and offset potentials is larger than the potential width at half peak ($\Delta E_{1/2}$) by about two times, i.e., $E_{off} - E_{on} \approx 2\Delta E_{1/2}$. When using linear scan cyclic voltammetry or cyclic voltammetry to measure the capacitance, Y-axis of the voltammogram curve is current density. By integrating the current peak in the potential range of $E_{off} - E_{on}$, charge quantity of the redox reaction (Q_{pc}) can be obtained, so that the intrinsic specific pseudocapacitance can be derived from the following equation:

$$C_{sp} = \frac{Q_{pc}}{m\left|E_{off} - E_{on}\right|} = \frac{Q_{pc}}{2m\left(\Delta E_{1/2}\right)}. \tag{2.125}$$

As a supercapacitor consists of electrode with both double-layer and pseudocapacitive materials, its total capacitance will be enlarged. Ideally, the total capacitance (C^T) of the electrode is summation of the double-layer capacitance (C_{dl}) and the pseudocapacitance (C_{pc}), expressed as:

$$C^T = C_{dl} + C_{pc}(E)$$

$$= C_{dl} + \frac{n^2 F^2}{RT} dC_{O_X}^0$$

$$\frac{\exp\left[\frac{nF}{RT}\left(E_{O_X/R_d}^0 - E\right) + g\frac{C_{O_X}}{C_{O_X}^0}\right]}{\left\{1 + \exp\left[\frac{nF}{RT}\left(E_{O_X/R_d}^0 - E\right) + g\frac{C_{O_X}}{C_{O_X}^0}\right]\right\}^2 + g\exp\left[\frac{nF}{RT}\left(E_{O_X/R_d}^0 - E\right) + g\frac{C_{O_X}}{C_{O_X}^0}\right]}. \tag{2.126}$$

Summary

In conventional electrical circuits, capacitors are essential elements. The principles and fundamental aspects of capacitors have been the bases to understand electrochemical supercapacitors. The fundamentals of electrochemical double-layer supercapacitors have been considered as the foundation of supercapacitor science and technology. For instance, the fundamental principles of supercapacitors, i.e., electrode–electrolyte double-layer theory resulted in the establishment of double-layer models and differential capacitances, which have been used to design and fabrication of supercapacitors. Calculation and relation of energy and power densities have been discussed, which are important for practical applications of the energy storage devices. The charging and discharging processes of supercapacitors could be well described mathematically by taking into account Equivalent Series Resistance (ESR) and faradic leakage resistance. All these aspects will always play significant roles in design and fabrication of supercapacitors.

Keywords: Supercapacitor, Capacitor, Electrode, Parallel-plate capacitor, Dielectric materials, Dielectric constant, Dielectric loss, Dielectric polarization, Charge, Discharge, Electric field, Electric potential, Capacitance, Energy storage, Electrochemical double-layer (ECDL), Double-layer potential, Electrolyte, Potential difference, Gouy-Chapman-Stern (GCS) model, Differential capacitance, Equivalent Series Resistance (ESR), Equivalent circuit, Energy density, Power density, Ragone plot

References

[1]   Hickey HV, Villines WM. Elements of Electronics: New York: McGraw Hill; 1970.
[2]   Halliday D, Resnick R, Walker J. Fundamentals of Physics, 8th Ed. New York: John Wiley & Sons; 2008.
[3]   von Hippel AR. Dielectrics and Waves: Cambridge, MA: MIT Press; 1954.
[4]   Ballou G. The Electrical Engineering Handbook: Boca Raton: CRC Press; 1993.
[5]   Bockris JO, Conway BE, Yeager E. Comprehensive Treatise of Electrochemistry. New York: Plenum Press; 1980.
[6]   Ban S, Zhang JJ, Zhang L, Tsay K, Song DT, Zou XF. Charging and discharging electrochemical supercapacitors in the presence of both parallel leakage process and electrochemical decomposition of solvent. Electrochimica Acta. 2013; 90: 542–9.
[7]   Linden D, Reddy T. Handbook of Batteries, 4th Ed. New York: McGraw Hill; 2010.
[8]   Ragone DV. Review of battery systems for electrically powered vehicles. SAE Transactions. 1968; 77: 131–9.
[9]   Conway BE, Birss V, Wojtowicz J. The role and utilization of pseudocapacitance for energy storage by supercapacitors. Journal of Power Sources. 1997; 66: 1–14.
[10] Bard AJ, Faulkner LR. Electrochemical Methods, Fundamentals, and Applications. New York: John Wiley & Sons; 1980.
[11] Hubbard AT, Anson FC. Linear potential sweep voltammetry in thin layers of solution. Analytical Chemistry. 1966; 38: 58–61.
[12] Gileadi E, Conway BE. Modern Aspects of Electrochemistry. London: Butterworth; 1965.

Carbon Based Supercapacitors

Ling Bing Kong,[1,*] *Wenxiu Que,*[2] *Lang Liu,*[3] *Freddy Yin Chiang Boey,*[1,a] *Zhichuan J. Xu,*[1,b] *Kun Zhou,*[4] *Sean Li,*[5] *Tianshu Zhang*[6] *and Chuanhu Wang*[7]

Overview

Initially, carbon materials, especially carbon blacks [1–3], were the main electrode materials of supercapacitors, with the discovery of new types of carbon materials, such as carbon nanotubes (CNTs) and graphene [2, 4–6], the electrode materials of supercapacitors have been largely widened. This chapter serves to provide an overview on the progress in development of carbon based electrode materials of supercapacitors, including carbon blacks, CNTs, graphene oxide and graphene, as well as their combinations.

Nanocarbons–Carbon Blacks (CAs)

Due to their cost-effectiveness, high chemical and thermal stability, stable reproducibility and tunable microstructures, nanocarbons, such as carbon blacks and activated carbons, have been extensively applied to supercapacitors with large scale production [1]. Carbon black, also known as acetylene black, channel black, furnace black, lamp black and thermal black, is usually produced through incomplete combustion of heavy petroleum products, such as FCC tar, coal tar, ethylene cracking tar and vegetable oils. Carbon

---

[1] School of Materials Science and Engineering, Nanyang Technological University, 50 Nanyang Avenue, Singapore 639798.
[a] Email: mycboey@ntu.edu.sg
[b] Email: xuzc@ntu.edu.sg
[2] Electronic Materials Research Laboratory and International Center for Dielectric Research, Xi'an Jiaotong University, Xi'an 710049, Shaanxi Province, People's Republic of China.
  Email: wxque@xjtu.edu.cn
[3] School of Chemistry and Chemical Engineering, Xinjiang University, Urumqi 830046, Xinjiang, People's Republic of China.
  Email: llyhs1973@sina.com
[4] School of Mechanical and Aerospace Engineering, Nanyang Technological University, 50 Nanyang Avenue, Singapore 639798.
  Email: kzhou@ntu.edu.sg
[5] School of Materials Science and Engineering, The University of New South Wales, Australia.
  Email: sean.li@unsw.edu.au
[6] Anhui Target Advanced Ceramics Technology Co. Ltd., Hefei, Anhui, People's Republic of China.
  Email: 13335516617@163.com
[7] Department of Material and Chemical Engineering, Bengbu Univresity, Bengbu, Peoples's Republic of China.
  Email: bbxywch@126.com
* Corresponding author: elbkong@ntu.edu.sg

black has a para-crystalline structure, with a high specific area. Activated carbon could have even higher specific surface area of $> 1000 \text{ m}^2 \cdot \text{g}^{-1}$, as well as very large pore volume of $> 0.5 \text{ cm}^3 \cdot \text{g}^{-1}$. The conventional activated carbons exhibited pores with a wide size distribution (PSD), ranging from micropores (< 2 nm) to macropores (> 50 nm). With the advancement in activation procedures, together with more available precursors, the microstructures of activated carbons could be well controlled, with presence of mesopores (2–50 nm). Due to their more versatility in processing and characterization as compared with carbon blacks, only activated carbons will be discussed in a detailed way here.

Processing and characterization of Activated Carbons (ACs)

Activated carbons can be prepared by using either physical or chemical methods. Physical methods have been conducted in various oxidizing gases, including air, O_2, CO_2, steam or their mixtures, while chemical activation uses chemicals, like KOH, NaOH, H_3PO_4 and $ZnCl_2$, and so on, as the activating agents. The physical activation usually involves two steps, i.e., (i) pyrolysis of precursors at temperatures of 400–900°C in inert atmosphere (e.g., N_2) and (ii) partial gasification with the presence of oxidizing gases at 350–1000°C to increase porosity and specific surface area. Chemical activation is a one-step process, in which the carbon precursors are treated by using an appropriate activating agent and then pyrolyzed at temperatures over 450–900°C. Furthermore, combination of the physical and chemical processes has been widely used to produce activated carbons with higher performances.

Similar to carbon blacks, activated carbons are also generated mainly from organic precursors. More recently, various biomasses, such as wood, sawdust, peat, coconut shells, fruit bones or rice husk, are being used to prepare activated carbons [7–18]. Meanwhile, coal, low temperature lignite coke, charcoal and biochar are used as carbonization precursors [19–21]. With the presence of newly developed nanocarbons, i.e., nanotubes, nanofibers and graphenes, activated carbons have been incorporated to obtain higher and more properties [22–26].

Physical methods

During physical activation processes, precursors are pyrolyzed in an inert atmosphere at 400–900°C to remove the volatile components and then partially gasified in oxidizing gases at 350–1000°C. The pyrolysis off-products trapped within the pores are burnt out, because of the active oxygen in the activating agents burns, so that some of the closed pores are opened. With further oxidation, the porosity is increased and microporous structure is gradually formed, during which CO and CO_2 are released, depending on the nature of the gases used and the activation temperature adopted. The most commonly used activating agents include CO_2, air and steam, owing to their wide availability and cost-effectiveness. Properties, such as porosity, surface chemistry and surface area, of the activated carbons prepared in this way are closely related to the properties of the precursors, characteristics of the oxidizing agents, reaction temperature and the degree of activation. Generally, the higher the activation temperature and the longer the activation time, the larger the porosity will be developed in the activated carbons. At the same time, a high porosity also leads to a wide Pore Size Distribution (PSD). Therefore, a tradeoff is usually adopted, when there is a specific requirement for a given application.

Compared with steam and CO_2, air (oxygen) has been more likely used as the oxidizing agent of physical activation of carbons, because of its high reactivity [27, 28]. However, it is because of the high reactivity, together with a strong exothermic effect, the reaction of carbon with oxygen is inevitable, when using air (oxygen) as the activating agent, resulting in a reduction in the production yield. As a result, controlling the reaction has been a challenge when using the physical activation process to produce activated carbons.

A comparison study has been conducted on the variation in pore structures of the physically activated carbons from several coal chars during the gasification in air and CO_2 [29]. After pyrolysis at 900–1150°C, gasification in air was carried out at 380°C, while that in CO_2 gasification was performed at 800°C. It was found that in air gasification, the surface area and pore volume initially increase rapidly and then slowly with carbon conversion, as shown in Fig. 3.1.

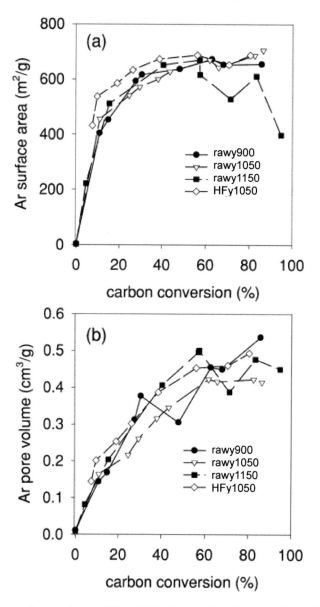

Fig. 3.1. Variations in pore surface area (pores < 250 nm) (a) and pore volume (pores < 250 nm) (b), as a function carbon conversion, for the Yarrabee coal chars gasified in air at 653 K, derived from Ar adsorption isotherms obtained at 87 K. Reproduced with permission from [29], Copyright © 2003, Elsevier.

Three peaks were observed on the distribution curves of pore surface area and pore volume, corresponding to pore size ranges of < 1 nm, 1–2 nm and 2–5 nm, respectively. Variations in surface area and pore volume as a function of carbon conversion, for Yarrabee coal chars gasified in air at 653 K, determined from Ar adsorption isotherms obtained at 87 K are illustrated in Figs. 3.2 and 3.3, respectively. Both the surface area and volume of the small micropores with pore diameters of < 1 nm were almost unchanged as a function of carbon conversion, when the conversion was proceeded to a certain level, while those in the ranges of 1–2 nm and 2–5 nm were increased with increasing conversion.

In comparison, with CO_2 gasification, both the surface area and volume of small micropores were increased significantly as the gasification proceeded. At the same time, those for the other pore size ranges

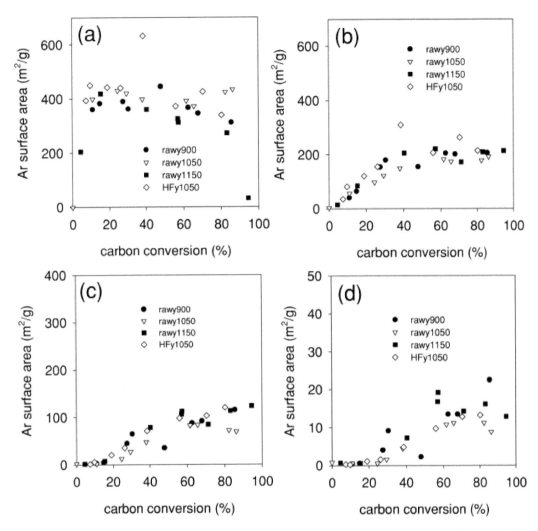

Fig. 3.2. Variations in surface area as a function of carbon conversion, for Yarrabee coal chars gasified in air at 653 K, determined from argon adsorption isotherms obtained at 87 K, over various pore size ranges: (a) 0–1 nm, (b) 1–2 nm, (c) 2–5 nm and (d) 5–250 nm. Reproduced with permission from [29], Copyright © 2003, Elsevier.

were also increased with increasing conversion. Carbon materials with surface area of up to 700 m²·g⁻¹ and pore volume of up to 0.55 cm³·g⁻¹ were achieved by using air gasification, while the values were up to 500 m²·g⁻¹ and 0.22 cm³·g⁻¹, respectively, when using CO_2 gasification.

It was observed that although the gasification rate in air could not be normalized with the total surface area, the experimental data could be fitted by using the random pore model. In air gasification, the fitted structural parameter matched the measured data for the partially gasified char without closed pores, whereas the fitted structural parameters were not matched with the measured results for other chars when the gasification was conducted in CO_2. The observation was attributed to the fact that the closed pores were not fully opened, since CO_2 is less reactive than O_2 in removing heavy molecules and functional groups from the chars. This hypothesis could also be used to explain the gradual increase in the surface area and pore volume of the small micropores during the CO_2 gasification, as compared with the rapid increase in the beginning of the air gasification. Nevertheless, further clarification is needed with more experiments.

It is believed that chemically adsorbed O_2 are able to migrate along the micropores so as to approach the active sites for reaction [30]. In other words, there was a process consisting of chemisorption within

32 *Nanomaterials for Supercapacitors*

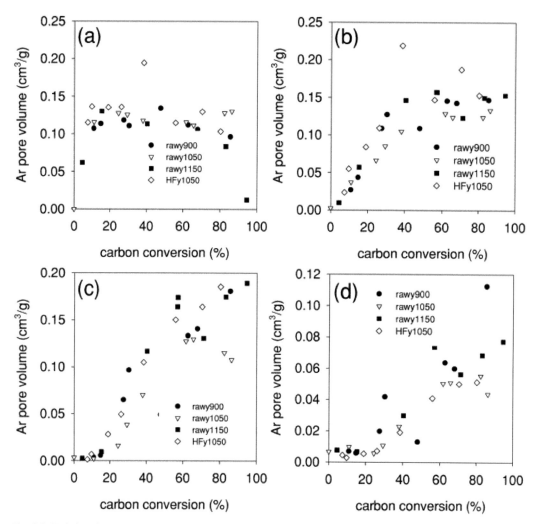

Fig. 3.3. Variations in pore volume as a function of carbon conversion, for Yarrabee coal chars gasified in air at 653 K, determined from argon adsorption isotherms obtained at 87 K, over various pore size ranges: (a) 0–1 nm, (b) 1–2 nm, (c) 2–5 nm and (d) 5–250 nm. Reproduced with permission from [29], Copyright © 2003, Elsevier.

micropores, migration of the activated oxygen towards the reactive sites and reaction at the sites, whereas no such pathway was available when using CO_2 gasification. A similar phenomenon was observed for a bituminous coal, which was ascribed to the difference in degree of penetration of the micropores by the two reactants, thus leading to different growth behaviors of the pores during the carbon burn-off [31, 32].

A similar method was reported to prepare activated carbons from almond tree pruning, with a surface area up to 560 $m^2 \cdot g^{-1}$ and a pore volume below 0.31 $cm^3 \cdot g^{-1}$ by using air gasification at 190–260°C [33]. Figure 3.4 shows a schematic of the experimental setup used to prepare the activated carbons. A stainless steel cylinder reactor was coupled into a tubular oven, equipped with a heating and temperature control system. A gas inlet system, a tar condensation and recovery system, together with a gas sample collector, were complemented in the setup. Both the specific surface area and pore volume were increased with increasing gasification temperature and time.

Carbides, such as TiC, have been used as the precursor to develop activated carbons, known as carbide-derived carbons (TiC-CDCs), with gasification in both air and CO_2 [34]. TiC was chlorinated at 600°C to form 600°C-TiC-CDC [35], while further chlorination was carried out at 1000°C and 1300°C

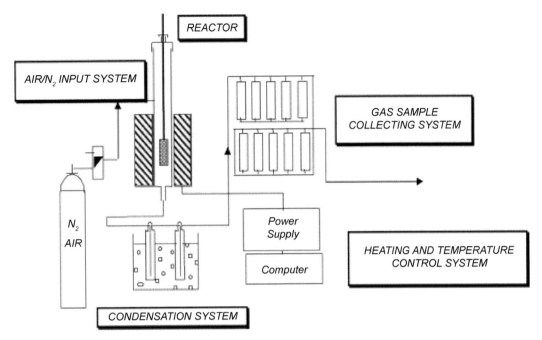

Fig. 3.4. Schematic of the experimental set-up. Reproduced with permission from [33], Copyright © 2003, Elsevier.

[36], thus forming samples of 600°C-TiC-CDC, 1000°C-TiC-CDC and 1300°C-TiC-CDC. At the desired temperatures, Cl_2 was introduced at a flow of 15 ml·min^{-1} to extract Ti through the reaction: TiC + 2Cl$_2$ (g) → TiCl$_4$ (g) + C (s).

Thermal analysis results indicated that oxidation of TiC-CDC in air was started above 350°C and accelerated at temperatures at 400°C and 450°C for the 600°C-TiC-CDC and 1000°C-TiC-CDC, respectively. Comparatively, the 1300°C-TiC-CDC required much higher oxidation temperatures of > 500°C, due to its lower reactivity. In addition, higher reaction temperature resulted in a large average pore size and a broader pore size distribution. Therefore, the 600°C-TiC-CDC and 1000°C-TiC-CDC samples were activated at an initial temperature of 430°C, so as to well control over the porosity. Figure 3.5 shows variations in porosity of the 1000°C-TiC-CDC sample after activation at 430°C in air for different time durations. BET Specific Surface Area (SSA) was increased with increasing activation time, as seen in Fig. 3.5(a). After activation for 6.5 hours, a maximum value of 1800 m^2·g^{-1} was achieved. As shown in Fig. 3.5(b), total pore volume was increased from 0.7 to 0.78, 0.84 and 0.87, corresponding to increment by 11.4, 20 and 24%, after activation for 3, 6.5 and 10 hours, respectively. Among them, volume of the micropore (< 1 nm) was increased from 0.34 to 0.36 cm^3·g^{-1} (+6%) after activation for 3 hours and stabilized at this value for longer activation times. The porosity was formed through a two-step process, (i) formation of new micropores at moderate conditions and (ii) widening of the micropores into larger ones, thus resulting in an increase in the total pore volume [37, 38].

The variations in BET-SSA and burn-off after activation for 3 and 6.5 hours at different temperatures are demonstrated in Fig. 3.5(c). The BET-SSA was increased with burn-off (weight loss) at low activation temperatures, reached the maximum value of 1800 m^2·g^{-1} after 6.5 h and 3 hours at 425°C and 475°C, respectively. After that, no further increase was observed. The corresponding variations in pore volume are illustrated in Fig. 3.5(d). The total pore volume was significantly increased with increasing activation temperature up to 55%, whereas the micropore volume had only a very slight increase of about 6% at low temperature. Moreover, the micropore was decreased at higher activation temperatures, due to the increase in size of the micropores, i.e., the formation of mesopores. In summary, activation in air at low temperatures (e.g., 430°C) for long activation times of > 6.5 hours could be used to large micropore

34 *Nanomaterials for Supercapacitors*

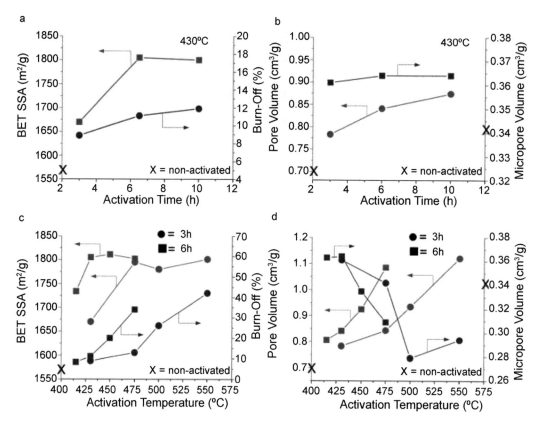

Fig. 3.5. Variations in SSA (a) and pore volume (b) of the 1000°C-TiC-CDC sample after activation at 430°C in air for different times. Similar trends were found for SSA (c) and pore volume (d) after activation for 3 hours (circles) and 6 hours (squares) between 400 and 550°C. Reproduced with permission from [34], Copyright © 2009, Elsevier.

volume and BET-SSA, with a weight loss of only 10%. The total pore volume was increased by 60% with burn-off. The optimum activation conditions in terms to high BET-SSA and low weight loss were 6.5 h at 430°C and 3 hours at 475°C.

Figure 3.6 shows pore-size distributions of the 1000°C-TiC-CDC sample after activation in air. The jaggedness in the PSDs was due to the numerical model rather than the real features of the pore structure. For 1000°C-TiC-CDC, oxidation at 430°C for 3 hours resulted in a small increase in the volume of pores that were smaller than 0.8 nm, as observe in Fig. 3.6(a). With increasing oxidation time, the fraction of small pores was decreased gradually, thus leading to a shift in the average pore-size from 1.10 of the non-activated sample to 1.20 and 1.25 nm, after activation at 430°C for 3 and 6 hours, respectively. For a given activation time, an increase in activation temperature resulted in a decrease in the pore volume for the pore sizes of < 0.8 nm, as revealed in Fig. 3.6(b), confirming formation of large pores (> 2 nm). The average pore size was increased from 1.10 for the non-activated sample to 1.22, 1.50 and 1.90 nm, after activation for 6.5 hours at 415, 450 and 475°C, respectively.

In order to further increase the micropore volume, the 600°C-TiC-CDC sample was activated by using CO_2. For the 600°C-TiC-CDC sample, the formation of graphitic carbon was minimized, due to the relative low chlorination temperature. Figure 3.7 shows variations in BET-SSA and porosity of the 600°C-TiC-CDC sample that was activated at different conditions. For a given activation time (e.g., 2 hours), the BET SSA was proportional to the value of burn-off and increased with increasing temperature from about 1300 $m^2 \cdot g^{-1}$ for the non-activated sample to > 3000 $m^2 \cdot g^{-1}$ after activation at 950°C, as depicted

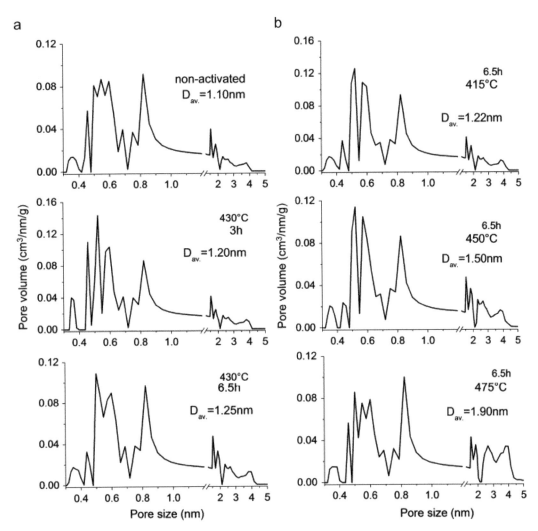

Fig. 3.6. Pore-size distribution and average pore diameter (D_{av}) of the 1000°C-TiC-CDC sample after activation in air for 3 and 6.5 hours at 430°C (a) and 6.5 hours at 415, 450 and 475°C (b). Reproduced with permission from [34], Copyright © 2009, Elsevier.

in Fig. 3.7(a). At activation at 875°C, the total pore volume (+28%) and micropore volume were increased by 28 and 18%, respectively, as compared with those of the non-activated 600°C-TiC-CDC sample, as demonstrated in Fig. 3.7(b). Meanwhile, the total pore volume was increased in a similar way with increasing activation temperature, having an increment by 150% at 950°C, whereas the micropore volume was decreased by about 8%, so that it became to be lower than that of the non-activated 600°C-TiC-CDC sample, as observed in Fig. 3.7(b).

Burn-off and BET SSA of the 600°C-TiC-CDC sample activated at 875 and 950°C for different activation time durations are shown in Fig. 3.7(c). Both were increased monotonically with activation time. After activation at 850°C for 8 hours, BET-SSA and burn-off reached their maximum values of 2700 $m^2 \cdot g^{-1}$ and 55 wt%, respectively, while these values were increased to 3100 $m^2 \cdot g^{-1}$ and 75 wt% if the activation was conducted at 950°C. Figure 3.7(d) shows the corresponding variations in the pore volume. The total pore volume had a similar variation trend of BET-SSA, increasing with activation time and reaching maximum values of 1.14 $cm^3 \cdot g^{-1}$ (12 hours at 875°C) and 1.34 $cm^3 \cdot g^{-1}$ (2 hours at 950°C). The micropore volume

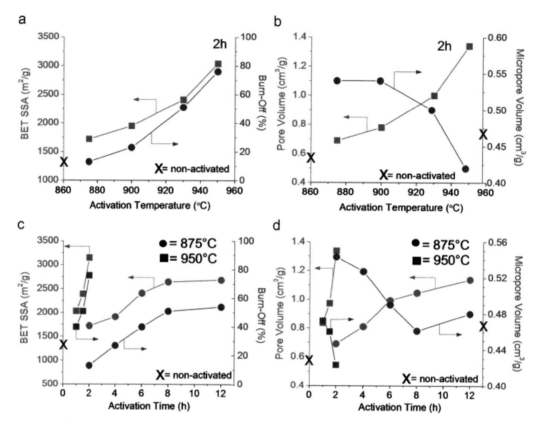

Fig. 3.7. BET-SSA and pore volume of the 600°C-TiC-CDC sample after oxidation in CO$_2$ as a function of activation temperature (a, b) and activation time (c, d). Reproduced with permission from [34], Copyright © 2009, Elsevier.

was then decreased with further increasing activation time, which was similar to that observed in the case of activation in air. Therefore, the two-step process of pore formation was also present. The optimum conditions for activation of the 600°C-TiC-CDC sample in CO$_2$, in terms of high BET-SSA, presence of micropores and low weight loss, were low temperature of 875°C and long activation time (> 8 hours). After oxidation at 875°C for 4 and 12 hours, the average pore size was increased from 0.73 to 0.96 and 1.51 nm, respectively. At a given activation temperature, the small pores were enlarged with increasing activation time when being oxidized in CO$_2$, so that the number of micropores was decreased while that of the mesopores were increased. After activation at 875, 925 and 950°C for 2 hours, the average pore sizes were 0.80, 1.35 and 1.52 nm, respectively, as compared with the initial 0.73 nm.

As stated earlier, steam and CO$_2$ are more effective than air as activating agents to prepare activated carbons, while it is well recognized that steam is more reactive than CO$_2$. Therefore, even lower temperatures are required to obtain activated carbon materials with similar performances [22, 37, 39]. A general trend is that all the surface area, pore volume and average pore diameter of the final activated carbons are increased with increasing degree of burn-off [29, 33, 34, 39–47]. However, it is also somehow contradictory regarding the evolution of pores. Studies have indicated activated carbons produced by using steam activation had a narrower pore size distribution and more pronounced micropore structure than those prepared by using activation agent of CO$_2$ [39, 41–43], while there have been also reports demonstrating that steam-activated carbons possessed relatively low volume of micropores but large external surface areas [40, 44–47]. Currently, activated carbons with surface areas of up to 1800 m^2·g^{-1} and pore volumes of up to 0.9 cm^3·g^{-1} could be readily achieved by using steam as the activating agent [22, 23, 26–28, 33, 34, 37, 39–47].

Activated carbons have been derived from scrap tyres through gasification by using steam and CO_2 under different experimental conditions (temperature and time) [39]. A cylindrical stainless steel atmospheric pressure reactor was used, with a ceramic furnace and power source heating system, gas-feed inlets and accessories to collect the liquid and gas samples. A heat exchanger at the top was used to keep the samples at room temperature before each run, with a suspension system to lower the basket containing the samples. For each run, 20 g scrap tyre was used, while the basket was placed in the cooling zone of the reactor, after N_2 flow at a rate of 100 $cm^3 \cdot min^{-1}$ was circulated for 1 hour to purge the air from the system. The heating system was turned on as the temperature reached 800°C, while the basket was lowered into the heating zone and kept at the temperature for 1 hour. After that, the heating system was turned off, so that the sample was cooled down to room temperature. About 43% carbon could be obtained from the tyre feedstock. A schematic diagram of the experimental setup is shown in Fig. 3.8.

Figure 3.9 shows carbon burn-off results as a function of activation time at different temperatures for the steam and carbon dioxide activation processes [39]. At a given temperature, the burn-off was always linearly proportional to the activation time. All the lines passed through the origin when extrapolated to the y-axis, which implied that the activation of the tyre char was conducted at a constant rate, during reaction between the graphitic carbonaceous structure of the char with the activation agent. This was because the carbonization temperature was sufficiently high (800°C), so that all volatiles in the raw material had been removed. Activation energies for carbon dioxide and steam procedures were 135 and 131 $kJ \cdot mol^{-1}$, respectively, suggesting that the gasification was chemically rather than mass-transfer controlled. It was found that the burn-off achieved by using carbon dioxide was averagely 69.5% of that obtained by using steam, suggesting that steam was more efficient than carbon dioxide in activating the tyre char. The lower reactivity of the carbon dioxide was attributed to lower diffusion rate due to the larger molecular size of carbon dioxide.

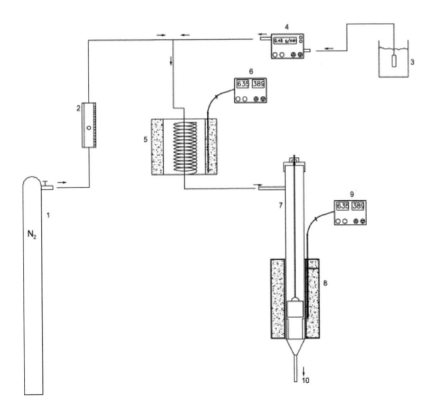

Fig. 3.8. Experimental set-up for carbon activation process: (1) N_2, (2) gas feeding system, (3) distilled water recipient, (4) peristaltic bomb, (5) evaporator, (6) temperature control and calefaction system, (7) reactor, (8) electric furnace, (9) temperature control and calefaction system and (10) gases out. Reproduced with permission from [39], Copyright © 2006, Elsevier.

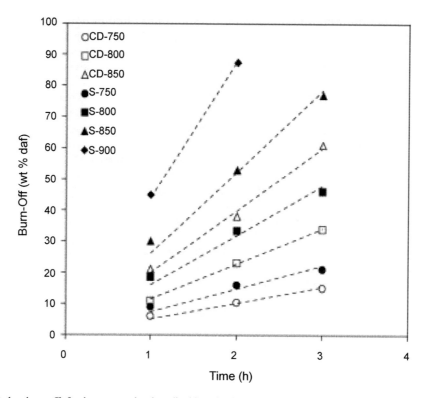

Fig. 3.9. Carbon burn-offs for the steam and carbon dioxide activation processes, as a function of activation time, at different temperatures. Reproduced with permission from [39], Copyright © 2006, Elsevier.

Chemical methods

In chemical activation processes, the mixtures of carbon precursor and activating agent were thermally treated at temperatures in the range of 450–900°C. As compared with physical activation, chemical activation has several advantages, including (i) one-step processing, (ii) lower pyrolysis temperatures, (iii) much higher carbon yield, (iv) very high specific surface area (3600 $m^2 \cdot g^{-1}$) and (v) controllable microporosity. The last two advantages are specifically important for the applications in supercapacitors, due to the requirements of large surface areas and microporosity suitable for the electrolyte ion diffusion [48–52].

Various reagents have been explored and evaluated for chemical activation to prepare activated carbons, such as $ZnCl_2$, H_3PO_4, $AlCl_3$, $MgCl_2$, KOH, NaOH and so on, among which $ZnCl_2$, H_3PO_4 and KOH are the most widely used up to now. $ZnCl_2$ and H_3PO_4 serve as dehydrating agents, while KOH acts as an oxidant. For example, if H_3PO_4 was used, the dehydration could be promoted, so that a lower activation temperature was required, as compared with the cases of normal thermal treatment [53]. Consequently, the formation of CO and CO_2 was started at a lower temperature. Dehydration of the carbon precursor during the heat treatment was triggered by the activating agent that was inside the precursor particles. At this dehydrating stage, cross-linking reactions, such as cyclization and condensation processes, could play an important role. Due to the dehydration, the size of the particles of the carbon precursors was decreased. At the same time, because the activating agent was inside the particles during the thermal treatment, it acted as a template to produce microporosity. Moreover, H_3PO_4 could react with organic species to form phosphate and polyphosphate bridges, which linked the biopolymer fragments together, thus preventing the materials from contracting as the temperature is increased.

A comparison study on the activation behaviors of H_3PO_4 and $ZnCl_2$ has been reported [54]. It was found that heterogeneity in microporosity occurred during the activation by using H_3PO_4, which was not related to the content of H_3PO_4, but instead, it was attributed to the presence of the mixture of phosphate molecules during the activation process, such as H_3PO_4, $H_4P_2O_5$, $H_{13}P_{11}O_{34}$, because they had different sizes. In comparison, due to its small size and without the presence of any other derivatives, $ZnCl_2$ resulted small micropores with uniform size distribution.

Granular activated carbons with a particle size of about 2 mm were obtained through chemical activation of peach stones by using $ZnCl_2$ as activation agent [54]. The degree of activation characterized by the amount of Zn (X_{Zn}) that was incorporated into the precursor, ranging from 0.20–1.2 g Zn/g of precursor. The carbonization was performed in N_2 flow at 500°C. Figure 3.10 shows porosity and yield of the activation process with $ZnCl_2$. The points at $X_{Zn} = 0$ were derived from the samples without impregnation, at which the porosity was very low because the carbonization was still incomplete. A small content of $ZnCl_2$ could facilitate the formation of pores, especially the micropores. At $X_{Zn} > 0.4$, porosity reached a pretty high level, after which the increase in porosity was slow down. However, the pore size distribution was narrowed. Although the increase in porosity was continued until $X_{Zn} = 1$, corresponding to an impregnation ratio (weight ratio of $ZnCl_2$/precursor) of 2/1, most of them were mesopores. This observation was confirmed by the evolution of the yield, as shown in Fig. 3.10. The formation of micropores is usually accompanied by an increase in yield, while the yield is decreased when mesopores begin to appear. This is the reason why the yield is increased along with the microporosity at low values of X_{Zn}.

Figure 3.11 shows isotherms of two granular carbon samples activated with $ZnCl_2$ in N_2 at 77 K. According to the shape of the isotherm of the sample after a low degree of activation ($X_{Zn} = 0.23$ g/g), only micropores were present, with a relatively uniform pore size distribution. Almost no mesopores were observed. As for the isotherm the sample with $X_{Zn} = 0.96$ g/g, mesopores started to be present, because a plateau at high relative pressures was present. The increase in porosity with increasing X_{Zn} implied that the formation of the pores was attributed to the presence of spaces left by $ZnCl_2$ after it was washed out. Especially for the samples with $X_{Zn} < 0.4$, the total volume of micropores and mesopores was almost the same as that of $ZnCl_2$ used in the reactions. In other words, the activation agent acted as template of the pores, as mentioned earlier. For those with $X_{Zn} > 0.4$, whereas the increase in microporosity slowed down and the increase in mesoporosity became dominant, the total volume of the pores was smaller than that of the activation agent. This observation was ascribed to the fact that the carbon particles reacted with the activation agent, as confirmed by the decrease in yield observed in the samples with $X_{Zn} > 0.4$.

The effect of carbonization on pore evolution of the impregnated precursor of over 500–800°C has been studied. It was revealed that the samples experienced a slight weight loss and particle shrinkage after complete carbonization. At the same time, there was a decrease in microporosity with increasing reaction temperature, especially at the low impregnation ratios. The results indicated that the effect of $ZnCl_2$ was most pronounced during the thermal degradation of the impregnated samples below 500°C, while at higher temperatures its reaction with the char was largely slowed down.

The degree of activation, given by the content of phosphorous incorporated into the particles, X_p, was in the range of 0.09–0.91g P/g precursor, with the activation to be conducted in N_2 flow at 450°C. Figure 3.12 shows relationships porosity as a function of impregnation ratio. A rapid formation of microporosity was observed at low concentrations, with predominant porosity up to $X_p = 0.3$, along with pore volumes up to 0.6 $cm^3 \cdot g^{-1}$. At higher concentrations, almost no micropores were developed, while mesopores were present instead. A similar trend in the variation of yield is well coupling with the development of microporosity/mesoporosity. The formation of micropores in the carbon samples with low values of X_p was evidenced by the shape of the adsorption isotherms. Gradually, the micropore size distributed was widened, followed by the formation of mesopores. For instance, the carbon sample activated with $X_p = 0.34$ g/g consisted of microporous with a relatively wide pore size distribution, while the one activated with $X_p = 0.91$ exhibited an even wider size distribution.

The volume occupied by the 'phosphoric acid' inside the precursor particles was estimated, as the decomposition product at 450°C was 82% P_4O_{10}–18% H_2O having a density of 2.06 $g \cdot cm^{-3}$. The volume obtained was in a good agreement with V_{mi} (n-butane), up to 0.4 $cm^3 \cdot g^{-1}$. This observation suggested that there was a wide range of concentrations, at which the shrinkage of the precursor particles during the heat treatment was not pronounced. For carbon samples with higher impregnation ratios, e.g., $X_p = 0.63$ and 0.91 g/g, the volume of the pores was larger than the volume of the activation agent. This was because

40 *Nanomaterials for Supercapacitors*

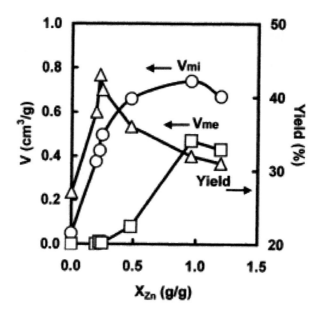

Fig. 3.10. Volume of micropores (V_{mi}), volume of mesopores (V_{me}) and yield of the carbon samples as a function of the degree of activation X_{Zn} (i.e., ratio of g Zn/g precursor). Reproduced with permission from [54], Copyright © 2004, Elsevier.

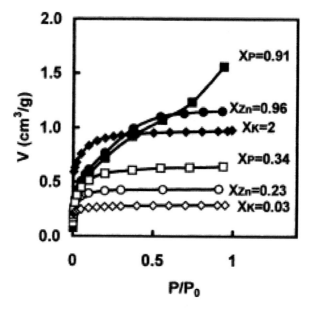

Fig. 3.11. Adsorption isotherms for N_2 at 77 K of representative carbons chemically activated with $ZnCl_2$ (X_{Zn}), KH_2PO_4 (X_P) and KOH (X_K). Reproduced with permission from [54], Copyright © 2004, Elsevier.

the intense chemical reaction between the highly concentrated H_3PO_4 and precursor particles led to the formation of mesopores and even macropores.

It was illustrated that KOH and NaOH had no dehydrating effect, while they acted only as oxidants [55–57]. There was a redox reaction, given by $6KOH + 2C \rightarrow 2K + 3H_2 + 2K_2CO_3$. The carbon was

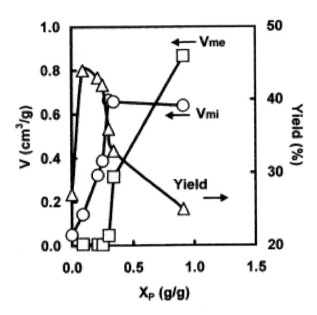

Fig. 3.12. Volume of micropores (V_{mi}), volume of mesopores (V_{me}) and yield of the carbon samples as a function of the degree of activation X_p (i.e., g P/g precursor). Reproduced with permission from [54], Copyright © 2004, Elsevier.

oxidized into carbonate ions and potassium compounds were formed due to the intercalation, both of which could be washed out during the consequent procedure. As a result, carbon framework with the presence of pores was created. In addition, the formation of CO_2 owing to the decomposition of K_2CO_3 at temperatures of > 700°C would further increase the porosity through carbon gasification [58]. Moreover, it has been demonstrated that a precursor with high reactivity corresponded to a low temperature to facilitate the gasification [59], thus resulting in high degree of gasification and high porosity.

With an increasing amount of activating agent, the porosity development, i.e., both specific surface area and pore volume, is accelerated, which also leads to a wide pore size distribution [60, 61]. At the early state, microporosity is dominant, while mesopores start to appear, as the level of the activating agent is gradually increased. The degree of the point at which the mesopores are present is closely related to the type of the activating agents. For instance, during the activation of peach stones with KOH, only the micropore size was widened, while $ZnCl_2$ caused widening in both the micropores and small mesopores, whereas the use of H_3PO_4 led to the formation of large mesopores and macropores, as discussed above [54].

A polypyrrole prepared with $FeCl_3$ as oxidant was chemically activated by heating the PPy-KOH mixture, with KOH/PPy weight ratios = 2 and 4, in N_2 at temperatures of 600–850°C [60]. After activation, the samples were thoroughly washed with HCl (10 wt%) to remove inorganic salts and then washed with distilled water to neutral pH. Activated carbon powders were obtained after drying at 120°C. The activated carbons were denoted as CP-x-y, where x was the KOH/PPy weight ratio and y was the activation temperature.

The polypyrrole had a sponge-like structure constructed by particles with an average diameter of about 1 μm, as shown in Fig. 3.13(a). Due to this special morphology, a large interfacial area was available for the polymer to react with KOH, so that the activation process was highly uniform, thus ensuring desired pore evolution. With different degree of activation, the activated carbons possessed different shapes. Therefore, after mid-stage activation, the samples exhibited a combined morphology, i.e., the sponge-like structure of polypyrrole precursor and highly vesiculated particles, as demonstrated in Fig. 3.13(c, d). As the activation was further progressed, highly activated carbons were obtained, which consisted of particles with irregular shapes, as observed in Fig. 3.13(d). The micropores were randomly distributed in the carbon particles, as illustrated in Fig. 3.13(e, f), for samples CP-4–600 and CP-4–850,

42 *Nanomaterials for Supercapacitors*

respectively. The micropores were slightly enlarged as the activation proceeded from the middle to near final stages.

Figure 3.14 shows N_2 sorption isotherms and PSDs of the activated carbons obtained at temperatures of 600–800°C with KOH/PPy = 2 [60]. It was found that the shape of the isotherms was varied, as the activation temperature was increased from 600°C to 800°C, implying that the porous structure of the activated carbons was changed, especially the enlargement of the pores. As shown in Fig. 3.14(b), the sample prepared at 600°C (CP-2–600) contained mainly micropores with size of about 1 nm and a narrow

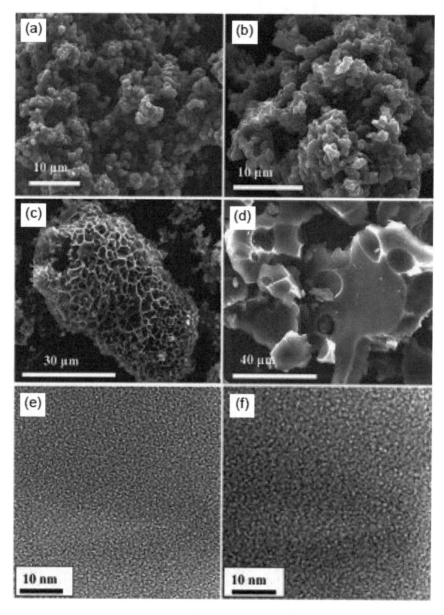

Fig. 3.13. SEM images of polypyrrole (a) and activated carbons CP-2–600 (b, c) and CP-4–850 (d). HRTEM images of CP-4–600 (e) and CP-4–850 (f). Reproduced with permission from [60], Copyright © 2011, WILEY-VCH Verlag GmbH & Co. KGaA, Weinheim.

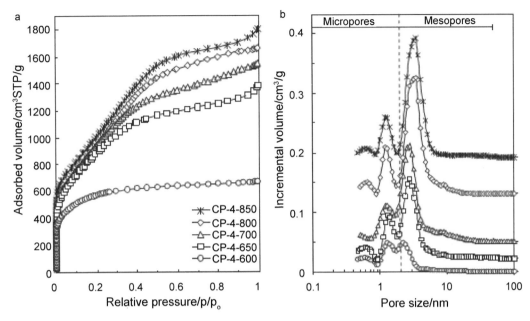

Fig. 3.14. (a) N_2 adsorption isotherms and (b) pore size distribution profiles of the activated carbons prepared with KOH/PPy = 2 at different activation temperatures. Reproduced with permission from [60], Copyright © 2011, WILEY-VCH Verlag GmbH & Co. KGaA, Weinheim.

size distribution. With increasing activation temperature, the size of the micropore was gradually increased. After activation at 800°C, a large portion of mesopores with a size of about 2.7 nm were observed in the sample CP-2-800.

According to the textural parameters of the porous carbons, both the BET surface area and the pore volume were increased enormously from 1700 $m^2 \cdot g^{-1}$ to 3410 $m^2 \cdot g^{-1}$ and from 0.88 $cm^3 \cdot g^{-1}$ to 1.94 $cm^3 \cdot g^{-1}$, respectively, as the temperature was increased from 600°C to 800°C. The pores were mainly micropores (< 2 nm), with a large fraction of them to be micropores of < 0.7 nm. It was also observed that most of the activated carbons obtained with KOH/PPy = 4 possessed ultrahigh surface areas, in the range of 3000–3500 $m^2 \cdot g^{-1}$, together with large pore volumes. For the samples prepared at 800 and 850°C, pore volumes could be as high as 2.6 $cm^3 \cdot g^{-1}$. In comparison with the samples activated at the mild activating conditions, i.e., KOH/PPy = 2, those obtained with KOH/PPy = 4 contained pores with relatively wider size distribution, which were characterized by the presence of two types of pores, (i) uniform supermicropores (1.2 nm) and (ii) mesopores. For the mesopores, the size was increased from 2.2 nm to 3.4 nm, as the activation temperature was raised from 600°C to 850°C.

It has been confirmed that there is an optimum activation temperature in terms of the porosity evolution, which is dependent on the type and properties of both the activating agents and the carbon precursors. Generally, the porosity is increased with increasing temperature first, reaches a maximum at a given temperature and then is decreased with further increase in temperature that is mainly caused by the shrinkage and collapse of the carbon structures [53, 58, 62–64]. For example, at temperatures of > 500°C, for chemical activation with $ZnCl_2$ or H_3PO_4, both weight loss and shrinkage of the carbons could occur, which resulted in reduction and narrowing in total porosity. The structural variation at high temperatures has been attributed to the fact that the cross-linking structures formed at low temperatures due to the activation of $ZnCl_2$ or H_3PO_4 lack thermal stability. As a result, breakdown and rearrangement of carbonaceous aggregates took place, thus leading to collapse and shrinkage of the pores [53, 62–64]. Comparatively, if KOH was used as the activation agent, the maximum porosity could survive at higher temperatures, e.g., 700–900°C, implying that the cross-linked carbon structures activated by KOH are more stable [58, 64].

A systematic study was reported on the preparation of porous carbons from various biomass resources, including cornstalks, rice straws, pine needles and pinecone hulls [65]. All the raw materials were washed, dried at 100°C and then thermally treated at 400–500°C for 3 hours in N_2. The samples were carbonized, which were denoted as CP1, CP2, CP3, CP4 cornstalks, rice straws, pine needles and pinecone hulls, respectively. About 1.0 g carbonized species was soaked for 1 day in 25 mL KOH solution with concentrations in the range of 4–24 M. After that, the powders were collected and activated at 750–800°C in N_2 for 2 hours. The activated products were washed with distilled water and treated with dilute HCl solution, leading carbon materials of C1, C2, C3 and C4, respectively, which were further washed with distilled water until the pH value reached 7.0 and finally dried at 100°C.

N_2 adsorption/desorption isotherms of the porous carbons derived from pinecone hulls, cornstalks, pine needles and rice straws are shown in Fig. 3.15(a). The isotherms belonged to type-I adsorption curve, which indicated that carbon materials contained mainly micropores. Figure 3.15(b) shows their Horvath–Kawazoe pore size distribution curves, demonstrating that the porous carbons had pores with diameters in the range of 1–2 nm, which was confirmed by TEM observation results.

Raman spectra of the porous carbons had two broad bands at 1580 cm^{-1} (G band) and 1360 cm^{-1} (D band), due to the C–C bond vibrations of carbon atoms with sp^2 electronic configuration of graphene layered structure and the disordered and imperfect structures of carbon materials, respectively. It implied that the degree of graphitization of the porous carbons was relatively low, even although the temperatures were as

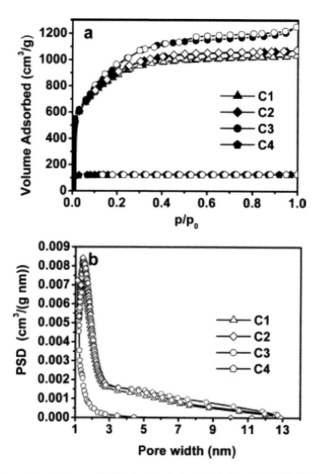

Fig. 3.15. N_2 adsorption/desorption isotherms at 77 K (a) and pore size distributions (b) of C1, C2, C3 and C4, with specific surface areas of 3200, 3315, 3545 and 397 m^{2}g^{-1}, respectively. Reproduced with permission from [65], Copyright © 2008, Elsevier.

high as 750–800°C. Similar results were observed by using IR spectra. It was concluded that most of the heteroatom elements, including O and H, in the biomasses had been removed after the carbonization and activation, while graphene sheets were formed by the carbon atoms. The graphene sheets were assembled to produce the pores in the carbonized products, in which there were tar matter, disordered carbons and other species. The pore structures were further developed after the activation treatment. Meanwhile, new pores were generated due to the oxidation of the reactive centers of the carbon matters. As a result, the carbonized products were transferred to porous carbon materials with high porosity and randomly distributed pores, as well as specific high surface areas.

Although the four biomass resources had similar compositions, the final porous carbon materials exhibited a different surface area, due the difference in texture of the raw materials. Figure 3.16 shows representative SEM images of the four carbonized samples. A large number of macro-pores and/or channels were observed in CP1 and CP2, as shown in Fig. 3.16(a, b). The presence of macro-pores or channels was an advantage for the absorption of KOH during the activation process, so that a full oxidation of the reactive centers to form micropores was realized. As a result, the final carbons had a huge surface area of > 3000 $m^2 g^{-1}$. The sample CP3 possessed much less macro-pores as compared with CP1 and CP2, so that the activation of CP3 was not complete as those of CP1 and CP2. As a consequence, the surface area of the porous carbon derived from CP3 was always less than 2500 $m^2 g^{-1}$. As shown in Fig. 3.16(d), CP4 contained the least macro-pores, which resulted in a relatively insufficient activation. Therefore, CP4 led to porous carbon material with surface area of ≤ 1000 $m^2 g^{-1}$.

The concentration of activating agent also played an important role in determining specific surface area of the biomass resources derived carbon materials through carbonization and activation. Figure 3.17(a) shows a specific surface area of the porous carbon derived from the rice straws as a function the concentration of activating solution, with values of > 3000 $m^2 g^{-1}$, over the concentrations of the KOH

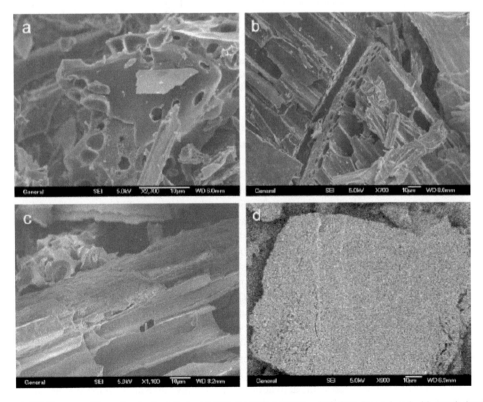

Fig. 3.16. SEM images of the carbonized products: (a) CP1, (b) CP2, (c) CP3 and (d) CP4. Reproduced with permission from [65], Copyright © 2008, Elsevier.

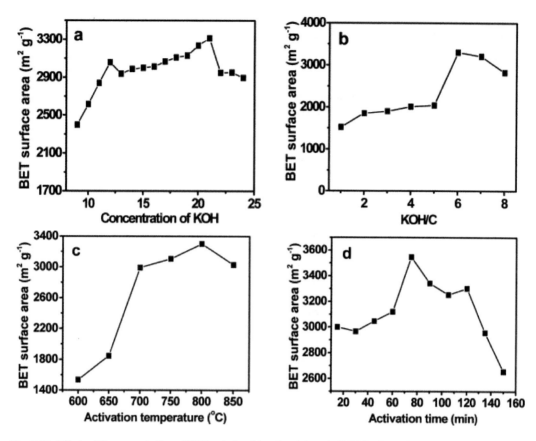

Fig. 3.17. Effects of the concentration of KOH solution (a) and weight ratio KOH/C (b) on the BET surface areas of the porous carbon materials derived from rice straws. Effects of activation temperature (c) and activation time (d) on BET surface areas of the porous carbon derived from pine needles. Reproduced with permission from [65], Copyright © 2008, Elsevier.

solution in the range of 12–21 M. However, further increase in the activation solution led to decrease in specific surface area of the porous carbons. Specific surface area of the sample derived from pine needles was also increased with the weight ratio of activating agent to carbonized product (KOH/C) as the values are < 6.0. However, as the weight ratio was over 6.0, the surface area would be decreased gradually, as seen in Fig. 3.17(b). With increasing concentration of KOH solution or KOH/C weight ratio, more KOH could be introduced into the carbonized products, thus leading to more efficient activation for produce more micropores. However, as the concentration was further increased, although more micropores could be formed, the porous structures of the carbons could be destroyed.

Similarly, specific surface area of the microporous carbon materials was also influenced by the activation time and temperature. BET surface areas as functions of activation time and temperature for the porous carbon prepared from pine needles are shown in Fig. 3.17(c) and (d), respectively. The surface area was increased with increasing temperature until 800°C. Further increase in activation temperature resulted in a slight decrease in the specific surface area, because too high temperature would also damage the porous structure of the carbons. Therefore, the optimal activation temperature was 800°C. As demonstrated in Fig. 3.17(d), the surface area was always above 3000 $m^2 g^{-1}$, if the activation time was < 2 hours, with a maximum value of > 3500 $m^2 g^{-1}$, after activation for about 1 hour. Too long activation time resulted in decrease in the surface area, which could also be attributed to the damage of the pore structures.

If $ZnCl_2$ and H_3PO_4 were used as the activation agents, the pore size distribution was not affected by the carbonization temperature, while volume fraction of micropores was in the range of 80–90% [64], which has been confirmed by various studies [62, 66–69]. In contrast, when using activation agent of KOH, the activation temperature exhibited a significant effect on PSD [64]. In other words, activation temperature could be a critical parameter in determining the PSD [70–73]. This observation has been attributed to the fact that the micropores were mainly produced through the release of volatiles in the temperature range of 500–700°C. As the temperature was above 700°C, CO_2 from K_2CO_3 formed during carbonization was released strongly, while carbon gasification could also occur at the same time. The direct consequence of these two processes was the opening up of the closed pores and enlargement of the present micropores. As a result, the pore size distribution would be widened. In addition, metallic K could be formed, which then intercalated the graphite-like laminar structure, further increasing the spaces of the adjacent carbon atomic layers, i.e., increasing the pore volume [73]. Other parameters that could influence the porosity of the porous carbons include (i) mixing procedure (solution or mechanical mixing), (ii) activation time, (iii) rate of gas flow, (iv) properties of the gas used during the heat treatment and (v) heating rate [70, 74, 75].

Among the various activating agents, KOH was able to result in activated carbons with surface areas of as high as 4000 $m^2 \cdot g^{-1}$, together with pore volumes of up to 2.7 $cm^3 \cdot g^{-1}$ [65, 70–72, 76–80]. Meanwhile a tunable and narrow pore size distribution could be achieved, through the controlling of the activating conditions (e.g., temperature and concentration of KOH) [80]. In contrast, the other activation agents would be able to produce carbons with the surface areas of 1500–2000 $m^2 \cdot g^{-1}$ and pore volumes of < 1.5 $cm^3 \cdot g^{-1}$, together with a relatively broad pore size distribution [62, 80, 81].

Hybrid methods

Hybrid method means that both the physical and chemical are used in one processing. Usually, a hybrid method is started with a chemical activation step, e.g., with H_3PO_4 or $ZnCl_2$, followed by a physical activation (e.g., with CO_2). It has widely been used to further strengthen the pore development and also tune the pore size distribution of the activated carbons. For example, by using CO_2 at 800°C, it was able to further activate the $ZnCl_2$ or KOH activated carbons derived from coconut shells and palm stones, which led to an increase in mesoporosity [82]. The two-step activated carbon materials possessed surface areas of up to 2400 $m^2 \cdot g^{-1}$ and mesoporosities of 14–94%.

Coconut shells and palm stones were dried at 110°C and crushed into samples with particle size of 1.0–2.0 mm. 20 g precursors were activated in solutions of either $ZnCl_2$ or KOH. CZ3 and CZ8 stood for coconut shell-based carbons activated by $ZnCl_2$ with weight ratios of 0.75 and 3.0, PZ1 and PZ2 denoted palm stone-based carbons activated by $ZnCl_2$, with weight ratios of 1.0 and 2.0, while CK26 and CK27 represented coconut shell-based carbons activated by KOH with weight ratios of 0.5, respectively. The mixtures were dehydrated at 110°C and then pyrolyzed in N_2 flow of 5 $l \cdot hours^{-1}$. The temperature was then increased to 800°C at a heating rate of 10°C·minutes^{-1}. Upon reaching 800°C, CO_2 of 20 $l \cdot h^{-1}$ was used to replace N_2. The activation was carried out for 2 hours. CK26 and CK27 were activated for 4 and 5 hours, respectively. After cooling, the activated samples were washed with deionized water to remove the undesired chemicals. Zinc or potassium cations were completely removed by thoroughly washing the samples with hydrochloric acid of about 0.1 $mol \cdot l^{-1}$, followed by washing with hot deionized water.

Figure 3.18 shows N_2 adsorption-desorption isotherms of the $ZnCl_2$-activated carbons [82]. The CZ3 had a steep Type I isotherm, with a nearly horizontal plateau at high relative pressures. This observation suggested that the sample contained micropores with a relatively narrow pore size distribution. A narrow hysteretic loop of type H4 was observed at the desorption branch, implying that it had a high microporosity and the micropores were slit-like pores. PZ1 displayed a Type I isotherm, with a slight steepness and a slightly sloped-up plateau. Comparatively, the nitrogen uptake of CZ8 was sustained over the entire range of pressure, leading to adsorption isotherms with a shape to be the combination of Types I and II, corresponding to the amount of nitrogen adsorbed at both the very low relative pressures (Type I) and higher pressures (Type II). Gradually, the hysteretic loops of desorption (H3 type) became more and more pronounced, suggesting the presence of mesopores. In addition, the nitrogen uptake of PZ2 was increased over the entire range of pressure, which led to an isotherm to be close to Type 2, confirming its mesoporous structure.

Figure 3.19 shows nitrogen isotherms of the KOH activated carbons [82]. The nitrogen adsorption of both the CK26 and CK27 was increased monotonically throughout the whole range of relative pressure,

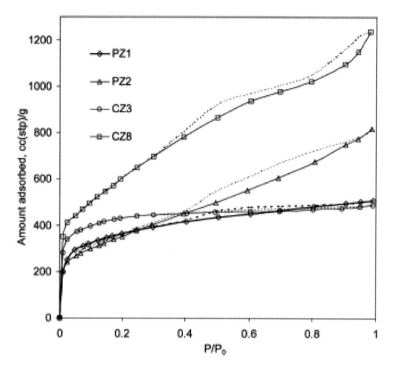

Fig. 3.18. N$_2$ adsorption–desorption isotherms of the ZnCl$_2$-activated carbons. Reproduced with permission from [82], Copyright © 2003, Elsevier.

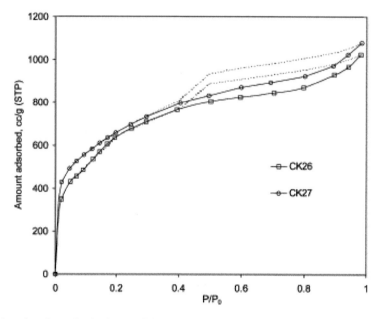

Fig. 3.19. N$_2$ adsorption–desorption isotherms of the KOH-activated carbons. Reproduced with permission from [82], Copyright © 2003, Elsevier.

indicating that their isotherms were a combination of Type I and Type II. In addition, both samples possessed an obvious hysteresis loop, which implied that both of them contained mainly micropores, together with mesopores.

It was found that CZ8 and CK26 and CK27 possessed a quite high surface area of > 2000 m²·g⁻¹. Total pore volume of all samples was > 0.5 cm³·g⁻¹ of general activated carbon. Specifically, CZ8, CK26 and CK27 had total pore volumes of 1.913, 1.584 and 1.665 cm³·g⁻¹, respectively. In comparison, although CZ3 had the lowest total pore volume, its volume fraction of micropore was relatively large, while CZ8 had high fraction of mesopores and total pore volume. Therefore, $ZnCl_2$/shell ratio led to the formation of mesoporous and widening of some micropores. In addition, CZ3 is a highly microporous, with a microporosity of 86% (1–V_{me}/V_{tot}), whereas CZ8 was highly mesoporous with a mesoporosity (V_{me}/V_{tot}) of 71%. CK26 and CK27 exhibited high fractions of both micropores and mesopores. Although PZ2 had a very small micropore volume of 0.075 cm³·g⁻¹, it possessed a relatively large mesopore volume of 1.188 cm³·g⁻¹. As a result, it demonstrated a significantly high mesoporosity of 94%.

Figure 3.20 shows pore size distributions of the activated carbons. The distribution curve of CZ3 was similar to that of typical activated carbon, with the pore volume mainly contributed by the pores with diameters of < 0.2 nm. PZ1 possessed a second small peak at about 3.7 nm, suggesting the presence

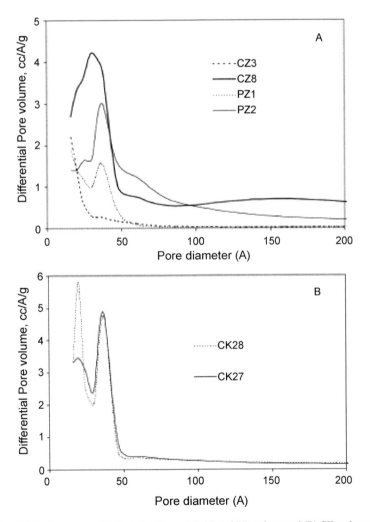

Fig. 3.20. Pore size distribution curves of activated carbons: (A) CZ and PZ carbons and (B) CK carbons. Reproduced with permission from [82], Copyright © 2003, Elsevier.

of mesopores. Both CZ8 and PZ2, derived from high chemical/carbon ratio, had maximum at 3.49 and 3.99 nm, respectively. CK26 and CK27 displayed a double-peak curve, with the two peaks at 2.0 and 3.7 nm, respectively. The peak of CK27 at 2.0 nm was lower, which indicated that the mesopores were more developed. This confirmed that high chemical/carbon ratio and prolonged activation could be used to enlarge the pores when synthesizing activated carbons. All the samples had a narrow pore size distribution, with a very small fraction of the pores to be > 5 nm. The average pore size was in the range of 2.43–3.99 nm, so that they were small mesopores.

Another example is the combination of chemical activation with H_3PO_4 or $ZnCl_2$ and CO_2 activation at 750°C, with coconut shell as the carbon source, which led to activated carbons with highly textural structures [83]. By using small amounts of chemical reagent, it was able to produce activated carbons with narrow microporosity without compromising their bulk densities, while the physical activation allowed to modify the pore structure of the primary pores that were developed after the chemical activation.

Figure 3.21 shows pore size distribution curves of the samples [83]. The three samples possessed similar total pore volume, although they were prepared by using different procedures. The distribution curves revealed that physical activation resulted in carbons with more uniform pores with a narrow size distribution, whereas chemical activation led to the formation of mesoporosity. It was found that chemical activation with relatively low levels of chemicals ensured the development of micropores with a narrow size distribution, while the bulk density of the samples was not decreased significantly, such as those with X_P and X_{Zn} below 0.15 and 0.32, respectively, as seen in Fig. 3.22. The fact that the granular carbons with well-developed and uniform distributed porosities was confirmed by the nitrogen adsorption–desorption isotherms. The adsorption capacity by mass unit and the bulk density were increased oppositely, the volumetric adsorption capacity had a maximum, near 85 V/V, no matter H_3PO_4 or $ZnCl_2$ was used in the chemical activation process. A similar procedure has been reported to produce activated carbons from an oil palm shell [84], together with various other examples which are available in the open literature and not described in a more detailed way here [47, 85, 86].

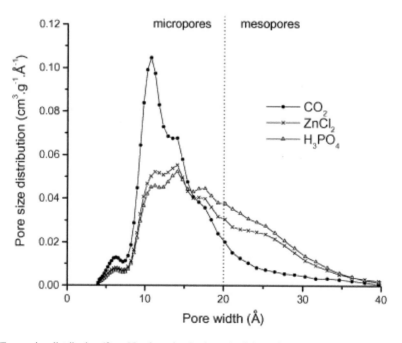

Fig. 3.21. DFT pore size distribution (from N_2 adsorption isotherms) of the activated carbons derived from coconut shell (2.00–2.83 mm) by physical activation with CO_2 (750°C for 110 h), chemical activation with H_3PO_4 (X_P = 0.30) and chemical activation with $ZnCl_2$ (X_{Zn} = 0.50). Reproduced with permission from [83], Copyright © 2008, Elsevier.

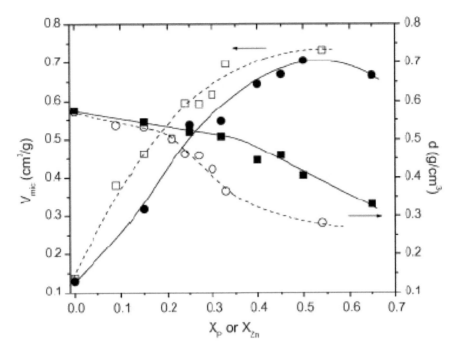

Fig. 3.22. Micropore volume and bulk density of the activated carbons prepared by using chemical activation of coconut shell (2.00–2.83 mm) with H_3PO_4 (open symbols—dashed lines) and $ZnCl_2$ (closed symbols—solid lines), as a function of levels of the chemicals used in the activation process. Reproduced with permission from [83], Copyright © 2008, Elsevier.

Monoliths of activated carbons

From the practical application point of view, it is necessary to increase the packing density, because activated carbons usually have a relatively low density. In this regard, activated carbons in the form of monoliths will be advantageous for practical applications as compared with loosely packed powders. If the carbon precursor and the preparation route are properly selected, activated carbon monoliths can be prepared directly without the use of any binders. This offers various advantages, including (i) less steps involved in the synthesis process, (ii) no loss of porosity due to the absence of binders and (iii) no loss in electrical resistivity, which are important for various applications in general and for electrochemical applications in particular [87, 88]. Monoliths of activated carbons could be readily prepared by using polymers or pitch binders [89–94]. Selected examples are discussed in detail as follows.

It was found that Expanded Natural Graphite (ENG) could be used both as thermal binder and as mechanical binder due to its thermoplastic polymeric characteristics [89]. Activated carbon powder was coated with a thermoplastic binder (PVA + PEG), which was mixed with ENG and then compressed into pellets or cylinders. The compacted samples were heated to 220°C, i.e., the melting point of the polymer. After that, the samples were cooled in air at given applied stresses. The solidification of the polymer ensured mechanical cohesion of the composites. The compression led to anisotropic behavior of the composites, i.e., parallel and orthogonal (or radial) to the uni-axial direction of compression. Due to the preferential orientation, the graphite layers were normal to the direction of uni-axial compression, so that the heat and mass transfer in the radial direction were largely enhanced. The composites were characterized with apparent density (ρ_a) and weight ratio to ENG (w_1).

Figure 3.23(a) shows apparent densities (ρ_a) as a function of applied stress (σ) of the composites with different weight ratios of ENG (w_1), as well as the two pure components, ENG and PX-21 activated carbon powder [89]. The apparent density was increased exponentially at low applied stresses, while the increase was slowed down at high stresses. Over the applied stress range of 0.1–2 MPa, the densification was mainly attributed to the activated carbon particles, in the form of clusters. As the applied stress was increased to > 2 MPa, the compaction of ENG, in the form of graphite layers that were coated on the activated carbon clusters, was responsible for the increase in density. Figure 3.24 shows a schematic diagram demonstrating the densification profile of the composites.

Pore size distribution of the activated carbon powder and the consolidated composites from the activated carbon were studied. Two peaks were observed on the distribution curves, corresponding to (i) the pores

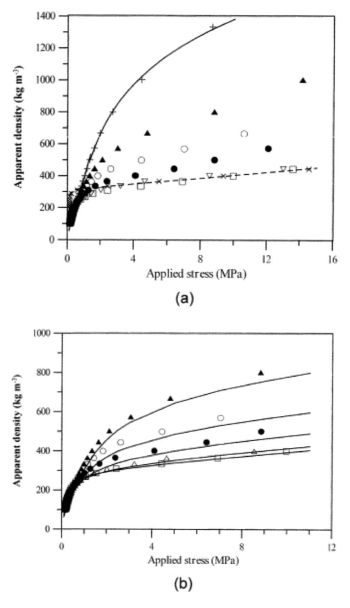

Fig. 3.23. (a) Consolidation related apparent density curves of the composites with different weight ratios of ENG: (×) 0%, (□) 10%, (▽) 20%, (•) 40%, (○) 60%, (▲) 80%, (+) 100%. The line (—) indicates equation (a) fitted curve for ENG's, while the line (- - -) indicate equation (b) fitted curve for PX-21's. Equation (a) is $\rho_1 = 367\sigma_1 + 19.6$ for $\sigma_1 < 1.4$ MPa and $\rho_1 = 446 \ln(\sigma_1) + 355.4$, whereas equation (b) is: $\rho_2 = 9.6\sigma_2 + 308$ for $\sigma_2 > 0.5$ MPa and $\rho_2 = 22.6 \ln(\sigma_2) + 326.4$. (b) Comparison between experimental results and current model (—). Reproduced with permission from [89], Copyright © 2001, Elsevier.

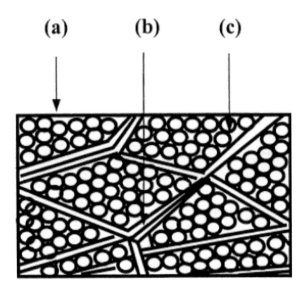

Fig. 3.24. Schematic diagram showing microstructure of the composites: (a) composite with apparent density, ρ_a (in kg of block per volume of block), (b) graphite layers with apparent density, ρ_1 (in kg of ENG per volume of ENG) and (c) confined PX-21 clusters with apparent density, ρ_2 (in kg of PX-21 per volume of PX-21). Reproduced with permission from [89], Copyright © 2001, Elsevier.

located in between the activated carbon particles (i.e., interparticle void) in the range of 0.5–5 μm for the activated carbon powder and (ii) those in the range of 0.1–1 μm for the activated carbon composites. The second peak was similar for all the samples, corresponding to the mesopores located inside the activated carbon particles.

Both the inter-particle void and the corresponding pore diameter were decreased significantly as the density of the activated carbon composite was increased, which was caused by the compression. Meanwhile, the compression had almost no effect on mesoporous volume and diameter of the corresponding mesopores. In other words, internal structure of the activated carbon particles was not influenced by the compression. Mesoporous volume was about 0.66 cm^3 g^{-1} while the corresponding mean diameter was 7 nm. Gradually, the microporous were merged into the mesoporous range. The total microporous and mesoporous volume was about 1.8 cm^3 g^{-1} for pores with diameters of < 20 nm. In comparison, the volume of the pores with diameters of < 5 nm was 1.623 cm^3 g^{-1}.

The effect of binder on properties of activated carbon monoliths has been studied [90]. To prepare the activated carbon monoliths, chemically activated carbon powders, obtained by using KOH as the activating agent and a Spanish anthracite as the precursor, were used as the starting materials [95]. Binders used to prepare the activated carbon monoliths include a Humic Acid-derived Sodium salt (HAS) from Acros organics, polyvinyl alcohol (PVA), a novolac Phenolic Resin (PR), a proprietary binder from Waterlink Sutcliffe Carbons (WSC), commercial Teflon (TF) and an adhesive cellulose-based binder (ADH). The binders were used to achieve monoliths with 15 wt% binder in the final products. The binder/activated carbon mixtures were dried in the form of powder, which were then pressed into pellets by using a die. The temperatures of the die were 135°C for HAS, PR and WSC, 200°C for PVA and Teflon, while room temperature process was used for ADH. The obtained pellet monoliths with HAS, PR, and WSC were pyrolyzed at 750°C for 2 hours to realize a complete carbonization. For the ADH monoliths, the pellet monoliths were dried at 200°C overnight.

Figure 3.25 shows nitrogen adsorption isotherms of the starting activated carbon powder (KUA31752), the activated carbon heated at 750°C (KUA31752HT) and the activated carbon monoliths with different binders [90]. All the samples exhibited a Type I isotherm, which is the characteristic of microporous solids. The presence of a very small fraction of mesopores was evidenced by the plateau isotherm. Therefore, the formation of mesoporosity was effectively prevented. In addition, all the monoliths had a relatively low

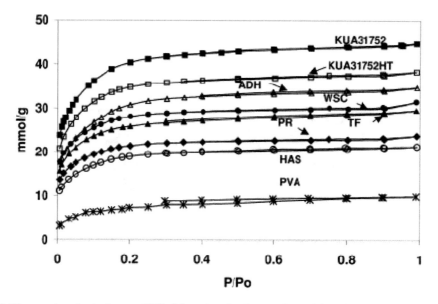

Fig. 3.25. Nitrogen adsorption isotherms at 77 K of the activated carbon powders and the activated carbon monoliths prepared with different binders. Reproduced with permission from [90], Copyright © 2002, Elsevier.

nitrogen adsorption capacity than the activated carbon raw material, because the composites contained 15 wt% binders that had no adsorption capacity. In fact, the monoliths were not ideal mixtures, since the decrease in the adsorption capacity was larger than 15%, which was attributed to at least two reasons. Firstly, the binders occupied spaces and could also block the pores in the monoliths. Secondly, the microporous structure of the activated carbon precursor could have been altered in the final monoliths.

Figure 3.26 shows SEM images of representative activated carbon monoliths [90]. The monolith with the binder of WSC possessed more compact microstructure, as seen in Fig. 3.26(c), so that it had the highest mechanical strength. In comparison, the monolith with TF exhibited a more open structure, which contained voids in between the particles, as demonstrated in Fig. 3.26(a). The sample with binder ADH was an intermediate case, as shown in Fig. 3.26(b).

Activated carbon monoliths have been prepared with precursors from olive stones through chemical activation with KOH, by using binder mixing method [91]. After demineralization, the carbon powder was impregnated with aqueous solution of KOH, leading to four samples with KOH/carbon weight ratios of 1/1, 2/1, 4/1 and 5/1. The suspensions were heated at 60°C for 12 hours and then completely dried at 110°C. After that, the samples were pyrolyzed at 300°C for 3 hours and the temperature was increased to 800°C, which they were dwelled for 2 hours in N_2 flow 300 cm^3·min^{-1}, at a heating rate of 10°C·minute^{-1}. The activated carbons were denoted as A1, A2, A4 and A5, corresponding to the KOH/carbon weight ratios of 1/1, 2/1, 4/1 and 5/1, respectively.

Another two activated carbon samples were prepared in a similar way to A2, but with different particle sizes in the demineralised carbonized powder. The activated carbons B and C were derived samples with particle size ranges of 0.63–0.80 mm and 0.08–0.15 mm, respectively. Activated carbon B was further steam-activated at 800°C to 22% burn-off, leading to sample BW. Two more samples, AC and H, were prepared from chemically activated carbons with particle size range of 1–2 mm, with a KOH/olive stone ratio of 2/1. The activation process was similar to that of A2 sample. It was observed that the precursor grains were broken during the impregnation, which resulted in very fine powders to ensure the homogeneity of the mixture between the raw material and the activating agent [54]. All activated carbon samples were washed with 0.1 N HCl and then distilled water to completely remove the chloride ions.

Figure 3.27 shows N_2 adsorption isotherms of the samples A1–A5. The samples all belonged to type I isotherm, with a plateau practically parallel to the p/p_0-axis, confirming their microporous microstructure,

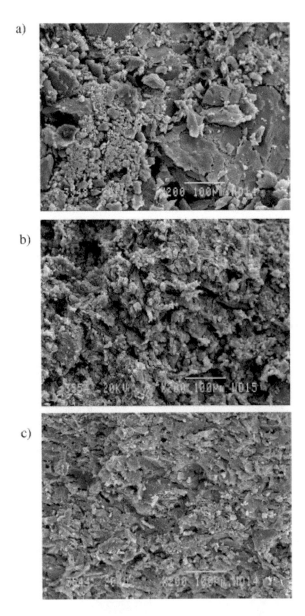

Fig. 3.26. SEM images of representative activated carbon monoliths with different binders: (a) TF, (b) ADH and (c) WSC. Reproduced with permission from [90], Copyright © 2002, Elsevier.

without or with very low level of mesoporosity. It was found that BET surface area was increased initially up to a KOH/carbon weight ratio of 2/1, beyond which the increase in S_{BET} was suddenly slowed down. Therefore, KOH/carbon weight ratio of 2/1 was considered to be the optimal ratio. Figure 3.28 shows SEM images of selected samples. There were large macropores on surface of the particles, typically observed when a raw material was carbonized to activated carbons, e.g., A2 and B. The sample H exhibited a different microstructure, due to its fine particles. A photograph of the monoliths from sample is shown in Fig. 3.29.

Activated carbon samples (R-1/2, R-1/3 and R-1/4), derived from corn grain precursor with varied ratios of the carbonized corn grain char to chemical activating agent, were used to prepare activated carbon monoliths [92]. Activated carbon monoliths were prepared by using different types, combinations and quantities of binders. Three binders, including polyvinyl alcohol (PVA), polyvinyl pyrrolidone (PVP)

56 *Nanomaterials for Supercapacitors*

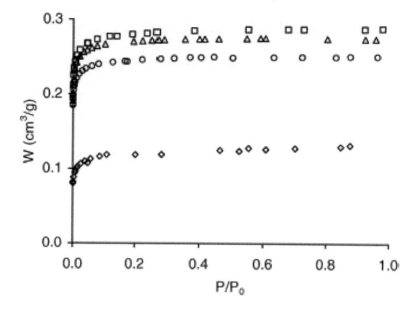

Fig. 3.27. N_2 adsorption isotherms at −196°C of the activated carbon samples: (◊) A1, (○) A2, (△) A4 and (□) A5. Reproduced with permission from [91], Copyright © 2006, Elsevier.

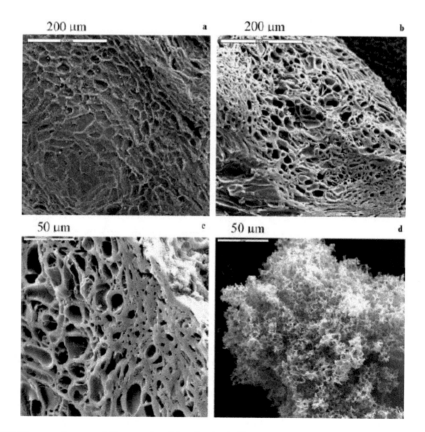

Fig. 3.28. SEM images of representative samples: (a) carbonized olive stones, (b) A2, (c) B and (d) H. Reproduced with permission from [91], Copyright © 2006, Elsevier.

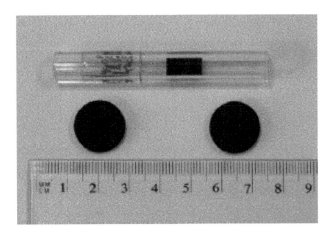

Fig. 3.29. Photographs of the activated carbon monoliths from sample H. Reproduced with permission from [91], Copyright © 2006, Elsevier.

and carboxymethyl cellulose sodium salt (CMC), were used to prepare carbon monoliths. Binder contents included PVA-10 wt%, PVP-10 wt%, CMC-5 wt% and CMC-10 wt% and binder combination of PVA/PVP-10%. To prepare monoliths, the activated carbon powder and binder were thoroughly mixed with a certain amount of water. After that, the water in the slurries was removed by thermal evaporation. The resultant mixtures were compressed in to pellets at 5 MPa at room temperature. The pellets were completely dried at 120°C, leading to monolith samples.

N_2 adsorption and desorption isotherms of the MR-1/4 carbon monoliths are shown in Fig. 3.30(a) [92]. It was found that, as compared with other MR-1/4 carbon monoliths, both MR-1/4_CMC-5% and MR-1/4_PVA/PVP-10% experienced less reduction in surface area and had relatively higher packing density. N_2 adsorption and desorption isotherms MR-1/3 (MR-1/3_CMC-5% and R-1/3_PVA/PVP-10%) and MR-1/2 (MR-1/2_CMC-5% and MR-1/2_PVA/PVP-10%) monoliths are shown in Fig. 3.30(b) and (c). Comparatively, CMC-5 wt% monolith showed relatively less reduction in surface area and exhibited a higher packing density, as compared with other samples.

With increasing ratio of the carbonized corn grain char to chemical activating agent, all the surface area, total pore volume, mesopore volume, micropore volume ($V_{DR}(N_2)$), average pore width were increased. The Pore Size Distribution (PSD) based on the DFT theory indicated that the PSD was broadened, as the ratio of carbonized corn grain char to chemical activating agent was increased, as shown in Fig. 3.31 [92]. In addition, the micropore size distributions (MPSDs) were evaluated according to the two different DR micropore volumes, which were derived from the N_2 and CO_2 adsorption isotherms. The smaller the difference between $V_{DR}(N_2)$ and $V_{DR}(CO_2)$, the narrower the MPSD would be, and vice versa. Therefore, the carbon monoliths, MR-1/2_CMC-5% and MR-1/3_CMC-5%, possessed relatively narrow and broad MPSDs, respectively.

Figure 3.32 shows SEM images of the carbonized corn grain char, powdered activated carbon (R-1/4) and two monoliths (i.e., MR-1/4_CMC-5% and MR-1/4_PVA/PVP-10%). By comparing the images of the carbonized corn grain char (Fig. 3.32(a)) and the powdered activated carbon (Fig. 3.32(b)), the formation of the pores due to the chemical activation could be readily identified. After the carbon powders were compressed into monoliths, the spacing in between the particles was significantly reduced, as seen in Fig. 3.32(c) and (d).

Binder-free carbon monoliths have also attracted much attention. A simple method was reported to prepare binder-free monoliths of activated carbon from compacted cellulose [96]. Firstly, cellulose microcrystal powder was pressed into monoliths, as shown schematically in Fig. 3.33. Figure 3.34(a) shows a photograph of the as-compacted pellet of the activated carbon powder. After that, the pellet was carbonized at 800°C. After carbonization, the monoliths were physically activated by using with CO_2 at 800–900°C, leading samples like that shown in Fig. 3.34(b).

58 *Nanomaterials for Supercapacitors*

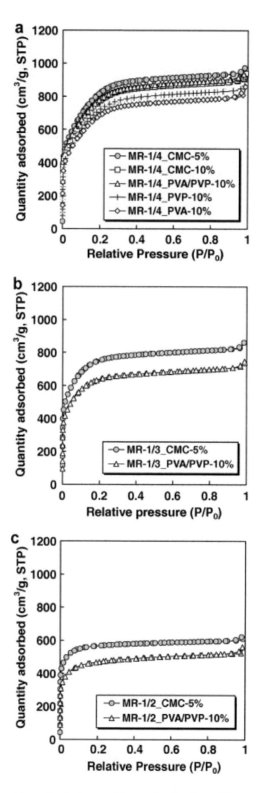

Fig. 3.30. Nitrogen adsorption and desorption isotherms of the powdered and monolith corn grain-derived activated carbon. Reproduced with permission from [92], Copyright © 2009, Elsevier.

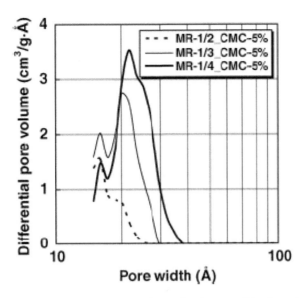

Fig. 3.31. DFT pore size distribution curves of the corn grain-based carbon monoliths. Reproduced with permission from [92], Copyright © 2009, Elsevier.

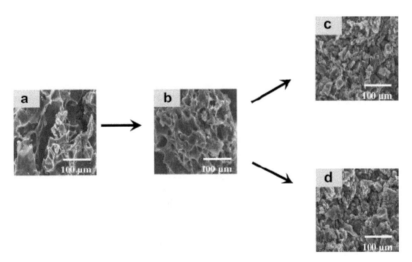

Fig. 3.32. SEM Images of (a) carbonized corn grain char, (b) powdered activated carbon, (c) carbon monolith with CMC-5% binder and (d) carbon monolith with PVA/PVP-10% binders. Reproduced with permission from [92], Copyright © 2009, Elsevier.

The carbon monoliths prepared in this way had bulk densities in the range of 0.56–0.99 g·cm^{-3}, which are higher than that of activated carbon powder by two–four times. The bulk density was increased with increasing compaction pressure, while the true density (1.9 g·cm^{-3}) was not affected by the compaction pressure. However, high compaction pressure was detrimental to porosity development and surface area. Both the pore volume and surface area were decreased as the compaction pressure was increased. For instance, the surface area and pore volume could be decreased from 1790 m^2·g^{-1} and 0.74 cm^3·g^{-1} to 1100 m^2·g^{-1} and 0.46 cm^3·g^{-1}, respectively. Figure 3.35 shows surface SEM images of the activated carbon powder (sample 1) and activated carbon pellets. The cellulose domains were not bond together, so that the morphology after the activation was retained in the samples. However, the carbonized cellulose microcrystals were highly packed.

60 *Nanomaterials for Supercapacitors*

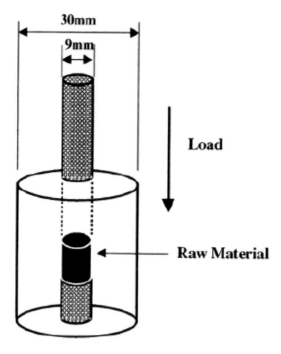

Fig. 3.33. Schematic illustration of the die used for compacting the precursor powders. Reproduced with permission from [96], Copyright © 2002, Elsevier.

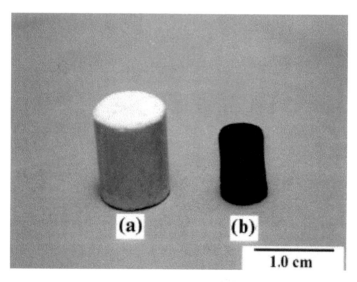

Fig. 3.34. Photographs of the raw material (a) and the activated carbon pellet (b). Reproduced with permission from [96], Copyright © 2002, Elsevier.

All samples had a Type I isotherm, confirming the microporous structure of the materials. The precursors were effectively activated with a high volume of micropore. The pore volume was linearly decreased with the compression pressure, while the variation in porous texture was saturated at $P = 49.0$ MPa. Therefore, the higher the compression pressure applied to the raw material, the more difficult it would

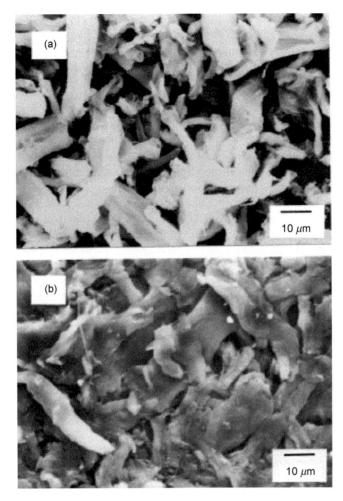

Fig. 3.35. SEM images of the activated carbon: (a) activated carbon powder and (b) activated carbon pellet. Reproduced with permission from [96], Copyright © 2002, Elsevier.

be to increase the activation efficiency. Meanwhile, the total pore volume, micropore volume and surface area were all increased with increasing activation time duration. In addition, the average pore diameter was independent on the compression pressure.

Binder-free activated carbon monoliths have also been prepared for various other precursors. For example, activated carbon disks were derived from olive stones through chemical activation with H_3PO_4 and $ZnCl_2$ [97, 98]. A hot compression with temperature of up to 300°C was used to compact the powder. Bulk density of the disk prepared in this way was higher than that of the samples from granular carbons, because of the considerable reduction in the inter-particle space. They were in the range of 0.57–0.98 $cm^3 \cdot g^{-1}$ and 0.48–0.81 $cm^3 \cdot g^{-1}$ for the H_3PO_4-activated and the $ZnCl_2$-activated carbons, respectively. On the other hand, the higher the compaction temperature, the higher the bulk density due to the removal of volatile matter during the pressing stage, which favors compaction.

A similar approach has been used to produce activated carbon monoliths with coconut shells and African palm stones as the precursor, by using $ZnCl_2$ and H_3PO_4 as the activating agents, [99–101]. By making use of the self-sintering characteristic, petroleum mesophase pitch have been developed into activated carbon monoliths, for different applications, including supercapacitors [102–104]. The monoliths were also used

62 *Nanomaterials for Supercapacitors*

as supercapacitor electrodes, demonstrating high specific capacitances in H_2SO_4, together with very low electrical resistivity, thus leading to very efficient energy storage device.

To prepare activated carbon monoliths for supercapacitor applications, a petroleum residue (ethylene–tar) was pyrolyzed at 440°C and 1.0 MPa for 4 hours to produce mesophase pitch (MP) [102]. The synthesis of the activated carbon binder-free monoliths (ACMs) involved several steps: (i) mixture of the MP with KOH through ball mill for 30 minutes, (ii) uniaxial compaction at room temperature to produce monoliths (Φ = 13 mm × 1 mm) at a pressure of 400 MPa, (iii) heat treatment at 800°C in N_2 for 2 hours, (iv) washing the monoliths with 1 M HCl solution and (v) washing with distilled water until pH = 7. The mixture of the carbon precursors and the activating agent could be consolidated due to the high plasticity. Finally, the monoliths were dried at 100–110°C for 24 hours. Three different monoliths, i.e., ACM-A, ACM-B and ACM-C, were prepared with ratios of KOH to MP to be 4:1, 2:1 and 1:1, respectively.

Figure 3.36 shows N_2 at 77 K adsorption isotherms of the three samples. They all belonged to Type I, with a well-defined plateau, indicating that the materials were essentially microporous. The broadening of the knee in the range of very low relative pressure implied that the microporosity of the material was widened, with increasing KOH/MP ratio, so that the characteristic of sample ACM-A was the most pronounced. ACM-A had the highest BET surface of 2650 $m^2 \cdot g^{-1}$, with a total pore volume measured at P/P_0 = 0.95 to be 1.27 $cm^3 \cdot g^{-1}$. Most of the pores were micropores, with an average pore width of 1.12 nm. ACM-C had the lowest BET surface of 400 $m^2 \cdot g^{-1}$, with a very low mesopore volume of 0.03 $cm^3 \cdot g^{-1}$ and a narrow pore width of 0.86 nm. ACM-B was close to ACM-C, but with relatively larger BET surface of 1100 $m^2 \cdot g^{-1}$ and larger pore volume of 0.49 $cm^3 \cdot g^{-1}$.

Figure 3.37 shows a representative cross-sectional SEM image of the monoliths, illustrating that they were a continuous mass without the presence of grain boundaries. There were voids with sizes of micrometers, which could be formed in the precursors reacting with the KOH particles, due to the boundaries of the MP grains, as observed in the figure. Therefore, due to the presence of the inter-particle voids, these activated carbon binder-free monoliths possessed a high accessibility owing to large fraction of microporosity.

Another strategy is to use self-adhesive properties of the pre-carbonized precursors, such as oil palm empty fruit bunches and rubber wood sawdust to prepare activated carbon binder-free monoliths [105, 106]. Usually, the pre-carbonized substances were firstly ball-milled to obtain self-adhesive grains, followed by activation and carbonization. A new type of porous carbon monolith with well-defined multi-length scale

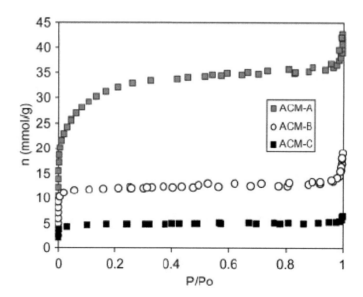

Fig. 3.36. N_2 Adsorption isotherms at 77 K of the three samples. Reproduced with permission from [102], Copyright © 2009, Elsevier.

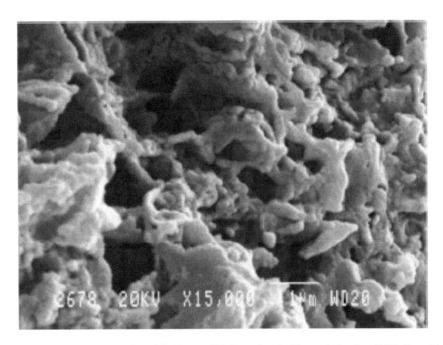

Fig. 3.37. Cross-sectional SEM image of the ACM-B monolith. Reproduced with permission from [102], Copyright © 2009, Elsevier.

pore structures synthesized porous carbon monoliths, through self-assembly of poly(benzoxazine-co-resol) combined with carbonization process [107].

In the experiment, resorcinol and Pluronic F127 were dissolved in a solvent mixture of ethanol and deionized water with magnetic stirring at 25°C [107]. After that, 1, 6-diaminohexane was added into the solution and stirred for 30 minutes at 25°C, so that the solution became pale yellow. Then, formalin (37 wt%) containing formaldehyde (denoted F) was quickly injected into the solution. The reaction system quickly (< 1 minute) turned to a white homogeneous emulsion. The reaction mixture was stirred at 25°C for 10 minutes. The white homogeneous emulsion was then sealed and heated at 90°C, which became solid within 15 minutes. The gel was then cured for 4 hours. The as-made polymer monolith (HPM) was first dried at 50°C for 24 hours and then pyrolyzed at 800°C for 2 hours in N_2, leading to a crack-free porous carbon monolith (HCM-DAH-1). By varying the amount of DAH, different polymer monoliths could be obtained. In all syntheses, the molar ratios of R/F127 and R/F were fixed to be 275:1 and 1:2, respectively. The mass ratio of water to ethanol was 1:1. The obtained monoliths were denoted HCM-DAH-x, where x was the multiples of the amount of the organic amine DAH used for sample HCM-DAH-1. For comparison, four other organic amines, including EDA, DMA, TMA and TEA, were used to prepare porous carbon monoliths in the same way. pH values of the reaction systems were adjusted to be the same as that of HCMDAH-1 by controlling the amount of the organic bases. The obtained carbon monoliths were denoted as HCM-y, where y = EDA, DMA, TMA, or TEA.

Surfactant Pluronic F127 was introduced into the reaction system to direct the formation of interconnected mesopores (cubic $Im3m$ symmetry) during the assembly of poly(benzoxazineco-resol). The resultant carbon monolith HCM-DAH-1 was characterized with various techniques, with results shown in Fig. 3.38. Figure 3.38(a) shows low-angle XRD pattern of HCM-DAH-1, with well-resolved reflections corresponding to a body-centered cubic symmetry ($Im3m$). The (110) peak corresponds to a d-spacing of 9.6 nm, suggesting a unit cell parameter of 11.1 nm. Figure 3.38(b) shows Small-Angle X-ray Scattering (SAXS) pattern of HCM-DAH-1, which was also related to a body centered cubic $Im3m$ symmetry.

Figure 3.38(c) shows N_2 isotherm and pore size distribution of HCM-DAH-1. Its isotherm belonged to type IV in shape, with a type H2 hysteresis loop, implying that the sample was a mesoporous material with

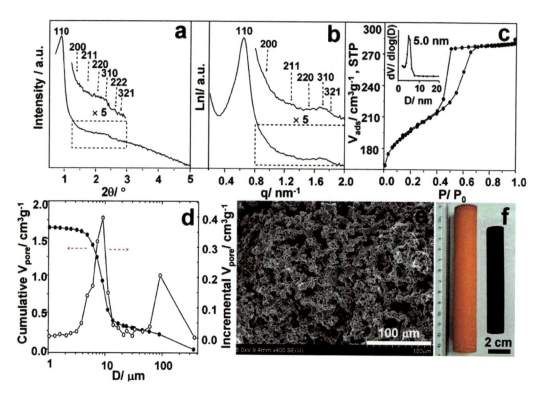

Fig. 3.38. Structural characterization results of the sample HCM-DAH-1: (a) low-angle XRD and (b) SAXS patterns, (c) N$_2$ isotherm and the corresponding PSD (inset), (d) Hg intrusion curve and differential PSD, (e) SEM image and (f) photograph of the as-synthesized polymer and the carbon monoliths. Reproduced with permission from [107], Copyright © 2011, American Chemical Society.

a 3D caged pore structure [108]. The PSDs shown as inset of Fig. 3.38(c) indicated that the mesopores were centered at a size of about 5.0 nm. It had a specific surface area and pore volume of 670 m^2·g^{-1} and 0.46 cm^3·g^{-1}, respectively. The macroporosity of HCM-DAH-1 was also confirmed by Hg intrusion analysis, as shown in Fig. 3.38(d). As the pore size as close to about 3.8 μm, a plateau was present, with a saturation Hg uptake of 1.67 cm^3·g^{-1}, which suggested that HCM-DAH-1 had rich and entirely interconnected macropores with a large volume fraction. The sizes of the macropore were centered at about 9.1 μm.

The formation of the continuous macropores was attributed to the polymerization-induced phase separation. Micrometer-range heterogeneity was first generated due to presence of gel and fluid phases. As the conjugate fluid phase (solvents) was removed, continuous macropore spaces were present [109]. Figure 3.38(e) shows SEM image of HCM-DAH-1, demonstrating the interconnected macropores appearing as a robust sponge-like branched skeleton. Figure 3.38(f) shows photographs of the polymer monolith (HPM-1) and the corresponding carbon monolith (HCM-DAH-1). The carbonization process caused a linear and volume shrinkages of ~ 30% and ~ 65%, respectively. The size of the monolith was only determined by shape and size of the reaction container. As the synthesis was scaled up by a factor of 5, all the properties were retained, indicating the potential of large scale production.

Representative TEM images of HCM-DAH-1 are shown in Fig. 3.39(a–c), confirming its regular mesoporous structure [107]. The three images were viewed along the [100], [110] and [111] directions, respectively, together with the FFT diffractograms, further confirming the interconnected cubic (*Im3m*) mesoporous structure [110, 111]. The cell parameter estimated from the TEM images was about 10.8 nm, which was comparable with that derived from the XRD pattern. HR-SEM images are shown

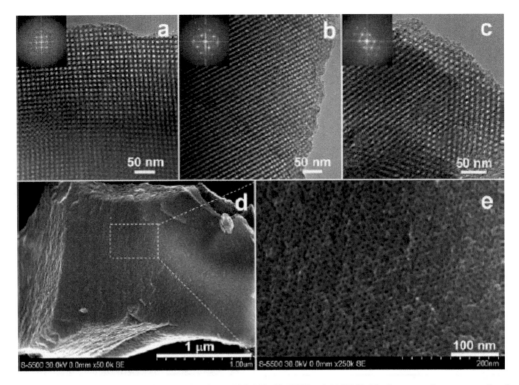

Fig. 3.39. TEM images viewed in different directions: (a) [100], (b) [110] and (c) [111]. The insets are the corresponding fast Fourier transform (FFT) diffractograms. (d, e) Low and high magnification SEM images of the carbon monolith HCM-DAH-1. Reproduced with permission from [107], Copyright © 2011, American Chemical Society.

Fig. 3.39(d–e), which further illustrated that HCM-DAH-1 possessed long-range ordering and large domain interconnected 3D cubic structure. According to the images, HCM-DAH-1 had unit cell parameter and pore size of 11.0 nm and 5.2 nm respectively, which were in good agreement with the results from other characterization techniques.

As mentioned earlier, the solution of HCM-DAH-1 became opaque white homogeneous emulsion within 1 minute, while EDA involved solution would become opaque yellow homogeneous emulsion. Meanwhile, the DMA involved system became transparent golden yellow in 5 minutes, with sol-gel transition in 30 minutes. For TMA and TEA, the solutions exhibited a similar reaction behavior. The solutions gradually became pale yellow in 5 minutes and then brown, which were solidified in about 60 minutes. The samples were then carbonized in the same way as that of HCM-DAH-1, thus leading to carbon monoliths HCM-EDA, HCM-DMA, HCM-TMA, and HCM-TEA.

Figure 3.40 shows SEM images and N_2 isotherms of the samples. It was found that HCM-EDA and HCM-DMA had a microstructure similar to that of HCM-DAH-1, in terms of the interconnected framework and macroporosity. The thickness of the struts was about 0.45 μm. The N_2 sorption isotherms of HCM-EDA and HCM-DMA were categorized to Type IV, characterized by a pronounced hysteresis loop in the relative pressure range of 0.4–0.7, implying the presence of well-developed mesopores. The mesopore sizes were centered at about 3.5 nm. Therefore, EDA and DMA also could be used to synthesize poly(benzoxazine-co-resol)s towards carbonaceous materials with hierarchical porosity. In contrast, the monoliths HCM-EDA and HCM-DMA exhibited very weak mechanical strength.

Figure 3.30(c) shows N_2 sorption isotherms of the samples, demonstrating microporous characteristics. When TMA and TEA were present in the polymerization process, mesostructures could not be formed in

66 Nanomaterials for Supercapacitors

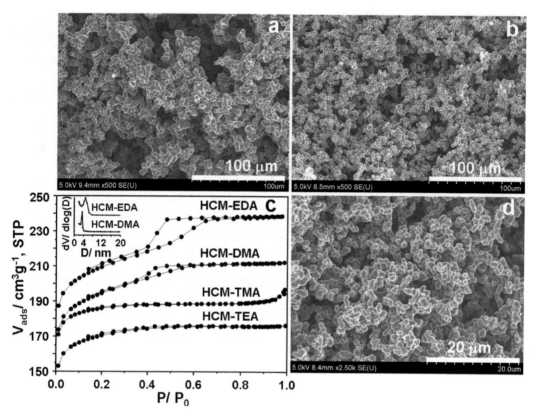

Fig. 3.40. SEM images of HCM-EDA (a), HCM-DMA (b) and HCM-TEA (d). (c) N$_2$ isotherms of HCM-EDA, HCM-DMA, HCM-TMA, and HCMTEA, together with PSDs of HCM-EDA and HCM-DMA (inset). The isotherms of HCM-EDA and HCM-DMA were vertically offset by 20 and 15 cm^3·g^{-1}, STP. Reproduced with permission from [107], Copyright © 2011, American Chemical Society.

the resultant carbons, i.e., HCM-TMA and HCM-TEA, respectively. This implied that the property of the organic amines was a critical factor that could influence the assembly of the items to form mesostructure. It was thus concluded that protic organic bases, such as DMA (secondary amine), EDA (primary amine) and DAH (primary amine), facilitated the formation of ordered mesopores, whereas aprotic organic bases, i.e., TMA and TEA, could only support the formation of micropores.

Supercapacitor performances of activated carbons

As stated previously, Electrochemical Double-Layer Capacitors (EDLCs) store energies through the charge separation at the interfaces between electrode and electrolyte. In this case, the charge separation distance is shortened to the dimensions very close to that of the ions in the electrolyte, while the porous electrodes could achieve a very large surface area, so that EDLCs are able to store much more energy than the conventional capacitors. At the same time, the storing process is an electrostatic behavior, which could offer high power densities, good reversibility and long cycle life, when compared with batteries.

Pseudocapacitors store energy through the fast reversible Faradaic charge transfer reactions between the electrolytes and electroactive species on surface of the electrodes. Redox reactions are usually observed for the Faradaic charge transfer, in which the oxidation states that some elements in the materials are varied in the way similar to that in battery systems. The amount of energy stored by pseudocapacitors is higher than that by EDLCs, but pseudocapacitors have relatively lower cyclability and power density.

Electrical Double Layer Capacitors (EDLCs)

Activated carbons have been widely used in commercial EDLCs, because of their high surface area, wide availability, low cost, high electrical conductivity and chemical stability. Most of the activated carbons used in commercial supercapacitors are powders that are derived from coconut shells. Although a wide range of new carbon materials, such as mesoporous templated carbons, zeolite templated carbons, carbon xerogels, carbon nanotubes and carbide derived carbons, have been studied as supercapacitor electrodes, activated carbons are the primary candidate.

Significant achievements have been made to establish the interrelationship between the nanoporous structures of activated carbons and their capacitance performances in various electrolytes. However, it is found that specific capacitance is not always linearly dependent on surface area, because of various factors, such as inaccessibility of the pores too small to the solvated ions, difference in electroadsorption behavior of the micropores, presence of external surface area and pseudocapacitance effects caused by the oxygen groups. For example, according to a study, in which electrochemical performance of 11 activated carbon microbeads and 23 activated carbon fibers in KOH aqueous solution were evaluated, no linear correlation between the total surface area and the specific capacitance was observed, while the specific capacitance demonstrated a strong dependence on the surface area of micropores and the external surface area [112].

Modeling results indicated that the double layer capacitance per unit surface area of the micropores was comparable with that of the carbon basal plane of 15–20 $mF \cdot cm^{-2}$, while the double layer capacitance per unit external surface area closely related to the porous structure and surface morphology of the carbon materials. By studying the electrochemical behavior of activated carbons with different surface areas and PSDs, it was concluded that the bigger the pores, the easier and faster that the ions could access electrochemically into the electrodes [113].

In a separate study, supercapacitor performances of activated milled mesophase carbon fibers and powder-type activated carbons in an organic electrolyte were examined [114]. The samples used for the study were comprised of activated carbon (AC and MP) powders, with different surface areas. The ACs were commercially available products, made by carbonizing and activating coconut shells in steam plus CO_2. The AC-series were denoted as AC-a, AC-b, AC-c, with BET specific surface areas of 959.7, 1499.9 and 1848.4 $m^2 \cdot g^{-1}$, respectively. Other groups of samples were commercial milled fiber-type activated carbon fibers (AC-mMPCF), obtained by carbonizing and activating mesophase carbon fiber (MPCF). Samples MP-a and MP-b had surface areas of 770.7 and 683.3 $m^2 \cdot g^{-1}$, respectively.

Figure 3.41 shows accessibilities of the AC-group and MP-group samples. The accessibility was calculated by dividing the Specific Capacitance (SC) by the BET surface areas, which was then converted into a percentage against the Maximum Specific Capacitance (MSC), i.e., the equation: Accessibility = (SC/S)/MSC × 100, so that the accessibility was a relative dimensionless variable. In this figure, accessibilities of most of the MP-group samples were very close to one and another. In contrast, the AC-series had a relatively wide range of accessibilities from < 50% to 70%. Therefore, MP-group samples had higher capacitance properties, including high conductivity, high specific capacitance and high accessibility, although they had a smaller specific surface area. As a result, it was concluded that structure of the pores was more important to the accessibility of the electrolyte ions than the specific surface area.

Similar studies on electrochemical performances of various activated carbons in H_2SO_4 indicated that the specific capacitance was dependent on two factors, i.e., total surface area and the concentration of CO-generating surface species [115]. Also, it has been widely accepted that the ion salvation shell was seriously distorted, so that partial subnanometer pores were destroyed, which resulted in an unexpected increase in capacitance of the carbon materials with pore sizes of < 1 nm [116]. Therefore, it is important to match pore sizes of the electrode materials to given electrolytes, while a high surface area alone is not sufficient to achieve high electrochemical performances. In other words, a narrow pore size distribution combined with a desired size of the pores will determine specific capacitance of a given electrode material.

A series of activated carbons with progressively varied nanotextural characteristics were prepared by carbonizing a bituminous coal at temperatures in the range of 520–1000°C and then activating with KOH at 700 and 800°C [117]. Surface area of the activated carbons was over a wide range of 800–3000 $m^2 \cdot g^{-1}$, with average micropore sizes over 0.65–1.51 nm, which were determined by the carbonization

68 Nanomaterials for Supercapacitors

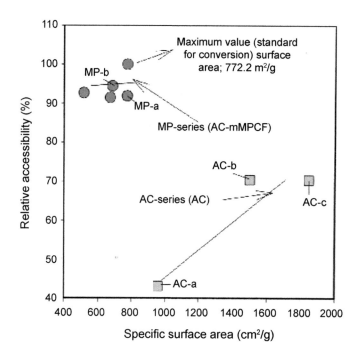

Fig. 3.41. Relative accessibilities per unit surface area against the maximum value of specific capacitance for the two groups of samples. Reproduced with permission from [114], Copyright © 2001, Cambridge University Press.

temperature. Electrochemical performances of the activated carbons were evaluated in solutions of H_2SO_4, KOH and $TEABF_4$/acetonitrile. Specific capacitances were 160–310 $F \cdot g^{-1}$, 124–286 $F \cdot g^{-1}$ and 120–180 $F \cdot g^{-1}$, respectively. Figure 3.42 shows specific capacitances of the group of samples activated at 800°C in 6 $mol \cdot L^{-1}$ KOH, as a representative. It was observed that the higher the surface area, the higher the specific capacitance would be. The specific capacitance per unit surface area was gradually increased with pore size as the pores were relatively small, which reached maximum at about 0.7 nm in aqueous media and 0.8 nm in organic electrolytes, respectively. In addition, the specific capacitance was saturated for the samples with high specific surface area, no matter what kind of electrolytes was used, because high surface is usually accompanied by large pores. The large pores were not efficient in participating in the formation of the double layer.

The porosity of activated carbons could be tuned controlling the synthesizing conditions. For instance, when preparing activated carbons by using hydrothermal method with carbonized substances or polypyrrole, the porosity could be well controlled by adjusting the activation temperature over 600–900°C or KOH/precursor weight ratio [60, 118]. These carbon materials had large surface areas in the range of 2000–3500 $m^2 \cdot g^{-1}$ and high pore volumes of 1–2.6 $cm^3 \cdot g^{-1}$. Additionally, the porous structure had also a close relation to the type of precursor. It was found that microporous carbons were obtained from the hydrothermally carbonized substances, while polypyrrole led to carbon materials with bimodal pore structures, i.e., micropores of 1.2–1.3 nm and mesopores of 2–3.4 nm.

Microporous carbons derived from natural precursors have been explored for supercapacitor applications, which exhibited high specific capacitance in one of the most common organic electrolytes, $MeEt_3NBF_4$/AN, together with rapid charging/discharging capability [119]. Various low-cost natural precursors were used to develop the microporous carbons, including cellulose, potato starch and eucalyptus wood saw dust. Hydrochars were produced from the precursors at 230–250°C, crushed into powders and then activated at 700–800°C. The obtained activated carbons were denoted as: AC-C700 and AC-C800 (from cellulose), AC-S700 (from starch), AC-W700 and AC-W800 (from wood), with numbers to be the activation temperature.

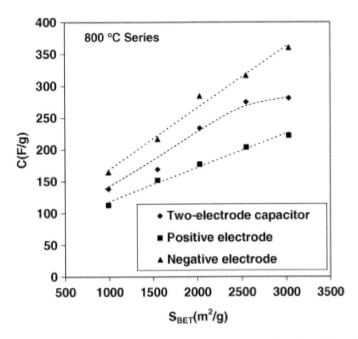

Fig. 3.42. Gravimetric capacitances of the series of carbons activated at 800°C in 6 mol·L^{-1} KOH. Values for the two-electrode capacitor and the positive and negative electrodes were obtained with the three-electrode cell using Hg/HgO as the reference electrode. Reproduced with permission from [117], Copyright © 2006, Elsevier.

The carbon materials exhibited promising electrochemical performances. The samples displayed retention of 64–85% of the capacitance, as the current density was increased from 0.6 to 20 A g^{-1}, as seen in Fig. 3.43(a). The samples AC-W800 and AC-C800 exhibited the highest capacitance retention, which was in a good agreement with the CV measurements. Due to the nearly entire microporosity, the activated carbon samples demonstrated pretty high capacitance of up to 175 F g^{-1} at a current density of as high as 20 A g^{-1}, as compared with the ACs with large mesopores. Therefore, it was further confirmed that the presence of mesopores in carbon electrodes is not beneficial to rapid ion transport, high capacity retention and high power characteristics.

Figure 3.43(b) shows specific capacitance measured at 0.1 A g^{-1} as a function of the surface area of micropores, indicating a linear relationship between the two parameters. This observation was supportive to the conclusion that larger mesopores contributed to lower capacitance per unit surface area [48, 117]. Specific capacitance of the activated carbons exhibited saturation at frequencies of < 0.04 Hz, implying that the ion adsorption could achieve near equilibrium in several seconds, which was faster than that of commercial devices (10–100 s). In addition, the AC-W800 and AC-C800 samples had the fastest frequency response, which was attributed to their slightly larger average pore size and higher capacitance retention at high sweep rates in the CV measurements or higher current densities in C-D tests, as illustrated in Fig. 3.43(a).

A straw based activated carbon was also evaluated as electrodes for organic electrolyte EDLCs [120]. Pretty high capacitance of 251.1 and 236.4 F·g^{-1} were achieved in the organic electrolytes, MeEt$_3$NBF$_4$/Propylene Carbonate (PC) and MeEt$_3$NBF$_4$/acetonitrile (AN), respectively. The pore structure and capacitance performances, as well as their effects on electrochemical performances of the devices, were studied. Accessible volume to solution was determined by monitor the mass loss of the carbon impregnated by AN, while minimal diameter of the effective pore was clarified.

Figure 3.44 shows cycle voltammetry curves of the EDLCs assembled with the activated carbons at different scan rates at room temperature. The results obtained in the electrolytes of MeEt$_3$NBF$_4$/PC and MeEt$_3$NBF$_4$/AN are shown in Fig. 3.44(a) and (b), respectively. All the CV curves at the cathode and the anode processes were highly symmetric, indicating that the activated carbons exhibited typical capacitive

70 *Nanomaterials for Supercapacitors*

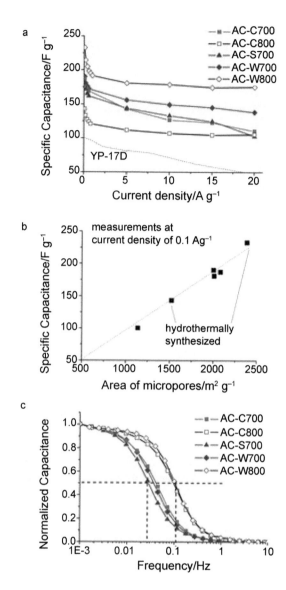

Fig. 3.43. Electrochemical characterization results of the hydrothermally synthesized carbon materials in 1 M tetraethylammonium tetrafluoroborate (TEABF$_4$) solution in acetonitrile (AN) at room temperature: (a) capacitance retention as a function of current density, (b) specific capacitance measured at 0.1 A·g^{-1} as a function of the NL-DFT specific surface area of the micropores and (c) frequency response. The performance of commercially available YP-17D activated carbon is included for comparison. Reproduced with permission from [119], Copyright © 2011, WILEY-VCH Verlag GmbH & Co. KGaA, Weinheim.

behaviors in the organic electrolytes, at both positive and negative potentials. The shape of the curves became more and more rectangular with decreasing scan rate, whereas the voltammetric currents were increased with increasing scan rate. This implied that, the lower the scan rates, the more the time the electrolytes could use to fill the micropores, so that more effectively the micropores were used to form double layer capacitances.

The specific capacitance in MeEt$_3$NBF$_4$/AN was higher than that in MeEt$_3$NBF$_4$/PC, which could be attributed to the difference in molecular size between the two solvents. Compared with AN, PC with larger molecular size was more difficult to access to the micropores. The resistance of the EDLCs in

Carbon Based Supercapacitors 71

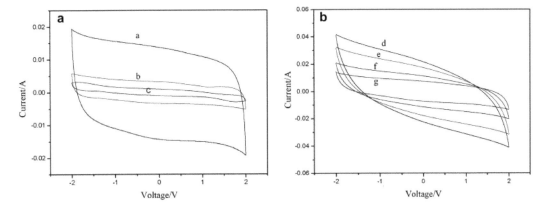

Fig. 3.44. Cycle voltammetry curves of EDLCs made by wheat straw based activated carbon at different scan rates at room temperature, (a) MeEt$_3$NBF$_4$/PC, 50 mV·s^{-1}, (b) MeEt$_3$NBF$_4$/PC, 10 mV·s^{-1}, (c) MeEt$_3$NBF$_4$/PC, 2 mV·s^{-1}, (d) MeEt$_3$NBF$_4$/AN, 300 mV·s^{-1}, (e) MeEt$_3$NBF$_4$/AN, 200 mV·s^{-1}, (f) MeEt$_3$NBF$_4$/AN, 100 mV·s^{-1} and (g) MeEt$_3$NBF$_4$/AN, 50 mV·s^{-1}. Reproduced with permission from [120], Copyright © 2011, Elsevier.

MeEt$_3$NBF$_4$/AN electrolyte was 0.62 Ω, while that of ones in MeEt$_3$NBF$_4$/PC electrolyte was 1.63 Ω, because MeEt$_3$NBF$_4$/AN had higher conductivity.

N$_2$ adsorption–desorption isotherms can provide important pore structure information of materials and can be applied more effectively to the substances with almost same size as that of N$_2$ molecule. If the substances have a size different from that of N$_2$ molecule, N$_2$ adsorption–desorption isotherm can only be used as guidance. Because the size of AN molecule is larger than that of N$_2$, not all the pores of the activated carbons that had been observed through the N$_2$ adsorption–desorption isotherm can be filled by the AN molecules. In other words, only partial surface had contribution to capacitance.

To identify the real contribution to capacitance, TGA was employed as a new approach to quantify the impregnated degree of the activated carbon by the electrolyte solution. TGA was used to monitor the mass loss due to evaporation of the liquid and desorption of the adsorbed substances. The total adsorbed volume could be the sum of volume of all the accessible pores by the solution. There were three stages that could be observed on the TGA curves, with the corresponding AN evaporation behaviors shown schematically in Fig. 3.45.

The first stage was characterized a rapid reduction in mass of the sample, due to the vaporization of AN outside of the pores, as illustrated in Fig. 3.45. At the second stage, the evaporation of the solvent was gradually slowed down, corresponding to the evaporation of solvent in between the carbon particles. Because of the larger resistance, solvent in between the carbon particles, the evaporation rate was much lower than that at the first stage. When second order derivative value vs. time was plotted, the maximal was the critical point marking the transition from the second stage to the third stage, at which the AN molecules in between the carbon particles were vaporized completely. After that, the AN molecules inside the pores started to vaporize. Therefore, there was a distinct difference in mass loss between the second stage and the third stage. At the third stage, the rate of mass loss was further decreased, the vaporization of the solvent inside the pores was a desorption process. In this case, the diffusion resistance was relatively large and evaporation propulsion was relatively low, because the smaller the pores, the lower the equilibrium vapor pressure would be.

According to the quantity of AN impregnated pores and the surfaces of the activated carbons at a given temperature (e.g., at 700°C), the desorption volume of AN per unit mass of the activated carbon at the third stage was estimated to be 1.02 mL·g^{-1}. From the N$_2$ adsorption of cumulative volume vs. pore width, the accessible volume of 1.02 mL·g^{-1} was the sum of all volumes of the pores with sizes of > 0.85 nm. Therefore, the minimal pore size to host a molecule of AN was 0.85 nm, so that larger pores could be impregnated by AN to contribute to double layer capacitance. The TGA result indicated that double layers could be effectively formed inside the pores with sizes of > 0.85 nm.

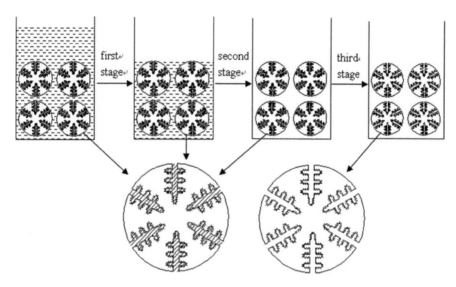

Fig. 3.45. Schematic diagram of mass loss process from the AN impregnated porous carbon. Reproduced with permission from [120], Copyright © 2011, Elsevier.

Several other studies have reported on KOH-activated carbons with outstanding performance (> 250 F·g^{-1}) in aqueous electrolytes [121–124]. There are also studies on KOH activated carbon-based electrodes with modest capacitance values (100–200 F·g^{-1}) in aqueous electrolytes [125–127]. For an organic electrolyte, the capacitance values are normally lower than 150 F·g^{-1} [128–130]. One specific example introduced here is on KOH-activated carbons in H$_2$SO$_4$ electrolyte [131]. The activated carbons were derived from various coal and pitch-derived carbonaceous materials at activation temperature of 800°C for 5 hours, with a KOH/precursor weight ratio = 4. The final activated carbons exhibited well-developed microporous structure, with surface areas in the range of 1900–3200 m^2·g^{-1} and pore volumes of 1.05–1.61 cm^3·g^{-1}. Specific capacitances in the range of 200–320 F·g^{-1} were achieved, comparable with the value of 240 F·g^{-1} of the commercially available activated carbon, PX-21.

Galvanostatic charge/discharge curves demonstrated a triangular shape, confirming typical capacitive characteristics of the carbon materials. At the same time, Ohmic drop with a certain degree of diffusion polarization in the micropores was observed during the switching time. The capacitive behavior was also confirmed by the impedance spectroscopy measurements. The carbon samples A-PM and A-C had capacitance values of 200 F·g^{-1} at 100 mHz and 100 F·g^{-1} at 1 Hz. Estimated time constants (RC) of all the capacitors were in the range of 2–3 s, which suggested that quick charge propagation was responsible for the performance of the capacitors made with the highly porous carbons. Figure 3.46 shows the results of voltammetry experiments at scan rates of 1–20 mV·s^{-1}, which could be described as $C = f(E)$. Over the scan rate range, the capacitance was decreased from 300 to 200 F·g^{-1}, due to the diffusion limitation in the highly microporous carbons. It proved that to achieve high current loads or high scan rates carbon materials with only mesoporosity are not sufficient.

Figure 3.47 shows durability of the capacitors by cycling in galvanostatic regime of 165 mA·g^{-1}. The results indicated that the carbon samples A-AC, A-PS and A-PM exhibited a pretty good cycleability. In comparison, samples A-CS and A-C experienced a capacitance loss of 100 F·g^{-1} after 2500 cycles. The loss of capacity during cycling was attributed to the self-discharge and leakage currents of the capacitors. The values of leakage current for the first group of samples were only 4.0 mA·g^{-1}, while self-discharge was about 50% after a duration of 20 hours. In contrast, the examples A-CS and A-C had much higher leakage currents of 8.2 and 12.8 mA·g^{-1}, respectively. Self-discharge approached 80% after 20 hours for these two carbons. Therefore, besides high values of capacitance, capacitance stability against cycling is equally important for practical applications.

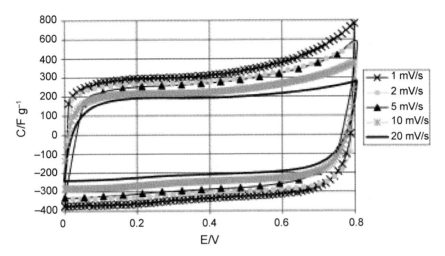

Fig. 3.46. Voltammetry characteristics of a capacitor built from the KOH activated carbon (A-PM) at different scan rates of voltage, in electrolytic solution of 1 mol·l^{-1} H$_2$SO$_4$. Reproduced with permission from [131], Copyright © 2004, Elsevier.

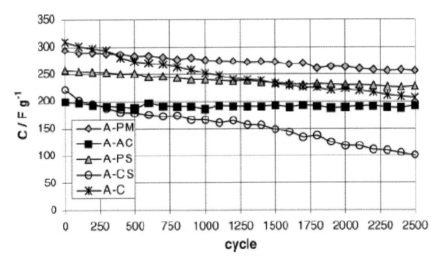

Fig. 3.47. Cycleabilities of the capacitors derived from the KOH activated carbons at current load of 165 mA·g^{-1}. Reproduced with permission from [131], Copyright © 2004, Elsevier.

Activated carbons prepared by using other chemical agents or physical activation have also been used as electrodes of supercapacitors demonstrating promising electrochemical performances. For instance, activated carbons obtained by using NaOH to activate anthracite, with a surface area of 2500 m^2·g^{-1}, exhibited capacitance values of up to 167 F·g^{-1} in organic electrolyte [132]. Another group of activated carbons from PVDC through NaOH activation, had surface areas in the range of 1829–2675 m^2·g^{-1} and specific capacitances in the range of 108–155 F·g^{-1} in an organic electrolyte of Et$_4$NBF$_4$/PC, PC = propylene carbonate [133].

On the other hand, Rufford et al. have measured specific capacitance up to 300 F·g^{-1} in H$_2$SO$_4$ for ZnCl$_2$-activated sugar cane bagasse and up to 134 F·g^{-1} in TEABF$_4$/acetonitrile for ZnCl$_2$-activated waste coffee [134, 135]. The carbonized Sugar cane Carbons (SCCs) were denoted as SCC-0, SCC-1, SCC-2 and SCC-3.5, with reference to the weight ratio of ZnCl$_2$ to bagasse [135]. Sample SCC-1-750 was prepared by the same method with a ZnCl$_2$ to bagasse ratio of 1 and a maximum activation temperature of 750°C. Similarly, if ionic liquid electrolytes (EMImBF$_4$ and EdMPNTf$_2$N) were used, activated carbons obtained through CO$_2$ activation of pyrolyzed sucrose could achieve specific capacitance of 172 F·g^{-1} [136].

The activated carbon prepared without using ZnCl$_2$ (SCC-0) had a narrow CV profile, with poor electrochemical performance, as evidenced by the small surface area, due to its low pore volume. In contrast, the activated carbons obtained with ZnCl$_2$ possessed well-developed pore structures, so that they demonstrated much higher performances than SCC-0. Their CV profiles had a rectangular shape, a characteristic of ideal electrochemical double-layer capacitance. The degree of rectangularity was decreased in the order of SCC-3.5 > SCC-2.0 > SCC-1-750 > SCC-1.

Charge–discharge profiles of the samples were symmetrical, with low iR voltage drops, illustrating their characteristic of typical electrochemical double-layer capacitor. Specific capacitances as a function of current load are shown in Fig. 3.48(a). At low current loads, SCC-1-750 exhibited the highest specific capacitance of > 300 F·g^{-1}, while SCC-2 and SCC-3.5 also possessed a specific capacitance of close to 300 F·g^{-1} at current loads of up to 0.25 A·g^{-1}. SCC-1 had a relatively specific capacitance of about 230 F·g^{-1}, i.e., because the level of the activation agent ZnCl$_2$ was too low. Figure 3.48(b) shows stabilities of SCC-1 and SCC-2 during charge–discharge cycling at a current density of about 5 A·g^{-1}. The SCC-1 experienced a drop in capacitance by 77% after 5000 cycles from the first cycle, but only 5% drop was observed after 10,000 cycles. SCC-2 had higher stability, with 83% retention in capacitance after 5000 cycles.

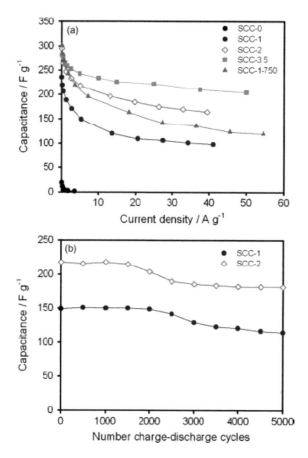

Fig. 3.48. (a) Specific capacitance as a function of current density of the activated carbons obtained from sugar cane bagasse. (b) Charge–discharge cycling stability of SCC-1 and SCC-2 at a current load of 20 mA, with the specific capacitance to be calculated from the discharge profiles of galvanostatic cycling curves measured with two-electrode cells. Reproduced with permission from [135], Copyright © 2011, Elsevier.

Figure 3.49 shows Ragone plots of the activated carbons derived from the sugar cane bagasse. Specific energies were estimated based on the active carbon mass, with values of up to 10 W·h·kg^{-1} that could be achieved by the SCC supercapacitor cells at low current loads. Once again, the one made with SCC-1-750 displayed the highest value, because the carbon sample had the highest specific capacitance. As specific power was increased, the effect of the mesopores in the carbon electrodes became more and more pronounced, evidenced by the stable performance of SCC-3.5, which retained to be 5.9 W·h·kg^{-1} at 10,000 W·kg^{-1}.

Carbon aerogels could be potential candidates as electrodes of supercapacitors, because of their high electrical conductivity, relatively high surface area, controllable pore structure and the possibility to be monoliths. Generally, carbon aerogels are mesoporous in structure, which is a problem as supercapacitor electrodes. To address this problem, a postsynthesis activation step has to be used in order to introduce microporosity. For example, activated carbon aerogels have been developed from resorcinol–formaldehyde in the flow of CO_2, leading to doubled surface area and increase in micropore volume from 0.02 to 0.61 cm^3·g^{-1} [137]. Therefore, the specific capacitance was increased by two times, reaching 100 F·g^{-1} in organic electrolyte of Et_4NBF_4/PC.

A CO_2-activation step was employed to activate carbon aerogels that were synthesized from cresol–formaldehyde, which resulted in an increase in surface area [138]. The surface area was enlarged from the original 245 m^2·g^{-1} to values in the range of 401–1418 m^2·g^{-1}, while pore volume was increased to 0.22–0.43 cm^3·g^{-1}, depending on the activation temperature (800–900°C) and time duration (1–3 hours). Due to the increase in porosity, the specific capacitance was increased correspondingly from 78 F·g^{-1} to 146 F·g^{-1} in 30% KOH electrolyte solution. In addition, the activated carbon aerogel showed a promising rate stability, with a small drop of 10% in capacitance, as the current density was increased from 1 to 20 mA·cm^{-2}.

Various soft-templated carbons have been activated to improve their electrochemical performances as electrodes of supercapacitors [139–141]. For instance, mesoporous carbon with a narrow pore size distribution centered at about 9 nm, which was synthesized through self-assembly of block copolymer and phloroglucinol-formaldehyde resin by using a soft-template method, was activated with CO_2 and KOH [139]. Activation conditions, including activation temperature, time and mass ratio of KOH/C, had an effect on textural properties of the activated mesoporous carbons. The activated mesoporous carbons possessed large BET specific surface areas of up to 2000 m^2·g^{-1} and pore volumes of up to 1.6 cm^3 g^{-1},

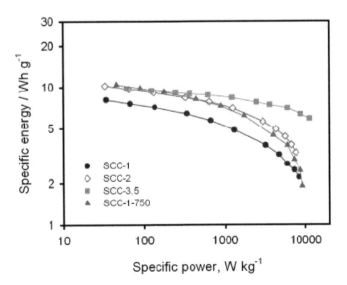

Fig. 3.49. Ragone plots of activated carbons obtained from sugar cane bagasse. Specific energy and specific power were estimated based on the mass of the active electrode material, while the masses of the electrolyte, current collectors and cell packaging were excluded. Reproduced with permission from [135], Copyright © 2011, Elsevier.

76 *Nanomaterials for Supercapacitors*

while the highly mesoporous structure was intact. When using CO_2 as the activation agent, a moderate to high extent of activation was necessary in order to increase BET surface area of the activated carbons by two–three times. In comparison, if KOH was used to activate the mesoporous carbons, a much smaller degree of activation could reach the same surface area. After activation, capacitance of the activated mesoporous carbons was significantly increased.

Pseudocapacitors

Heteroatom induced pseudocapacitance

Activated carbons usually have oxygen-containing surface functional groups, which lead to pseudocapacitance effects, especially in aqueous electrolytes, so as to increase specific capacitance and the amount of energy of the supercapacitors. Thus, with the presence of carbonyl- or quinone-type groups (CO desorbing groups in TPD experiments), the following equilibrium reaction may occur in the carbon electrode:

$$> C_xO + H^+ => C_xO//H^+, \tag{3.1}$$

where $> C_xO//H^+$ stands for a proton adsorbed by a carbonyl or quinone-type site caused by the attraction of ion-dipole. This special adsorption process could offer an excessive specific double layer capacitance, because of the local changes in electronic charge density. Also, as the negative electrode is charged, a strong bond could be formed between the carbonyl- or quinone-type groups and protons, because electrons are transferred across the double layer through the following reaction:

$$> C_xO + H^+ + e => C_xOH. \tag{3.2}$$

When discharging, the reaction takes place backwards. It is because of this redox reaction that a pseudocapacitance is added to the double-layer capacitance, thus leading an increase in total capacitance of the materials [142].

In addition, the presence of oxygen-containing functional groups could largely enhance wettability of the carbon electrode, because they are electrochemically inert. As a result, the specific capacitance could be increased, because of the improvement in pore accessibility and the utilization of surfaces of the materials [143].

Figure 3.50 shows the potential distribution at the basal layer of graphite at negative bias potential. The differential capacitance of a semiconductor interface was composed of three series components, i.e., (i) the capacitance of the space charge layer within the semiconductor, (ii) the capacitance of the compact double-layer and (iii) the capacitance of the diffuse ionic layer of the electrolyte. For an intrinsic semiconductor, the charge-carrier (hole or electron) densities are very low, which is similar to the case of an electrolyte solution at a very low concentration. As a consequence, the charge carriers would be away from the interface and extended into the bulk of the electrode over a large distance, which was inversely related to the charge carrier density. The charge distribution distance inside the basal orientation of the electrode was of the same order of magnitude as that the of diffusion layer as the electrolyte solutions are diluted.

However, due to their high polarity, oxygen groups, including carboxyl, anhydride and lactone (CO_2-desorbing groups), could hinder the ionic species to move about inside the electrode, which results in an increase in the resistance and thus a decrease in capacitance at high current densities [144]. The specific capacitance of activated carbons at low current densities has been well correlated with both the amount of CO-desorbing groups and the total surface area, while the variation in the specific capacitance as a function of current density could be linked with the amount of CO_2-desorbing groups and the average pore width (L_0) [115]. Similarly, a close relationship between the specific capacitance of activated carbons and the amount of surface oxygen groups desorbing as CO has been observed for activated carbons [59]. Therefore, it has been widely accepted that oxygen-containing groups in activated carbons have significant influence on their pseudocapacitance [145–149].

Electrochemical behavior of activated carbons with different oxygen contents and their contribution to pseudocapacitance were studied [145]. In this study, a mesophase-derived activated carbon was thermally

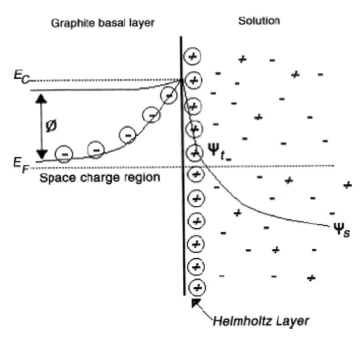

Fig. 3.50. Potential profile at the graphite basal layer/electrolyte interface at negative bias potential. E_c and E_F are potentials of the conductive band and Fermi-level, respectively. Reproduced with permission from [143], Copyright © 2002, Elsevier.

treated at 600°C and 1000°C, in order to modify the texture and surface chemistry, leading to samples of AC-600 and AC-1000, respectively. Electrochemical behavior of the samples was studied in two- and three-electrode cells in either acidic (H_2SO_4) or basic (KOH) media.

The activated carbons obtained after chemical activation with KOH exhibited a high microporosity, as evidenced by the large volume of adsorbed N_2 and the shape of the Type I isotherm, as demonstrated in Fig. 3.51. The textural parameters were derived from the isotherms. The total pore volume of this carbon was 0.85 cm^3·g^{-1}, with 88% of the pores in the regime of microporosity and an average pore diameter of 0.97 nm. The micropore volume estimated from the CO_2 isotherms was significantly lower than that obtained from the N_2 isotherms, which implied that the contribution of smaller micropores could be ignored. At the same time, the contribution of mesopores was very less pronounced, with a value of about 0.10 cm^3·g^{-1}. Due to the high volume of micropores, the samples had large BET surface areas.

After the activated carbon was thermally treated at 600°C, the total pore volume was slightly decreased, as seen in Fig. 3.51, while the type of porosity remained to be the same as that of the as-derived sample. As the heating temperature was raised to 1000°C, the texture of the activated carbon was changed more significantly, while both the total pore volume and average micropore size were decreased, due to the textural reorganization of the activated carbon. It was found that the amount of oxygen present in the samples was decreased very significantly after the sample was thermally treated at 600° and 1000°C.

XPS analysis results indicated that the oxygen was mainly attached at the surface of carbons. The decrease in the oxygen content was accompanied by a significant increment in pH value, varying from 2.8 of the as-derived AC to 7.3 of AC-1000. This observation implied that the AC contained carboxylic groups that could be removed through thermal annealing. Figure 3.52 shows TPD profiles of the samples, illustrating the variation of CO_2 and CO. CO_2 was formed at low temperatures, due to the decomposition of the carboxylic groups, lactones or anhydrides [150]. CO was formed at higher temperatures, because of the decomposition of the basic or neutral groups, such as phenols, ethers and carbonyl groups. It was observed that the amount of CO was larger than that of CO_2. After thermal treatment at 600°C, most of the acidic groups were removed, while the groups related to CO were still retained. As the temperature was increased to 1000°C, most of the oxygenated functional groups were eliminated.

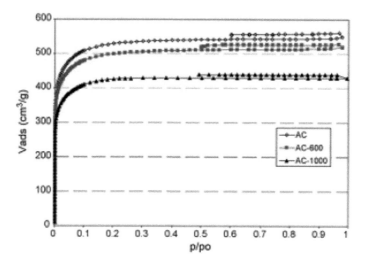

Fig. 3.51. N_2 adsorption/desorption isotherms of the activated carbons. Reproduced with permission from [145], Copyright © 2007, Elsevier.

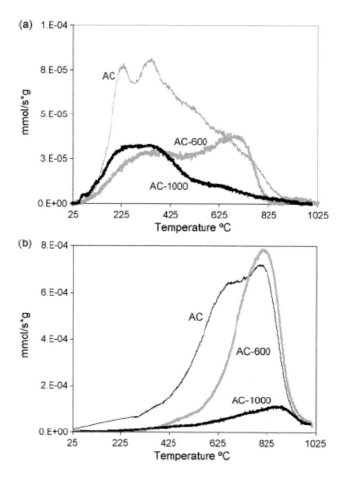

Fig. 3.52. TPD profiles of the activated carbons: (a) CO_2 evolution and (b) CO evolution. Reproduced with permission from [145], Copyright © 2007, Elsevier.

Figure 3.53 shows specific capacitance values of the three samples as a function of current density, with the data measured by using a two-electrode cell in H_2SO_4 electrolyte solution. The original AC sample had the highest specific capacitance, due to its largest pore volume and surface area. Thermal treatment resulted in a decrease in capacitance, especially for the sample annealed at 1000°C. The decrease in capacitance was much larger than expected according to the corresponding variation in specific surface area and pore volume. The additional effect could be readily related to the variation in the oxygenated surface functional groups, which contributed to Faradaic charge transfer reactions, i.e., pseudocapacitance, as mentioned above.

It was observed that the decrease in capacitance from AC to AC-600 was almost proportional to the reduction in microporous surface area, which was about 8%, as seen in Fig. 3.53. This implied that the elimination of oxygen due to the annealing at 600°C that corresponded to the groups to form CO_2 had no contribution to the total capacitance. However, specific capacitance of the AC-1000 sample was lower than that of the AC by 36%, while the reduction in microporous surface area was only 14%. Therefore, the removal of CO-desorbing complexes contributed to the pseudo-faradaic reactions, so that the reduction in capacitance was much larger than the reduction in specific surface area [151]. In other words, pseudocapacitance contributed to the total capacitance of the AC sample by 22% in H_2SO_4 electrolyte. A similar effect was observed in KOH electrolyte for the samples.

Oxidation treatment, either chemically (with HNO_3 or H_2SO_4) or electrochemically, has also been employed to produce oxygen-containing functional groups on activated carbons, so as to enhance electrochemical performances of the materials [152–154]. For instance, an activated carbon was oxidized in HNO_3 for 3 hours at room temperature, followed by thermally treating at different temperatures in the range of 250–900°C for 1 hour, so that samples containing different amounts of surface oxygen complexes could be developed [153]. The treatments were conducted in N_2, H_2 and air for comparison. Steady cyclic voltammograms were achieved, which was attributed to the redox processes associated with the CO-type groups corresponding to the broad peak below 0.6 V RHE and carboxylic anhydrides leading to the peak at about 0.63 V RHE. The surface oxygen groups exerted effects on total capacitance of the porous carbons in at least two aspects. Firstly, they contributed to the capacitance through the Faradic processes involving reactions with one-two electron transfer. Also, the surface oxygen groups or the dangling bonds produced after decomposition of the materials in inert atmosphere, increased surface wettability of the carbon materials by the aqueous electrolyte, enhanced on the other hand the double layer capacitance.

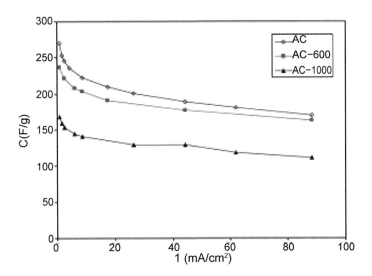

Fig. 3.53. Variation in specific discharge capacitance of the activated carbons as a function of current density in 1 M H_2SO_4 aqueous solution, with voltage window = 0.8 V. Reproduced with permission from [145], Copyright © 2007, Elsevier.

80 *Nanomaterials for Supercapacitors*

Nitrogen functionalization has been acknowledged to be another effective technique to introduce heteroatom to carbon materials, in order to provide Faradaic pseudocapacitive reactions. Various strategies have been employed to realize such functionalization, including the direct use of N-rich carbon precursor, such as polypyrrole, polyacrylonitrile and waste coffees [60, 146, 155], introduction of nitrogen containing groups to modify the surface of activated carbons, use of melamine or urea combined with heat-treatment [148, 156], and thermal treatment with ammonia [157, 158] or ammoxidation (i.e., air/ammonia mixture) [159, 160].

The properties of the nitrogen functionalized surfaces can be well controlled by controlling the temperature of thermal treatment. It has been recognized that amides, aromatic amines and protonated amides, with functional groups to be external to the aromatic ring structure, are predominant at low temperatures of 400–700°C, while pyridine, pyrrole, aromatic amines, quaternary nitrogen and protonated pyridine, in which nitrogen is within the aromatic ring structure either with a delocalized charge or without charge, dominate at high temperatures of > 700°C [157, 161].

A strong dependence of capacitance on the chemistry of N-containing surface groups of a series of activated carbons, which were oxidized and then treated with melamine and urea, followed by carbonization at 950°C in an inert atmosphere [156]. The results indicated that a direct relationship was established between the number of basic groups quantified from wet titration and the gravimetric capacitance, especially at high current densities, and between the normalized capacitance per unit volume of micropores and the population of quaternary nitrogen and pyridine-N-oxide inside the pores. It could be concluded that pseudocapacitive interactions occurred on negatively charged pyrrolic-N and pyridinic-N, while the positive charge on quaternary-N and pyridine-N-oxide were responsible for the electron transport through the carbon atoms. As a result, the N-functionalized activated carbons exhibited specific capacitances in the range of 260 and 330 $F \cdot g^{-1}$ in H_2SO_4 electrolyte, for melamine- and urea-treated samples, respectively.

Activated carbons were derived from wood (BAX-1500) were oxidized with 50% HNO_3 for 4 hours and then thoroughly washed with water to remove excessive acid and water-soluble products of oxidation [156]. To introduce nitrogen functional groups, the original and oxidized carbon powders were treated with urea or melamine suspension by stirring at room temperature for 5 hours. Then, the mixtures were boiled to evaporate the alcohol to obtain carbon samples after drying at 120°C. The samples impregnated with urea or melamine were heated in N_2 at 950°C for 0.5 hours. After that, the samples were washed with boiling water to remove excessive urea or melamine decomposition products. The carbons after treatment denoted as BAX, with U or M to show the modification with urea or melamine, respectively, while the preoxidized samples were labeled with O. For example, BAX-MO stood for BAX-1500 preoxidized, treated with melamine and heated at 950°C.

The nitrogen content could be up to 8% for BAX-MO. Even more nitrogen could be introduced into the carbon structure if the samples were oxidized with HNO_3 before the impregnation, while at the same time melamine was used as the source of nitrogen. In this case, the content of nitrogen on the surface was slightly lower than that inside the bulk interior. Meanwhile, it was found that the relative content of phosphorus on the surface of the materials was also increased after the treatment.

The differences in the contents of nitrogen between the samples treated with melamine and urea could be related to the difference in content of nitrogen in the precursors. For example, melamine and urea contain 67 and 47% nitrogen, respectively, thus exhibiting different enhancing effects to retain the nitrogen-containing groups. Because BAX was prepared at relatively low temperatures, it was very susceptible to chemical treatment [53, 162]. Comparatively, the contents of oxygen and phosphorus were decreased by about 50%, due to the decomposition of the functional groups on surface. For the samples treated with urea, no matter whether there was a pretreatment process, the total contents of oxygen and phosphorus were almost the same, implying that the oxygen containing functional groups introduced during the oxidation were not very stable against the thermal treatment at 950°C.

The oxygen atoms could be redistributed in the materials, so as to form more basic surface groups [163]. For the samples treated with melamine, surface preoxidation could significantly increase the total amount of nitrogen incorporated into the activated carbons. If the content of phosphorus was assumed to be constant, the oxygen containing groups in the oxidized samples were more stable than those in the untreated samples. The sample BAX-M possessed the lowest contents of oxygen and phosphorus. It was found that a relatively high content of nitrogen was accompanied by the lowest content of hydrogen, implying that the majority of nitrogen was incorporated in the format of pyridine or quaternary. Noting the chemical formula of melamine and its conversion to melamine resins at high temperatures, this observation was within the expectation [164]. While the oxygen containing groups attracted melamine and urea, they also contributed to the surface reactions, as demonstrated in Figs. 3.54 and 3.55 [156].

Therefore, although urea had less nitrogen that could be introduced, it could be accommodated in relatively smaller pores. In comparison, the treatment of melamine resulted in the formation of bulky resin, which only occurred in pores with sufficiently large sizes. As a result, the effect of the latter on the pore size distribution was much more pronounced. Once the functional groups were attached to the sites at the entrances to the micropores with a relatively small size, they would block the adsorbate and ions of the electrolytes. Although the melamine treated samples contained more nitrogen that the urea treated ones, it was highly possible that some of the nitrogen was present as the quaternary nitrogen and thus inactive in the acid–base interactions. Furthermore, the distribution of nitrogen could be different and 'patches' could be formed, as melamine was used as the source of nitrogen. Oxygen atoms inside the materials promoted the incorporation of nitrogen onto the surface, by accepting hydrogen from the amines to form water molecules, as a result of surface reactions, as illustrated in Figs. 3.54 and 3.55 [156].

Supercapacitors based on electrodes made with N-doped activated carbons have been widely reported [146, 148, 155, 156, 159, 160, 165–175]. Although the quaternary and pyridinic-N-oxide nitrogen groups could improve the capacitive behavior of activated carbons, because of the positive charge that increased the electron transfer at high current densities, the most important functional groups boosting the energy storage performance of the related supercapacitors were pyrrolic and pyridinic nitrogen, together with quinone oxygen [148].

Nitrogen functionalized porous carbons were prepared by activating a nitrogen-enriched carbon with KOH [166]. The nitrogen-enriched carbon was synthesized with the aid of pores of a mesoporous silica. It was observed that the pore volume was increased with increasing mass ratio of KOH/carbon. The optimized mass ratio was 5, which led to samples with the highest pore volume of $2.68 \ ml \cdot g^{-1}$. The porous carbon samples demonstrated promising performances as electrodes of supercapacitors. A capacitance of as high as $318 \ F \cdot g^{-1}$ was achieved in 6 M KOH electrolyte, with the sample derived by using the mass ratio of KOH/carbon to be 3.

Ordered mesoporous nitrogen-containing carbon (MCN-1) was obtained by using the mesoporous silica SBA-15 as the template [176]. Ethylene diamine and carbon tetrachloride were mixed with calcined SBA-15, under magnetic stirring for 2 hours at room temperature, so that the reactants would enter the pores of the SBA-15 channels. After that, the mixture was refluxed and stirred at 363 K for 6 hours for pre-polymerization. Then, the solid mixture was dried and carbonized at 873 K for 6 hours. The silica template SBA-15 was removed by using dilute hydrofluoric acid solution. The MCN-1 sample was collected,

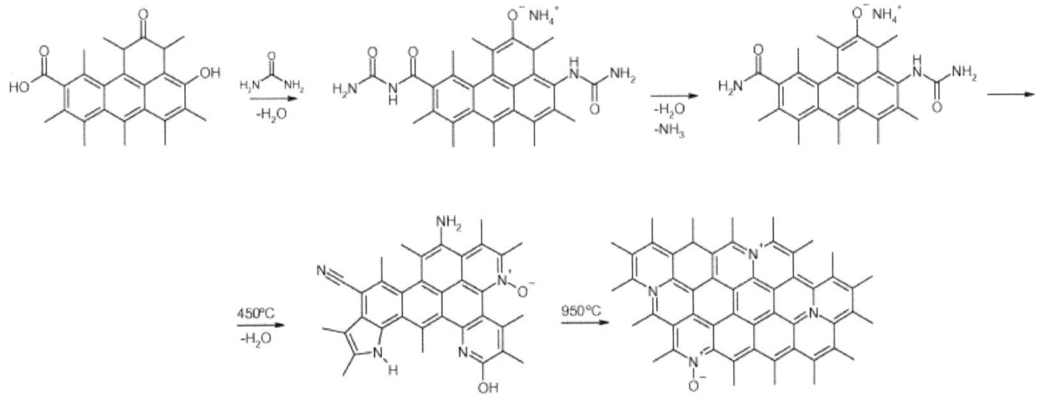

Fig. 3.54. Possible chemical reactions of urea with surface functional groups and thermal transformations. Reproduced with permission from [156], Copyright © 2008, Elsevier.

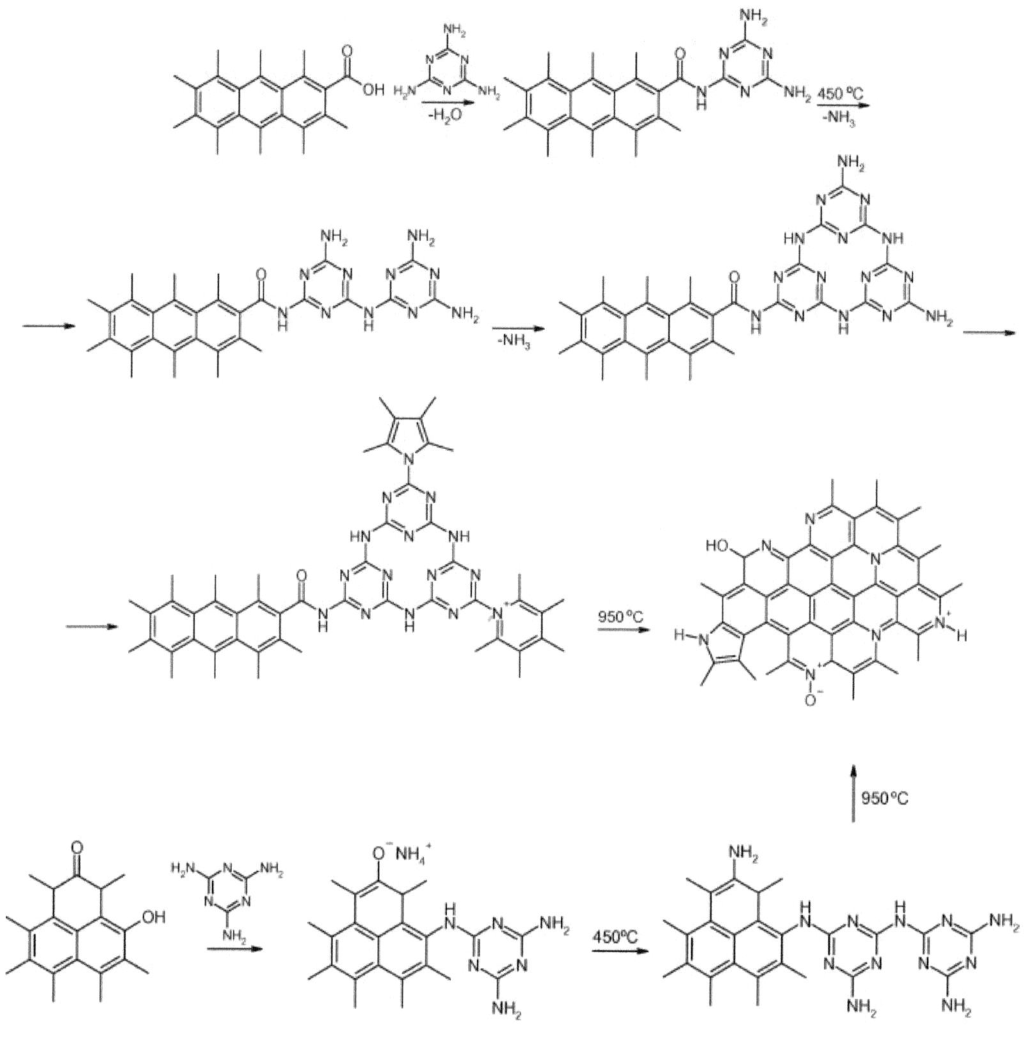

82 *Nanomaterials for Supercapacitors*

Fig. 3.55. Possible chemical reactions of melamine with surface functional groups and thermal transformations. Reproduced with permission from [156], Copyright © 2008, Elsevier.

washed and dried. To activate the MCN-1, it was mixed with KOH and heated at 1023 K for 1 hour in the flow of Ar. The samples were washed with hydrochloric acid solution and de-ionized water thoroughly and were dried at 373 K. The samples were denoted as MCN-1-ACn, in which AC means activated carbon and n was the mass ratio of KOH/MCN-1. Ordered mesoporous carbon CMK-3 was also synthesized by using SBA-15 as the template and sucrose as the carbon source [177]. After the same KOH activation, the samples were denoted as CMK-3-ACn.

Figure 3.56(a) shows nitrogen adsorption/desorption isotherms of the mesoporous MCN-1 and the activated samples MCN-1-Can, while their mesopore and micropore size distributions are illustrated in Fig. 3.56(b) and (c), respectively. All the MCN-1 and MCN-1-ACn samples contained both mesopores and micropores. The nitrogen adsorption/desorption values of the MCN-1-ACn samples were largely increased, as mass ratio of KOH/MCN-1 was increased, as seen in Fig. 3.56(a). The volume ratios of the mesopores to micropores were altered correspondingly. The sorption isotherms of MCN-1 and MCN-1-AC5 belonged to Type IV. The isotherms of MCN-1-AC1 and MCN-1-AC3 were very close to Type I, because

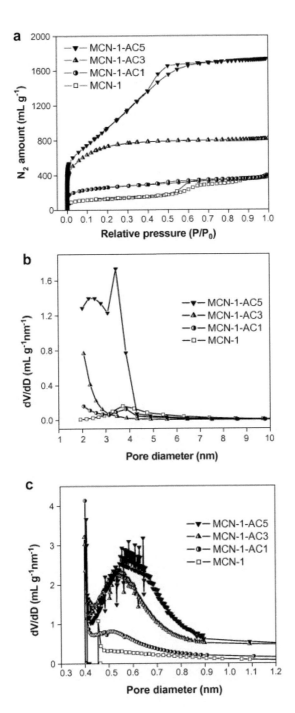

Fig. 3.56. Nitrogen adsorption/desorption isotherms (a), together with mesopore (b) and micropore size distributions (c) of the KOH-activated nitrogen-containing carbons. Reproduced with permission from [166], Copyright © 2009, Elsevier.

of the larger volume fraction of micropores. Accordingly, the volume distribution ratios of mesopore to micropore were 0.76 and 0.34 for MCN-1-AC1 and MCN-1-AC3, respectively.

The specific surface area, micropore volume, mesopore volume and total pore volume were all increased with increasing mass ratios of KOH/MCN-1. Specifically, the specific surface area and micropore volume were increased significantly from 916 to 2416 m²·g⁻¹ and 0.34 to 0.93 ml·g⁻¹, as the mass ratio of KOH/MCN-1 was increased from 1 to 3. The total pore volume of MCN-1-AC5 was as high as 2.68 ml·g⁻¹, together with a S_{BET} value of 2833 m²·g⁻¹, because of the formation of large amounts of mesopores (1.36 ml·g⁻¹) and micropores (1.32 ml·g⁻¹).

Figure 3.57 shows HRTEM images of the KOH activated samples, demonstrating the porous characteristics of the structures. As shown in Fig. 3.57(a), MCN-1 possessed long-range ordered mesopores, with average pore diameters of about 4 nm, which was confirmed by low-angle XRD results. In comparison, the long-range ordered mesopore domains were absent in MCN-1-AC1. However, short-range ordered mesopore domains were still retained in MCN-1-AC1, as seen in Fig. 3.57(b). As the mass ratio of KOH/MCN-1 was increased to 3 and 5, the pore structures became completely disordered. As a result, no low-angle XRD peaks were present in these two samples.

Figure 3.58(a) shows specific capacitance of the samples, with values quite high in the basic electrolyte. The values of MCN-1-AC3 and MCN-1-AC5 were 328 and 318 F·g⁻¹, respectively, at a scan rate of 2 mV·s⁻¹. As the scan rate was increased to 50 mV·s⁻¹, 75 and 81% specific capacitances were retained. The slightly higher specific capacitance retention ratio of the MCN-1-AC5 could be ascribed to its higher volume fraction of mesopores, mainly because the mesopores could provide fast ion channels, so as to improve the ion mobility at high scan rates [178–180].

Figure 3.58(b) shows Ragone plots of the samples, demonstrating the relationship between the output and the energy density, important parameters in terms of practical applications. Energy densities of MCN-1-AC3 and MCN-1-AC5 were pretty high, reaching about 8 W·h·kg⁻¹ at a power density of as high as 1000 W·kg⁻¹, ensuring their potential application as the electrodes of high energy density supercapacitors at high outputs [181]. Specific capacitances per unit micropore volume (C_{MV}) and per total pore volume

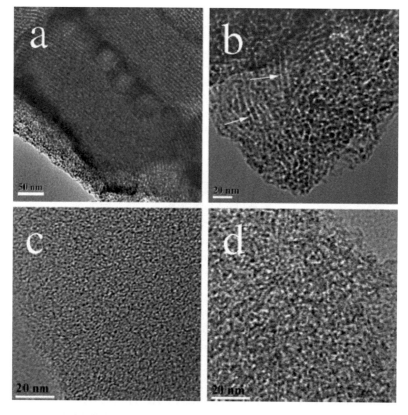

Fig. 3.57. HRTEM images of the samples: (a) pristine MCN-1, (b) MCN-1-AC1, (c) MCN-1-AC3 and (d) MCN-1-AC5, with scale bars to be 50, 20, 20 and 20 nm, respectively. Reproduced with permission from [166], Copyright © 2009, Elsevier.

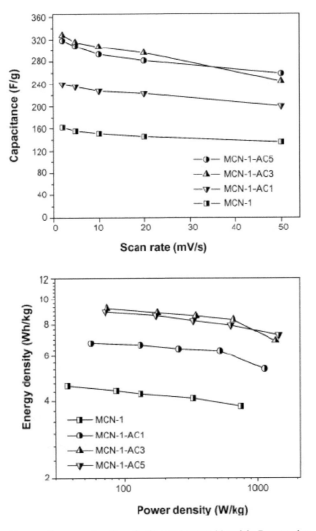

Fig. 3.58. Specific gravimetric capacitance as a function of voltage scan rates (a) and the Ragone plots showing the relationship between energy density and power density (b) for the KOH-activated nitrogen-containing carbons. Reproduced with permission from [166], Copyright © 2009, Elsevier.

(C_{TV}) of the samples were also compared. Generally, the values of capacitance were decreased as the processing parameters were increased. For example, from MCN-1 to MCN-1-AC1, to MCN-1-AC3, both the pore size and pore connectivity were decreased, as observed in Fig. 3.57. Such a decrease resulted in a hindering effect on the transport of the electrolyte ions, so that the efficiency of the pores was decreased accordingly. From MCN-1-AC3 to MCN-1-AC5, although both the pore size and pore connectivity were slightly increased, the specific capacitances were decreased, which could be attributed to the fact that the efficiency of the micropores of MCN-1-AC3 centered at 0.65 nm might be higher than that of the micropores of MCN-1-AC5 centered at 0.68 nm.

Activated carbons with pseudocapacitive effect caused by sulfur-containing functional groups, such as sulfone, sulfide or sulfonic groups, have been reported [174, 182–184]. One example is discussed as follows. In N_2, sublimed sulfur and CTAB were mixed, together with acetonitrile and benzene, which was then hydrothermally treated at 400°C for 16 hours [174]. After reaction, the precipitates were collected by centrifugation, followed by thorough washing with diluted hydrochloric acid, distilled water and ethanol absolute. The resulting powder samples after drying were denoted as NS-PCMSs-SU, in which the −SU was used to represent the first two letters of the sulfur sources. Similarly, NS-PCMSs-TH and NS-PCMSs-TM were also prepared by using thiourea and TMTD as the source of sulfur, respectively. The samples were further activated with KOH in Ar at 750°C for 1 hour, in order to enlarge the specific surface area and create more active sites [185]. The weight ratio of sample to KOH was 2:1. The final samples were denoted as NS-PCMSs-SU750, NS-PCMSs-TH750 and NS-PCMSs-TM750, respectively.

Figure 3.59 shows TEM images of the as-obtained NS-PCMSs without activation. Diameter of the NS-PCMSs was about 4 μm. Figure 3.60 shows SEM images of the as-synthesized NS-PCMSs and the corresponding samples activated with KOH. The unactivated samples had particle diameter in the range of 3–6 μm, while their morphology was uniform and nearly perfect. However, after activation with KOH, cracks were observed on surface of the samples, as illustrated Fig. 3.60(b, d and f). The presence of the cracks was the direct result of the etching effect of KOH, which led to collapse of the carbon microspheres. The corresponding EDS elemental mapping images are shown in the insets of Fig. 3.60(b, d and f), which indicated that the distributions of S and N in the activated NS-PCMSs were quite homogeneous, implying that both S and N had been incorporated into the NS-PCMSs.

N_2 adsorption-desorption isotherms indicated that all the NS-PCMSs exhibited a hierarchical pore structure, with relatively lower specific surface areas in the range of 397.4–413.8 $m^2 \cdot g^{-1}$. X-ray Photoelectron Spectroscopy (XPS) results revealed that both S and N were well incorporated into the carbon frameworks. Electrochemical characterization results demonstrated that the NS-PCMSs had superior specific capacitance of 242–295 $F \cdot g^{-1}$ at 0.1 $A \cdot g^{-1}$, excellent rate capability with specific capacitance to 216–247 $F \cdot g^{-1}$ at 10 $A \cdot g^{-1}$ and good cyclic stability without almost no decay after 10000 cycles.

Composites pseudocapacitors

Pseudocapacitance has also been realized in activated carbon based supercapacitors by using composite structures consisting activated carbons and electroactive items, such as metal oxides or conducting polymers. Metal oxides include Ru_2O, NiO, MnO_2, Fe_3O_4, $NiCo_2O_4$, etc., while conducting polymers could be polyaniline, polypyrrole, polythiophene and their derivatives. Usually, metal oxides are nonconductive, with very low electrical conductivity. When metal oxides are embedded in the activated carbon matrix, the whole composites will have sufficiently high conductivity to achieve high electrochemical performances of the electrodes. As the activated carbons are combined with conducting polymers, the carbon matrix has a special function to buffer the volume variation of the polymers caused by the charge/discharge processes, so as to improve the cycling performance. At the same time, the presence of conducting polymers provide additional conductivity to the electrodes, which is also beneficial to the electrochemical behaviors.

Ruthenium oxide (RuO_2) has the highest pseudocapacitive effect among all oxides. Due to its very high cost, it is impossible to use RuO_2 for large scale applications. To address this issue, one strategy is to minimize the amount and maximize its efficiency, which is realized by dispersing it over high surface area materials. Therefore, activated carbons are candidates for such a purpose [186–191]. For example,

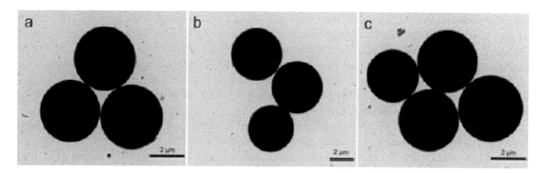

Fig. 3.59. TEM images of different carbons: (a) NS-PCMSs-SU, (b) NS-PCMSs-TM and (c) NS-PCMSs-TH. Reproduced with permission from [174], Copyright © 2016, Elsevier.

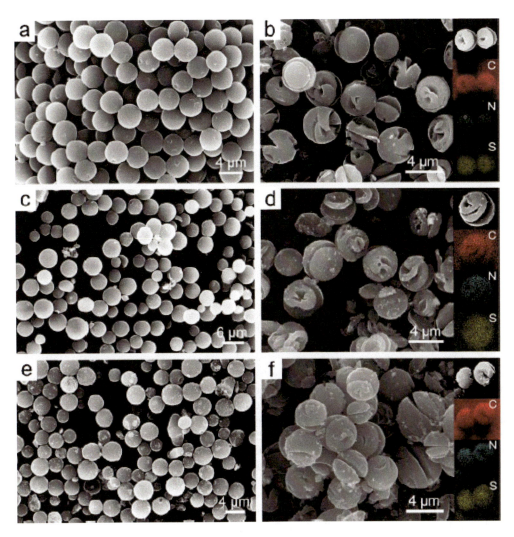

Fig. 3.60. SEM images of the samples: (a) NS-PCMSs-SU, (b) NS-PCMSs-SU-750, (c) NS-PCMSs-TM, (d) NS-PCMSs-TM-750, (e) NS-PCMSs-TH and (f) NS-PCMSs-TH-750, with the insets showing the corresponding SEM elemental-mapping images of carbon, nitrogen and sulfur. Reproduced with permission from [174], Copyright © 2016, Elsevier.

it was found that specific capacitance of activated carbon could be increased by 20% to 308 F·g^{-1}, if 7.1 wt% Ru was loaded [188]. In this study, ruthenium oxide was loaded into various types of activated carbon by dispersing the activated carbon powders in an aqueous solution of RuCl$_3$ followed by a neutralization process.

In a separate study, varying levels of RuO$_2$ were loaded on activated carbons by using an electroless deposition process [187]. The electroless deposition of Ru on activated carbon was conducted in an alkaline bath. Carbon particles were immersed in an aqueous bath of RuCl$_3$, sodium hypophosphite, di-ammonium hydrogen citrate and ammonium oxalate. pH value and bath temperature were controlled to prevent precipitation of Ru. The amount of carbon used for deposition was kept constant, while the amount of RuCl$_3$ in the bath was varied in order to control the final loading of Ru in the composites. Composites with 5, 9, 15 and 20 wt% RuO$_2$ loaded on carbon were prepared to study the effect of RuO$_2$ on electrochemical performance of the materials.

Figure 3.61 shows capacitances of the composite electrodes as a function of discharge current density. It was observed that capacitance of the 20 wt% composite was decreased with increasing current densities

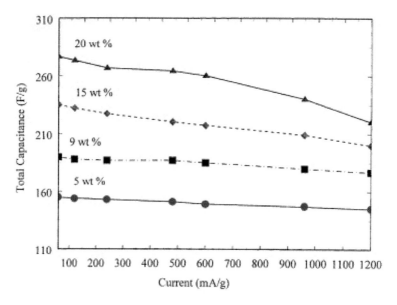

Fig. 3.61. Total capacitance as a function of current density for the RuO$_2$–AC composite electrodes. Reproduced with permission from [187], Copyright © 2001, Elsevier.

faster than that of the 5 wt% composite. With 5 wt% loading of RuO$_2$, the capacitance remained almost constant over the whole range of the discharge current. In comparison, for the 20 wt% RuO$_2$ sample, the specific capacitance started decreased quickly as the discharge current was higher than 250 mA·g^{-1}. SEM results indicated that as the RuO$_2$ loading was increased, large clusters of Ru nanoparticles were formed on surface of the carbon particles. Therefore, the path of proton transfer was increased. In this case, mass transportation was limited at high current densities, thus leading to the decrease in capacitance.

As stated earlier, manganese oxides are among the most promising transition metal oxides for supercapacitor applications, because of their cost-effectiveness, environmentally friendliness, ideal capacitive behavior and high operating safety. However, pure manganese oxides are nonconductive materials, with very low electrical conductivity, so that their cycling stability and rate capability cannot meet the requirements of practical applications. One of the ways to address this problem is to incorporate manganese oxides with activated carbons [192–198].

A simple chemical coprecipitation method was employed to coat MnO$_2$ nanostructures onto activated mesocarbon microbeads (a-MCMB) [193]. Characterization results indicated that the nano-MnO$_2$ had a structure like a short fiber, with a length of about 200 nm and a width of about 60 nm. The composite exhibited promising performance as the electrode materials of supercapacitors. Electrochemical results revealed that the composite electrode demonstrated a maximum specific capacitance of 183 F·g^{-1} in 1 M LiPF$_6$ (EC + DMC) organic electrolyte, with operating voltages of up to 3.0 V. Specific capacitance base on the mass of MnO$_2$ was 475 F·g^{-1}, corresponding to an energy density of 106 W·h·kg^{-1}.

The MCMB had an average diameter of 25 μm and a specific surface area of 32.6 m^2·g^{-1}. It was activated with KOH, with mass ratio of the MCMB to KOH to be 5:1. The mixture was first added with certain amount of deionized water and then heated 850°C for 1 hour in the flow of Ar. The activated sample, a-MCMB, was washed and dried. To load MnO$_2$, the a-MCMB was added to 0.05 M KMnO$_4$ solution for 15 minutes with stirring, so that KMnO$_4$ was fully absorbed onto surface of the a-MCMB particles. After that, 0.075 M Mn(Ac)$_2$·4H$_2$O solution was added drop by drop into the mixed solution by stirring. The whole mixed solution was vigorously stirred for 8 hours to trigger the precipitation reaction. The precipitates were filtered and washed with distilled water and ethanol and finally dried in vacuum at room temperature. For comparison, pure MnO$_2$ without a-MCMB and MCMB/MnO$_2$ composite with non-activated MCMB were also prepared in a similar way. The presence of MnO$_2$ was confirmed by XRD results.

SEM images of the a-MCMB and pure MnO$_2$ are shown in Fig. 3.62(a) and (b), respectively. The a-MCMB had a spherical morphology with a relatively rough surface. Cracks were observed occasionally on the spheres, which was attributed to the etching effect of KOH at high temperatures [199]. The pure MnO$_2$ without the presence of a-MCMB possessed agglomerates with sizes in a wide range of 0.1–10 µm, as seen in Fig. 3.62(b). SEM images of the AMMO (a-MCMB/nano-MnO$_2$ composite) at different magnifications are shown in Fig. 3.62(c) and (d). The AMMO possessed typical composite microstructure, with the MnO$_2$ nanoparticles to be deposited over the spherical a-MCMB particles, as illustrated in Fig. 3.62(c). The morphologies of the MnO$_2$ on the a-MCMB were significantly different from the agglomerated pure MnO$_2$. During the formation of MnO$_2$, the reaction between Mn (II) and Mn (VII) through chemical coprecipitation easily led to agglomeration, due to the absence of surfactants. In this regard, the a-MCMB acted as dispersing agent to support the nano-MnO$_2$ without the requirement of surfactants.

Figure 3.63 shows a schematic diagram describing the possible formation mechanism of the composites, consisting of nano-MnO$_2$ on the surface of the a-MCMB particles. After the activation reaction, the MCMB became a-MCMB with high-adsorption capacity, due to high specific surface area and formation of various oxygen functional groups. As a result, when the a-MCMB particles were added into the KMnO$_4$ solution, the Mn^{7+} in the solution were adsorbed onto the a-MCMB particles, thus forming mixture of a-MCMB + KMnO$_4$, so that the a-MCMB served as a carrier for the reactants. As the Mn(Ac)$_2$·4H$_2$O was added into the solution drop by drop, Mn^{2+} ions were oxidized to the Mn^{4+} ions on the surface of the a-MCMB particles, while the resulting Mn^{4+} interacted with water to form MnO$_2$ nanoparticles.

With increasing concentration of Mn^{2+}, the initially formed MnO$_2$ acted as nucleation centers, thus leading to more and more MnO$_2$ that were deposited onto surface of the a-MCMB particles. As a result,

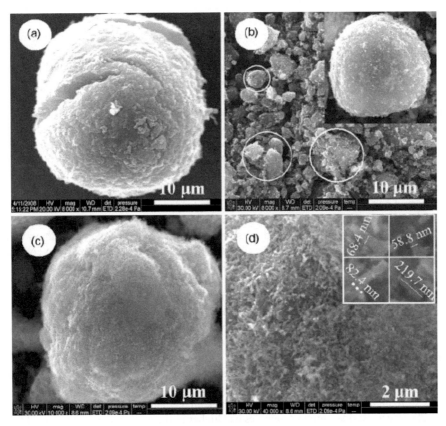

Fig. 3.62. SEM images of (a) pure a-MCMB, (b) pure MnO$_2$ with the insets showing the MCMB/MnO$_2$ composites at the top right corner, (c) AMMO at 10,000 × magnification and (d) AMMO at 80,000 × magnification, with the inset at the top right corner showing the dimensions of the particles. Reproduced with permission from [193], Copyright © 2010, Elsevier.

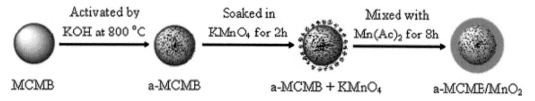

Fig. 3.63. Schematic diagram demonstrating the growth process of the nano-MnO$_2$ onto the a-MCMB particles. Reproduced with permission from [193], Copyright © 2010, Elsevier.

MnO$_2$ nanostructures were gradually formed along the broad spherical-surface of the a-MCMB particles. In this regard, the a-MCMB particles acted as a substrate to support the formation of the carbon/MnO$_2$ composite, which had several advantages as compared with the carbonaceous materials with lower specific surface area and irregular shape. Firstly, the high specific surface area of the a-MCMB sample was a critical parameter to ensure the effective adsorption of the reactant particles. Secondly, the nearly perfect spherical a-MCMB particles provided a broad surface for the growth of the nano-MnO$_2$ at a large scale. Lastly, the oxygen functional groups on the surface of the a-MCMB particles promoted the dispersion and attachment of the nano-MnO$_2$ in the composite [200].

Nickel/cobalt oxide/hydroxide is another electroactive species commonly used in pseudocapacitors [201–205]. For instance, a nickel hydroxide/activated carbon composite was prepared by using a simple chemical precipitation method [201]. With the presence of 2–6% Ni(OH)$_2$, the specific capacitance was increased by about 23.3% to 292–314 F·g^{-1}, as compared with that of the pure activated carbon (255 F·g^{-1}). However, if the content of Ni(OH)$_2$ was increased by 8–10%, the capacitance was decreased to 261–302 F·g^{-1}. In addition, the composite electrodes had promising electrochemical performances and high charge–discharge stability.

Among all conducting polymers, polyaniline (PANI) is the most promising candidate used to prepare activated carbon/conducting polymer composites, because of its high conductivity and high stability and relatively low cost [206–210]. For example, an electrodeposition method was used to deposit PANI onto microporous activated carbon fabric, resulting in electrode materials with specific capacitances of up to 320 F·g^{-1} in H$_2$SO$_4$ [206]. Activated Carbon Fabrics (ACF) made from polyacrylonitrile (PAN), with a BET surface area and pore volume of 1200 m^2·g^{-1} and 0.59 cm^3·g^{-1}, respectively, were used as the precursor. The ACF was pretreated at 900°C for 20 minutes in N$_2$ flow to remove the surface oxides and volatile impurities.

Aqueous aniline solutions with concentrations of 10–40 mM were prepared in N$_2$ purge. The thermally treated ACF and the aniline solutions in glass-stoppered flask were put in a 25°C shaker bath at a shaker rate of 100 rpm. After reaching equilibrium, the aniline-loaded ACF samples were taken out from the solution. The aniline-loaded ACFs were adhered to a stainless steel current collector to serve as the working electrode for electrochemical polymerization of the loaded aniline. Polymerization was carried out in 1 M H$_2$SO$_4$, using Pt and Ag/AgCl as the counter and reference electrodes, respectively. A potential of 0.85 V was applied to the ACF electrode for PANI deposition, which was denoted as sequence A. For comparison, polymerization experiments in a conventional way, i.e., sequence N, were conducted in 1 M H$_2$SO$_4$ solutions containing different amounts of aniline monomer.

Figure 3.64 shows discharge capacitance measured at 1 mA as a function of the content of PANI. In both groups, the capacitance was increased initially with the amount of PANI, reached a maximum and then decreases as the amount of PANI was further increased. The decrease in capacitance at high contents of PANI could be attributed to the variation of Open Circuit Potential (OCP). The potential window of a constituting electrode in the two-electrode capacitor is near the OCP of the electrode. Experimental results indicated that an increase in the amount of PANI led to a decrease in OCP. The capacitances of group A capacitors were higher than those of group N ones, which could be related to the higher OCP of the sequence A electrodes. Besides, group A samples had higher intrinsic capacitance, because of their lower resistance for electrolyte migration.

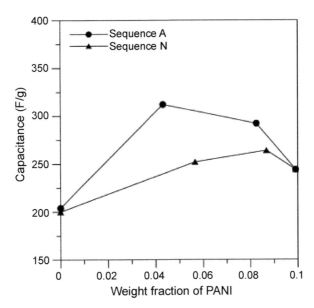

Fig. 3.64. Discharge capacitances measured at 1 mA over 0.7–0 V of the supercapacitors assembled with sequences A and N electrodes. Reproduced with permission from [206], Copyright © 2003, Elsevier.

Hybrid supercapacitors

Hybrid supercapacitors have been acknowledged to be the next generation of energy storage devices that have higher energy densities for practical applications [167, 211–217]. The positive and negative electrodes in a hybrid supercapacitor are made of different materials, with one of them always belonging to EDLC-type based on carbon materials. Hybrid supercapacitors can be classified into two groups, (i) asymmetrical supercapacitors, i.e., EDLC-type electrode plus pseudocapacitance-type electrode and (ii) lithium-ion capacitors or battery-like capacitors, i.e., EDLC-type electrode plus battery-type electrode. Due to the large difference in performance between the two types of electrodes, their mass balance has been a critical requirement to fully utilize each electrode to achieve optimal cycle capability [218].

Asymmetrical supercapacitors

Asymmetrical capacitors consist of an EDLC-type electrode based on carbon materials and a pseudocapacitance-type electrode made of metal oxides/nitrides, conductive polymers or carbon/pseudocapacitive composites [219–221]. Asymmetrical capacitors have the advantages of both fast redox reversible reactions of pseudocapacitors and ultra-fast electrostatic ion adsorption and robust cycle life of carbon materials. In most cases, the negative electrode is composed of an activated carbon, whereas the positive electrode comprises the pseudocapacitive material. This particular configuration allows a substantial enlargement of the working voltage window when working with aqueous electrolytes (sometimes well beyond the thermodynamic limit of 1.2 V) as a result of the high overpotentials for H_2 and O_2 evolution at the carbon-based negative electrode and the pseudocapacitance-type positive electrode, respectively. This enhances the energy density of the device even more.

For instance, CoAl double hydroxide has been used as a positive electrode to assemble hybrid supercapacitors, together with activated carbons in 6 mol·L^{-1} KOH electrolyte solution [222]. Operational voltage of the AC/nanostructured CoAl double hydroxide could be as high as 1.5 V, corresponding to energy density of 15.5 Wh·kg^{-1}, power density of 0.3 kW·kg^{-1} and 90% retention of the initial capacity after 1000 cycles. Commercially available activated carbon was used in the experiment. Cobalt aluminum double hydroxide was synthesized by using a simple wet-chemical route. 0.2 mol·L^{-1} Co(NO$_3$)$_2$·6H$_2$O and

0.1 mol·L^{-1} Al(NO$_3$)$_3$·7H$_2$O solutions, with mole ratio of Co/Al to be 2:1, were mixed and co-precipitated in 2 mol·L^{-1} NaOH solution containing 2 mol·L^{-1} Na$_2$CO$_3$ at 40°C. After that, the suspension was stirred at 70°C for 48 hours. The precipitate was then collected through centrifugation and thoroughly washed with distilled water. The precipitate was finally dried at 80°C for 12 hours.

XRD results indicated that the CoAl double hydroxide had a composition to be very close to Ni$_6$Al$_2$(OH)$_{16}$(CO$_3$·OH)·4H$_2$O, so that the obtained materials were of layered structure. Figure 3.65 shows SEM and TEM images of the CoAl double hydroxide, which was in the form of aggregation comprising of nanoparticles with sizes in the range of 60–70 nm. The CoAl double hydroxide had a composition of Co$_{0.67}$Al$_{0.33}$(CO$_3$)$_{0.165}$(OH)$_2$·nH$_2$O ([Co$_{1-x}$Al$_x$(OH)$_2$]$^{x+}$·[CO$_3^{2-}$]$_{x/2}$·[H$_2$O]$_z$, $x = 0.33$).

Another example is stabilized Al-substituted α-Ni(OH)$_2$ that was prepared by using a chemical coprecipitation method [223]. Optimized composition was 7.5% Al-substituted α-Ni(OH)$_2$, which demonstrated a high specific capacitance of 2.08×10^3 F·g^{-1} and excellent rate capability, due to the high stability of Al-substituted α-Ni(OH)$_2$ in alkaline electrolyte solutions. Asymmetric type supercapacitor of pseudo/electric double-layer was constructed with the α-Ni(OH)$_2$ and activated carbon as the positive and negative electrodes. It exhibited a maximum specific capacitance of 127 F·g^{-1} and specific energy of 42 W·h·kg^{-1}, within the voltage window of 0.4–1.6 V. The hybrid supercapacitor demonstrated pretty good electrochemical stability, with 82% retention of the initial capacitance over 1,000 cycles.

To prepare the Al-substituted α-Ni(OH)$_2$, two metal salts solution with a total concentration of 1.5 M total concentration, containing NiCl$_2$·6H$_2$O and Al$_2$(SO$_4$)$_3$, were mixed under stirring. Molar ratios of NiCl$_2$·6H$_2$O/Al$_2$(SO$_4$)$_3$ included 1:0.031, 1:0.042, 1:0.053, 1:0.067 and 1:0.083. The mixed solution was slowly adjusted to pH = 9 by the addition of 5 wt% NH$_4$OH at 10°C, where the NH$_4$OH solution was added dropwise at a time interval of 5 seconds. The suspension was stirred at the temperature for another 3 hours. After that, the precipitate was collected through filtration and thoroughly washed with distilled water, followed by drying at 100°C in air for 6 hours. Commercial AC was used as the negative electrode, which had a specific surface area of 2,000 m^2·g^{-1} and diameters of 5–10 μm.

XRD results of the aged samples indicated that the 7.5 and 9.16% Al-substituted Ni(OH)$_2$ experienced almost no changes in terms of the structure of α-Ni(OH)$_2$ after ageing at room temperature for 15 days in 6 M KOH. All the samples exhibited a phase composition of α-Ni(OH)$_2$·0.75H$_2$O. However, there was no Al, the sample would be entirely converted to β-phase. For the sample of 4.72% Al-substituted α-Ni(OH)$_2$ sample, although no peaks of β-phase were observed, the peaks of α-phase were widened and weakened after ageing for 15 days. Therefore, 7.5% Al was the minimum content in order to achieve high crystallinity and structural stability of α-Ni(OH)$_2$ in alkaline medium.

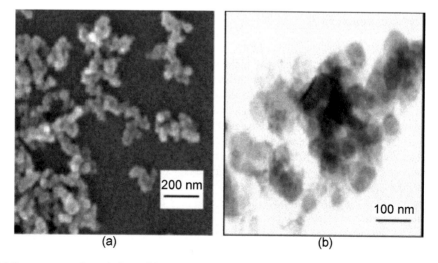

Fig. 3.65. Microstructure and morphology of the CoAl double hydroxide: (a) SEM and (b) TEM images. Reproduced with permission from [222], Copyright © 2006, Elsevier.

Another group of asymmetric supercapacitors that are worth mentioning is those based on manganese oxides with activated carbons [224, 225]. For example, a study was reported on the effect of electrodes mass ratio of MnO_2/activated carbon supercapacitors on the potential window of the two electrodes, in order to optimize the operating voltage [225]. It was found that the theoretical mass ratio of $R = 2$, which was calculated by taking into account the equivalent charge passing across both the electrodes, was underestimated. Therefore, optimized R values should be in the range of 2.5–3 instead, so that the extreme potential for each electrode was close to the stability limits of the electrolyte and active materials, thus leading to the maximum voltage. For the galvanostatic cycling up to 2 V, $R = 2.5$ resulted in the highest performance. Specific capacitance was increased from 100 to 113 $F \cdot g^{-1}$ after 2000 cycles, which was slightly declined after 6000 cycles and eventually stabilized at 100 $F \cdot g^{-1}$. SEM observation indicated that the manganese electrode experienced a significant morphological variation after thousands of cycles, which was attributed to oxidation and dissolution of Mn(IV) at high potentials.

Similarly, asymmetric supercapacitor with high energy density has been reported by using graphene/MnO_2 composite as positive electrode and activated carbon nanofibers (ACN) as negative electrode in a neutral aqueous of Na_2SO_4 electrolyte [226]. The asymmetric supercapacitor constructed with optimized conditions exhibited a stable cycling reversibility within the voltage window of 0–1.8 V, with maximum energy density of 51.1 $Wh \cdot kg^{-1}$, which is much higher than that of the MnO_2//DWCNT counterparts with an energy density of 29.1 $Wh \cdot kg^{-1}$. In addition, the graphene/MnO_2//ACN asymmetric supercapacitor demonstrated excellent cycling durability, with 97% specific capacitance retention after 1000 cycles.

The graphene/MnO_2 composites were prepared through redox reaction between graphene and potassium permanganate with microwave irradiation [226]. In the experiment, graphene was ultrasonically treated for 1 hour, after which $KMnO_4$ powder was added into the suspension. The resultant suspension was heated in a household microwave oven (2450 MHz, 700 W) for 5 minutes. As a result, a black deposit was obtained, which was collected through filtration, followed by thorough washing with distilled water and alcohol and drying at 100°C for 12 hours in vacuum.

In a similar way, asymmetric capacitors made of AC//conducting polymers, also exhibited enhanced cyclic stability as compared with the symmetric capacitors derived from conducting polymers only [227–229]. For instance, an AC//doped poly(3-methylthiophene) supercapacitor was fabricated, with over 10,000 cycles in an organic electrolyte at 20 $mA \cdot cm^{-2}$ at cut off potentials of 1.0–3.0 V [227]. Both DLCS and the C//pMeT supercapacitors were constructed with two composite electrodes that were separated by two or four sheets of microporous polytetrafluoroethylene (PTFE) separators, with ~ 25 μm in thickness and 80% porosity, while propylene carbonate (PC)-tetraethylammonium tetrafluoborate ($TEABF_4$) with a concentration of 1.0 $mol \cdot L^{-1}$ as the electrolyte. The composite carbon electrodes were made of 90% commercially available activated carbon, whereas the composite polymer electrodes were prepared with 55 and 80% poly(3-methylthiophene) or pMeT. The composite electrodes with densities of 7–18 $mg \cdot cm^{-2}$ were fabricated by mixing active material carbon conducting additive and binder, carboxy methylcellulose (Cmc) and Teflon in the form of a paste. The paste was then laminated on the current collector made of stainless steel grid and dried at 80°C in vacuum.

A prototype of AC//poly(4-fluorophenyl-3-thiophene) was fabricated, which could be sustained for 8000 cycles over the voltage range 1.0–3.0 V in the organic electrolyte. Although half of the initial capacitance was lost after 100 cycles, a constant value was retained after that, offering an energy density of 3.3 $Wh \cdot kg^{-1}$ [228]. The coupling reaction of 3-bromothiophene and 4-fluorophenylmagnesium bromide in tetrahydrofuran (THF) resulted in the monomer, in which $NiCl_2$ (diphenylphosphinopropane) was used as the catalyst, with a yield of > 90 wt%. The monomer 4-FPT was polymerized through the direct oxidation reaction in chloroform or dichloromethane with $FeCl_3$ as oxidant, having a yield of > 80 wt%.

Lithium-ion supercapacitors

The second group of hybrid supercapacitors is called lithium-ion capacitors or battery-type capacitors, consisting of a battery-type electrode (e.g., lithium intercalation compounds) and an EDLC-type electrode (i.e., mainly various carbon materials). Such hybrid supercapacitors possessed various advantages, such as rapid charging rate, high cycle stability and relatively high energy density. It has been recognized that

94 *Nanomaterials for Supercapacitors*

because the intercalation-deintercalation of Li^+ in the battery-type electrode takes place at a shallower State-Of-Charge (SOC), as compared with the Li-ion batteries working with the same electrode, higher cycling stability and safety are expected. In this case, activated carbons served as electrode materials, acting as either positive or negative electrodes, which is dependent on redox potential of the Li-intercalation compounds [218]. Various Li-ion hybrid supercapacitors have been developed and explored, with several examples discussed as follows [230–237].

The very first hybrid battery/supercapacitor cell was assembled with activated carbon as positive electrode and nanostructured $Li_4Ti_5O_{12}$ as negative electrode, with $LiPF_6$ EC:DMC as electrolyte in which PF_6 anions were adsorbed on surface of the ACs, while Li^+ cations intercalated into the lattice of $Li_4Ti_5O_{12}$, thus leading to $Li_xTi_5O_{12}$ [230]. With the presence of intercalation compounds acting as the negative electrode, the electrode voltage could be readily pinned at potentials that are sufficiently negative with respect to SHE, so as to attain high gravimetric and volumetric energy densities. The cell had a sloping voltage profile decreasing from 3 to 1.5 V, due to the flat two phase intercalation voltage profile of $Li_4Ti_5O_{12}$ and the typical linear voltage profile of carbon EDLC, with 90% capacity utilization at a charge/discharge rate of 10C and only 10–15% capacity loss after 5000 cycles.

Although lithium metal reference electrode can be used to accurately characterize three-electrode electrochemical behaviors in Li^+ electrolytes, it is quite complicated to fabricate Li reference electrodes, while they cannot reused in most cases [230]. In comparison, quasi-reference Ag metal electrodes that have been widely used in aqueous electrochemical cells can be used to replace Li reference electrodes, although it has a voltage drift of 15 mV. Three-electrode cells were prepared, which consisted of Teflon Swagelok™ bodies and seals with oxidation and reduction resistant stainless steel plungers. The third electrode was made of a flattened metal wire that was inserted into the cell. The reference electrode was placed between two sheets of borosilicate glass fiber separators.

The three-electrode characterization was used to calibrate and verify accuracy of the Ag quasi-reference electrode [230]. The Ag quasi-reference electrode and Li metal reference electrodes were assembled into a three-electrode cell, with $LiCoO_2$ and mesocarbon microbeads (MCMB) 25–28 graphite to be the positive and negative electrodes, respectively. The Li and Ag reference electrodes exhibited almost the same behavior. By using the Ag quasi-reference electrode and the Li reference electrode, Solid Electrolyte Interface (SEI) was formed on graphite at –2.2 and 0.8 V, for the Ag quasi-reference and Li/Li^+ reference cell, respectively. It was found that very small voltage drift during cycling was observed for the Ag alkali quasi-reference electrode. Therefore, the Ag quasi-reference can be satisfactorily used to replace the Li/Li^+ reference electrodes. The Ag quasi-reference electrode potential could be calibrated to be about 3.0 V vs. Li/Li^+, which is very close to that of the standard hydrogen electrode (SHE).

Another example is to replace $Li_4Ti_5O_{12}$ with TiO_2–B as the negative electrode, to construct asymmetric supercapacitors with AC as the positive electrode and $LiPF_6$ EC:DMC as electrolyte [231]. With this device, energy densities of 45–80 $Wh \cdot kg^{-1}$ and power densities of 0.2–0.4 $kW \cdot kg^{-1}$ were achieved. The maximum cell voltage could be in the range of 2.75–3.5 V. To synthesize the titanate powder, $K_2Ti_4O_9$ as the starting material was prepared by thermally treating the mixture of KNO_3 and TiO_2, with a molar ratio of 1:2, at 1000°C for 6 hours [238]. The treated sample was crushed into a powder, which was then hydrolyzed for 3 days in HNO_3 with concentrations of < 0.5 M. After filtering, the powder was calcined at 500°C in air for 15 hours.

Figure 3.66 shows XRD pattern of the TiO_2 (B) sample after hydrolysis and thermal treatment. It was of a monoclinic crystal structure (space group C 2/m), lattice parameters of $a = 1.2163(5)$ nm, $b = 0.3735(2)$ nm, $c = 0.6451(2)$ nm and $\beta = 107.29(5)$. The structure was constructed with TiO_6 octahedra arranged in the perovskite ReO_3 subunits assembled in the same plane (a, c) through edge-sharing. An idealized 3D framework was formed due to the edge-sharing subunits. There were parallel infinite channels along the [010] axis, which provided with intercalation sites at which Li^+ ions could be accommodated without significant structural distortion [239]. Figure 3.67 shows SEM images of the TiO_2 (B) powder, which consisted of needle shaped grains, with sizes at the scale of several micrometers in length and a prismatic section of 0.3 μm × 0.1 μm. EDX analysis revealed that less than 1 at % K per TiO_2 unit formula was retained in the sample.

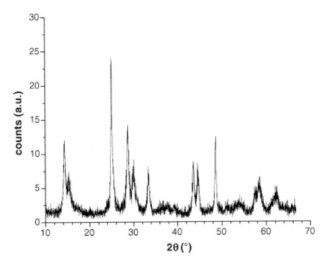

Fig. 3.66. XRD pattern of the TiO$_2$ (B) powder. Reproduced with permission from [231], Copyright © 2006, Elsevier.

Fig. 3.67. SEM images of the starting TiO$_2$ (B) powder: (a) general view of the grains size and shape and (b) detail of a particle confirming the 'bi-dimensional' microstructure, with bars of 1 μm. Reproduced with permission from [231], Copyright © 2006, Elsevier.

Specifically, long term stability of the hybrid device was observed at different cell voltages, as revealed in Fig. 3.68. The hybrid cell based on the TiO$_2$ (B) electrode could reach its nominal energy density after only a few cycles. It was known as activation process, which was related to the faradic behavior of the TiO$_2$ (B) electrode because lithium intercalation was restricted. This step could be the key factor to achieve high capacity and long term cycling behavior for the hybrid cells. At cell voltage of 3.5 V and high cycling rate, the system delivered a relatively low specific energy of 30 Wh·kg^{-1}. After 600 cycles, the energy density was decreased to Wh·kg^{-1}. This could be avoided in packaged devices.

A non-aqueous (LiPF$_6$ in EC:DEC) Li-ion capacitor with AC as the positive electrode and sub-micron sized LiCrTiO$_4$ particles as the anode was reported [232]. Electrochemical Li-insertion properties were evaluated with half-cell configurations (Li/LiCrTiO$_4$) by using both the galvanostatic and potentiostatic modes over 1–2.5 V vs. Li. Reversible insertion of about one mole of lithium (~ 155 mA h·g^{-1}) was at a low current rate of 15 mA g^{-1}, offering an excellent cycling performance. A non-aqueous Li-ion electrochemical hybrid capacitor (Li-HEC) was assembled with an optimized mass loading of Activated Carbon (AC) as the cathode and synthesized LiCrTiO$_4$ as the anode in 1 M LiPF$_6$ in ethylene carbonate–diethyl carbonate solution, which was cycled over 1–3 V at ambient conditions. The Li-HEC delivered maximum specific energy density and power density of 23 W h·kg^{-1} and 4 kW·kg^{-1}, respectively.

A similar non-aqueous (LiPF$_6$ in EC:DMC) cell was constructed with activated graphitized carbon as the positive electrode and AC/Li$_4$Ti$_5$O$_{12}$ composite as the positive electrode [233]. A symmetric sandwich-type SC cell with identical activated carbon electrodes and different organic electrolytes was also assembled for comparison. Four types of carbons, with different specific surface areas in the range of 1,000–1,600 m^2·g^{-1} and texture parameters, were tested. At the same time, three types of organic electrolytes, including Et$_4$NBF$_4$–Propylene Carbonate (PC), LiBF$_4$–PC and LiPF$_6$–DMC/EC for the symmetric SC cells, were evaluated and compared. Specific capacitance of the symmetric SC was up to 70 F·g^{-1}, while that of the asymmetric SC cells could reach 150 F·g^{-1}. Both types of supercapacitors exhibited promising cycling behaviors.

Because aqueous electrolytes can be used for Li-ion capacitors, the fabrication and packaging process could be extremely simplified, as compared with unlike Li-ion batteries. In this regard, an asymmetric cell was assembled with AC as the negative electrode and LiMn$_2$O$_4$ as the positive electrode in a mild Li$_2$SO$_4$ aqueous electrolyte [234]. The charge/discharge processes involved only the transport of Li ions between the two electrodes, similar to those the conventional Li-ion batteries. As a result, the electrolyte acted as an ionic conductor, so that it was not consumed during the charge/discharge processes. Therefore, the problem of electrolyte depletion during the charge process of the conventional Li-ion capacitors could

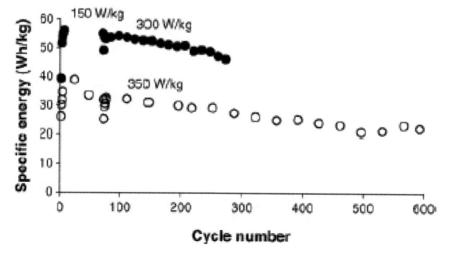

Fig. 3.68. Long term stability of the TiO$_2$ (B)/PICACTIF hybrid cells at constant cycling powers, cell voltage of 2.75 V (black dots) and cell voltage of 3.5 V (open dots). Reproduced with permission from [231], Copyright © 2006, Elsevier.

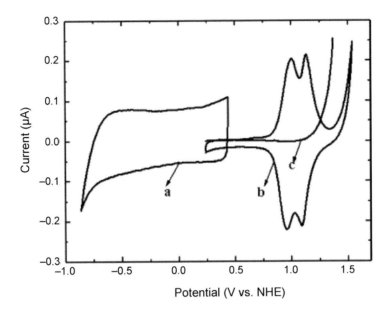

Fig. 3.69. CV curves of (a) AC, (b) LiMn$_2$O$_4$ and (c) carbon black at a scan rate of 10 mV·s^{-1}. Reproduced with permission from [234], Copyright © 2005, Elsevier.

be effectively prevented [235, 236]. The hybrid cell exhibited a sloping voltage profile over the voltage window of 0.8–1.8 V, offering a specific energy of 35 W·h·kg^{-1}. Moreover, the cell demonstrated promising cycling behavior, with < 5% capacity loss after 20,000 cycles at 10 C charge/discharge rate.

In this case of lithium doped Li$_{1.1}$Mn$_{1.9}$O$_4$ as the positive electrode (LiMn$_2$O$_4$), the excess lithium ion not only stabilized its crystal structure, but also modified the voltage profile from a flat plateau to a sloping plateau. This offered an opportunity to control the charge cut-off voltage of the lithium doped spinel as compared with the stoichiometric LiMn$_2$O$_4$. The intercalation profile of LiMn$_2$O$_4$ and capacitance profile of the activated carbon in 1 M Li$_2$SO$_4$ solution were examined by using micro-electrode. Figure 3.69 shows cyclic voltagramms of LiMn$_2$O$_4$ and activated carbon. LiMn$_2$O$_4$ exhibited two pairs of redox peaks, at 1.0 and 1.13 V (oxidation potential vs. NHE), corresponding to the insertion/extraction reactions of LiMn$_2$O$_4$. It was also observed that Li-ion could be extracted from the host lattice before the evolution of O$_2$ (1.3 V vs. NHE). Due to its poor electrical conductivity, LiMn$_2$O$_4$ electrode should be incorporated with various carbon materials. Therefore, it was necessary to study the evolution overpotential for O$_2$ in the carbon black. O$_2$ was evolved at about 1.2 V vs. NHE, which was lower than that of the LiMn$_2$O$_4$ electrode. There was a high over-potential for O$_2$ evolution on the surface of LiMn$_2$O$_4$. The AC electrode in the mild electrolyte solution had a rectangular shaped curve within the potential range of 0–0.6 V (vs. NHE), corresponding to the nonfaradic reversible reaction of Li-ion on surface of an activated carbons.

Carbon Nanotubes (CNTs) Based Supercapacitors

CNTs have been explored extensively as electrodes of supercapacitors, because of their various attractive characteristics, such as high specific surface area, relatively low density, high chemical stability and high electrical conductivity [4, 240–242]. According to the number of layers of the walls, CNTs include Single-Walled (SWCNTs), Double-Walled (DWCNTs) and Multi-Walled (MWCNTs), which could be used directly as electrodes of supercapacitors, with electrochemical behaviors similar to those of the activated carbons as discussed before. Moreover, more and more work has been carried out to use CNTs to develop nanocomposites with conductive polymers (CNT–polymer) and hybrids with metallic oxides (CNT–oxides), which formed a new type of nanostructured carbon based electrodes for supercapacitor applications. CNT-

98 *Nanomaterials for Supercapacitors*

based nanocomposites and nanohybrids have various advantages as supercapacitor electrodes [243–245], including (i) strong interaction between the electroactive materials and CNTs, (ii) creation of open porous network structure due to the entanglement of nanotubes so as to boost the transportation of the electrolyte ions towards the active surfaces and (iii) mechanical resiliency to buffer the volumetric changes during the charge and discharge process so as to ensure cycling stability.

CNT-based materials

CNT-based materials have been extensively studied as electrodes of supercapacitors [246–249]. A supercapacitor electrode has been fabricated by using SWCNTs [246]. The monolithic chemical composition of the SWCNT electrode could withstand durable operation with a range of 4 V. In this case, there were no surface functional groups, conducting agents and binders. Moreover, the monolithic SWCNTs exhibited a fiber-like structure, which contributed to the life stability, due to the mechanical integrity of the electrodes. Over the voltage range of up to 4 V, the electrode demonstrated high energy and power performances. A high specific capacitance of 160 $F \cdot g^{-1}$ was achieved, which was attributed to the combination of high specific surface area and strong electrochemical doping effect, thus leading to enhanced energy storage capability. The strong mechanical strength of the SWCNT electrode enhanced the ion transport, which resulted in high power capability. Furthermore, a dual functionality, i.e., as both electrode and current-collector, was realized by using the SWCNT electrode, because of the excellent electronic conductivity.

The electrode was made of vertically aligned SWCNTs (forests) that were synthesized by using a water-assisted chemical vapor deposition (CVD) [250]. The SWCNT forests, with volumetric occupancy of 3–4% and mass density of 0.03 $g \, cm^{-3}$ were achieved by using iron catalysts with an average diameter of 1 nm sputtered on single crystal Si substrates from ethylene as the carbon source with the presence of water. The forests had a high carbon purity of > 99.98%, due to the requirement of minimal amount of catalyst. The one-dimensional fiber-like nature of SWCNTs allowed for assembling into sheets, similar to the paper derived from cellulous fibers, without using any binders. Figure 3.70(a) shows a schematic diagram illustrating fabrication processes of the electrode from the SWCNTs. During the fabrication process, SWCNT forest was flattened and densified into a highly densely packed sheet, with a density of 0.5 $g \, cm^{-3}$, an area of 1 cm^2 and a thickness of about 100 μm [251].

AC sheet electrodes, with a density of 0.6 $g \, cm^{-3}$, an area of 1 cm^2 and a thickness of 100 μm, were also prepared for comparison, by mixing and kneading AC powder (YP17: 1640 $m^2 \, g^{-1}$), together with poly(tetrafluoroethylene) (PTFE) as a binder and carbon black as a conducting additive. There was an obvious difference between the aligned well-defined structure of SWCNT electrodes and the random irregular structure of AC electrodes, as demonstrated in Fig. 3.70(b) and (c). The SWCNT electrode was of a mesoporous structure, with a maximum pore size of 10 nm and a specific surface area of 1250 $m^2 \, g^{-1}$. The high purity of the SWCNTs was further evidenced by IR spectroscopy and TGA results. Supercapacitors were assembled with both the AC and SWCNT electrodes, as illustrated in Fig. 3.70(d). The SWCNT had a higher electrical conductivity of 21 $S \, cm^{-1}$ than the AC electrodes of 0.3 $S \, cm^{-1}$, so that the SWCNT sheets could serve as both the electrodes and current-collectors. It was expected that electrodes with relatively low densities and small thickness would result in high power capability, due to enhanced ion transport.

It was found that specific capacitance of the SWCNT electrodes with MCC was 160 $F \cdot g^{-1}$, which exhibited almost no variation with discharge rate. It was much higher than that of the AC electrodes (100 $F \cdot g^{-1}$) which however was decreased with increasing discharge rate. Also, capacitance of the SWCNTs electrode was significantly increased from 68 $F \cdot g^{-1}$ at 2 V to 160 $F \cdot g^{-1}$ at 4 V, which could be attributed to electrochemical doping [252]. This implied that the SWCNT electrodes could be used high voltage operation. The near constant capacitance of the SWCNT electrodes with increasing discharge rate indicated their fast ion transport, thus leading to high power capability. Although the capacitance of the SWCNT electrode without MCC was lower than that of the one with MCC, the value was higher than that of the AC electrodes, which was readily ascribed to the high electrical conductivity of the SWCNT electrode.

A compact-designed supercapacitor was assembled by using large-scaled free-standing and flexible SWCNT films as both the anode and cathode [247]. A prototype of the processing procedures was developed to obtain the uniform spreading of the SWCNT films onto the separators serving as both electrodes and

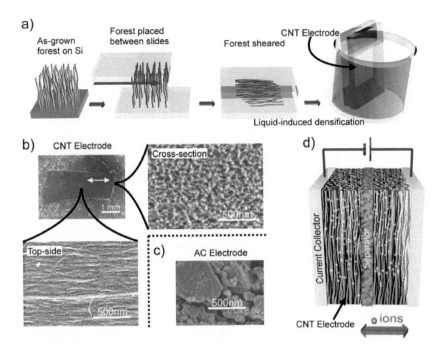

Fig. 3.70. Electrode fabrication and characterization: (a) fabrication of the SWCNT electrode, (b) SEM image of the SWCNT electrode demonstrating ordered pore structure and alignment, (c) SEM image of the AC with random pore structure and (d) representative cell structure. Reproduced with permission from [246], Copyright © 2010, WILEY-VCH Verlag GmbH & Co. KGaA, Weinheim.

charge collectors without metallic current collectors, leading to a simplified and lightweight architecture. The area of SWCNT film on a separator can be scaled up and its thickness can be extended. High energy and power densities (43.7 Wh kg^{-1} and 197.3 kW kg^{-1}, respectively) were achieved from the prepared SWCNT film-based compact-designed supercapacitors with small equivalent series resistance. The specific capacitance of this kind of compact-designed SWCNT film supercapacitor is about 35 F·g^{-1}. These results clearly show the potential application of free-standing SWCNT film in compact-designed supercapacitor with enhanced performance and significantly improved energy and power densities.

The SWCNT films were synthesized by using a floating chemical vapor deposition method [248]. To prepare electrodes of supercapacitors in a rolled configuration, the SWCNT films and separators were cut to small pieces, as illustrated in Fig. 3.71(a). The obtained SWCNT films and a tailored separator were soaked in ethanol, as demonstrated in Fig. 3.71(b), in which the SWCNT films were spread out and attached onto the separator, in the way of end to end or/and layer by layer. In the case of end to end contact, the neighboring SWCNT films were overlapped at the ends, as seen in Fig. 3.71(c), effective contact between the adjacent SWCNT films would be ensured. After that, the ethanol was evaporated, so that the SWCNT films were strongly attached to the separator. Two pieces of separators coated with the SWCNT films were stacked and rolled up, to form compact-designed supercapacitors, which was filled with 1 M LiClO$_4$ in a mixture of ethylene carbonate (EC), diethyl carbonate (DEC), and dimethylene carbonate (DMC), with a volume ratio of EC/DEC/DMC = 1:1:1, as the electrolyte, as demonstrated in Fig. 3.7(d) and (e).

Figure 3.72(a) shows a photograph of the directly grown SWCNT film, with a large-scale of > 50 cm^2 and a homogeneous thickness of about 400 nm. The free-standing SWCNT film had a tensile strength of 250 MPa, making it sufficiently strong to be tailored and processed. Also, because SWCNT film was directly grown without any additional treatment, it had a very low sheet resistance of 5–50 Ω·□$^{-1}$, for the thickness in the range of 500–100 nm.

Figure 3.72(b) shows SEM image of the free-standing SWCNT film, illustrating a nanoporous architecture with almost no impurity, Fig. 3.72(c) depicts SEM image of the SWCNT bundles peeled off

100 Nanomaterials for Supercapacitors

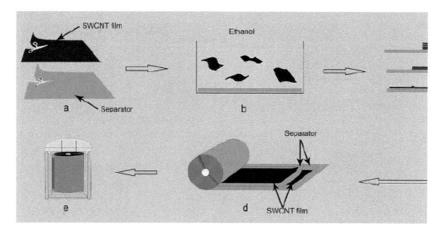

Fig. 3.71. Schematic diagram describing the assembling of the compact-designed supercapacitor with free-standing flexible SWCNT films. Reproduced with permission from [247], Copyright © 2011, The Royal Society of Chemistry.

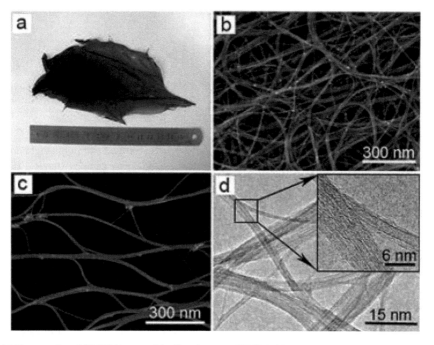

Fig. 3.72. (a) Photograph and (b) SEM image of the directly grown SWCNT film. (c) SEM image of a single layer of bundles peeled off from a thick SWCNT film. (d) TEM image of the directly grown SWCNT film. Reproduced with permission from [247], Copyright © 2011, The Royal Society of Chemistry.

from a thick film, presenting a continuous 2D reticulate structure. Because of this unique architecture, the directly grown SWCNT films possessed higher strength and higher conductivity than those post-deposited CNT films. In the films, the sidewall of the SWCNTs was nearly free of amorphous carbon, as revealed in Fig. 3.72(d), which was beneficial to the improvement of specific capacitance as the SWCNT film was used electrodes [253].

Figure 3.73(a) shows a photograph of a separator coated with the SWCNT films, on which the SWCNT films were well adhered onto the separator, so that films could not be peeled off, after the evaporation of

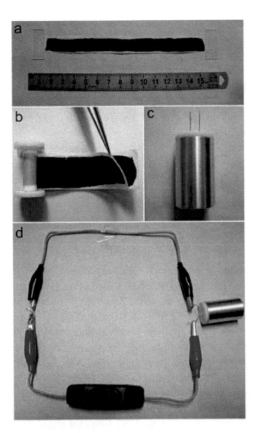

Fig. 3.73. Photograph of (a) the SWCNT films spread out on a separator, (b) rolled design of the separator with SWNCT films and (c) the final compact-designed supercapacitor. (d) Photograph of the compact-designed supercapacitor used to power a red light-emitting diode (LED). Reproduced with permission from [247], Copyright © 2011, The Royal Society of Chemistry.

ethanol and drying in air. Due their high electrical conductivity, the directly grown SWCNT films could be used as current collectors to replace the metallic counterparts. Figure 3.73(b) shows the photograph of the fabrication procedure to roll up the two pieces of SWCNT film/separator. The compact-designed supercapacitor was demonstrated in Fig. 3.73(c). When fully charged, the SWCNT film-based compact-designed supercapacitor could light up a light-emitting diode, as seen Fig. 3.73(c). Due to the slow kinetics of carbon oxidation, SWCNT had higher electrochemical stability than ACs, which, together with the high conductivity and excellent mechanical strength, made the free-standing SWCNT films to be promising candidate as electrodes for supercapacitor applications.

CV curves of a compact-designed supercapacitor at different scan rates possessed a typical rectangular shape, suggesting that the SWCNT film had excellent capacitance behavior with very rapid current response on voltage reversal with low ESR [254]. Specific capacitance of the compact-designed SWCNT film supercapacitor was about 35 F·g^{-1} for the two-electrode cell configuration, which corresponded to 140 F·g^{-1} for the three-electrode cell configuration.

A MWCNT film has been developed by using the electrophoretic deposition (EPD) method, which could be employed as electrodes of supercapacitors with high power density [255]. The EPD method has various advantages, such as short fabrication time, simple apparatus requirement, enhanced adherence of the coatings and potential large scale production. Due to the very small ESR value and fast diffusion of the electrolyte, the MWCNT-based electrode displayed rectangular CV curves, even at relatively high scan rates, with a power density of 20 kW kg^{-1}.

MWCNTs synthesized by using CVD method were used as the starting materials, after purification before the EPD experiment. To proceed the EPD, 6 mg MWCNTs were dispersed in 60 ml absolute ethanol through strong ultrasonication. To have surface charges on the carbon nanotubes, 10^{-5}–10^{-4} mol $Mg(NO_3)_2 \cdot 6H_2O$ were added into the suspension as electrolyte to facilitate EPD. Two nickel foils with a diameter of 5 mm were used as the EPD electrodes, which were put into the suspension with a parallel configuration. DC voltages of 40–50 V were applied to the EPD electrodes, so that the charged carbon nanotubes were migrated towards the cathode and deposited onto it. After EPD, the films were thermally annealed at 500°C for 30 minutes in H_2.

A spray deposition technique has been used to deposit carboxylic acid- and ester-functionalized SWCNTs and MWCNTs on the polymer substrate as electrodes of supercapacitors [256]. It was found that the ester-functionalized CNTs (E-MWCNT/E-SWCNT) exhibited an EDLC capacitive behavior, while the acid-functionalized ones (A-MWCNT/A-SWCNT) demonstrated a pseudocapacitance feature. Both the pure CNTs (P-MWCNT/P-SWCNT) and the ester-modified CNT films (E-MWCNT/E-SWCNT) displayed almost perfectly symmetrical charge–discharge curves. The A-MWCNT and A-SWCNT have specific capacitance of 77 and 155 F g^{-1} at a scan rate of 100 mV s^{-1}, respectively, both of which were higher than those of the P-MWCNT (23 F·g^{-1}) and P-SWCNT (38 F·g^{-1}).

After purification, the carboxylic and ester functionalized SWCNT and MWCNT were suspended in 0.1 wt% Sodium Dodecyl Benzene Sulfonate (SDBS) aqueous solution, deionized water and ethanol through high shear mixing, with a concentration of 1 mg·ml^{-1}. The mixtures were sonicated by using a probe at 600 W and 20 kHz for 30 minutes in ice bath. The SWCNT and MWCNT suspensions were deposited onto various substrates, including glass slides, polyethylene naphthalate (PEN) films, and stainless steel and Al foils, by using a spray apparatus. During spraying, the dispersions were syringe pumped at a flow rate of 1 ml·min^{-1}, while the substrates were heated at temperatures in the range of 70–130°C, depending on the substrates used. The substrates were moved in x–y plane, over a constant spray distance of 13 cm, at a linear velocity of 70 mm·s^{-1} during the spraying process. A layer of 1% (w/v) branched poly(ethyleneimine) (PEI-b, M_w = 70000) aqueous solution was pre-sprayed onto stainless steel substrates as current collectors, in order to ensure the adhesion and stability of the CNT films for electrochemical characterization.

A stretchable supercapacitor was developed by using buckled SWCNT macrofilms [257]. The initial specific capacitances of the stretchable supercapacitor were 54 F·g^{-1} and 52 F·g^{-1}, for the states of non-strain and 30% strain, respectively. Moreover, the values of the stretchable supercapacitors showed almost no decay after up to 1000 charge–discharge cycles. PDMS substrates were prepared by mixing silicone-elastomer base and the curing agent at a weight ratio of 10:1, casting into a glass container and baking at 100°C for 45 minutes. Rectangular slabs of 1.5 × 6 cm^2 were cut from the polymerized samples, which were rinsed with isopropyl alcohol (IPA) to remove contaminations and then dried with N_2 blowing. The prestrained PDMS substrate was subjected to a flood exposure with UV light, with 185-nm and 254-nm radiation for 150 s. Ozone was generated due to the 185-nm radiation, which was dissociated to O_2 molecules and O atoms after the 254-nm radiation, so that chemically activated surface was created.

The SWCNT films were grown by using a floating CVD method, from a mixture of ferrocene and sulfur with a weight ratio 10:1 in Ar/H_2. Ferrocene served as the carbon feedstock/catalyst, while sulfur acted as an additive to promote the growth of the SWCNT films. The as-synthesized SWCNT films possessed a hydrophobic surface. It became hydrophilic after the post purification, i.e., oxidation either by thermally annealing in air at 450°C for 1 hour or soaking in 30% H_2O_2 solution for 72 hours, followed by rinsing with 37% HCl acid, due to the reduction in the amount of impurities and the increase in the quantity of the functional groups, such as O = C–OH. The hydrophilic behavior of the surface ensured strong interaction of the SWCNT films with the activated PDMS.

Two-electrode configuration supercapacitor was assembled for electrochemical characterization. Tetraethylammonium tetrafluoroborate was dissolved in propylene carbonate to form a 1 M solution that was used as the electrolyte. Two PDMS sheets with buckled SWCNT films were pressed together, sandwiching a filter paper which was soaked in the electrolyte. The SWCNT films also served as the current collectors. Two thin copper strips were placed at the edge of each SWCNT-film electrode to ensure appropriate electrical contact. Resistance of the device was measured by placing two copper strips at the edges of the SWCNT films attached to the PDMS substrate. Tensile strain was applied by holding both sides of the PDMS sheet, while maintaining the electrical contact between the copper and SWCNT film during the presence of the strain.

The stretchable supercapacitors experienced almost no change in electrochemical performances under strains, including specific capacitance, power density and energy density. The buckled SWCNT macrofilms were prepared by removing the PDMS substrates with 30% prestrain. Pristine SWCNT macrofilms were also examined as electrodes for comparison. CV curves of the stretchable supercapacitor measured at different scan rates are shown in Fig. 3.74(a) and (b). The CV curves possessed a rectangular shape, indicating the ideal capacitive behavior of the electrodes, even at a scan rate of as high as $1000 \ mV \cdot s^{-1}$, as seen in Fig. 3.74(b). With the application of 30% strain, CVs of the stretchable supercapacitors were almost not changed. Galvanostatic charge–discharge behaviors at a constant current density of $1 \ A \cdot g^{-1}$ up to 1000 cycles are depicted in Fig. 3.74(c–e). The initial specific capacitance was $54 \ F \cdot g^{-1}$ for the stretchable supercapacitor at zero strain (top panel in Fig. 3.74(f)), while it was slightly decreased to $52 \ F \cdot g^{-1}$ at 30% applied (middle panel in Fig. 3.74(f)), both of which were very close to that ($50 \ F \cdot g^{-1}$) of the pristine SWCNT macrofilm as electrodes (bottom panel in Fig. 3.74(f)). The stretchable supercapacitors display a very stable cycling stability, with the specific capacitance to be unchanged after charge–discharge cycles of up to 1000. Additionally, both the energy and power densities of the stretchable supercapacitors were also kept to be constant at strains, illustrating their high mechanical flexibility.

An omnidirectionally stretchable high-performance supercapacitor has been fabricated by using isotropic buckled CNT films [258]. The omnidirectional stretchability of the CNT film was stemmed from the continuous isotropic buckled architecture of the CNTs. Also, the stretchable film exhibited a high electrical conductivity, with a sheet resistance of $2.75 \ \Omega \cdot \square^{-1}$, together with strong mechanical strength. The omnidirectionally stretchable supercapacitor was assembled by using the buckled CNT film as electrode and H_2SO_4/poly(vinyl alcohol) (PVA) as electrolyte. The supercapacitor could withstand an omnidirectional stretch by 200% strain, making it promising for applications in flexible energy storage devices. To further enhance electrochemical performance of the supercapacitor, conductive and pseudocapacitive PANI was employed to wrap the electrodes. It was deposited onto the surface of the isotropic buckled acid treated CNT film with a weight percentage of 116% by using electrochemical deposition.

Freestanding highly conductive films, with a thickness of 2 μm and density of $807 \ mg \cdot cm^{-3}$, were prepared from randomly oriented CNTs that were synthesized by using the floating catalyst chemical vapor deposition method (FCCVD) [259]. Figure 3.75 shows surface SEM images of the CNT films. Buckled CNT films with different omnidirectional stretched degrees in the range of 50–200% were achieved, as demonstrated in Fig. 3.76(a–d), by simply controlling the prestrain level of the substrate. The capability of a buckled film to withstand the external strain is closely related to the liner density of the buckles and the corresponding wavelengths, i.e., smaller and denser the buckled structure, the larger the strain that the buckled film could withstand [260]. The buckled CNT film was isotropic and uniform, if it was separated from the substrate in all directions simultaneously. Otherwise, if the substrate was removed, in the x-direction first and then in other directions, the buckles of the film would be anisotropic and non uniform.

The supercapacitor exhibited a specific capacitance of $9.52 \ mF \cdot cm^{-2}$ at the scan rate of $50 \ mV \cdot s^{-1}$. To further increase the specific capacitance and energy density of the stretchable supercapacitors, three types of electrodes were examined, including isotropic buckled acid treated CNT film, CNT@PANI film and acid treated CNT@PANI film. It was found that the acid treated CNT@PANI had the highest areal specific capacitance of $1147.12 \ mF \cdot cm^{-2}$ and energy density of $50.98 \ \mu Wh \cdot cm^{-2}$, at a scan rate of $10 \ mV \cdot s^{-1}$. The supercapacitors could retain unchanged capacitive performance at uniaxial, biaxial and omnidirectional enlongational strains of up to 200%, making the supercapacitors to be potential for a wide range of applications.

CNT–polymer composites

As mentioned above, the combination of conductive polymers and CNTs has been an effective way that can be used to improve electrochemical performances of supercapacitors [261–264]. A group of composite films, consisting of conductive polymer PPy and SWCNTs, were employed as electrodes of supercapacitors [261]. The composite films were prepared through electro-polymerization of the homogenous mixtures of

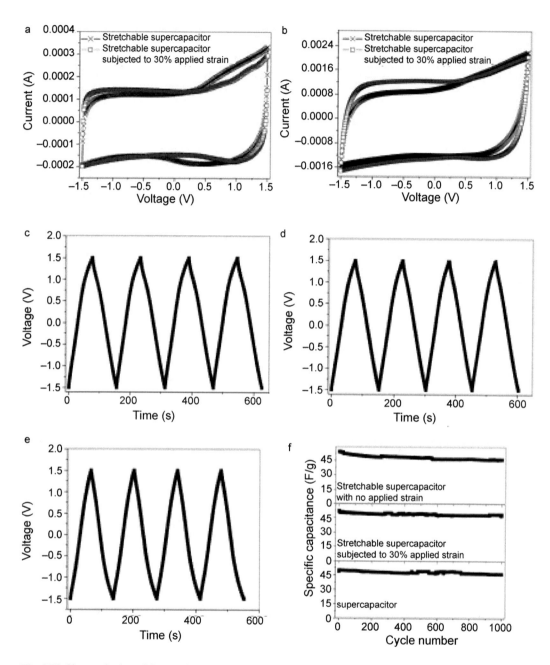

Fig. 3.74. Characterization of the stretchable supercapacitors subjected to 0 and 30% tensile strain, together with those of supercapacitors derived from the pristine SWCNT films. (a, b) Cyclic voltammograms of the stretchable supercapacitors measured at 100 mV·s^{-1} and 1000 mV·s^{-1}, respectively. (c–e) Charge–discharge cycles measured at a constant current density of 1 A·g^{-1} at 0 and 30% strains for the supercapacitors made of the two types of SWCNT films, respectively. (f) Long charge–discharge cycling at a constant current density of 1 A·g^{-1}. Reproduced with permission from [257], Copyright © 2009, WILEY-VCH Verlag GmbH & Co. KGaA, Weinheim.

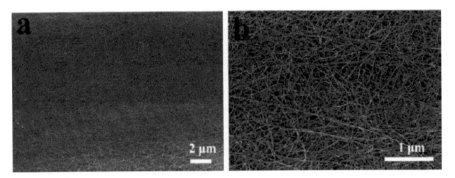

Fig. 3.75. SEM images of the CNT film: (a) low magnification and (b) high magnification. Reproduced with permission from [258], Copyright © 2016, American Chemical Society.

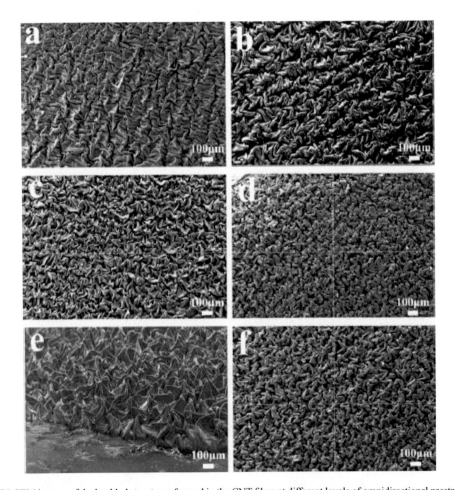

Fig. 3.76. SEM images of the buckled structures formed in the CNT films at different levels of omnidirectional prestrains: (a) 50%, (b) 100%, (c) 150% and (d) 200%. (e) Side view of the buckled CNT film formed at a prestrain of 200%. (f) Buckled structure (omnidirectional prestrain = 200%) after 1000 uniaxial stretching cycles at a strain of 200%. Reproduced with permission from [258], Copyright © 2016, American Chemical Society.

pyrrole and SWCNTs. The PPy/SWCNTs and PPy/functionalized SWCNTs composites could reach specific capacitances of 144 F·g^{-1} and 200 F·g^{-1}, respectively, at a scan rate of 200 mV·s^{-1}. Both values were higher than that of PPy, which was attributed to the fact that the nanocomposites possessed mesoporous structures that were beneficial to electrochemical reaction, due to the presence of the SWCNTs. In addition, charge–discharge behavior of the PPy/functionalized SWCNTs composite film was further enhanced as compared with that of the PPy/SWCNTs film, because the former possessed a higher degree of dispersion of CNTs. The increased electrochemical properties of the composites as compared with pure PPy was ascribed to the significant increase in electrical conductivity and the minimized volume variation in the discharging states. Figure 3.77 shows SEM images of the composite films. For the PPy/SWCNTs sample, two CNTs were uniformly coated with PPy, as seen in Fig. 3.77(a). As for the PPy/functionalized SWCNTs, there were many shorter SWCNTs coated by PPy, because the SWCNTs were broken during the functionalized process, as demonstrated in Fig. 3.77(b).

An electrochemically deposited nanoporous composite film, comprising of MWCNTs and PPy, was developed for electrodes of supercapacitors [262]. The low-frequency capacitances of the films with different thicknesses indicated that the devices possessed specific capacitances per mass (C_{mass}) and geometric area (C_{area}) were 192 F·g^{-1} and 1.0 F·cm^{-2}, respectively. CV and EIS results revealed that the supercapacitors demonstrated excellent charge storage and transfer capabilities, because the films possessed high surface area, high electrical conductivity and enhanced electrolyte accessibility, due to their nanoporous structures.

The composite films were electrolytically deposited onto a graphite plate that acted as the working electrode in a three-electrode configuration single compartment electrochemical cell. The polymerization electrolyte was aqueous solution, containing 0.5 M pyrrole monomer and 0.4 wt% oxidized suspension of MWCNTs, with an outer diameter of about 10 nm and lengths of 0.2–2.5 μm. The MWCNTs were oxidized to create functional groups, such as hydroxyl, carbonyl and carboxylic groups, so that they could be suspended in polar solvents, including distilled water. The pyrrole monomer was anodically polymerized on the working electrode, at a constant current of 3.0 mA·cm^{-2} or a constant potential of 0.7 V, against a saturated calomel reference electrode (SCE). Due to their negative charges, the MWCNTs acted as both the supporting electrolyte during the polymerization and the dopant in the PPy.

Figure 3.78 shows SEM images of the composite films, where the PPy was uniformly coated on the individual MWCNTs, so that the coated tubes were bonded together to form a nanoporous 3D network. The PPy coating had a thickness of up to 250 nm, depending on the polymerization conditions. Thermal analysis results revealed that the content of MWCNTs was about 60 wt%. After coating with PPy, the MWCNTs were detangled in such a way that they were preferentially aligned parallel to the plane of the film, while maintaining random orientation within the plane. Due to this special structure, the composite film could improve ionic conductivity, increased capacitance and boost rate of response. Firstly, due to the large surface area caused by the PPy, the porous morphology of the composite film ensured smooth electrolyte access. Secondly, because the PPy was coated on individual MWCNT as a thin layer, the ion

Fig. 3.77. SEM images of the composite films: (a) PPy/SWCNTs and (b) PPy/functionalized SWCNTs. Reproduced with permission from [261], Copyright © 2007, Elsevier.

Carbon Based Supercapacitors 107

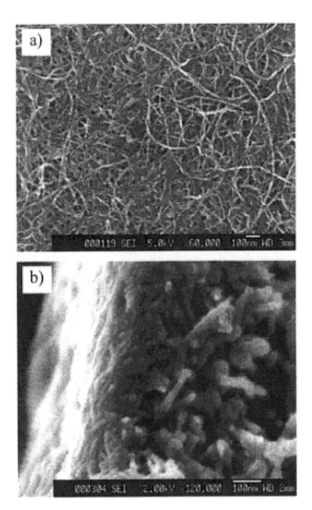

Fig. 3.78. SEM images of the PPy-coated MWCNTs in the composite films: (a) surface and (b) fractured cross-section. Reproduced with permission from [262], Copyright © 2002, American Chemical Society.

intercalation distance was effectively decreased to nanometer scale. Furthermore, the homogeneous disperse of the MWCNTs increased electrical conductivity of the composite film.

A microwave hydrothermal method was used to functionalize CNTs with a thorn-like organometallic methyl orange–iron(III) chloride (MO–FeCl$_3$) complex [265]. The complex acted as an oxidant, so that PPy nanoparticles could be attached directly onto the CNTs through the polymerization of pyrrole. Galvanostatic charge/discharge for the PPy/CNT composite, conducted at a current density of 1.0 mA·cm^{-2} over the potentials in the range of −0.5–0.5 V, revealed that the pseudocapacitance behavior of the composite. The PPy/CNT composite exhibited a maximum specific capacitance of was 304 F·g^{-1}, which was ascribed to the large accessible surface area of the functionalized CNTs to the electrolyte. As a result, the electrochemical redox reaction of PPy could be effectively used.

Pyrrole was distilled, while CNTs were purified by refluxing in concentrated nitric acid. In typical experiment, purified CNTs and methyl orange (MO, (CH$_3$)$_2$NC$_6$H$_4$-N=NC$_6$H$_4$SO$_3$Na) were dispersed in deionized water and sonicated for 30 minutes, in order for MO to be adsorbed onto surface of the CNTs. During stirring, FeCl$_3$ was added into the suspension, which was subject to microwave hydrothermal reaction for 30 minutes. The mixture with the CNT/MO–FeCl$_3$ composite was sonicated for 30 minutes, after which pyrrole monomer solution was then dropped into the solution. After reaction for 3 minutes in a microwave oven, a dark precipitate was obtained, which was filtered and thoroughly washed with deionized water and ethanol, followed by drying at 50°C overnight in vacuum, leading to PPy/CNT composite powder. The complex acted as both the morphology-guiding agent and oxidant for the formation of the nanocomposite, so that polypyrrole nanoparticles could be well incorporated with the CNTs during polymerization of pyrrole.

Figure 3.79 shows representative TEM images of the functionalized CNTs obtained by using the microwave hydrothermal and conventional solution methods, together with the PPy/CNT composite [265]. After microwave hydrothermal reaction, surface of the CNT treated became highly rough, as seen in Fig. 3.79(a), which was obviously different from the smooth surface of the pristine CNTs. Surface of the CNTs was uniformly distributed with thorn-like items, as illustrated in Fig. 3.79(b). In comparison, the CNTs functionalized through the conventional solution route had very few thorns, as revealed in Fig. 3.79(c), because most of them were not attached onto the CNTs. It has been reported that MO–FeCl$_3$ could be attached onto the surface of acid-treated CNTs through electrostatic interaction between the negatively charged CNTs and the positively charged MO [266]. Due to the weakness of the interaction, dissociation easily occurred under mild conditions. In this regard, microwave hydrothermal process had boosted the association between the CNTs and the MO–FeCl$_3$, which led to the strong absorption of the

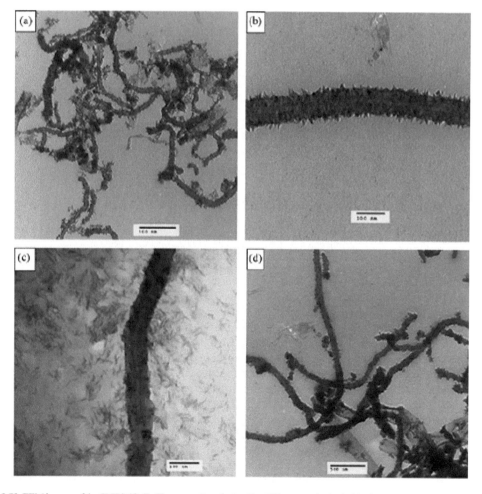

Fig. 3.79. TEM images of the CNT/MO–FeCl$_3$ composites obtained by different methods: (a, b) microwave hydrothermal method and (c) conventional solution method. (d) TEM image of the PPy/CNT composite at magnification of 50,000. Reproduced with permission from [265], Copyright © 2010, Elsevier.

Carbon Based Supercapacitors 109

MO–FeCl$_3$ onto surface of the CNTs. It was anticipated that Fe^{3+} moieties on the CNT/MO–FeCl$_3$ could have acted as an initiator to trigger the oxidation polymerization, which was experimentally confirmed, as demonstrated in Fig. 3.79(d).

MO contains both amine and sulfonate groups at its two sides, where the amine side could be positive charged when amine is transferred to ammonium in acidic solution. As a result, because of the electrostatic interaction between the ammonium and –COO–, MO could be readily attached onto the CNTs. With the presence of FeCl$_3$, CNT–MO–FeCl$_3$ complexes could thus be formed, due to the interaction between the negative –SO$_3^-$ and the positive Fe^{3+} groups [267]. The presence of Fe, O and S in the CNT/MO–FeCl$_3$ was confirmed by EDS spectrum, indicating the formation of the CNT/MO–FeCl$_3$ composite.

The second group of CNT-conductive polymer nanocomposites involves another important conductive, polymer, polyaniline or PANI [268–271]. MWCNT/PANI composite films were prepared by using an *in-situ* electrochemical polymerization process, with an aniline solution at different levels of MWCNTs [268]. It was found that the MWCNT/PANI films exhibited much higher specific capacitance, higher power density, more stable cyclic behavior, than pure PANI film, as the electrodes of supercapacitors. The MWCNT/PANI composite film with 0.8 wt% demonstrated the maximized specific capacitance of 500 F·g^{-1}. MWCNTs were dispersed in 10 ml of aniline with concentrations of 0, 0.2, 0.4 and 0.8 wt%. The mixtures were then reflexed for 5 hours in the dark. After cooling, MWCNT–aniline solutions were developed after filtration through 0.1-μm-diameter membrane.

Electropolymerization of aniline with different contents of MWCNTs to form the MWCNT/PANI films was conducted in aqueous solutions of 0.5 M H$_2$SO$_4$ + 0.325 M aniline. Continuous potential scanning at 100 mV·s^{-1} over the potential range between −0.2 and −1.2 V resulted in a thin uniform green coating on SS sheet electrode. Peaks at 0.3 V, 0.5 V and 0.9 V corresponded to formation of the radical cations, oxidation of head-to-tail dimmer and conversion from emeraldine to pernigraniline structure, respectively. Meanwhile, intensities of all the peaks were increased with increasing number of potential cycles, implying that the films were highly conductive and electroactive.

Surface morphology and microstructure of pure PANI and the MWCNT/PANI composite films were examined by using SEM. The particles of the pure PANI film were highly aggregated, thus resulting in a rough surface. However, the presence of MWCNTs could alter the morphology of PANI. For example, with the presence of 0.8 wt% MWCNTs, the film was formed with fibrous morphology, having a diameter of about 100 nm. Due to the fiber structure, the film became highly porous. In this case, the highly dispersed MWCNTs acted as condensation nuclei during the polymerization of aniline, so that PANI was attached onto the MWCNTs, thus forming the fiber structure. Such a fiber structure could provide a conductive path and network for high conductivity and offer more active sites for faradic reaction, which led to high specific capacitance.

PANI/SWCNT composites have been fabricated by using a similar method, i.e., electrochemical polymerization of aniline onto SWCNTs, with their electrochemical performance being examined in 1 M H$_2$SO$_4$ electrolyte [269]. The PANI/SWCNT composites exhibited much higher specific capacitance, specific energy and specific power than either pure PANI or SWCNTs. Optimized specific capacitance, specific power and specific energy were 485 F·g^{-1}, 228 Wh·kg^{-1} and 2250 W·kg^{-1}, for the sample with 73 wt% PANI. The PANI/SWCNT composites also demonstrated higher long cyclic stability.

The SWCNTs had a purity > 90%, with small amount of metal impurities. The morphology of the PANI/SWCNT composites had a close relationship with their compositions. Figure 3.80 shows a schematic diagram demonstrating microstructure of the PANI/SWCNT composites. At low contents of PANI, the polymer was tightly coated onto the SWCNTs, as demonstrated in Fig. 3.80(a). With increasing content of PANI up to 70–75 wt%, the thickness of the polymer layer was increased gradually, as seen in Fig. 3.80(b). At the same time, morphology of the surface became mesoporous at these levels of PANI. As the content of PANI was further increased, some of the polymer would be deposited separately, instead of coating on the SWCNTs, i.e., it could be present either near the surface of the composite or inside the mesopores, as revealed in Fig. 3.80(c). Such microstructural variations had a direct influence on capacitive performances of the PANI/SWCNT composites.

A well-constructed CNT mesh/PANI composite film was developed by alternatively cross-stacking superaligned CNT sheets and *in-situ* coating PANI from chemical solution, which was used as the electrode

110 Nanomaterials for Supercapacitors

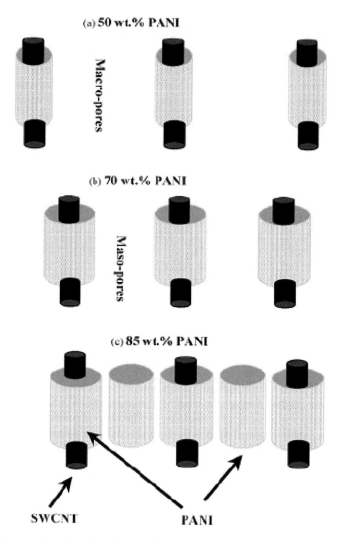

Fig. 3.80. Schematic diagrams of morphologies of the PANI/SWCNT composites with different contents of PANI: (a) 50 wt%, (b) 70 wt% and (c) 85 wt%. Reproduced with permission from [269], Copyright © 2006, Elsevier.

of flexible supercapacitors [271]. The composite exhibited a uniform and oriented structure in both the plane and thickness direction, with a thickness of only 0.8 μm. Because thickness of the electrode could be readily controlled, the effect of thickness of the electrode thickness on electrochemical performance of the supercapacitors could be quantitatively evaluated. It was found that the thickness of the electrode should be ≤ 5 μm, in order to achieve optimized specific capacitance, energy density and power density. An optimal power density of 9.0 kW·kg^{-1} had been obtained, which much higher than those of the CNT/PANI composites fabricated by using other methods. The composite could find potential applications in flexible energy storage devices.

The CNT sheet could be continuously drawn from superaligned CNT arrays, with a height of 300 μm grown on an 8 in. Si wafers, with a width of 20 cm and a length of 200 m. The CNT sheet was extremely thin, with an average thickness of 10s of nanometers, a mass per unit area of 1.5 μg·cm^{-2}, a sheet resistance of 700–1500 Ω·□$^{-1}$ and a high degree of orientation. A well-constructed CNT mesh could be prepared, by alternatively cross-stacking layers of the CNT sheets. The well-constructed CNT meshes contained layer number of 5–200.

PANI was coated on the CNT meshes by using an *in-situ* wet-chemical method. At the first step, the well-constructed CNT meshes were soaked with ethanol and then immersed in 40 mL aqueous solution having 0.04 mol HCL and 0.002 mol aniline monomers to conduct an infiltration for 10 minutes. After that, 40 mL precooled aqueous solution with 0.002 mol ammonium persulfate as an oxidant for the polymerization was dropped slowly in the solution. The mixed solution was then kept at 0°C for 24 hours to complete the reaction, leading to well-constructed CNT mesh/PANI composite film, after washing with deionized water, ethanol and acetone, followed by drying at 80°C in vacuum for 24 hours. Samples with thicknesses of 0.8, 1.7, 3.5, 5.2, 8.7, 17.4, and 34.6 μm were prepared in the same way. The content of PANI was estimated to about 80 wt% in the composite films. Randomly-constructed CNT networks were also prepared to obtain CNT/PANI composite film for comparison.

Supercapacitors were constructed with a sandwich-like configuration, i.e., electrode–separator–electrode, which were encased in a stainless button mold. The CNT/PANI composite films were used as the electrodes, while filter papers were employed as the separator and 1 M H_2SO_4 aqueous solution served as the electrolyte. Representative photographs of the well-constructed CNT mesh/PANI composite films, with a dimension of 3 × 3 cm and thicknesses of 0.8 and 17.4 μm, are shown in Fig. 3.81(a) and (b), respectively. The 0.8 μm thick composite was translucent and highly flexible, whereas the 17.4 μm sample had a smooth surface and was highly elastic. The freestanding films were attributed to the van der Waals interactions among the CNTs.

Plane view SEM images of the well-constructed CNT mesh/PANI composite film are depicted in Fig. 3.81(c, d), confirming that it was highly uniform, strongly textured and extremely porous in the plane direction. The superaligned microstructure was changed after the coating of PANI. The PANI was uniformly deposited on surface of the CNTs, with the presence of large amount of pores. Figure 3.81(e) shows a representative cross-sectional SEM image of the well-constructed CNT mesh/PANI lamina with a thickness of 17.4 μm, which indicated that the composite also possessed a uniform, oriented and porous structure in the thickness direction. The adjacent CNT sheets were in close contact, with uniform coating of PANI.

Figure 3.81(f) shows a SEM image of the randomly-constructed CNT network/PANI composite film. It had a disordered microstructure, with pores having varied size and different morphology. Therefore,

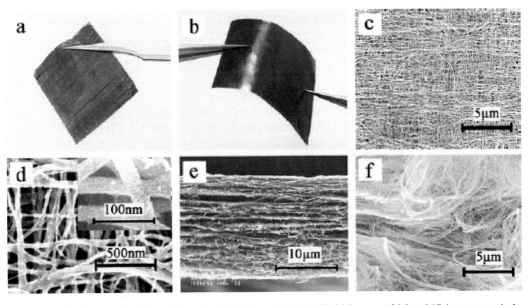

Fig. 3.81. (a, b) Photographs of the well-constructed CNT mesh/PANI with thicknesses of 0.8 and 17.4 μm, respectively. (c, d) Plane view SEM images of the well-constructed CNT mesh/PANI at magnifications of 5 μm, 500 nm and 100 nm. (e) Side view SEM image of the well-constructed CNT mesh/PANI with a thickness of 17.4 μm. (f) Plane view SEM images of the randomly-constructed CNT network/PANI with magnification of 5 μm. Reproduced with permission from [271], Copyright © 2012, American Chemical Society.

the well-constructed composite film was much more uniform than the randomly-constructed composite sample. In addition, the latter had a porosity of 70%, which was lower than that of the ordered sample.

The effect of thickness on electrochemical performance of the composite film could be explained in terms of effective electrical potential–potential distribution in the electrode was different for different thicknesses, which resulted in the difference in specific capacitance [271]. On one hand, pseudocapacitance is generated due to the fast faradaic reaction, once an appropriate potential was present. On the other hand, as the PANI molecules attract charged ions from the electrolyte through the Coulomb force, it is also necessary to have a potential. For the given device configuration, i.e., the electrode–separator–electrode structure, the total potential difference across the thickness of the device was the charge–discharge voltage, which was 0.8 V in this case.

According to Ohm's law, the current that flows through a resistor is proportional to the potential difference and inversely proportional to the resistance. For the supercapacitor in this study, there was also potential drop over the equivalent series resistance of the electrode. As a result, the potential over the electrode decreased from the side close to the collector to that close to the separator. In this case, if the electrode layer was too thick, e.g., ≥ 5 μm, the potential difference would be insufficient for the PANI molecules near the center of the separator (or center of the device) to attract electrolyte ions. This is the reason why the effective pseudocapacitance contributed by PANI was decreased with increasing thickness of the electrode. Because specific capacitance of the CNT/PANI composite electrodes was mainly contributed by the pseudocapacitance of PANI, the specific capacitance was thus decreased as the thickness of the electrode was increased above a critical value.

Poly(3,4-ethylenedioxythiophene) (PEDOT)/functionalized SWCNTs (F-SWCNT) composite, as the third group of composite electrode materials, exhibited promising electrochemical performances [272]. It was found that both the PEDOT and the composite demonstrated CV curves with a nearly rectangular shape, indicating their EDLC capacitive behavior. The PEDOT/F-SWCNT composite had a specific capacitance of 210 $F \cdot g^{-1}$, which was much higher than that of PEDOT (120 $F \cdot g^{-1}$), at a scan rate of 10 $mV \cdot s^{-1}$. The enhanced electrochemical capacitive performance observed for the composite was ascribed to the synergistic effect of the PEDOT and F-MWCNT. In addition, the composite could retain 86% of the initial capacitance, as the scan rate was increased from 10 $mV \cdot s^{-1}$ to 200 $mV \cdot s^{-1}$. The fast charge–discharge capability of the composite was closely related to the large number of meso-tunnels of the composite. After 1000 cycles, the composite electrode could retain 92% of the initial capacitance value, with a discharging capacity of nearly 100%.

To prepare the composite, EDOT was distilled before use and stored at $-10°C$ in N_2 atmosphere. SWCNTs, with a purity of > 90 wt%, outer diameter of < 8 nm and lengths of 0.5–2 μm, were used as the starting material. The SWCNTs were functionalized by treatment in concentrated H_2SO_4/HNO_3 solution and sonicating in a water bath at temperatures of 60–70°C.

More recently, conducting PEDOT conformally coated aligned carbon nanotubes (PEDOT/A-CNTs) were prepared to be used as electrodes of supercapacitors [273]. The PEDOT/A-CNTs with 5% volume fraction (V_f) of A-CNTs had a specific volumetric capacitance of 84.0 $F \cdot cm^{-3}$, which was much higher than those of the non-coated and non-densified A-CNTs (1% V_f). Specific energy and power densities of 11.8 $Wh \cdot l^{-1}$ and 34.0 $kW \cdot l^{-1}$ were achieved by using the PEDOT/A-CNTs electrodes, together with high capacitance retention as compared with those of the PEDOT coated random CNTs.

Aligned multi-walled carbon nanotubes (A-CNTs) were synthesized by using a thermal catalytic CVD method, with a thin catalyst layer of Fe/Al_2O_3 at 750 C [274]. The as-grown CNT forests had a volume fraction (V_f) of about 1%, with densities of 10^9–10^{10} $CNTs \cdot cm^{-2}$. The A-CNTs with 2–3 walls exhibited an average diameter of 8 nm and a spacing between the aligned CNTs of about 80 nm. To coat the A-CNT forests with PEDOT by using the oxidative chemical vapor deposition process (oCVD) [275], iron (III) chloride was used as the oxidant that was heated $in\text{-}situ$ at 300 C to sublime. Liquid EDOT monomer was heated in a monomer jar connected the chamber to 150°C and flowed into the vacuum chamber. A conformal PEDOT coating was formed over the A-CNT forest inside the oCVD chambers. The coating reaction was carried out at 70°C and a pressure of 50 mTorr. The PEDOT films were electrically conductive due to the doping with the chloride ions from the oxidant iron (III) chloride, so that no further doping steps were needed.

CNTs–metallic oxide hybrids

Hybrids of CNTs with various metallic oxides have been developed for the applications as electrodes of supercapacitors. Due to their high electrical conductivity, the presence of CNTs could significantly improve electrochemical performances of the supercapacitors with the nanohybrid electrodes. Almost all the oxides studied as active materials of supercapacitors have been involved in such nanohybrids. Representative examples for each metallic oxides, such as RuO_2 [276–278], transitional metal oxides and other oxides, will be presented and discussed as follows.

RuO_2/multiwalled carbon nanotube (MWCNT) nanocomposites have been developed as electrode of supercapacitors [277]. It was found that electrostatic charge storage capability and pseudofaradaic reactions of the RuO_2 nanoparticles could be influenced by surface functionality of the MWCNTs, due to the increased hydrophilicity. Such increased hydrophilicity promoted the access of the solvated ions to the electrode/electrolyte interface, thus increasing the number of the faradaic reactive site in the RuO_2 nanoparticles. The RuO_2/MWCNT (pristine) nanocomposites exhibited specific capacitances of 70 $F \cdot g^{-1}$ and 500 $F \cdot g^{-1}$, for the combined loading and the mass of RuO_2, respectively. After functionalization, the RuO_2/MWCNT (hydrophilic) nanocomposites possessed specific capacitances 120 $F \cdot g^{-1}$ and 900 $F \cdot g^{-1}$, respectively.

$RuCl_3 \cdot nH_2O$ was reacted with sodium ethoxide to form $Ru(OC_2H_5)_3$. This reaction was conducted in ethanol at 78°C in a dried nitrogen atmosphere for 3 hours. After the filtration of precipitated NaCl, solution of ruthenium ethoxide was obtained. The MWCNT powder was treated in boiled 70% nitric acid solution for 1 hour, followed by washing with distilled water and drying at 90°C for 24 hours. To prepare the RuO_2/MWCNT nanocomposites, Ru-ethoxide solution was mixed with MWCNTs (both pristine and acid-treated), which were then heated in air condition to obtain amorphous ruthenium oxide. After thermal annealing at 200°C for 10 hours, RuO_2/MWCNT nanocomposites were produced, which were thoroughly washed with distilled water and ethanol.

A simple yet effective method was reported to deposit Ru oxide on MWCNTs, through spontaneous reduction of Ru(VI) and Ru(VII) [278]. Both purified and acid functionalized nanotubes, p-MWCNTs and a-MWCNTs, were used to prepare the composites for comparison. The Ru oxide/p-MWCNT and Ru oxide/a-MWCNT composites demonstrated specific capacitances of 213 ± 16 $F \cdot g^{-1}$ and 184 ± 11 $F \cdot g^{-1}$, respectively. In terms of the Ru oxide, the specific capacitances were as high as 704 ± 62 $F \cdot g^{-1}$ and 803 ± 72 $F \cdot g^{-1}$, respectively. Current vs. potential curves exhibited capacitance peaks at about +0.5 V vs. Ag/AgCl. In addition, the Ru oxide/p-MWCNT composite was stable up to 20,000 charge/discharge cycles.

MWCNTs, with a diameter of 10 nm diameter, lengths of 0.1–10 μm and purity of ~ 90% C, were purified to remove the residual metals of Fe and Co. 0.100 M aqueous NaOH(aq) solution, $KRuO_4$ and concentrated acids (hydrochloric, sulphuric, and nitric) were used as the starting materials. MWCNT powder (6.6 g) was suspended in 200 mL mixture of H_2SO_4 and HNO_3 with a volume ratio of 3:1. After the addition of 200 mL distilled water, the suspension was stirred overnight on at 60°C, followed by filtration, thorough washing with distilled water, and drying at 70°C.

To prepare Ru(VI) solutions, Ru oxide was mixed with an excessive amount of NaOH (4.6 g), which was then heated at 550°C. After fusion for 20 minutes, Na_2O_2 (0.95 g) was added, which was heated at 550°C for another 10 minutes. After cooling down, the sample was dissolved in deionized water (200 mL). The solution was filtered through glass fiber paper. p-MWCNT (0.125 g) was added to 50 mL of Ru(VI) solution, which was stirred until the supernatant solution became colorless. The Ru oxide/p-MWCNT was collected after filtration and washing with deionized water. The composite was aged in 1 mL of deionized water for 2 weeks.

To prepare the Ru oxide/a-MWCNT composite, 0.2 g Ru oxide, 6.6 g NaOH and 1.8 g Na_2O_2 were used to prepare 200 mL Ru(VI) solution. a-MWCNT powder (0.25 g) was suspended in 100 mL Ru(VI) solution for 1 week. Then, 2.4 M HCl was slowly added until pH value approached about 7, which was heated at 70–80°C for 5 minutes. The Ru oxide/a-MWCNT composite sample was collected through filtration, followed by thorough washing with distilled water. The Ru oxide/a-MWCNT composite was gelatinous, which was different from the powdered form of the Ru oxide/p-MWCNT. Ru oxide/p-MWCNT composite was also derived from the reaction of MWCNT with $KRuO_4$. The p-MWCNT was first dispersed in water and then mixed with 0.05 M $KRuO_4$ solution in 0.1 KOH.

Figure 3.82 shows TEM images of the Ru oxide/p-MWCNT and Ru oxide/a-MWCNT composites samples, in which the small sized dark spots distributed on surface of the CNTs were Ru oxide nanoparticles. The degree of dispersion and the monodispersity of the Ru oxide nanoparticles was closely related to the degree of surface functionality of the MWCNTs. The Ru oxide/p-MWCNT composite possessed higher uniformity and narrower particle size distribution of 1–3 nm, as demonstrated in Fig. 3.82(a), as compared with those of the Ru oxide/a-MWCNT composite (Fig. 3.82(b)).

The TEM images also revealed that the two composites had a slight difference in morphology of the MWCNTs. As the CNTs were intentionally functionalized with acids, apparent thickening and smearing of the nanotube surface were observed, which was ascribed to the degree of surface functionalization, so that the composite based on the a-MWCNT was more like a coating. Also, a higher initial degree of surface functionalization would result in more agglomeration of the Ru oxide nanoparticles, i.e., the presence of the acid groups would have hindered the reduction and uniform distribution of the Ru on surface of the MWCNTs.

A core–shell templated approach was devised to prepare hollow structured RuO_2 nanoparticles supported by carbon or carbon nanotubes, $hRuO_2$/C or $hRuO_2$/CNT nanocomposites, which were employed as electrode materials of supercapacitors [279]. Ag–Ru nanoparticles with core-shell structures were synthesized first, which were then loaded on the carbon or carbon nanotube supports. After that, the Ag cores were etched out with saturated aqueous Na_2S solution, resulting in hollow structured Ru nanoparticles that were loaded on the carbon or carbon nanotube support. The Ru nanoparticles were converted into RuO_2 by using thermal treatment in air, during which the hollow structure was well retained. The $hRuO_2$/C and $hRuO_2$/CNT nanocomposites had specific capacitances of 817.1 and 819.9 $F \cdot g^{-1}$, respectively, at a current density of 0.2 $A \cdot g^{-1}$, in H_2SO_4 based electrolyte. The specific capacitances could be as high as 805.8 and 770.2 $F \cdot g^{-1}$, respectively, as the current density was increased to 0.5 $A \cdot g^{-1}$, demonstrating high cycling stability.

To synthesize core–shell Ag–Ru nanoparticles, 17 mg $AgNO_3$ was dissolved in 20 mL oleylamine. The solution was heated at 150°C for 1 hour, so that Ag^+ ions were reduced by oleylamine, which was also used as a surface stabilizer. After that, 34 mg $RuCl_3$ was added quickly, while the temperature was increased to 240°C and kept for 3 hours, in order to reduce the Ru^{3+} ions. Once the reaction was finished, core–shell Ag–Ru nanoparticles were obtained, which were purified by precipitation with methanol, centrifugation, washing with methanol and re-dispersed in 20 mL toluene. To load the core–shell Ag–Ru nanoparticles,

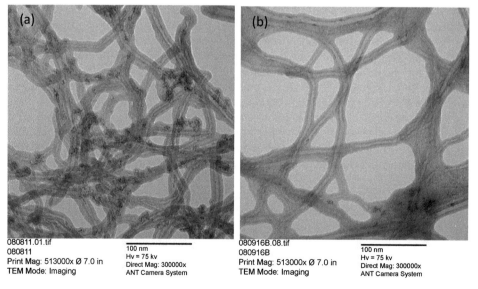

Fig. 3.82. TEM images of the composites: (a) 25.1% Ru oxide/p-MWCNT and (b) 15.6% RuO_2/a-MWCNT. Reproduced with permission from [278], Copyright © 2009, Elsevier.

50 mg carbon powder or CNT was added to toluene suspension of the core–shell Ag–Ru nanoparticles, which was stirred for 24 hours. The core–shell nanoparticles on C or CNT were denoted as Ag–Ru/C or Ag–Ru/CNT, which were collected by centrifugation. The obtained samples were then redispersed in 20 mL acetic acid with ultrasonication. The resulting mixtures were refluxed in acetic acid at 120°C for 3 hours to remove the oleylamine on surface of the particles.

In order to etch out Ag from the core–shell Ag–Ru nanoparticles, 20 mg Ag–Ru/C or Ag–Ru/CNT was added into 20 mL saturated Na$_2$S solution, which was aged for 24 hours at room temperature with vigorous stirring, leading to the formation of hRu/C or hRu/CNT. The samples were collected by centrifugation, washed thoroughly with water and dried in vacuum at room temperature. The hRu/C and hRu/CNT samples were the annealed at 350°C for 6 hours in air to oxidize Ru into RuO$_2$ to form hRuO$_2$/C and hRuO$_2$/CNT nanocomposites. For comparison, Ag–Ru/C and Ag–Ru/CNT were also treated to prepare Ag–RuO$_2$/C and Ag–RuO$_2$/CNT nanocomposites.

RuO$_2$ nanoparticles with a hollow structure were expected to have excellent electrochemical performances. It was found that the inside-out diffusion of Ag in core–shell metallic nanoparticles with Ag as the core could be boosted by etching the multiply twinned structure of decahedral Ag seeds with O$_2$, through the formation of Ag$^+$ ions [280]. With this mechanism, hollow structured Ru nanoparticles supported on carbon or carbon nanotube were synthesized.

Figure 3.83 shows representative TEM images of the samples with different compositions. During the core–shell Ag–Ru/C and Ag–Ru/CNT nanoparticles were aged in saturated aqueous Na$_2$S solution, Ag$^+$ ions

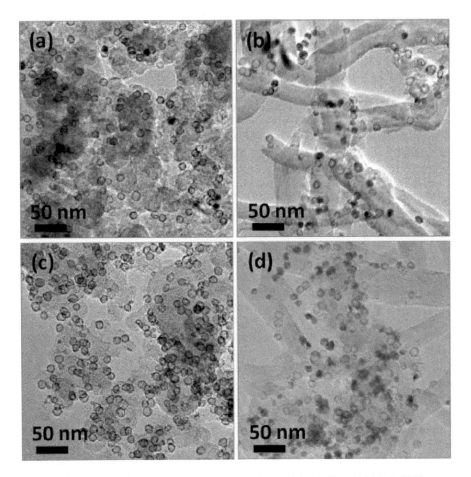

Fig. 3.83. TEM images of as-prepared samples: (a) hRu/C, (b) hRu/CNT, (c) hRuO$_2$/C and (d) hRuO$_2$/CNT nanocomposites. Reproduced with permission from [279], Copyright © 2015, Elsevier.

116 *Nanomaterials for Supercapacitors*

formed due to the O_2 etching diffused out through the discontinuous Ru shell, to react with Na_2S to form Ag_2S. Because Ag_2S was soluble in the saturated Na_2S solution, Ag in the core of the nanoparticles was removed, thus resulting in hRu/C and hRu/CNT nanoparticles without variation in size, as demonstrated in Fig. 3.83(a) and (b), respectively. Hollow structures of the two types of nanoparticles were also confirmed by EDX analysis results. The hRu/C and hRu/CNT nanocomposites were converted into $hRuO_2/C$ and $hRuO_2/CNT$ nanocomposites at a high temperature in air. Both XPS spectra and XRD patterns indicated that RuO_2 was the dominant phase after the thermal treatment. As revealed in Fig. 3.83(c) and (d), the resultant RuO_2 nanoparticles in the nanocomposites exhibited almost the same morphology and size as the original samples.

A thick-film CNT/V_2O_5 nanowire composite was developed as an anode to fabricate asymmetric supercapacitors, together with commercial activated carbon as a cathode, with an organic electrolyte [281]. The nanocomposite electrode exhibits promising rate capability, high specific capacitance and high cycling stability, with an optimal energy density of 40 $Wh \cdot kg^{-1}$ at a power density of 210 $W \cdot kg^{-1}$, corresponding to a maximum power density of 20 $kW \cdot kg^{-1}$. The overall energy and power performance of the asymmetric cell was higher than the general Electric Double-Layer Capacitors (EDLCs). Noting that the improvement in energy density had not compromised the power density, the supercapacitor could find practical device applications in various fields. Therefore, nanocomposite could be explored for promising high energy and high power electrochemical storage devices. The CNT/V_2O_5 nanocomposite was prepared by using a one-pot hydrothermal synthetic process with aqueous vanadium oxide precursors in the presence of pretreated hydrophilic CNTs [282].

A representative SEM image of the nanocomposite with 18 wt% CNTs is shown in Fig. 3.84(A), revealing that it possessed a continuous fibrous structure. Electrical conductivity of the intertwined networks of the CNTs and the nanowires was about 3.0 $S \cdot cm^{-1}$, which was much higher than that (0.037 $S \cdot cm^{-1}$) of the V_2O_5 nanowires. Figure 3.84(B) shows a representative TEM image of the V_2O_5 nanowires, which had a diameter of about 50 nm. The High-Resolution TEM (HRTEM) image shown as the inset indicated that the nanowires exhibited a layered structure. The small dimension of the nanowires was expected to be beneficial to the diffusion of Li^+ ions. Furthermore, nitrogen sorption isotherms and higher resolution SEM images of the etched composite film, as seen in the inset of Fig. 3.84(A), demonstrated that the composite had a hierarchically porous structure, containing both large and small pores, with the small pores contributing a surface area of 125 $m^2 \cdot g^{-1}$ to the composites. The large pores would allow for rapid transport of the electrolyte, whereas the small pores could enlarge surface area of the materials to provide with more active sites for the electrochemical reactions.

As CNTs were employed to be incorporated with V_2O_5, they could significantly boost electrochemical performance of V_2O_5 [283]. In this case, the CNTs acted as a support to facilitate the uniform distribution of V_2O_5. Ti layer with a thickness of 200 nm as the conducting layer was deposited on Si substrates by using sputtering. To grow CNT arrays, Ni film with a thickness of 10 nm was deposited on Si substrates as catalysts by using e-beam evaporation. Then, CNTs were grown by using an Inductively Coupled Plasma Chemical-Vapor Deposition (ICPCVD) method [284]. V_2O_5 was deposited by using arc-ion plating in Ar gas flow for a deposition time of 30 minutes at arc current of 40 A [285].

It was observed that the pristine CNTs had a well-aligned morphology, while V_2O_5 films exhibited a relatively dense microstructure. After the CNT arrays were coated with V_2O_5 films, surface morphology of the CNTs were largely altered. V_2O_5 particles were uniformly distributed on the CNTs to form nanocomposites. Such nanocomposites possessed a high surface area, which was thus beneficial to the charging–discharging capacitive behavior when they were used as electrodes of supercapacitors.

Figure 3.85 shows CV curves of bare V_2O_5, pristine CNT arrays and V_2O_5–CNT nanocomposites, measured at a scanning rate of 600 $mV \cdot s^{-1}$ in 2 M KCl and 1 M $KHSO_4$. As seen in Fig. 3.85(a), the CV curve of bare V_2O_5 was almost flat, implying its very weak capacitive behavior, because V_2O_5 has a very low electrical conductivity, while its dense microstructure was not favorable for pseudocapacitive reaction. CV curve of the pristine CNTs demonstrated a typical electrical double-layer capacitance behavior, as illustrated in Fig. 3.85(b), although the specific capacitance was not very high. In contrast, upon the formation of V_2O_5–CNT nanocomposites, the CV profile was significantly changed, with an increase by a nearly three

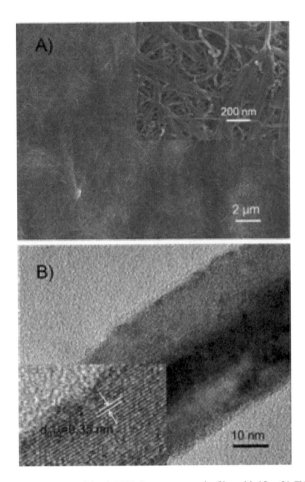

Fig. 3.84. (A) Representative SEM images of the CNT/V$_2$O$_5$ nanocomposite film with 18 wt% CNTs before and after (inset) etching with 1 wt% HF solution. (B) TEM and HRTEM (inset) images of a representative V$_2$O$_5$ nanowire, demonstrating its layered crystalline structure. Reproduced with permission from [281], Copyright © 2011, WILEY-VCH Verlag GmbH & Co. KGaA, Weinheim.

orders of magnitude, as demonstrated in Fig. 3.85(c). CV curves of the V$_2$O$_5$–CNT nanocomposites at the different of scan rates are depicted in Fig. 3.86(a), which all exhibited an ideal square capacitive profile. Figure 3.86(b) shows capacitance of the V$_2$O$_5$–CNT nanocomposites as a function of scanning rate, with a typical decreasing trend with the scanning rate.

As one of the most important electrode materials of supercapacitors, MnO$_2$ has been extensively and intensively studied. Therefore, it is natural to incorporate with CNTs, considering the high electrical conductivity of CNTs. For instance, a flexible paper-based supercapacitor has been fabricated by using the combination of CNTs and MnO$_2$ [286]. The CNT papers were prepared by using a drop–dry method from CVD synthesized CNT arrays, on which MnO$_2$ nanoparticles were deposited by using an electrochemical deposition method. The MnO$_2$/CNT paper supercapacitors exhibited an optimal specific capacitance of 540 F·g^{-1}, corresponding to specific energy and power densities of 20 Wh·kg^{-1} and 1.5 kW·kg^{-1}, respectively, at a current density of 5 A·g^{-1} in aqueous solution of 0.1 M Na$_2$SO$_4$.

CNTs were synthesized by using a water-assisted CVD method, in which a 10 nm Al and 1 nm Fe were e-beam evaporated on Si substrates with a thin layer of SiO$_2$. Ethylene (C$_2$H$_4$), H$_2$ and Ar were used as the carbon source and carrier gases to synthesize CNT arrays, at flow rates of 50, 100 and 125 sccm, respectively. At the same time, a small amount of water was introduced through the flow of Ar at a flow rate of 0.75 sccm. The reaction were conducted at 800°C for 5–10 minutes.

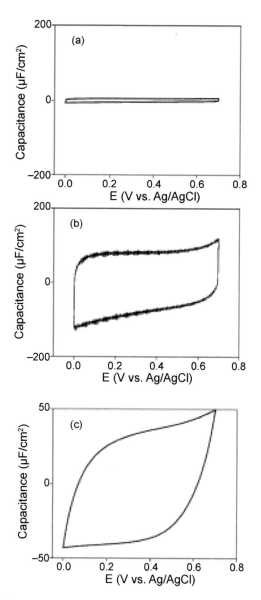

Fig. 3.85. CV curves of the samples: (a) bare V_2O_5 films, (b) pristine CNTs and (c) V_2O_5–CNT nanocomposites. Reproduced with permission from [283], Copyright © 2008, American Chemical Society.

CNT ink was prepared by dispersing 20 mg CNTs and 20 mg sodium dodecylbenzenesulfonate (SDBS) surfactant in 20 mL DI water, with the aid of bath-sonicated for 5 minutes and bar-sonicated for 20 minutes, followed by centrifugation at 3000 rpm for 5 minutes, in order to remove the surfactant in the supernatant. The CNT precipitates were dried by putting them on poly(tetrafluoroethylene) (PTFE) filter paper with pore size of about 1 μm and then washed with DI water. After that, the CNTs were redispersed ultrasonically in 20 mL DI water to form CNT ink, which was used coat papers. Typically, 0.2 mL CNT suspension was casted on a piece of paper with an area of 1 cm × 1 cm, followed by drying at 80°C for 2 hours. MnO_2 were deposited on the CNT/paper by sweeping voltage in the range of 0–1.2 V for 100 cycles in aqueous solution containing 0.1 M Na_2SO_4 and 0.1 M $Mn(CH_3COO)_2$. After drying at 80°C for 1 hour, MnO_2/CNT papers were obtained.

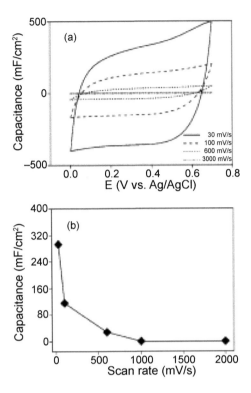

Fig. 3.86. (a) CV curves of the V$_2$O$_5$–CNT nanocomposites measured at different scan rates. (b) Capacitance of the V$_2$O$_5$–CNT nanocomposites as a function of the scan rate. Reproduced with permission from [283], Copyright © 2008, American Chemical Society.

A composite consisting of manganese oxide and MWCNTs was prepared by simply mixing manganese oxide and MWCNTs in ethanol [287]. A nonaqueous hybrid supercapacitor was constructed by using the manganese oxide/MWNTs (M/M) composite as a positive electrode and MWCNTs as a negative electrode. The asymmetric hybrid capacitor exhibited higher electrochemical performances than its symmetric counterparts. The hybrid composite demonstrated a specific capacitor of 56 F·g^{-1} at the current density of 1 mA·cm^{-2}, much higher than that of the electrode based on MWCNTs only. Meanwhile, an energy density of 32.91 Wh·kg^{-1} was achieved at the current density of mA·cm^{-2}, in electrolyte solution of 1.0 M LiClO$_4$.

The nanocrystalline manganese oxide particles had a needle-like morphology, with lengths of 100–300 nm and diameters of 20–40 nm. Such nanostructures were favorable to ionic charge transport during the electrochemical reactions. The M/M composite contained spherical agglomerates of manganese oxide nanoparticles, which are randomly mixed with the MWCNTs. XRD results indicated that the manganese oxide consisted of α-MnO$_2$ and MnO$_{1.88}$.

Figure 3.87 shows cycle performances of the three capacitors at the current density of 10 mA·cm^{-2}. Comparatively, the hybrid capacitor had both higher capacitance and higher cycling stability, which suggested that electrochemical performance of MWCNTs could be enhanced by making into M/M composite, because the MWCNTs network ensured a sufficiently high conductivity of the electrodes and also had a contribution to charge storage due to the EDLC effect. Meanwhile, the tubular structure of MWCNTs could have increased the filling factor to boost discharge capacitance, whereas their linear structure could facilitate a rapid insertion and removal of Li$^+$ ions [288].

Au-MnO$_2$/CNT hybrid coaxial nanotube arrays, used as electrodes of supercapacitors, were fabricated with the combination of electrodeposition, vacuum infiltration and CVD methods, with porous AAO templates [289]. Due to the highly conductive CNT core, the coaxial hybrid structure could facilitate enhanced electronic transport to the MnO$_2$ shell and possessed strong adhered interface between Au and

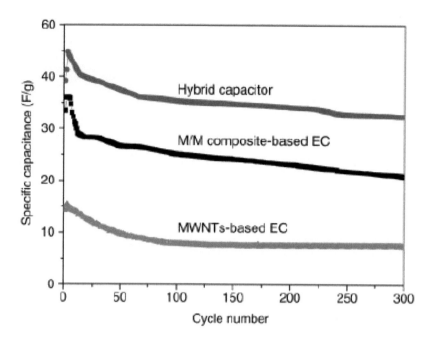

Fig. 3.87. Cycle performances of the three supercapacitors at the current density of 10 mA·cm^{-2}. Reproduced with permission from [287], Copyright © 2005, Elsevier.

the MnO$_2$/CNT segments. Such a nanoscale contact with the electrode and the current collector resulted in a very low contact resistance. The Au-MnO$_2$/CNT hybrid coaxial nanotube electrodes exhibited promising supercapacitive behavior, with an optimal specific capacitance of 68 F·g^{-1}, together with a power density and an energy density of 33 kW·kg^{-1} and 4.5 Wh kg^{-1}, respectively, owing to the dual storage mechanisms of EDLC and pseudocapacitance.

Figure 3.88 shows a schematic diagram to demonstrate experimental procedure for the growth of Au-MnO$_2$/CNT coaxial arrays [289]. A layer of Ag with a thickness of about 100 nm was coated onto one side of AAO template by using thermal evaporation method, which would be used as the working electrode for the consequent electrochemical deposition. Au was then deposited into the nanopores by using the electrodeposition process to form nanowires (AuNWs), with the standard three-electrode potentiostat configuration. The reference electrode was Ag/AgCl, while a piece of platinum wire was employed as the counter electrode. Above the AuNWs, MnO$_2$ nanotubes were deposited by using a vacuum infiltration method, with manganese nitrate solution. After that, the MnO$_2$ infiltrated AAO was calcined at 300°C for 10 hours, which led to Au-MnO$_2$-nanotube configuration. To fabricate Au-MnO$_2$/CNT coaxial arrays, CNTs were grown inside the MnO$_2$ tubes. The CNT arrays were obtained at 650°C in Ar flow of 100 sccm, followed by introducing a mixture of C$_2$H$_2$ (20%) and Ar (80%) at a flow rate of 20 sccm for 1 hour. Finally, the AAO template was removed with 3 M NaOH aqueous solution.

Figure 3.89 depicts representative SEM images of the Au-MnO$_2$/CNT hybrid nanostructures [289]. In the uniform coaxial nanostructure, two segments could be readily observed, as demonstrated in Fig. 3.89(a, b). The coaxial structure of the MnO$_2$ shell and the CNT core is clearly revealed in Fig. 3.89(c, d). The Au-MnO$_2$/CNT hybrid nanostructures, including shell thickness and nanotube length, could be simply maintained by controlling the infiltration time. Due to the presence of the Au segments, the conduction from the electrode to the current collector was totally ensured, while the CNT cores enhanced electrical conductivity, mechanical stability and electrochemical performance, when the nanostructures were used as electrodes.

Composites of amorphous MnO$_2$ (a-MnO$_2$) and SWCNTs (weight ratios of 5–40 wt% SWCNTs) were fabricated, with a focus on the long cycle performance at a high charge–discharge current of A·g^{-1} [290].

Carbon Based Supercapacitors 121

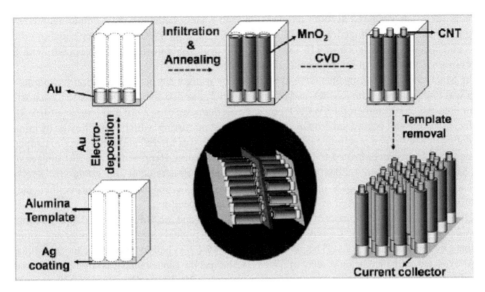

Fig. 3.88. Schematic diagram showing procedure to grow the Au-MnO$_2$/CNT hybrid coaxial nanotube arrays inside an AAO template, by using the combination of electrodeposition, vacuum infiltration and CVD techniques. Reproduced with permission from [289], Copyright © 2009, American Chemical Society.

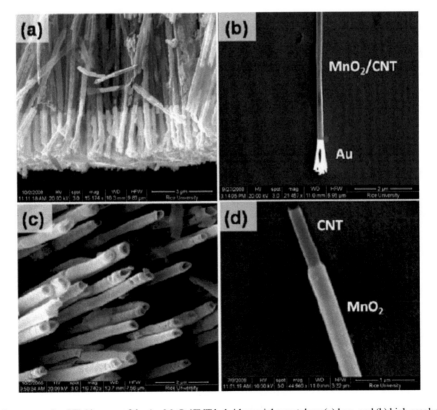

Fig. 3.89. Representative SEM images of the Au-MnO$_2$/CNT hybrid coaxial nanotubes: (a) low- and (b) high-resolution SEM images showing the Au-MnO$_2$/CNT, (c) low- and (d) high-resolution SEM images showing the MnO$_2$ shell and CNT core. Reproduced with permission from [289], Copyright © 2009, American Chemical Society.

The composites were prepared at room temperature with KMnO$_4$, ethanol and commercial SWCNTs as starting materials. The optimized composition was MnO$_2$:20 wt% SWCNT, with couloumbic efficiency of 75% and specific capacitance of 110 F·g^{-1} after 750 cycles. Meanwhile, the composite with 5 wt% SWCNTs showed the highest initial specific capacitance.

The MnO$_2$:SWCNT composites were synthesized by using a simple precipitation technique. MnO$_2$ powder was first prepared from KMnO$_4$ and ethanol. KMnO$_4$ was dissolved in DI water to form a saturated solution, into which a commercial SWCNT power (with diameter of < 2 nm and lengths of 0.5–40 μm). After that, 10 ml ethanol was added drop by drop under constant stirring, which was followed by immediate precipitation of MnO$_2$. The precipitate of MnO$_2$ and SWCNTs was filtered and dried at room temperature. The valence of Mn was confirmed to be +4 by using XPS spectra. Representative TEM images of pure MnO$_2$, SWCNTs and the MnO$_2$:20 wt% SWCNT composite are illustrated in Fig. 3.90(a–c). The SWCNT and a-MnO$_2$ were homogeneously incorporated, which would be beneficial to improve their electrical conductivity when they were used as the electrodes of supercapacitors.

Figure 3.90(d) shows CV curves of pure a-MnO$_2$, SWCNTs and the MnO$_2$:20 wt% SWCNT composite, measured at a scan rate of 2 mV·s^{-1}, with all samples having a typical capacitive behavior and completely reversible electrochemical response. The pure a-MnO$_2$ and SWCNTs had specific capacitance of 202 and 22 F·g^{-1}. As expected, with increasing content of SWCNTs, the specific capacitance was slightly decreased, with values of 195, 180, 162 and 151 F·g^{-1} for the composites with 5, 10, 20 and 40 wt% SWCNTs, respectively.

A flexible nanostructured hybrid supercapacitor was developed with α-Fe$_2$O$_3$/MWCNT composite, which was fabricated by using a spray deposition method [291]. The composite exhibited specific energy density of 50 Wh·kg^{-1}, corresponding to a specific power density of 1.0 kW·kg^{-1}, over a potential range

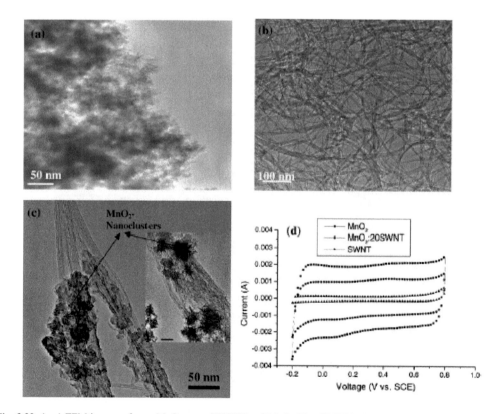

Fig. 3.90. (a–c) TEM images of pure MnO$_2$, pure SWCNT and MnO$_2$:20 wt% SWCNT composite. Inset of (c) is a high magnification image of the composite, with scale bar = 20 nm. (d) CV curves of the pure MnO$_2$, the pure SWCNT and the MnO$_2$:20 wt% SWCNT composite at a scan rate of 2 mV·s^{-1}. Reproduced with permission from [290], Copyright © 2006, Elsevier.

of 0–2.8 V. The high electrochemical performance of the hybrid supercapacitor could be readily ascribed to the effectively decreased internal resistance, as well as the improvement in both the ion diffusion and the structural integrity of the films.

Commercial CVD grown MWCNTs with diameter of 10–12 nm and BET surface area of 250–300 $m^2 \cdot g^{-1}$ were refluxed with 9 M HNO_3 for 24 hours for purification and carboxylation, followed by filtration and thorough washing with DI water. α-Fe_2O_3 nanospheres were derived from $FeCl_3 \cdot 6H_2O$ aqueous solution, with the presence of cetrimonium bromide (CTAB), which was heated to 100°C for 15 hours, followed by filtration, washing and drying. α-Fe_2O_3/MWCNT suspensions were prepared by dispersing MWCNTs in water with the aid of ultrasonication, which was then mixed with $FeCl_3 \cdot 6H_2O$ and CTAB, with weight ratios of α-Fe_2O_3 to MWCNT to be 60:1, 9:1, 3:1 and 1:1. Film electrodes were obtained by spraying suspensions of α-Fe_2O_3 and α-Fe_2O_3/MWCNT in ethanol with a concentration of 1 mg·mL^{-1} on Cu foils.

The formation of α-Fe_2O_3 in the composite films was confirmed by XRD results. Figure 3.91 shows representative SEM and TEM images of the samples. As demonstrated in Fig. 3.91(a), the purified MWCNT film consisted of randomly oriented MWCNTs, forming an entangled and interconnected nanoporous structure, with ionic/electronic conducting channels [292]. The α-Fe_2O_3 particles had a spherical morphology, with a narrowly distributed diameter of about 100 nm, as seen in Fig. 3.91(b). The nanospheres possessed a loose packing, thus leading to a large surface area but weak mechanical strength. After incorporation with MWCNTs, the α-Fe_2O_3 nanoparticles were uniformly distributed within the continuous MWCNT network, as demonstrated in Fig. 3.91(c). Both the morphology and size of the α-Fe_2O_3 nanoparticles remained the same after the incorporation of the MWCNTs. Figure 3.91(d) shows

Fig. 3.91. (a) Top surface SEM image of the spray deposited films: (a) purified MWCNTs, (b) α-Fe_2O_3 and (c) α-Fe_2O_3/MWCNT (3:1 w/w) composite. (d) TEM image of the α-Fe_2O_3/MWCNT (3:1 w/w) composite. The inset in (b) shows the α-Fe_2O_3 nanospheres at higher magnification, while that in (d) shows the electron diffraction pattern from the α-Fe_2O_3 nanospheres in the composite. Reproduced with permission from [291], Copyright © 2009, Royal Society of Chemistry.

124 *Nanomaterials for Supercapacitors*

TEM image of an α-Fe$_2$O$_3$ nanosphere that was attached to a MWCNT. In general, TEM investigations confirmed that the α-Fe$_2$O$_3$ nanospheres and MWNTs were well mixed with numerous points of α-Fe$_2$O$_3$-MWNT contact. The Fe$_2$O$_3$ was of rhombohedral crystal structure, as evidenced by the electron diffraction pattern in the inset of Fig. 3.91(d).

Similarly, a flexible supercapacitor was assembled with a 3D α-Fe$_2$O$_3$/carbon nanotube (CNT@Fe$_2$O$_3$) sponge electrode [293]. The sponge electrode possessed a porous hierarchical structure, which was made of a compressible conductive CNT network that was coated with a layer of Fe$_2$O$_3$ nanohorns. The hybrid sponges exhibited an optimal specific capacitance of 300 F·g^{-1}, with an equivalent series resistance of about 1.5 Ω. The flexible supercapacitor could withstand large deformations, while maintaining normal functions and high reliability. For example, 90% of the initial specific capacitance could be retained under a compressive strain of 70%, corresponding to a decrease in volume by 70%. Stable operation could be maintained without variation in specific capacitance after 1000 compression cycles at a strain of 50%.

The CNT sponges were synthesized by using a CVD method [294]. To prepare α-Fe$_2$O$_3$/CNT composite, the CNT sponges were cut into rectangular blocks and exposed to UV for 1 hour, which were then soaked in 30 mL FeCl$_3$ ethanol/aqueous (volume ratio of 1:1) solution with concentrations of 0.01–1.0 M, followed by hydrothermal reaction at 100°C for 2 hours. After that, the CNT@FeOOH sponges were thoroughly washed with deionized water and freeze-dried. Then, the composite samples were calcined in air at 300°C for 3 hours to form α-Fe$_2$O$_3$/CNT sponges.

Figure 3.92(a) shows a schematic diagram of the fabrication process of the α-Fe$_2$O$_3$/CNT composites, in which porous Fe$_2$O$_3$ nanoparticles were uniformly distributed inside the CNT sponge, while the original 3D network was well retained [293]. The CNT sponges were treated with UV/O$_3$ to increase their hydrophilicity, due to the formation of oxygen-containing groups and other carbon hexagonal rings, which provided with anchoring sites or nucleation sites for the attachment of FeOOH [295]. Therefore, hierarchical CNT@FeOOH sponges were obtained after the hydrothermal treatment, with the FeOOH nanoparticles to be homogeneously distributed on surface of the CNTs. Final thermal calcination led to the formation of α-Fe$_2$O$_3$/CNT composite sponges, without losing the porosity and flexibility of the original CNT sponges.

A photograph of a CNT sponge is shown as the inset in Fig. 3.92(b), which possessed a 3D porous network structure. Such a porous structure could be used as a host to deposit active materials with high efficiency, because it allowed an easy infiltration of precursor solutions. As expected, FeOOH nanospindles, with lengths of 20–30 nm and a diameter of 10 nm, were uniformly deposited inside the CNTs sponges, as demonstrated in Fig. 3.92(c). After calcination treatment in air, Fe$_2$O$_3$ nanohorns were derived from the FeOOH nanospindles, as seen in Fig. 3.92(d). Due to the strong mechanical strength of the skeleton formed by the CNTs, the α-Fe$_2$O$_3$/CNT composite well retained the original structure, as illustrated as the inset in Fig. 3.92(d). It was found that the Fe$_2$O$_3$ nanoparticles were highly porous, while their size and morphology remained almost unchanged, as compared with their FeOOH counterparts, as revealed in Fig. 3.92(e, f). The presence of the porous structure was attributed to the reduction in volume from FeOOH to denser α-Fe$_2$O$_3$.

NiOx/CNT electrodes have been developed by incorporating Ni(OH)$_2$ with carbon nanotube (CNT) films [296]. The incorporation of Ni(OH)$_2$ was realized by using electrochemical deposition on CNT films as substrates, followed by thermal annealing at 300°C. X-ray photoelectron spectroscopy results revealed that the pseudocapacitive behavior of the NiO$_x$/CNT electrodes in 1 M KOH solution was resulted from the redox reactions of NiO$_x$/NiO$_x$OH and Ni (OH)$_2$/NiOOH. The sample with 8.9 wt% NiO$_x$ exhibited a specific capacitance of as high as 1701 F·g^{-1} based on the content of NiO$_x$, accompanied by a high rate capability. The promising electrochemical performance of the composite electrodes was readily attributed to the 3D nanoporous network structure, in which a thin NiO$_x$ layer was uniformly coated on the CNT film substrate. Meanwhile, the NiO$_x$/CNT electrode with 36.6 wt% NiO$_x$ demonstrated optimal geometric and volumetric capacitances of 127 mF·cm^{-2} and 254 F·cc^{-1}, respectively, corresponding to a specific capacitance of 671 F·g^{-1}. As compared with that of the 8.9 wt% sample, the specific capacitance was decreased quite significantly, which was explained by considering the utilization efficiency of the oxide. The high loading level of oxide blocked the channels of the 3D network, so that the ionic transportation became less effective.

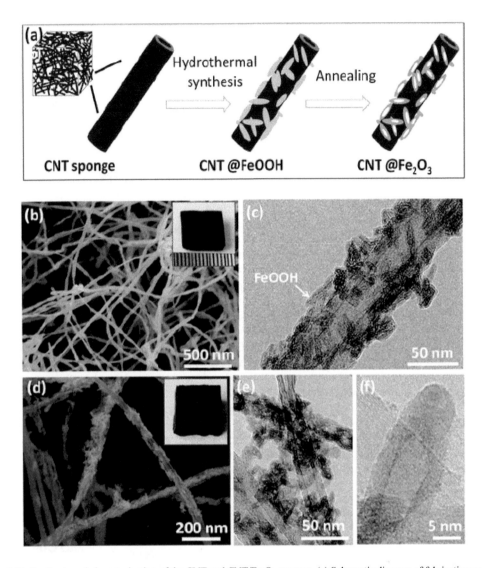

Fig. 3.92. Synthesis and characterization of the CNT and CNT/Fe$_2$O$_3$ sponges. (a) Schematic diagram of fabrication process of the CNT@Fe$_2$O$_3$ sponges. (b) SEM image of the as-grown CNT sponge, with the inset showing a photograph of the sponge with a dimension of 11 × 10 × 3 mm. (c) TEM image of the CNT@FeOOH showing uniformly distributed FeOOH nanospindles on surface of the CNTs. (d) SEM image of the CNT@Fe$_2$O$_3$ sponge, with the inset showing a photograph of the CNT@Fe$_2$O$_3$ sponge block. (e, f) TEM images of the CNT@Fe$_2$O$_3$ showing the hollow structure of the rice-like Fe$_2$O$_3$ nanohorns. Reproduced with permission from [291], Copyright © 2015, Royal Society of Chemistry.

A simple hydrothermal method was reported to synthesize NiO/MWCNT composites for the applications as electrodes of supercapacitors [297]. Morphologies of the NiO nanoparticles could be well controlled, from 2D flake to mesoporous sphere, by adjusting the concentration of Sodium Dodecyl Sulfate (SDS). It was found that electrochemical performance of the NiO/MWCNT composites was dependent on the morphology and distribution of the NiO nanoparticles. The mesoporous spherical NiO nanoparticles exhibited the highest specific capacitance of 1329 F·g^{-1}, together with a stable cycling capability up to 1000 cycles in 1 M KOH electrolyte solution at a high current density of 84 A·g^{-1}.

To synthesize the NiO/MWCNT composites, 1 g commercial MWCNTs powder, 1 g NiCl$_2$·6H$_2$O and 40 g urea were dispersed in 50 mL DI water, followed by stirring for 3 hours, into which surfactant

SDS was added with different contents. Hydrothermal synthesis was conducted at 80°C for 4 hours. After that, the products were filtered, washed and dried at 60°C for 12 hours in vacuum, which thus led to NiO/MWCNT composites after calcination at 300°C for 6 hours in air. Three samples, NiO/MWCNT-0, NiO/MWCNT-10 and NiO/MWCNT-20, were obtained, corresponding to the contents of SDS of 0, 10 and 20 g, respectively.

Without the presence of MWCNTs and SDS, the NiO particles had an average size at micrometer scale, which were agglomerates consisting of nanosheets. With the presence of MWCNTs, the NiO particles with a uniform sheet-like morphology were distributed on the MWCNTs. Once SDS was present, spherical NiO particles were formed. With increasing content of SDS, the NiO nanoparticles were varied in morphology from nanodisk to nanofilament. The variation in morphology of the NiO nanoparticles was closely related to the attractive interactions between the dodecyl sulfate anions and the Ni^{2+} ions. TEM observation indicated that no morphology change was observed before and after the calcination of the samples at 300°C.

CNT–ZnO nanocomposites were used as an electrodes to assemble supercapacitors with a gel polymer as the electrolyte solution [298]. The CNT films were first fabricated using screen-printing method on CuNi alloy substrate, on which ZnO nanoparticles were deposited by ultrasonic spray pyrolysis (USP) process for different time durations to control over the loading level. Specific capacitance of the CNT-based electrode was 92.7 $F·g^{-1}$, while it was increased to 126.3 $F·g^{-1}$ after the incorporation of CNT with deposition time of 5 minutes (i.e., CNT–ZnO-5 composite). However, further increase in the content of ZnO, the value of specific capacitance was significantly decreased. Because ZnO has lower conductivity than CNTs, the presence of too much ZnO would result in high resistance, especially the surface of the CNTs was entirely covered by ZnO nanoparticles.

More recently, a single precursor, from Metal–Organic Frameworks (MOFs) (zeolitic imidazolate framework, ZIF-8) and CNTs, was synthesized, which could be used as a template to further develop ZnO quantum dots (QDs)/carbon/CNTs and porous N-doped carbon/CNTs, as electrodes of supercapacitors [299]. Figure 3.93 shows a schematic diagram illustrating synthetic route of the samples. The ZnO QDs/carbon/CNTs exhibited a specific capacitance of 185 $F·g^{-1}$ at the current density of 0.5 $A·g^{-1}$, while that of the porous N-doped carbon/CNTs was 250 $F·g^{-1}$ at 1 $A·g^{-1}$. Moreover, when the two types of composites were used to assemble all-solid-state asymmetric supercapacitor (ASC), with the ZnO QDs/carbon/CNTs as the positive electrode and the porous N-doped carbon/CNTs as the negative electrode, the device had a working potential of 1.7 V, which led to a maximized energy density of 23.6 $Wh·kg^{-1}$ and a corresponding power density of 16.9 $kW·kg^{-1}$.

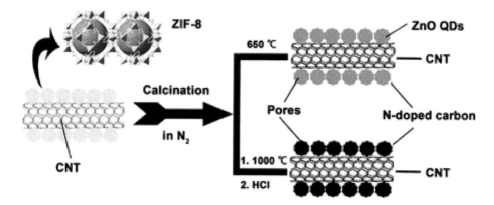

Fig. 3.93. Schematic diagram demonstrating synthesis route of the ZnO QDs/carbon/CNTs and porous N-doped carbon/CNTs (ZIF = zeolitic imidazolate framework, QDs = quantum dots and CNTs = carbon nanotubes). Reproduced with permission from [299], Copyright © 2016, Royal Society of Chemistry.

Graphene based Electrodes

Graphene is a newly emerged special type of carbon material, with one-atom thick 2D honeycomb nanostructure, which is a hot research topic for various potential applications. Specifically, it has also been extensively studied as electrodes of electrochemical supercapacitors, both experimentally and theoretically, due to its low mass density, high electrical conductivity, strong mechanical strength, high chemical stability, excellent thermal conductivity and high specific surface area [300–310]. Here recent progress in the development of supercapacitor electrodes, based on graphene and graphene-based composites, is presented and discussed, by using examples of each type of materials.

Graphene/graphene oxide

A graphene based electrode of supercapacitors was reported, with noble metal films in between the interlayer of graphene and Al foil as the current collector [308]. The graphene was derived from graphite powder, by using a combined process of chemical exfoliation and mild thermal reduction. The graphene layers were coated on Au/Pd by using a drop casting method, which were used as the working electrodes of supercapacitors, exhibiting a specific capacitance of 50 $F \cdot g^{-1}$ at the current density of 300 $mV \cdot s^{-1}$. The rectangular-shaped CV curve was well retained even at a scan rate of as high as 3000 $mV \cdot s^{-1}$. The initial specific capacitance was retained by about 82% up to 1500 cycles. Such graphene based electrodes possessed a power density and an energy density at a current density of 33.3 $A \cdot g^{-1}$ to be 40 $kW \cdot kg^{-1}$ and 40 $Wh \cdot kg^{-1}$, respectively.

Graphene nanosheet has been deposited into Ni foams as electrodes of supercapacitors, by using electrophoretic deposition (EPD) method [309]. The EPD graphene based electrodes demonstrated a non-rectangular CV curve, indicating their pseudocapacitive profile, which was ascribed to the adsorption/desorption of the positively charged ions at negative potentials. The initial capacitance of the graphene based electrodes was retained by 89 and 60% at 20 $mV \cdot s^{-1}$ and 100 $mV \cdot s^{-1}$, respectively. The low electrical conductivity of the graphene prepared in this way could be readily related to the adsorption of the oxidized product of p-phenylenediamine (OPPD). Also, due to the lack of porosity of the graphene, the migration of electrolyte ions could be effectively promoted, thus resulting in inefficient charge propagation.

In a more recent study, reduced Graphene Oxide/Nickel Foam (rGO/NF) composite was prepared for the applications as electrodes of high-performance supercapacitors, by using flame-induced reduction method, started with dry Graphene Oxide (GO) coated on nickel foam [311]. The flame-reduced rGO exhibited a disordered cross-linking network with randomly distributed pores, so as to ensure fast transportation of ions during the charge-discharge reaction. The rGO/NF electrode demonstrated higher electrochemical performance, as compared with that based on rGO films. A specific capacitance of 228.6 $F \cdot g^{-1}$ at a current density of 1 $A \cdot g^{-1}$, high rate capability of > 81% retention at 32 $A \cdot g^{-1}$ and high cycling stability with only 95% capacitance retention after 10,000 cycles at a high scan rate of 1000 $mV \cdot s^{-1}$ were achieved by using the rGO/NF based electrode.

To prepare normal rGO paper based electrode, GO was dispersed in DI water, with a concentration of 10.0 $mg \cdot mL^{-1}$, which was used to obtain GO paper through vacuum membrane filtration, followed by flame reduction. The rGO paper electrode was characterized with a symmetrical two-electrode cell configuration, in 6 M KOH electrolyte. To develop rGO/NF electrode, the same GO suspension was filled into a piece of NF, followed by drying at 55°C for 12 hours, with the content of GO to tailored by controlling the volume of the GO suspension. The dried rGO/NF composites then reduced similarly by using the flame reduction method. The loading level of rGO was about 0.35 $mg \cdot cm^{-2}$. Figure 3.94(a) shows a schematic diagram describing fabrication process of the rGO/NF composites, while symmetrical supercapacitor was assembled by stacking two electrodes in a face-to-face way with the rGO/NF composites, followed by the application of an appropriate pressure, as seen in Fig. 3.94(b).

As shown in Fig. 3.95(a–c), after the flame reduction, the sample experienced a change in color from brown (gray) GO/NF to black rGO/NF. The samples all possessed a well-retained 3D structure, as demonstrated in Fig. 3.95(d–g). After drying, the inner walls of the 3D foam scaffold were uniformly covered by the GO layer, as seen in Fig. 3.95(e). The flame reduction process had almost no effect on

128 *Nanomaterials for Supercapacitors*

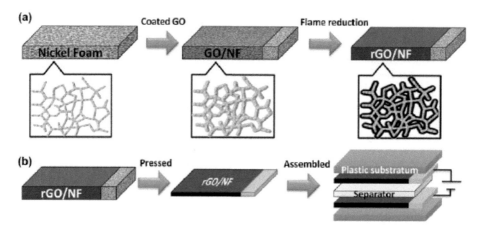

Fig. 3.94. (a) Schematic diagram showing fabrication process of the flame-reduced rGO/NF electrode, together with local magnified models of the corresponding parts. (b) Assembling and configuration of solid-state supercapacitors based on the rGO/NF composite electrodes. Reproduced with permission from [311], Copyright © 2016, Royal Society of Chemistry.

Fig. 3.95. (a–c) Photographs of NF, GO/NF composite and rGO/NF composite, respectively. (d) SEM image of NF. (e, f) Cross-sectional SEM images of the GO/NF and rGO/NF composites, respectively. (g) Magnified SEM image of the rGO/NF composite. Reproduced with permission from [311], Copyright © 2016, Elsevier.

microstructure and morphology of the graphene nanosheets, as observed in Fig. 3.95(f). It was found that a cross-linked rGO foam was formed during the flame reduction, owing to the rapid increase in spacing in between the rGO nanosheets. Figure 3.95(g) indicated that the rGO layer on the NF was present as a disordered cross-linking network, accompanied by randomly distributed pores with pore sizes from sub-micrometer to micrometers, which offered accessible paths to maintain fast transportation of ions when the sample was used as an electrode of supercapacitors.

A new group of carbon material, called Chemically Modified Graphene (CMG), with 1-atom thick nanosheets of carbon, was developed to be used as electrodes of supercapacitors [312]. Such CMG based electrodes could achieve specific capacitances of 135 and 99 $m^2 \cdot g^{-1}$ in aqueous and organic electrolytes, respectively. Due to the high electrical conductivity, the CMG materials could be operated over a wide range of scan rates of voltage. The CMG was synthesized by suspending GO nanosheets in water and then reducing with hydrazine hydrate. The reduced graphene nanosheets were agglomerated into particles with diameters of 15–25 µm, with C/O and C/N atomic ratios to be 11.5 and 23.0, respectively.

Figure 3.96(a) shows surface SEM image of the CMG agglomerates, while a TEM image revealing detailed profile of the individual graphene nanosheets is illustrated in Fig. 3.96(b) [312]. The graphene nanosheets at the surface of the agglomerates could be directly in contact with the electrolyte. The CMG powder had a BET specific surface area of 705 $m^2 \cdot g^{-1}$. Figure 3.96(c) depicts surface SEM images of the as-prepared electrode based on the CMG powder. A two-electrode ultracapacitor cell was assembled to characterize electrochemical performance of the CMG based electrode, as shown schematically in Fig. 3.96(d).

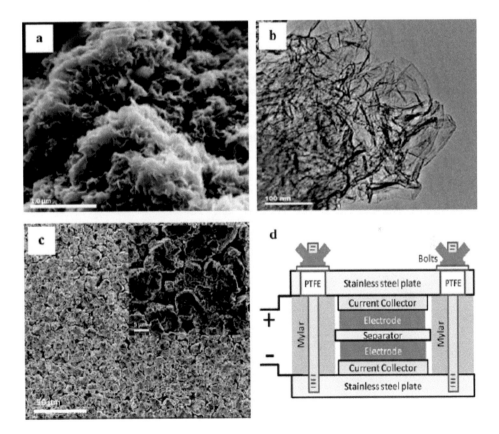

Fig. 3.96. (a) Surface SEM image of the CMG particles. (b) TEM image showing individual graphene nanosheets extending from surface of the CMG particle. (c) Low and high (inset) magnification surface SEM images of the CMG electrode. (d) Schematic of supercapacitors based on the CMG. Reproduced with permission from [312], Copyright © 2008, American Chemical Society.

Various synthetic routes have been reported in the open literature to prepare graphene, graphene oxide and reduced graphene oxide, used as electrode of supercapacitors, including low temperature exfoliation [313, 314], gas-solid reduction process [315], mild solvothermal or hydrothermal synthetic routes [316–319], graphene with functionalization [320–322] and KOH activation of graphene [323, 324]. These newly developed graphene based materials have been demonstrated to have enhanced electrochemical performances, in terms of specific capacitance, cycling stability and high rate capability and high energy densities.

For instance, a hydrothermal reaction method was used to synthesize three-dimensional nitrogen-doped graphene hydrogels (3D NG), which demonstrated high mechanical strength and promising electrochemical properties [318]. An optimal specific capacitance of 387.2 $F \cdot g^{-1}$ at 1 $A \cdot g^{-1}$ in 6 M KOH, was retained by 90% at a high current density of 5 $A \cdot g^{-1}$ after 5500 cycles. The electrochemical performance of the 3D NGs was attributed to the introduction of the pyrrolic and pyridinic nitrogen atoms into the graphene nanosheets, which enhanced the pseudo-capacitive effect of the materials.

GO powder was dispersed in DI water to form suspension with a concentration of 3 $mg \cdot mL^{-1}$, with the dispersion to be promoted by adding about 0.1 mL ammonia (25 wt%) solution for every 5 mL GO suspension. Ammonia, together with urea, was also used a source of nitrogen that was incorporated with GO during the hydrothermal reaction. The GO dispersion (60 mL) and urea were mixed with mass ratios of 3:n (n = 0, 1, 3, 4). Hydrothermal reaction was then conducted at 180°C for 24 hours, to obtain GO samples, denoted as NG-3:n (n = 0, 1, 3, 4). The NG samples had strong mechanical strength, so that the hydrogel cylinders with a diameter of 1.8 cm and a height of 2.0 cm was sunk in water and could be handled freely, as seen in Fig. 3.97(a–c). Mechanical strength of the NG cylinders was further demonstrated in Fig. 3.97(d–e).

The GO hydrogels had an interconnected porous 3D microstructure after freeze drying, as demonstrated in Fig. 3.98 [318]. Without the addition of urea, the sample exhibited a compact stacking microstructure, with the presence of a high porosity, as revealed in Fig. 3.98(a). With the presence of urea of relatively low mass ratio of GO to urea (3:1, 3:3), the NG hydrogels possessed microstructures characterized by interconnected 3D mesoporous networks, with pore sizes in the range between 10s and 100s of nanometer, in which the wall of the pores was made of ultrathin closely stacked graphene nanosheets, as illustrated in Fig. 3.98(b–c). It was found that too high GO/urea mass ratio (3:4), the NG nanosheets would be re-stacked, due to the high content of nitrogen, as observed in Fig. 3.98(d).

Fig. 3.97. Photographs of the NG hydrogels demonstrating their physical integrity and mechanical strength: (a) NG hydrogels, (b) NG hydrogel sunk in water and (c–e) mechanical strength demonstration. Reproduced with permission from [318], Copyright © 2016, Elsevier.

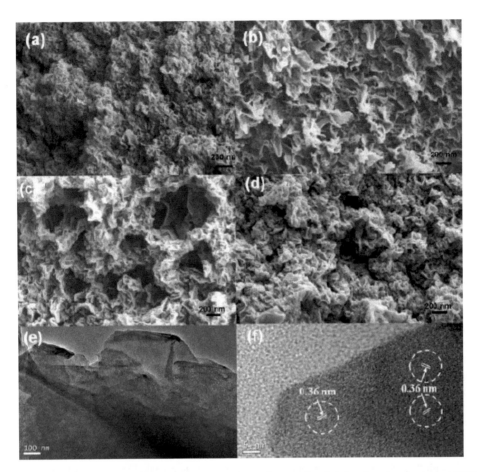

Fig. 3.98. (a–d) SEM images of the 3D NG-3:*n* (*n* = 0, 1, 3, 4) samples revealing their interior microstructures. (e) TEM and (f) HRTEM of the 3D NG-3:3 sample. Reproduced with permission from [318], Copyright © 2016, Elsevier.

Figure 3.98(e) shows TEM images of the 3D NG-3:3 sample, depicting its interior microstructural characteristics, in which graphene nanosheets were staggered by one another, thus forming the 3D cross-linked network structure [318]. Figure 3.98(f) shows a HRTEM of the 3D NG-3:3 sample. There were clear fringes with an inter-planar spacing of 0.36 nm, corresponding to the *d* of the (002) plane of graphene, which was in a good agreement with the XRD results. Therefore, it was expected that the 3D interconnected network of the NG samples would be highly beneficial to electron transfer during the charge–discharge process, thus resulting in high electrochemical performances, when they were used as electrodes of supercapacitors.

Graphene–conductive polymer nanocomposites

Both PANI and PPy have been used to fabricate nanocomposites with graphene to achieve electrodes of supercapacitors with enhanced electrochemical performances. For example, a Graphene/PANI Composite Paper (GPCP) was developed through an *in-situ* anodic electropolymerization process, which exhibited an optimal specific capacitance of 233 $F \cdot g^{-1}$ than was higher than that of the graphene counterpart [325]. A similar graphene/PANI nanocomposite was prepared by using a chemical precipitation method, with specific capacitances in the range of 300–500 $F \cdot g^{-1}$ at the current density of 0.1 $A \cdot g^{-1}$, together with high cyclability [326]. Graphene/PANI composite films were derived from CMG and PANI nanofibers by using the vacuum filtration process [327]. The PANI-NFs exhibited a high specific capacitance of 214 $F \cdot g^{-1}$, due to the concurrent effect of the PANI-NFs and the CMG. A sulfonated graphene/PANI nanocomposite (SGEP) was developed by using a morphology-controlled method combined with liquid/liquid interfacial polymerization, which demonstrated a specific capacitance of 763 F·g^{-1}, with 96% retention of the value after 100 cycles [328].

More specifically, an *in-situ* polymerization strategy was employed to produce GNS/PANI composite [329]. The CV shape of the GNS/PANI composite demonstrated pseudocapacitance behavior attributed towards the redox transition of PANI. The substantially high capacitance value (978 F·g^{-1}) of the composite material was perhaps due to the homogeneous dispersion of PANI over the GNS sheets with large surface area. The dispersion of PANI particles on GNS reduced the diffusion and migration length of the electrolyte and in the process the electrochemical utilization of PANI was improved.

GNS was obtained by reducing graphene oxide with hydrazine hydrate [329]. To synthesize the GNS/PANI composite, 0.25 M aniline monomer, with 1 M HCl as the solvent, was dispersed in 135 mL GNS suspension, with the aid of sonication for 30 minutes. The mixture was then cooled to 2°C in an ice-water bath, into which an equal volume of 0.25 M ammonium persulfate (APS) solution was added and then the temperature was maintained at 0–4°C for 4 hours. After that, the GNS/PANI composites were thoroughly washed with distilled water and ethanol, followed by vacuum drying at 80°C for 12 hours. The final product had a mass content of PANI to be 85%.

Figure 3.99 shows XRD patterns of the GNS, PANI and GNS/PANI composite [329]. Two diffraction peaks were observed on the XRD pattern at $2\theta = 24.5°$ and $42.8°$ corresponded to the (002) and (100) of graphite-like structure. Pure PANI exhibited crystalline peaks at $2\theta = 15.3°$, $20.7°$ and $25.2°$, being attributed to its (011), (020) and (200) crystal planes of the form of emeraldine salt. The XRD patterns of the GNS/PANI composite were almost the same as that of pure PANI, implying that the GNS was entirely incorporated with the PANI particles.

Figure 3.100 shows a schematic diagram demonstrating formation mechanism of the GNS/PANI composite [329]. As an electron acceptor and donor, the GNS and aniline would interact to form a weak charge-transfer complex [330]. During the mixing of aniline monomers and the GNS suspension, aniline monomers would be adsorbed onto surfaces of the GNS through the electrostatic attraction. Without the presence of GNS, PANI particles with a size of about 40 nm were polymerized to form worm-like agglomerates with a diameter of about 170 nm and lengths of 1–2 μm, as demonstrated in Fig. 3.101(a). The GNS, with thicknesses of 2–3 nm and $S_{BET} = 267$ m^2·g^{-1}, as revealed in Fig. 3.101 (b), served as a supporter with numerous active sites for the nucleation of PANI, so that uniform PANI particles were

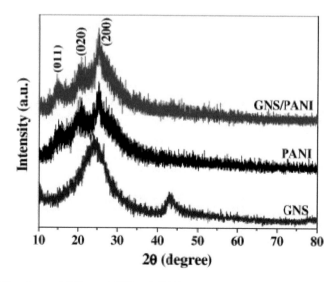

Fig. 3.99. Typical XRD patterns of GNS, pure PANI and GNS/PANI composite. Reproduced with permission from [329], Copyright © 2010, Elsevier.

Carbon Based Supercapacitors 133

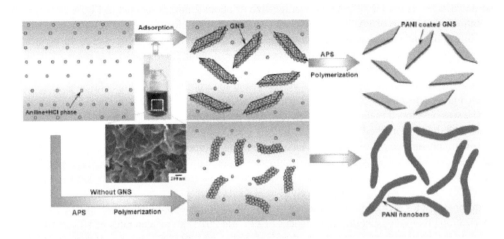

Fig. 3.100. Schematic diagram showing synthetic processing of the GNS/PANI composites. Reproduced with permission from [329], Copyright © 2010, Elsevier.

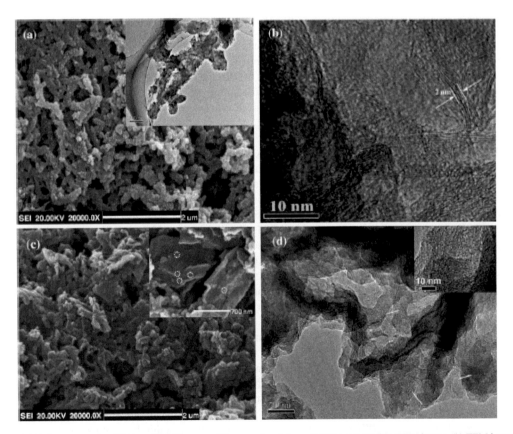

Fig. 3.101. Representative SEM and TEM images of the samples: (a) pure PANI inset exhibits TEM image, (b) TEM image of GNs, (c) SEM and (d) TEM images of THE GNS/PANI composite (with inset showing a high-magnification TEM image). Reproduced with permission from [329], Copyright © 2010, Elsevier.

134 *Nanomaterials for Supercapacitors*

coated on surfaces of the GNS, with an average size of about 2 nm, as labeled in Fig. 3.101(d). In this case, both the dispersion of PANI and the contact of PANI with electrolyte were effectively enhanced, thus leading to the enhancement in electrochemical performance of the nanocomposite as the electrode for supercapacitors. PANI agglomerates with an average size of about 20 nm were also present on surfaces of the GNS, as indicated by the circles in Fig. 3.101(c).

PPy is the second conducive polymer that has been incorporated with graphene or graphene oxide to obtain high performance electrodes of supercapacitors [331–336]. For instance, a graphene/PPy composite was developed to have high supercapacitive performance, by depositing graphene layers on Ti substrate with the EPD method [331]. Graphene/PPy composite was then formed by electropolymerizing Py monomers on surface of the graphene nanosheets. The composite based electrodes demonstrated an optimal area capacitance of 151 $mF \cdot cm^{-2}$, volume capacitance of 151 $F \cdot cm^{-3}$ and specific capacitance of as high as 1510 $F \cdot g^{-1}$, at the scan rate of 10 $mV \cdot s^{-1}$, corresponding to energy density and power density of 5.7 $Wh \cdot kg^{-1}$ and 3.0 $kW \cdot kg^{-1}$, respectively.

Similar examples include graphene nanosheet (GNS)/PPy nanocomposites prepared by using *in-situ* polymerization possessing specific capacitances of 417 and 267 $F \cdot g^{-1}$ at scan rates of 10 and 100 $mV s^{-1}$, respectively, together with excellent electrochemical cyclic stability [332], *in-situ* polymerized GNS/PPy nanocomposites from pyrrole monomer in the presence of graphene at acidic conditions, with a specific capacitance of 482 $F g^{-1}$ at a current density of 0.5 $A g^{-1}$ [333] and nanocomposite films of sulfonated graphene and PPy electrochemically deposited from aqueous solution containing pyrrole monomer, sulfonated graphene sheets, and DBSA, with high electrochemical performances [334].

More specifically, a nanoarchitecture composed of nanostructured polypyrrole (PPy) and electrically conductive graphene nanosheets with a multilayered configuration has been fabricated to be used as electrodes for supercapacitor applications [335]. In this nanoarchitecture, aligned monolayers of the graphene nanosheets were interspersed within the fibrous network of PPy, which served as current collectors inside the multilayer bulk configuration, when it was used as the electrodes of supercapacitors. Thanks to the presence of both the fibrous network of PPy and the highly conductive graphene nanosheets, both the ionic and electronic transports were significantly enhanced. A low equivalent series resistance was observed, due to the interlayer charge transport between the large graphene basal plane and the π conjugated polymer chain. The electronic charge transport related to the graphene nanosheets would increase electrochemical cyclic stability of the nanocomposite electrodes. A specific capacitance of nanostructured composite electrode exhibits 165 $F \cdot g^{-1}$ was achieved at a discharge current density of 1 $A \cdot g^{-1}$ after 1000 cycles.

For chemical polymerization to prepare PPy nanowires, 0.91 g surfactant CTAB was mixed with 0.3 mL pyrrole monomer in 125 mL 0.2 M aqueous HCl solution, with constant stirring at 0–5°C for 2 hours [335]. After that, ammonium persulfate (APS) as an oxidizing agent that was dissolved in 10 mL 0.2 M HCl solution was added to the mixture of CTAB and pyrrole, which was then kept at 0–5°C for 24 hours with constant stirring. The black precipitate was collected with filtration and thorough washing. To prepare monolayer nanowire films, 0.1 g nanowire was dispersed in water with aid of ultrasonication, into which chloroform was added as the second immiscible phase. The mixture was further sonicated so that the nanowires were cumulated at the liquid–liquid interface. Emulsified droplets were generated after shaking, so as to float to the surface of the suspension, where the chloroform was evaporated, thus leading to a thin layer of nanowires there.

The PPy nanowires obtained in this way had an average diameter in the range of 40–60 nm [335]. The fibrous nanowires possessed a specific surface area of 91 $m^2 \cdot g^{-1}$ and a total pore volume of 0.376 $cm^3 \cdot g^{-1}$. Self-assembly of hydrophobic organic nanowires at the liquid–liquid interface was demonstrated earlier. The PPy nanowire film was hydrophobic with a contact angle of about 69°, so that the nanowires could be adsorbed at the liquid–liquid interface and transported to the surface of the suspension, resulting in the thin layer.

In order to disperse the PPy nanowires on graphene nanosheets, a thin layer of nanowires was first transferred to a substrate with a pre-coated monolayer of graphene nanosheets [335]. A tight adhesion was formed between the PPy nanowires and the graphene nanosheets due to the wetting effect water. As the water continuously evaporated, the nanowires were collapsed into a thin layer, caused by the strong capillary force. Finally, the PPy nanowire layer and the graphene layer were assemble into a bilayered nanostructure, owing to the strong attraction of van der Waal's force between the highly aromatic graphene basal plane and the π conjugated polymer film. Representative SEM images are shown in Fig. 3.102(a, b), confirming that highly dispersed network of PPy nanowires were well deposited on the graphene nanosheet. Multilayer deposition led to the formation of thin films with fibrous microstructure of PPy nanowires, as illustrated in Fig. 3.102(c, d).

Figure 3.103 shows SEM images of the graphene nanosheets that were deposited on the PPy nanowire layers [335]. It was observed that the graphene nanosheets were highly dispersed in such a way that an edge shared interconnected network was generated. Such a network offered a conductive conduit inside the electrodes, thus leading to an enhanced electrochemical performance. Besides the inter-connection of the graphene nanosheets near the edges, as seen in Fig. 3.103(b–d), they also strongly interacted with the polypyrrole nanowires through the strong van der Waal's force as mentioned earlier. In addition, the large sized graphene nanosheets were tightly adhered to surface of the current collector in the top section, so as to act as a series of current collectors in multilayer configuration. Multilayered deposition of the two nanostructured layers have rendered binder free electrodes of supercapacitors.

More recently, holey graphene/PPy hybrid aerogels (HGPAs) with 3D hierarchical structures were prepared by freeze-drying holey graphene/PPy hydrogels, which were derived from Holey Graphene (HG) nanosheets and PPy nanoparticles, in order to develop high performance electrodes of supercapacitors [336]. In the HGPAs interconnected 3D porous network, PPy nanoparticles were uniformly distributed in the aerogels, so that the holey graphene nanosheets were effectively separated without the occurrence of restacking. Due to the unique hierarchical porous structure and synergistic effect of the PPy nanoparticles and HG nanosheets, the HGPA aerogels exhibited promising electrochemical performances as electrodes

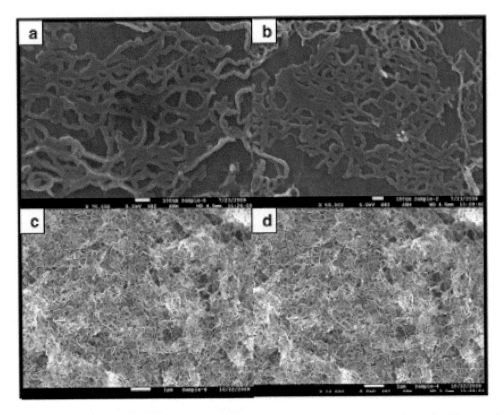

Fig. 3.102. (a, b) Morphology and microstructure of the PPy nanowires, showing a highly dispersed network of PPy nanowires with an average diameter of 40–60 nm to be deposited on the large basal plane of graphene nanosheets. (c, d) Highly fibrous morphology of the PPy nanowires after multilayer deposition. Reproduced with permission from [335], Copyright © 2010, American Chemical Society.

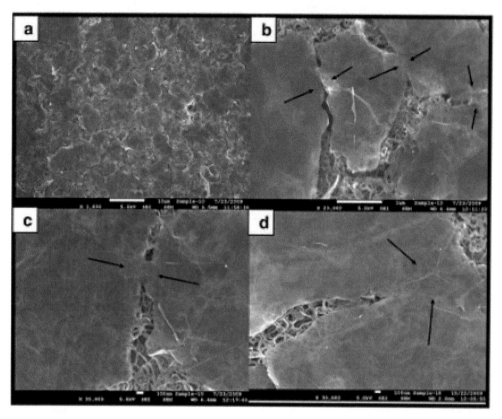

Fig. 3.103. Morphology and microstructure of the graphene nanosheets deposited on PPy nanowires: (a) low magnification image demonstrating the uniform monolayer coverage of the graphene nanosheets deposited on the fibrous network of PPy nanowire. (b–d) High magnification images illustrating the dispersed interconnected network of graphene nanosheets near their edges (indicated by the arrows) of the samples. Reproduced with permission from [335], Copyright © 2010, American Chemical Society.

of supercapacitors. For instance, the sample with a PPy/HGO mass ratio of 0.75 possessed the optimized specific capacitance of 418 F·g^{-1} at a current density of 0.5 A·g^{-1}, together with high rate capability (80%) at current densities over 0.5–20 A·g^{-1} and excellent cycling performance (74%) after 2000 cycles in 1.0 M KOH aqueous solution. It was found that PPy nanoparticles with relatively larger sizes were more suitable as electrodes in terms of electrochemical performance.

To prepare Holey Graphene Oxide (HGO), 100 mg GO powder was dispersed in 50 mL ultrapure water to form homogeneous GO suspension with a concentration of 2 mg·mL^{-1}, with the aid of ultrasonication. 50 mL GO suspension was then mixed with 5 mL 30% H$_2$O$_2$ solution, followed by refluxing at 100°C for 4 hours. The precipitate was thoroughly washed with DI water and then dispersed in water to obtain HGO suspension with a concentration of 2 mg·mL^{-1}. As a comparison, a similar treatment was applied to GO suspension without the presence of H$_2$O$_2$. To obtain PPy nanoparticles with average size of 150 nm, 0.2 g PVA was dissolved in 40 mL ultrapure water, into which 2.48 g FeCl$_3$·6H$_2$O was added to form a solution. Then, 280 μL Py monomer was added into the solution, where polymerization of Py was triggered at 5°C. After reaction for 4 hours, the PPy precipitate was collected with aid of centrifugation, followed by thorough washing with hot water and vacuum drying. By controlling the content of PVA (3, 1 and 0.008 g), PPy nanoparticles with average sizes of 40, 70 and 250 nm could be obtained.

To synthesize the nanocomposites, 12 mg PPy with average size of 150 nm was added to 8 mL HGO aqueous suspension with constant stirring and ultrasonication each for 1 hour. 200 μL 2 M VC aqueous solution was mixed with the HGO-PPy suspension. The mixed suspension was then sealed in a glass vial and heated at 80°C for 5 hours, which resulted in black HG/PPy hydrogels. After thorough washing with ultrapure water, the samples were freeze-dried for 15 hours, thus leading to 3D holey graphene/polypyrrole hybrid aerogels, denoted as HGPA-0.75. By controlling the PPy/HGO mass ratio, samples with different contents of PPy were obtained, denoted as HGPA-X, with X = 0.25, 0.5, 0.75 and 1.0. For the HGPA-0.75 hybrid aerogels, PPy nanoparticles with different sizes were used. In addition, the graphene/PPy hybrid aerogel (GPA) and pure holey graphene aerogel (HGA) were also prepared for comparison.

Figure 3.104 shows a schematic diagram demonstrating formation process of the HGPAs [336]. Due to the high dispersity of the HGO, HGO-PPy suspension was readily derived, as the PPy nanoparticles were mixed with the HGO suspension. Uniform distribution of the PPy nanoparticles in the HGO suspension ensured the formation of the HGPA with the unique macroscopic architecture. During the heating process, HGO was reduced into HG, while the PPy nanoparticles were entrapped in between the HG nanosheets, because of the π–π interaction and hydrogen bonding between the HG nanosheets and the aromatic PPy rings, thus resulting in the HG/PPy hydrogels. After freeze-drying, HGPA with 3D hierarchical structure was achieved.

Figure 3.105 shows SEM images of the samples obtained at different stages [336]. The PPy nanoparticles had a spherical morphology with uniform size distribution and an average diameter of about 150 nm, as seen in Fig. 3.105(a). The HGO nanosheets possessed a plate-like morphology that was

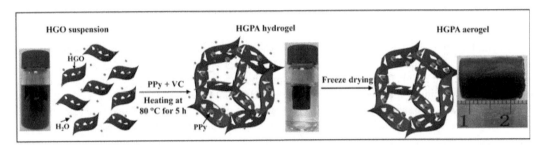

Fig. 3.104. Schematic diagram illustrating formation process of the HGPA, together with corresponding photographs of the samples at different stages. Reproduced with permission from [336], Copyright © 2016, Elsevier.

Fig. 3.105. SEM images of the samples obtained at different stages: (a) PPy, (b) HGO, (c) HGA, (d) HGPA-0.75 and (e) GPA-0.75. Reproduced with permission from [336], Copyright © 2016, Elsevier.

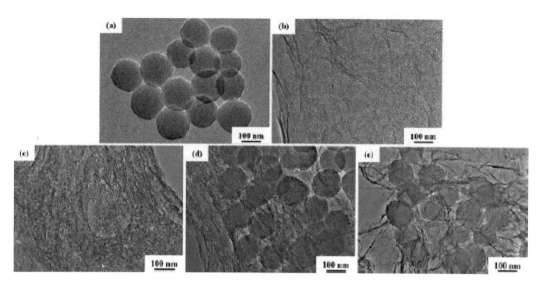

Fig. 3.106. TEM images of the samples obtained at different stages: (a) PPy, (b) HGO, (c) HGA, (d) HGPA-0.75 and (e) GPA-0.75. Reproduced with permission from [336], Copyright © 2016, Elsevier.

curved and loosely packed, as illustrated in Fig. 3.105(b). After reduction reaction, the HGO nanosheets were reduced to HGA, forming an interconnected 3D macroporous network through self-assembling. The porous structure had a wide range of pore sizes from sub-micrometer to several micrometers, with walls of the pores containing single layer or a few layers of HG nanosheets, as revealed in Fig. 3.105(c). As for the HGPAs, e.g., the sample with a PPy/HGO mass ratio of 0.75, 3D porous network was formed, as illustrated in Fig. 3.105(d). Due to the presence of the PPy nanoparticles in between the HG nanosheets, their restacking was effectively prevented. As shown in Fig. 3.105(e), the sample GPA-0.75 had a similar morphology.

Figure 3.106 shows TEM images of the samples obtained at different stages, demonstrating similar microstructures to those revealed by the SEM images [336]. The PPy nanoparticles were nearly monodisperse, as illustrated in Fig. 3.106(a), whereas the HGO possessed a density of in-plane nanopores that were uniformly distributed on the HGO nanosheets, as observed in Fig. 3.106(b). As seen in Fig. 3.106(c), the HGA consisted of either single layer or few layered nanosheets, forming a hierarchical porous structure. Specifically, in the sample HGPA-0.75, PPy nanoparticles were wrapped by the HG nanosheets, with in-plane nanopores being clearly visible in Fig. 3.106(d). In contrast, the GPA-0.75 had no nanopores on graphene nanosheets, as shown in Fig. 3.106(e).

Graphene–metal oxide hybrids

RuO_2 was used to fabricate graphene-metal oxide hybrids as electrodes of supercapacitors [337]. For example, a RuO_2/graphene nanosheet composite (ROGSCs) was prepared by using a sol–gel process [337]. The ROGSCs with 38.3 wt% Ru loading showed optimized specific capacitance of 570 $F \cdot g^{-1}$, which was much higher than that of graphene (148 $F\ g^{-1}$). An energy density obtained of 20.1 $Wh \cdot kg^{-1}$ was observed for the materials, corresponding to a power density of 50 $W \cdot kg^{-1}$ at a current density of 0.1 $A \cdot g^{-1}$. The specific capacitance was retained by 97.9% after 100 cycles.

Figure 3.107 shows a schematic diagram describing the formation process of the ROGSCs [337]. Graphene nanosheets (GSs) of ≤ 3 layers, prepared through chemical exfoliation of natural graphite powder, were dispersed in mixture of ethanol and water, so that the stacked nanosheets were separated, onto which RuO_2 nanoparticles could be loaded, whereas water served as an aqueous medium of NaOH in order to form ruthenium hydroxide through the reaction of $RuCl_3$ and NaOH. The ruthenium hydroxide NPs were

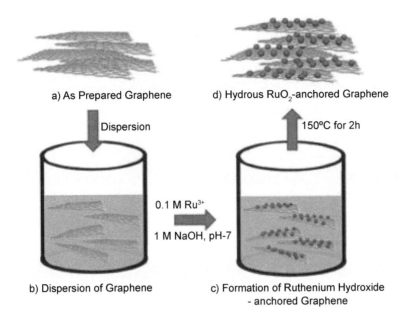

Fig. 3.107. Schematic diagram demonstrating fabrication process of the ROGSCs. Reproduced with permission from [337], Copyright © 2010, WILEY-VCH Verlag GmbH & Co. KGaA, Weinheim.

attached onto the surface of the GSs due to strong chemical interactions and van der Waals interactions. After calcination at 150°C, ruthenium hydroxide NPs were converted to hydrous amorphous RuO_2 NPs. Because of the presence of ruthenium hydroxide NPs, agglomeration of the GSs during the calcination process was effectively prevented, which in turn ensured a high dispersal of the RuO_2 NPs. As a result, the ROGSCs were also highly porous, with a relatively large surface area, which was beneficial to the charge-storage behavior of the nanocomposites, as compared with either GSs or RuO_2.

Figure 3.108 shows representative SEM and TEM images of the ROGSCs as well as GSs [337]. As demonstrated in Fig. 3.108(a), the solvent-dispersed GSs were near transparent with the presence of visible wrinkles. In the ROGSCs, RuO_2 NPs with sizes in the range of 5–20 nm were uniformly distributed on surface of the GSs, as illustrated in Fig. 3.108(b), which was attributed to the presence of the oxygen-containing functional groups on the surface of the GSs, as stated earlier. Also, the pure GSs were randomly distributed and prone to restacking, as seen in Fig. 3.108(c). In comparison, due to the spacing effect of the RuO_2 nanoparticles, the restacking of GSs in the ROGSCs was avoided, as revealed in Fig. 108(d).

Figure 3.109 shows CV results of the electrodes based on GSs, RuO_2 and ROGSCs (Ru, 38.3 wt%), measured in 1 M H_2SO_4 at potentials from −0.2 to 0.8 V vs. Saturated Calomel Electrode (SCE). As seen in Fig. 3.109(a), the GSs based electrode possessed a typical EDLC behavior. In contrast, both the electrodes based on RuO_2 and ROGSCs exhibited a broad redox peaks, suggesting the presence of a typical pseudocapacitive behavior. Specific capacitances of GSs, RuO_2 and ROGSCs based electrodes were 148 $F \cdot g^{-1}$, 606 $F \cdot g^{-1}$ and 570 $F \cdot g^{-1}$, respectively, at 1 $mV \cdot s^{-1}$, as illustrated in Fig. 3.109(b). It was observed that the ROGSCs based electrode had a much higher specific capacitance than the GS one. In addition, the ROGSCs based electrode also had a higher rate capability. The largely enhanced electrochemical performance of the ROGSC electrode could be attributed to the electrochemical Faradaic reactions of the fine RuO_2 in nanocomposites.

The maximum literature data on graphene-oxide hybrids for electrodes of supercapacitors are based on manganese oxides [338–360]. For example, a hydrothermal method was explored to synthesize MnO_2 nanorods/graphene nanocomposites [339]. Electrochemical properties of the nanocomposite based electrodes were closely related to the content of the MnO_2 nanorods. An optimized specific capacitance

140 *Nanomaterials for Supercapacitors*

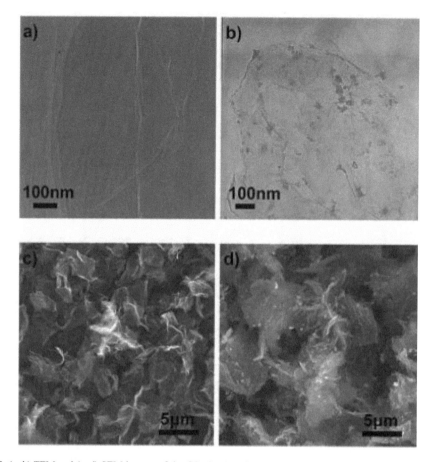

Fig. 3.108. (a, b) TEM and (c, d) SEM images of the GSs (a, c) and ROGSCs (b, d) with 38.3 wt% Ru. Reproduced with permission from [337], Copyright © 2010, WILEY-VCH Verlag GmbH & Co. KGaA, Weinheim.

of 218 F·g^{-1} was achieved, when the electrodes were measured at the scan rate of 5 mV·s^{-1} in 1 M Na$_2$SO$_4$ aqueous solution. Correspondingly, the MnO$_2$ nanorods/graphene nanocomposites possessed an optimal energy density of 16 Wh·kg^{-1} at power density of 95 W·kg^{-1}. In addition, a high capacitance retention with a loss of < 6% was observed after 1000 cycles.

GO exfoliated from graphite was dispersed in 200 mL DI water, into which 8 g glucose and 2 mL ammonia solution (25 wt%) were added and stirred for 5 minutes [339]. The mixture was the heated at 95°C for 1 hour, followed by cooling down. After that, the black dispersion was filtered, thoroughly washed with distilled water and dried at 70°C for 6 hours, thus leading to graphene powder. To synthesize MnO$_2$ nanorods/graphene composites, 0.05 g of graphene was dispersed in 80 mL of distilled water. Then, MnSO$_4$·H$_2$O solution and KMnO$_4$ solution with different concentrations (0.025, 0.05, 0.125 and 0.25 M) were added and stirred for 1 hour to form suspension for hydrothermal reaction, corresponding to final samples denoted as MnO$_2$-1/G, MnO$_2$-2/G, MnO$_2$-5/G and MnO$_2$-10/G, respectively. A sample of MnO$_2$ nanorods without graphene was also prepared for comparison. Hydrothermal reaction was conducted at 150°C for 24 hours. The products were collected through centrifugation, followed by thorough washing with distilled water and then drying at 80°C. XRD results indicated that the MnO$_2$ nanorods were well-crystallized tetragonal phase α-MnO$_2$. The nanorod morphology was confirmed by SEM and TEM observation results.

CV curves of the electrodes based on pure MnO$_2$ and the four MnO$_2$ nanorods/graphene composites were measured at the scan rate of 5–50 mV·s^{-1} [339]. The mirror-image and rectangular CV curves

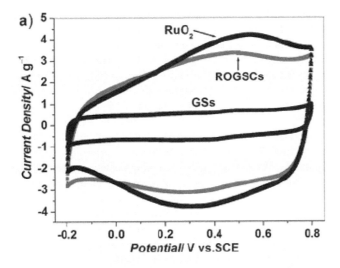

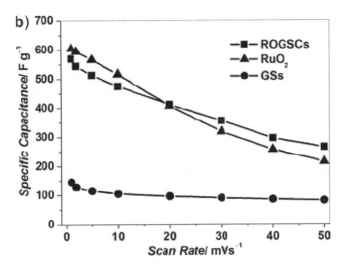

Fig. 3.109. (a) CV curves of the GSs, RuO$_2$, and ROGSCs (Ru, 38.3 wt%) based electrodes measured at 5 mV s^{-1} at over potentials from −0.2 to 0.8 V vs. SCE in 1 M H$_2$SO$_4$. (b) C_{sp} of the GSs, RuO$_2$, and ROGSCs (Ru, 38.3 wt%) based electrodes as a function of scan rate. Reproduced with permission from [337], Copyright © 2010, WILEY-VCH Verlag GmbH & Co. KGaA, Weinheim.

indicated ideal capacitive behavior of all the samples. The electrode based on MnO$_2$-5/G had the largest area, implying its high specific capacitance, due to the synergistic effect of MnO$_2$ and graphene. At high scan rate of 50 mV·s^{-1}, specific capacitances of the MnO$_2$, MnO$_2$-1/G and MnO$_2$-2/G were greatly reduced, although their CV curves were still nearly rectangular. The low capacitances could be attributed to the low content of MnO$_2$ in the nanocomposites. In comparison, the area of the CV curve of the MnO$_2$-5/G was increased, even though the curve was slightly deviated from the rectangular shape.

In another study, a Nitrogen-doped Graphene/MnO$_2$ nanosheet (NGMn) composite was fabricated by using a hydrothermal synthetic route [349]. The nitrogen-doped graphene not only was used as a template to guide the growth of layered δ-MnO$_2$, but also served to enhance electrical conductivity of the composite. The NGMn composite possessed an optimal specific capacitance of 305 F·g^{-1} at the scan rate of 5 mV·s^{-1}, which was employed as the cathode to fabricate flexible solid-state asymmetric supercapacitors, together with activated carbon as the anode and PVA−LiCl gel as the electrolyte. The device demonstrated an optimized energy density of 3.5 mWh·cm^{-3} and a power density of 0.019 W·cm^{-3}, with an operating voltage of 1.8 V. The specific capacitance could be retained by > 90% after 1500 cycles.

To obtain chemically reduced Graphene (G), 0.8 g NaHSO$_3$ was dispersed in 200 mL GO solution of 0.4 mg·mL^{-1}, which was then heated at 95°C for 3 hours, followed by filtration and thorough washing with DI water. The G powder was re-dispersed in DI with the aid of ultrasonic vibration to form a suspension, which was hydrothermally reacted with ammonia solution (12.5−14% v/v%) at 120°C for 6 hours, in order to obtain Nitrogen-doped Graphene (NG). NGMn composite was fabricated by dissolving 0.1048 g KMnO$_4$ into 70 mL NG suspension, which was hydrothermally treated at 140°C for 50 minutes. The NGMn composite was collected through filtration, followed by thorough washing with DI water and drying at 80°C in air. Graphene/MnO$_2$ nanosheet (GMn) composite was also prepared for comparison.

Figure 3.110 shows representative SEM images of the NGMn composite [349]. The NG nanosheets possessed a curly and interconnected structure, with MnO$_2$ nanoparticles to be attached on surface of the NG, as illustrated in Fig. 3.110(a). The corrugated nanosheet morphology was demonstrated in Fig. 3.110(b). The interconnected NG nanosheets ensured high conductivity of the composite. XRD indicated that the manganese oxide in the composite was birnessite-type δ-MnO$_2$. The presence of N in the graphene nanosheets was confirmed by EDX analysis results.

Figure 3.111 shows TEM images and SAED pattern of the NGMn sample [349]. The well distribution of the layer-structured MnO$_2$ nanosheet on the surface of NG was confirmed by the TEM images, as seen in Fig. 3.111(a, b). The MnO$_2$ nanosheets were polycrystalline, evidenced by the presence of domains and grain boundaries. The lattice spacing of the MnO$_2$ nanosheets was 0.24 nm, as shown in Fig. 3.111(c), corresponding to the (11$\bar{1}$) plane of birnessite δ-MnO$_2$. As demonstrated in Fig. 3.111(d), the SAED pattern had three faint continuous diffraction rings, which could be indexed to the (001), (110), (11$\bar{1}$) and (31$\bar{2}$) planes, confirming the polycrystalline characteristics of the manganese oxide in the composite.

Various other oxides, e.g., SnO$_2$, have also been explored to form nanocomposites with graphene as electrodes of electrochemical supercapacitors with improved performances [361–364].

A solvent-based synthesis method was reported to prepare SnO$_2$/Graphene (SnO$_2$/G) nanocomposites by employing the oxidation–reduction reaction between Graphene Oxide (GO) and SnCl$_2$·2H$_2$O [363]. The SnO$_2$/graphene nanocomposites exhibited an optimal specific capacitance of 363.3 m^2·g^{-1}, which was much higher than that of the pure graphene (68.4 m^2·g^{-1}). The promising electrochemical performance of the nanocomposites was attributed to the synergistic effects of the high electrical conductivity of the graphene and the capacitive effect of the SnO$_2$ nanoparticles.

To synthesize the nanocomposite, GO was dispersed in absolute ethanol with a concentration of 1 mg·mL^{-1}, so as to obtain homogeneous suspension after stirring for 30 minutes. After that, 5 mM SnCl$_2$·2H$_2$O was added into the suspension with constant stirring for 30 minutes. Hydrothermal reaction of the suspension was carried out at 180°C for 24 hours. The nanocomposite hydrogels were collected through

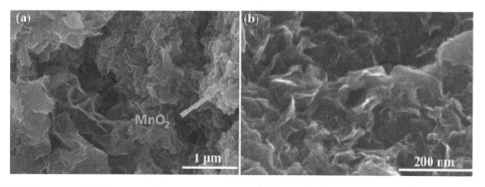

Fig. 3.110. (a) Low- and (b) high-magnification SEM images of the NGMn composites. Reproduced with permission from [349], Copyright © 2016, American Chemical Society.

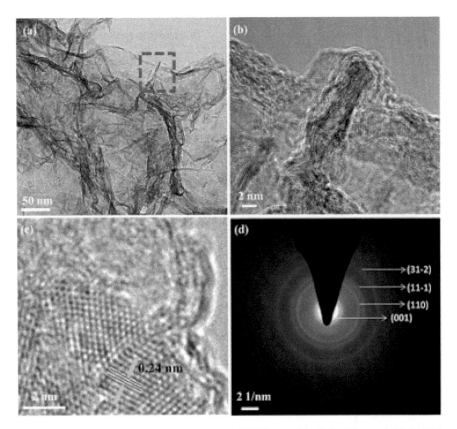

Fig. 3.111. (a) TEM images of the NGMn composite. (b) Zoom-in TEM image of the dashed line area in (a). (c) High-resolution TEM image of the NGMn composite, showing MnO_2 lattices. (d) Selected area electron diffraction (SAED) pattern of the NGMn composite. Reproduced with permission from [349], Copyright © 2016, American Chemical Society.

centrifugation, followed by thorough washing with distilled water and ethanol. SnO_2/G nanocomposite powder was obtained after drying at 60°C. Pure graphene hydrogel denoted as HGO was also prepared for comparison.

Figure 3.112 shows XRD patterns of the SnO_2/G nanocomposite, as well as GO, HGO, and SnO_2 [363]. The diffraction peak of GO at 9.8° in Fig. 3.112(a) was absent in that of HGO in Fig. 3.112(b), which meant that the GO was reduced. The diffraction peaks of SnO_2 observed in Fig. 3.112(c, d) belonged to rutile phase of SnO_2 with tetragonal structure. The broadened diffraction peaks of the SnO_2 and SnO_2/G samples were attributed to the nanocrystalline nature of the SnO_2 particles. An impurity peak at 13.4° was present in the patterns of HGO, SnO_2/G and SnO_2, owing to the sample holder of Perspex.

Figure 3.113 shows representative SEM images of the samples [363]. The HGO had a creased and layered morphology, as seen in Fig. 3.113(a). As demonstrated in Fig. 3.113(b), the SnO_2 nanoparticles possessed an average size of 95 nm. The SnO_2/G nanocomposite contained well-defined interlinked 3D graphene nanosheets, which led to a porous loose sponge-like structure. The hydrothermal treatment facilitated overlapping and coalescence of the graphene nanosheets, so that cross-links were developed and thus framework of the graphene hydrogel was obtained. In addition, the large conjugated basal planes of the graphene nanosheets ensured mechanical strength and integrity of the hydrogels.

SEM images of the SnO_2/G nanocomposite samples derived from precursors with concentrations of 25 mM, 5 mM and 2.5 mM are shown in Fig. 3.113(d–f) [363]. The sample from the high concentration precursor of 25 mM exhibited a certain degree of aggregations, as observed in Fig. 3.113(d), while too low

144 *Nanomaterials for Supercapacitors*

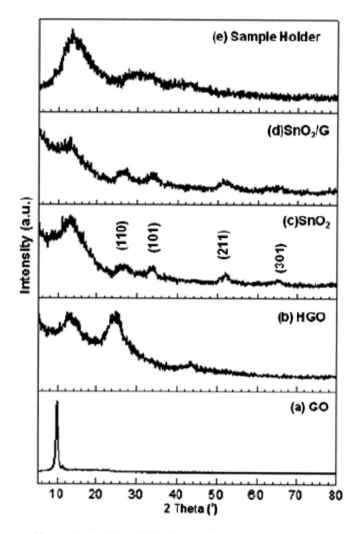

Fig. 3.112. XRD patterns of the samples: (a) GO, (b) HGO, (c) SnO$_2$, (d) SnO$_2$/G and (e) sample holder (Perspex). Reproduced with permission from [363], Copyright © 2013, Elsevier.

concentration of precursor led to less content of SnO$_2$, as illustrated in Fig. 3.113(f). Therefore, 5 mM of Sn^{2+} was an optimized concentration, where the SnO$_2$ nanoparticles were uniformly distributed on surface of the graphene nanosheets, as revealed in Fig. 3.113(e). Meanwhile, average size of the SnO$_2$ nanoparticles in the nanocomposite was 10 nm, which was smaller than that of pure SnO$_2$ nanoparticles by nearly one order of magnitude. In this case, the graphene nanosheets served as a support for the distribution of the SnO$_2$ nanoparticles, so that their aggregation was effectively avoided.

A possible mechanism was proposed to explain formation process of the SnO$_2$/G 3D nanocomposites [363]. When the negatively charged GO nanosheets were mixed with positively charged Sn^{2+}, the Sn^{2+} ions were adsorbed onto surface of the GO nanosheets, through the electrostatic interaction between the oppositely charged ions. Therefore, nucleation sites were formed. During the hydrothermal reaction, the GO nanosheets with Sn^{2+} ions were randomly orientated inside the vehicle, whereas cross-linking among the nanosheets was most likely established. At the same time, the GO served as an oxidizing agent, so that the Sn^{2+} was oxidized to SnO$_2$ and while the GO was reduced to graphene. As a result, a 3D graphene matrix uniformly distributed with SnO$_2$ nanoparticles was achieved.

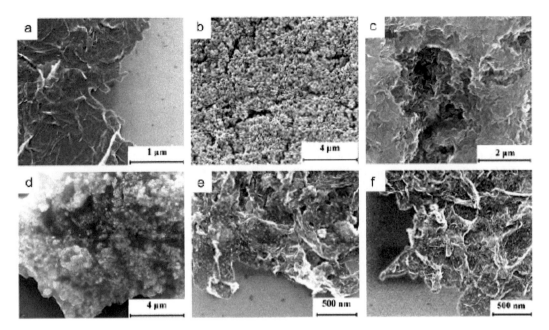

Fig. 3.113. SEM images of (a) HGO, (b) SnO$_2$ nanoparticles and (c) low magnification image for SnO$_2$/G nanocomposite. High magnification SEM images of the SnO$_2$/G nanocomposites from solutions with different contents of Sn^{2+} cation: (d) 25 mM, (e) 5 mM and (f) 2.5 mM. Reproduced with permission from [363], Copyright © 2013, Elsevier.

Graphene–CNTs hybrids

Combination of nanocarbons, including graphene, CNTs, carbon nanofibers (CNFs) and ACs, could be an effective approach to form hybrids with enhanced performances through the utilization of the advantages of each component. Two examples regarding graphene-CNTs hybrids are presented and discussed below.

An electrochemical reduction method was used to prepare binder-free composites of reduced graphene oxide (ecrGO) and multiwalled carbon nanotubes (MWCNTs), as electrodes of supercapacitors [365]. The ecrGO/MWCNTs composites possessed a higher specific capacitance than ecrGO, due to the intercalation of the MWCNTs into the ecrGO sheets, which increased surface areas of the nanocomposites. The composites could have a wide range of mass ratios of GO to MWCNTs, from 10:1 to 1:10. The ecrGO/MWCNTs composite with a GO:MWCNTs = 5:1 exhibited an optimized specific capacitance of 165 F·g^{-1} at the current density of 1 A·g^{-1}, with 93% retention after 4000 cycles of charge/discharge.

Figure 3.114 shows a schematic diagram illustrating formation process of the ecrGO-MWCNTs nanocomposites [365]. Commercial MWCNTs, with lengths of 1–12 μm, outside diameters of 13–18 nm and purity of > 99 wt%, and GO suspension with a concentration of 5 g·L^{-1}, containing 79% carbon and 20% oxygen, were used as the starting materials. The GO suspension was diluted to 1 g·L^{-1} with DI water with the aid of ultrasonication. MWCNTs powder was then added into the GO suspension to form GO/MWCNTs precursor, with mass ratios of GO:MWCNTs = 1:0, 10:1, 5:1, 1:1, 1:5 and 1:10. A Pt plate, used as a current collector, was covered with the GO/MWCNTs precursor suspension, followed by drying at 150°C. The mass loading of GO/MWCNTs was about 1.8 mg·cm^{-2}. Electrochemical reduction of the GO/MWCNTs composite was conducted, with cyclic voltammetry form −1.2 V to 0 V in 0.5 M NaCl solution at a scan rate of 50 mV·s^{-1}. A three-electrode configuration was employed, where Pt wire and Ag/AgCl were used as the counter electrode and reference electrode, respectively.

An *in-situ* partial unzipping method was developed to prepare CNT carbon nanotube/graphene (CNT/G) nanocomposite sponge, onto which PPy was coated through electro-polymerization, thus leading to CNT/graphene/polypyrrole (CNT/G/PPy) ternary composite sponge [366]. The degree of unzipping of the CNT could be controlled by adjusting the reaction time duration. An optimal specific capacitance of 225 F·g^{-1}

was achieved, together with a capacitance retention of 90.6% after 1000 cycles. The CNT/G/PPy sponge exhibited a high flexibility under compression.

Figure 3.115 shows the fabrication process of the CNT/G sponges [366]. CNT sponge was first synthesized by using a CVD method. Then, the CNT sponge blocks were immersed in $KMnO_4/H_2SO_4$ solution with a concentration of 0.03 g·mL^{-1} at 60°C for 1, 2 and 3 hours, so that the CNTs were partially unzipped into GO. The as-obtained CNT/GO sponge was soaked in the DI water to dilute the concentrated sulfuric acid and remove the MnO_2 particles formed during the reaction. After being thoroughly washed and cleaned, the CNT/GO sponge was reduced with 45 wt% HI solution. After another round of thorough cleaning and washing, the samples were freeze-dried to obtain CNT/G sponge. Micro-morphologies of the CNTs, CNT/GO and CNT/G are shown in Fig. 3.115(d–f). Electrochemical polymerization with three-electrode configuration was employed to coat PPy onto surface of the CNT/G sponges.

The contents of PPy in the composite sponges were 28.6, 38.0 and 50.6 wt%, as the polymerization time durations were 5, 10 and 15 minutes. Figure 3.116 shows SEM images of the composite sponge samples. Obviously, thickness of the PPy layer was gradually increased with increasing content of PPy. At a low content of 28.6 wt%, the CNT/G/PPy sponge had a morphology (Fig. 3.116(a)) similar to that of the CNT/G sponge. Figure 3.116(b) revealed that the contact regions of the CNT/G networks were coated with PPy. The presence of the PPy could significantly increase electrical conductivity of the composite sponge. Figure 3.116(c) shows an image of the CNT/G core networks and the PPy coating. The CNT/G networks were entirely covered by the PPy layer, where each CNT/G/PPy component had a thickness of about 80 nm, as the content of PPy was increased to 50.6 wt%, as illustrated in Fig. 3.116(d).

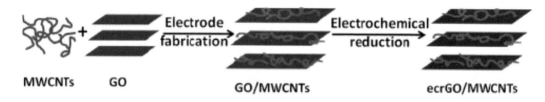

Fig. 3.114. Schematic diagram showing formation process of the ecrGO-MWCNTs nanocomposites. Reproduced with permission from [365], Copyright © 2016, Elsevier.

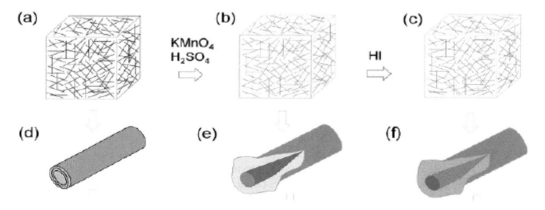

Fig. 3.115. Schematic diagram demonstrating fabrication process from a CNT sponge to a CNT/G composite sponge. (a, d) Pristine CNT sponge. (b, e) CNT/GO sponge immersed in $KMnO_4/H_2SO_4$. (c, f) CNT/G sponge in deionized water. Reproduced with permission from [366], Copyright © 2015, Elsevier.

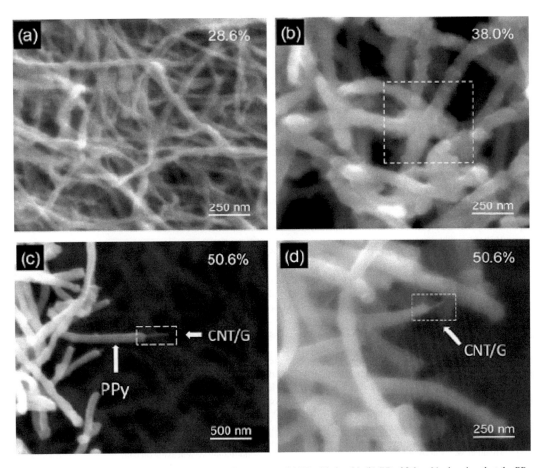

Fig. 3.116. SEM images of the CNT/G/PPy composite sponges: (a) PPy 28.6 wt%, (b) PPy 38.0 wt%, showing that the PPy coating interlinks the CNT/G networks at inter-contacts and (c, d) PPy 50.6 wt%. Reproduced with permission from [366], Copyright © 2015, Elsevier.

Summary

Activated carbons, with high apparent surface area of up to 4000 $m^2 \cdot g^{-1}$ and various morphologies, such as powders, fibers, cloths, monoliths, have been extensively studied as electrodes of supercapacitors. Although electrodes based on activated carbons have enhanced the energy storage capabilities of supercapacitors, the cycling stability and power density are at levels of EDLCs. Issues still remain such as material fabrication and fundamental electrochemistry to be clarified to further improve the performances of this group of supercapacitors. Moreover, beyond that of specific surface area, the influence of size of pores and pore size distribution on electrochemical behavior of the conventional carbon materials has not been well clarified, with very little information in the open literature, which should be a research topic for the future in this field.

As new types of nanocarbon materials, carbon nanotubes (CNTs) and graphene, have also been employed to develop electrodes of supercapacitors. As compared with the conventional carbon materials, these new nanocarbons have several advantages, such as higher electrical conductivity, more freedom in material processing and easier incorporation with other components. For example, both CNTs and graphene can be made into 1D fibers, 2D papers and 3D hierarchical structures, which served as a host or support to integrate nanoparticles of metal oxides, thus forming pseudocapacitors. In addition, CNT-graphene hybrids form a new family of nanocarbon materials, which combine the advantages of the two components. However, the hybrids that have been widely reported in the open literature are usually fabricated through

148 *Nanomaterials for Supercapacitors*

physical mixing, which cannot fully utilize their advantages. Future study should focus on design and fabrication of CNT-graphene hybrids with strong chemical interaction. It will be more challenging when other components such as metal oxides are incorporated to form multi-component hybrids.

Keywords: Carbon black, Active carbon, Activation, Carbon nanotubes (CNTs), Single-walled carbon nanotubes (SWCNTs), Multi-walled carbon nanotubes (MWCNTs), CNT fibers, CNT papers, CNT-MnO_2, CNT-Fe_2O_3, CNT-CoO, CNT-NiO, Graphene, Graphene oxide, Reduced graphene oxide, Graphene fibers, Graphene papers, G-MnO_2, G-Fe_3O_4, Specific surface area, Nanopore, Mesopore, Micropore, Pore size, Pore size distribution, Supercapacitor, Specific capacitance, Rate stability, Power density, Charge, Discharge

References

[1] Sevilla M, Mokaya R. Energy storage applications of activated carbons: supercapacitors and hydrogen storage. Energy & Environmental Science. 2014; 7: 1250–80.

[2] Zhang LL, Zhao XS. Carbon-based materials as supercapacitor electrodes. Chemical Society Reviews. 2009; 38: 2520–31.

[3] Pandolfo AG, Hollenkamp AF. Carbon properties and their role in supercapacitors. Journal of Power Sources. 2006; 157: 11–27.

[4] Bose S, Kuila T, Mishra AK, Rajasekar R, Kim NH, Lee JH. Carbon-based nanostructured materials and their composites as supercapacitor electrodes. Journal of Materials Chemistry. 2012; 22: 767–84.

[5] Zhai YP, Dou YQ, Zhao DY, Fulvio PF, Mayes RT, Dai S. Carbon materials for chemical capacitive energy storage. Advanced Materials. 2011; 23: 4828–50.

[6] Zhang LL, Zhou RF, Zhao XS. Graphene-based materials as supercapacitor electrodes. Journal of Materials Chemistry. 2010; 20: 5983–92.

[7] Brown PA, Gill SA, Allen SJ. Metal removal from wastewater using peat. Water Research. 2000; 34: 3907–16.

[8] Dias JM, Alvim-Ferraz MCM, Almeida MF, Rivera-Utrilla J, Sanchez-Polo M. Waste materials for activated carbon preparation and its use in aqueous-phase treatment: A review. Journal of Environmental Management. 2007; 85: 833–46.

[9] Foo KY, Hameed BH. Utilization of rice husk ash as novel adsorbent: A judicious recycling of the colloidal agricultural waste. Advances in Colloid and Interface Science. 2009; 152: 39–47.

[10] Mohamed AR, Mohammadi M, Darzi GN. Preparation of carbon molecular sieve from lignocellulosic biomass: A review. Renewable & Sustainable Energy Reviews. 2010; 14: 1591–9.

[11] Abdullah MO, Tan IAW, Lim LS. Automobile adsorption air-conditioning system using oil palm biomass-based activated carbon: A review. Renewable & Sustainable Energy Reviews. 2011; 15: 2061–72.

[12] Abioye AM, Ani FN. Recent development in the production of activated carbon electrodes from agricultural waste biomass for supercapacitors: A review. Renewable & Sustainable Energy Reviews. 2015; 52: 1282–93.

[13] Bagheri S, Julkapli NM, Abd Hamid SB. Functionalized activated carbon derived from biomass for photocatalysis applications perspective. International Journal of Photoenergy. 2015: 218743.

[14] Jain A, Balasubramanian R, Srinivasan MP. Hydrothermal conversion of biomass waste to activated carbon with high porosity: A review. Chemical Engineering Journal. 2016; 283: 789–805.

[15] Liu M, Chen Y, Chen K, Zhang N, Zhao XQ, Zhao FH et al. Biomass-derived activated carbon for rechargeable lithium-sulfur batteries. Bioresources. 2015; 10: 155–68.

[16] Rafatullah M, Ahmad T, Ghazali A, Sulaiman O, Danish M, Hashim R. Oil palm biomass as a precursor of activated carbons: A review. Critical Reviews in Environmental Science and Technology. 2013; 43: 1117–61.

[17] Sethupathi S, Bashir MJK, Akbar ZA, Mohamed AR. Biomass-based palm shell activated carbon and palm shell carbon molecular sieve as gas separation adsorbents. Waste Management & Research. 2015; 33: 303–12.

[18] Skodras G, Diamantopouiou I, Zabaniotou A, Stavropoulos G, Sakellaropoulos GP. Enhanced mercury adsorption in activated carbons from biomass materials and waste tires. Fuel Processing Technology. 2007; 88: 749–58.

[19] Andresen JM, Burgess CE, Pappano PJ, Schobert HH. New directions for non-fuel uses of anthracites. Fuel Processing Technology. 2004; 85: 1373–92.

[20] Azargohar R, Dalai AK. Biochar as a precursor of activated carbon. Applied Biochemistry and Biotechnology. 2006; 131: 762–73.

[21] Liu DD, Wu ZS, Ge XY, Cravotto G, Wu ZL, Yan YJ. Comparative study of naphthalene adsorption on activated carbon prepared by microwave-assisted synthesis from different typical coals in Xinjiang. Journal of the Taiwan Institute of Chemical Engineers. 2016; 59: 563–8.

[22] Ryu SK, Jin H, Gondy D, Pusset N, Ehrburger P. Activation of carbon-fibers by steam and carbon-dioxide. Carbon. 1993; 31: 841–2.

[23] Sevilla M, Alam N, Mokaya R. Enhancement of hydrogen storage capacity of zeolite-templated carbons by chemical activation. Journal of Physical Chemistry C. 2010; 114: 11314–9.

[24] Frackowiak E, Delpeux S, Jurewicz K, Szostak K, Cazorla-Amoros D, Beguin F. Enhanced capacitance of carbon nanotubes through chemical activation. Chemical Physics Letters. 2002; 361: 35–41.

[25] Raymundo-Pinero E, Cazorla-Amoros A, Delpeux S, Frackowiak E, Szostak K, Beguin F. High surface area carbon nanotubes prepared by chemical activation. Carbon. 2002; 40: 1614–7.

[26] Yoon SH, Lim S, Song Y, Ota Y, Qiao WM, Tanaka A et al. KOH activation of carbon nanofibers. Carbon. 2004; 42: 1723–9.

[27] Rodriguezreinoso F, Molinasabio M. Activated carbons from lignocellulosic materials by chemical and or physical activation—An overview. Carbon. 1992; 30: 1111–8.

[28] Dawson EA, Parkes GMB, Barnes PA, Chinn MJ. An investigation of the porosity of carbons prepared by constant rate activation in air. Carbon. 2003; 41: 571–8.

[29] Feng B, Bhatia SK. Variation of the pore structure of coal chars during gasification. Carbon. 2003; 41: 507–23.

[30] Salatino P, Senneca O, Masi S. Gasification of a coal char by oxygen and carbon dioxide. Carbon. 1998; 36: 443–52.

[31] Ballal G, Zygourakis K. Evolution of pore surface-area during noncatalytic gas solid reactions 2. Experimental results and model validation. Industrial & Engineering Chemistry Research. 1987; 26: 1787–96.

[32] Ballal G, Zygourakis K. Evolution of pore surface-area during noncatalytic gas solid reactions 1. Model development. Industrial & Engineering Chemistry Research. 1987; 26: 911–21.

[33] Ganan J, Gonzalez JF, Gonzalez-Garcia CM, Ramiro A, Sabio E, Roman S. Air-activated. carbons from almond tree pruning: Preparation and characterization. Applied Surface Science. 2006; 252: 5988–92.

[34] Osswald S, Portet C, Gogotsi Y, Laudisio G, Singer JP, Fischer JE et al. Porosity control in nanoporous carbide-derived carbon by oxidation in air and carbon dioxide. Journal of Solid State Chemistry. 2009; 182: 1733–41.

[35] Dash R, Chmiola J, Yushin G, Gogotsi Y, Laudisio G, Singer J et al. Titanium carbide derived nanoporous carbon for energy-related applications. Carbon. 2006; 44: 2489–97.

[36] Kravchik AE, Kukushkina JA, Sokolov VV, Tereshchenko GF. Structure of nanoporous carbon produced from boron carbide. Carbon. 2006; 44: 3263–8.

[37] Rodriguezreinoso F, Molinasabio M, Gonzalez MT. The use of steam and CO_2 as activating agents in the preparation of activated carbons. Carbon. 1995; 33: 15–23.

[38] Navarro MV, Seaton NA, Mastral AM, Murillo R. Analysis of the evolution of the pore size distribution and the pore network connectivity of a porous carbon during activation. Carbon. 2006; 44: 2281–8.

[39] Gonzalez JF, Encinar JM, Gonzalez-Garcia CM, Sabio E, Ramiro A, Canito JL et al. Preparation of activated carbons from used tyres by gasification with steam and carbon dioxide. Applied Surface Science. 2006; 252: 5999–6004.

[40] Johns MM, Marshall WE, Toles CA. The effect of activation method on the properties of pecan shell-activated carbons. Journal of Chemical Technology and Biotechnology. 1999; 74: 1037–44.

[41] Chang CF, Chang CY, Tsai WT. Effects of burn-off and activation temperature on preparation of activated carbon from corn cob agrowaste by CO_2 and steam. Journal of Colloid and Interface Science. 2000; 232: 45–9.

[42] San Miguel G, Fowler GD, Sollars CJ. A study of the characteristics of activated carbons produced by steam and carbon dioxide activation of waste tyre rubber. Carbon. 2003; 41: 1009–16.

[43] Linares-Solano A, de Lecea CSM, Cazorla-Amoros D, Martin-Gullon I. Porosity development during CO_2 and steam activation in a fluidized bed reactor. Energy & Fuels. 2000; 14: 142–9.

[44] Roman S, Gonzalez JF, Gonzalez-Garcia CM, Zamora F. Control of pore development during CO_2 and steam activation of olive stones. Fuel Processing Technology. 2008; 89: 715–20.

[45] Gonzalez JF, Roman S, Gonzalez-Garcia CM, Valente Nabais JM, Luis Ortiz A. Porosity development in activated carbons prepared from walnut shells by carbon dioxide or steam activation. Industrial & Engineering Chemistry Research. 2009; 48: 7474–81.

[46] Singh A, Lal D. Preparation and characterization of activated carbon spheres from polystyrene sulphonate beads by steam and carbon dioxide activation. Journal of Applied Polymer Science. 2010; 115: 2409–15.

[47] MolinaSabio M, Gonzalez MT, RodriguezReinoso F, SepulvedaEscribano A. Effect of steam and carbon dioxide activation in the micropore size distribution of activated carbon. Carbon. 1996; 34: 505–9.

[48] Chmiola J, Yushin G, Gogotsi Y, Portet C, Simon P, Taberna PL. Anomalous increase in carbon capacitance at pore sizes less than 1 nanometer. Science. 2006; 313: 1760–3.

[49] Janes A, Lust E. Electrochemical characteristics of nanoporous carbide-derived carbon materials in various nonaqueous electrolyte solutions. Journal of the Electrochemical Society. 2006; 153: A113–A6.

[50] Janes A, Permann L, Arulepp M, Lust E. Electrochemical characteristics of nanoporous carbide-derived carbon materials in non-aqueous electrolyte solutions. Electrochemistry Communications. 2004; 6: 313–8.

[51] Chmiola J, Yushin G, Dash R, Gogotsi Y. Effect of pore size and surface area of carbide derived carbons on specific capacitance. Journal of Power Sources. 2006; 158: 765–72.

[52] Chmiola J, Largeot C, Taberna PL, Simon P, Gogotsi Y. Desolvation of ions in subnanometer pores and its effect on capacitance and double-layer theory. Angewandte Chemie-International Edition. 2008; 47: 3392–5.

[53] Jagtoyen M, Derbyshire F. Activated carbons from yellow poplar and white oak by H_3PO_4 activation. Carbon. 1998; 36: 1085–97.

150 *Nanomaterials for Supercapacitors*

[54] Molina-Sabio M, Rodriguez-Reinoso F. Role of chemical activation in the development of carbon porosity. Colloids and Surfaces A-Physicochemical and Engineering Aspects. 2004; 241: 15–25.

[55] Lillo-Rodenas MA, Cazorla-Amoros D, Linares-Solano A. Understanding chemical reactions between carbons and NaOH and KOH—An insight into the chemical activation mechanism. Carbon. 2003; 41: 267–75.

[56] Lillo-Rodenas MA, Juan-Juan J, Cazorla-Amoros D, Linares-Solano A. About reactions occurring during chemical activation with hydroxides. Carbon. 2004; 42: 1371–5.

[57] Lozano-Castello D, Calo JM, Cazorla-Amoros D, Linares-Solano A. Carbon activation with KOH as explored by temperature programmed techniques, and the effects of hydrogen. Carbon. 2007; 45: 2529–36.

[58] Teng HS, Hsu LY. High-porosity carbons prepared from bituminous coal with potassium hydroxide activation. Industrial & Engineering Chemistry Research. 1999; 38: 2947–53.

[59] Bleda-Martinez MJ, Macia-Agullo JA, Lozano-Castello D, Morallon E, Cazorla-Amoros D, Linares-Solano A. Role of surface chemistry on electric double layer capacitance of carbon materials. Carbon. 2005; 43: 2677–84.

[60] Sevilla M, Valle-Vigon P, Fuertes AB. N-doped polypyrrole-based porous carbons for CO_2 capture. Advanced Functional Materials. 2011; 21: 2781–7.

[61] Macia-Agullo JA, Moore BC, Cazorla-Amoros D, Linares-Solano A. Activation of coal tar pitch carbon fibres: Physical activation vs. chemical activation. Carbon. 2004; 42: 1367–70.

[62] Teng HS, Yeh TS, Hsu LY. Preparation of activated carbon from bituminous coal with phosphoric acid activation. Carbon. 1998; 36: 1387–95.

[63] Suarez-Garcia F, Martinez-Alonso A, Tascon JMD. Pyrolysis of apple pulp: chemical activation with phosphoric acid. Journal of Analytical and Applied Pyrolysis. 2002; 63: 283–301.

[64] Hsu LY, Teng I. Influence of different chemical reagents on the preparation of activated carbons from bituminous coal. Fuel Processing Technology. 2000; 64: 155–66.

[65] Zhang F, Li GD, Chen JS. Effects of raw material texture and activation manner on surface area of porous carbons derived from biomass resources. Journal of Colloid and Interface Science. 2008; 327: 108–14.

[66] Puziy AM, Poddubnaya OI, Martinez-Alonso A, Suarez-Garcia F, Tascon JMD. Synthetic carbons activated with phosphoric acid—I. Surface chemistry and ion binding properties. Carbon. 2002; 40: 1493–505.

[67] Puziy AM, Poddubnaya OI, Martinez-Alonso A, Suarez-Garcia F, Tascon JMD. Synthetic carbons activated with phosphoric acid—II. Porous structure. Carbon. 2002; 40: 1507–19.

[68] Puziy AM, Poddubnaya OI, Martinez-Alonso A, Suarez-Garcia F, Tascon JMD. Synthetic carbons activated with phosphoric acid III. Carbons prepared in air. Carbon. 2003; 41: 1181–91.

[69] Tsai WT, Chang CY, Lee SL. A low cost adsorbent from agricultural waste corn cob by zinc chloride activation. Bioresource Technology. 1998; 64: 211–7.

[70] IllanGomez MJ, GarciaGarcia A, deLecea CSM, LinaresSolano A. Activated carbons from Spanish coals. 2. Chemical activation. Energy & Fuels. 1996; 10: 1108–14.

[71] Eletskii PM, Yakovlev VA, Fenelonov VB, Parmon VN. Texture and adsorptive properties of microporous amorphous carbon materials prepared by the chemical activation of carbonized high-ash biomass. Kinetics and Catalysis. 2008; 49: 708–19.

[72] Niu JJ, Wang JN. Effect of temperature on chemical activation of carbon nanotubes. Solid State Sciences. 2008; 10: 1189–93.

[73] Ahmadpour A, Do DD. The preparation of activated carbon from macadamia nutshell by chemical activation. Carbon. 1997; 35: 1723–32.

[74] Lozano-Castello D, Cazorla-Amoros D, Linares-Solano A. Can highly activated carbons be prepared with a homogeneous micropore size distribution? Fuel Processing Technology. 2002; 77: 325–30.

[75] Ip AWM, Barford JP, McKay G. Production and comparison of high surface area bamboo derived active carbons. Bioresource Technology. 2008; 99: 8909–16.

[76] Qiao WM, Ling LC, Zha QF, Liu L. Preparation of a pitch-based activated carbon with a high specific surface area. Journal of Materials Science. 1997; 32: 4447–53.

[77] Lozano-Castello D, Lillo-Rodenas MA, Cazorla-Amoros D, Linares-Solano A. Preparation of activated carbons from Spanish anthracite I. Activation by KOH. Carbon. 2001; 39: 741–9.

[78] Guo YP, Yang SF, Yu KF, Zhao JZ, Wang ZC, Xu HD. The preparation and mechanism studies of rice husk based porous carbon. Materials Chemistry and Physics. 2002; 74: 320–3.

[79] Zhang F, Ma H, Chen J, Li GD, Zhang Y, Chen JS. Preparation and gas storage of high surface area microporous carbon derived from biomass source cornstalks. Bioresource Technology. 2008; 99: 4803–8.

[80] Suetcue H, Demiral H. Production of granular activated carbons from loquat stones by chemical activation. Journal of Analytical and Applied Pyrolysis. 2009; 84: 47–52.

[81] Yavuz R, Akyildiz H, Karatepe N, Cetinkaya E. Influence of preparation conditions on porous structures of olive stone activated by H3PO4. Fuel Processing Technology. 2010; 91: 80–7.

[82] Hu ZH, Guo HM, Srinivasan MP, Ni YM. A simple method for developing mesoporosity in activated carbon. Separation and Purification Technology. 2003; 31: 47–52.

Carbon Based Supercapacitors 151

[83] Prauchner MJ, Rodriguez-Reinoso F. Preparation of granular activated carbons for adsorption of natural gas. Microporous and Mesoporous Materials. 2008; 109: 581–4.

[84] Arami-Niya A, Daud WMAW, Mjalli FS. Comparative study of the textural characteristics of oil palm shell activated carbon produced by chemical and physical activation for methane adsorption. Chemical Engineering Research and Design. 2011; 89: 657–64.

[85] Caturla F, Molinasabio M, Rodriguezreinoso F. Preparation of activated carbon by chemical activation with $ZnCl_2$. Carbon. 1991; 29: 999–1007.

[86] Wu FC, Tseng RL. Preparation of highly porous carbon from fir wood by KOH etching and CO_2 gasification for adsorption of dyes and phenols from water. Journal of Colloid and Interface Science. 2006; 294: 21–30.

[87] Yates M, Blanco J, Avila P, Martin MP. Honeycomb monoliths of activated carbons for effluent gas purification. Microporous and Mesoporous Materials. 2000; 37: 201–8.

[88] Gamby J, Taberna PL, Simon P, Fauvarque JF, Chesneau M. Studies and characterisations of various activated carbons used for carbon/carbon supercapacitors. Journal of Power Sources. 2001; 101: 109–16.

[89] Biloe S, Goetz V, Mauran S. Characterization of adsorbent composite blocks for methane storage. Carbon. 2001; 39: 1653–62.

[90] Lozano-Castello D, Cazorla-Amoros D, Linares-Solana A, Quinn DF. Activated carbon monoliths for methane storage: influence of binder. Carbon. 2002; 40: 2817–25.

[91] Ubago-Perez R, Carrasco-Marin F, Fairen-Jimenez D, Moreno-Castilla C. Granular and monolithic activated carbons from KOH-activation of olive stones. Microporous and Mesoporous Materials. 2006; 92: 64–70.

[92] Balathanigaimani MS, Shim WG, Lee JW, Moon H. Adsorption of methane on novel corn grain-based carbon monoliths. Microporous and Mesoporous Materials. 2009; 119: 47–52.

[93] Giraldo L, Moreno-Pirajan JC. Synthesis of activated carbon honeycomb monoliths under different conditions for the adsorption of methane. Adsorption Science & Technology. 2009; 27: 255–65.

[94] Machnikowski J, Kierzek K, Lis K, Machnikowska H, Czepirski L. Tailoring porosity development in monolithic adsorbents made of KOH-activated pitch coke and furfuryl alcohol binder for methane storage. Energy & Fuels. 2010; 24: 3410–4.

[95] Lozano-Castello D, Cazorla-Amoros D, Linares-Solano A, Quinn DF. Influence of pore size distribution on methane storage at relatively low pressure: preparation of activated carbon with optimum pore size. Carbon. 2002; 40: 989–1002.

[96] Inomata K, Kanazawa K, Urabe Y, Hosono H, Araki T. Natural gas storage in activated carbon pellets without a binder. Carbon. 2002; 40: 87–93.

[97] Molina-Sabio A, Almansa C, Rodriguez-Reinoso F. Phosphoric acid activated carbon discs for methane adsorption. Carbon. 2003; 41: 2113–9.

[98] Almansa C, Molina-Sabio M, Rodriguez-Reinoso F. Adsorption of methane into $ZnCl_2$-activated carbon derived discs. Microporous and Mesoporous Materials. 2004; 76: 185–91.

[99] Paola Vargas-Delgadillo D, Giraldo L, Carlos Moreno-Pirajan J. Preparation and characterization of activated carbon monoliths with potential application as phenol adsorbents. European Journal of Chemistry. 2010; 7: 531–9.

[100] Garcia Blanco AA, Alexandre de Oliveira JC, Lopez R, Moreno-Pirajan JC, Giraldo L, Zgrablich G et al. A study of the pore size distribution for activated carbon monoliths and their relationship with the storage of methane and hydrogen. Colloids and Surfaces A-Physicochemical and Engineering Aspects. 2010; 357: 74–83.

[101] Paola Vargas D, Giraldo L, Silvestre-Albero J, Carlos Moreno-Pirajan J. CO_2 adsorption on binderless activated carbon monoliths. Adsorption-Journal of the International Adsorption Society. 2011; 17: 497–504.

[102] Ruiz V, Blanco C, Santamaria R, Ramos-Fernandez JM, Martinez-Escandell M, Sepulveda-Escribano A et al. An activated carbon monolith as an electrode material for supercapacitors. Carbon. 2009; 47: 195–200.

[103] Ramos-Fernandez JM, Martinez-Escandell M, Rodriguez-Reinoso F. Production of binderless activated carbon monoliths by KOH activation of carbon mesophase materials. Carbon. 2008; 46: 384–6.

[104] Wahby A, Ramos-Fernandez JM, Martinez-Escandell M, Sepulveda-Escribano A, Silvestre-Albero J, Rodriguez-Reinoso F. High-surface-area carbon molecular sieves for selective CO_2 adsorption. ChemSusChem. 2010; 3: 974–81.

[105] Deraman M, Omar R, Zakaria S, Mustapa IR, Talib M, Alias N et al. Electrical and mechanical properties of carbon pellets from acid (HNO_3) treated self-adhesive carbon grain from oil palm empty fruit bunch. Journal of Materials Science. 2002; 37: 3329–35.

[106] Taer E, Deraman M, Talib IA, Awitdrus A, Hashmi SA, Umar AA. Preparation of a highly porous binderless activated carbon monolith from rubber wood sawdust by a multi-step activation process for application in supercapacitors. International Journal of Electrochemical Science. 2011; 6: 3301–15.

[107] Hao GP, Li WC, Qian D, Wang GH, Zhang WP, Zhang T et al. Structurally designed synthesis of mechanically stable poly(benzoxazine-co-resol)-based porous carbon monoliths and their application as high-performance CO_2 capture sorbents. Journal of the American Chemical Society. 2011; 133: 11378–88.

[108] Matos JR, Kruk M, Mercuri LP, Jaroniec M, Zhao L, Kamiyama T et al. Ordered mesoporous silica with large cage-like pores: Structural identification and pore connectivity design by controlling the synthesis temperature and time. Journal of the American Chemical Society. 2003; 125: 821–9.

152 *Nanomaterials for Supercapacitors*

[109] Yu CZ, Tian BZ, Fan J, Stucky GD, Zhao DY. Nonionic block copolymer synthesis of large-pore cubic mesoporous single crystals by use of inorganic salts. Journal of the American Chemical Society. 2002; 124: 4556–7.

[110] Sakamoto Y, Kaneda M, Terasaki O, Zhao DY, Kim JM, Stucky G et al. Direct imaging of the pores and cages of three-dimensional mesoporous materials. Nature. 2000; 408: 449–53.

[111] Meng Y, Gu D, Zhang FQ, Shi YF, Cheng L, Feng D et al. A family of highly ordered mesoporous polymer resin and carbon structures from organic-organic self-assembly. Chemistry of Materials. 2006; 18: 4447–64.

[112] Shi H. Activated carbons and double layer capacitance. Electrochimica Acta. 1996; 41: 1633–9.

[113] Qu DY, Shi H. Studies of activated carbons used in double-layer capacitors. Journal of Power Sources. 1998; 74: 99–107.

[114] Endo M, Kim YJ, Maeda T, Koshiba K, Katayam K, Dresselhaus MS. Morphological effect on the electrochemical behavior of electric double-layer capacitors. Journal of Materials Research. 2001; 16: 3402–10.

[115] Centeno TA, Stoeckli F. The role of textural characteristics and oxygen-containing surface groups in the supercapacitor performances of activated carbons. Electrochimica Acta. 2006; 52: 560–6.

[116] Huang JS, Sumpter BG, Meunier V. A universal model for nanoporous carbon supercapacitors applicable to diverse pore regimes, carbon materials, and electrolytes. Chemistry-A European Journal. 2008; 14: 6614–26.

[117] Raymundo-Pinero E, Kierzek K, Machnikowski J, Beguin F. Relationship between the nanoporous texture of activated carbons and their capacitance properties in different electrolytes. Carbon. 2006; 44: 2498–507.

[118] Sevilla M, Fuertes AB, Mokaya R. High density hydrogen storage in superactivated carbons from hydrothermally carbonized renewable organic materials. Energy & Environmental Science. 2011; 4: 1400–10.

[119] Wei L, Sevilla M, Fuertes AB, Mokaya R, Yushin G. Hydrothermal carbonization of abundant renewable natural organic chemicals for high-performance supercapacitor electrodes. Advanced Energy Materials. 2011; 1: 356–61.

[120] Li XL, Han CL, Chen XY, Shi CW. Preparation and performance of straw based activated carbon for supercapacitor in non-aqueous electrolytes. Microporous and Mesoporous Materials. 2010; 131: 303–9.

[121] Toupin M, Belanger D, Hill IR, Quinn D. Performance of experimental carbon blacks in aqueous supercapacitors. Journal of Power Sources. 2005; 140: 203–10.

[122] Zhu YD, Hu HQ, Li WC, Zhang XY. Resorcinol-formaldehyde based porous carbon as an electrode material for supercapacitors. Carbon. 2007; 45: 160–5.

[123] He XJ, Geng YJ, Qiu JS, Zheng MD, Long SA, Zhang XY. Effect of activation time on the properties of activated carbons prepared by microwave-assisted activation for electric double layer capacitors. Carbon. 2010; 48: 1662–9.

[124] Li FH, Chi WD, Shen ZM, Wu YX, Liu YF, Liu H. Activation of mesocarbon microbeads with different textures and their application for supercapacitor. Fuel Processing Technology. 2010; 91: 17–24.

[125] Wang XF, Wang DZ, Liang J. Performance of electric double layer capacitors using active carbons prepared from petroleum coke by KOH and vapor re-etching. Journal of Materials Science & Technology. 2003; 19: 265–9.

[126] Olivares-Marin M, Fernandez JA, Lazaro MJ, Fernandez-Gonzalez C, Macias-Garcia A, Gomez-Serrano V et al. Cherry stones as precursor of activated carbons for supercapacitors. Materials Chemistry and Physics. 2009; 114: 323–7.

[127] Lee SG, Park KH, Shim WG, Balathanigaimani MS, Moon H. Performance of electrochemical double layer capacitors using highly porous activated carbons prepared from beer lees. Journal of Industrial and Engineering Chemistry. 2011; 17: 450–4.

[128] Ruiz V, Pandolfo AG. Polyfurfuryl alcohol derived activated carbons for high power electrical double layer capacitors. Electrochimica Acta. 2010; 55: 7495–500.

[129] Zhao XY, Cao JP, Morishita K, Ozaki J, Takarada T. Electric double-layer capacitors from activated carbon derived from black liquor. Energy & Fuels. 2010; 24: 1889–93.

[130] Zhai DY, Li BH, Du HD, Wang G, Kang FY. The effect of pre-carbonization of mesophase pitch-based activated carbons on their electrochemical performance for electric double-layer capacitors. Journal of Solid State Electrochemistry. 2011; 15: 787–94.

[131] Kierzek K, Frackowiak E, Lota G, Gryglewicz G, Machnikowski J. Electrochemical capacitors based on highly porous carbons prepared by KOH activation. Electrochimica Acta. 2004; 49: 515–23.

[132] Zhang CX, Zhang R, Xing BL, Cheng G, Xie YB, Qiao WM et al. Effect of pore structure on the electrochemical performance of coal-based activated carbons in non-aqueous electrolyte. New Carbon Materials. 2010; 25: 129–33.

[133] Xu B, Wu FC, Mu DB, Dai LL, Cao GP, Zhang H et al. Activated carbon prepared from PVDC by NaOH activation as electrode materials for high performance EDLCs with non-aqueous electrolyte. International Journal of Hydrogen Energy. 2010; 35: 632–7.

[134] Rufford TE, Hulicova-Jurcakova D, Fiset E, Zhu ZH, Lu GQ. Double-layer capacitance of waste coffee ground activated carbons in an organic electrolyte. Electrochemistry Communications. 2009; 11: 974–7.

[135] Rufford TE, Hulicova-Jurcakova D, Khosla K, Zhu ZH, Lu GQ. Microstructure and electrochemical double-layer capacitance of carbon electrodes prepared by zinc chloride activation of sugar cane bagasse. Journal of Power Sources. 2010; 195: 912–8.

[136] Wei L, Yushin G. Electrical double layer capacitors with sucrose derived carbon electrodes in ionic liquid electrolytes. Journal of Power Sources. 2011; 196: 4072–9.

[137] Fang B, Wei YZ, Maruyama K, Kumagai M. High capacity supercapacitors based on modified activated carbon aerogel. Journal of Applied Electrochemistry. 2005; 35: 229–33.

Carbon Based Supercapacitors 153

[138] Zhu YD, Hu HQ, Li WC, Zhang XY. Cresol-formaldehyde based carbon aerogel as electrode material for electrochemical capacitor. Journal of Power Sources. 2006; 162: 738–42.

[139] Wang XQ, Lee JS, Tsouris C, DePaoli DW, Dai S. Preparation of activated mesoporous carbons for electrosorption of ions from aqueous solutions. Journal of Materials Chemistry. 2010; 20: 4602–8.

[140] Lv YY, Zhang F, Dou YQ, Zhai YP, Wang JX, Liu HJ et al. A comprehensive study on KOH activation of ordered mesoporous carbons and their supercapacitor application. Journal of Materials Chemistry. 2012; 22: 93–9.

[141] Jin J, Tanaka S, Egashira Y, Nishiyama N. KOH activation of ordered mesoporous carbons prepared by a soft-templating method and their enhanced electrochemical properties. Carbon. 2010; 48: 1985–9.

[142] Hsieh CT, Teng H. Influence of oxygen treatment on electric double-layer capacitance of activated carbon fabrics. Carbon. 2002; 40: 667–74.

[143] Qu DY. Studies of the activated carbons used in double-layer supercapacitors. Journal of Power Sources. 2002; 109: 403–11.

[144] Nian YR, Teng HS. Influence of surface oxides on the impedance behavior of carbonbased electrochemical capacitors. Journal of Electroanalytical Chemistry. 2003; 540: 119–27.

[145] Ruiz V, Blanco C, Raymundo-Pinero E, Khomenko V, Beguin F, Santamaria R. Effects of thermal treatment of activated carbon on the electrochemical behaviour in supercapacitors. Electrochimica Acta. 2007; 52: 4969–73.

[146] Rufford TE, Hulicova-Jurcakova D, Zhu ZH, Lu GQ. Nanoporous carbon electrode from waste coffee beans for high performance supercapacitors. Electrochemistry Communications. 2008; 10: 1594–7.

[147] Balathanigaimani MS, Shim WG, Lee MJ, Kim C, Lee JW, Moon H. Highly porous electrodes from novel corn grains-based activated carbons for electrical double layer capacitors. Electrochemistry Communications. 2008; 10: 868–71.

[148] Hulicova-Jurcakova D, Seredych M, Lu GQ, Bandosz TJ. Combined effect of nitrogen- and oxygen-containing functional groups of microporous activated carbon on its electrochemical performance in supercapacitors. Advanced Functional Materials. 2009; 19: 438–47.

[149] Tashima D, Yamamoto E, Kai N, Fujikawa D, Sakai G, Otsubo M et al. Double layer capacitance of high surface area carbon nanospheres derived from resorcinol-formaldehyde polymers. Carbon. 2011; 49: 4848–57.

[150] Tremblay G, Vastola FJ, Walker PL. Thermal desorption analysis of oxygen-surface complexes on carbon. Carbon. 1978; 16: 35–9.

[151] Nian YR, Teng HS. Nitric acid modification of activated carbon electrodes for improvement of electrochemical capacitance. Journal of the Electrochemical Society. 2002; 149: A1008–A14.

[152] Bleda-Martinez MJ, Lozano-Castello D, Morallon E, Cazorla-Amoros D, Linares-Solano A. Chemical and electrochemical characterization of porous carbon materials. Carbon. 2006; 44: 2642–51.

[153] Mysyk R, Raymundo-Pinero E, Anouti M, Lemordant D, Beguin F. Pseudo-capacitance of nanoporous carbons in pyrrolidinium-based protic ionic liquids. Electrochemistry Communications. 2010; 12: 414–7.

[154] Milczarek G, Ciszewski A, Stepniak I. Oxygen-doped activated carbon fiber cloth as electrode material for electrochemical capacitor. Journal of Power Sources. 2011; 196: 7882–5.

[155] Ra EJ, Raymundo-Pinero E, Lee YH, Beguin F. High power supercapacitors using polyacrylonitrile-based carbon nanofiber paper. Carbon. 2009; 47: 2984–92.

[156] Seredych M, Hulicova-Jurcakova D, Lu GQ, Bandosz TJ. Surface functional groups of carbons and the effects of their chemical character, density and accessibility to ions on electrochemical performance. Carbon. 2008; 46: 1475–88.

[157] Mangun CL, Benak KR, Economy J, Foster KL. Surface chemistry, pore sizes and adsorption properties of activated carbon fibers and precursors treated with ammonia. Carbon. 2001; 39: 1809–20.

[158] Chen WF, Cannon FS, Rangel-Mendez JR. Ammonia-tailoring of GAC to enhance perchlorate removal. I: Characterization of NH_3 thermally tailored GACs. Carbon. 2005; 43: 573–80.

[159] Jurewicz K, Babel K, Ziolkowski A, Wachowska H. Ammoxidation of active carbons for improvement of supercapacitor characteristics. Electrochimica Acta. 2003; 48: 1491–8.

[160] Pietrzak R, Jurewicz K, Nowicki P, Babel K, Wachowska H. Nitrogen-enriched bituminous coal-based active carbons as materials for supercapacitors. Fuel. 2010; 89: 3457–67.

[161] Pels JR, Kapteijn F, Moulijn JA, Zhu Q, Thomas KM. Evolution of nitrogen functionalities in carbonaceous materials during pyrolysis. Carbon. 1995; 33: 1641–53.

[162] Adib F, Bagreev A, Bandosz TJ. Adsorption/oxidation of hydrogen sulfide on nitrogen-containing activated carbons. Langmuir. 2000; 16: 1980–6.

[163] Biniak S, Szymanski G, Siedlewski J, Swiatkowski A. The characterization of activated carbons with oxygen and nitrogen surface groups. Carbon. 1997; 35: 1799–810.

[164] Devallencourt C, Saiter JM, Fafet A, Ubrich E. Thermogravimetry Fourier-transform infrared coupling investigations to study the thermal-stability of melamine-formaldehyde resin. Thermochimica Acta. 1995; 259: 143–51.

[165] Kim YJ, Abe Y, Yanaglura T, Park KC, Shimizu M, Iwazaki T et al. Easy preparation of nitrogen-enriched carbon materials from peptides of silk fibroins and their use to produce a high volumetric energy density in supercapacitors. Carbon. 2007; 45: 2116–25.

[166] Jiang JH, Gao QM, Xia KS, Hu J. Enhanced electrical capacitance of porous carbons by nitrogen enrichment and control of the pore structure. Microporous and Mesoporous Materials. 2009; 118: 28–34.

[167] Li B, Dai F, Xiao QF, Yang L, Shen JM, Zhang CM et al. Nitrogen-doped activated carbon for a high energy hybrid supercapacitor. Energy & Environmental Science. 2016; 9: 102–6.

[168] Lu WJ, Liu MX, Miao L, Zhu DZ, Wang X, Duan H et al. Nitrogen-containing ultramicroporous carbon nanospheres for high performance supercapacitor electrodes. Electrochimica Acta. 2016; 205: 132–41.

[169] Ma GF, Zhang ZG, Peng H, Sun KJ, Ran FT, Lei ZQ. Facile preparation of nitrogen-doped porous carbon for high performance symmetric supercapacitor. Journal of Solid State Electrochemistry. 2016; 20: 1613–23.

[170] Ramakrishnan P, Shanmugam S. Nitrogen-doped carbon nanofoam derived from amino acid chelate complex for supercapacitor applications. Journal of Power Sources. 2016; 316: 60–71.

[171] Sahu V, Grover S, Singh G, Sharma RK. Nitrogen-doped carbon nanosheets for high-performance liquid as well as solid state supercapacitor cells. RSC Advances. 2016; 6: 35014–23.

[172] Tian XD, Zhao N, Song Y, Wang K, Xu DF, Li X et al. Synthesis of nitrogen-doped electrospun carbon nanofibers with superior performance as efficient supercapacitor electrodes in alkaline solution. Electrochimica Acta. 2015; 185: 40–51.

[173] Zhan CZ, Xu Q, Yu XL, Liang QH, Bai Y, Huang ZH et al. Nitrogen-rich hierarchical porous hollow carbon nanofibers for high-performance supercapacitor electrodes. RSC Advances. 2016; 6: 41473–6.

[174] Zhang JC, Zhou JS, Wang D, Hou L, Gao FM. Nitrogen and sulfur codoped porous carbon microsphere: a high performance electrode in supercapacitor. Electrochimica Acta. 2016; 191: 933–9.

[175] Zhang ZJ, Chen XY. Nitrogen-doped nanoporous carbon materials derived from folic acid: Simply introducing redox additive of p-phenylenediamine into KOH electrolyte for greatly improving the supercapacitor performance. Journal of Electroanalytical Chemistry. 2016; 764: 45–55.

[176] Vinu A, Ariga K, Mori T, Nakanishi T, Hishita S, Golberg D et al. Preparation and characterization of well-ordered hexagonal mesoporous carbon nitride. Advanced Materials. 2005; 17: 1648–52.

[177] Ryoo R, Joo SH, Jun S. Synthesis of highly ordered carbon molecular sieves via template-mediated structural transformation. Journal of Physical Chemistry B. 1999; 103: 7743–6.

[178] Fuertes AB, Lota G, Centeno TA, Frackowiak E. Templated mesoporous carbons for supercapacitor application. Electrochimica Acta. 2005; 50: 2799–805.

[179] Yoon SH, Oh SM, Lee CW, Ryu JH. Pore structure tuning of mesoporous carbon prepared by direct templating method for application to high rate supercapacitor electrodes. Journal of Electroanalytical Chemistry. 2011; 650: 187–95.

[180] Frackowiak E, Lota G, Machnikowski J, Vix-Guterl C, Beguin F. Optimisation of supercapacitors using carbons with controlled nanotexture and nitrogen content. Electrochimica Acta. 2006; 51: 2209–14.

[181] Xing W, Qiao SZ, Ding RG, Li F, Lu GQ, Yan ZF et al. Superior electric double layer capacitors using ordered mesoporous carbons. Carbon. 2006; 44: 216–24.

[182] Chen WZ, Shi JJ, Zhu TS, Wang Q, Qiao JL, Zhang JJ. Preparation of nitrogen and sulfur dual-doped mesoporous carbon for supercapacitor electrodes with long cycle stability. Electrochimica Acta. 2015; 177: 327–34.

[183] Yi HT, Zhu YQ, Chen XY, Zhang ZJ. Nitrogen and sulfur co-doped nanoporous carbon material derived from p-nitrobenzenamine within several minutes and the supercapacitor application. Journal of Alloys and Compounds. 2015; 649: 851–8.

[184] Zhou Y, Ma RG, Candelaria SL, Wang JC, Liu Q, Uchaker E et al. Phosphorus/sulfur co-doped porous carbon with enhanced specific capacitance for supercapacitor and improved catalytic activity for oxygen reduction reaction. Journal of Power Sources. 2016; 314: 39–48.

[185] Jiang JG, Chen H, Wang ZC, Bao LK, Qiang YW, Guan SY et al. Nitrogen-doped hierarchical porous carbon microsphere through KOH activation for supercapacitors. Journal of Colloid and Interface Science. 2015; 452: 54–61.

[186] Nanaumi T, Ohsawa Y, Kobayakawa K, Sato Y. High energy electrochemical capacitor materials prepared by loading ruthenium oxide on activated carbon. Electrochemistry. 2002; 70: 681–5.

[187] Ramani M, Haran BS, White RE, Popov BN, Arsov L. Studies on activated carbon capacitor materials loaded with different amounts of ruthenium oxide. Journal of Power Sources. 2001; 93: 209–14.

[188] Sato Y, Yomogida K, Nanaumi T, Kobayakawa K, Ohsawa Y, Kawai M. Electrochemical behavior of activated-carbon capacitor materials loaded with ruthenium oxide. Electrochemical and Solid State Letters. 2000; 3: 113–6.

[189] Zhang JR, Jiang DC, Chen B, Zhu JJ, Jiang LP, Fang HQ. Preparation and electrochemistry of hydrous ruthenium oxide/active carbon electrode materials for supercapacitor. Journal of the Electrochemical Society. 2001; 148: A1362–A7.

[190] Chen WC, Hu CC, Wang CC, Min CK. Electrochemical characterization of activated carbon-ruthenium oxide nanoparticles composites for supercapacitors. Journal of Power Sources. 2004; 125: 292–8.

[191] Egashira M, Matsuno Y, Yoshimoto N, Morita M. Pseudo-capacitance of composite electrode of ruthenium oxide with porous carbon in non-aqueous electrolyte containing imidazolium salt. Journal of Power Sources. 2010; 195: 3036–40.

[192] Lee YJ, Park HW, Park SY, Song IK. Electrochemical properties of Mn-doped activated carbon aerogel as electrode material for supercapacitor. Current Applied Physics. 2012; 12: 233–7.

[193] Li ZS, Wang HQ, Huang YG, Li QY, Wang XY. Manganese dioxide-coated activated mesocarbon microbeads for supercapacitors in organic electrolyte. Colloids and Surfaces A-Physicochemical and Engineering Aspects. 2010; 366: 104–9.

[194] Wang HQ, Li ZS, Yang JH, Li QY, Zhong XX. A novel activated mesocarbon microbead(aMCMB)/Mn_3O_4 composite for electrochemical capacitors in organic electrolyte. Journal of Power Sources. 2009; 194: 1218–21.

[195] Bello A, Fashedemi OO, Barzegar F, Madito MJ, Momodu DY, Masikhwa TM et al. Microwave synthesis: Characterization and electrochemical properties of amorphous activated carbon-MnO_2 nanocomposite electrodes. Journal of Alloys and Compounds. 2016; 681: 293–300.

[196] Huang TF, Qiu ZH, Wu DW, Hu ZB. Bamboo-based activated carbon @ MnO_2 nanocomposites for flexible high-performance supercapacitor electrode materials. International Journal of Electrochemical Science. 2015; 10: 6312–23.

[197] Ramirez-Castro C, Crosnier O, Athouel L, Retoux R, Belanger D, Brousse T. Electrochemical performance of carbon/MnO_2 nanocomposites prepared via molecular bridging as supercapacitor electrode materials. Journal of the Electrochemical Society. 2015; 162: A5179–A84.

[198] Zhao Y, Meng YN, Jiang P. Carbon@MnO_2 core-shell nanospheres for flexible high-performance supercapacitor electrode materials. Journal of Power Sources. 2014; 259: 219–26.

[199] Zhang Y, Feng H, Wu XB, Wang LZ, Zhang AQ, Xia TC et al. Progress of electrochemical capacitor electrode materials: A review. International Journal of Hydrogen Energy. 2009; 34: 4889–99.

[200] Sharma RK, Oh HS, Shul YG, Kim HS. Carbon-supported, nano-structured, manganese oxide composite electrode for electrochemical supercapacitor. Journal of Power Sources. 2007; 173: 1024–8.

[201] Huang QH, Wang XY, Li J, Dai CL, Gamboa S, Sebastian PJ. Nickel hydroxide/activated carbon composite electrodes for electrochemical capacitors. Journal of Power Sources. 2007; 164: 425–9.

[202] Gong SL, Cao Q, Jin LE, Zhong CG, Zhang XH. Electrodeposition of three-dimensional $Ni(OH)_2$ nanoflakes on partially crystallized activated carbon for high-performance supercapacitors. Journal of Solid State Electrochemistry. 2016; 20: 619–28.

[203] Sui LP, Tang SH, Chen YD, Dai Z, Huangfu HX, Zhu ZT et al. An asymmetric supercapacitor with good electrochemical performances based on $Ni(OH)_2$/AC/CNT and AC. Electrochimica Acta. 2015; 182: 1159–65.

[204] Tang YF, Liu YY, Yu SX, Guo WC, Mu SC, Wang HC et al. Template-free hydrothermal synthesis of nickel cobalt hydroxide nanoflowers with high performance for asymmetric supercapacitor. Electrochimica Acta. 2015; 161: 279–89.

[205] Xie LJ, Sun GH, Xie LF, Su FY, Li XM, Liu Z et al. A high energy density asymmetric supercapacitor based on a CoNi-layered double hydroxide and activated carbon. New Carbon Materials. 2016; 31: 37–45.

[206] Lin YR, Teng HS. A novel method for carbon modification with minute polyaniline deposition to enhance the capacitance of porous carbon electrodes. Carbon. 2003; 41: 2865–71.

[207] Wang Q, Li JL, Gao F, Li WS, Wu KZ, Wang XD. Activated carbon coated with polyaniline as an electrode material in supercapacitors. New Carbon Materials. 2008; 23: 275–80.

[208] Bleda-Martinez MJ, Morallon E, Cazorla-Amoros D. Polyaniline/porous carbon electrodes by chemical polymerisation: Effect of carbon surface chemistry. Electrochimica Acta. 2007; 52: 4962–8.

[209] Chen WC, Wen TC, Teng HS. Polyaniline-deposited porous carbon electrode for supercapacitor. Electrochimica Acta. 2003; 48: 641–9.

[210] Tamai H, Hakoda M, Shiono T, Yasuda H. Preparation of polyaniline coated activated carbon and their electrode performance for supercapacitor. Journal of Materials Science. 2007; 42: 1293–8.

[211] Alguail AA, Al-Eggiely AH, Gvozdenovic MM, Jugovic BZ, Grgur BN. Battery type hybrid supercapacitor based on polypyrrole and lead-lead sulfate. Journal of Power Sources. 2016; 313: 240–6.

[212] Ke QQ, Zheng MR, Liu HJ, Guan C, Mao L, Wang J. 3D TiO_2@$Ni(OH)_2$ core-shell arrays with tunable nanostructure for hybrid supercapacitor application. Scientific Reports. 2015; 5: 13940.

[213] Lee JH, Kim HK, Baek E, Pecht M, Lee SH, Lee YH. Improved performance of cylindrical hybrid supercapacitor using activated carbon/niobium doped hydrogen titanate. Journal of Power Sources. 2016; 301: 348–54.

[214] Ma T, Yang HX, Lu L. Development of hybrid battery-supercapacitor energy storage for remote area renewable energy systems. Applied Energy. 2015; 153: 56–62.

[215] Shellikeri A, Hung I, Gan ZH, Zheng JP. In situ NMR tracks real-time Li ion movement in hybrid supercapacitor-battery cevice. Journal of Physical Chemistry C. 2016; 120: 6314–23.

[216] Srikanth VVSS, Ramana GV, Kumar PS. Perspectives on state-of-the-art carbon nanotube/polyaniline and graphene/polyaniline composites for hybrid supercapacitor electrodes. Journal of Nanoscience and Nanotechnology. 2016; 16: 2418–24.

[217] Zhang LJ, Hui KN, Hui KS, Lee HW. High-performance hybrid supercapacitor with 3D hierarchical porous flower-like layered double hydroxide grown on nickel foam as binder-free electrode. Journal of Power Sources. 2016; 318: 76–85.

[218] Plitz I, DuPasquier A, Badway F, Gural J, Pereira N, Gmitter A et al. The design of alternative nonaqueous high power chemistries. Applied Physics A-Materials Science & Processing. 2006; 82: 615–26.

[219] Algharaibeh Z, Pickup PG. An asymmetric supercapacitor with anthraquinone and dihydroxybenzene modified carbon fabric electrodes. Electrochemistry Communications. 2011; 13: 147–9.

[220] Qu DY, Wang LL, Zheng D, Xiao L, Deng BH, Qu DY. An asymmetric supercapacitor with highly dispersed nano-Bi_2O_3 and active carbon electrodes. Journal of Power Sources. 2014; 269: 129–35.

[221] Qu QT, Li L, Tian S, Guo WL, Wu YP, Holze R. A cheap asymmetric supercapacitor with high energy at high power: Activated carbon//$K_{0.27}MnO_2 \cdot 0.6H_2O$. Journal of Power Sources. 2010; 195: 2789–94.

156 *Nanomaterials for Supercapacitors*

[222] Wang YG, Cheng L, Xia YY. Electrochemical profile of nano-particle CoAl double hydroxide/active carbon supercapacitor using KOH electrolyte solution. Journal of Power Sources. 2006; 153: 191–6.

[223] Lang JW, Kong LB, Liu M, Luo YC, Kang L. Asymmetric supercapacitors based on stabilized a-Ni(OH)$_2$ and activated carbon. Journal of Solid State Electrochemistry. 2010; 14: 1533–9.

[224] Khomenko V, Raymundo-Pinero E, Beguin F. Optimisation of an asymmetric manganese oxide/activated carbon capacitor working at 2 V in aqueous medium. Journal of Power Sources. 2006; 153: 183–90.

[225] Demarconnay L, Raymundo-Pinero E, Beguin F. Adjustment of electrodes potential window in an asymmetric carbon/MnO$_2$ supercapacitor. Journal of Power Sources. 2011; 196: 580–6.

[226] Fan ZJ, Yan J, Wei T, Zhi LJ, Ning GQ, Li TY et al. Asymmetric supercapacitors based on graphene/MnO$_2$ and activated carbon nanofiber electrodes with high power and energy density. Advanced Functional Materials. 2011; 21: 2366–75.

[227] Di Fabio A, Giorgi A, Mastragostino M, Soavi F. Carbon-poly(3-methylthiophene) hybrid supercapacitors. Journal of the Electrochemical Society. 2001; 148: A845–A50.

[228] Laforgue A, Simon P, Fauvarque JF, Sarrau JF, Lailler P. Hybrid supercapacitors based on activated carbons and conducting polymers. Journal of the Electrochemical Society. 2001; 148: A1130–A4.

[229] Laforgue A, Simon P, Fauvarque JF, Mastragostino M, Soavi F, Sarrau JF et al. Activated carbon/conducting polymer hybrid supercapacitors. Journal of the Electrochemical Society. 2003; 150: A645–A51.

[230] Amatucci GG, Badway F, Du Pasquier A, Zheng T. An asymmetric hybrid nonaqueous energy storage cell. Journal of the Electrochemical Society. 2001; 148: A930–A9.

[231] Brousse T, Marchand R, Taberna PL, Simon P. TiO$_2$ (B)/activated carbon non-aqueous hybrid system for energy storage. Journal of Power Sources. 2006; 158: 571–7.

[232] Aravindan V, Chuiling W, Madhavi S. High power lithium-ion hybrid electrochemical capacitors using spinel LiCrTiO$_4$ as insertion electrode. Journal of Materials Chemistry. 2012; 22: 16026–31.

[233] Mladenov M, Alexandrova K, Petrov NV, Tsyntsarski B, Kovacheva D, Saliyski N et al. Synthesis and electrochemical properties of activated carbons and Li$_4$Ti$_5$O$_{12}$ as electrode materials for supercapacitors. Journal of Solid State Electrochemistry. 2013; 17: 2101–8.

[234] Wang YG, Xia YY. A new concept hybrid electrochemical surpercapacitor: Carbon/LiMn$_2$O$_4$ aqueous system. Electrochemistry Communications. 2005; 7: 1138–42.

[235] Zheng JP. The limitations of energy density of battery/double-layer capacitor asymmetric cells. Journal of the Electrochemical Society. 2003; 150: A484–A92.

[236] Zheng JP. High energy density electrochemical capacitors without consumption of electrolyte. Journal of the Electrochemical Society. 2009; 156: A500–A5.

[237] Yan JW, Sun YY, Jiang L, Tian Y, Xue R, Hao LX et al. Electrochemical performance of lithium ion capacitors using aqueous electrolyte at high temperature. Journal of Renewable and Sustainable Energy. 2013; 5: 021404.

[238] Marchand R, Brohan L, Tournoux M. TiO$_2$ (B) a new form of titanium-dioxide and the potassium octatitanate K$_2$Ti$_8$O$_{17}$. Materials Research Bulletin. 1980; 15: 1129–33.

[239] Nuspl G, Yoshizawa K, Yamabe T. Lithium intercalation in TiO$_2$ modifications. Journal of Materials Chemistry. 1997; 7: 2529–36.

[240] Saito R, Dresselhaus G, Dresselhaus MS. Electronic-structure of double-layer graphene tubules. Journal of Applied Physics. 1993; 73: 494–500.

[241] Issi JP, Langer L, Heremans J, Olk CH. Electronic-properties of carbon nanotubes-experimental results. Carbon. 1995; 33: 941–8.

[242] Frackowiak E, Jurewicz K, Delpeux S, Beguin F. Nanotubular materials for supercapacitors. Journal of Power Sources. 2001; 97-8: 822–5.

[243] Arabale G, Wagh D, Kulkarni M, Mulla IS, Vernekar SP, Vijayamohanan K et al. Enhanced supercapacitance of multiwalled carbon nanotubes functionalized with ruthenium oxide. Chemical Physics Letters. 2003; 376: 207–13.

[244] Lee CY, Tsai HM, Chuang HJ, Li SY, Lin P, Tseng TY. Characteristics and electrochemical performance of supercapacitors with manganese oxide-carbon nanotube nanocomposite electrodes. Journal of the Electrochemical Society. 2005; 152: A716–A20.

[245] Su LH, Zhang XG, Yuan CZ, Gao B. Symmetric self-hybrid supercapacitor consisting of multiwall carbon nanotubes and coal layered double hydroxides. Journal of the Electrochemical Society. 2008; 155: A110–A4.

[246] Izadi-Najafabadi A, Yasuda S, Kobashi K, Yamada T, Futaba DN, Hatori H et al. Extracting the full potential of single-walled carbon nanotubes as durable supercapacitor electrodes operable at 4 V with high power and energy density. Advanced Materials. 2010; 22: E235–E41.

[247] Niu ZQ, Zhou WY, Chen J, Feng GX, Li H, Ma WJ et al. Compact-designed supercapacitors using free-standing single-walled carbon nanotube films. Energy & Environmental Science. 2011; 4: 1440–6.

[248] Ma WJ, Song L, Yang R, Zhang TH, Zhao YC, Sun LF et al. Directly synthesized strong, highly conducting, transparent single-walled carbon nanotube films. Nano Letters. 2007; 7: 2307–11.

[249] Shah R, Zhang XF, Talapatra S. Electrochemical double layer capacitor electrodes using aligned carbon nanotubes grown directly on metals. Nanotechnology. 2009; 20.

[250] Hata K, Futaba DN, Mizuno K, Namai T, Yumura M, Iijima S. Water-assisted highly efficient synthesis of impurity-free single-walled carbon nanotubes. Science. 2004; 306: 1362–4.

[251] Futaba DN, Hata K, Yamada T, Hiraoka T, Hayamizu Y, Kakudate Y et al. Shape-engineerable and highly densely packed single-walled carbon nanotubes and their application as super-capacitor electrodes. Nature Materials. 2006; 5: 987–94.

[252] Kimizuka O, Tanaike O, Yamashita J, Hiraoka T, Futaba DN, Hata K et al. Electrochemical doping of pure single-walled carbon nanotubes used as supercapacitor electrodes. Carbon. 2008; 46: 1999–2001.

[253] Toth S, Fule M, Veres M, Selman JR, Arcon D, Pocsik I et al. Influence of amorphous carbon nano-clusters on the capacity of carbon black electrodes. Thin Solid Films. 2005; 482: 207–10.

[254] Kaempgen M, Chan CK, Ma J, Cui Y, Gruner G. Printable thin film supercapacitors using single-walled carbon nanotubes. Nano Letters. 2009; 9: 1872–6.

[255] Du CS, Pan N. High power density supercapacitor electrodes of carbon nanotube films by electrophoretic deposition. Nanotechnology. 2006; 17: 5314–8.

[256] Zhao X, Chu BTT, Ballesteros B, Wang WL, Johnston C, Sykes JM et al. Spray deposition of steam treated and functionalized single-walled and multi-walled carbon nanotube films for supercapacitors. Nanotechnology. 2009; 20.

[257] Yu CJ, Masarapu C, Rong JP, Wei BQ, Jiang HQ. Stretchable supercapacitors based on buckled single-walled carbon nanotube macrofilms. Advanced Materials. 2009; 21: 4793–7.

[258] Yu JL, Lu WB, Pei SP, Gong K, Wang LY, Meng LH et al. Omnidirectionally stretchable high-performance supercapacitor based on isotropic buckled carbon nanotube films. ACS Nano. 2016; 10: 5204–11.

[259] Li YL, Kinloch IA, Windle AH. Direct spinning of carbon nanotube fibers from chemical vapor deposition synthesis. Science. 2004; 304: 276–8.

[260] Niu ZQ, Dong HB, Zhu BW, Li JZ, Hng HH, Zhou WY et al. Highly stretchable, integrated supercapacitors based on single-walled carbon nanotube films with continuous reticulate architecture. Advanced Materials. 2013; 25: 1058–64.

[261] Wang J, Xu YL, Chen X, Sun XF. Capacitance properties of single wall carbon nanotube/polypyrrole composite films. Composites Science and Technology. 2007; 67: 2981–5.

[262] Hughes M, Chen GZ, Shaffer MSP, Fray DJ, Windle AH. Electrochemical capacitance of a nanoporous composite of carbon nanotubes and polypyrrole. Chemistry of Materials. 2002; 14: 1610–3.

[263] Chen GZ, Shaffer MSP, Coleby D, Dixon G, Zhou WZ, Fray DJ et al. Carbon nanotube and polypyrrole composites: Coating and doping. Advanced Materials. 2000; 12: 522–6.

[264] An KH, Jeon KK, Heo JK, Lim SC, Bae DJ, Lee YH. High-capacitance supercapacitor using a nanocomposite electrode of single-walled carbon nanotube and polypyrrole. Journal of the Electrochemical Society. 2002; 149: A1058–A62.

[265] Mi HY, Zhang XG, Xu YL, Xiao F. Synthesis, characterization and electrochemical behavior of polypyrrole/carbon nanotube composites using organometallic-functionalized carbon nanotubes. Applied Surface Science. 2010; 256: 2284–8.

[266] Guo P, Wang XS, Guo HC. TiO_2/Na-HZSM-5 nano-composite photocatalyst: Reversible adsorption by acid sites promotes photocatalytic decomposition of methyl orange. Applied Catalysis B-Environmental. 2009; 90: 677–87.

[267] Huang HP, Feng XM, Zhu JJ. Synthesis, characterization and application in electrocatalysis of polyaniline/Au composite nanotubes. Nanotechnology. 2008; 19.

[268] Zhang J, Kong LB, Wang B, Luo YC, Kang L. *In-situ* electrochemical polymerization of multi-walled carbon nanotube/polyaniline composite films for electrochemical supercapacitors. Synthetic Metals. 2009; 159: 260–6.

[269] Gupta V, Miura N. Polyaniline/single-wall carbon nanotube (PANI/SWCNT) composites for high performance supercapacitors. Electrochimica Acta. 2006; 52: 1721–6.

[270] Meng CZ, Liu CH, Chen LZ, Hu CH, Fan SS. Highly flexible and all-solid-state paper like polymer supercapacitors. Nano Letters. 2010; 10: 4025–31.

[271] Yin YL, Liu CH, Fan SS. Well-constructed CNT mesh/PANI nanoporous electrode and its thickness effect on the supercapacitor properties. Journal of Physical Chemistry C. 2012; 116: 26185–9.

[272] Wang J, Xu YL, Sun XF, Li XF, Du XF. Electrochemical capacitance of the composite of poly (3,4-ethylenedioxythiophene) and functionalized single-walled carbon nanotubes. Journal of Solid State Electrochemistry. 2008; 12: 947–52.

[273] Ghaffari M, Kosolwattana S, Zhou Y, Lachman N, Lin M, Bhattacharya D et al. Hybrid supercapacitor materials from poly(3,4-ethylenedioxythiophene) conformally coated aligned carbon nanotubes. Electrochimica Acta. 2013; 112: 522–8.

[274] Wardle BL, Saito DS, Garcia EJ, Hart AJ, de Villoria RG, Verploegen EA. Fabrication and characterization of ultrahigh-volume-fraction aligned carbon nanotube-polymer composites. Advanced Materials. 2008; 20: 2707–14.

[275] Vaddiraju S, Cebeci H, Gleason KK, Wardle BL. Hierarchical multifunctional composites by conformally coating aligned carbon nanotube arrays with conducting polymer. ACS Applied Materials & Interfaces. 2009; 1: 2565–72.

[276] Lo AY, Jheng Y, Huang TC, Tseng CM. Study on RuO_2/CMK-3/CNTs composites for high power and high energy density supercapacitor. Applied Energy. 2015; 153: 15–21.

[277] Park JH, Ko JM, Park OO. Carbon nanotube/RuO_2 nanocomposite electrodes for supercapacitors. Journal of the Electrochemical Society. 2003; 150: A864–A7.

158 *Nanomaterials for Supercapacitors*

[278] Liu XR, Huber TA, Kopac MC, Pickup PG. Ru oxide/carbon nanotube composites for supercapacitors prepared by spontaneous reduction of Ru(VI) and Ru(VII). Electrochimica Acta. 2009; 54: 7141–7.

[279] Wang PF, Xu YX, Liu H, Chen YF, Yang J, Tan QQ. Carbon/carbon nanotube-supported RuO_2 nanoparticles with a hollow interior as excellent electrode materials for supercapacitors. Nano Energy. 2015; 15: 116–24.

[280] Liu H, Qu JL, Chen YF, Li JQ, Ye F, Lee JY et al. Hollow and cage-bell structured nanomaterials of noble metals. Journal of the American Chemical Society. 2012; 134: 11602–10.

[281] Chen Z, Augustyn V, Wen J, Zhang YW, Shen MQ, Dunn B et al. High-performance supercapacitors based on intertwined CNT/V_2O_5 nanowire nanocomposites. Advanced Materials. 2011; 23: 791–5.

[282] Chen Z, Qin YC, Weng D, Xiao QF, Peng YT, Wang XL et al. Design and synthesis of hierarchical nanowire composites for electrochemical energy storage. Advanced Functional Materials. 2009; 19: 3420–6.

[283] Fang WC. Synthesis and electrochemical characterization of vanadium oxide/carbon nanotube composites for supercapacitors. Journal of Physical Chemistry C. 2008; 112: 11552–5.

[284] Weng CH, Leou KC, Wei HW, Juang ZY, Wei MT, Tung CH et al. Structural transformation and field emission enhancement of carbon nanofibers by energetic argon plasma post-treatment. Applied Physics Letters. 2004; 85: 4732–4.

[285] Leu MS, Chen BF, Chen SY, Lee YW, Lih WC. Properties of (Ti,Al)N coatings deposited by the magnetic filter cathodic arc. Surface & Coatings Technology. 2000; 133: 319–24.

[286] Kang YJ, Kim BW, Chung HG, Kim W. Fabrication and characterization of flexible and high capacitance supercapacitors based on $MnO_2/CNT/papers$. Synthetic Metals. 2010; 160: 2510–4.

[287] Wang GX, Zhang BL, Yu ZL, Qu MZ. Manganese oxide/MWNTs composite electrodes for supercapacitors. Solid State Ionics. 2005; 176: 1169–74.

[288] Honda K, Yoshimura M, Kawakita K, Fujishima A, Sakamoto Y, Yasui K et al. Electrochemical characterization of carbon nanotube/nanohoneycomb diamond composite electrodes for a hybrid anode of Li-ion battery and super capacitor. Journal of the Electrochemical Society. 2004; 151: A532–A41.

[289] Reddy ALM, Shaijumon MM, Gowda SR, Ajayan PM. Multisegmented Au-MnO_2/carbon nanotube hybrid coaxial arrays for high-power supercapacitor applications. Journal of Physical Chemistry C. 2010; 114: 658–63.

[290] Subramanian V, Zhu HW, Wei BQ. Synthesis and electrochemical characterizations of amorphous manganese oxide and single walled carbon nanotube composites as supercapacitor electrode materials. Electrochemistry Communications. 2006; 8: 827–32.

[291] Zhao X, Johnston C, Grant PS. A novel hybrid supercapacitor with a carbon nanotube cathode and an iron oxide/carbon nanotube composite anode. Journal of Materials Chemistry. 2009; 19: 8755–60.

[292] Lee SW, Kim BS, Chen S, Shao-Horn Y, Hammond PT. Layer-by-layer assembly of all carbon nanotube ultrathin films for electrochemical applications. Journal of the American Chemical Society. 2009; 131: 671–9.

[293] Cheng XP, Gui XC, Lin ZQ, Zheng YJ, Liu M, Zhan RZ et al. Three-dimensional α-Fe_2O_3/carbon nanotube sponges as flexible supercapacitor electrodes. Journal of Materials Chemistry A. 2015; 3: 20927–34.

[294] Gui XC, Wei JQ, Wang KL, Cao AY, Zhu HW, Jia Y et al. Carbon nanotube sponges. Advanced Materials. 2010; 22: 617–21.

[295] Wang ZY, Luan DY, Madhavi S, Hu Y, Lou XW. Assembling carbon-coated α-Fe_2O_3 hollow nanohorns on the CNT backbone for superior lithium storage capability. Energy & Environmental Science. 2012; 5: 5252–6.

[296] Nam KW, Kim KH, Lee ES, Yoon WS, Yang XQ, Ki KB. Pseudocapacitive properties of electrochemically prepared nickel oxides on 3-dimensional carbon nanotube film substrates. Journal of Power Sources. 2008; 182: 642–52.

[297] Lin P, She QJ, Hong BL, Liu XJ, Shi YN, Shi Z et al. The nickel oxide/CNT composites with high capacitance for supercapacitor. Journal of the Electrochemical Society. 2010; 157: A818–A23.

[298] Zhang YP, Sun XW, Pan LK, Li HB, Sun Z, Sun CQ et al. Carbon nanotube-zinc oxide electrode and gel polymer electrolyte for electrochemical supercapacitors. Journal of Alloys and Compounds. 2009; 480: L17–L9.

[299] Zhang YD, Lin BP, Wang JC, Tian JH, Sun Y, Zhang XQ et al. All-solid-state asymmetric supercapacitors based on ZnO quantum dots/carbon/CNT and porous N-doped carbon/CNT electrodes derived from a single ZIF-8/CNT template. Journal of Materials Chemistry A. 2016; 4: 10282–93.

[300] Brownson DAC, Banks CE. Fabricating graphene supercapacitors: highlighting the impact of surfactants and moieties. Chemical Communications. 2012; 48: 1425–7.

[301] Chen W, Rakhi RB, Alshareef HN. Capacitance enhancement of polyaniline coated curved-graphene supercapacitors in a redox-active electrolyte. Nanoscale. 2013; 5: 4134–8.

[302] Le LT, Ervin MH, Qiu HW, Fuchs BE, Lee WY. Graphene supercapacitor electrodes fabricated by inkjet printing and thermal reduction of graphene oxide. Electrochemistry Communications. 2011; 13: 355–8.

[303] Li WK, Yang YJ. The reduction of graphene oxide by elemental copper and its application in the fabrication of graphene supercapacitor. Journal of Solid State Electrochemistry. 2014; 18: 1621–6.

[304] Polat EO, Kocabas C. Broadband optical modulators based on graphene supercapacitors. Nano Letters. 2013; 13: 5851–7.

[305] Skinner B, Fogler MM, Shklovskii BI. Model of large volumetric capacitance in graphene supercapacitors based on ion clustering. Physical Review B. 2011; 84.

[306] Xiong SX, Shi YJ, Chu J, Gong M, Wu BH, Wang XQ. Preparation of high-performance covalently bonded polyaniline nanorods/graphene supercapacitor electrode materials using interfacial copolymerization approach. Electrochimica Acta. 2014; 127: 139–45.

[307] Yoo JJ, Balakrishnan K, Huang JS, Meunier V, Sumpter BG, Srivastava A et al. Ultrathin planar graphene supercapacitors. Nano Letters. 2011; 11: 1423–7.

[308] Ku KH, Kim BW, Chung HG, Kim W. Characterization of graphene-based supercapacitors fabricated on Al foils using Au or Pd thin films as interlayers. Synthetic Metals. 2010; 160: 2613–7.

[309] Chen Y, Zhang X, Yu P, Ma YW. Electrophoretic deposition of graphene nanosheets on nickel foams for electrochemical capacitors. Journal of Power Sources. 2010; 195: 3031–5.

[310] Liu CG, Yu ZN, Neff D, Zhamu A, Jang BZ. Graphene-based supercapacitor with an ultrahigh energy density. Nano Letters. 2010; 10: 4863–8.

[311] Zhang JZ, Yang WR, Liu JQ. Facile fabrication of supercapacitors with high rate capability using graphene/nickel foam electrode. Electrochimica Acta. 2016; 209: 85–94.

[312] Stoller MD, Park SJ, Zhu YW, An JH, Ruoff RS. Graphene-based ultracapacitors. Nano Letters. 2008; 8: 3498–502.

[313] Lv W, Tang DM, He YB, You CH, Shi ZQ, Chen XC et al. Low-temperature exfoliated graphenes: Vacuum-promoted exfoliation and electrochemical energy storage. ACS Nano. 2009; 3: 3730–6.

[314] Singh C, Mishra AK, Paul A. Highly conducting reduced graphene synthesis via low temperature chemically assisted exfoliation and energy storage application. Journal of Materials Chemistry A. 2015; 3: 18557–63.

[315] Wang Y, Shi ZQ, Huang Y, Ma YF, Wang CY, Chen MM et al. Supercapacitor devices based on graphene materials. Journal of Physical Chemistry C. 2009; 113: 13103–7.

[316] Lin ZY, Liu Y, Yao YG, Hildreth OJ, Li Z, Moon K et al. Superior capacitance of functionalized graphene. Journal of Physical Chemistry C. 2011; 115: 7120–5.

[317] Jia Z, Li CY, Liu DQ, Jiang LX. Direct hydrothermal reduction of graphene oxide based papers obtained from tape casting for supercapacitor applications. RSC Advances. 2015; 5: 81030–7.

[318] Liao YQ, Huang YL, Shu D, Zhong YY, Hao JN, He C et al. Three-dimensional nitrogen-doped graphene hydrogels prepared via hydrothermal synthesis as high-performance supercapacitor materials. Electrochimica Acta. 2016; 194: 136–42.

[319] Sari FNI, Ting JM. One step microwaved-assisted hydrothermal synthesis of nitrogen doped graphene for high performance of supercapacitor. Applied Surface Science. 2015; 355: 419–28.

[320] Du QL, Zheng MB, Zhang LF, Wang YW, Chen JH, Xue LP et al. Preparation of functionalized graphene sheets by a low-temperature thermal exfoliation approach and their electrochemical supercapacitive behaviors. Electrochimica Acta. 2010; 55: 3897–903.

[321] Campbell PG, Merrill MD, Wood BC, Montalvo E, Worsley MA, Baumann TF et al. Battery/supercapacitor hybrid via non-covalent functionalization of graphene macro-assemblies. Journal of Materials Chemistry A. 2014; 2: 17764–70.

[322] Jana M, Saha S, Khanra P, Samanta P, Koo H, Murmu NC et al. Non-covalent functionalization of reduced graphene oxide using sulfanilic acid azocromotrop and its application as a supercapacitor electrode material. Journal of Materials Chemistry A. 2015; 3: 7323–31.

[323] Li YM, van Zijll M, Chiang S, Pan N. KOH modified graphene nanosheets for supercapacitor electrodes. Journal of Power Sources. 2011; 196: 6003–6.

[324] Zheng B, Chen TW, Xiao FN, Bao WJ, Xia XH. KOH-activated nitrogen-doped graphene by means of thermal annealing for supercapacitor. Journal of Solid State Electrochemistry. 2013; 17: 1809–14.

[325] Wang DW, Li F, Zhao JP, Ren WC, Chen ZG, Tan J et al. Fabrication of graphene/polyaniline composite paper via *in situ* anodic electropolymerization for high-performance flexible electrode. ACS Nano. 2009; 3: 1745–52.

[326] Gomez H, Ram MK, Alvi F, Villalba P, Stefanakos E, Kumar A. Graphene-conducting polymer nanocomposite as novel electrode for supercapacitors. Journal of Power Sources. 2011; 196: 4102–8.

[327] Wu Q, Xu YX, Yao ZY, Liu AR, Shi GQ. Supercapacitors based on flexible graphene/polyaniline nanofiber composite films. ACS Nano. 2010; 4: 1963–70.

[328] Hao QL, Wang HL, Yang XJ, Lu LD, Wang X. Morphology-controlled fabrication of sulfonated graphene/polyaniline nanocomposites by liquid/liquid interfacial polymerization and investigation of their electrochemical properties. Nano Research. 2011; 4: 323–33.

[329] Yan J, Wei T, Shao B, Fan ZJ, Qian WZ, Zhang ML et al. Preparation of a graphene nanosheet/polyaniline composite with high specific capacitance. Carbon. 2010; 48: 487–93.

[330] Sun Y, Wilson SR, Schuster DI. High dissolution and strong light emission of carbon nanotubes in aromatic amine solvents. Journal of the American Chemical Society. 2001; 123: 5348–9.

[331] Mini PA, Balakrishnan A, Nair SV, Subramanian KRV. Highly super capacitive electrodes made of graphene/poly(pyrrole). Chemical Communications. 2011; 47: 5753–5.

[332] Bose S, Kim NH, Kuila T, Lau KT, Lee JH. Electrochemical performance of a graphene-polypyrrole nanocomposite as a supercapacitor electrode. Nanotechnology. 2011; 22.

[333] Zhang DC, Zhang X, Chen Y, Yu P, Wang CH, Ma YW. Enhanced capacitance and rate capability of graphene/polypyrrole composite as electrode material for supercapacitors. Journal of Power Sources. 2011; 196: 5990–6.

160   *Nanomaterials for Supercapacitors*

[334] Liu AR, Li C, Bai H, Shi GQ. Electrochemical deposition of polypyrrole/sulfonated graphene composite films. Journal of Physical Chemistry C. 2010; 114: 22783–9.

[335] Biswas S, Drzal LT. Multilayered nanoarchitecture of graphene nanosheets and polypyrrole nanowires for high performance supercapacitor electrodes. Chemistry of Materials. 2010; 22: 5667–71.

[336] He YB, Bai YL, Yang XF, Zhang JY, Kang LP, Xu H et al. Holey graphene/polypyrrole nanoparticle hybrid aerogels with three-dimensional hierarchical porous structure for high performance supercapacitor. Journal of Power Sources. 2016; 317: 10–8.

[337] Wu ZS, Wang DW, Ren W, Zhao J, Zhou G, Li F et al. Anchoring hydrous RuO_2 on graphene sheets for high-performance electrochemical capacitors. Advanced Functional Materials. 2010; 20: 3595–602.

[338] Chen CY, Fan CY, Lee MT, Chang JK. Tightly connected MnO_2-graphene with tunable energy density and power density for supercapacitor applications. Journal of Materials Chemistry. 2012; 22: 7697–700.

[339] Deng SX, Sun D, Wu CH, Wang H, Liu JB, Sun YX et al. Synthesis and electrochemical properties of MnO_2 nanorods/graphene composites for supercapacitor applications. Electrochimica Acta. 2013; 111: 707–12.

[340] Dong XC, Wang XW, Wang L, Song H, Li XG, Wang LH et al. Synthesis of a MnO_2-graphene foam hybrid with controlled MnO_2 particle shape and its use as a supercapacitor electrode. Carbon. 2012; 50: 4865–70.

[341] Gao HC, Xiao F, Ching CB, Duan HW. High-performance asymmetric supercapacitor based on graphene hydrogel and nanostructured MnO_2. ACS Applied Materials & Interfaces. 2012; 4: 2801–10.

[342] Ge J, Yao HB, Hu W, Yu XF, Yan YX, Mao LB et al. Facile dip coating processed graphene/MnO_2 nanostructured sponges as high performance supercapacitor electrodes. Nano Energy. 2013; 2: 505–13.

[343] Ghasemi S, Hosseinzadeh R, Jafari M. MnO_2 nanoparticles decorated on electrophoretically deposited graphene nanosheets for high performance supercapacitor. International Journal of Hydrogen Energy. 2015; 40: 1037–46.

[344] Gui DY, Chen W, Liu CL, Liu JH. Graphene-like membrane supported MnO_2 nanospheres for supercapacitor. Journal of Materials Science-Materials in Electronics. 2016; 27: 5121–7.

[345] Hao JN, Zhong YY, Liao YQ, Shu D, Kang ZX, Zou XP et al. Face-to-face self-assembly graphene/MnO_2 nanocomposites for supercapacitor applications using electrochemically exfoliated graphene. Electrochimica Acta. 2015; 167: 412–20.

[346] He YM, Chen WJ, Li XD, Zhang ZX, Fu JC, Zhao CH et al. Freestanding three-dimensional graphene/MnO_2 composite networks as ultra light and flexible supercapacitor electrodes. ACS Nano. 2013; 7: 174–82.

[347] Li ZM, An YF, Hu ZG, An N, Zhang YD, Guo BS et al. Preparation of a two-dimensional flexible MnO_2/graphene thin film and its application in a supercapacitor. Journal of Materials Chemistry A. 2016; 4: 10618–26.

[348] Liu CL, Gui DY, Liu JH. Process dependent graphene-wrapped plate-like MnO_2 nanospheres for high performance supercapacitor. Chemical Physics Letters. 2014; 614: 123–8.

[349] Liu YC, Miao XF, Fang JH, Zhang XX, Chen SJ, Li W et al. Layered-MnO_2 nanosheet grown on nitrogen-doped graphene template as a composite cathode for flexible solid-state asymmetric supercapacitor. ACS Applied Materials & Interfaces. 2016; 8: 5251–60.

[350] Mao L, Zhang K, Chan HSO, Wu JS. Nanostructured MnO_2/graphene composites for supercapacitor electrodes: the effect of morphology, crystallinity and composition. Journal of Materials Chemistry. 2012; 22: 1845–51.

[351] Naderi HR, Norouzi P, Ganjali MR. Electrochemical study of a novel high performance supercapacitor based on MnO_2/nitrogen-doped graphene nanocomposite. Applied Surface Science. 2016; 366: 552–60.

[352] Patil UM, Sohn JS, Kulkarni SB, Park HG, Jung Y, Gurav KV et al. A facile synthesis of hierarchical alpha-MnO_2 nanofibers on 3D-graphene foam for supercapacitor application. Materials Letters. 2014; 119: 135–9.

[353] Sun SM, Wang PY, Wang S, Wu Q, Fang SM. Fabrication of MnO_2/nanoporous 3D graphene for supercapacitor electrodes. Materials Letters. 2015; 145: 141–4.

[354] Xie XY, Zhang C, Wu MB, Tao Y, Lv W, Yang QH. Porous MnO_2 for use in a high performance supercapacitor: replication of a 3D graphene network as a reactive template. Chemical Communications. 2013; 49: 11092–4.

[355] Xiong CY, Li TH, Khan M, Li H, Zhao TK. A three-dimensional MnO_2/graphene hybrid as a binder-free supercapacitor electrode. Rsc Advances. 2015; 5: 85613–9.

[356] Yu GH, Hu LB, Liu NA, Wang HL, Vosgueritchian M, Yang Y et al. Enhancing the supercapacitor performance of graphene/MnO_2 nanostructured electrodes by conductive wrapping. Nano Letters. 2011; 11: 4438–42.

[357] Zhang JY, Yang XF, He YB, Bai YL, Kang LP, Xu H et al. δ-MnO_2/holey graphene hybrid fiber for all-solid-state supercapacitor. Journal of Materials Chemistry A. 2016; 4: 9088–96.

[358] Zhang Y, Yao QQ, Gao HL, Wang LX, Wang LZ, Zhang AQ et al. Synthesis and electrochemical properties of hollow-porous MnO_2-graphene micro-nano spheres for supercapacitor applications. Powder Technology. 2014; 267: 268–72.

[359] Zhao Y, Meng YN, Wu HP, Wang Y, Wei ZX, Li XJ et al. *In situ* anchoring uniform MnO_2 nanosheets on three-dimensional macroporous graphene thin-films for supercapacitor electrodes. RSC Advances. 2015; 5: 90307–12.

[360] Zheng YH, Pann WX, Zhengn DY, Sun CX. Fabrication of functionalized graphene-based MnO_2 nanoflower through electrodeposition for high-performance supercapacitor electrodes. Journal of the Electrochemical Society. 2016; 163: D230–D8.

[361] Chen G, Lu QF, Zhao HB. SnO_2-decorated graphene/polyaniline nanocomposite for a high-performance supercapacitor electrode. Journal of Materials Science & Technology. 2015; 31: 1101–7.

[362] Jin YH, Jia MQ. Design and synthesis of nanostructured graphene-SnO_2-polyaniline ternary composite and their excellent supercapacitor performance. Colloids and Surfaces A-Physicochemical and Engineering Aspects. 2015; 464: 17–25.

[363] Lim SP, Huang NM, Lim HN. Solvothermal synthesis of SnO_2/graphene nanocomposites for supercapacitor application. Ceramics International. 2013; 39: 6647–55.

[364] Wang WJ, Hao QL, Lei W, Xia XF, Wang X. Graphene/SnO_2/polypyrrole ternary nanocomposites as supercapacitor electrode materials. RSC Advances. 2012; 2: 10268–74.

[365] Yang Q, Pang SK, Yung KC. Electrochemically reduced graphene oxide/carbon nanotubes composites as binder-free supercapacitor electrodes. Journal of Power Sources. 2016; 311: 144–52.

[366] Zhang YY, Zhen Z, Zhang ZL, Lao JC, Wei JQ, Wang KL et al. *In-situ* synthesis of carbon nanotube/graphene composite sponge and its application as compressible supercapacitor electrode. Electrochimica Acta. 2015; 157: 134–41.

4

Oxide Based Supercapacitors
I-Manganese Oxides

Ling Bing Kong,[1], Wenxiu Que,[2] Lang Liu,[3] Freddy Yin Chiang Boey,[1,a] Zhichuan J. Xu,[1,b] Kun Zhou,[4] Sean Li,[5] Tianshu Zhang[6] and Chuanhu Wang[7]*

Overview

Manganese oxide is among those that have pseudocapacitive (Faradic) behavior in aqueous solutions, while charge storage mechanism of manganese oxide electrodes has been well established [1, 2]. Pseudocapacitive reactions have been observed both on the surface and in bulk of the electrodes. Surface Faradaic reaction is due to the surface adsorption of electrolyte cations (C^+ = H^+, Li^+, Na^+ and K^+) onto surface of the manganese oxide, given by [3, 4]:

$$(MnO_2)_{surface} + C^+ + e^- \leftrightarrow (MnOOC)_{surface}. \qquad (4.1)$$

---

[1] School of Materials Science and Engineering, Nanyang Technological University, 50 Nanyang Avenue, Singapore 639798.
[a] Email: mycboey@ntu.edu.sg
[b] Email: xuzc@ntu.edu.sg
[2] Electronic Materials Research Laboratory and International Center for Dielectric Research, Xi'an Jiaotong University, Xi'an 710049, Shaanxi Province, People's Republic of China.
 Email: wxque@xjtu.edu.cn
[3] School of Chemistry and Chemical Engineering, Xinjiang University, Urumqi 830046, Xinjiang, People's Republic of China.
 Email: llyhs1973@sina.com
[4] School of Mechanical and Aerospace Engineering, Nanyang Technological University, 50 Nanyang Avenue, Singapore 639798.
 Email: kzhou@ntu.edu.sg
[5] School of Materials Science and Engineering, The University of New South Wales, Australia.
 Email: sean.li@unsw.edu.au
[6] Anhui Target Advanced Ceramics Technology Co. Ltd., Hefei, Anhui, People's Republic of China.
 Email: 13335516617@163.com
[7] Department of Material and Chemical Engineering, Bengbu Univresity, Bengbu, Peoples's Republic of China.
 Email: bbxywch@126.com
* Corresponding author: elbkong@ntu.edu.sg

Bulk Faradaic reaction is related to intercalation or deintercalation of the electrolyte cations inside the manganese oxide, expressed as [3, 4]:

$$MnO_2 + C^+ + e^- \leftrightarrow MnOOC. \tag{4.2}$$

Both charge storage mechanisms involve the redox reaction between III and IV oxidation states of Mn. As compared with the RuO_2, hydrated manganese oxides have lower specific capacitances, usually in the range of 100–200 $F \cdot g^{-1}$ in alkali salt solutions. Moreover, the currently available MnO_2-based supercapacitors have encountered serious limitations, such as intrinsic problem of MnO_2, low specific capacitance, low energy density, structural instability, weak long-term cyclability, and low rate-capacity. For practical applications, the electrode materials should have high reversible capacitance, structural flexibility, long-time stability, fast cation diffusion at high charge–discharge rates, cost-effectiveness and environmental friendliness [5].

Mn is a multivalent transition metal element, with several stable oxides, including MnO, Mn_3O_4, Mn_2O_3 and MnO_2 [6], as well as various types of crystal structures, such as bixbyite, psilomelane, pyrolusite, ramsdellite, nsutite and phyllomanganate [7–9]. Due to the diverse crystal structures, defect chemistry, morphology, porosity and textures, manganese oxides possess rich electrochemical properties, which can be considered when they are used as electrodes of supercapacitors [6, 10–14].

One of the most serious problems of bulk MnO_2 electrodes is the low intrinsic electrical conductivity of MnO_2. By using an ultrathin film of MnO_2 for electrochemical capacitors, although the low conductivity problem can be addressed, the total capacity of the device is largely limited due to the low mass loading. Therefore, in order to increase the electrical conductivity of MnO_2, various elements have been studied as dopants. One way is to mix MnO_2 electrodes with other transition metal oxides, such as NiO, CuO, Fe_2O_3, V_2O_5, CoO, MoO_3 and RuO_2 [15–18]. The other strategy is to modify MnO_2 electrodes by chemically doping with small amounts of other elements, such as Al, Sn and Pb [19, 20]. It has been demonstrated that defect chemistry can be used to increase the electrical conductivity and thus electrochemical performance of MnO_2 electrodes.

An additional effective approach is to incorporate conductive layers, such as carbon nanofoams, templated mesoporous carbon, graphene nanosheets and nanotube assemblies, with richly porous structure and high specific surface area, so that MnO_2 electrodes could be deposited with high mass loadings. The basic idea is that the carbon substrates serve as a highly conductive network, while the interconnected pores provide a continuous pathway for electrolyte diffusion. Therefore, the MnO_2 electrodes would have shortened solid-state transport paths for ion transportation and reaction [21]. Various methods have been used to deposit nanostructured MnO_2 onto carbon structures. For example, MnO_2 nanoparticles can be deposited on carbon nanotubes (CNTs) by using a variety of ways, such as physical mixing of the components [22], thermal decomposition [23] and chemical deposition [24, 25] and electrochemical deposition [26, 27].

Besides electrical conductivity, structural stability, flexibility and electrochemical dissolution of active materials have also been issues of MnO_2 electrodes, because they readily resulted in poor long term cyclability. It has been found that the problem of electrochemical dissolution can be avoided by using coatings. For instance, a self-limited growth process based on electro-polymerization of o-phenylenediamine was proposed to develop such coatings [28, 29]. The polymer layer confined the oxide nanoscale network, which isolated the MnO_2 nanoparticles from the electrolyte, thus avoiding chemical dissolution. However, although the underlying MnO_2 was still electrochemically accessible, there was a significant reduction in conductivity of the whole electrodes [29]. In this regard, conductive polymers, polyaniline, polypyrrole and polythiophene and their derivatives, have been attempted to address this problem [30–33].

The strategies used to develop MnO_2 based electrodes can be classified into three categories [34–36]: (i) chemical and structural modification of manganese oxide materials to create more electrochemically active sites for the redox reaction between Mn^{3+} and Mn^{4+}, (ii) shortening the transport path for both electrons and cations by using porous carbon architectures, with high surface area and high electrical conductivity and (iii) increasing structural stability and flexibility, together with decreasing electrochemical dissolution of active materials by using conductive polymer coatings on the active materials.

164 *Nanomaterials for Supercapacitors*

Manganese Oxide Based Electrodes

Electrode fabrication and characterization

Powder electrodes of manganese oxide can be either crystalline or amorphous. Electrochemical properties of amorphous hydrated manganese dioxide powders were first studied in 1999, which were derived from the reaction between $KMnO_4$ and $Mn(CH_3COO)_2$ in aqueous solution [1, 2]:

$$KMnO_4 + 3/2Mn\,(CH_3COO)_2 \rightarrow 5/2MnO_2. \tag{4.3}$$

The amorphous hydrated MnO_2 powder electrodes demonstrated ideal capacitive behavior in aqueous solutions of KCl, NaCl and LiCl. Subsequently, hydrated MnO_2 powders were obtained by reducing $KMnO_4$ with various reducing agents, such as $MnSO_4$ [4, 12], potassium borohydride [37], sodium dithionite [37], sodium hypophosphite and hydrochloric acid [37], aniline [38] and ethylene glycol [39]. Besides aqueous solution processing, the reduction of $KMnO_4$ has also been conducted in organic solvents, such as AOT/iso-octane [40], interface of H_2O/CCl_4 and ferrocene/chloroform [13, 41].

Specifically, surfactant sodium bis(2-ethylhexyl) sulfosuccinate (AOT) act as both dispersant and reduce agent. $KMnO_4$ aqueous solution that is dispersed in iso-octane with the presence of AOT to form nano-droplets in the water phase, in which then $KMnO_4$ was reduced by AOT [40]. The product was denoted as s-MnO_2. Chemical co-precipitation method was also used to prepare MnO_2. A 0.1 $mol\cdot L^{-1}$ $KMnO_4$ solution was mixed with 0.15 $mol\cdot L^{-1}$ $Mn(CH_3COO)_2$, with a molar ratio of $KMnO_4$:$Mn(CH_3COO)_2$ = 2:3. A dark brown precipitate was immediately obtained due to the reaction of Eq. (4.3). The sample was denoted as co-MnO_2.

Two steps might be involved in the formation of MnO_2 precipitation. Firstly, $KMnO_4$ aqueous solution was dispersed in iso-octane (oil phase) with the presence of surfactant AOT to form nano-droplets in the water phase. Due to its strong oxidation capability, $KMnO_4$ oxidized the sulfonates, while it was reduced subsequently by AOT to produce MnO_2 precipitation. At the same time, the growth of MnO_2 particles was restricted by the extra AOT, thus leading to MnO_2 nanoparticles.

Figure 4.1 shows SEM and TEM images of the s-MnO_2 and co-MnO_2 powders [40]. The co-MnO_2 had a rod-like morphology, with a diameter of about 12 nm and length of about 100 nm, as shown in Fig. 4.1(c, d). The s-MnO_2 powder consisted of agglomerates of small spherical particles with an individual particle size of about 4 nm, as seen in Fig. 4.1(a, b), much smaller than the dimension of co-MnO_2. Figure 4.2 shows XRD patterns of the two samples. They both had only several broad peaks, suggesting their amorphous state. Broad peaks at about $2\theta = 37.0°$ and $65.3°$ implied that the samples were partially crystalline, corresponding to the peaks of α-MnO_2.

SBET values of the s-MnO_2 and co-MnO_2 samples were 145.7 and 217.3 $m^2\cdot g^{-1}$, respectively. Figure 4.3 shows pore size distribution profiles of the s-MnO_2 and co-MnO_2 samples. According to the IUPAC nomenclature, micropores are < 2 nm in diameter, mesopores are in the range of 2–50 nm, and macropores are > 50 nm. As seen in Fig. 4.3, the s-MnO_2 exhibited a broad size distribution, over 2–100 nm, with a mean width of about 10 nm in the mesopore and macropore regimes. According to TEM images, the primary particles of s-MnO_2 had an average diameter of 4 nm and there were no pores inside the primary particles. As a result, it could be concluded that both the mesopores and macropores of the s-MnO_2 sample were formed due to the agglomeration of the primary and secondary particles. Its specific surface area was mainly attributed to the surface of its primary particles. The co-MnO_2 powder possessed two narrow peaks with mean pore widths of 1.3 and 4 nm. Because of its large particle size, the co-MnO_2 powder exhibited a porous structure with the presence of many small pores. Therefore, the high specific surface area of co-MnO_2 was mainly contributed by the inside surface of the pores in the particles.

The other two processes involved Interfacial Reactions taking place at the aqueous/organic interfaces have been used to prepare MnO_2. One example is the formation of MnO_2 at the interface of aqueous $KMnO_4$/ferrocene in chloroform. By using sodium dodecylsulfate (SDS), micelles containing $KMnO_4$ were formed at the aqueous/organic interface [13]. Figure 4.4 shows a schematic diagram used to describe the formation of MnO_2. With the present of SDS, micelles were formed, with a function of colloid nanoreactors. As the

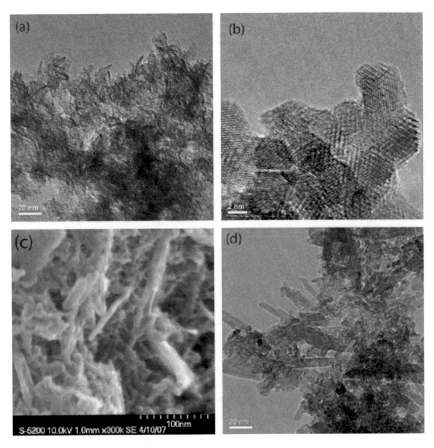

Fig. 4.1. SEM and TEM images of the s-MnO$_2$ (a, b) and co-MnO$_2$ (c and d) powders. Reproduced with permission from [40], Copyright © 2008, Elsevier.

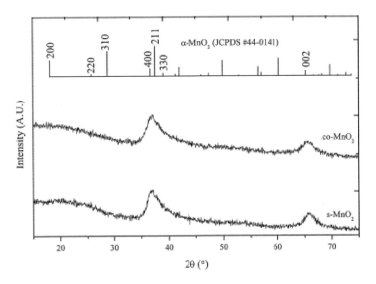

Fig. 4.2. XRD patterns of the s-MnO$_2$ and co-MnO$_2$ powders. Reproduced with permission from [40], Copyright © 2008, Elsevier.

166 *Nanomaterials for Supercapacitors*

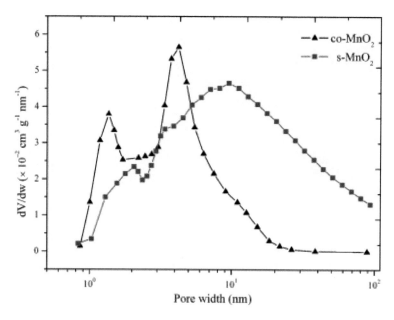

Fig. 4.3. Pore size distribution profiles of the co-MnO$_2$ and s-MnO$_2$ powders. Reproduced with permission from [40], Copyright © 2008, Elsevier.

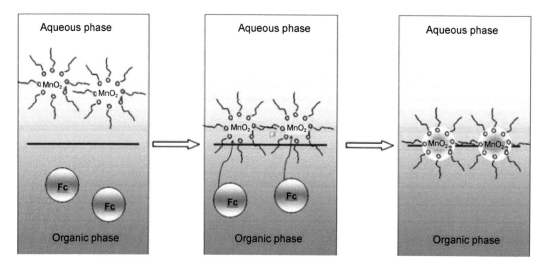

Fig. 4.4. Schematic graph of the formation of MnO$_2$ at interface of potassium permanganate in water/ferrocene in chloroform. Reproduced with permission from [13], Copyright © 2007, Elsevier.

ferrocene molecules were in contact with permanganate at the core of the micelles, redox reaction occurred at the interface, leading the formation of MnO$_2$ particles inside the micelles.

The concentration of KMnO$_4$ was 0.01 M, while the concentrations of SDS were 3, 8, 16, 40, 80, 120, 160 and 320 mM. 25 ml solution was added on surface 25 ml 0.03 M ferrocene/CHCl$_3$ solution, resulting in a static organic/aqueous interface. After a short while, dark brown precipitates were formed at the interface. As the precipitates became dark and dark, the color of the aqueous phase became gradually lighter. Finally, a dark brown precipitate layer was formed at the liquid/liquid interface. The organic phase was then taken away, in order to collect the precipitates through centrifugation. The precipitates were thoroughly washed with distilled water and ethanol. The samples were dried at 70°C in a vacuum for 10 hours, which were denoted as sample A, B, C, D, E, F, G and H, corresponding to the SDS contents used in the experiment. XRD results indicated that the samples of MnO_2 were amorphous.

Figure 4.5 shows SEM images of sample E after reaction for different times, i.e., 3.5, 4.5, 6.5 and 24 hours, respectively. Particle size of the MnO_2 powders was increased with increasing reaction time. Because the sample reacted for 4.5 hours possessed the highest specific capacitance of 260 $F \cdot g^{-1}$, 4.5 hours was used as the optimal reaction time. Morphology of the MnO_2 powders was also affected by the concentration of SDS. MnO_2 was aggregated into spherical particles. The particle size was decreased from 300 to 30 nm, as the concentration of SDS was increased from 3 to 320 mM. The samples prepared at low concentration of SDS showed more compact morphologies. BET surface area was decreased gradually from 264 $m^2 \cdot g^{-1}$ of sample A to 173 $m^2 \cdot g^{-1}$ of sample H. Samples A, E and H possessed an adsorption average pore diameter of 7.4, 13.1 and 16.7 nm, respectively. As a result, sample A had the largest surface area and it contained the largest particles. This implied that the surface area was mainly contributed by the pores inside the MnO_2 particles. Sample A had a narrow pore size distribution centered at about 5 nm, sample E displayed a pore size distribution centered at 5 and 15 nm, whereas the pore size of sample H was centered at 5 and 30 nm. The number of the micelles was increased while the micelle size was decreased, as the concentration of SDS was increased, which thus resulted in particles with smaller sizes. Meanwhile, the distance between the adjacent micelles was increased as the concentrations of SDS increased, due to the increase in charge repulsion. In other words, the particle size was determined by the micelle size, whereas the pore size was dependent on the distance between adjacent micelles.

Electrochemical properties of the hydrated MnO_2 powder electrodes are evaluated by using Cyclic Voltammetry (CV), galvanostatic charge–discharge and Electrochemical Impedance Spectroscopy (EIS).

Fig. 4.5. SEM images of MnO_2 prepared at different reaction times: (a) 3.5 hours; (b) 4.5 hours; (c) 6.5 hours; and (d) 24 hours. Reproduced with permission from [13], Copyright © 2007, Elsevier.

Specific Capacitance (C) with a unit of $F \cdot g^{-1}$ is determined with the voltammetric charge (Q) derived from the cyclic voltammograms or the galvanostatic charge–discharge curves. Q is divided by the mass of the active materials in the electrodes (m) and the width of the potential window ΔE, given by:

$$C = Q/(m\Delta E). \tag{4.4}$$

Figure 4.6 shows CV curves of sample E, at scan rates of 2, 5, to 10 $mV \cdot s^{-1}$, over voltage range of 0.0–0.8 V, vs. SCE in 0.5 M K_2SO_4 solution. The typical quasi-rectangular curves implied that the materials possessed promising EDLC behaviors. Furthermore, the quasi-rectangular symmetric CV curves could be readily retained over the voltage range of 0–0.8 V, as the scan rate was increased from 2 to 10 $mV \cdot s^{-1}$, demonstrating the excellent capacitive ability of the amorphous MnO_2 at high current densities.

In a separate study, nanostructured MnO_2 (nanoMnO_2) powder was synthesized by mixing $KMnO_4$ with ethylene glycol in a beaker just under ambient conditions [39]. Permanganate as a strong oxidizing agent reacted with ethylene glycol as a strong reducing agent, which was an exothermic redox reaction to form nanoMnO_2 and oxidized products of glycol, together with certain amount of aldehydes. After the reaction, the powder was washed thoroughly with distilled water and ethanol, followed by drying in air at 60°C overnight. The sample was heated at temperatures of 200–600°C.

The hydrated MnO_2 powders were mostly amorphous or had poor crystallinity, as shown in Fig. 4.7. The as-prepared hydrated MnO_2 powders was amorphous, as illustrated in Fig. 4.7(a) [38, 39]. The amorphous state of the hydrated MnO_2 powder retained until 300°C, as seen in Fig. 4.7(b and c), while α-MnO_2 was obtained after the sample was thermally treated at temperatures of > 400°C, as revealed in Fig. 4.7(d–f) [38]. Similar structural evolutions as a result of thermal annealing have been observed in the open literature. For example, decomposition of α-MnO_2 led to the formation of α-Mn_2O_3 (bixbyite-C, Ia3 space group) at 400°C, which was then well-crystallized as the major phase at 600°C [12]. Similarly, as-prepared α-MnO_2 was converted to α-Mn_2O_3 at temperatures of 500–800°C and then to Mn_3O_4 at 900°C [42].

SEM analysis results indicated that the nanoMnO_2 were nearly spherical nanoparticles, with diameters of 5–13 nm. The primary particles of the nanoMnO_2 with secondary pores due to the interparticle space were agglomerated in part to form secondary particles with an average size of about 30 nm, which assembled into 1D nanorods with lengths of 200–300 nm and diameter of about 40 nm, after heat treatment at 600°C.

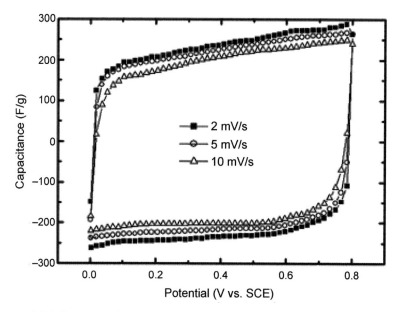

Fig. 4.6. CV curves of MnO_2 in 0.5 M K_2SO_4 electrolyte at different scan rates: (a) 2 $mV \cdot s^{-1}$, (b) 5 $mV \cdot s^{-1}$ and (c) 10 $mV \cdot s^{-1}$. Reproduced with permission from [13], Copyright © 2007, Elsevier.

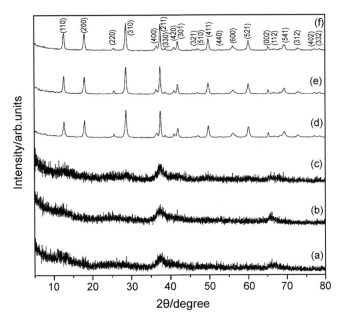

Fig. 4.7. XRD patterns of the as-prepared and thermally treated samples at different temperatures: (a) nanoMnO$_2$, (b) 200°C, (c) 300°C, (d) 400°C, (e) 500°C and (f) 600°C. Reproduced with permission from [39], Copyright © 2009, American Chemical Society.

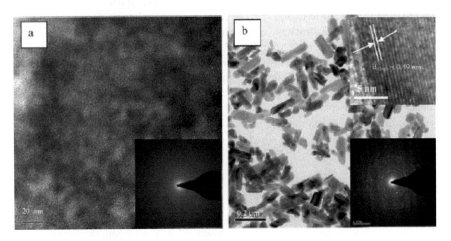

Fig. 4.8. TEM images of the nanoMnO (a) and 600°C-annealed sample (b). Corresponding SAED patterns are shown as insets. HRTEM image of nanorod MnO$_2$ is shown as the inset in (b), with lattice spacing of 0.49 nm corresponding to d$_{(200)}$. Reproduced with permission from [39], Copyright © 2009, American Chemical Society.

Representative TEM image of the nanoMnO$_2$ is shown in Fig. 4.8(a), revealing the particles did not have well-defined shapes. At the same time, SAED demonstrated a diffused pattern, suggesting the characteristic of an amorphous state. After thermal annealing, rod shape of high crystal quality was confirmed by the well-defined circles in the SAED pattern and the lattice fringes seen in HRTEM image of single nanorod, with an interlayer spacing of 0.49 nm, corresponding to d$_{(200)}$ of α-MnO$_2$, as demonstrated in Fig. 4.8(b) and the insets.

170 *Nanomaterials for Supercapacitors*

BET isotherm indicated that the nanoMnO$_2$ contained secondary mesopores formed in between the particles, which was in a good agreement with SEM observations. The sample had a BET surface area of 230 m$^2\cdot$g^{-1}, with a narrow pore size distribution and an average pore size of about 14.5 nm, which suggested that the pores were derived from the agglomerates or compacts of the uniform spherical particles. The presence of the secondary pores would be beneficial to bring about enhanced electrochemical performance of the nanostructured materials. However, the surface area was decreased with increasing annealing temperature.

Electrochemical performance of the nanoMnO$_2$ was examined in 0.1 M Na$_2$SO$_4$ solution. Figure 4.9(a) shows Cyclic Voltammograms (CV) curves of the nanoMnO$_2$ for the first, 1200th and 2000th cycles. A rectangular shape demonstrated the characteristic ideal capacitive behavior of the materials. The three CV nearly overlapped one another, suggesting the high stability of the nanoMnO$_2$ electrode, after cycling up to 2000 times. Figure 4.9(b) shows Galvanostatic charge-discharge cycling data of the nanoMnO$_2$ electrode for the first few cycles at a current density of 0.5 mA$\cdot$cm^{-2}. The variation in potential as the function of time was linear during both the charging and discharging processes, which is an important requirement of electrode materials.

The discharge capacitance obtained according to the data in Fig. 4.9(b) was 250 F$\cdot$g^{-1}. The variation in specific capacitance as a function of cycling up to 2000 cycles is illustrated in Fig. 4.9(c). A high electrochemical stability with a loss of on < 8% loss in capacitance was observed after 2000 charge-discharge cycles. It was also demonstrated that the specific capacitance was decreased as the sample was heated at 200 and 300°C, which was attributed to loss of water molecules and the decrease in surface area. Higher thermal annealing temperature led to serious decrease in specific capacitance. High surface area, secondary pores in between particles and appropriate water content of the nanoMnO$_2$ had been acknowledged as the main factors that were responsible for the capacitance stability against long time cycling is strongly attributed to the nature possessing. There was only a very slight decrease in surface area and specific capacitance as the annealing temperature was up to 300°C. In contrast, if the temperature was higher than this temperature, both the surface area and specific capacitance were largely decreased.

Besides variation in crystal structure, morphology and chemistry of the hydrated MnO$_2$ powders are also changed as a result of thermal annealing. Amorphous MnO$_2$ most likely appeared as clustered granules with sizes ranging from several to 10s of nanometers [38, 39]. After thermal annealing at 300°C, nanorods started to appear, whereas well-defined nanorods with lengths of 500–750 nm and diameters of 50–100 nm were formed as the annealing temperature was increased to 500°C and 600°C [38, 39]. Similar morphological change from spherical nanoparticles to nanorods was also observed in hydrated MnO$_2$ powders synthesized by using microemulsion method [42]. The as-prepared MnO$_2$ powders contained large amount of hydrate MnOOH and residual structural water, which could be reduced or removed through further thermal annealing [4, 39].

Crystalline MnO$_2$ powder electrodes

As stated earlier, manganese oxides have several types of crystal structures that can be present under different conditions, with structural frameworks composed of MnO$_6$ octahedra sharing vertices and edges. Through the stacking of the MnO$_6$ octahedra, 1D, 2D and 3D tunnel structures can be formed. The different crystal structures can be described by the size of the tunnel determined by the number of octahedra subunits ($n \times m$). The tunnels can be filled with either water molecules or cations, including Li$^+$, Na$^+$, K$^+$ and Mg^{2+}, which make crystalline manganese oxides have unique electrochemical properties in mild aqueous electrolytes [5].

It has been reported that MnO$_2$ powders, with various crystal structures, including α-, β-, δ-, γ- and λ-type crystal structures, could be synthesized by using co-precipitation and sol–gel techniques under different synthesis conditions [6]. The crystalline structure has a strong influence on electrochemical capacitances of the materials, simply because the degree and capability of intercalation of the cations are closely related to the size of the tunnels. For example, if birnessite δ-MnO$_2$ with a 2D tunnel structure was doped with potassium, high specific capacitance values of up to 110 F$\cdot$g^{-1} could be achieved, even the BET surface area was just moderate (17 m$^2\cdot$g^{-1}). β- and γ-MnO$_2$ had 1D tunnel structures, so that

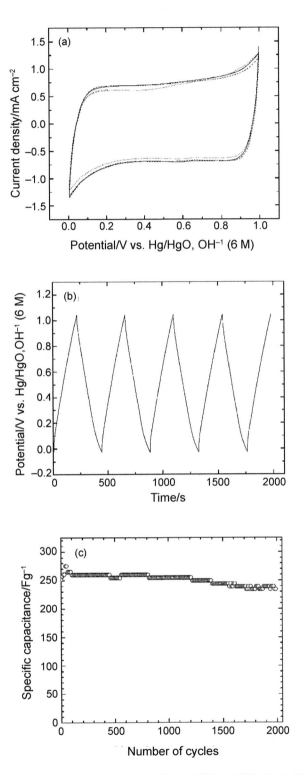

Fig. 4.9. Cyclic voltammograms in 0.1 M Na$_2$SO$_4$ at 10 mV·s^{-1} of first (solid line), 1200th (dashed line), and 2000th (solid line) cycle (a), charge-discharge curves of first few cycles at a constant current density 0.5 mA·cm^{-2} (b), and specific capacitance as a function of cycle number (c) for the nanoMnO$_2$. Reproduced with permission from [39], Copyright © 2009, American Chemical Society.

they exhibited only pseudo-Faradaic surface capacitance, which was thus strongly dependent on BET surface area of the crystalline materials. 3D tunnel structures such as λ-MnO$_2$ demonstrated intermediate electrochemical performance between birnessite and 1D tunnel structures. Examples of different types of manganese oxides are present as follows.

Crystalline δ-MnO$_2$ (birnessite) could be synthesized by using a sol-gel method [43]. A pre-synthesized xerogel without any washing was annealed at 600°C for 10 hours, leading to a powder, known as birnessite. The precursor was dispersed in water with stirring for 16 hours, known as H$_2$O washed birnessite, whereas it also was dispersed in 1 M H$_2$SO$_4$, known as H$_2$SO$_4$ washed birnessite. Other precursors, such as NaMnO$_4$, could be used to replace KMnO$_4$, which resulted in similar results.

Crystalline λ-MnO$_2$ and γ-MnO$_2$ could also be prepared by using the sol-gel process. A precursor of LiMn$_2$O$_4$ was first prepared by using the Pechini technique [44], in which a polymer matrix was used as a template. LiNO$_3$ and Mn(NO$_3$)$_2$·6H$_2$O were first dissolved in a mixture of citric acid and ethylene glycol, with a molar ratio of 1:4, which led to the formation of a transparent solution heating at 90°C for 20 minutes. After that, esterification and removing the ethylene glycol excess were facilitated by raising the temperature to 140°C. The temperature was then increased to 180°C to heat the solution in vacuum, followed by thermal annealing at 250°C in air, which resulted in ultrafine powder of LiMn$_2$O$_4$. The LiMn$_2$O$_4$ powder was hydrolyzed in 0.5 M HCl at 25°C for 24 hours, leading to the formation of λ-MnO$_2$. If it was treated in 0.5 M HCl at 95°C, γ-MnO$_2$ was formed.

Similarly, a series of MnO$_2$ allotropic phases were prepared, in order to study the effects of crystallographic forms of MnO$_2$ on their electrochemical performances [14]. 1D structures, such as pyrolusite, ramsdellite, cryptomelane, Ni-doped todorokite (Ni–todorokite) and OMS-5, as well as 2D and 3D structures of birnessite and spinel, were prepared and systematically compared. Figure 4.10 shows schematic diagrams of representative MnO$_2$ forms with different crystallographic structures. As mentioned earlier, MnO$_2$ crystal structures of different allotropic forms are constructed with MnO$_6$ octahedra building blocks [45]. The MnO$_6$ units are assembled by sharing corners, edges or faces to build up the crystal structures. In the MnO$_2$ unit cells, channel cavities with different sizes could be created, by controlling the number of the MnO$_6$ units [46]. For example, the birnessite compound, as shown in Fig. 4.10(f), could be produced through the following redox reaction between MnO$_4^-$ and Mn^{2+} under basic conditions:

$$2Mn^{7+} + 3Mn^{2+} \rightarrow 5Mn^{4+}. \tag{4.5}$$

The products were hydrated MnO$_2$ with a layered structure, with sodium ions present in between the layers [45]. The 2D structure was stabilized due to the presence of the Na$^+$ ions and H$_2$O molecules.

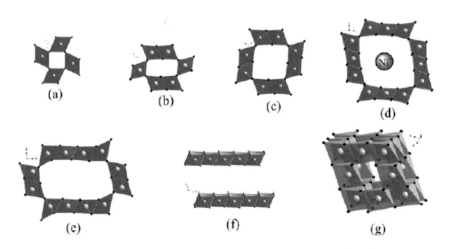

Fig. 4.10. Schematic diagrams showing the crystallographic structures of MnO$_2$: (a) pyrolusite, (b) bramsdellite, (c) cryptomelane, (d) Ni-todorokite, (e) OMS-5, (f) birnessite and (g) spinel. Reproduced with permission from [14], Copyright © 2009, American Chemical Society.

MnO$_2$ pyrolusite of Fig. 4.10(a) could be derived from lithiated birnessite by using a hydrothermal treatment. During the hydrothermal reaction, the layered sheets were collapsed, so that H$_2$O molecules were driven out the interlayered space, which resulted in a tunnel-type structure [46]. In acidic solutions, due to the presence of protons, the interactions between Na$^+$ and H$_2$O were largely weakened. At the same time, Mn^{3+} ions could be partially oxidized to Mn^{4+}. Therefore, the number of cations used for charge balance in between the layers was decreased, so that small 1 × 1 tunnels were formed. If the reaction systems were basic, the interactions between Na$^+$ and H$_2$O were much stronger. Consequently, the hydrated sodium ions with a larger size led to larger tunnel sizes, thus forming OMS-5 with 2 × 4 tunnels, as shown in Fig. 4.10(e).

Figure 4.10(d) shows a schematic of the Ni-todorokite, whose large tunnel cavity is attributed to the precursor of Ni-buserite [14]. The MnO phase has a basal spacing of 10 Å, which is larger than that of Ni-birnessite (7 Å). The larger basal spacing was attributed to the presence of a second interlayer, which was caused by H$_2$O molecules, as the birnessite was transferred to buserite during the preparation process. In addition, Ni$^+$ ions also acted as a stabilizer to maintain the layered structure. After the Ni-buserite was hydrothermally processed, the layers were collapsed, so that a 3 × 3 tunneled structure was obtained [47, 48]. It was found that the doping metal ions were inserted in the bulk crystal as the buserite structure was transferred to the todorokite on [49]. Furthermore, the ignition method has been used transfer the layered structure into tunneled structure, where the preparation of the cryptomelane (Fig. 4.10(c)) is an example. The ignition could result in an increase in pH value of the solution, while the potassium ions could be partially depleted after washing with water, which lead to the presence of 2 × 2 tunnels. MnO$_2$ ramsdellite (Fig. 4.10(b)) could be obtained through the oxidation of MnSO$_4$ with (NH4)$_2$S$_2$O$_8$, whereas MnO$_2$ spinel, e.g., Li$_{0.2}$Mn$_2$O$_4$, could be synthesized through the delithiation of LiMn$_2$O$_4$.

Figure 4.11 shows XRD patterns of the MnO$_2$ powders with different structures [14]. According to Fig. 4.11(a), the birnessite phase of MnO$_2$ had a monoclinic type crystal structure, with space group C2/m and unit cell parameters of $a = 5.174$ Å, $b = 2.850$ Å, $c = 7.336$ Å and $\beta = 103.18°$ [50]. The broadened diffraction peaks indicated that the sample exhibited a relatively poor crystallinity. For (Fig. 4.11c) powders, the XRD patterns of Fig. 4.11(b) and (c) demonstrated the pure phases of cryptomelane and pyrolusite, respectively, corresponding to a tetragonal crystal structure [51]. High crystallinity of the two

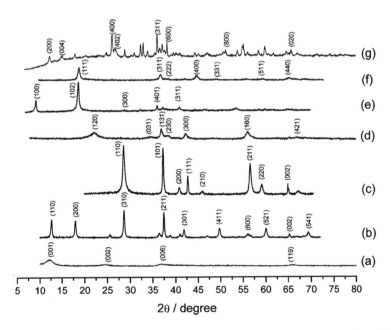

Fig. 4.11. XRD patterns of the MnO$_2$ with different crystal structures: (a) birnessite, (b) cryptomelane, (c) pyrolusite, (d) ramsdellite, (e) Ni-todorokite, (f) spinel and (g) OMS-5. Reproduced with permission from [14], Copyright © 2009, American Chemical Society.

174 *Nanomaterials for Supercapacitors*

samples was confirmed by the sharp diffraction peaks. The tetragonal cryptomelane MnO_2 possessed a 2×2 tunnel structure, with space group *I4/m* and cell parameters $a = 9.815$ Å and $c = 2.847$ Å [12]. The MnO_2 pyrolusite possessed cell parameters of $a = 4.38$ Å and $c = 2.85$ Å, with space group *P42/mnm* [52].

The broad diffraction peaks of the pattern for the ramsdellite as shown in Fig. 4.11(d) were attributed to the presence of stacking faults in the structure of the sample [6, 53]. It belonged to the orthorhombic crystal system, with unit cell parameters of $a = 9.273$ Å, $b = 2.866$ Å and $c = 4.533$ Å, as well as space group *Pnma* [6]. According to Fig. 4.11(e), the Ni-todorokite MnO_2 exhibited a 3×3 tunnel structure, with a monoclinic crystal structure, having space group *P2/m* and unit cell parameters of $a = 9.757$ Å, $b = 2.842$ Å, $c = 9.560$ Å and $\beta = 94.07°$ [47, 48, 54].

The spinel sample had a 3D microstructure, with a cubic crystal structure and a *Fd3m* space group, as well as lattice constant $a = 8.03$ Å [52, 55]. According to the XRD pattern of OMS-5, hydrothermal reaction under basic conditions led to MnO_2 with a 2×4 tunnel structure [46, 56, 57]. Splitting of diffraction peaks was observed, suggesting that the sample was a mixture of poorly crystallized microdomains that possessed different cell parameters [12]. The MnO_2 phase belonged to monoclinic crystal system, with a *C2/m* space group and cell parameters of $a = 14.434$ Å, $b = 2.849$ Å, $c = 23.976$ Å and $\beta = 98.18°$.

Figure 4.12 shows SEM images of MnO_2 powders with different crystal structures [14]. A strong dependence of morphology of the powders on the preparation method, i.e., crystal structures, could be clearly observed. 1D group samples, including pyrolusite, ramsdellite and OMS-5, as shown in Fig. 4.12(a), (b) and (c), respectively, appeared as nanorods. The pyrolusite sample contained uniform nanorods, with a diameter of about 50 nm and a length of 300 nm. The ramsdellite powder consisted of spherical particles with an urchinlike morphology, which were assembled by the nanorods. The particles had diameters in the range of 3–8 μm, in which the nanorods possessed diameters of 70–100 nm and lengths of 400–500 nm, as demonstrated in the inset of Fig. 4.12(b). The nanorods of OMS-5 were about 50 nm in diameter and 500–700 nm in length, as seen in Fig. 4.12(c). The cryptomelane sample consisted of nanoparticles with a relatively wide particle size distribution, whereas the Ni-todorokite sample had a fragmented platelet-like morphology, as illustrated in Fig. 4.12(d) and (e), respectively. The birnessite powder consisted of densely packed particles, as demonstrated in Fig. 4.12(f). The spinel powder was characterized by interconnected nanofibers, with a diameter of 10 nm and a length of 200 nm.

The MnO_2 powders with different crystal structures exhibited specific surface areas ranging from 19 to 156 $m^2 \cdot g^{-1}$ from OMS-5 to spinel phases. The specific surface areas were increased in the order of OMS-5 < cryptomelane < Ni-todorokite < pyrolusite < birnessite < ramsdellite < spinel. The surface structure was mainly attributed to the micropores, which contributed to the total area by about 50% for the cryptomelane, Ni-todorokite and pyrolusite, and about 63% for the OMS-5, ramsdellite and spinel. The birnessite powder had the highest fraction of micropores of 82% and pore volume of 0.0045 $m^3 \cdot g^{-1}$, due to its large interlayer spacing. The spinel MnO_2 exhibited a pretty high pore volume of 0.003 $m^3 \cdot g^{-1}$, which could be attributed to its 3D interconnected tunnel structures. The average pore volume of the rest of the MnO_2 sample (1D structure) was 0.0013 $m^3 \cdot g^{-1}$.

All the MnO_2 materials demonstrated a relatively low electrical conductivity, while different crystal structures led to different conductivity, which could be varied by four orders of magnitude [14]. However, it was difficult to link the electrical conductivity of the MnO_2 materials with their microstructures. Electrical conductivity is an important factor that determines electrochemical storage mechanism and performance of the materials. Comparatively, ionic conductivity was in the range of 0.004–0.02 $S \cdot cm^{-1}$, with an increasing trend with channel size and structural variation from 1D to 3D, because larger channels or more opened structures would allow fast ion diffusion.

Specific capacitance, ionic conductivity and BET surface areas of the MnO_2 materials are summarized in Fig. 4.13 [14]. It was found that the specific capacitance of the different MnO_2 phases was not well matched with the BET surface area. For, the OMS-5 phase possessed the lowest surface area as stated earlier, while it offered the highest capacitance. In contrast, the ramsdellite had a relatively high surface area, but its electrochemical performance was not as good as expected. Generally, if the capacitance is attributed to double-layer charging or adsorption of cations at the material surface, the capacitance will be proportional to the BET surface area [52]. When a material has a faradic behavior, the electrolyte accesses

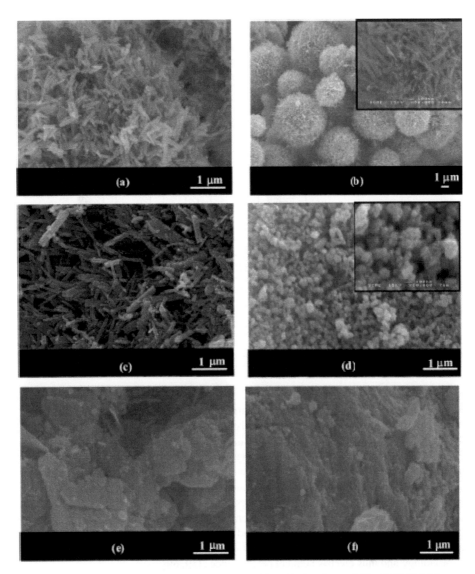

Fig. 4.12. SEM images of the MnO$_2$ with different crystal structures: (a) pyrolusite, (b) ramsdellite, (c) OMS-5, (d) cryptomelane, (e) Ni-todorokite and (f) birnessite. Reproduced with permission from [14], Copyright © 2009, American Chemical Society.

bulk of the material for the charge-storage process, so that the capacitance is not dependent on the BET surface area [6].

It was observed that channel size the 1D MnO$_2$ structures did not have a direct effect on the micropore surface areas and the total specific areas, mainly because the micropore surface area was included by the BET measurement, due to the fact that the channel sizes of MnO$_2$ pyrolusite and ramsdellite were too narrow to adsorb krypton gas [14]. Therefore, the surface area was underestimated. In addition, the krypton could also be blocked by the ionic species from the precursor salts of the MnO$_2$ powders, or the structure frameworks had been stabilized. According to the relationship between ionic conductivities of the materials and the specific capacitance of the devices, it was concluded that the electrochemical behavior of the MnO$_2$-based electrodes was controlled by the electrolyte accessibility and diffusion through the material bulk instead of the measured surface area.

176 *Nanomaterials for Supercapacitors*

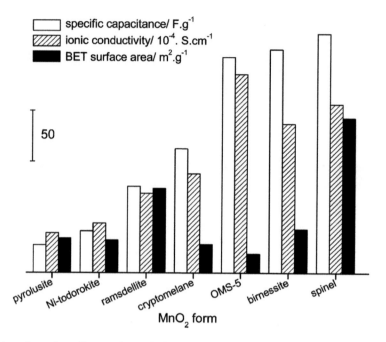

Fig. 4.13. Relative values of specific capacitance, ionic conductivity and BET surface area of the as-synthesized MnO$_2$ materials. Reproduced with permission from [14], Copyright © 2009, American Chemical Society.

Ionic conductivity has a strong effect on the charge transfer at the electrolyte/MnO$_2$ interface during the redox processes, in which Mn^{n+} cations were involved. It also governed the charge-storage mechanisms of the MnO$_2$ materials. Therefore, the specific capacitance could have been mainly attributed to the ionic conductivity. Nevertheless, a large BET surface area is still desirable, because a large surface area is beneficial to wetting behavior of the materials in contact with the electrolytes.

Hydrothermal and solvothermal methods have been widely used to synthesize various materials with controllable nanoarchitectures, including nanoparticles, nanorods, nanowires, nano-urchins and nanotubes. This can be realized by properly controlling the reaction temperature, reaction time, precursor concentration, type of precursor, filling level, solvents, minerals and additives. A hydrothermal route based on aqueous solutions of MnSO$_4$·H$_2$O and KMnO$_4$ was reported to prepare MnO$_2$ nanostructured materials [58], with tunable morphology and thus electrochemical performances [59, 60].

Thoroughly mixed aqueous solutions of KMnO$_4$ and hydrated MnSO$_4$ were thermally treated at 140°C [59]. The reaction time was changed in the range of 1–18 hours in order to optimize the material with desired electrochemical performances. After reaction, the precipitates were filtered and washed thoroughly with distilled water until the pH of the washed water was 7, in order to completely remove the unreacted materials. The precipitated MnO$_2$ powders were dried at 100°C in air. For comparison, amorphous MnO$_2$ sample was precipitated by leaving the same the starting materials in a beaker overnight, so as to check the structural evolution of the MnO$_2$ nanorods from room temperature to the state after the time-dependent hydrothermal reaction. Figure 4.14 shows a schematic diagram describing the structural evolution of the MnO$_2$ nanorods, together with the respective TEM images [59].

Once the hydrated MnSO$_4$ and KMnO$_4$ were mixed, precipitation of MnO$_2$ occurred quickly. TEM image indicated that the MnO$_2$ precipitated at room temperature had a flowerlike morphology constructed with nanowhiskers. After hydrothermal reaction at 140°C for 1 hour, the individual whiskers showed a nanostructured surface with a plate-like morphology, as revealed by the SEM results shown in Fig. 4.15. The nanowhiskers were heavily agglomerated with an increased density, which were dissembled as the hydrothermal reaction time was increased. As the hydrothermal reaction time was further increased, the size of the individual nanowhiskers also increased. At the same time, a rod-like structure at nanoscale

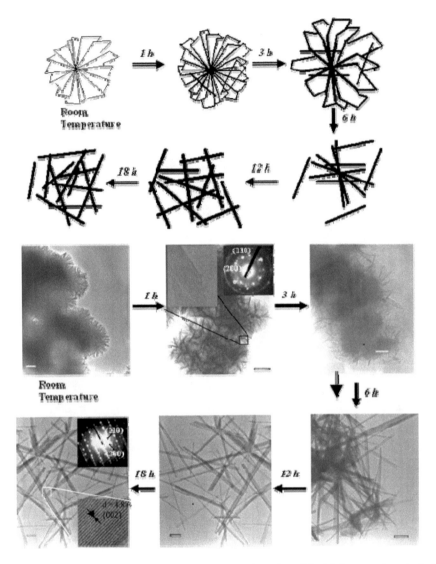

Fig. 4.14. Schematic diagram and TEM images to demonstrate the formation of the MnO_2 nanorods under hydrothermal conditions. Insets of 1 hour and 18 hours TEM images show the electron diffraction and HRTEM of the MnO_2 synthesized for the respective hydrothermal reaction time. Reproduced with permission from [59], Copyright © 2005, American Chemical Society.

was formed, as seen in Fig. 4.14. Therefore, the nanorods were derived from the nanowhiskers that were formed at room temperature. After hydrothermal reaction, rod-like architectures were developed, in which the size of the individual nanorods varied with the hydrothermal reaction time. The Electron Diffraction (ED) of the sample after hydrothermal reaction for 1 hour displayed the nucleation of the nanocrystalline particles. While lattice fringes of the 1 hour sample were still not clearly evolved, the 18 hours sample was well crystallized, as evidenced by the ED and HRTEM results. The lattice fringes possessed *d*-spacing of 4.8 Å, corresponding to the (002) plane of MnO_2. It was found that the hydrothermal reaction time had a strong influence on pore structure and surface area of the MnO_2 nanostructures and hence electrochemical properties of the materials.

Figure 4.15 shows SEM images of the MnO_2 samples hydrothermally reacted for different time durations [59]. The nanostructure was evolved gradually as the hydrothermal reaction time was increased.

178 *Nanomaterials for Supercapacitors*

Fig. 4.15. SEM images of the MnO$_2$ nanostructures synthesized for different hydrothermal dwell times. Reproduced with permission from [59], Copyright © 2005, American Chemical Society.

The room temperature precipitated sample consisted of heavily agglomerated nanowhiskers of MnO$_2$, which were changed to nanostructured plate-like morphology after reaction for 1 hour. As the reaction time was increased to 3 hours and 6 hours, nanoarchitecture with few rods and nanostructured plate-like morphology were developed, respectively. If the hydrothermal reaction time was further raised to 12 hours and 18 hours, the samples contained large amount of individual nanorods. It was concluded that hydrothermal reaction is a versatile technique that can be used to manipulate the morphology of the synthesized nanomaterials.

Figure 4.16 shows XRD patterns of the MnO$_2$ samples treated for different times [59]. The sample precipitated at room temperature was amorphous MnO$_2$, which was the precursor of α-MnO$_2$. After hydrothermally treated at 140°C for 1 hour, nanocrystalline MnO$_2$ was obtained, which could be indexed to α-MnO$_2$. The amorphous phase was still present, evidenced by the broad peaks. The crystallinity was gradually increased with increasing hydrothermal reaction time from 1 hour to 18 hours, because the peaks became sharper and sharper, which was also confirmed by the TEM, ED and HRTEM results. After hydrothermal reaction for ≥ 12 hours, single phase α-MnO$_2$, with a tetragonal crystal structure and space group *I*4/*m* was obtained.

The 1 hour hydrothermally treated sample had a BET surface area of 100 m^2·g^{-1}. As the hydrothermal reaction time was increased to 6 hours and 12 hours, the surface areas were increased to 132 m^2·g^{-1} and 150

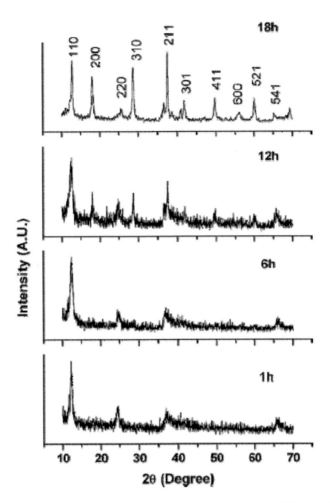

Fig. 4.16. XRD patterns of the MnO$_2$ nanostructures synthesized hydrothermally at 140°C for different time durations. Reproduced with permission from [59], Copyright © 2005, American Chemical Society.

m^2·g^{-1}, respectively [59]. The surface area was more dependent on the structure of the pores that facilitated the N$_2$ adsorption-desorption. As stated earlier, the samples with shorter hydrothermal reaction time possessed more agglomerated nanowhiskers of MnO$_2$, which were converted to the nanorod architecture after the reaction time was increased. As a result, the pore volume was increased. However, the MnO$_2$ obtained with different reaction times were mesoporous, with pores diameters to be about 8 nm.

The MnO$_2$ sample with hydrothermal reaction time of 6 hours demonstrate the highest capacitance with a value of 168 F·g^{-1}, as compared with 140 F·g^{-1} for the sample with reaction time of 1 hour. However, further increase in reaction time led a decrease in capacitance to 118 F·g^{-1} and 72 F·g^{-1} for 12 hours and 18 hours, respectively. The sample with reaction time of 1 hour at 140°C still contained amorphous MnO$_2$ and had serious particle agglomeration, which blocked the electrolyte for the electrochemical reactions. Prolonged reaction time led to loosening of the agglomeration and hence higher capacitance, due to the improved electrolyte accessibility. The specific capacitance of the materials with respect to the reaction time is closely related to the variation in crystallinity, specific surface area, coexistence of more metal ions such as K$^+$ and the loss of the chemically and physically adsorbed water from MnO$_2$ [10, 61–65].

Another example is to use hydrothermal process, with KMnO$_4$, sulfuric acid and Cu scraps, to synthesize α-MnO$_2$ hollow spheres and hollow urchins [66]. The hollow sphere or urchin structured α-MnO$_2$ materials possess a highly loose, mesoporous cluster structure consisting of thin plates or nanowires and exhibit enhanced rate capacity and cyclability. $KMnO_4$ was dissolved in deionized water, into which sulfuric acid (98 wt%) was slowly dropped. After stirring for 15 minutes, Cu scraps was added into the solution. Then, the mixture was transferred into a stainless steel autoclave for hydrothermal reaction at 110°C for different time durations. After reaction, the precipitates were collected, washed with distilled water and dried at 80°C for 12 hours.

Figure 4.17 shows XRD patterns of the samples after hydrothermal reaction for different time durations [66]. As seen in Fig. 4.17(a), after hydrothermal reaction for 3 hours, four weak peaks were present, which could be indexed to α-MnO_2. As the hydrothermal reaction time was increased, intensities of all the peaks were increased, which matched very well with the standard XRD pattern of α-MnO_2. There were no impurities in the samples. α-MnO_2 is constructed from double chains of octahedral [MnO_6], forming 2 × 2 tunnels, which makes it to be promising electrode material in batteries and supercapacitors.

The 3 hours α-MnO_2 sample had a panoramic morphology, as shown in Fig. 4.18(a), with spherical particles having diameters in the range of 1–3 µm [66]. Figure 4.18(b) shows high-magnification SEM

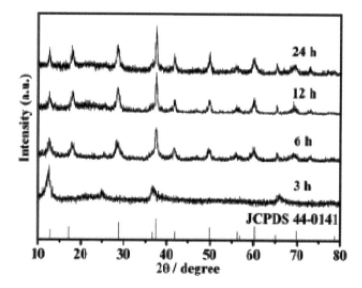

Fig. 4.17. XRD patterns for the standard values of JCPDS 44-0141 and the samples obtained at 110°C for different reaction times. Reproduced with permission from [66], Copyright © 2007, American Chemical Society.

Fig. 4.18. (a) Low-magnification and (b) high-magnification FESEM images of the products obtained by hydrothermal reaction for 3 hours. Reproduced with permission from [66], Copyright © 2007, American Chemical Society.

image of the 3 hours sample, indicating that the particles were formed by many interleaving thin plates. After hydrothermal reaction for 6 hours, size of the aggregates was increased, while some urchin-like α-MnO$_2$ was present. On the surface of the urchin-like spheres, numerous compactly growing nanowires could be observed, which implied that the thin plates were gradually changed into nanowires after hydrothermal reaction for 6 hours. It was also found that the as-obtained α-MnO$_2$ spheres had possibly a hollow structure.

Figure 4.19 shows SEM images of the samples after hydrothermal reaction for 12 hours [66]. The samples were composed of hollow urchins with cavities. High-magnification SEM images indicated that the shell of the hollow urchins was formed by densely aligned nanorods with uniform diameters and lengths, as revealed in Fig. 4.19(c) and (d). Therefore, reaction time was a key to the formation of the hollow-structured α-MnO$_2$. If the hydrothermal reaction was prolonged to 24 hours, the urchin structure disappeared completely. Instead, only nanorods were obtained.

The TEM observation results confirmed that the α-MnO$_2$ after hydrothermal reaction for 12 hours consisted of intercrossed nanorods. In HRTEM image, clear lattice fringes were observed, confirming the single crystal characteristic of the α-MnO$_2$ phase. The interplay lattice spacing of 0.69 nm corresponded to the spacing of (110) crystal planes. The evolution of crystallinity and morphology of the α-MnO$_2$ samples could be explained according to the Ostwald ripening process [67, 68]. At the initial stage of reaction, nuclei were formed in a short time through the Ostwald ripening process, from which crystals grew slowly. Then, both the size and density of the aggregates were continuously increased, resulting in spheres with a solid core. After that, interior cavities were gradually formed through the core evacuation process, because of the higher surface energies. Further increase in the hydrothermal reaction time led to complete damage of the urchin structures, while the size of the individual nanorods was increased. As stated earlier, 12 hours was sufficient for the α-MnO$_2$ hollow urchins to reach their equilibrium size.

In order to further understand the forming mechanism of the different structured α-MnO$_2$, the samples obtained from different reaction times were analyzed carefully by TEM investigations. As shown in

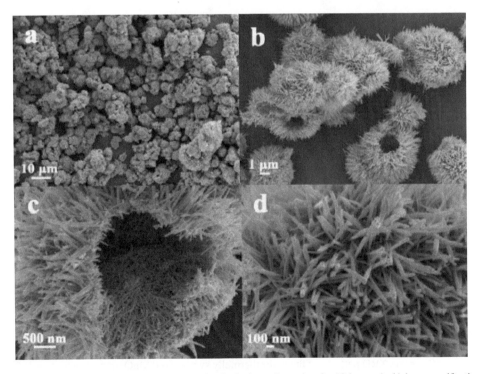

Fig. 4.19. Morphologies of the products obtained after hydrothermal reaction for 12 hours: (a, b) low magnification and (c, d) profile images of the α-MnO$_2$ hollow urchins. Reproduced with permission from [66], Copyright © 2007, American Chemical Society.

182 *Nanomaterials for Supercapacitors*

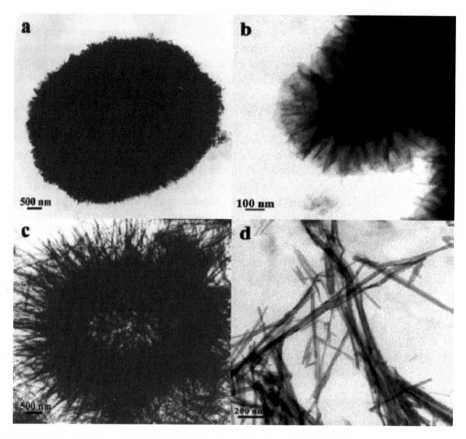

Fig. 4.20. TEM of the α-MnO$_2$ sample obtained for different reaction times: (a) 3, (b) 6, (c) 12, and (d) 24 hours. Reproduced with permission from [66], Copyright © 2007, American Chemical Society.

Fig. 4.7, several obvious evolution stages could be clearly observed. In the initial stage (shorter reaction time, 3 hours), only a close-grained sphere is observed; after hydrothermal reaction for 6 hours, the surface of the α-MnO$_2$ sphere changes to flower-like nanostructure which consists of nanoflakes and nanowires. When the reaction time was prolonged to 12 hours, an interior cavity sphere is easily observed; after reaction for 24 hours, the sphere structures disappear completely and only nanorods can be observed. The results of TEM investigation further confirm the above Ostwald ripening forming mechanism of the different structured α-MnO$_2$.

Similar morphologies were also observed in α-MnO$_2$ prepared from KMnO$_4$ and nitric acid-containing reactants, and in α-MnO$_2$ and ε-MnO$_2$ nanostructures prepared at a low temperature (110°C) [69, 70]. For instance, a simple method was developed to synthesize α-MnO$_2$ nanostructures through the hydrothermal decomposition of KMnO$_4$ in nitric acid. It was found that the hydrothermal reaction time exhibited a significant role in determining morphology and phase composition of the final products and thus electrochemical properties of the materials used as electrodes of supercapacitors [69]. To synthesize the α-MnO$_2$ nanostructures, 6.7 g KMnO$_4$ and 4.2 mL concentrated HNO$_3$ were dissolved in 30 mL distilled water to form the reaction solution. The solution was reacted in a Teflon-lined autoclave at 120°C for different time durations. The reaction time was changed from 1 to 12 hours so as to obtain materials with optimized electrochemical performances.

Figure 4.21 shows XRD patterns of the hydrothermally synthesized MnO$_2$ nanostructures at 120°C for different time durations. After hydrothermal for 1 hour, the sample exhibited broad peaks, implying that it was a mixture of amorphous and nanocrystalline phases. After reaction for 3 hours, diffraction peaks

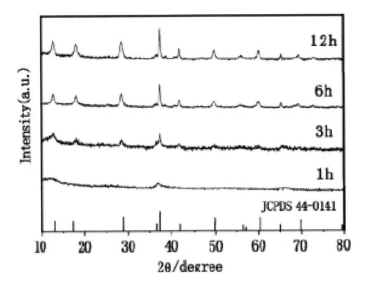

Fig. 4.21. XRD patterns of the MnO$_2$ nanoparticles synthesized hydrothermally at 120°C for different times. Reproduced with permission from [69], Copyright © 2009, Elsevier.

became clearer, but they are still quite weak, suggesting the incomplete crystallization. The peaks were found to belong to α-MnO$_2$. As the reaction time was further increased, all the peaks were significantly sharpened, which were in a good agreement with the standard XRD pattern. The 6 hours sample was pure α-MnO$_2$, with a body-centered tetragonal crystal structure, having space group I4/m and lattice constants of a = 0.9785 nm and c = 0.2863 nm. No impurities were detected in the XRD measurement, demonstrating high purity and crystallinity of the samples.

Figure 4.22 shows TEM images of the α-MnO$_2$ samples, from which mechanisms for phase formation and microstructure development could be deduced. As shown in Fig. 4.22(a–e), evolution stages of the microstructure evolution could be clearly identified. The 1 hour sample consisted of solid nanospheres, with interleaving nanowhiskers. After hydrothermal reaction for 3 hours, the nanowhisker sphere structure was varied to flower-like nanostructure, comprising of nanorods on surface of the particles. As the reaction time was increased to 9 hours, spheres with an interior cavity were developed. The spheres consisted of nanorods, while the flower-like aggregates were absent, as observed in Fig. 4.22(d). After reaction for 12 hours, the sphere structure was disassembled, with the presence of entirely nanorods. Figure 4.22(f) shows a representative HRTEM image of an individual α-MnO$_2$ nanorod from the sample after hydrothermal reaction for 12 hours. The rod had a diameter of 25 nm and lengths of 1–2 mm. A lattice spacing of 0.31 nm was observed, corresponding to the distance between two adjacent (110) crystal planes (inset image in Fig. 4.22(f)). SAED pattern indicated that the sample was of a single crystal structure, as seen in the inset of Fig. 4.22(f).

The time dependent evolution of crystallinity and morphology of the α-MnO$_2$ nanostructures could be explained in terms of Ostwald ripening [67, 71]. The consequence of hydrothermal treatment of KMnO$_4$ aqueous solution led to its decomposition into MnO$_2$ with the presence of O$_2$. Because decomposition reaction was very fast, a large number of nuclei were formed within a short time period, followed by a relatively slow crystal growth into spherical structure with a solid core initially. This stage was over a time duration of several hours, after which an interior cavity was gradually developed, due to the core evacuation process related to the surface energies. As the hydrothermal reaction time was further increased, dissolution–recrystallization process was started, during which the amorphous items in the agglomerates were dissolved, resulting in the formation of the nanorods. The larger nanorods continued to grow at the expense of the smaller ones, so that urchin structures were destroyed, thus leading to the presence of isolated nanorods.

184 *Nanomaterials for Supercapacitors*

Fig. 4.22. TEM images of the MnO$_2$ nanostructures after hydrothermal reaction for different times: (a) 1 hour, (b) 3 hours, (c) 6 hours, (d) 9 hours, (e) 12 hours and (f) individual MnO$_2$ nanorods prepared for 12 hours. Inset of (f) shows ED and HRTEM of the MnO$_2$ sample prepared for 12 hours. Reproduced with permission from [69], Copyright © 2009, Elsevier.

The α-MnO$_2$ sample after reaction at 120°C for 6 hours had specific capacitances of 152, 1381 and 114 F·g^{-1} at 2.5, 5 and 10 mA, respectively, demonstrating that it was a promising candidate of electrode materials for supercapacitors. Figure 4.23 shows variation of specific capacitance over 100 cycles of the sample. After 100 cycles, the capacitance was retained to be about 87% of that for the first cycle, suggesting that, within the voltage window of 0.2–0.8 V, the charge and discharge processes had not induced obvious structural or microstructural variation of the α-MnO$_2$ electrodes during the pseudocapacitive reactions. With this long-term stability, the α-MnO$_2$ nanostructures synthesized in this way could be used as electrodes of supercapacitors with advanced performances.

A hydrothermal method was reported to synthesize lamellar birnessite δ-MnO$_2$ structures with different interlayer spacings, which could be simply controlled by adjusting pH value of the initial reaction solutions [70]. As the precursor, α-NaMnO$_2$, was prepared by using the conventional solid-state process [72]. Na$_2$CO$_3$ and Mn$_2$O$_3$ were mixed and calcined at 650°C for 40 hours in Ar flow to form α-NaMnO$_2$.

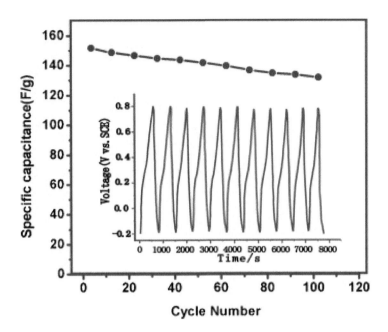

Fig. 4.23. Cycle life of the α-MnO$_2$ electrode at current of 500 mA·g^{-1} in 1 M Na$_2$SO$_4$ over 0.2–0.8 V, with the inset showing charge–discharge curves of the α-MnO$_2$ electrode at a constant current of 500 mA·g^{-1}. Reproduced with permission from [69], Copyright © 2009, Elsevier.

Birnessite-type MnO$_2$ was synthesized by using hydrothermal reaction. 1 g α-NaMnO$_2$ was dispersed in 30 mL distilled water at pH values of 0.63, 2.81 and 8.24, which were adjusted with 10% HNO$_3$ solution. One more solution had a pH value of 12.43, which was obtained without the presence of HNO$_3$. Hydrothermal reaction was conducted at 120°C for 12 hours.

Figure 4.24 shows XRD patterns of the MnO$_2$ samples synthesized from solutions with different pH values. The MnO$_2$ samples obtained from the solutions with pH values of 2.81, 8.24 and 12.43 were monoclinic phase with birnessite-type structure. The MnO$_2$ sample derived from the pH = 0.63 solution consisted of two phases, which are α-MnO$_2$ and γ-MnO$_2$. Figure 4.25 shows SEM images of α-NaMnO$_2$ and the MnO$_2$ samples. The α-NaMnO$_2$ contained circular cylinders with thicknesses of 1–3 mm and diameters of 1.5–4 mm, as seen in Fig. 4.25(a). The MnO$_2$ samples from the solutions with pH values of 12.43, 8.24 and 2.81 possessed a lamellar structure, with diameters of 2–4 mm, as revealed in Fig. 4.25(b–e), whereas the MnO$_2$ synthesized at pH = 0.63 had a rod-like structure, as illustrated in Fig. 4.25(f).

A mechanism was proposed to explain formation of the MnO$_2$ lamellar structure. During the hydrothermal reaction, α-NaMnO$_2$ was disproportionate through the following equation, due to instability [73]. Some of the Mn^{3+} ions were transformed into Mn^{4+}, while MnO$_2$ lamellar structure was formed with MnO$_6$ octahedra with the intercalation of water molecule and exchange of Na$^+$ with protons.

$$2Mn^{3+} \rightarrow Mn^{2+} + Mn^{4+}. \tag{4.6}$$

In the solution with high pH values, ionization of MnOH groups could take place, while Na$^+$ ions were exchanged with protons, so that the concentration of Na$^+$ was increased in the van der Waals gap [74]. As a result, the electrostatic attraction in between the adjacent layers was enhanced, because the total numbers of positive charged Na$^+$ ions and the net negative charge of the MnO$_6$ octahedral layers were increased. In this case, the interlayer spacing was reduced, as the pH value of the starting solution was increased. For example, at pH value of 0.63, due to the intercalation of water molecules and the weak electrostatic interaction in between the adjacent MnO$_6$ octahedra sheets, the lamellar MnO$_2$ curled and collapsed into items with rod-like morphology, as illustrated in Fig. 4.25(f).

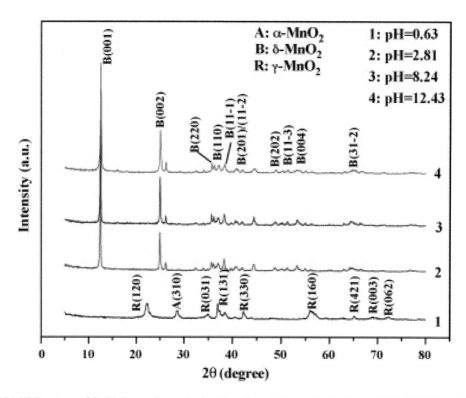

Fig. 4.24. XRD patterns of the MnO$_2$ samples synthesized from the solutions with pH values of 0.63, 2.81, 8.24 and 12.43. Reproduced with permission from [70], Copyright © 2010, Elsevier.

Initial specific capacitance of the MnO$_2$ sample from the starting solution with pH of 2.81 was 242.1 F·g^{-1} at a low current density of 2 mA·cm^{-2} in 2 M (NH$_4$)$_2$SO$_4$ aqueous electrolyte, which was decreased gradually with increasing current density. Additionally, a maximum power density of 10.4 kW·kg^{-1} was achieved at the minimum energy density of 6.8 Wh·kg^{-1}. Therefore, it was suggested that the birnessite-type MnO$_2$ synthesized by using the hydrothermal method could be a promising candidate as electrode materials for high rate supercapacitors.

A modified hydrothermal method was employed to synthesize Mn$_3$O$_4$ and MnOOH single crystals from the precursors of Mn(CH$_3$COO)$_2$·4H$_2$O and K$_2$S$_2$O$_8$ [75]. Before the hydrothermal reaction, the precursor solution was saturated with O$_2$ by blowing pure O$_2$ bubbling or adding 10 mM K$_2$S$_2$O$_8$ into the solution, which was the sealed and thermally treated at 120°C for 12 hours. The precursor solution experienced a color change gradually from transparent to brown during the heating process, followed by the formation of precipitation. After the hydrothermal reaction was finished, all products were collected with centrifugation, washing and drying.

Figure 4.26 shows XRD patterns of the Mn$_3$O$_4$ and MnOOH samples. The diffraction peaks of pattern (1) were matched well with those of the tetragonal hausmannite structure (space group *I*41/*amd*) of Mn$_3$O$_4$, with lattice parameters of a = b = 5.762 Å and c = 9.470 Å. The single phase characteristic of the sample was confirmed by the Raman spectrum. For pattern (2), all the diffraction peaks belonged to the monoclinic MnOOH, with space group *P*21/*c*14 and lattice parameters of a = 5.3 Å, b = 5.278 Å and c = 5.307 Å. Therefore, it was concluded that crystalline structure of the resultant oxides was strongly dependent on the type of the oxidants that were used in the precursor solutions. Also, the simple low-temperature hydrothermal synthesis method directly led to hausmannite and MnOOH single crystals.

Figure 4.27 shows representative SEM and HRTEM images of the two powders. As shown in Fig. 4.27(A) and (B), the Mn$_3$O$_4$ and MnOOH single crystals had high purity and could be produced at large

Fig. 4.25. SEM images of (a) pristine α-NaMnO$_2$ and (b–f) MnO$_2$ synthesized from the solutions with different pH values and (d) the enlargement of the dashed segment in (c). Reproduced with permission from [70], Copyright © 2010, Elsevier.

quantity. HRTEM and electron diffraction pattern indicated the single-crystalline characteristic of MnOOH, as observed in Fig. 4.27 (C). The direction of the zone of electron beam, [$\bar{1}0\bar{1}$], was clearly defined from the directions of the (020) and (20$\bar{2}$) diffraction dots. From the HRTEM image, lattice spacings of 2.6 and 2.2 Å, corresponding to the inter-layer distance of (020) and (20$\bar{2}$) planes, respectively. The HRTEM and electron diffraction pattern (Fig. 4.27 (D)) confirmed the Mn$_3$O$_4$ synthesized in this way was single crystal. The direction of the electron zone, [$\bar{1}3\bar{1}$], was defined from the direction of the (112) and (10$\bar{1}$) diffraction dots. The HRTEM image gave rise to lattice spacings of 4.9 and 3.1 Å, corresponding to the inter-layer distances of (101) and (112) planes, respectively.

With potentiodynamic (CV) activation for 200 cycles over 0–1.0 V in 1 M Na$_2$SO$_4$ at 25 mV·s^{-1}, the activated Mn$_3$O$_4$ displayed a relatively higher capacitance of about 170 F·g^{-1} at 500 mV·s^{-1}, with high power capability and excellent stability, making it a potential candidate as electrodes of supercapacitors. The activated Mn$_3$O$_4$ possessed an ideal capacitive response, as compared with the potentiodynamically activated MnOOH.

188 *Nanomaterials for Supercapacitors*

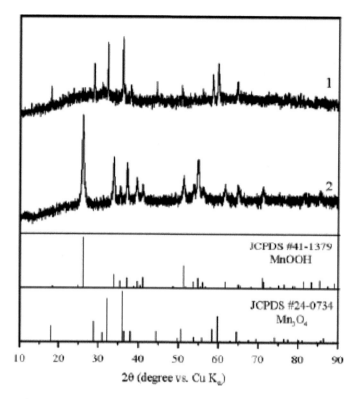

Fig. 4.26. XRD patterns of as-prepared powders: (1) Mn$_3$O$_4$ and (2) MnOOH. Reproduced with permission from [75], Copyright © 2008, American Chemical Society.

A simple method was reported to prepare single crystalline Mn$_3$O$_4$ nano-octahedrons with a very narrow size distribution, by using an ethylenediaminetetraacetic acid disodium salt (EDTA-2Na) aided hydrothermal synthesis [76]. The average side length of square base of the octahedrons was 160 nm. The Mn$_3$O$_4$ nano-octahedrons exhibited a specific capacitance of 322 F·g^{-1}, making them promising electrode materials for supercapacitors.

To synthesize the Mn$_3$O$_4$ nano-octahedrons, 2 ml 0.5 M EDTA-2Na and 5 ml 0.2 M KMnO$_4$ aqueous solution were mixed, with the mixture to be filled in a 50 ml Teflon autoclave, which was then added with distilled water to 70% of the total volume. pH value of the solution was adjusted to be 6.0 with 2 M HNO$_3$ aqueous solution under vigorous magnetic stirring. The Teflon-lined autoclave was sealed tightly and heated at 180°C for 4 hours. After cooling down, the precipitates were collected by filtration and thoroughly washed with distilled water and absolute ethanol, which was finally dried at 60°C for 6 hours.

Representative SEM images of the samples are shown in Fig. 4.28(A–C), demonstrating the general morphology of the Mn$_3$O$_4$ nano-octahedrons [76]. Every octahedron was constructed by two inverted pyramids connected at the square base, thus having eight triangular facets. The edges between the facets were very sharp, while all the facets had a very smooth surface without any obvious defects, as illustrated in Fig. 4.28(C). Average side length of the square base of the octahedrons was 160 nm. Figure 4.28(D) shows XRD pattern of the Mn$_3$O$_4$ nano-octahedrons, exhibiting strong and sharp diffraction peaks, which implied that the sample had been highly crystallized. The sample was a mixture of two manganese oxides, hausmannite Mn$_3$O$_4$ as major phase and pyrolusite MnO$_2$ as minor phase. No other impurities were observed in the XRD pattern.

Figure 4.29(A) shows a TEM image of the Mn$_3$O$_4$ nano-octahedrons, which possessed a uniform tetragonal projected shape, in a good agreement with the SEM results [76]. Figure 4.29(B) shows a representative high-magnification TEM image of an individual nano-octahedron, together with the

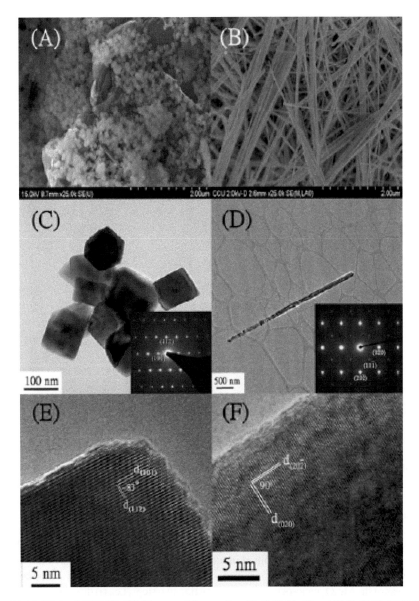

Fig. 4.27. (A, B) FE-SEM and (C-F) HR-TEM images of the powders: (A, C, E) Mn$_3$O$_4$ and (B, D, E) MnOOH. The insets are the corresponding SAED patterns. Reproduced with permission from [75], Copyright © 2008, American Chemical Society.

corresponding edge outlines, as demonstrated in Fig. 4.29(C). Clearly, a high geometric symmetry of the octahedron with smooth surfaces was observed. EDS spectrum is shown in Fig. 4.28(D), indicating that the crystals contained Mn and O, with a chemical formula of Mn$_3$O$_{4.3}$. Therefore, there were 85% Mn$_3$O$_4$ and 15% MnO$_2$, consistent with the XRD pattern shown in Fig. 4.28(D). HRTEM images of two typical vertexes of the octahedron are shown in Fig. 4.29(F, G). The inter-planar spacing of 0.25 nm and 0.27 nm corresponded to the distance of the (211) and (103) planes, respectively. Figure 4.29(H) shows SAED pattern of an individual Mn$_3$O$_4$ nano-octahedron in Fig. 4.29(E), demonstrating its single crystalline nature.

It was found that morphology of the Mn$_3$O$_4$ was dependent on various experimental parameters, including pH value of the reaction solution, type of precursor and hydrothermal reaction temperature. Especially, pH value was a key factor to facilitate the formation of uniform Mn$_3$O$_4$ nano-octahedrons.

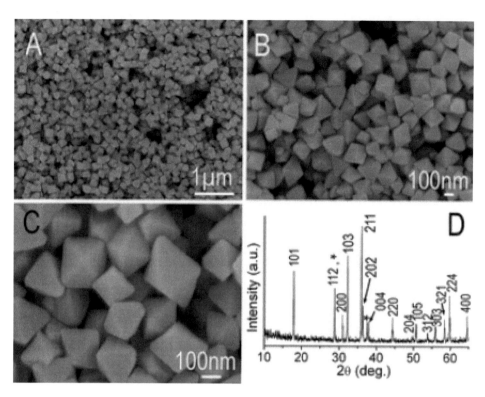

Fig. 4.28. (A–C) SEM images of the Mn$_3$O$_4$ nano-octahedrons at different magnifications and (D) the corresponding XRD pattern. Reproduced with permission from [76], Copyright © 2010, The Royal Society of Chemistry.

Without the pH value adjustment with HNO$_3$, disordered nanostructures would be formed. If the pH value was < 3.0, no precipitation was obtained. Figure 4.30(A) shows SEM image of the sample after reaction for 1.5 hours, which contained nanorods with an average diameter of 40 nm. As the reaction time was increased to 2 hours, nano-octahedrons with high geometric symmetry were present, as seen in Fig. 4.30(B). Further increase in reaction time to 3 hours led to the formation of well-crystallized nano-octahedrons, while all the nanorods were absent, as revealed in Fig. 4.30(C). Hydrothermal reaction of 4 hours was the optimized time, because reaction time to 8 hours resulted in truncated octahedrons, as illustrated in Fig. 4.30(D).

According to the time-dependent morphology evolution, a mechanism was proposed to describe formation process of the Mn$_3$O$_4$ nano-octahedrons, consisting two steps, i.e., (i) dissolution–recrystallization and (ii) Ostwald ripening. Initially, a large amount of nuclei were formed, which grew into nanorods, due to the requirement of minimizing overall energy of the whole system. In fact, the nanorods were unstable and thus in intermediate phases. After reaction for 2 hours, some of the nanorods were dissolved, while nanooctahedrons were developed. As the reaction time was increased to 3 hours, all the nanorods were converted to nano-octahedrons, which implied that the dissolution of the nanorods and the growth of the nano-octahedrons took place at the same time. Further increase in the reaction time led to the formation of well-defined and uniform nano-octahedrons with increased size. Finally, truncated octahedrons were present due to the Ostwald ripening.

Low temperature reduction process is another important method that has been widely used to synthesize crystalline MnO$_2$ powders, with various shapes and morphologies [77–84]. Selected examples will be presented and discussed as follows. For instance, rod-shaped MnO$_2$ powders, consisting of both α- and γ-MnO$_2$, have been synthesized under high viscosity conditions at a relative low temperature of 65°C [77]. The MnO$_2$ had lengths of 250–350 nm and diameters of 10–15 nm, with an aspect ratio of > 20:1. CV behaviors of the MnO$_2$-based supercapacitors were characterized in 2 mol·L^{-1} (NH$_4$)$_2$SO$_4$ aqueous

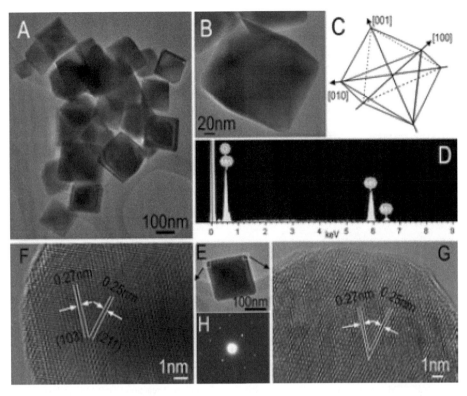

Fig. 4.29. (A–D) TEM images and EDS spectrum of the Mn$_3$O$_4$ nano-octahedrons. (B) Representative high-magnification TEM image of an individual nano-octahedron with smooth surfaces, together with the corresponding edge outlines illustrated in (C). (F, G) HRTEM images taken from the white box areas from (E) and (H) ED pattern. Reproduced with permission from [76], Copyright © 2010, The Royal Society of Chemistry.

solution as the electrolyte. A maximum Specific Capacitance (SC) of 398 F·g^{-1} was achieved according to the galvanostatic charge/discharge curves in the 2 mol·L^{-1} (NH$_4$)$_2$SO$_4$ aqueous solution of electrolyte at a constant current of 10 mA. The SC value was only slightly decreased to 328 F·g^{-1} at a high current of 50 mA, which suggested that such rod-shaped MnO$_2$ could be used for high power supercapacitors. The electrolyte exhibited a resistance of 0.21 Ω·cm^2 according to EIS. Long cycle-life and high coulombic efficiency of the rod-shaped MnO$_2$ electrode were demonstrated by charging and discharging for 1000 cycles.

The Polyethylene Glycol (PG) and polyacrylamide (PAM) used in the experiment had molecular weights of 6000 and 14,000,000, respectively. Starting solutions, including KMnO$_4$ (0.5 mol·L^{-1}), MnCl$_2$ (2 mol·L^{-1}), PG (1 wt%) and PAM (0.1 wt%), were prepared first. 150 mL KMnO$_4$ and 10 mL PG solutions were mixed, with pH value was adjusted to 6–8 with NaOH solution (10 wt%). After that, PAM solution was added to control the viscosity of the mixture to 40–50 MPa·s at 65°C. Finally, 60 mL MnCl$_2$ solution was slowly introduced, followed by ageing for 12 hours. The precipitate was filtered and thoroughly washed with hot deionized water. The filtration process was repeated, followed by treatment in steam for 3 hours. Eventually, black powder MnO$_2$ was obtained after drying in vacuum at 80°C.

A room temperature precipitation route was reported to synthesize rod-shaped MnO$_2$ with δ-type structure from the solution of MnSO$_4$ + K$_2$S$_2$O$_8$ [78]. Electrochemical performance of MnO$_2$ nanorods was studied in 0.5 mol·L^{-1} aqueous electrolyte solutions, including Li$_2$SO$_4$, Na$_2$SO$_4$ and K$_2$SO$_4$. It was found that, at slow scan rates, the nanorods exhibited the highest capacitance of 201 F·g^{-1} in Li$_2$SO$_4$, which was attributed to the fact an additional capacitance was created due to the reversible intercalation/deintercalation of Li$^+$ in the solid phase, in addition to the capacitance caused by the absorption/desorption reaction. In comparison, at fast scan rates, the largest capacitance was achieved in K$_2$SO$_4$, mainly because

192 *Nanomaterials for Supercapacitors*

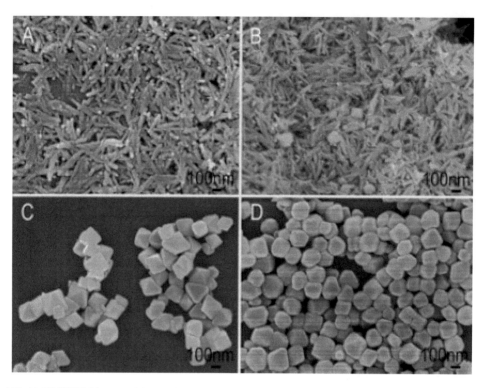

Fig. 4.30. (A–D) FESEM images of the Mn$_3$O$_4$ nano-octahedrons for different reaction times: (A) 1.5 hours, (B) 2 hours, (C) 3 hours and (D) 8 hours. Reproduced with permission from [76], Copyright © 2010, The Royal Society of Chemistry.

K$^+$ has the smallest hydration radius, highest ionic conductivity and lowest Equivalent Series Resistance (ESR). After incorporation with activated carbon to form asymmetric supercapacitor (AC)/K$_2$SO$_4$/MnO$_2$, it possessed an energy density of 17 Wh·kg^{-1} at 2 kW·kg^{-1}, according to the reversible cycle between 0 and 1.8 V. The value was higher than those achieved in its counterparts of AC/K$_2$SO$_4$/AC and AC/Li$_2$SO$_4$/LiMn$_2$O$_4$. Furthermore, the asymmetric supercapacitor demonstrated superb cycling behavior with < 6% capacitance loss after 23,000 cycles at 10 C rate.

To synthesize the MnO$_2$ nanorods, 200 mL 0.1 mol·L^{-1} MnSO$_4$ solution was first mixed with 200 mL 0.1 mol·L^{-1} K$_2$S$_2$O$_8$ solution. After that, 100 mL 1.2 mol·L^{-1} NaOH was added dropwise, so that a precipitate dark brown in color was quickly generated, which was then filtered, thoroughly washed with deionized water and dried at 60°C for 24 hours. The obtained MnO$_2$ powder had a BET specific surface area of about 135 m^2·g^{-1} measured. At the same time, δ-MnO$_2$ microcrystals with a high crystallinity were obtained the thermal decomposition of KMnO$_4$ at 800°C for comparison.

Figure 4.31 shows XRD pattern and TEM images of the MnO$_2$ sample, demonstrating its crystalline structure and morphology. The diffraction peaks at about 12.5° and 25.0° corresponded to (001) and (002), respectively, confirming the δ-type crystal structure of the MnO$_2$, with a layered structure [85]. TEM image revealed that the MnO$_2$ sample contained nanorods, with diameters of < 10 nm. ICP elemental analysis results indicated that the as-prepared products had a small level of Na$^+$ ions, with a Na/Mn mole ratio to be about 1/10, while almost no other elements, such as S and K, were detected.

A low temperature ambient atmosphere method was used to synthesize monodisperse manganese oxide with flowerlike nanostructures [79]. It was observed that reaction temperature had a significant effect on morphology of the MnO$_2$ nanostructures. In addition, the sample prepared at 40°C for 8 hours exhibited a large specific surface area, uniform pore size distribution and promising capacitive performance and thus could be potential candidates as electrode materials for supercapacitor applications.

Oxide Based Supercapacitors: I-Manganese Oxides 193

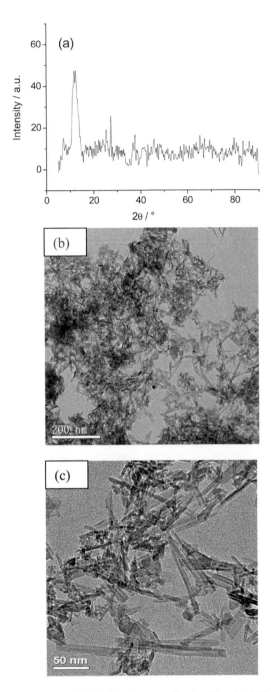

Fig. 4.31. XRD pattern (a) and TEM images of the MnO$_2$ at low magnification (b) and at high magnification (c). Reproduced with permission from [78], Copyright © 2009, American Chemical Society.

To synthesize the MnO$_2$ nanostructures, KMnO$_4$ was dissolved in deionized water, into which formamide (HCONH$_2$) was added with constant stirring. The mixed solution was further stirred vigorously for 5 minutes before it was put into a water bath at 40°C for 8 hours. Black precipitate was formed and collected through centrifugation. The collected samples were thoroughly washed with deionized water

194 *Nanomaterials for Supercapacitors*

and alcohol to remove the possible residuals. After drying in vacuum at 60°C for 10 hours, samples were ready to be characterized.

XRD results indicated that the samples were MnO_2 with a relatively low crystallinity. Figure 4.32 shows TEM and HRTEM images together with SAED pattern of the sample reacted at 40°C for 8 hours. As demonstrated in Fig. 4.32(a), the powder had flowerlike nanostructures, with an average size of 35 nm and uniform size distribution. Figure 4.32(b) shows a higher magnification TEM image of the MnO_2 nanostructures, with the presence of a large number of lamellar platelets. It was also found that many nanoparticles with small size were attached on some of the nanoflowers, as confirmed by the HRTEM image in Fig. 4.32(c). SAED pattern of the MnO_2 nanoparticles revealed broadened and diffused polycrystalline rings, as illustrated in Fig. 4.32(d), confirming the less crystallized nanoflowers, in a good agreement with the XRD result.

Figure 4.33 shows representative TEM images of the samples collected at different reaction times, i.e., 1/12 and 1/6 hours, which served to understand the formation mechanism of the MnO_2 flowerlike nanostructures. After reaction form 1/12 hours, the solution had almost no change in color, so that a very small amount of precipitate could be collected. The small nanostructures consisted of many tiny nanoparticles, as seen in Fig. 4.33(a). As the reaction time was increased to 1/6 hours, the solution was still purple-red in color, while the quantity of precipitate was similarly low. However, size of the nanostructures was increased to about 30 nm, with flowerlike morphology consisting of a number of lamellar platelets, as demonstrated in Fig. 4.33(b). After reaction for 8 hours, the solution became yellow-brown in color, with a large amount of MnO_2 precipitate to be present. In addition, the monodisperse nanostructures had an average size of about 35 nm. A large quantity of small nanoparticles were attached onto the flowerlike structures, as stated earlier.

Reaction temperature also posed influences on morphology and size of the MnO_2 nanostructures. It was observed that, as the reaction temperature was increased from 40 to 99°C, the MnO_2 experienced a change in morphology from monodisperse nanoflowers to the coexistence of nanowires and flowerlike nanostructures. The sample reacted at 80°C already contained several nanowires, which implied that nanowires began to be developed from the flowerlike nanostructures, as seen in Fig. 4.34(a). As the reaction temperature was increased to 99°C, the fraction of nanowires was largely increased, as demonstrated in Fig. 4.34(c). Therefore, with increasing reaction temperature, the nanostructure was evolved from flowerlike nanostructures to nanowires. High temperature meant more energy, so that oriented growth of nuclei was promoted. The diffusion of the smaller MnO_2 nanoparticles at low temperatures was slower than that at high temperatures, so that it was more likely that some of the smaller MnO_2 nanostructures tended to aggregate [86]. As a result, the monodisperse flowerlike MnO_2 nanostructures can be obtained at low temperature. However, at relatively high temperatures, thermal disturbance would increase the rate of mass transport, including diffusion, migration and convection [87]. In this case, the nuclei tended to grow along the low-energy direction on the surface of the flowerlike nanostructures, thus leading to the formation of the MnO_2 nanowires. XRD patterns of the samples synthesized at 80 and 99°C, are shown in Fig. 4.34(b) and (d), respectively. Both patterns confirmed the poor crystallinity of the MnO_2 nanostructures, in a good agreement with the samples after reaction at 40°C, as illustrated in Fig. 4.32(a).

It was found that electrochemical discharge capacity when using the MnO_2 nanostructures as electrode materials was sensitive to morphology. The monodisperse MnO_2 flowerlike nanostructures discharged with a high capacity of 121.5 $F \cdot g^{-1}$ in the first charge-discharge process, due to their high surface area of 225.93 $m^2 \cdot g^{-1}$ and mesoporous structure. The present electrode materials based on the monodisperse MnO_2 nanoflowers also exhibited a good cycle performance at a high current density of 1000 $mA \cdot g^{-1}$. Therefore, monodisperse MnO_2 nanoflowers could be used to assemble MnO_2/AC supercapacitors.

Low-cost layered manganese oxides with the rancieite type of structure have been synthesized through the reduction of $KMnO_4$ or $NaMnO_4$ in acidic aqueous medium [80]. The products could be modified through the successive proton- and alkali-ion exchange reactions, which resulted samples with high surface area of up to 200 $m^2 \cdot g^{-1}$. When using $KMnO_4$ as the precursor, α-MnO_2 phase was obtained. It was found that both the morphology and size of the MnO_2 particles exhibited a direct influence on capacitance of the supercapacitors, with the K-containing rancieite-type compounds demonstrating promising cycleability.

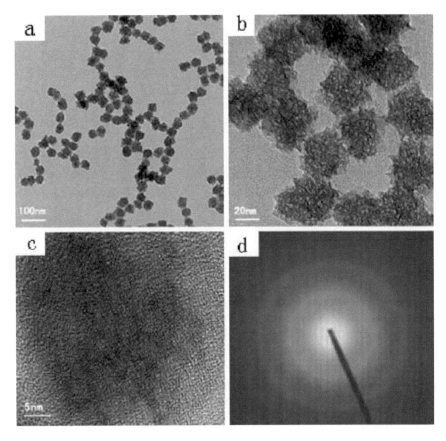

Fig. 4.32. (a) Low-magnification TEM image, (b) high magnification TEM image, (c) HRTEM image and (d) SAED pattern of the monodisperse MnO_2 flowerlike nanostructures after reaction at 40°C for 8 hours. Reproduced with permission from [79], Copyright © 2009, American Chemical Society.

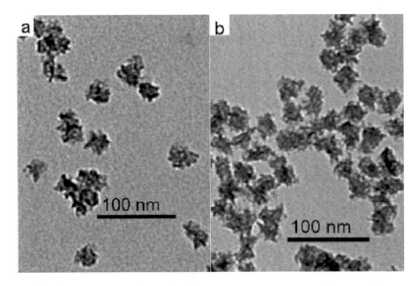

Fig. 4.33. TEM images of the monodisperse MnO_2 3D flowerlike nanostructures after reaction at 40°C for different redox times: (a) 1/12 hour and (b) 1/6 hour. Reproduced with permission from [79], Copyright © 2009, American Chemical Society.

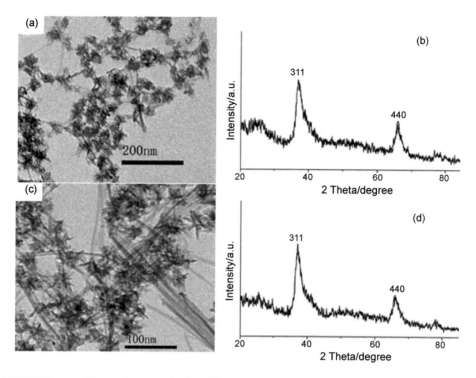

Fig. 4.34. TEM images of the samples synthesized at different reaction temperatures for 8 hours: (a) 80°C and (c) 99°C. XRD patterns of the samples at different reaction temperatures: (b) 80°C and (d) 99°C. Reproduced with permission from [79], Copyright © 2009, American Chemical Society.

MnO_2 powders with different morphologies and sizes have been synthesized by using various wet-chemical methods, such as chemical co-precipitation, hydrothermal synthesis, sovothermal reaction, sol-gel and solution combustion techniques, in order to identify electrode materials for supercapacitors with desired electrochemical performances in mild aqueous electrolytes. For both the amorphous and crystalline MnO_2 powders, specific capacitance had a wide range of values. The charge storage process is determined by various factors, such as porosity, morphology, defect chemistry, crystal structure and residual water content. Because there is no standard to follow, it is difficult to compare the performance of the electrodes made of different MnO_2 powders. In addition, because MnO_2 powders have a serious problem of low electronic conductivity, it is still a challenge to maintain high charge–discharge rates.

Thin film MnO$_2$ electrodes

MnO_2 electrodes in the forms of thin films or coatings have been studied for potential applications as micro-scale energy storage devices. Generally, a thin layer of manganese oxides is deposited on a current collector as the electrodes. Various methods, such as sol–gel coating [88–91], electrodeposition [92], electrophoresis deposition [93, 94] and sputtering-electrochemical oxidation [11]. Selected examples for each method will be described as follows.

Sol–gel deposition

When using the sol–gel process to deposit MnO_2 thin film electrodes for supercapacitors, the first step is to prepare stable MnO_2 colloidal sols. Various MnO_2 sols have been developed for such purposes, including reduction of Mn(VII) (potassium permanganate) with Mn(II) (manganous perchlorate) in alkaline aqueous medium [10], reduction of tetrapropylammonium permanganate with 2-butanol [95], introduction of solid fumaric acid to NaMnO$_4$ [96], mixture of manganese acetate with citric acid containing n-propyl alcohol [97], reaction of KMnO$_4$ with H$_2$SO$_4$ [98, 99], and reduction of KMnO$_4$ with polyacrylamide (PAM) or polyvinyl alcohol (PVA) [100]. After that, MnO$_2$ thin films could be deposited by using dip-coating or casting onto desired substrates, which were then calcined at high temperatures to achieve further crystallization.

One example is the sol-gel synthesis of manganese oxide thin films used electrodes of pseudocapacitive with manganese acetate as the precursor [97]. It was found that the sol-gel derived manganese oxide films consisted of both Mn$_2$O$_3$ and Mn$_3$O$_4$, after thermal annealing at temperatures of > 300°C. The manganese oxide coatings possessed a porous structure caused by the burnout of organic components, with specific capacitances of 53.2, 230.5, 185.6 and 189.9 F·g^{-1} after annealing at 250, 300, 350 and 400°C, respectively. After 300 charging–discharging cycles, the manganese oxide electrodes annealed at 300°C had a specific capacitance retain of 78.6%.

Graphite substrates were used to deposit the manganese oxide coatings. The manganese oxide precursor sols were obtained by dissolving manganese acetate (Mn(CH$_3$COO)$_3$·2H$_2$O) in citric acid containing n-propyl alcohol at room temperature, with a molar ratio of manganese acetate to citric acid to be 1:2. After the solution was stirred for 12 hours at room temperature, ammonium hydroxide was added to adjust pH value to be about 9. After that, transparent viscous sols were developed. To deposit the manganese films, the graphite substrates were slowly dipped into sols. The sol coated substrates were dried at room temperature for 15 minutes, heated at 80°C for 1 hour and then annealed in air at different temperatures.

Figure 4.35 shows XRD patterns of the sol–gel derived manganese oxide coatings thermally annealed at different temperatures [97]. As the annealing temperature was lower than 200°C, amorphous-like structure was observed, without the presence of crystalline phase. Once the annealing temperature was increased to 250°C, diffraction peaks of Mn$_3$O$_4$ and Mn$_2$O$_3$ were detected. With increasing annealing temperature, peak intensity of Mn$_2$O$_3$ was increased, while that of Mn$_3$O$_4$ was decreased, suggesting that Mn$_3$O$_4$ was gradually converted to Mn$_2$O$_3$ during the annealing process. In other words, Mn$_2$O$_3$ phase was more stable at high temperatures [101]. After annealing at 500°C, there was still Mn$_3$O$_4$. It was found that if the annealing temperature was higher than 500°C, voltammetric properties of the materials would be seriously degraded.

CV curve of the 250°C-annealed coating was close to a rectangular shape, with a mirror-image characteristic, implying its reversibility and ideal pseudo-capacitive behavior. However, it had the lowest

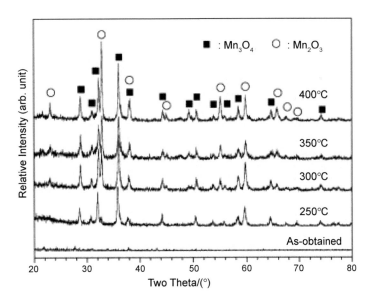

Fig. 4.35. XRD patterns of the sol–gel derived manganese oxide coatings annealed at various temperatures. Reproduced with permission from [97], Copyright © 2007, Elsevier.

specific capacitance of 53.2 F·g^{-1}. Comparatively, CV curves of the 300, 350 and 400°C annealed coatings were somehow deviated from rectangular shapes, whereas their specific capacitance values were much higher than that of the 250°C-annealed sample. Among all the samples, the 300°C-annealed manganese coating demonstrated the highest electrochemical performance. According to the CV curves measured at scan rates of 25, 50 and 100 mV·s^{-1}, specific capacitance were 230.5, 172.3 and 133.1 F·g^{-1}, respectively.

Figure 4.36 shows SEM images of the 300°C-annealed coating before and after 300 cycles of CV testing [97]. After annealing at 300°C, the film sample exhibited a porous structure, as demonstrated in Fig. 4.36(a). The pores were believed to be attributed to the evaporation and burnout of the organic components involved during the sample synthesis process. After 300 CV cycles, a fibrous-like microstructure was developed, as illustrated in Fig. 4.36(b). The 350 and 400°C-annealed coatings displayed similar microstructure. In comparison, the 250°C-annealed film had a relatively smooth surface with limited small pores. After 300 CV cycles, the pores were enlarged, but no fibrous-like microstructure was observed. Because of the mixture phase of Mn$_3$O$_4$ and Mn$_2$O$_3$ and relatively smooth surface with small number of

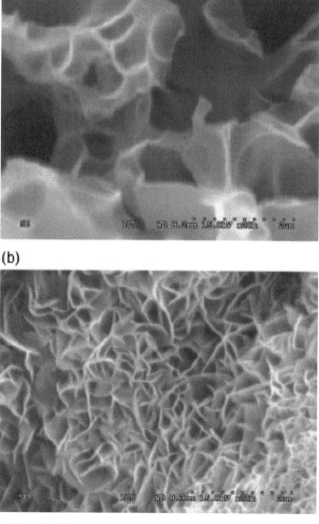

Fig. 4.36. SEM images of the sol–gel derived manganese oxide coatings annealed at 300°C before (a) and after (b) CV measurement. Reproduced with permission from [97], Copyright © 2007, Elsevier.

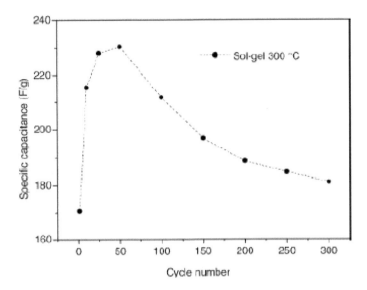

Fig. 4.37. Variation in specific capacitance with respect to CV cycle number for the sol–gel manganese oxide coatings annealed at 300°C. Reproduced with permission from [97], Copyright © 2007, Elsevier.

pores, the effective interaction between the electrolyte ions and the electrode was limited, thus leading to the lowest specific capacitance of the 250°C-annealed sample.

Figure 4.37 shows variation in specific capacitance as a function of CV testing cycles [97]. The specific capacitance was increased initially from 170 F·g^{-1} to a maximum of 230.5 F·g^{-1} after 50 cycles, which was attributed to the fact that surface of the 300°C-annealed was activated at the early stage. After that, the sample experiences a gradual decrease in specific capacitance with increasing number of CV cycle. The specific capacitance approached 181.1 F·g^{-1} after 300 cycles, which was about 78.6% of the maximum value. Synchrotron X-ray absorption spectroscopy results indicated that the relatively low valence states of Mn$^{2.67+}$ (Mn$_3$O$_4$) and Mn^{3+} (Mn$_2$O$_3$) were oxidized into high valences of Mn^{4+} after the CV testing.

Anodic electrodeposition

Anodic electrodeposition works due to the oriented diffusion of charged reactive species through an electrolyte at given electric fields, so that the charged species are oxidized on the deposition surface that also acts as an electrode. To anodically electrodeposit MnO$_2$ thin film, electro-oxidation of Mn(II) species takes place on surface of the anode, given by:

$$Mn^{2+} + 2H_2O \rightarrow MnO_2 + 4H^+ + 2e^-. \qquad (4.7)$$

Desired electrode materials of supercapacitors should have 3D, mesoporous and ordered/periodic architectures, in order for electrolytes and reactants to diffuse in. In this respect, strong efforts have been made to control the morphology of MnO$_2$ electrodes, so as to create as many as possible accessible electroactive sites and shorter cation diffusion lengths. When using anodic electrodeposition method, morphology controlled growth can be realized by controlling the deposition parameters, filling degree of the template membranes and using etched nanoporous substrates. Anodic electrodeposition through galvanostatic or potentiostatic modes usually leads to MnO$_2$ electrodes with surface morphology in the form of a 3D porous fibrous network structure [102–110].

The thin films of carambola-like γ-MnO$_2$ nanoflakes with thickness of about 20 nm and width of 200 nm were deposited on nickel sheets by using the combination of potentiostatic and cyclic voltammetric electrodeposition techniques [107]. The MnO$_2$ nanomaterials were used as the active material of the positive electrode for supercapacitors, with a high specific capacitance of 240 F·g^{-1} at a current density of 1 mA·cm⁻². The γ-MnO₂ nanostructured thin films were electrodeposited at room temperature with the three electrode configuration. Before electrodeposition, nickel sheets used as working electrodes were polished to a smooth surface finish and then washed with distilled water and ethanol, followed by drying under an air stream. A Saturated Calomel Electrode (SCE) was used as the reference electrode, while graphite rod was used as the counter electrode. The electrolyte was composed of 0.1 M Na$_2$SO$_4$ and 0.1 M Mn(CH$_3$COO)$_2$ at pH = 6.0.

The potentiostatic method to deposit MnO$_2$ thin films was conducted at 0.6 V (SCE) for 15 minutes. The MnO$_2$ nanoflakes were prepared by cycling for 15 minutes in combination of cyclic voltammetric technique over the potential range between +0.60 V and +0.30 V (SCE) at a rate of 250 mV·s⁻¹ for 30 s and PS technique at the potential of 0.6 V (SCE) for 1.5 minutes. The MnO$_2$ nanoflakes with a mass densities of 0.05 mg·cm⁻², 0.1 mg·cm⁻² and 0.2 mg·cm⁻² were obtained with the combination of CV and PS techniques for 5 minutes, 15 minutes and 30 minutes, respectively.

Figure 4.38 shows SEM and TEM images of the MnO$_2$ nanostructures with different morphologies. The MnO$_2$ nanoflakes had similar structures to that of carambolas, with star-shaped ridged profile, as

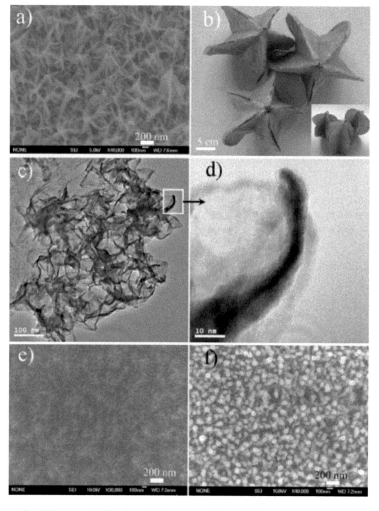

Fig. 4.38. Representative SEM (a) and TEM (c and d) images of the γ-MnO$_2$ samples deposited by using the combined potentiostatic and cyclic voltammetric electrodeposition techniques. SEM images of the γ-MnO$_2$ prepared by using the potentiostatic (e) and cyclic voltammetric (f) techniques. (b) Photograph of a fruit called carambola native to Southeast Asia. Reproduced with permission from [107], Copyright © 2006, Elsevier.

demonstrated in Fig. 4.38(b). TEM images shown in Fig. 4.38(c) and (d) were obtained with samples prepared by directly electrodepositing MnO$_2$ onto carbon-coated copper grid using combination of CV and PS techniques. As illustrated in Fig. 4.38(c), agglomerated nanoflakes with a thickness of 10 nm were developed. HRTEM image revealed that the MnO$_2$ nanostructures were not well crystallized, as seen in Fig. 4.38(d). SEM image of the MnO$_2$ thin film electrodeposited by using the PS technique is shown in Fig. 4.38(e), exhibiting crossed needle-like nanostructures with diameters of 5–10 nm and lengths of 50–100 nm. A top view image of the oriented MnO$_2$ nanorod arrays electrodeposited by using the CV technique is shown in Fig. 4.38(f), with the nanorods having diameters of 50–70 nm.

Figure 4.39 shows XRD patterns of the films having nanoneedles, nanorods and nanoflakes, which indicated that γ-MnO$_2$ was formed when the different electrodeposition techniques were employed. Diffraction peaks of (110), (021) and (061) could be readily identified, confirming the γ-MnO$_2$ of the as-prepared samples. In comparison, both the nanoflake and nanoneedle films possessed higher crystallinity than the nanorod films.

Formation mechanisms of the γ-MnO$_2$ nanorods and nanoflakes during the electrodeposition by using different methods were studied. SEM images of the γ-MnO$_2$ nanostructures deposited by using the CV technique electrodepositing for 0.5 minute, 1.5 minutes and 15 minutes, are shown in Fig. 4.40(a–c), respectively. As illustrated in Fig. 4.40(a), after a short time of deposition, spherical morphology was developed first, with the spheres being closely packed on the surface of the nickel substrate. As the reaction time was slightly increased to 1.5 minutes, surface roughness of the film was significantly increased, as seen in Fig. 4.40(b). Finally, uniform nanorods were obtained, after 15 minutes deposition, as revealed in Fig. 4.40(c). SEM images of the electrodeposited films by using the combined CV and PS techniques for 2 minutes, 6 minutes and 15 minutes, are shown in Fig. 4.40(d–f), respectively. As demonstrated in Fig. 4.49(d), both needle-like and flake-like nanostructures were formed. With increasing deposition time, the needle-like morphology was absent, whereas the flakes were enlarged, thus leading to the formation of the network structures, as illustrated in Fig. 4.40(f).

Another example is electrochemical deposition of amorphous nano-structured manganese oxide onto a stainless-steel electrode [108]. It was found that morphological and capacitive characteristics of the hydrous manganese oxide were strongly dependent on the electrochemical deposition parameters. A highest specific capacitance of 410 F·g^{-1} and specific power of 54 kW·kg^{-1} were achieved at a scanning

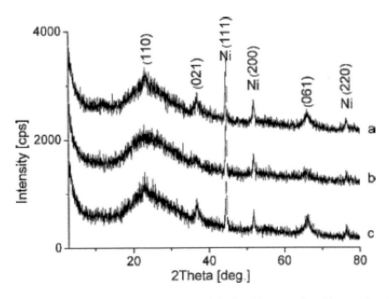

Fig. 4.39. XRD patterns of the γ-MnO$_2$ films with different morphologies: (a) nanoneedles, (b) nanorods and (c) nanoflakes, which were electrodeposited on the Ni substrates by using potentiostatic, CV and combined potentiostatic and CV techniques, respectively. Reproduced with permission from [107], Copyright © 2006, Elsevier.

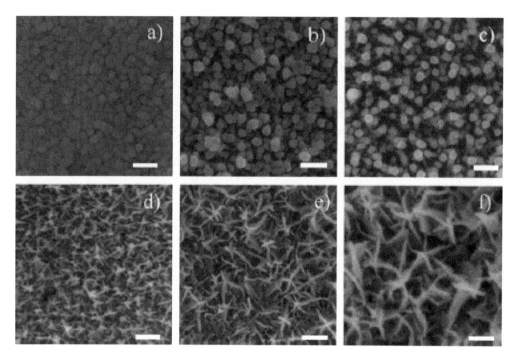

Fig. 4.40. SEM images of the γ-MnO$_2$ films electrodeposited by using CV technique for different time durations: (a) 0.5 minute, (b) 1.5 minutes and (c) 15 minutes. SEM images of the γ-MnO$_2$ films electrodeposited by using the combination of CV and PS techniques for different time durations: (d) 2 minutes, (e) 6 minutes and (f) 15 minutes. All scale bars = 200 nm. Reproduced with permission from [107], Copyright © 2006, Elsevier.

rate of 400 mV·s^{-1}. The specific capacitance could be well retained up to 10,000 cycles, demonstrating excellent cyclic stability of the amorphous hydrous manganese oxide as electrodes of supercapacitors.

Commercially available Stainless-Steel (SS) sheets (grade 304, 0.2 mm thick) were cut into pieces with sizes of 1 cm × 10 cm, which were then polished with emery paper to a rough finish, followed by washing and air-drying. H$_2$SO$_4$, MnSO$_4$·5H$_2$O and Na$_2$SO$_4$ were used as raw materials. An electrochemical cell was assembled with three-electrode configuration, in which the SS strip, a platinum (Pt) plate and a Saturated-Calomel Electrode (SCE) were employed as the working, counter and reference electrode, respectively. Manganese oxide was electrochemically deposited onto the tip area (1 cm × 1 cm) of the SS strip, in an electrolyte solution of 0.15 M H$_2$SO$_4$ mixed with 2 M MnSO$_4$·5H$_2$O, under galvanostatic, potentiostatic and potentiodynamic conditions. After deposition, the electrodes were washed with distilled water and then dried at 40°C for one day.

The deposition conditions are shown in Fig. 4.41(a–c) [108]. pH value of the electrolyte solution was controlled to be 1.2 ± 0.1. Galvanostatic deposition was carried out in the current range of 1.25–5 mA, as illustrated in Fig. 4.41(a). Potentiostatic deposition was conducted at different potentials, as demonstrated in Fig. 4.41(b). Potentiodynamic deposition was processed at different scan rates, as seen in Fig. 4.41(c). According to the potential–pH diagram of Mn, to anodically deposit MnO$_2$ from Mn^{2+}, the experiment must be conducted in the potential range of 1.1–1.4 V at pH ≈ 1.2. As observed in Fig. 4.41(c), the onset potential to oxidize Mn^{2+} was well within this range. For the galvanostatic deposition, a plateau was present in the specific potential range. At current density of ≥ 3 mA·cm^{-2}, the plateau was above the 1.4 V range, implying that Mn^{2+} had been over oxidized. The constant potential as a function of time also revealed a stabilizing current density of about 1 mA·cm^{-2}, as observed in Fig. 4.41(b).

Furthermore, two distinct types of decays were observed in Fig. 4.41(b). The first rapid decay could be attributed to the decomposition of the unstable Mn hydrous oxide, such as MnO$_4^-$, which was formed due to the anodic polarization, whereas the second gradual decay was closely related to the redox potential of

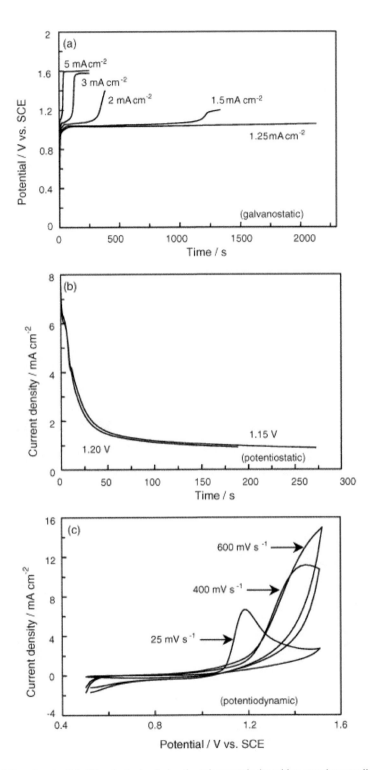

Fig. 4.41. (a) Variations in potential of the electrodes during the galvanostatic deposition at various anodic current densities, (b) variations in the anodic current density of the electrodes during the potentiostatic deposition at various potentials and (c) CV curves of the potentiodynamic deposition at various scan rates, in 2 M MnSO$_4$ + 0.15 M H$_2$SO$_4$ solution at 25 ± 1°C. Reproduced with permission from [108], Copyright © 2006, Elsevier.

MnO$_2$/Mn$_2$O$_3$ [102]. All the curves were similar one another at different potential ranges, which suggested that they might have similar oxidation states, while the electrochemical properties would be mainly dependent on the structure of the materials and the porosity inside the structure at the nanoscale level.

XRD patterns of the manganese oxide samples annealed at different temperatures had been measured. At room temperature, the manganese oxide was of an amorphous structure, which was remained even after annealing at 400°C. However, as the annealing temperature was increased to 500°C, crystalline Mn$_2$O$_3$ was developed. Further increase in annealing temperature resulted in enhanced crystallinity of the manganese oxide.

Figure 4.42 shows SEM images of the films deposited under different conditions. Both the morphology and porosity were dependent on the processing parameters. The nano-sized manganese oxide possessed a wire-like morphology, with diameters of 10–15 nm and lengths of 30–70 nm. Capacitive behavior of the materials was strongly influenced by their porosity and dimension, both of which were closely related to the deposition conditions. The manganese oxide nano-structure exhibited a relatively low density and low meso-porosity.

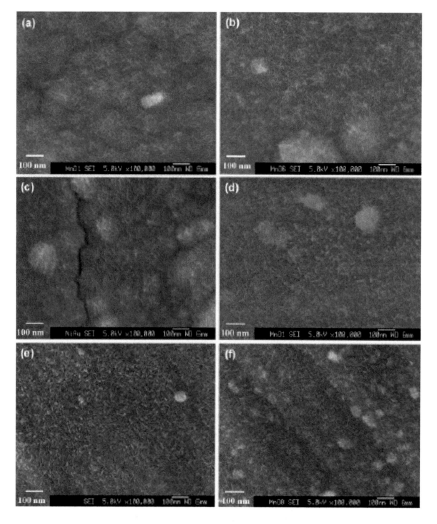

Fig. 4.42. SEM images of the manganese oxide films deposited with different methods under different conditions: (a) 1.5 mA·cm^{-2} (galvanostatic), (b) 1.15 V (potentiostatic), (c) 25 mV·s^{-1} (potentiodynamic), (d) 100 mV·s^{-1}, (e) 400 mV·s^{-1} and (f) 600 mV·s^{-1}. Reproduced with permission from [108], Copyright © 2006, Elsevier.

After potentiodynamic deposition at 400 and 600 mV·s⁻¹, both the density and the meso-porosity of the nano-structure deposits were comparatively high. Specifically, the size of the nano-structure at deposition scan rate of 400 mV·s⁻¹ was smaller than that of the sample deposited at 600 mV·s⁻¹, as demonstrated in Fig. 4.42(e) and (f), respectively. It was therefore concluded that morphology of the manganese oxide deposits could be maintained by controlling the electrochemical deposition parameters.

Cross-sectional SEM images of the sample deposited at 400 mV·s⁻¹ revealed that thickness of the manganese oxide layer was about 10 μm. The SEM images also indicated that the sample was highly porous throughout the thickness, which was observed for all samples, although a quantitative evaluation of porosity according to the cross-sectional SEM images was inaccurate. It was expected that electrolyte could readily penetrate the deposit because of the porosity, so as to improve electrochemical performance of the materials.

Figure 4.43 shows CV curves of the manganese oxide/SS electrodes over the potential range of 0–1.0 V for the samples deposited with different methods. All the curves were of a near-rectangular shape, with symmetrical anodic and cathodic halves, which suggested that the manganese oxide films exhibited near ideal capacitive behavior. The slight distortion from ideal behavior could be attributed to the impedance of the electrodes, which is actually a typical feature of redox-capacitors. The galvanostatic at 1.5 mA·cm⁻² and potentiostatic at 1.15 V samples had specific capacitances of 174 and 153 F·g⁻¹, respectively. As the scan rate of the potentiodynamic deposition was increased from 25 to 600 mV·s⁻¹, the specific capacitance was varied accordingly from 74 to 333 F·g⁻¹. Therefore, different deposition method led to materials with different microstructure and thus electrochemical properties.

MnO_2 nanocrystals with three types of crystal structures, including hexagonal ε-MnO_2 (complex-free), defective rock salt MnO_2 (ethylenediaminetetraacetic acid) and defective antifluorite MnO_2 (citrate), have been synthesized by using an electrodeposition method [109]. Capacitive behaviors of the MnO_2 nanocrystals exhibited a close relationship with their crystal structures. MnO_2 with defective rock salt and antifluorite structures possessed higher capacitive performances than ε-MnO_2. The difference in electrochemical performance between the MnO_2 nanostructures with different crystal structures could be readily understood based on the crystal chemistry. It was observed that specific capacitance of the defective rock salt and antifluorite MnO_2 was not dependent on scanning rate, which could be attributed

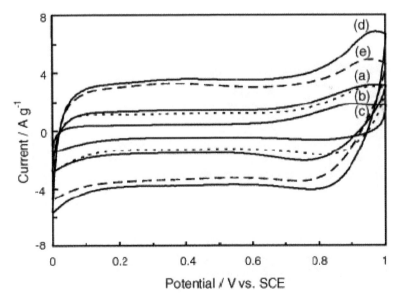

Fig. 4.43. CV curves of the manganese oxide films deposited with different methods under different conditions: (a) 1.5 mA·cm⁻² (galvanostatic), (b) 1.15 V (potentiostatic), (c) 25 mV·s⁻¹ (potentiodynamic), (d) 400 mV·s⁻¹ and (e) 600 mV·s⁻¹. Reproduced with permission from [108], Copyright © 2006, Elsevier.

206 *Nanomaterials for Supercapacitors*

to the fact that physicochemical activities, e.g., phase transformations or morphology variations, took place at high cycling rates.

To obtain the three types of MnO_2, three groups of solutions were used for anodic electrodeposition, including (i) 0.3 M $MnSO_4$, (ii) 0.3 M $MnSO_4$ + 0.2 M EDTA disodium salt and (iii) 0.3 M $MnSO_4$ + 0.3 M sodium citrate. Deposition current density, pH value of electrolyte and electrolyte temperature were 100 mA/cm², 7.0, and 70°C, respectively. The electrodeposition was conducted under vigorous stirring. Quantity of the MnO_2 nanocrystals on the Pt electrodes was controlled to be 0.20–0.25 mg/cm² based on surface area of the electrode. After electrodeposition was finished, the working electrodes were rinsed with deionized water, dried at 100°C for 60 minutes in air and then stored in a vacuum desiccator for electrochemical characterization.

Figure 4.44(a) shows a representative SEM image of the MnO_2 nanocrystals derived from the 0.3 M $MnSO_4$ solution without the use of any complexing agents. The sample exhibited a typical fibrous morphology of electrochemically MnO_2. A bright field (BF) TEM image is depicted in Fig. 4.44(b), indicating that the soft agglomeration of the nanofibers. Figure 4.44(c) shows a Dark Field (DF) TEM image of the sample, confirming the nanocrystalline characteristics of the MnO_2 coating. The oxide grains had a diameter of < 20 nm. Figure 4.44(d) shows corresponding SAD pattern for the MnO_2 coating, in which five characteristic diffraction rings at *d*-spacings of 0.420, 0.243, 0.212, 0.164 and 0.140 nm were observed. With these data, it could be confirmed that the MnO_2 nanocrystals had a hexagonal ε-MnO_2 crystal structure, with space group *P6₃/mmc* and lattice parameters of a = 0.280 nm and c = 0.445 nm.

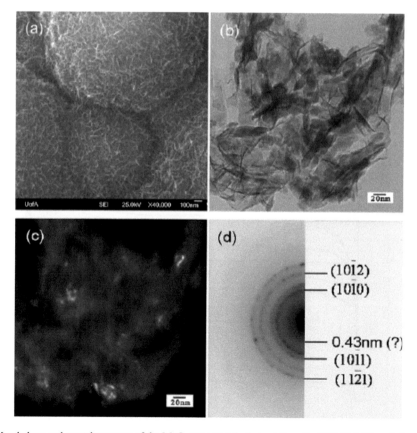

Fig. 4.44. Morphology and crystal structure of the MnO_2 nanocrystals derived from the 0.3 M $MnSO_4$ solution without the presence of complexing agents: (a) SEM image, (b) TEM BF image, (c) TEM DF image and (d) SAD pattern. Reproduced with permission from [109], Copyright © 2008, American Chemical Society.

Figure 4.45 shows representative SEM and TEM images, together with SAD pattern, of the oxide nanocrystals obtained from the 0.3 M MnSO$_4$ + 0.2 M EDTA solution. As seen in Fig. 4.45(a), that the MnO$_2$ coating was comprised of entangled nanoscale fibers. The special morphology of the sample was further confirmed by the BF TEM image, as illustrated in Fig. 4.45(b). Figure 4.45(c) shows a DF TEM image of the sample, indicating that the MnO$_2$ crystals consisted of equiaxed nanocrystals with a diameter of < 10 nm. SAED pattern revealed its Face-Centered Cubic (FCC) symmetry. Combined with simulated electron diffraction patterns based on several possible FCC-type model structures, the MnO$_2$ had a defective rock salt-type structure [111]. The metastable structure was constructed with an FCC array of oxygen anions, with Mn cations to randomly occupy the octahedral interstices.

Figure 4.46 shows the morphology and crystal structure of the MnO$_2$ nanocrystals prepared from the solution of 0.3 M sodium citrate. As illustrated in Fig. 4.46(a), the sample had a rough surface containing agglomerated MnO$_2$ nanoparticles, which exhibited a denser microstructure than the previous two samples. Representative TEM images of the sample are shown in Fig. 4.46(b) and (c), further demonstrating that the sample possessed nanocrystalline grains with a diameter of < 5 nm. The weak and diffused SAD ring pattern, as shown in Fig. 4.46(d), indicated its incomplete crystallinity. d-spacing values revealed that it also possessed an FCC crystal structure. The peak intensities were decreased in the order of [111] > [222] > [200]. Together with simulated patterns from several FCC-type model structures, it was of a defective antifluorite-type structure, in which the Mn cations randomly occupied the eight tetrahedral interstices in the oxygen unit cell.

In summary, with the presence of complexing agents, including ethylenediaminetetraacetic acid (EDTA), disodium salt and sodium citrate salt, crystal structure of the MnO$_2$ nanocrystals synthesized by

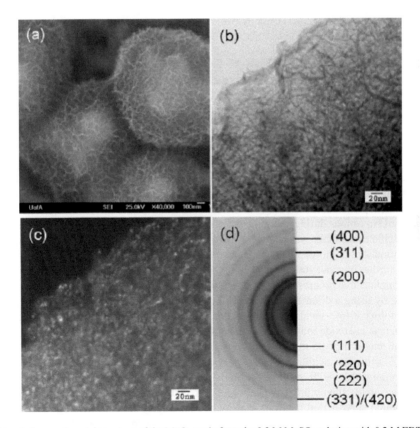

Fig. 4.45. Morphology and crystal structure of the MnO$_2$ made from the 0.3 M MnSO$_4$ solution with 0.2 M EDTA: (a) SEM image, (b) TEM BF image, (c) TEM DF image and (d) SAD pattern. Reproduced with permission from [109], Copyright © 2008, American Chemical Society.

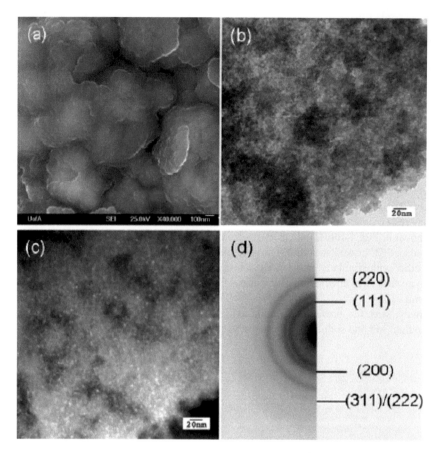

Fig. 4.46. Morphology and crystal structure of MnO$_2$ nanocrystals prepared from the 0.3 M MnSO$_4$ solution with 0.3 M sodium citrate: (a) SEM image, (b) TEM BF image, (c) TEM DF image and (d) SAD pattern. Reproduced with permission from [109], Copyright © 2008, American Chemical Society.

using the anodic electrodeposition could be well controlled. The MnO$_2$ nanocrystals with different crystal structures exhibited different electrochemical behaviors, which could be understood in terms of octahedral vacancies. The vacancies had a lower energy barrier in antifluorite MnO$_2$, thus allowing faster intercalation/deintercalation of Na cations during charging and discharging. For the defective rock salt and antifluorite MnO$_2$, their specific capacitance was not decreased with increasing scanning rate, which could be related to phase transformations or morphology changes at high cycling rates. These interesting findings could be used a guide to further improve the electrochemical performances of MnO$_2$ electrode materials.

Another template-free process has been developed by preparing MnO$_2$ based electrode materials for supercapacitors, by using a diluted electrolyte [112, 113] or applying cyclic voltammetry [103], which could be in the forms of free-standing micro- and nano-scale fibers, rods and interconnected nanosheets. For example, manganese oxide rods with high porosity and diameters of 1–1.5 μm were electrochemically prepared through anodic deposition under galvanostatic control, from a diluted solution of Mn(CH$_3$COO)$_2$, onto Au coated Si substrates, without using any surfactants, catalysts or templates [112]. Electrodes made of the rod-like manganese oxide exhibited a high specific capacitance at low current densities, due to their large surface areas. The optimized value of specific capacitance was 185 F·g^{-1}, which was obtained by using the sample deposited at a current density of 5 mA·cm^{-2}. A 75% retention of specific capacity was observed for all samples after 250 charge–discharge cycles in an aqueous solution of 0.5 M Na$_2$SO$_4$.

The rod-like manganese oxide coatings were anodically electrodeposited from two dilutes solutions, i.e., (i) 0.01 M MnSO$_4$ and (ii) 0.01 M Mn(CH$_3$COO)$_2$, onto Si substrates coated with Au of 500 nm in thickness, in a hot water bath under galvanostatic control, at current densities of 5–30 mA·cm^{-2}. A Pt plate was used as counter electrode, which was placed vertically 20 mm away from the vertical Au coated Si substrate, serving as the working electrode. The electrolyte solution had a pH value of 6.5 and temperature of 60°C. All deposition experiments were conducted for 10 minutes, under vigorous stirring. After electrodeposition, the working electrodes were rinsed with deionized water and then dried at 100°C for 60 minutes in air to remove the residual water. The specific deposit mass was controlled to be in the range of 0.15–0.20 mg·cm^{-2}.

Composition, morphology and crystal structure of the manganese oxide coatings could be well maintained by controlling the deposition parameters, including the use of complexing agents, current density of deposition, voltage and pH value of the solutions [112]. Figure 4.47 shows SEM images of the manganese oxide coatings derived from the solutions of 0.01 M MnSO$_4$ and 0.01 M Mn(CH$_3$COO)$_2$ (acetate served as complexing agent). As shown in Fig. 4.47(a) and (c), the two samples exhibited different morphologies. The coating derived from the solution of 0.01 M MnSO$_4$ was continuous with a certain level of porosity, as seen in Fig. 4.47(a) and (b). The one obtained from the solution of 0.01 M Mn(CH$_3$COO)$_2$ consisted of free-standing rods with a length of 10 μm and diameter of < 1.5 μm, as illustrated in Fig. 4.47(c) and (d).

Under galvanostatic control, the manganese oxide might be grown on energetically favorable sites, thus leading to a highly porous structure. The rod-like structures would result in a large specific surface area, which was expected to be able to provide efficient contact between the active materials and the electrolyte, so as to create more active sites for electrochemical reactions. As a result, such manganese oxide coatings would have sufficiently high conductivity for electronic transport, which would lead to increased capacitance and enhanced cycling rate capability. Furthermore, the smaller the diameter of rods, the larger the specific surface area would be available for electrolyte access. In addition, rod-like structures

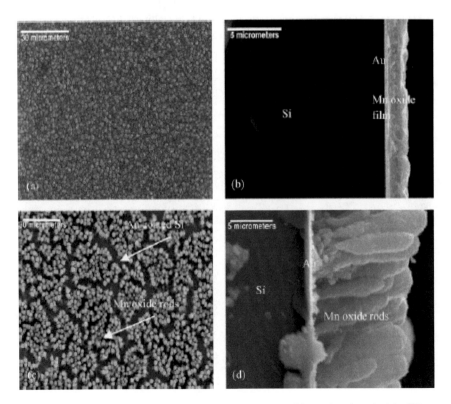

Fig. 4.47. Surface and cross-sectional SEM images of the manganese oxide coating deposited in different solutions (pH = 6.5): (a, b) 0.01 M MnSO$_4$ at i = 30 mA·cm^{-2} and (c, d) 0.01 M Mn(CH$_3$COO)$_2$ at i = 5 mA·cm^{-2}. Reproduced with permission from [112], Copyright © 2010, Elsevier.

implied short diffusion path lengths for both ions and electrons, while sufficiently high porosity would be beneficial to electrolyte penetration and thus offered high charge/discharge rates [33, 114].

The samples from the solution of Mn (CH$_3$COO)$_2$ displayed a more favorable morphology, with the manganese oxide rods that could be controlled by simply adjusting the current density of deposition. Figure 4.48 shows SEM images of the sampled deposited with the solution of 0.01 M Mn (CH$_3$COO)$_2$. It was found that the manganese oxide coatings deposited at relatively low deposition current densities consisted of uniform vertically arranged rods, as demonstrated in Fig. 4.48(a–e). At higher deposition current densities, the rods were agglomerated, as revealed in Fig. 4.48(c) and (f). Comparatively, the more uniform the microstructure, the higher the electrochemical performance would be.

Figure 4.49 shows TEM images of the individual manganese oxide rods in order to reveal more details of the microstructure. Figure 4.49(a) depicts a representative Bright Field (BF) image of two rods, deposited at 5 mA·cm^{-2}, demonstrating a fibrous surface morphology. Figure 4.49(b) shows an SAED pattern and a Dark Field (DF) image of the circled region in Fig. 4.49(a). The diffraction pattern indicated that the rods was of a polycrystalline structure, while the DF image revealed that the manganese oxide grains had a diameter of <10 nm. The SAED pattern could be indexed to NaCl-type FCC crystal structure, with space group $fm\overline{3}m$ and lattice parameter of 0.445 nm.

As mentioned earlier, manganese oxide rods deposited at low deposition current densities exhibited higher specific capacitances, with the optimal current density to be 5 mA·cm^{-2}. For the electrodes derived from the solution of 0.01 M Mn (CH$_3$COO)$_2$ at different deposition current densities, the capacitance could be retained 75%, after cycling at 20 mV s^{-1} for 250 cycles. The morphology of the samples deposited at lower current densities has no significant change after cycling, whereas the rods prepared at high current densities became a petal-like morphology. The variation in morphology was caused by the partial dissolution of the manganese oxide during the cycling. This was because the charge/discharge process of MnO$_2$ experienced a redox reaction between the Mn(III) and Mn(IV).

Anodic Al Oxide (AAO) templates and Lyotropic Liquid Crystalline (LLC) phases have been employed to template the deposition of MnO$_2$ for electrodes of supercapacitors [115–118]. MnO$_2$ nanostructures

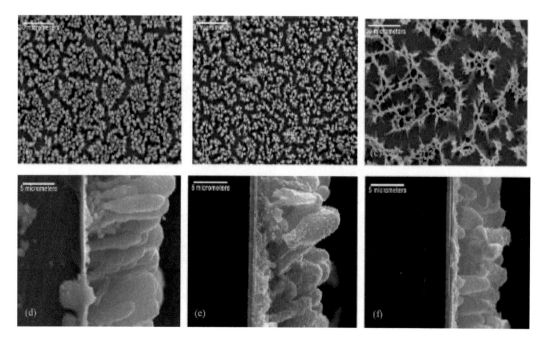

Fig. 4.48. Surface and cross-sectional SEM images of Mn oxide deposits prepared in the solution of 0.01 M Mn(CH$_3$COO)$_2$ at different current densities (pH = 6.5): (a, d) i = 5 mA·cm^{-2}, (b, e) i = 15 mA·cm^{-2} and (c, f) i = 30 mA·cm^{-2}. Reproduced with permission from [112], Copyright © 2010, Elsevier.

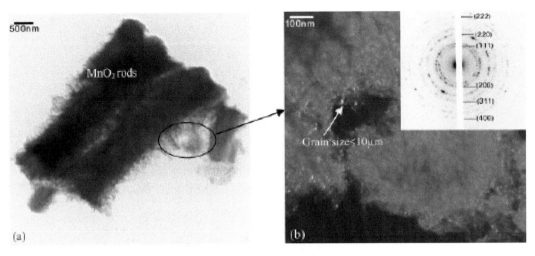

Fig. 4.49. Morphology and crystal structure of the manganese oxide deposited in the solution of 0.01 M Mn(CH$_3$COO)$_2$ at $i = 5$ mA·cm^{-2}. (a) TEM BF image and (b) TEM DF image and SAED pattern. Reproduced with permission from [112], Copyright © 2010, Elsevier.

could be in the forms with oriented nanofibrous, nanotubular and mesoporous ravine-like morphologies. One example is the growth of high surface area and ordered MnO$_2$ nanowire array through AAO template on Ti/Si substrate, which were used as electrodes of supercapacitors [115]. Electrochemical performances of the ordered MnO$_2$ nanowire arrays were characterized by using cyclic voltammetry in 0.5 M Na$_2$SO$_4$ aqueous solution. The nanowire array thin films acted as the working electrode. A specific capacitance of about 254 F·g^{-1} was achieved by using an electrode made of the ordered MnO$_2$ nanowire arrays.

Highly pure (99.999%) Al film with a thickness of about 3.0 μm was deposited on p-type silicon substrate coated with a Ti film of about 300 nm by using radio frequency sputtering. Th anodization process was conducted in 0.3 M oxalic acid solution at room temperature at 40 V, as the pore diameter was about 40 nm. The resulting alumina film was etched out with 0.4 M H$_3$PO$_4$ plus 0.2 M H$_2$Cr$_2$O$_4$ at 40°C for 20 minutes. The remaining aluminum was re-anodized under the same conditions until the metal film was entirely oxidized. The anodization was continuously processed for 90–120 minutes, so as to remove the barrier layer.

An aqueous solution of 0.5 M Mn (CH$_3$COO)$_2$ was used as the electrolyte to deposit the MnO$_2$ nanowire array thin films. Electrodeposition was performed at room temperature, with a three-electrode potentiostatic control, consisting of a DC electrodeposition system with a saturated calomel electrode as reference electrode, a 1.0 cm × 1.0 cm platinum plate as the counter electrode and the AAO/Ti/Si substrate as working electrode. The electrolysis was conducted at 1.0 V for 2 hours. After electrodeposition was finished, the samples were soaked in 0.5 M NaOH solution for 1 hour to remove the AAO on the Ti/Si substrate. To measure electrochemical performance, the nanowire arrays on the Ti/Si substrate were used as a working electrode, platinum foil and saturated calomel electrodes were employed as the counter and reference electrodes, respectively.

Figure 4.50 shows SEM images of the samples after the two-step anodization. The template exhibited parallel pores with a relatively narrow pore size distribution The pore array possessed an average pore diameter about 40 nm, an interspace of about 60 nm and a pore density of about 10^{10} cm^{-2}, as revealed in Fig. 4.50(a). As demonstrated in Fig. 4.50(b), the Ti adhesion layer had a thickness of about 300 nm, which was not uniform due to the fracture of the sample for SEM observation. The pores were partly open to the Ti layer, due to the short time of oxidation. The oxide barrier layer could be entirely removed by increasing the oxidation time.

Figure 4.51 shows a representative surface SEM image of the MnO$_2$ nanowire arrays. Clusters were formed from the Ti/Si substrate, offering a high surface area that was beneficial to the performance of

212 *Nanomaterials for Supercapacitors*

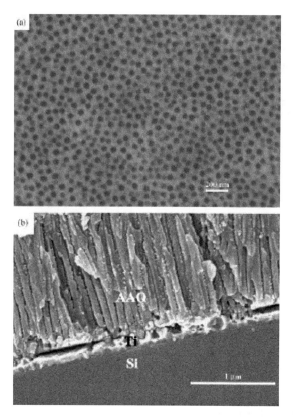

Fig. 4.50. SEM images of the alumina template on Ti/Si substrate prepared by using a two-step anodization process in 0.3 M oxalic acid solution at room temperature: (a) top view and (b) fractured side view. Reproduced with permission from [115], Copyright © 2006, Elsevier.

Fig. 4.51. Surface SEM image of the MnO_2 nanowire arrays grown on the AAO/Ti/Si substrate. Reproduced with permission from [115], Copyright © 2006, Elsevier.

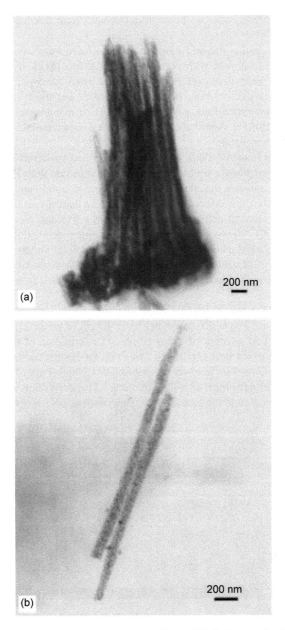

Fig. 4.52. TEM images of selected MnO$_2$ nanowires: (a) truss nanowires and (b) single nanowires. Reproduced with permission from [115], Copyright © 2006, Elsevier.

electrode. The clusters were derived, because the nanowires were uncovered from the framework of the porous anodic alumina template as an incomplete freestanding form. After the porous anodic alumina template was etched away, the nanowires inside the template were released gradually, which tended to aggregate to minimize the total free energy of the system. The nanowires were abundant, uniform and highly ordered over a large area, with a pore filling rate of above 90%. The diameter and the length of the nanowires are determined from Fig. 4.52 shows representative TEM images of some MnO$_2$ nanowires, which exhibited a diameter of about 40 nm and a length of over 1 μm. The diameter of the MnO$_2$ nanowires was comparable with the pore diameters of the AAO template. XRD patterns indicated that the samples contained main peaks of α-MnO$_2$, whereas the samples thermally treated at temperatures of < 400°C were amorphous.

Highly ordered MnO$_2$ nanotube and nanowire arrays have been similarly prepared by using electrochemical deposition technique with porous AAO templates [118]. It was found that the MnO$_2$ nanotube array had higher capacitive performance than the MnO$_2$ nanowire array. Besides, the MnO$_2$ nanotube array electrode also demonstrated better rate capability and higher cycling stability. To deposit the MnO$_2$ nanotube and nanowire arrays, a solution with 0.1 M manganese sulfate and 0.1 M sodium sulfate was used as the electrolyte. Alumina membranes with a pore diameter of 200 nm and a thickness of 60 μm were used as the templates.

Working electrode was made of a thin layer of Pt, which was sputtered on one side of the alumina membrane. The Pt-coated membrane was adhered onto a Pt substrate with Pt paste, which was heated at 700°C for 30 minutes to enhance the adhesion between the two components after drying of the paste. Parafilm was used to cover the exposed area of the Pt substrate, so that only surface of the alumina template was exposed. The electrochemical cell was constructed with a working electrode, a counter electrode of carbon rod and a Ag|AgCl reference electrode. Electrochemical deposition of MnO$_2$ was carried out under galvanostatic conditions at a constant current of 2 mA·cm^{-2}. By simply controlling the deposition time, either MnO$_2$ nanotubes or nanowires could be obtained, with deposition times of 10 minutes and 60 minutes respectively. The alumina template was completely removed from the deposited samples by soaking in 1 M KOH solution.

Figure 4.53 shows SEM images of the MnO$_2$ nanotube and nanowire arrays at different magnifications after removal of the template. The MnO$_2$ nanotubes had a length of about 2 μm, as seen in Fig. 4.53(c), while the nanowires exhibited a length of about 10 μm. The formation of the nanotube structure was closely related to the shape of the base electrodes. The Pt on the bottom surface of the alumina template and a portion of the inside channel surface were responsible for the presence of a ring morphology, so that MnO$_2$ was initially deposited in the form tube-like structure. As the deposition was proceeded continuously, the tubes would be filled, leading to the formation of the wire-like MnO$_2$ arrays, after a prolonged period of deposition.

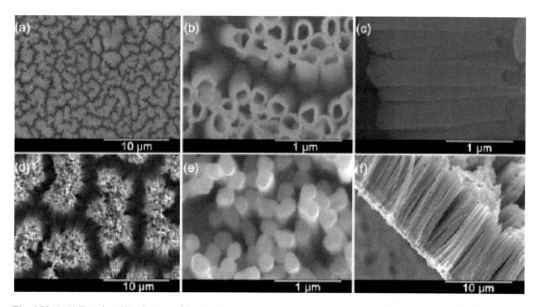

Fig. 4.53. (a, b) Top view SEM images of the MnO$_2$ nanotube arrays. (c) Cross-sectional SEM images of the MnO$_2$ nanotube arrays. (d, e) Top view SEM images of the MnO$_2$ nanowire arrays. (f) Cross-sectional SEM image of the MnO$_2$ nanowire arrays. Reproduced with permission from [118], Copyright © 2010, Elsevier.

Figure 4.54 shows representative TEM images of the MnO$_2$ nanotubes and nanowires. As illustrated in Fig. 4.54(a), it was an image of an open-ended MnO$_2$ tube at low magnification, with the tubular structure to be evidenced by the fact that the center area was brighter than the edges. Electron Diffraction pattern (ED) indicated that the as-electrodeposited MnO$_2$ tube was amorphous. The tube had inner diameters of about 120–130 nm and wall thicknesses of about 40–50 nm, as revealed in Fig. 4.54(b). Figure 4.54(c) shows a low magnification TEM image of a broken MnO$_2$ nanowire, confirming its amorphous nature. Figure 4.54(d) shows a TEM image of a MnO$_2$ nanowire, which had a diameter of about 200 nm, compatible to the pore diameter of the alumina template.

Electrochemical properties of the electrodes made of the MnO$_2$ nanotube and nanowire arrays were characterized through CV measurements in 1 M Na$_2$SO$_4$ electrolyte. The CV curves of the nanotube array were nearly rectangular in shape, with a mirror-image profile, so that the electrode exhibited good reversibility and ideal capacitive behavior. In comparison, CV curves of nanowire electrodes were distorted from the rectangular shape, indicating their poor capacitive effect. Specific capacitances for the MnO$_2$ nanotube and nanowire array electrodes were 320 and 101 F·g^{-1}, respectively. The superior capacitive behavior of the MnO$_2$ nanotube arrays could be understood from their nanotubular structure, which led to shorter diffusion paths for both the cations and electrons. For the MnO$_2$ nanotube array electrode, the initial capacitance could be well retained by 70%, as the current density was increased by 10 times, implying its high rate capability. In addition, about 81% of the initial capacitance was retained after 2000 cycles, suggesting its high cycling stability.

Fig. 4.54. (a, b) TEM images of the open-ended MnO$_2$ nanotube, with the inset in (a) showing the corresponding ED pattern. (c, d) TEM images of a broken MnO$_2$ nanowire, with the inset in (c) demonstrating the corresponding ED pattern. Reproduced with permission from [118], Copyright © 2010, Elsevier.

A nanoporous nickel (Ni) substrate was prepared through the selective dissolution of copper (Cu) from a Ni–Cu alloy layer [119], in which the etching of Cu and codeposition of Ni/Cu could be conducted in same solution. On the nanoporous Ni substrates, highly disperse fibrous manganese (Mn) oxide could be obtained through anodic deposition. Such manganese oxide could be used as electrode of supercapacitors. The electrochemical procedure to produce high-porosity structure was extremely simple and efficient. Pseudocapacitive behavior of the manganese oxide electrode evaluated by using cyclic voltammetry in 0.1 M Na_2SO_4 solution demonstrated that a specific capacitance of 502 $F·g^{-1}$ could be readily achieved. The capacitance value was retained 93% after 500 charge–discharge cycles, which was attributed to the nanoporous structure of the Mn oxide.

Ni–Cu alloy films were electrodeposited from a plating solution containing 1 M $NiSO_4$, 0.01 M $CuSO_4$ and 0.5 M H_3BO_3, with a pH of 4. The deposition was conducted at 25°C with a three-electrode cell, which consisted of a platinum counter electrode and a saturated calomel reference electrode (SCE, +0.241 V vs. SHE). Glass slides coated with Indium-doped Tin Oxide (ITO) with exposed area of 1.4 cm^2 served as the working electrode. The alloy films were deposited at a constant potential, while the total cathodic passed charge was controlled to be about 1 C. Cu was then selectively dissolved from the alloy in the same solution by applying an anodic potential of 0.5 V, until a cutoff current density of 10 $\mu A·cm^{-2}$ was reached. As a result, porous Ni electrode was developed. After that, anodic deposition of Mn oxide on the porous Ni substrate was carried out in a 0.5 M Mn $(CH_3COO)_2$ aqueous solution. The applied anodic potential was 0.5 V (vs. SCE), so that a total passed charge of 50 mC was achieved. Mn oxide of the same quantity was also deposited on a flat Ni substrate in order for comparison.

Figure 4.55(a) shows a surface SEM image of the alloy film deposited at −0.85 V, with a Ni/Cu atomic ratio of about 50/50. Figure 4.55(b) shows SEM image of the alloy film after it was etched at 0.5 V. The selective dissolution of Cu from the alloy resulted in porous Ni, which had pore sizes in the range of 50–150 nm and a pore density of about 10^{12} cm^{-2}. Figure 4.55(c) depicts a SEM image of the Mn oxide anodically deposited on the nanoporous Ni substrate. The Mn oxide exhibited a fibrous morphology, with nanosized diameter, which was mostly dispersed on surface of the electrode, with small amount of them to be filled into the nanopores. Therefore, the Mn oxide electrode had a high surface area, due to the nanoporous Ni substrate, as schematically demonstrated in Fig. 4.56.

The Mn oxide on the flat Ni substrates had a specific pseudocapacitance of 271 $F·g^{-1}$, the sample on the nanoporous Ni substrate exhibited a specific capacitance of 502 $F·g^{-1}$. The high surface area of the conductive Ni substrate offered more active sites for pseudocapacitive reaction and thus largely boosted the electrochemical reactivity of the Mn oxide. Also, due to the high porosity structure of the Ni substrate, the fibrous Mn oxide nanoparticles were uniformly dispersed, which significantly increased the accessibility of electrolyte and thus enhanced the transportation of ions within the electrode. As a result, higher electrochemical performance was achieved.

Similarly, high-porosity Mn oxide has also been deposited on micro-etched Duplex Stainless Steel (DSS) for electrode applications of supercapacitors [120]. It was found that as the γ phase was selectively dissolved, α phase network would be formed, and vice versa. A surface with a concave–convex morphology could be realized by controlling the extent of selective dissolution, so as to increase the surface area.

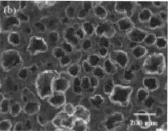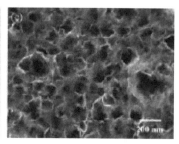

Fig. 4.55. SEM images of the samples: (a) as-deposited Ni–Cu alloy film, (b) nanoporous Ni film and (c) high-porosity Mn oxide electrode. Reproduced with permission from [119], Copyright © 2008, Elsevier.

Oxide Based Supercapacitors: I-Manganese Oxides 217

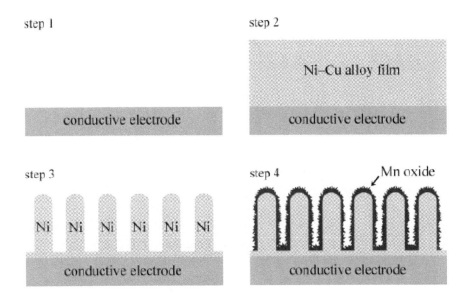

Fig. 4.56. Schematic diagram demonstrating the electrochemical deposition process of the high-porosity Mn oxide electrode. (Step 1) Conductive materials served as electrode substrates. (Step 2) Ni and Cu were codeposited on the substrate to form alloy films. (Step 3) Cu was selectively dissolved to leave Ni layer with a nanoporous structure. (Step 4) Fibrous Mn oxide was deposited on the porous Ni through anodic deposition. Reproduced with permission from [119], Copyright © 2008, Elsevier.

Manganese oxide layers were anodically deposited onto the etched dual phase steel substrate which also acted as the current collector. Cyclic Voltammetry (CV) results indicated that specific capacitance was increased as the etching depth of either the γ or α phase was increased, with the selective dissolution of α phase demonstrating more pronounced effect on the specific capacitance.

Figure 4.57 shows schematic diagram of deposition of the manganese oxide coating onto the stainless steel substrate with a concave/convex feature. A mixed solution of 2 M H_2SO_4 + 0.5 M HCl was used as the electrolyte, with etching potentials to be determined according to the potentiodynamic polarization curve of 2205 DSS. The potentiodynamic polarization curve in the active-to-passive transition region of 2205 DSS in mixed solution had been measured, with the polarization curve that could be dissociated into two sub-curves, one for the γ phase and the other for α phase, respectively. Their characteristic peak potentials, i.e., −260 mV and −320 mV, with respect to a saturated calomel electrode, SCE, were used for the potentiostatic etching of the γ and α phases, respectively. The etching time was in the range of 1–4 hours. The samples were denoted as A1, A2 and A4 for the 2205 DSS etched to selectively dissolve the γ phase for 1, 2 and 4 hours, respectively. Accordingly, sample A0 was not etched. The samples with α phase to be selectively dissolved were denoted as F0, F1, F2 and F4.

Manganese oxide was deposited onto the etched DSS substrates by anodic electroplating in 0.25 M $MnSO_4$ at room temperature. Anodic deposition was conducted with a three-electrode cell at a constant potential. During the deposition, the substrates acted as the anode, whereas the counter electrode and reference electrode were a Pt sheet and an SCE, respectively. A total charge of 0.2 C was applied at a constant potential of 0.9 V. Figure 4.68(a) shows representative SEM images of the steel samples after the γ phase was selectively dissolved for various time durations. As the γ phase was selectively dissolved, α phase was left, which extruded on the surface. With increasing etching time, the depth of the dissolved γ phase was increased, as illustrated in Fig. 4.58(b).

It was noted that α phase was also dissolved at a negligible rate. The etching depth could be estimated from the cross-sectional SEM images. For selective dissolution of the γ phase, the etch depth was increased almost linearly with increasing time, with an average rate of about 1.9 μm·h^{-1}. A depth of about 8 μm was achieved after etching for 4 hours. Surface area of the sample after 4 hours etching was increased by about 200%. Similar results were observed if α phase was selectively dissolved. The average dissolution

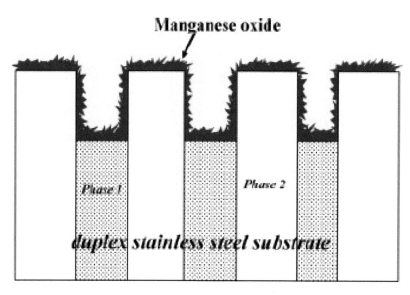

Fig. 4.57. Schematic diagram showing cross-sectional feature of the selectively etched 2205 DSS coated with manganese oxide. Reproduced with permission from [119], Copyright © 2008, Elsevier.

rate of the α phase was higher than that of the γ phase, with a value of about 3.1 μm·h^{-1} at −320 mV. The difference in etching rate between γ and α phases was attributed to the difference between chemical compositions and applied etching potentials.

Specific capacitances of the samples were measured potentials in the range of 0–0.8 V. At a scan rate of 5 mV·s^{-1}, capacitance of the F4 sample was 209 F·g^{-1}, which was higher than that of the F0 sample (132 F·g^{-1}) by nearly 60%. At higher scan rates, e.g., 100 mV·s^{-1}, the increase in specific capacitance for the F4 electrode was about 166%. Because the rate of selective dissolution of α phase was higher than that of γ phase at their respective peak potentials, the increase in surface area was higher for electrodes F1, F2 and F4 than for A1, A2 and A4, respectively. Therefore, selective dissolution of α phase was preferred in terms of enhancement of specific capacitance of the materials.

Cathodic electrodeposition

There are two electrochemical processes that can be used for cathodic deposition of manganese oxide thin films, according to the reactions at surface of the cathode. One is known as electrogeneration of base, involving either reactions that consume H$^+$ ions or electrolysis of water. Due to such reduction reactions, pH value of the electrolyte near the cathode could be increased, because of either the consumption of H$^+$ ions or the production of OH$^-$ ions, both of which compete with the reduction reaction of metal ions, thus becoming dominant in the reaction process. In this case, the reduction of metallic ions to metals is prevented. Instead, the metallic ions are deposited as hydroxides on the cathode, with reaction equation to be given by:

$$Mn^{2+} + 2OH^- = Mn(OH)_2\downarrow. \qquad (4.8)$$

The manganese hydroxide is dehydrated into manganese oxides after thermal annealing at appropriate temperatures. For example, smooth amorphous MnO$_x$ films have been deposited by using this method from polyethylenimine (PEI)– or chitosan–MnCl$_2$ solutions [92]. Mn$_3$O$_4$ films with a porous/nanoflake hierarchical architecture can be deposited from manganese acetate-containing solutions [121].

The second cathodic process was conducted through the electro-reduction of Mn(VII) on surface of the cathode, with reaction equation to be expressed as:

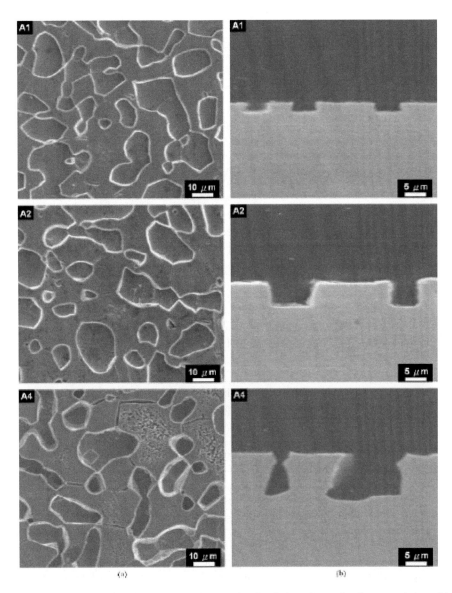

Fig. 4.58. (a) Surface SEM images of the 2205 DSS after selective dissolution of austenite phase at peak potentials for 1, 2 and 4 hours. (b) The corresponding cross-sectional images demonstrating the increase in etching depth with increasing etching time. Reproduced with permission from [119], Copyright © 2008, Elsevier.

$MnO_4^- + 2H_2O + 3e^- \rightarrow MnO_2 + 4OH^-$. (4.9)

Through reaction, nanostructured manganese dioxide films can be deposited, by using galvanostatic, pulse and reverse pulse electrodeposition [122–124]. It was found that concentration of the active MnO_4^- ions has a direct influence on deposition rate, composition and microstructure of the films.

Manganese oxide films were deposited by a cathodic electrodeposition from $KMnO_4$ aqueous solutions, which could be used as electrodes of supercapacitors with promising electrochemical performances [122]. At a given current density, quantity of the manganese oxide could be well controlled by adjusting the deposition time. It was found that nanostructured manganese oxide, $K_xMnO_{2+y}(H_2O)_z$ ($x = 0.35 \pm 0.04$), was formed as a result of the deposition, as shown in Fig. 4.59. The manganese oxide films were highly

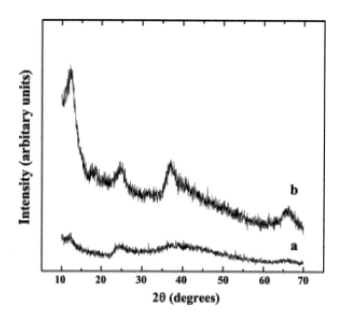

Fig. 4.59. XRD patterns of manganese oxide films: (a) as-prepared and (g) after thermal annealing at 200°C for 1 hour. Reproduced with permission from [122], Copyright © 2007, Elsevier.

porous, with pore sizes in the range of 100–200 nm. The manganese oxide films in 0.1 M Na_2SO_4 solution exhibited typical capacitive behavior within the voltage window of 0–1.0 V vs. SCE. Optimized specific capacitance of 353 $F \cdot g^{-1}$ was achieved for the 45 $\mu g \cdot cm^{-2}$ film at a scan rate of 2 $mV \cdot s^{-1}$.

In addition, it was also observed that the specific capacitance was decreased with increasing scan rate and film thickness. Heat treatment at 200°C would decrease porosity and enhance crystallinity of the manganese oxide films. In comparison with the as-prepared films, the samples thermally treated at 200°C had lower specific capacitance values at scan rates of < 10 $mV \cdot s^{-1}$ and higher values at scan rates of > 10 $mV \cdot s^{-1}$. Electrodeposition was carried out in 20 mM $KMnO_4$ aqueous solution at ambient temperature. The films were deposited on stainless steel foils by using the galvanostatic method at a current density of 2 $mA \cdot cm^{-2}$, with deposition time be varied up to 9 minutes. The electrochemical cell consisted of a cathodic substrate and a platinum counter-electrode.

Figure 4.60 shows SEM of the manganese oxide films before and after thermal annealing at 200°C. The as-prepared films possessed a nanostructured and highly porous three-dimensional network, as illustrated in Fig. 4.60(a). High magnification SEM image revealed that the pores had sizes in the range of 100–200 nm, as seen in Fig. 4.60(b). After thermal annealing, the open porosity was reduced, as observed in Fig. 4.60(c). As depicted in Fig. 4.60(d), nanowhiskers with an average length of about 100 nm were present on the surface of the films, which would be beneficial to ion transport and redox reactions for charge storage.

Nanostructured manganese dioxide films have also been deposited by using deposition methods, including galvanostatic, pulse and reverse pulse electrodeposition, in $KMnO_4$ solutions with concentrations in the range of 0.01–0.1 M [123]. The deposition was quantitatively monitored by using a Quartz Crystal Microbalance (QCM). It was found that both the deposition rate and microstructure of the films were closely related to the deposition parameters. Microstructure of the films could be varied from dense to porous. Due to the poor conductivity of manganese dioxide and to avoid cracking of the films, thickness of the dense films was controlled to be about 0.1 μm. Crack-free porous films with thicknesses of 1–2 μm could be readily deposited by using galvanostatic or reverse pulse deposition in 0.02 M $KMnO_4$ solution. Films with high porosity would be expected to demonstrate high electrochemical performances. Specifically, the porous nanostructured films deposited by using the reverse pulse exhibited higher specific capacitance, as

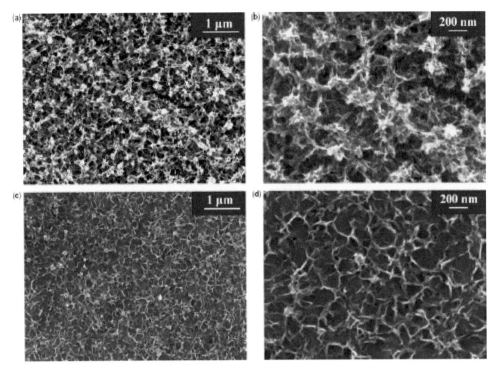

Fig. 4.60. SEM images of the manganese oxide films at different magnifications: (a, b) as-prepared and (c, d) after thermal annealing at 200°C. Reproduced with permission from [122], Copyright © 2007, Elsevier.

compared with those by using the galvanostatic films. An optimized specific capacitance of 279 F·g^{-1} was achieved within the voltage window of 1 V in 0.1 M Na$_2$SO$_4$ solutions at a scan rate of 2 mV·s^{-1}.

Microstructures of the films deposited by using galvanostatic method in the 0.02 M and 0.1 M KMnO$_4$ solutions were compared. The samples derived from the 0.02 M solutions possessed a porous microstructure, with pore sizes of 100–150 nm. Surface of the samples was characterized by the presence of nanowhiskers with a length of about 100 nm. Comparatively, the samples deposited in the 0.1 M KMnO$_4$ solution exhibited a relatively dense microstructure, whereas cracks were observed as thickness of the films was $\geq$ 0.1 μm. Therefore, films with porous microstructures were preferred in terms of the prevention of cracks and achieving high electrochemical performances.

MnO$_2$ films were also deposited by using the pulse deposition method [123]. For pulse electrodeposition, a series of pulses with constant current density were applied during the deposition process, in the 0.02 M KMnO$_4$ solution. Figure 4.61 shows deposit mass as a function of the deposition time, demonstrating a linear increase in mass during the time within which the current was 'on'. The mass was kept unchanged as the current was turned 'off'. Films deposited in this way exhibited a dense microstructure, as confirmed by SEM results. The films deposited by the using pulse deposition method in 0.02 M KMnO$_4$ solution, with 'on' time of 20 seconds and 'off' time of 10 seconds, possessed a microstructure similar to that of films derived from the 0.1 M KMnO$_4$ solutions.

Obviously, the concentration of MnO$_4^-$ at the electrode surface was decreased during the 'on' time due to the consumption, while it was gradually increased during the 'off' time due to diffusion caused by the concentration gradient. It was expected that such a depletion of MnO$_4^-$ at the electrode surface should have a significant influence on the microstructure of the films. In this case, because of the presence of the concentration gradient of MnO$_4^-$, the diffusion rate of the MnO$_4^-$ ions towards the cathode was largely promoted, so as to achieve a higher deposition rate of MnO$_2$. Accordingly, the higher the deposition rate, the more porous the films would be.

The third method used to deposit the manganese oxide films was reverse pulse deposition, in which the films were deposited onto cathodic at constant current densities of 1–2 mA·cm^{-2} for 2–5 minutes first and then currents of 0.5–1 mA·cm^{-2} with opposite polarity for 0.5–1 minute. Figure 4.62 shows variation in mass of the films and deposition pattern as a function of time. It was observed that the mass of the film was slightly decreased at the current of opposite polarity. SEM results indicated that the films deposited by using this method were crack-free and highly porous.

Capacitive behaviors of the films were examined over the potential range of 0–1.0 V vs. SCE. CV curves in 0.1 M Na$_2$SO$_4$ solution revealed that the manganese oxide electrodes demonstrated capacitive-like current–potential behavior, evidenced by the box shaped CVs. No redox peaks over the potential range of 0–1.0 V. The films deposited by using reverse pulse had a higher specific capacitance than the galvanostatic films. Figure 4.63 shows specific capacitances of the two samples, as a function of scan rate.

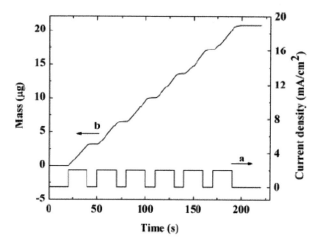

Fig. 4.61. (a) Current density and (b) deposit mass as a function of deposition time of the samples deposited in 0.02 M KMnO$_4$ solution by using pulse deposition. Reproduced with permission from [123], Copyright © 2007, Elsevier.

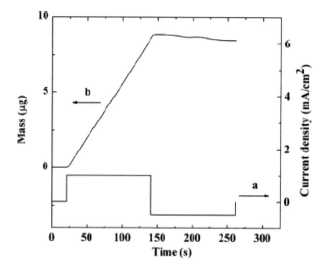

Fig. 4.62. (a) Current density and (b) deposit mass versus deposition time of the samples deposited in 0.02 M KMnO$_4$ solution by using the reverse pulse deposition. Reproduced with permission from [123], Copyright © 2007, Elsevier.

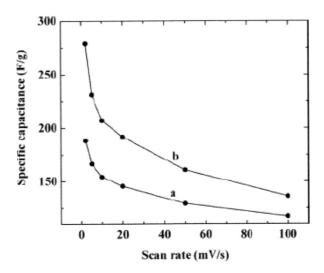

Fig. 4.63. Specific capacitances of the films of 80 μg·cm^{-2} as function of scan rate deposited by using different methods in 0.02 M KMnO$_4$ solution: (a) galvanostatically and (b) reverse pulse deposition. Reproduced with permission from [123], Copyright © 2007, Elsevier.

Electrochemical oxidation of metallic Mn

A special approach has been developed to synthesize manganese oxides, by using Physical Vapor Deposition (PVD), combined with a glancing vapor incidence angle (GLAD) to generate chevron-type porous Mn metallic structure, followed by electrochemically oxidation [125–128]. For instance, manganese oxide films with pseudocapacitive behaviors were obtained through the anodic oxidation of Mn metallic films that were pre-deposited by using sputtering [126]. The Mn films with thicknesses in the range of 20–200 nm were deposited on Pt coated Si substrates in Ar, which were electrochemically oxidized into porous dendritic structure that exhibited promising pseudocapacitive performances. An optimal specific capacitance of 700 F·g^{-1} at the current density of 160 μA·cm^{-2}. Besides, CV data at the scan rate of 5 mV·s^{-1} corresponded specific capacitances of 400–450 F·g^{-1}. Material loading of films was in the range of 25–75 μg·cm^{-2}.

Mn thin films were deposited by using magnetron sputtering, in vacuum with a base pressure of 7 × 10^{-5} Pa and high purity as the sputtering gas. The Mn target was prepared by sintering Mn powder. During sputtering, the chamber pressure was set to be 0.9 Pa, while the magnetron was operated at a power of 200 W. At the typical gun voltage of 416 V, a deposition rate of about 10 nm·min^{-1} could be achieved. The Si substrates were coated with 300 nm thick Pt. Nominal film thicknesses were set in the range of 20–200 nm. The Mn metallic films were then electrochemically oxidized in Na$_2$SO$_4$ electrolyte solution.

Figure 4.64 shows SEM images of the samples with various thicknesses, which were all oxidized at 160 μA·cm^{-2}. A highly convoluted dendritic structure was observed in the first three samples, which became coarser and coarser gradually, as the nominal film thickness was increased from 20 nm to 100 nm. However, further increase in film thickness to 200 nm resulted in almost no change in surface morphology. In addition, as the film thickness was increased, surface area of the samples was also increased.

It was discovered that a three-layered structure was formed, during the electrochemical oxidation process of the Mn films at a relatively anodic current in the solution of 1 M Na$_2$SO$_4$ [127]. Initially, the base layer of Mn/MnO with an undisturbed zigzag morphology was oxidized to Mn$_3$O$_4$. Meanwhile, a second layer was then formed just on surface of the zigzag layer, in which both crystalline Mn$_3$O$_4$ and amorphous phase were present. Finally, an amorphous layer of hydrated MnO$_2$ was achieved on surface the films, which was the only layer that contributed to the capacitive behavior of the materials. Porosity of the samples before oxidation was not important, because similar porous structure could be obtained from fully dense films.

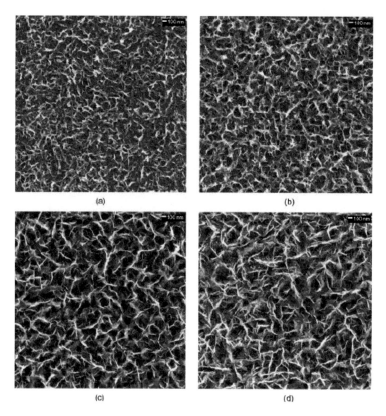

Fig. 4.64. SEM images of the Mn films after electrochemical oxidation, with different nominal thicknesses: (a) 20 nm, (b) 50 nm, (c) 100 nm and (d) 200 nm. The white bars = 100 nm. Reproduced with permission from [126], Copyright © 2004, Elsevier.

Figure 4.65 shows SEM images of the samples at the initial step, demonstrating that the zigzag layer was still not affected. After oxidation for a short while, surface of the films was almost unchanged, as seen in Fig. 4.65(b). Once the oxidation time was sufficiently long, two additional layers were formed, as demonstrated in Fig. 4.65(c). Just on top of the zigzag layer, a dense layer of disturbed material could be observed, on which a relatively thick and highly porous layer was present. The layers were denoted as dense intermediate layer and porous surface layer, respectively, for convenient discussion.

Figure 4.66 shows cross-sectional TEM images of the samples, confirming the structure of porous surface layer plus denser Mn zigzag layer. After oxidation for 1150 seconds and 1750 seconds, the porous surface layer had thicknesses of 250 nm and > 500 nm, respectively. EDS analysis results indicated that the porous consisted mainly of Mn and O, implying that original zigzag layer had been partly restructured so as to form the new layer. SAED patterns revealed that the crystalline phase in the porous surface layer was Mn_3O_4, without the presence of MnO, confirming that this layer contained crystalline Mn_3O_4 and amorphous phase.

Figure 4.67 shows high magnification TEM images of the bottom layers of the electrochemically oxidized structures, together with the corresponding diffraction patterns. After oxidation for 1150 seconds, the zigzag layer was transferred to a mixture of MnO and Mn_3O_4, as revealed in Fig. 4.67(a–c). Because the reflections of MnO in some areas were quite weak, it was implied that the zigzag layer consisted mainly of Mn_3O_4, as illustrated in Fig. 4.67(b). However, MnO was found in some areas in the zigzag layer, as demonstrated in Fig. 4.67(c), which meant that the oxidation was not uniform in the sample. As the oxidation time was increased to 1750 seconds, both the zigzag layer and the dense intermediate layer were still present, as seen in Fig. 4.67(d). The dense intermediate layer had a thickness of 80 nm, consisting of MnO and Mn_3O_4, as shown in Fig. 4.67(e). The residual part of the zigzag layer was single phase MnO,

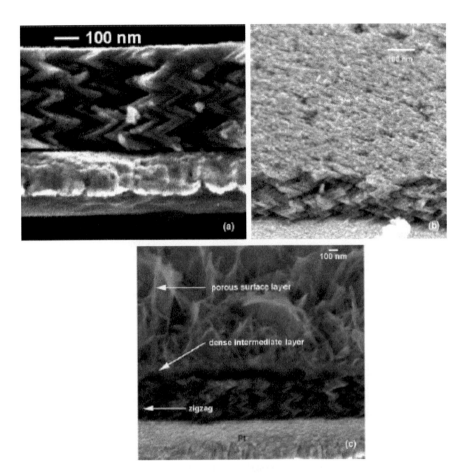

Fig. 4.65. SEM images of the samples: (a) as-deposited film, (b) partially oxidized for 1150 seconds and (c) partially oxidized for 1700 seconds. Reproduced with permission from [127], Copyright © 2005, Elsevier.

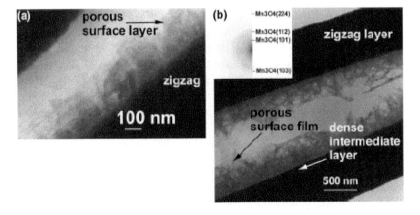

Fig. 4.66. (a) Cross-sectional TEM BF image of the sample oxidized for 1150 seconds. (b) Cross-sectional TEM BF image of the sample oxidized for 1750 seconds, with the inset to diffraction pattern showing certain degree of crystallinity, indexed as single phase Mn_3O_4. Reproduced with permission from [127], Copyright © 2005, Elsevier.

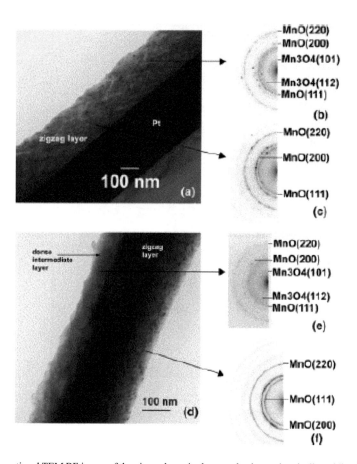

Fig. 4.67. (a) Cross-sectional TEM BF image of the zigzag layer in the sample electrochemically oxidized for 1150 seconds. (b) SAD pattern of zigzag layer indexed as MnO and Mn_3O_4. (c) SAD pattern of zigzag layer indexed as MnO. (d) Cross-sectional TEM BF image of the zigzag layer in sample electrochemically oxidized for 1750 seconds. (e) SAD pattern of the dense intermediate layer indexed as a mixture of MnO and Mn_3O_4. (f) SAD pattern of the zigzag layer indexed as single phase MnO. Reproduced with permission from [127], Copyright © 2005, Elsevier.

with a thickness of about 200 nm, as observed in Fig. 4.67(f). It was therefore could be concluded that Mn in the zigzag layer was first oxidized to MnO and then Mn_3O_4.

In addition, it was observed that after the completion of the oxidation process, the zigzag film was still not intact, while the continuous oxidation of the structure had almost no effect on the porous surface layer. Experimental results indicated that as long as the starting materials are Mn or the mixture of Mn/MnO, they can be electrochemically oxidized into porous structures that exhibit capacitive behaviors. Otherwise, as the valence of the starting materials is too high, e.g., $\geq 3^+$, electrochemical oxidation will not result in films with porous surface, so that no capacitive performance is available.

An electrochemical deposition process was also developed to deposit metallic manganese films from a butylmethylpyrrolidinium bis(trifluoromethylsulfony)imide (BMP–NTf_2) ionic liquid [129–131]. The electrodeposited Mn films were anodized in Na_2SO_4 aqueous solution by using various electrochemical methods, such as potentiostatic, galvanostatic and Cyclic Voltammetry (CV), so that Mn was converted to manganese oxides with different physical properties. The Mn oxide anodized under the CV condition had the largest surface area, highest hydrous state and lowest Mn valence state, thus leading to the highest specific capacitance.

Specifically, when electrodepositing manganese (Mn) in BMP–NTf_2 ionic liquid, both crystal structures and surface morphologies of the deposited Mn films were influenced by applied potentials (from −1.8

V to −2.2 V) and temperatures (from 50°C to 110°C) [129]. Experimental results revealed that although the obtained Mn films were amorphous, their morphologies were strongly dependent on the deposition conditions. The manganese oxide anodized from the Mn deposited at −1.8 V and 50°C had the highest specific capacitance of 402 F·g^{-1} at a scan rate of 5 mV·s^{-1}, with 94% retention after 500 CV cycles.

Mn cations were introduced into the BMP–NTf$_2$ ionic liquid by anodically dissolving a manganese block to a concentration of 0.05 M. Mn thin films were cathodically deposited on nickel (Ni) substrates at 50–110°C in the ionic liquid that was constantly stirred with a magnetic paddle. The electrodeposition process was conducted in a purified N$_2$ environment in a glove, in which the moisture and oxygen content were controlled to be < 1 ppm. A three-electrode electrochemical cell controlled by an AUTOLAB potentiostat was used for the deposition experiments. Ni foil with a thickness of 120 μm and an exposed area of 1 cm^2 was etched in 1 M H$_2$SO$_4$ solution at 85°C, followed by washing with pure water in an ultrasonic bath and subsequent drying, which was used as the working electrode. A Pt wire as the reference electrode was placed in a fritted glass tube that contained the BMP–NTf$_2$ ionic liquid with a ferrocene/ferrocenium couple (Fc/Fc$^+$ = 50/50 mol%, +0.55 V vs. SHE), thus building up a stable redox potential [132]. A Mn block was employed as the counter electrode, which would compensate for the consumption of Mn cations in the ionic liquid due to the electrodeposition. Cathodic potentials of −1.8 V, −2.0 V, and −2.2 V (vs. Fc/Fc$^+$) were applied, to yield a total passed charge of 0.3 C·cm^{-2}. After deposition, the Mn films were thoroughly washed with CH$_2$Cl$_2$ solution and dried in air. The Mn films were electrochemically oxidized in 0.1 M Na$_2$SO$_4$ aqueous solution at 25°C.

Figure 4.68 shows SEM images of the samples deposited at different potentials and temperatures. As demonstrated in Fig. 4.68(a), the Mn film deposited at −1.8 V in the ionic liquid at 50°C consisted of

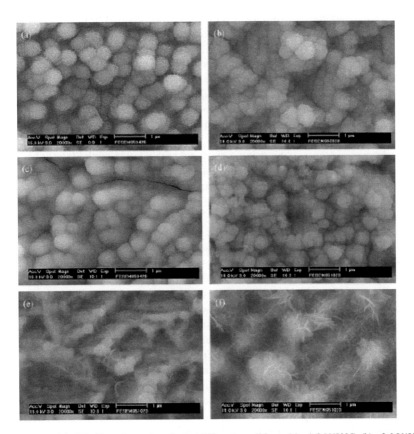

Fig. 4.68. SEM images of the Mn films electrodeposited at different conditions: (a) −1.8 V/50°C, (b) −2.0 V/50°C, (c) −2.2 V/50°C, (d) −1.8 V/70°C, (e) −1.8 V/90°C and (f) −1.8 V/110°C. Reproduced with permission from [129], Copyright © 2008, Elsevier.

228 *Nanomaterials for Supercapacitors*

spherical particles, with an average diameter of 400 nm. With close inspection, it was observed that the particles contained a large number of sub-grains, with sizes of a few nanometers. For the Mn films deposited at −2.0 V and −2.2 V, the fine surface morphology was absent, as seen in Fig. 4.68(b) and (c). Therefore, it could be concluded that the more negative the applied potential, the denser the microstructure the Mn films would have. A low the deposition rate and a less negative potential were the necessary conditions to result in Mn films to exhibit unique microstructure with nanosized profiles. SEM images of the Mn films, deposited at a potential of −1.8 V in the ionic liquids at 70°C, 90°C and 110°C, are illustrated in Fig. 4.68(d–f). With increasing deposition temperature, i.e., increasing deposition rate, surface morphology of the Mn films become from less uniform to fibrous structure.

Electrophoretic deposition (EPD)

Electrophoretic deposition (EPD) is another electro-deposition technique that has been widely used to deposit various coatings. EPD is realized due to the migration of charged particles of an electrode and deposition on it at an external electric field in a specific suspension towards. Bath compositions of EPD are comprised of different additives, in order to achieve stabilization and charge the inorganic particles in the suspensions. The deposition of MnO_2 films by using EPD usually consists of two steps, (i) preparation of a stable suspension of charged MnO_2 particles through the reduction of $KMnO_4$ in aqueous solutions with certain reducing agents and (ii) deposition at certain potentiostatic or galvanostatic conditions [93, 94, 133–135]. Deposition parameters that can be used control over microstructures and properties of MnO_2 films include voltage, current density, solution concentration, pH value, deposition temperature and type of additives.

A cathodic electrophoretic deposition (EPD) method was reported to deposit manganese dioxide films, with the addition of Phosphate Ester (PE) as an effective charging agent, which also stabilized the manganese dioxide nanoparticles in the suspensions [135]. Both the PE concentration and deposition voltage has significant influence on deposition efficiency of the films. Highly porous nanostructured films with thickness in the range of 0.5–20 μm were deposited by using EPD for applications as electrodes of supercapacitors. Cyclic voltammetry and chronopotentiometry results of the films in 0.1 M Na_2SO_4 solutions demonstrated their capacitive behavior in the voltage window of 1 V. An optimal specific capacitance of $377\ F\cdot g^{-1}$ was observed at a scan rate of $2\ mV\cdot s^{-1}$, which decreased with increasing thickness of the films and increasing scan rate from 2 to $100\ mV\cdot s^{-1}$.

MnO_2 powder was synthesized by using a chemical precipitation method described in a previous investigation [135]. EPD was carried out in the manganese dioxide suspensions in ethanol with solid contents of 5–$50\ g\cdot L^{-1}$ and PE of 0.1–$2\ g\cdot L^{-1}$. Manganese dioxide films were deposited on stainless steel foils and graphite substrates at constant voltages of 10–100 V, for 1–5 minutes. After deposition, the samples were dried in air at room temperature for 72 hours. XRD pattern indicated that the manganese dioxide was mainly amorphous, with weak diffraction peaks corresponding to birnessite, with a composition of $A_xMnO_{2+y}(H_2O)_z$ [136, 137]. The average oxidation state of Mn was 3.6–3.8, suggesting the presence of Mn^{4+} with minor level of Mn^{3+}.

It was found that, as the PE content was in the range of 0–$1\ g\cdot L^{-1}$, the deposition yield was increased with increasing content of PE, which could be ascribed to enhanced stability of suspensions and increased charge of the manganese dioxide nanoparticles. However, if the level of PE was further increased, the ionic strength of the suspensions would be increased, thus leading to a decrease in thickness of the electrical double layer on the manganese dioxide nanoparticles, so that stability of the suspensions would be decreased. As a result, the particles would be precipitated, thus resulting in a decrease in concentration of the suspensions and hence decrease in deposition rate. Figure 4.69 shows representative SEM images of the manganese oxide films, demonstrating a porous crack free surface, with pore sizes in the range of 10–100 nm. The films exhibited a uniform microstructure, with sizes of the nanoparticles in the range of 20–40 nm. The presence of the porosity was responsible for the crack free microstructure of the films, which was also beneficial to their electrochemical properties.

As discussed above, four types of deposition methods have been used to develop manganese oxide films for electrodes of supercapacitors, which include (i) sol-gel process, (ii) electrochemical deposition,

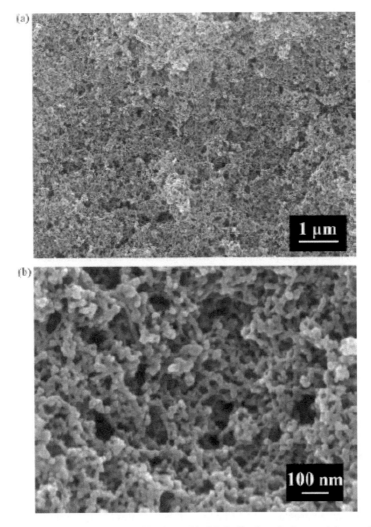

Fig. 4.69. SEM images at low (a) and high (b) magnifications of the MnO_2 film deposited on a stainless steel substrate, derived from 10 g·L^{-1} MnO_2 suspension, containing PE with a concentration of 0.2 g·L^{-1}, at a deposition voltage of 20 V. Reproduced with permission from [135], Copyright © 2008, Elsevier.

(iii) electrochemical oxidation of metallic Mn films and (iv) electrophoretic deposition. The sol-gel process requires relatively high temperatures to form the desired manganese oxide, which could be limited to the substrates that can withstand the high temperatures. Comparatively, electrochemical deposition and electrophoretic deposition can be conducted at room temperature, so that there is almost no limit to the substrates, whereas even plastic substrates could be used to deposit the manganese oxide films.

MnO_2 Based Mixed Oxides

Due to its low electrical conductivity, MnO_2 based supercapacitor electrodes have encountered limitations in specific capacitance and power density, caused by the high charge-transfer resistance [29]. Noting their semiconducting characteristics, transition metallic oxides could be introduced to MnO_2, so as to increase its electrical conductivity, through the introduction of defects and charge carriers. Various oxides have been explored to incorporate with MnO_2 to achieve desired electrochemical performances.

230 *Nanomaterials for Supercapacitors*

Mn–Fe mixed oxides

Mn-Fe oxide, $MnFe_2O_4$, also known as Mn-ferrite, has been demonstrated to have promising pseudocapacitive behavior in aqueous NaCl solutions [138]. $MnFe_2O_4$ fine powder was prepared by using a chemical coprecipitation method from $MnSO_4$ and $FeCl_3$, followed by calcination at different temperatures in N_2. Interestingly, as the Mn^{2+} ions occupied the tetrahedral sites in the spinel structure, high specific capacitance could be obtained. In addition, the calcination temperature had to be > 200°C for the powder to be crystallized to achieve high pseudocapacitance, otherwise, the samples would be amorphous, so that the capacitance was minimized.

More recently, nanocarbon materials have been incorporated to further increase its electrochemical performances [139–142]. For example, a simple method has been reported to fabricate high performance flexible electrodes of supercapacitors by dropping $MnFe_2O_4$/graphene hybrid inks onto flexible graphite sheets that were used as both current collectors and substrates, followed by drying with an infrared lamp [140]. The $MnFe_2O_4$/graphene hybrid inks were prepared by immobilizing the $MnFe_2O_4$ microspheres on graphene nanosheets through a solvothermal synthetic route. The $MnFe_2O_4$/graphene demonstrated a specific capacitance of 300 $F \cdot g^{-1}$ at the current density of 0.3 $A \cdot g^{-1}$. Furthermore, $MnFe_2O_4$/graphene hybrid could be used to develop supercapacitors with significantly high electrochemical performance. The supercapacitors had been made as a sandwich structure, consisting of two pieces of $MnFe_2O_4$/graphene hybrids modified electrodes that were separated by polyvinyl alcohol (PVA)-H_2SO_4 gel electrolyte. It was found that the flexible supercapacitors with a thickness of 227 μm demonstrated an optimal specific capacitance of 120 $F \cdot g^{-1}$ at the current density of 0.1 $A \cdot g^{-1}$, with excellent cycle performance by retaining 105% capacitance after 5000 cycles.

For one-pot solvothermal synthesis of $MnFe_2O_4$/graphene hybrids [143], 60.0 mg GO, 32.4 mg (0.12 mmol) $FeCl_3 \cdot 6H_2O$ and 11.8 mg (0.06 mmol) $MnCl_2 \cdot 4H_2O$ were dispersed in 40 mL ethylene glycol after ultra-sonication treatment for 2 hours. After that, 2.16 g NaAc and 1.0 g polyethylene glycol were added in, which was then stirred for 30 minutes. The mixture was finally solvothermally treated at 200°C for 10 hours to obtain $MnFe_2O_4$/graphene hybrids. $MnFe_2O_4$/graphene hybrid inks were prepared by mixing 2.0 mg $MnFe_2O_4$/graphene hybrid powder with 1 mL ethanol, thus leading to a concentration of 2.0 $mg \cdot mL^{-1}$, which were used to fabricate the $MnFe_2O_4$/graphene hybrids modified electrode, by simply dispersing 100 μL hybrids inks (2.0 mg mL^{-1}) onto flexible graphite sheet (GTS), followed by drying with an infrared lamp. PVA-H_2SO_4 gel, as the solid electrolyte, was developed by mixing 1.0 g H_2SO_4 with 10 mL DI water, into which 1.0 g PVA powder was added. After heating at 90°C under vigorous stirring for a certain time, clear electrolyte solution was formed, which could be used to fabricate solid-state supercapacitors.

Figure 4.70 shows representative SEM images of the samples [140]. As shown in Fig. 4.79(A), the solvothermal graphene (s-graphene) sample exhibited aggregation of crumpled graphene nanosheets. Meanwhile, the pure $MnFe_2O_4$ powder exhibited a monodisperse microsphere morphology, as seen in Fig. 4.70(B). In comparison, in the $MnFe_2O_4$/graphene hybrid sample, the $MnFe_2O_4$ microspheres were embedded inside thin graphene nanosheets with a quite uniform distribution, as illustrated in Fig. 4.70(C). High magnification image (Fig. 4.70(D)) further revealed that the $MnFe_2O_4$ microspheres were tightly wrapped with flake-like graphene nanosheets. With this configuration, agglomeration of the $MnFe_2O_4$ microspheres was effectively avoided. XRD results confirmed that the $MnFe_2O_4$ in the hybrids had been well crystallized, with cubic spinel structure.

Electrochemical co-deposition is another way to combine iron oxides with manganese oxides to form mixed oxide electrodes is [144]. In this case, Mn–Fe binary oxides, were prepared by using anodic deposition in mixed solutions containing both manganese acetate and iron chloride. Properties (crystal structure, surface morphology and chemical state) and electrochemical behaviors of the mixed oxide could be controlled by adjusting the annealing temperature. The optimal annealing temperature was 100°C, which resulted in mixed oxide with a specific capacitance of 280 $F \cdot g^{-1}$ at the scanning rate of 5 $mV \cdot s^{-1}$. The thermal annealing process had also effect on cyclic stability of the mixed oxide electrode.

TG/DTA results of the deposited binary oxide revealed there was a sharp weight loss at about 100°C, due to the evaporation of the adsorbed water. At 200°C, the weight loss was slowed down, while an exothermic peak was present, which was attributed to the decomposition of organic items derived from

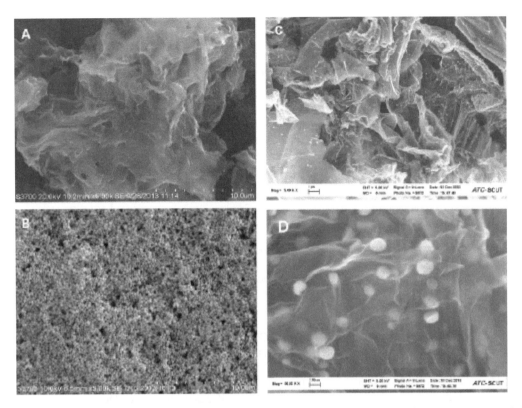

Fig. 4.70. SEM images of the samples: (A) solvothermal graphene (s-graphene), (B) MnFe$_2$O$_4$ microspheres, (C) MnFe$_2$O$_4$/graphene hybrids and (D) MnFe$_2$O$_4$/graphene hybrids at high magnification. Reproduced with permission from [140], Copyright © 2014, Elsevier.

the acetate plating solutions. A nearly constant weight over 200–500°C was closely related to the loss of structural water and further oxidation of the samples. A relative sudden weight loss at about 500°C, accompanied with an exothermic peak, was caused by the release of oxygen from the Mn oxide [145], is clearly discernible. Above 500°C, the weight loss seemed to be stopped. In addition, as the annealing temperature was < 400°C, all samples exhibited a poor crystallinity. At temperatures of > 500°C, crystalline (Mn–Fe)$_2$O$_3$ solid solution was formed.

Mn–Co mixed oxides

Co oxides formed the second group of candidates to be incorporated with Mn oxides, in order to develop high performance electrodes of supercapacitors [146]. For instance, hydrous manganese–cobalt oxides, (Mn+Co)O$_x$·nH$_2$O, were anodically deposited onto graphite substrates in 0.01 M MnCl$_2$ + 0.09 M CoCl$_2$ solutions with different pH values [146]. The binary hydrous mixed oxides displayed an optimal specific capacitance of 125 F·g^{-1} at 25 mV·s^{-1}. The plating solutions were adjusted to various pH values (7, 6, 5, 4, 3, and 2.5), by using either 0.1 M HCl or 0.1 M NaOH solutions. Anodic deposition was carried out at 1.0 V, with the total passed charge to be 0.3 C·cm^{-2}. As the pH value was ≥ 8, (Mn+Co) O$_x$·nH$_2$O could not be deposited, because of the formation of hydroxide gels. It was found that both the number and depth of the cracks were decreased with decreasing pH value of the plating solutions.

Figure 4.71 shows CV curves of the (Mn+Co) O$_x$·nH$_2$O deposited in the solutions with different pH values. At the negative sweeps, all curves had a shoulder at about 0.5 V, without the presence of pronounced redox peak, demonstrating typical rectangular and symmetric profiles. It was found that the voltammetric currents over the potential range from −0.1 and 1 V were almost independent on the upper and lower potential

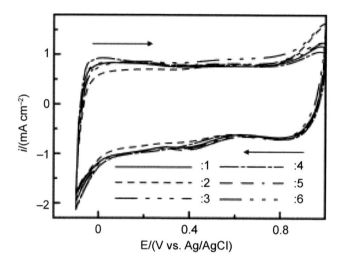

Fig. 4.71. CV curves of the (Mn+Co) O$_x$·nH$_2$O deposits obtained in 0.01 M MnCl$_2$·4H$_2$O + 0.09 M CoCl$_2$·6H$_2$O with different pH values: (1) 7, (2) 6, (3) 5, (4) 4, (5) 3 and (6) 2.5. Reproduced with permission from [146], Copyright © 2005, Elsevier.

limits of CV, which implied that the (Mn+Co)O$_x$·nH$_2$O possessed strong electrochemical reversibility over the potential range. Also, the value of pH of the plating solution had effect on electrochemical behaviors of the (Mn+Co) O$_x$·nH$_2$O, as the pH value was ≤ 7.

Figure 4.72 shows SEM images of the (Mn+Co) O$_x$·nH$_2$O samples deposited in the solutions with different pH values, characterized by uniform distribution of grains and cracks. However, the number and depth of the cracks were gradually decreased, as the pH value of the plating solutions was decreased. Specifically, the sample of pH = 2.5 contained only very narrow cracks, as illustrated in Fig. 4.72F(d). In other words, pH value of plating solutions could be used to control over morphologies of (Mn+Co) O$_x$·nH$_2$O samples. Further analysis results indicated that the content of Co in the (Mn+Co) O$_x$·nH$_2$O was decreased with decreasing pH value of the plating solutions, which could be linked to the variation trend in morphology of the samples.

Further studies demonstrated that the incorporation of Co could effectively prevent the irreversible anodic dissolution of the deposited oxides during cyclic voltammetric scans in KCl aqueous electrolyte [147]. As a result, electrochemical stability of the oxide electrodes was significantly improved. Furthermore, the presence of a suitable amount of Co oxide led to an enhancement in high rate charge–discharge capability of the binary oxide electrodes. However, as the content of Co was too high, e.g., > 15 wt%, oxide blocks were formed, while the fibrous morphology of the low Co content samples was absent, so that a large decrease in specific capacitance of the mixed oxide was observed.

Binary Mn–Co oxides were electroplated onto graphite substrates by using an anodic deposition method at room temperature. Co content in the deposited oxide was controlled by adjusting concentration of Co (CH$_3$COO)$_2$ in 0.25 M Mn (CH$_3$COO)$_2$ plating solutions. The deposition experiment was conducted with a three-electrode configuration. The graphite plate served as the anode, while a platinum sheet was employed as the counter electrode. Meanwhile, a Saturated Calomel Electrode (SCE) acted as the reference electrode. At a constant potential of 0.8 V, the deposition was carried out until reaching a total passed charge of 1.5 C·cm^{-2}. The oxide electrodes after drying had masses in the range of 1.1–1.2 mg.

Figure 4.73 shows XRD patterns of the samples deposited in the Mn (CH$_3$COO)$_2$ plating solutions with different contents of Co (CH$_3$COO)$_2$ [147]. All samples had almost the same diffraction pattern, which means that the incorporation of Co had no significant effect on crystal structure of the manganese oxide. At the same time, the diffraction peaks of the mixed oxides were relatively weak, as compared with those of the graphite substrate, which implied that their crystal sizes were at nanometer scale.

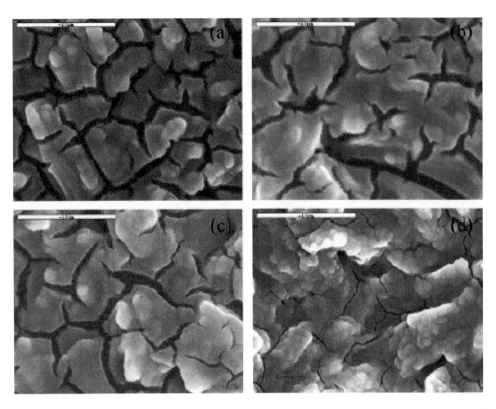

Fig. 4.72. SEM images of the (Mn+Co) O$_x$·nH$_2$O deposits prepared in 0.01 M MnCl$_2$·4H$_2$O + 0.09 M CoCl$_2$·6H$_2$O with pH values: (a) 6, (b) 4, (c) 3 and (d) 2.5. Reproduced with permission from [146], Copyright © 2005, Elsevier.

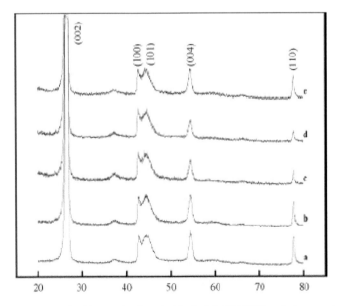

Fig. 4.73. XRD patterns of the mixed oxide samples deposited in the Mn(CH$_3$COO)$_2$ solution containing different contents of Co(CH$_3$COO)$_2$: (a) 0, (b) 0.05 M, (c) 0.1 M, (d) 0.15 M and (e) 0.2 M. Reproduced with permission from [147], Copyright © 2008, Elsevier.

Figure 4.74 shows SEM images of the mixed oxide samples [147]. As illustrated in Fig. 4.74(a), the Mn–Co oxide deposited in the mixed solution of 0.25 M Mn(CH$_3$COO)$_2$ and 0.10 M Co(CH$_3$COO)$_2$, exhibited a granular morphology, together with the presence of cracks, which could be caused by a shrinkage effect of the samples during the drying process. The pure Mn oxide was comprised of nanosized fibers with a 3D network microstructure, as revealed in Fig. 4.74(b). As the content of Co (CH$_3$COO)$_2$ in the plating solutions was increased, the nanofibers were gradually shortened and thickened, with a tendency of agglomeration, as demonstrated in Fig. 4.74(c)–(e). For the sample deposited in the solution containing 0.2 M Co (CH$_3$COO)$_2$, the fibrous structure was absent, as seen in Fig. 4.74(f).

At a relatively low potential scan rate of 5 mV·s^{-1}, the pure Mn oxide had a specific capacitance of 209 F·g^{-1}. After incorporation of Co, the specific capacitance was slightly decreased, with 195, 191 and 200 F·g^{-1} for the samples of Mn005Co, Mn010Co and Mn015Co, respectively. However, at high CV scan rates, the presence of Co had a positive effect on electrochemical performance. For example, at a scan rate of 150 mV·s^{-1}, the pure Mn oxide exhibited a specific capacitance of 124 F·g^{-1}, while those of the binary oxides, Mn005Co, Mn010Co and Mn015Co were 133, 136 and 144 F·g^{-1}, respectively. Therefore, the introduction of Co oxide led to mixed oxide electrodes with enhanced electrochemical performance at high power density.

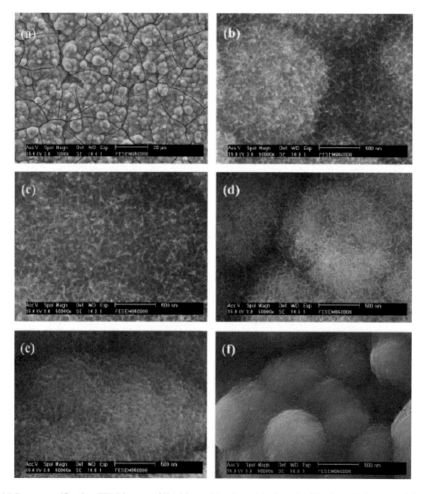

Fig. 4.74. (a) Low-magnification SEM image of the binary Mn–Co oxide deposited in the solution containing 0.1 M Co (CH$_3$COO)$_2$. (b)–(f) High-magnification SEM images of the oxides deposited in the Mn(CH$_3$COO)$_2$ plating solutions containing 0, 0.05, 0.1, 0.15 and 0.2 M Co (CH$_3$COO)$_2$, respectively. Reproduced with permission from [147], Copyright © 2008, Elsevier.

Mn–Ni mixed oxides

Mn–Ni mixed oxides, as electrodes of supercapacitors, were prepared by reducing $KMnO_4$ with Ni(II) acetate–manganese acetate [19]. With the introduction of 20% NiO, the Mn-Ni mixed oxide exhibited a specific capacitance 210 $F·g^{-1}$, which was higher than that of pure MnO_2 (166 $F·g^{-1}$). The mixed oxides also displayed higher rate capacities. The enhanced electrochemical properties of the mixed oxides could be ascribed to the enlarged surface area, because of the presence of micropores. To prepare the Mn–Ni mixed oxides, $KMnO_4$ was dissolved in distilled water to aqueous solution with a concentration of 0.17 M, while Mn $(CH_3CO_2)_2$ and Ni $(CH_3CO_2)_2$ were used to form mixed solution with nominal ratios of Mn/X. After that, the two solutions were mixed, under vigorous stirring at room temperature for 6 hours. Then, the precipitate in the form of a dark powder was filtered and thoroughly washed with distilled water, followed by calcination at appropriate temperatures.

XRD results indicated that the sample $MnNi_{0.25}O_x$ experienced a sudden transition from amorphous to crystalline state with increasing annealing temperature. At temperatures of $\leq 400°C$, the sample was amorphous, whereas it was well crystallized as the temperature was raised to 500°C, evidenced by the presence of the sharp diffraction peaks. The XRD pattern of the $MnNi_{0.25}O_x$ after annealing at 500°C could be indexed to Mn_2O_3 and $MnNiO_3$, both of which were of no electrochemical activity. Therefore, annealing temperature could not be $\geq 500°C$, in order to maintain high electrochemical performance of the mixed oxides.

A group of nanostructured Mn–Ni–Co Oxide composites (MNCO) were synthesized through the thermal decomposition of a precursor prepared by using a chemical co-precipitation from Mn, Ni and Co salts [148]. It was found that the MNCO based electrode demonstrated promising electrochemical performances, with an optimal specific capacitance of 1260 $F·g^{-1}$, over the potential range from -0.1 to 0.4 V, vs. Saturated Calomel Electrode (SCE) in 6 $mol·L^{-1}$ KOH electrolyte. To prepare the mixed oxide MNCO, Ni, Co and Mn chloride salts, with a nominal molar ratio of Ni:Co:Mn $-$ 1:1:1, were dissolved in distilled water to form a solution, into which 5% ammonia hydroxide solution was added slowly until the pH = 9.5. The precipitates collected and thoroughly washed with DI water. After drying, the powder sample was calcined in air at 200°C for 3 hours.

XRD results indicated that the MNCO powder was a mixture of Mn_3O_4, NiO and Co_3O_4, with a composition of $Mn_{0.325}Ni_{0.313}Co_{0.362}O_x$, close to the nominal composition. Figure 4.75 shows of the MNCO powder. As seen in Fig. 4.75(a), the powder consisted of honeycomb-like particles, with surface to have filament-like protrusions and an average particle size of about 1 μm. High magnification SEM image revealed that the MNCO particles were composed of networks with solid rods and pores of about 200 nm, as illustrated in Fig. 4.75(b). Such a nanostructured morphology would be beneficial to electrodes with high electrochemical performances, because the interconnected pores provided channels for the electrolyte to transport into the mixed oxide powders, thus enlarging the liquid–solid interfacial area for OH^- ions to insert and extract. Meanwhile, the nanostructured particles had a high density of active sites for fluid/solid reactions, while the fibrous open-weaved morphology made it relatively easy for the reactive fluid to penetrate the oxides. As a result, this MNCO mixed oxide could be used as electrodes of supercapacitors with high electrochemical performances.

Other mixed oxides

Various other metallic oxides, such as RuO_2, V_2O_5, PbO, MoO_3 and Al_2O_3, have also been incorporated with manganese oxides, in order to develop electrodes of supercapacitors, with enhanced electrochemical performance [18–20, 149–151]. Selected examples of each oxide are presented as follows.

Ruthenium–manganese oxides, $Ru_nMn_{1-n}O_x$, were synthesized by using an oxidative co-precipitation from the neutral mixed aqueous solution of Mn(VII) (potassium permanganate), Mn(II) (manganese acetate) and Ru(III) (ruthenium chloride), at room temperature, through the following reactions [149]:

$$2MnO_4^- + 3Mn^{2+} + 2H_2O \rightarrow 5MnO_2 + 4H^+, \tag{4.10}$$

$$Ru^{3+} + 2H_2O \rightarrow Ru(OH)_3 + 3H^+, \tag{4.11}$$

236 Nanomaterials for Supercapacitors

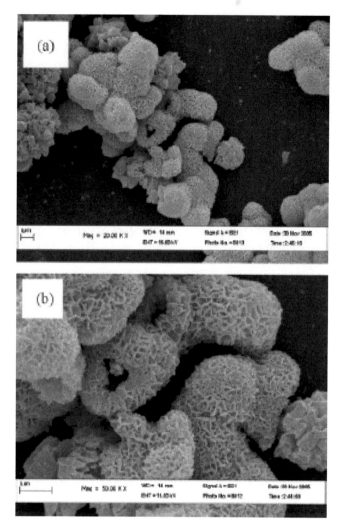

Fig. 4.75. SEM images of MNCO powder at different magnifications: (a) 20,000× and (b) 50,000×. Reproduced with permission from [148], Copyright © 2008, Elsevier.

$$4Ru(OH)_3 + O_2 \rightarrow 4RuO_2 + 6H_2O. \tag{4.12}$$

Figure 4.76 shows XRD patterns of the $Ru_nMn_{1-n}O_x$ ($n = 0.1$) powders calcined at different temperatures [149]. The patterns 1–3 corresponded to samples calcined at 80°C, 120°C and 170°C, respectively, with only one weak peak at $2\theta = 37.4°$. However, after calcining at 350°C, the sample (pattern 4) was well crystallized with a crystal structure of α-MnO_2. No diffraction peaks from ruthenium oxide were observed, due to its relatively high crystallization temperature and the low content in the sample. Therefore, the $Ru_nMn_{1-n}O_x$ mixed oxide prepared in this way was amorphous as the calcination temperature was $\leq 170°C$.

CV curves of the $Ru_nMn_{1-n}O_x$ ($n = 0.1$) mixed oxide electrodes had been compared with pure MnO_2 electrode over the potential range of 0–1 V vs. Ag/AgCl electrode. The CV curve of the $Ru_nMn_{1-n}O_x$ possessed an almost ideal rectangular profile, without the presence of sharp redox peaks within the potential window. The electrochemical charge–discharge currents were quite stable during the course of potential scanning. The $Ru_nMn_{1-n}O_x$ mixed oxide based electrodes showed typical pseudocapacitive behavior. In comparison, the $Ru_nMn_{1-n}O_x$ electrode demonstrated a near mirror symmetric rectangular CV curve, together

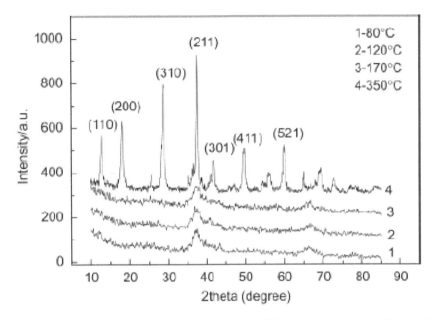

Fig. 4.76. XRD patterns of the $Ru_nMn_{1-n}O_x$ ($n = 0.1$) powders calcined at different temperatures. Reproduced with permission from [149], Copyright © 2009, Elsevier.

with fast current responses on voltage reversal at the end of potential window, whereas pure MnO_2 electrode exhibited a distorted rectangular CV curve, due to the delay in current responses. Therefore, $Ru_nMn_{1-n}O_x$ had higher electrochemical kinetic reversibility than pure MnO_2, because the doping of ruthenium oxide with quasi-metallic conductivity enhanced the electrochemical performance of the mixed oxides.

It was found that specific capacitance of the $Ru_nMn_{1-n}O_x$ based electrodes was increased with increasing n value for $n < 0.1$, accompanied a sharp increases for $n > 0.1$. This implied that doping with ruthenium could greatly enhance the electrochemical performance of manganese oxides. However, ruthenium is very expensive, an optimized composition was selected to be $n = 0.1$. The $Ru_nMn_{1-n}O_x$ ($n = 0.1$) calcined at 170°C had a specific capacitance of 264 $F \cdot g^{-1}$ after 200 cycles.

A co-electrospinning process was used to prepare MnO_x–RuO_2 composite fiber mats [150]. The electrospinning process was carried out with two precursor solutions, i.e., (i) Mn acetylacetonate + PVAc and (ii) $RuCl_3$ + PVAc. During the electrospinning experiment, the two nanofibers were ejected separately from the RuO_2/PVAc and MnO_x/PVAc solutions at an equal feeding rate. Because the substrate was moved back and forth continuously, the two nanofibers were stacked alternatively, so as to form the layer-by-layer fiber mats.

SEM observation results revealed that the as-spun MnO_x/PVAc and RuO_2/PVAc composite fibers had uniform diameters in the range of 400–700 nm. After calcination at 300 and 400°C, the MnO_x–RuO_2 fiber mats appeared as a network of fibers with two different morphologies, where RuO_2 fibers were circular, while the MnO_x fibers possessed a flattened-belt shape. The circular RuO_2 fibers had an average diameter of 400 nm, whereas the flattened MnO_x fibers possessed a thickness of about 100 nm. The MnO_x–RuO_2 composite fiber mats showed a high porosity, which would be beneficial for the penetration of electrolyte into the electrode for electrochemical reactions. XRD results indicated that the MnO_x–RuO_2 composite fibers had been well crystallized after thermal calcination at 300 and 400°C. The co-electrospun MnO_x–RuO_2 fiber mats calcined at 300°C demonstrated the optimized specific capacitance of 208.7 $F \cdot g^{-1}$ at the scan rate of 10 $mV \cdot s^{-1}$.

Mn–Mo mixed oxide thin films were deposited on platinum substrate by using anodic deposition method, through cycling the electrode potential from 0 to +1.0 V vs. Ag/AgCl, in Mn(II) aqueous solutions with certain amounts of molybdate anions (MoO_4^{2-}) [18]. During the deposition process, the Mn(II) cations were electro-oxidized to form MnO_2, while the MoO_4^{2-} ions were incorporated into the MnO_2 phase, due to protonation and dehydration. CV characterization results indicated that the Mn–Mo mixed oxide electrode in 0.5 M Na_2SO_4 aqueous solution demonstrated pseudocapacitive behavior, with higher specific capacitance and rate capability, as compared with those of the pure Mn oxide electrode. The enhancement in electrochemical performance was readily attributed to the increased electrical conductivity of the mixed oxide films.

In the first anodic scan, the oxidation of Mn^{2+} occurred on surface of the Pt electrode, whereas the reaction from the second cycle would be accumulated on the deposits. Without the presence of MoO_4^{2-}, the formation of the film blocked the electrons to transfer to the electrode surface. With the presence of MoO_4^{2-}, the films had higher conductivity. Figure 4.77 shows SEM images of the as-deposited films on Pt substrate after 36 potential cycles in 2 mM $MnSO_4$ solutions with and without 20 mM Na_2MoO_4. The film derived from the Mo-containing solution demonstrated a uniform and dense morphology. In comparison, the film from the solution without MoO_4^{2-} was less dense, with the presence of spherical grains having diameters in the range of 0.4–0.5 μm.

Figure 4.78 shows CV curves of the mixed Mn–Mo oxide and pure Mn oxide electrodes at different scan rates. The Mn–Mo oxide film exhibited a higher specific current than the pure Mn oxide. At the same time, the rectangular shaped CV curves of the Mn–Mo oxide were well retained at high potential scan rates.

Fig. 4.77. SEM images of the films derived from the 2 mM Mn_sO_4 solutions: (a) with 20 mM Na_2MoO_4 and (b) without Na_2MoO_4. The deposition was carried out by cycling the electrode potential between 0 and +1.0 V for 1 hour (36 cycles) at the scan rate of 20 mV·s^{-1}. Reproduced with permission from [18], Copyright © 2005, American Chemical Society.

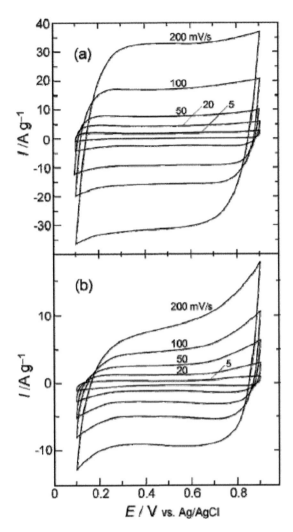

Fig. 4.78. CV curves of (a) Mn–Mo and (b) Mn oxide films at different scan rates in 0.5 M Na_2SO_4 solution. Reproduced with permission from [18], Copyright © 2005, American Chemical Society.

According to IR and electrical resistivity data, the improved charge-storage performance of the Mn–Mo oxide was readily ascribed to increase in electrical conductivity of the film electrode. Comparatively, the pure Mn oxide film experienced a distortion in rectangularity of the CV curves, as the scan rate was higher than 20 mV·s^{-1}. As the scan rate was increased from 5 to 200 mV·s^{-1}, specific capacitance values were decreased from 191 to 132 F·g^{-1} and from 83 to 38 F·g^{-1}, for the Mn–Mo and pure Mn oxides, respectively.

The last example here is aluminium (Al), which has been shown to be able to enhance electrochemical performance of manganese oxides [20]. The Al-doped MnO_2 oxides were prepared by using a high energy ball-milling method. The synthesis was started from a mixture of pure aluminium and manganese dioxide powders, with certain atomic ratios. The ball milling was conducted with a stainless steel vessel with steel balls as the milling media, at a ball-to-powder weight ratio of 20:1 and 250 rpm for time durations of up to 50 hours. Electrode made of the $Al_{0.05}/Mn_{0.95}O_2$ exhibited the optima specific capacitance. The enhanced electrochemical performance was attributed to the increase in electrical conductivity.

Figure 4.79 shows specific capacitances of the $Al_x/Mn_{1-x}O_2$ ($0 \leq x \leq 0.1$) samples at different scan rates. As compared with the pure MnO_2 electrode, the Al-doped MnO_2 electrodes had higher specific

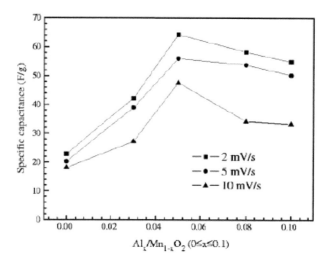

Fig. 4.79. Specific capacitances of the Al$_x$/Mn$_{1-x}$O$_2$ ($0 \le x \le 0.1$) samples at different scan rates. Reproduced with permission from [20], Copyright © 2008, Springer.

capacitance. At the optimized content of Al, i.e., Al$_{0.05}$/Mn$_{0.95}$O$_2$, the specific capacitances were 64.3 F·g^{-1}, 56.0 F·g^{-1} and 47.6 F·g^{-1}, at the scan rates of 2 mV·s^{-1}, 5 mV·s^{-1} and 10 mV·s^{-1}, respectively.

MnO$_2$–Polymer Nanocomposites

MnO$_2$ based electrodes have encountered another problem, i.e., relatively poor electrochemical cyclability, due to the dissolution of the active materials during the electrochemical cycling has been well recognized in some investigations, resulting in capacitance loss [3, 110]. To address this problem, it has been suggested to establish a barrier to protect the MnO$_2$ particles, by using a coating layer that could effectively block the dissolution of MnO$_2$ while allowing the electrolyte to access. Mechanical failure of the electrode materials, due to the repeated expansion/contraction in volume during the cycling is another important issue [152]. It has been demonstrated that conductive polymers, including polyaniline, polypyrrole and polythiophene, as well as their derivatives, can be employed to form MnO$_2$–polymer composites, in order to solve the two problems [31–33].

MnO$_2$–polyaniline (PANI)

Due to its simple synthetic process, high electrical conductivity and high chemical stability, PANI has been employed to improve electrochemical performances of manganese oxide based electrodes of supercapacitors [153–159]. For example, MnO$_2$–PANI composite electrode was fabricated by using a two-step electrochemical route, in which nanostructured MnO$_2$ was potentiodynamically deposited on a PANI matrix that was synthesized by using an electrochemical method [30]. In terms of thickness of the PANI film and the loading level of MnO$_2$, an optimized specific capacitance of 715 F·g^{-1} was achieved, with an energy density of 200 Wh·kg^{-1} at a charge–discharge current density of 5 mA·cm^{-2}.

In a separate study, PANI coating was electrochemically polymerized on MnO$_2$ nanoparticles to form PANI–MnO$_2$ composites [153]. To promote the interaction between MnO$_2$ and PANI, the MnO$_2$ nanoparticles were modified with triethoxysilylmethyl N-substituted aniline (ND42) before the electropolymerization of PANI, thus leading to PANI–ND–MnO$_2$ composite films, which were found to exhibit a higher specific capacitance than the PANI–MnO$_2$ film. Such an enhancement was attributed to the presence of the coupling reagent.

MnO$_2$ nanoparticles were synthesized by using a modified wet-chemical route, started from a mixed solution of MnSO$_4$ (1.0 M) and KMnO$_4$ (0.5 M) solution at 70°C for 4 hours [153]. The resultant precipitates were thoroughly washed with distilled water and ethanol, followed by vacuum drying at 110°C for 5 hours. The dried powders were calcined at 300°C for 3 hours, which were acidified with 2.0 M H$_2$SO$_4$ at 90°C for 2 hours. After washing and drying, the MnO$_2$ nanoparticles were modified with silane coupling agent, ND42, as schematically demonstrated in Fig. 4.80. To increase the content of –OH on surface of the nanoparticles, 0.2 g MnO$_2$ powder was heated in 80 ml boiling water for 1 hour. The hydroxylated nanoparticles were dispersed in 50 ml mixture of water and ethanol (volume ratio of 30:70), with the addition of 0.2 ml ND42 that pre-hydrated in small quantity of water, which was sonicated at 65°C for 30 minutes and then constantly stirred at 80°C for 4 hours. After thorough washing and drying, surface modified MnO$_2$ nanoparticles, ND-MnO$_2$, were obtained.

Saturated Calomel Electrode (SCE) was a reference electrode, while platinum wire was employed as a counter electrode, during the electrochemical experiments [153]. Aniline and ND-MnO$_2$ nanoparticles were electro-co-polymerized on a carbon cloth, in electrolyte solution with 0.2 g·L^{-1} ND-MnO$_2$ and 0.1 M aniline, together with 0.5 M H$_2$SO$_4$ and 0.6 g·L^{-1} (NaPO$_3$)$_6$, as seen in Fig. 4.80. Fifty successive cyclic voltammetric scans were conducted from -0.21 to 0.91 V at a scan rate of 10 mV·s^{-1} for the co-polymerization. Figure 4.81(a) shows TEM image of the as-synthesized MnO$_2$ nanoparticles, which exhibited narrowly distributed diameters of 15–20 nm and lengths of 10–80 nm. Figure 4.81(b) shows XRD pattern of the MnO$_2$ nanoparticles, demonstrating two sets of diffraction patterns, corresponding to tetragonal α-MnO$_2$ and orthorhombic γ-MnO$_2$.

SEM characterization results indicated that, with the presence of MnO$_2$ nanoparticles, the 1D growth behavior of PANI was boosted tremendously in the PANI–MnO$_2$ and PANI-ND-MnO$_2$ films [153]. As a result, thickness of the nanorods was decreased, so that the surface area and internal porosity of the

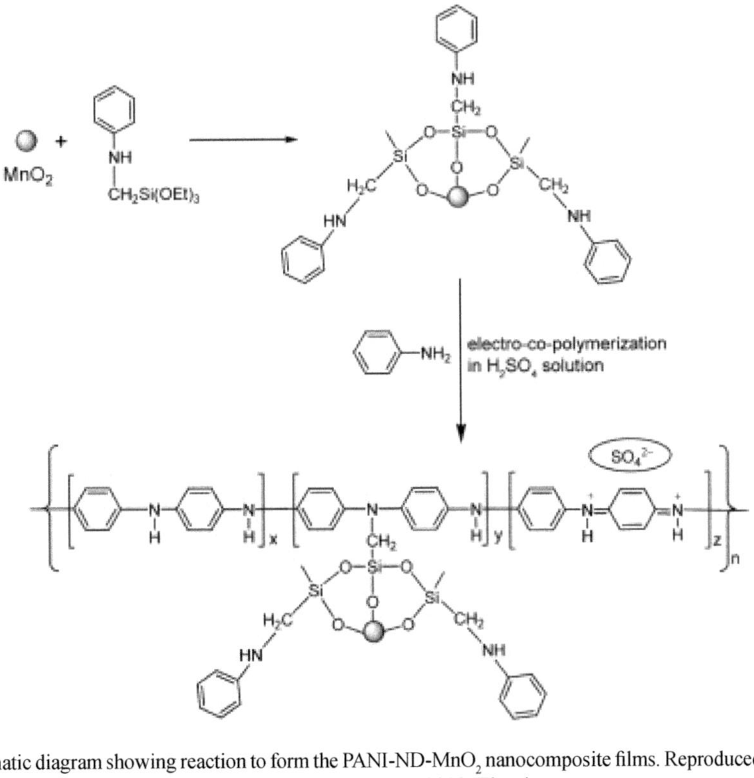

Fig. 4.80. Schematic diagram showing reaction to form the PANI-ND-MnO$_2$ nanocomposite films. Reproduced with permission from [153], Copyright © 2010, Elsevier.

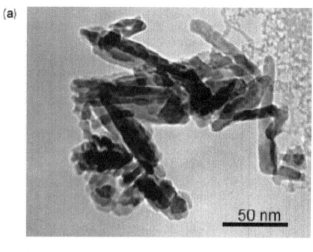

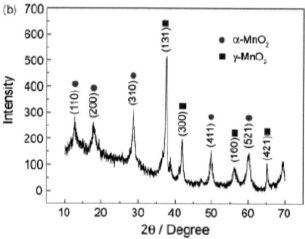

Fig. 4.81. (a) Bright-field TEM image and (b) XRD pattern of the as-synthesized MnO$_2$ nanoparticles. Reproduced with permission from [153], Copyright © 2010, Elsevier.

composite films were increased. For the three films of PANI, PANI–MnO$_2$ and PANI-ND-MnO$_2$, average diameters of the PANI rods were 500 nm, 200 nm and 150 nm, while their corresponding lengths are 1 μm, 0.7 μm and 0.4 μm, respectively. The ratios of surface area to volume of the individual PANI rod in the three composite films were 0.01 nm^{-1}, 0.013 nm^{-1} and 0.018 nm^{-1}, respectively, which implied that the PANI-ND-MnO$_2$ composite film had the largest surface area. Furthermore, because the MnO$_2$ nanoparticles were positively charged in the acidic co-polymerization solution, they thus electrostatically repulsed one and other, so that the entangling of the PANI chains was effectively prevented. As a consequence, the PANI-ND-MnO$_2$ composite film exhibited most promising pseudocapacitive perfomances.

An electrochemical co-deposition method was developed to deposit MnO$_2$–PANI composites for the application as electrodes of supercapacitors [154]. The MnO$_2$–PANI composites were electrochemically co-deposited on carbon cloth from solutions of aniline and MnSO$_4$ with potential cycling in the range of 0.2–1.45 V (vs. SCE). The co-deposition of PANI together with MnO$_2$ led to hybrid films with fibrous structures and large surface areas. The hybrid films exhibited pseudocapacitive behavior over the potential range of 0–0.65 V vs. SCE in an acidic aqueous solution of 1.0 M NaNO$_3$ with pH = 1. A specific capacitance of 532 F·g^{-1} at the discharging current density of 2.4 mA·cm^{-2}, with a coulombic efficiency of 97.5%, as well as a capacitance retention of 76% after 1200 cycles. For electrodepositions of the PANI–MnO$_2$ hybrid films, PM25, PM50, PM100, PM200 and PM400, solutions of 0.4 M aniline that contained 25, 50, 100, 200 and 400 mM Mn^{2+}, as well as 0.5 M H$_2$SO$_4$, were used, with PM50 to be the optimal composition in terms of electrochemical performances.

Other methods, including chemical polymerization [155, 156] and exchange reaction [157], have also been employed to prepare MnO$_2$–PANI composites. For instance, a MnO$_2$–PANI composite was deposited on a porous carbon electrode by oxidizing an aniline thin film in KMnO$_4$ solution [155]. At a rate of 50 mV·s^{-1}, the carbon/MnO$_2$–PANI composite electrode and the MnO$_2$–PANI composite film possessed a specific capacitance of 500 F·g^{-1}, which could be retained by 61% after 5000 cycles, because the dissolution of MnO$_2$ into the electrolyte was slowed down tremendously.

MnO$_2$–polypyrrole (PPy)

PPy is the second type of conducive polymers used for such purposes [160–165]. An electrochemical deposition method was used to synthesize MnO$_2$–PPy nanocomposite thin film, in which MnO$_2$ nanoparticles were embedded in PPy matrix [160]. Due to the long polymer chains of PPy, the MnO$_2$ nanoparticles could be homogeneously distributed in the matrix. Therefore, the porous PPy matrix offered a high active surface area for the MnO$_2$ nanoparticles to be attached, while the MnO$_2$ nanoparticles covering the polymer improved stability of the whole nanocomposites because the PPy polymer chains were strongly interlinked. Electrodes based on the MnO$_2$–PPy nanocomposite thin film exhibited promising electrochemical performance, with a specific capacitance of as high as 620 F·g^{-1}, as compared with those of MnO$_2$ (225 F·g^{-1}) and PPy (250 F·g^{-1}), as demonstrated in Fig. 4.82. XPS analysis results revealed that the MnO$_2$ was present in the form of hydrated manganese oxide with mixed oxidation states from Mn(II) to Mn(IV).

Figure 4.83 shows SEM images of the PPy, hydrous manganese oxide and the MnO$_2$–PPy nanocomposite obtained at same deposition conditions [160]. The electro-polymerized PPy film exhibited a high porosity, as shown in Fig. 4.83(a). As demonstrated in Fig. 4.83(b), the manganese oxide exhibited a dense microstructure, with agglomerated clusters resembled cauliflower. In comparison, the MnO$_2$–PPy nanocomposite had a morphology that was similar to that of PPy, which suggested that PPy had played a dominant role in determining microstructure of the nanocomposite. The uniformly distributed MnO$_2$

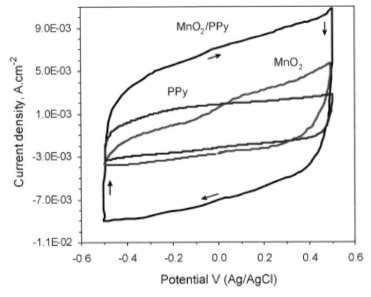

Fig. 4.82. CV curves of the samples in 0.5 M Na$_2$SO$_4$ at 50 mV·s^{-1}: (a) PPy, (b) hydrous MnO$_2$ and (c) MnO$_2$–PPy nanocomposite. Reproduced with permission from [160], Copyright © 2008, Elsevier.

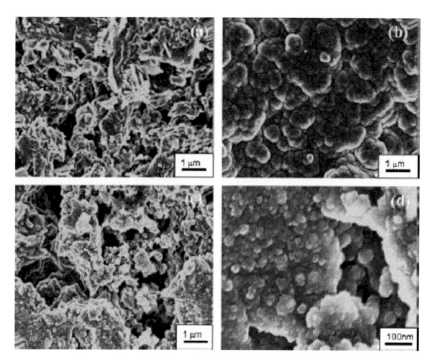

Fig. 4.83. SEM images of the films grown under identical conditions: (a) PPy, (b) MnO$_2$, (c) MnO$_2$–PPy nanocomposite and (d) MnO$_2$–PPy nanocomposite at high magnification. Reproduced with permission from [160], Copyright © 2008, Elsevier.

nanoparticles had an average particle size of 100 nm, which were covered by PPy, as seen in Fig. 4.83(d). The large surface area of the MnO$_2$–PPy nanocomposite was directly responsible for its highly enhanced electrochemical performance.

During the electrochemical co-deposition process, the porous PPy matrix offered a large surface to host the MnO$_2$ nanoparticles. Meanwhile, due to the blocking effect of the PPy chains, nucleation the MnO$_2$ nanoparticles was promoted, whereas their growth was inhibited. As a result, size of the MnO$_2$ nanoparticles was effectively controlled. Also, the presence of MnO$_2$ nanoparticles on the PPy scaffold improved cross-linking behavior of the PPy chains, resulting in a reduction in density of the chain defects. Furthermore, the rigid MnO$_2$ nanoparticle clusters in the PPy matrix essentially offered a structural stability of the PPy chains during the redox cycling. With a pseudo-capacitive behavior, the performance of MnO$_2$ is closely related to the number of available atoms at the surface atoms. In this respect, the MnO$_2$–PPy nanocomposite consisted of PPy framework with large surface to host the MnO$_2$ nanoparticles with sufficiently small size, which was believed to greatly contribute to the overall electrochemical performance of the nanocomposites.

Another example is the preparation of PPy–MnO$_2$ nanocomposite with Poly(4-styrenesulfonic acid) (PSS) dispersed with MWCNTs as a support [162]. Both electrical conductivity and mechanical integrity of the nanocomposites as electrodes of supercapacitors were greatly enhanced, due to the strong interaction between PPy and MnO$_2$, thus leading to high electrochemical performances. Due to the surface functionalities of –SO$_3^-$ groups on the MWCNT-PSS support, the MnO$_2$ nanoparticles were deposited inside the PPy matrix, in a highly ordered way, with a molecular level dispersion. As electrode of supercapacitors, the MWCNT-PSS/PPy–MnO$_2$ nanocomposite offered a specific capacitance of 268 F·g^{-1} at the current density of 5 mV·s^{-1}. The specific capacitance was retained by 93% as the current density was increased from 5 mV·s^{-1} to 100 mV·s^{-1}, while only 10% reduction was observed after 5000 CV cycles. The specific capacitance of the nanocomposite was 412 F·g^{-1} in 0.5 M Na$_2$SO$_4$ electrolyte, in terms of PPy–MnO$_2$.

In synthetic experiments, 10 mg MWCNTs, with diameters of 10–20 nm and lengths of 5–20 μm, were dispersed in 10 ml DI water containing 0.3 ml 18 wt% PSS [162]. Thus, the MWCNTs were wrapped by the PSS with the aid of ultrasonication, so as to be greatly stabilized, due to the charge repulsion effect. As a result, pyrrole and metal ions were attracted by the negatively charged PSS layer on the MWCNTs, forming the templates for the formation of the nanostructure. Because the large number of surface–SO_3^- functional groups interacted with the nuclei, the materials would be grown in just one direction to form 1D structure, which prevented the particles from being agglomerated. PPy–MnO_2 nanocomposite was formed by oxidizing pyrrole with $KMnO_4$.

The positive effect of PSS on the dispersion of MWCNTs was confirmed by TEM results, as demonstrated in Fig. 4.84(a, b) [162]. Without the presence of PSS, there was very weak interaction between PPy and MWCNTs, as seen in Fig. 4.84(a), whereas the MWCNTs (MWCNT-PSS) dispersed with PSS were wrapped with 30 nm thick PPy film, thus forming homogeneous MWCNT-PSS/PPy nanocomposite, as illustrated in Fig. 4.84(b). Such a morphology allowed the incorporation of MnO_2 into the PPy matrix to form MWCNT-PSS/PPy–MnO_2 nanocomposite, as evidenced in Fig. 4.84(c, d). The unique microstructure and morphology of the nanocomposite were also confirmed by other analysis results. The absence of agglomerations would be responsible for the high electrochemical performance of the nanocomposites when they were used as the electrodes of supercapacitors.

More recently, a ternary polypyrrole/manganese dioxide/carbon nanotubes (PPy/MnO_2/CNTs) composite, with tube-in-tube nanostructures, was synthesized by using an *in-situ* chemical oxidation polymerization method [164]. In this case, pyrrole was oxidized in the presence of the inorganic matrix of MnO_2 and CNTs, with a complex of methyl orange (MO)/$FeCl_3$ as a reactive self-degradable soft template. In the tube-in-tube nanostructures of the PPy/MnO_2/CNTs composite, the inner tubules were CNTs, while the outer tubules were template-synthesized PPy. MnO_2 nanoparticles were either sandwiched in the space between the inner and outer tubules or directly attached onto walls of the PPy tubes. Such composites exhibited promising electrochemical performances, with a specific capacitance of 402.7 $F \cdot g^{-1}$ at a current density of 1 $A \cdot g^{-1}$, as well as a high stability of up to 1000 cycles.

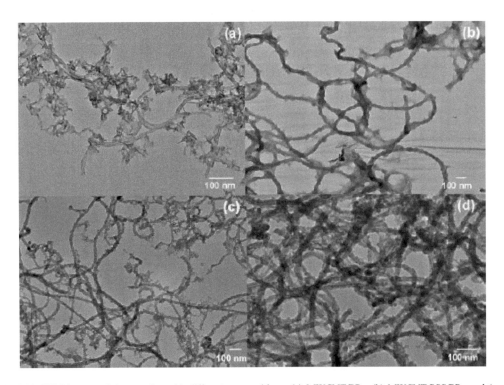

Fig. 4.84. TEM images of the samples with different compositions: (a) MWCNT/PPy, (b) MWCNT-PSS/PPy and (c, d) MWCNT-PSS/PPy–MnO_2 nanocomposite films. Reproduced with permission from [162], Copyright © 2010, Elsevier.

The MnO$_2$/CNTs composite was obtained by using the hydrothermal method [164]. The hydrothermal reaction solution was prepared with KMnO$_4$ and MnSO$_4$·H$_2$O, which were dissolved in distilled water with a molar ratio of 2:3 under vigorous stirring for 10 minutes. After that, CNT powder was added with a molar ratio of CNTs to KMnO$_4$ to be 5:9, with the aid of ultrasonication in order to promote the dispersion of CNTs. The suspension was then hydrothermally reacted at 140°C for 10 hours. The precipitates were thoroughly washed with distilled water and anhydrous alcohol, followed by vacuum drying at 60°C for 24 hours.

To form oxidant solution, FeCl$_3$ was dissolved in Methyl Orange (MO) solution with the aid of ultrasonication, which led to the presence of precipitate [164]. Then, the MnO$_2$/CNTs composite powder with an appropriate amount was added to the suspension, into which pyrrole monomer was introduced with strong ultrasonication. To facilitate the reaction, the mixture was kept at static condition for 24 hours at the temperatures between −5°C and 0°C. The composite products, with a nominal composition of MnO$_2$, PPy and CNTs to be about 57.0, 41.2 and 1.8%, respectively, were thoroughly washed, followed by vacuum drying at 50°C.

TEM observation results indicated that most MnO$_2$ particles were attached on surface of the CNTs, with a small fraction to be freely present in the sample [164]. For the PPy/MnO$_2$/CNTs composite, PPy tubes were formed due to the template effect of the MnO$_2$/CNTs composite, thus resulting in a tube-in-tube nanostructure. Thin tubes were formed by CNTs, with inside and outside diameters to be 20 nm and 40 nm, respectively, whereas PPy tubes were relatively thick, with diameters in the range of 90–250 nm. The MnO$_2$ particles in the composites were present in two forms, with (i) majority to be sandwiched between the CNTs and PPy tubes and (ii) minority being latched on walls of the PPy tubes. In addition, the MnO$_2$ nanoparticles had an irregular spherical morphology, with radius in the range of 15–30 nm. Both the SEM and TEM images disclose that the tube-in-tube nanostructured PPy/MnO$_2$/CNTs was successfully prepared.

The formation processes of the MnO$_2$/CNTs composite, PPy nanotubes and tube-in-tube PPy/MnO$_2$/CNTs composite described are schematically demonstrated in Fig. 4.85 [164]. Initially, most MnO$_2$ particles with relatively small sizes were attached onto the CNTs during the hydrothermal reaction. However, those that were too large could not be held by the CNTs, so as to be present as free particles, as seen in

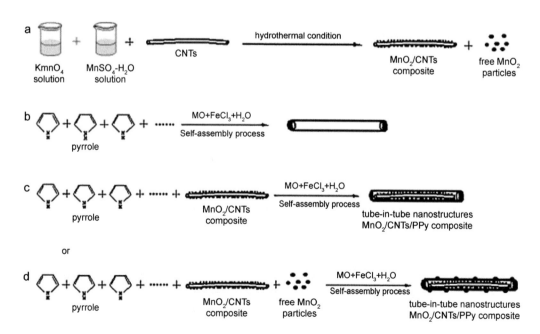

Fig. 4.85. Schematic diagram demonstrating formation process of the samples with different compositions or configurations: (a) MnO$_2$/CNTs composite, (b) PPy nanotubes and (c, d) tube-in-tube PPy/MnO$_2$/CNTs composite. Reproduced with permission from [164], Copyright © 2013, Elsevier.

Fig. 4.85(a). Then, the surface of the MnO_2/CNTs composite was wrapped by the pyrrole monomers, which underwent *in-situ* polymerization, so that the PPy/MnO_2/CNTs composite with tube-in-tube nanostructure was obtained, as illustrated in Fig. 4.85(c). As observed in Fig. 4.85(d), the freely-present large MnO_2 particles would be attached onto the outside walls of the PPy tubes.

MnO_2–polythiophene (PTh)

The third group of conductive polymers is polythiophene (PTh) and its derivatives, which have high electrical conductivity, high chemical/physical stability and strong mechanical flexibility, to form MnO_2–PTh composites with MnO_2, in order to further enhance the electrochemical performances as electrodes of supercapacitors [31, 33, 166, 167]. Selected examples are discussed representatively as follows.

A mesoporous nanocomposite consisting of polythiophene and MnO_2, as electrodes of supercapacitors with improved electrochemical properties, was fabricated by using a modified interfacial method [166]. The nanocomposite possessed a uniform hierarchical microstructure with submicron-spheres assembled from ultrathin nanosheet with diameters of < 10 nm, due to the ability of the method to prevent the overgrowth of nuclei. An optimal specific capacitances of 282 $F·g^{-1}$ was achieved at the current density of 1 $A·g^{-1}$. The nanocomposite electrode exhibited promising cycle performance, with a 97.3% retention of the initial specific capacitance after 1000 cycles at a charge/discharge rate of 2 $A·g^{-1}$. As the charge/discharge rate was increased from 1 $A·g^{-1}$ to 10 $A·g^{-1}$, the specific capacitance would be retained by 76.6%, demonstrating a high-power capability of the nanocomposite electrode. The polythiophene played an important role in both increasing electrical conductivity of the nanocomposite and preventing the dissolution of manganese oxides during the charge–discharge cycles.

To fabricate the nanocomposite, 2 mL thiophene was dissolved in 100 ml dichloromethane (CH_2Cl_2), while 0.1 g $KMnO_4$ was dissolved in 100 ml DI water, with pH = 2 adjusted with diluted HCl solution [166]. The aqueous solution was pipetted in to the organic solution in a very low manner, so that a static organic/inorganic interface was developed. The reaction system was then kept at stable temperature of 4°C, in order to facilitate the oxidative polymerization of thiophene and the reduction of MnO_4^- to manganese oxides at the organic/inorganic interface, which led to the formation of dark purplish-brown powder after reaction for about 1 hour.

Representative SEM images of the nanocomposite are depicted in Fig. 4.86(A, B) [166]. As shown in Fig. 4.86(A), the nanocomposite contained spherical particles with diameters in the range of 500–800 nm. The spherical particles were made of radial nanosheets that were less than 10 nm in thickness, as seen in Fig. 4.86(B). The hierarchical microstructure of the nanocomposite was further confirmed by TEM image, as illustrated in Fig. 4.87(C).

The sample possessed typical type IV isotherm with a hysteresis loop of type H3, according to the N_2 adsorption/desorption isotherm, demonstrating its disordered mesoporous structure, which could be attributed to the presence of the ultrathin nanosheets. A high BET surface area of 110 $m^2·g^{-1}$ was observed for the nanocomposite. Also, it had a relatively narrow pore size distribution, with diameter to be centered at about 4 nm. Such kind of nanoporous matrix added benefit to the nanocomposite, in terms of ion transfer and the access of electrolyte and ions into the electrodes. After thermal annealing 400°C for 1 hour, the submicron-sphere/nanosheet hierarchical microstructure was essentially retained, as revealed in Fig. 4.86(D).

Similar to PTh, various PTh derivatives have also been explored to fabricate nanocomposites with MnO_2 to enhance electrochemical performances. For instance, a composite was developed by using an electrodeposition method, with manganese oxide on titanium substrates that were modified with poly(3-methylthiophene) (PMeT) [31]. The PMeT polymer films were deposited by using galvanostatic deposition conducted at 2 $mA·cm^{-2}$, with deposition charges in the range of 250–1500 $mC·cm^{-2}$. The nanocomposites were characterized as electrodes of supercapacitors in 1 $mol·L^{-1}$ Na_2SO_4 aqueous solution. Due to the presence of the PMeT polymer layers, electrochemical properties of the manganese oxide were largely boosted. As compared with that of Ti/MnO_2 (122 $F·g^{-1}$), specific capacitances of the Ti/$PMeT_{250}$/MnO_2 and Ti/$PMeT_{1500}$/MnO_2 were 218 and 66 $F·g^{-1}$, corresponding to 381 and 153 $F·g^{-1}$, respectively, when normalized in terms of the mass of the oxide mass. It was found that the presence of the polymer layers

248 *Nanomaterials for Supercapacitors*

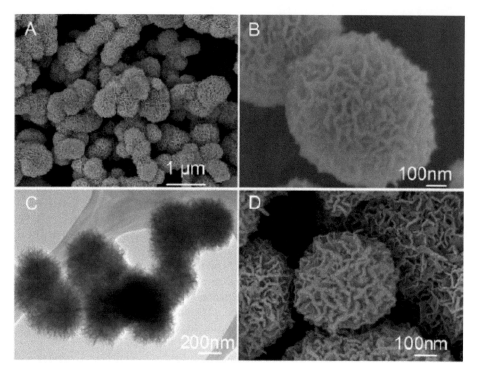

Fig. 4.86. Representative SEM (A, B and D) and TEM (C) images of the PTh/MnO$_2$ nanocomposite before (A, B and C) and after (D) thermal annealing at 400°C for 1 hour. Reproduced with permission from [166], Copyright © 2011, Elsevier.

altered the morphology of the manganese oxide, which also had a contribution to the improvement in electrochemical properties of the nanocomposites.

MnO$_2$–poly(o-phenylenediamine) (PoPD)

Manganese oxide based electrodes usually encounter the problem of reduction-dissolution during the electrochemical reaction, if mildly acidic and near-neutral electrolytes are used, due to the formation of soluble Mn(II) [168]. As a consequence, they could be used more likely in neutral or basic aqueous electrolytes. In this case, their electrochemical performances, such as specific capacitance and rate capacity, would be damaged, because of the presence of the insertion cations, such as Li$^+$, Na$^+$ and K$^+$, which are more sluggish than H$^+$ insertion. To address this important issue, a self-limiting electropolymerization method was proposed, in which nanostructured MnO$_2$ ambigels were coated with a thin layer of conformal poly(o-phenylenediamine) (PoPD) [28, 29]. The self-limited PoPD thin films, with a thickness of < 10 nm, were electrodeposited on planar Indium-Tin Oxide (ITO) electrodes, from an aqueous borate buffered solution with pH = 9. Because the ultrathin PoPD coating acted as a barrier to protect the MnO$_2$ nanoparticles, thus effectively reducing their electrochemical dissolution behavior, so that the PoPD–MnO$_2$ nanocomposite based electrodes could be operated in acid aqueous electrolytes.

More recently, a composite consisting of MnO$_2$ and PoPD was fabricated, by a one-step co-precipitation method, through the reaction of KMnO$_4$ with o-phenylenediamine in an acidic solution at room temperature [169]. It was found that morphology and electrochemical performance of the nanocomposites were influenced by the content of Cetyl Trimethyl Ammonium Bromide (CTAB) and Sodium Dodecyl Sulfate (SDS). An optimal specific capacitance of 262.2 F·g^{-1} was achieved with the potential window of 0–0.9 V, vs. saturated calomel electrode, in 1.0 mol·L^{-1} KNO$_3$ solution.

To synthesize the MnO$_2$/PoPD composite without surfactant, 20 mL 0.1 mol·L^{-1} KMnO$_4$ was slowed added into 20 mL 0.15 mol·L^{-1} oPD containing 0.01 mol·L^{-1} HCl, with continuous stirring for 6 hours [169]. MnO$_2$/PoPD composite samples were obtained after filtration, washing and drying. A similar process was used to prepared composites including CTAB and SDS. CTAB was added into 0.15 mol·L^{-1} oPD containing 0.01 mol·L^{-1} HCl, so that CTAB concentrations were 1.25, 2.50, 5.00, 7.50, 10.0 and 20.0 g·L^{-1}. Same compositions were also employed for SDS. Twenty mL solution was mixed with 20 mL 0.1 mol·L^{-1} KMnO$_4$, under vigorous stirring for 6 hours. The samples of CTAB were denoted as C1, C2, C3, C4, C5 C6, while those for SDS were S1, S2, S3, S4, S5 and S6, respectively. The surfactant-free sample was denoted as A0.

Figure 4.87 shows XRD patterns of the samples A0, C2 and S2 [169]. The three samples exhibited a similar XRD pattern, which suggested that the surfactant CTAB and SDS had no obvious effect on phase composition of the MnO$_2$/PoPD nanocomposites. However, they were all of an amorphous-like characteristic, as evidenced by the relatively low peak intensity and broad peaks. The weak diffraction peaks indicated that the patterns could be indexed as α-MnO$_2$.

Figure 4.88 shows SEM images of the samples A0, S2 and C2, at different magnifications [169]. Particle sizes of the samples were in the range of 0.5–2 μm. The surfactant-free sample had serious agglomeration, which was effectively avoided, with the presence of surfactants. For instance, the sample with 2.5 g·L^{-1} SDS an average size of about 1 μm, with relatively narrow size distribution. Furthermore, by including 2.5 g·L^{-1} CTAB, the agglomeration was almost completed prevented, while the particle size was increased to about 2 μm. This observation was readily attributed to the interaction between the surfactants and oPD, by forming complexes on which the manganese dioxide particles were grown and PoPD was developed.

Under the optimal condition, including 2.5 g·L^{-1}, specific capacitances of C2 and S2 were 262.2 F·g^{-1} and 246.3 F·g^{-1}, at the scan rate of 5 mV·s^{-1}. Figure 4.89 shows specific capacitances of the supercapacitors based on the three composites as a function of cycle number [169]. The value of specific capacitance was decreased relatively quickly during the first 100 cycles, after which values were gradually leveled off. After 1000 cycles of charge–discharge, the specific capacitances of C2, S2 and A0 were retained by 83.9, 80.8

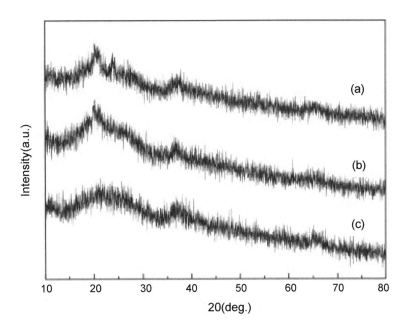

Fig. 4.87. Representative XRD patterns of the samples: (a) A0, (b) C2 and (c) S2. Reproduced with permission from [169], Copyright © 2013, Elsevier.

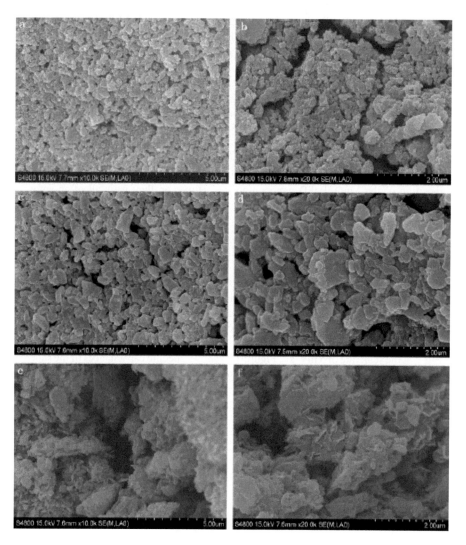

Fig. 4.88. Representative SEM images of the samples: (a, b) A0, (c, d) S2 and (e, f) C2. Reproduced with permission from [169], Copyright © 2013, Elsevier.

and 63.1%, respectively. In this case, the sample C2 had the highest specific capacitance and most stable cycling behavior, because the presence of CTAB led to a more stable structure, which could withstand the impact due to the charge–discharge cycles.

MnO_2–Nanocarbon Hybrids

MnO_2–carbon nanotubes (CNTs)

CNTs have been extensively studied as a nanostructured carbon, with 1D tubular structures, having unique physicochemical properties, including high electrical conductivity, strong mechanical strength, high chemical stability and high surface areas. Because CNTs are intrinsically carbon materials, it is not surprising that they have a relatively low specific capacitance when they are used as electrodes of supercapacitors. Similar to conductive polymers, CNTs can be incorporated with MnO_2, as a conductive component and structural stabilizer, to form hybrids that could be used as electrodes of supercapacitors with

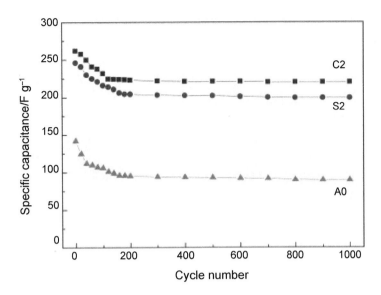

Fig. 4.89. Cycling behaviors of the samples C2, S2 and A0 in 1 mol·L^{-1} KNO$_3$ aqueous solution, at the current density of 500 mA·g^{-1}. Reproduced with permission from [169], Copyright © 2013, Elsevier.

enhanced electrochemical performances. Various strategies, such as chemical co-precipitation [170–177], thermal decomposition [178], electrophoretic deposition (EPD) and electrochemical deposition, have been employed to fabricate hybrids of MnO$_2$ and CNTs.

As the most widely used method, chemical co-precipitation to synthesize MnO$_2$–CNTs hybrids can be realized in different ways, such as reduction of KMnO$_4$ [170–173], hydrothermal oxidation of manganese acetate (Mn(CH$_3$COO)$_2$) [174] and oxidization of MnSO$_4$ with (NH$_4$)$_2$S$_2$O$_8$ and (NH$_4$)$_2$S$_2$O$_4$ [175]. For instance, spherical nanostructured MnO$_2$/MWCNTs hybrids were obtained by using a co-precipitation method combined with microwave irradiation [175]. The MWCNTs served as a support of the MnO$_2$ nanoplates so that they were not converted into nanorods. Electrodes based on the nanohybrids exhibited specific capacitances of 298, 213 and 198 F·g^{-1} at the current densities of 2, 10 and 20 mA·cm^{-2}, respectively.

MWCNTs were purified with a mixture of 3:1 (v:v) H$_2$SO$_4$/HNO$_3$ at 140°C for 2 hours, followed by filtration and thorough washing [175]. To prepare MnO$_2$–CNTs hybrids, 100 mg purified CNTs were dispersed in 65 ml distilled water with the aid of ultrasonication for 0.5 hour, into which 2.30 g MnSO$_4$·H$_2$O, 3.57 g (NH$_4$)$_2$S$_2$O$_8$ and 5.24 g (NH$_4$)$_2$S$_2$O$_4$ were added, followed by stirring for 4 hours at room temperature. After that, the mixture was irradiated with a commercial microwave oven at 700 W for 16 minutes. For comparison, the mixture without the addition of CNTs was similarly irradiated for 8 minutes and 16 minutes. All the samples were thoroughly washed with DI water and ethanol, followed by drying at 60°C in air.

SEM images of the MnO$_2$ powders without the presence of CNTs are shown in Fig. 4.90(a) and (b) [175]. It was found that the 8 minutes pure MnO$_2$ powder had a spherical structured morphology, with diameters in the range of 1–2 μm. The spheres possessed a sub-structure of plate-like MnO$_2$. As the irradiation time was increased to 16 minutes, the lamellar structure was collapsed into MnO$_2$ nanorods, with diameters of 10–30 nm and lengths of 1–2 μm, as shown in Fig. 4.90(b). Figure 4.90(c) shows SEM image of the purified MWCNTs, which were cut into short segments, with lengths in the range of 0.5–1 μm. These short CNTs were entangled to form agglomerates with an average diameter of 1 μm. In comparison, MnO$_2$–MWCNTs hybrids possessed spheres consisting of MnO$_2$ nanoplates and CNTs, as seen in Fig. 4.90(d).

Figure 4.91 shows XRD patterns of the samples with and without the presence of MWCNTs [175]. The MnO$_2$ powder obtained after microwave irradiation for 8 minutes contained two phases of MnO$_2$, with α-MnO$_2$ to be majority and γ-MnO$_2$ as minority. The diffraction peaks were relatively broad, indicating its nanocrystalline characteristics. As the irradiation time was prolonged to 16 minutes, the diffraction peaks

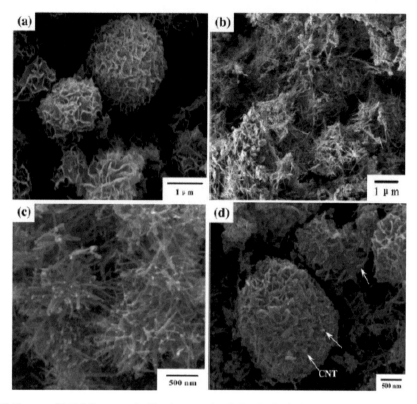

Fig. 4.90. SEM images of (a) MnO$_2$ prepared with microwave irradiation for 8 minutes, (b) MnO$_2$ prepared with microwave irradiation for 16 minutes, (c) purified MWCNTs, and (d) MnO$_2$–MWCNTs prepared with microwave irradiation for 16 minutes. Reproduced with permission from [175], Copyright © 2008, Elsevier.

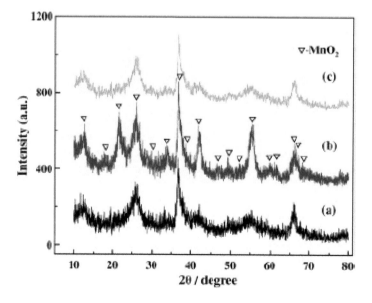

Fig. 4.91. XRD patterns of (a) MnO$_2$ prepared with microwave irradiation for 8 minutes, (b) MnO$_2$ prepared with microwave irradiation for 16 minutes and (c) MnO$_2$–MWCNTs prepared with microwave irradiation for 16 minutes. Reproduced with permission from [175], Copyright © 2008, Elsevier.

of MnO$_2$ were significantly sharpened and strengthened, as illustrate in Fig. 4.91(b). With the presence of MWCNTs, crystallization of MnO$_2$ was weakened, evidenced by the broadened diffraction peaks of the MnO$_2$–MWCNTs sample, as demonstrated in Fig. 4.91(c).

Figure 4.92 shows a schematic diagram describing formation mechanism of the MnO$_2$–MWCNTs hybrids [175]. Initially, Mn^{2+} ions in the solution were adsorbed by the MWCNTs through electrostatic attraction due to the presence carboxyl groups on the CNT surfaces, which were oxidized by (NH$_4$)$_2$SO$_4$ to form amorphous MnO$_2$ during the microwave irradiation. After that, the amorphous MnO$_2$ was converted into layer structures. With increasing irradiation time, the layer structured MnO$_2$ was enlarged, thus leading to the formation of interlacement structures of MnO$_2$ nanoplates and CNTs. In this case, entangled CNT agglomerates served as a support to prevent the nanoplates from collapsing, as described in Fig. 4.92(b–c). In contrast, without the presence of CNTs, the lamellar structure would be collapsed into MnO$_2$ nanorods, as observed in Fig. 4.90(b).

Furthermore, CNTs could act as a reducing agent to reduce KMnO$_4$ in an aqueous solution, which resulted in heterogeneous nucleation of MnO$_2$, where the CNTs also served as a template to guide the growth of the MnO$_2$ nanoparticles [176, 177]. Various parameters, such as composition of solutions, pH value, reaction temperature and reaction time duration, can be used to control morphology and composition of the MnO$_2$–CNT nanocomposites. For example, in a study to deposit MnO$_2$ on CNTs, a thin and uniform layer of MnO$_2$ was formed on CNTs at an initial pH = 7, while MnO$_2$ nanorods were obtained at an initial pH value of 1 [176].

Figure 4.93 shows SEM images of the CNTs with and without the presence of the MnO$_2$ nanoparticles [176]. The MnO$_2$ nanoparticles were deposited at 70°C from 200 ml KMnO$_4$ aqueous solution with a concentration of 0.1 M and different initial pH values, together with 1.0 g CNTs. As shown in Fig. 4.93(a), the CNTs and thicknesses in the range of 20–30 nm. With the presence of MnO$_2$ nanoparticles, CNTs were slightly thickened to diameters of 30–50 nm, confirming that the CNTs served as a template to guide the heterogeneous precipitation of the MnO$_2$ nanoparticles, as demonstrated in Fig. 4.93(b d). The MnO$_2$ layers became coarser and coarser, as the pH value of the initial solution was gradually increased. This observation implied that the morphology of the MnO$_2$ particles could be determined by the reduction rate of MnO$_4^-$ ions to MnO$_2$ by the CNTs. It was found that the time required to remove ions from the solutions was decreased with decreasing pH value, which suggested that the reduction rate was increased with decreasing pH value of the initial solution.

MnO$_2$–CNT nanocomposites used as electrodes of supercapacitors have also been developed by using the thermal decomposition method. For instance, a thermal decomposition method was employed to achieve Mn$_3$O$_4$ nanoparticles that were attached onto carbon nanotube arrays (CNTA) [178]. With the deposition of the Mn$_3$O$_4$ nanoparticles, the CNTAs became hydrophilic from hydrophobic, while their

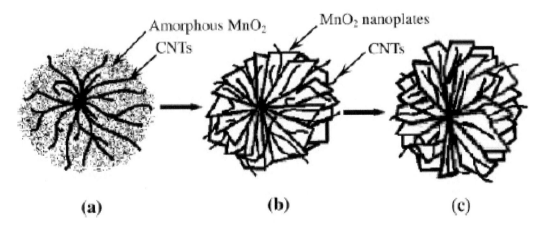

Fig. 4.92. Schematic diagram showing formation process of the MnO$_2$–MWCNTs hybrids. Reproduced with permission from [175], Copyright © 2008, Elsevier.

254　*Nanomaterials for Supercapacitors*

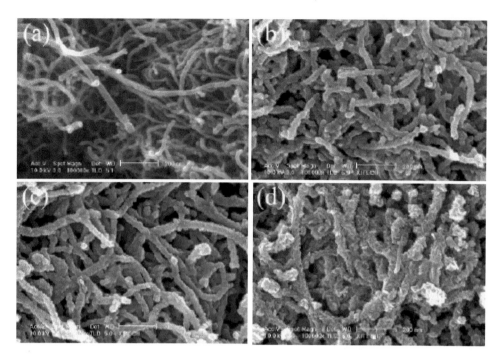

Fig. 4.93. SEM images of (a) pristine CNTs and the MnO$_2$/CNT nanocomposites prepared in the 200 ml aqueous solution of 0.1 M KMnO$_4$ containing 1.0 g CNTs at 70°C with different pH values in the initial solutions: (b) pH = 7, (c) pH = 2.5 and (d) pH = 1. Reproduced with permission from [176], Copyright © 2007, Elsevier.

original alignment and integrity were well retained. Due to its hydrophilic characteristics, the Mn$_3$O$_4$/CNTA nanocomposite exhibited enhanced performance as electrodes of supercapacitors, as compared with CNTA counterpart without the manganese oxide. The Mn$_3$O$_4$/CNTA nanocomposite based electrode had an optimized specific capacitance of 143 F·g^{-1}, corresponding to a high area-normalized capacitance of 1.70 F·cm^{-2}. When normalized to the Mn$_3$O$_4$ nanoparticles, the specific capacitance was 292 F·g^{-1}. In contrast, specific capacitance of the pure CNTA based electrode was as low as 1–2 F·g^{-1}.

The CNTAs were synthesized by using a Catalytic Chemical Vapor Deposition (CCVD) on (100) cut *p*-type Si wafers that were coated with a thin layer of 30 nm Al$_2$O$_3$ and a catalyst film of 3 nm Fe with DC magnetron sputtering [178]. To grow MWCNT arrays, a single-zone quartz tube furnace with an inner diameter of 5 in was first evacuated to ≤ 0.1 Torr. After purging with Ar for 1 hour, the furnace was heated to 750°C and kept for 60 minutes at Ar flow rate of 200 sccm Ar and H$_2$ flow rate of 400 sccm. Then, C$_2$H$_4$ was introduced at a flow rate of 400 sccm for 30 minutes. Finally, the flows of H$_2$ and C$_2$H$_4$ were stopped, while the furnace was cooled down to < 100°C with Ar purging. The CNTA obtained in this way had a length of as long as 0.9 mm.

To prepare Mn$_3$O$_4$/CNTA composites, the CNTAs were peeled off from the substrate, which were then immersed for 3 minutes in an ethanol solution of Mn (CH$_3$COO)$_2$·4H$_2$O [178]. Then, the CNTA sample was quickly taken out, onto which 10 drops of the same solution were dropped, with each drop for 1 minute. After that, the CNTA was dried at room temperature in air for 2 hours and at 100°C in a furnace for 1 hour, followed by calcination at 300°C for 2 hours, thus leading to Mn$_3$O$_4$/CNTA composite. One side of the composite sample was coated with a thin layer of Au by using sputtering, which was attached to a double-sided conductive tape, while the other side was sealed with a green Mask-it. The solvent was denatured alcohol consisting of 85% ethanol and 15% methanol. 0.25 M ethanol solution of Mn (CH$_3$COO)$_2$·4H$_2$O was used to obtain Mn$_3$O$_4$/CNTA(0.84), whereas 0.1 M ethanol solution of Mn(CH$_3$COO)$_2$·4H$_2$O resulted in sample Mn$_3$O$_4$/CNTA(0.33). The pure CNTAs were also studied as a comparison.

Figure 4.94 shows a schematic diagram to demonstrate formation process of the Mn$_3$O$_4$/CNTA composites [178]. The MWCNTs of the CNTAs had an average diameter of 11.5 nm, with wall number of 5–7. Height and density of the CNTAs were 0.9 mm and 0.069 g·cm^{-3}, respectively, corresponding to 92.3% unfilled space. The CNTAs were hydrophobic, with a contact angle of 138°. It was found that non-aqueous solution Mn(CH$_3$COO)$_2$, i.e., ethanol solution, was an important factor for the CNTAs to absorb the precursor of manganese oxide. The calcination at 300°C in air for 2 hours triggered the conversion of absorbed Mn(II) to Mn$_3$O$_4$ nanoparticles, so that Mn$_3$O$_4$/CNTA composites were achieved with the Mn$_3$O$_4$ nanoparticles to be uniformly distributed on the CNTA framework.

Figure 4.95 shows TEM images and EDS spectra of the composite samples, to reveal their morphology, composition and crystal structure [178]. A representative TEM image of a strip of Mn$_3$O$_4$/CNTA(0.33) at center region of the composite is shown in Fig. 4.95(a). Obviously, the CNT walls were decorated with Mn$_3$O$_4$ nanoparticles. These nanoparticles are well dispersed along the CNTs. As illustrated in Fig. 4.95(b), C, O, Mn and Cu (from Cu grid) were detected, indicating phase pure characteristics of the composites. Selected Area Diffraction (SAD) pattern of the manganese oxide nanoparticles is shown in Fig. 4.95(c), confirming the presence of CNTs and tetragonal hausmannite-Mn$_3$O$_4$ phase. A TEM image of a strip of the Mn$_3$O$_4$/CNTA(0.84) composite is shown in Fig. 4.95(e), in which a similar morphology of the Mn$_3$O$_4$ nanoparticles was observed at high loading level. However, the nanoparticles had a certain degree of aggregation in the Mn$_3$O$_4$/CNTA(0.84) sample. The two samples had narrow size distribution, with an average sizes of 6.0 nm and 7.3 nm for the Mn$_3$O$_4$/CNTA(0.33) and Mn$_3$O$_4$/CNTA(0.84), respectively, as observed in Fig. 4.95(d) and (f).

Figure 4.96 shows surface and cross-sectional SEM images of the composites calcined at 300°C for 2 hours [178]. The volume of the CNTA was decreased by 7% after dipping with the ethanol solution and calcination. In the composites, the original morphology of the CNTA with aligned macropores or macrochannels was retained, as demonstrated in Fig. 4.96(b) and (d). The aligned macrochannels would be of benefit to electrochemical performance of the composites used as electrodes of supercapacitors. In

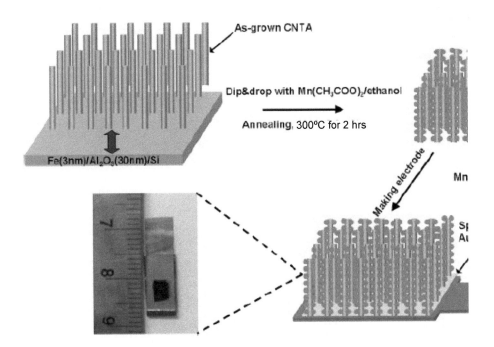

Fig. 4.94. Schematic diagram demonstrating fabrication process of the Mn$_3$O$_4$/CNTA nanocomposite based electrode. The inset image is a photograph of the nanocomposite based electrode before sealed with green Mask-it. Reproduced with permission from [178], Copyright © 2011, Elsevier.

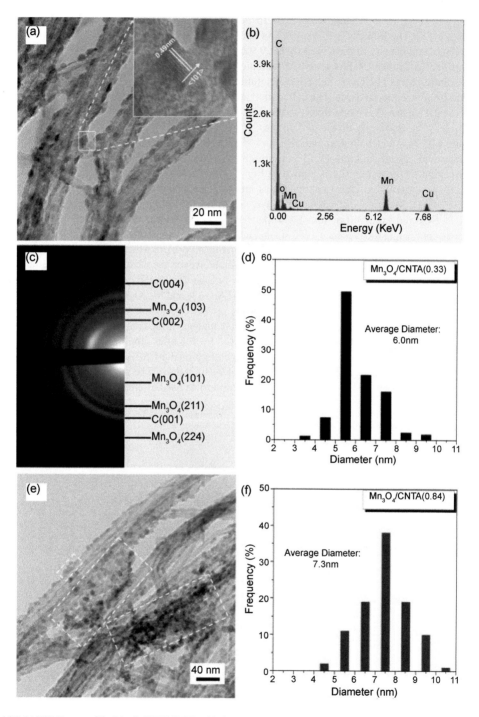

Fig. 4.95. (a) TEM image of the Mn$_3$O$_4$/CNTA(0.33) with the inset showing a HRTEM image of the Mn$_3$O$_4$ nanoparticles. (b) EDS and (c) SAD patterns of the Mn$_3$O$_4$/CNTA(0.33) in (a). (d) Size distribution of the Mn$_3$O$_4$ nanoparticle deposited onto the CNTA in Mn$_3$O$_4$/CNTA(0.33). (e) TEM image of Mn$_3$O$_4$/CNTA(0.84). (f) Size distribution of the Mn$_3$O$_4$ nanoparticle on the CNTA in Mn$_3$O$_4$/CNTA(0.84). Reproduced with permission from [178], Copyright © 2011, Elsevier.

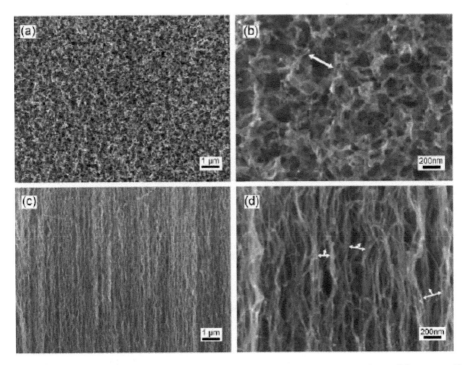

Fig. 4.96. (a, b) Surface and (c, d) cross-sectional SEM images of the Mn$_3$O$_4$/CNTA nanocomposites at different magnifications. Reproduced with permission from [178], Copyright © 2011, Elsevier.

addition, EDS mapping result indicated that the Mn$_3$O$_4$ nanoparticles were uniformly distributed in the composites.

Electrical deposition methods, including electrophoretic deposition (EPD) and electrochemical deposition, have been an alternative way to deposit MnO$_2$ on CNTs to obtain MnO$_2$–CNTs nanocomposites used as electrodes of supercapacitors. Electrophoretic deposition is to use electric field to align and thus deposit the charged particles of both MnO$_2$ and CNTs onto certain substrates [179, 180]. Parameters that can be used to control the composition, microstructure, thickness and properties of the films, including MnO$_2$/CNTs ratio in the suspensions, deposition time and voltage/current.

For instance, a cathodic EPD approach was reported to deposit MnO$_2$–MWCNT nanocomposite films, in which dopamine (DA) was used as the charging additive, in order to stabilize both MnO$_2$ nanoparticles and MWCNTs in the suspensions [179]. It was found that the deposition efficiency was affected by the concentration of DA. The EPD deposited MnO$_2$–MWCNT nanocomposite films had a porous microstructure, with electrochemical properties to be characterized in 0.5 M Na$_2$SO$_4$ solutions over voltage window of 0–0.9 V. An optimal specific capacitance of 650 F·g^{-1} was achieved at the scan rate of 2 mV·s^{-1}.

During electrochemical deposition, MnO$_2$ is usually deposited on CNTs through redox reactions, so as to form MnO$_2$–MWCNT nanocomposites [181, 182]. For instance, composites consisting of manganese oxide nanoflower-like particles and carbon nanotube array (CNTA), with hierarchical porous structure, large surface area, and high electrical conductivity, were fabricated by using the electrodeposition method on vertically aligned CNTA framework [181]. The binder-free manganese oxide/CNTA composite based electrode exhibited 50.8% capacity retention at 77 mA·g^{-1}, high capacitances of 199 F·g^{-1} and 305 F·cm^{-3} and only a 3% capacity loss after 20,000 charge/discharge cycles.

Figure 4.97 shows a schematic diagram to demonstrate the fabrication process of the nanocomposites [181]. To prepare the nanocomposites, vertically aligned CNTAs were firstly grown on a Ta foil by using

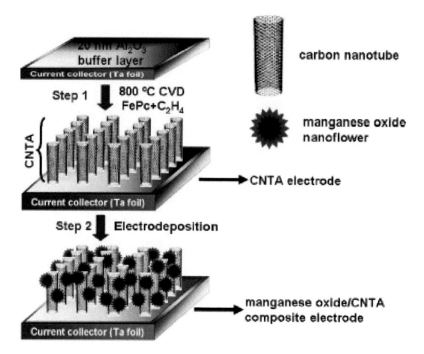

Fig. 4.97. Schematic diagram to present fabrication process of the manganese oxide/CNTA composites. Reproduced with permission from [181], Copyright © 2008, American Chemical Society.

a CVD method at 800°C, onto which manganese oxide was electrodeposited by using a potentiodynamic method. The CVD derived CNTA possessed a regular pore structure, with large pore sizes, as seen in Fig. 4.98(a, b), which ensured full adsorption of the precursor solutions by the CNTA. Also, due to the high electrical conductivity of the CNTAs, the manganese oxide could have been uniformly deposited onto the CNTA framework. In this case, Ta foils were used as the substrates, because Ta is sufficiently stable in acidic $MnSO_4$ precursor solutions, so that the loading level of manganese oxide could be accurately measured.

Figure 4.98 shows representative SEM and TEM images of the nanocomposites together with their components [181]. The inset in Fig. 4.98(a) shows photographs of the bare (left) and CNTA-covered (right) Ta foils. The CNTAs had a thickness of 35 μm, as illustrated in Fig. 4.98(a). As revealed in Fig. 4.98(b), the CNTA was comprised of densely packed and aligned CNTs. In the nanocomposites, the manganese oxide particles possessed an average diameter of 150 nm and were homogeneously distributed on the CNTAs, as observed in Fig. 4.98(c). EDX pattern of the composite indicated the presence of manganese oxide, as illustrated as the inset in Fig. 4.98(c). TEM image of Fig. 4.98(d) confirmed the surfboard-shaped nanosheets, constructing a dandelion-like flower morphology. The surfboard-shaped 'petal' had length and thickness of 50 and 3 nm, respectively.

Electrochemical performances of the electrodes based on the manganese oxide/CNTA composites are presented in Fig. 4.99 [181]. CV curves of the manganese oxide/CNTA composite possessed a rectangular shape at a high scan rate of 200 mV·s^{-1}, as seen in Fig. 4.99(a), owing to the highly capacitive nature and fast ion response. Electrochemical Impedance Spectroscopy (EIS) revealed the equivalent series resistance of manganese oxide/CNTA based electrode to be as low as 1.66 Ω. The manganese oxide/CNTA composite based electrode exhibited an optimized specific capacitance of 199 F·g^{-1} at a low current density, much higher than that (27 F·g^{-1}) of the CNTA, as demonstrated in Fig. 4.99(b). The high performances of the composite based electrodes could be attributed to the special microstructures of the materials, as illustrated in Fig. 4.99(d).

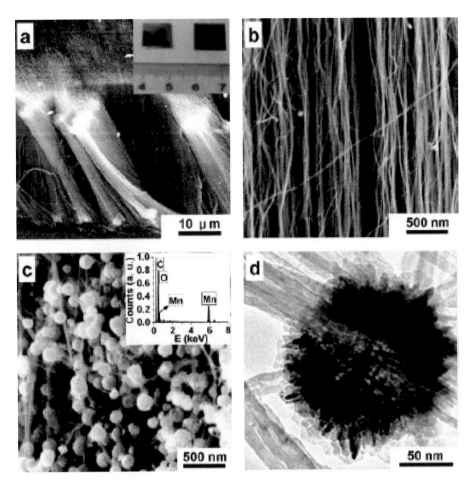

Fig. 4.98. Morphology and microstructure of the CNTA and the manganese oxide/CNTA composites. (a, b) SEM images at different magnifications of the original CNTA and a photograph (inset of (a)) of the original (left) and CNTA-covered (right) Ta foils. (c) SEM image of the manganese oxide/CNTA composite and its EDX pattern (inset). (d) TEM image of a manganese oxide nanoflower. Reproduced with permission from [181], Copyright © 2008, American Chemical Society.

MnO_2–graphene nanosheets

Graphene is a newly emerged nanocarbon material, with various special properties, which is mainly synthesized through chemical exfoliation of graphite. MnO_2–graphene nanosheet composites as electrodes of supercapacitors have been discussed in the previous chapter. For the purpose of completeness, as brief description on this group of nanocomposites will be presented here [183–192]. For example, a hydrothermal synthesis method was used to fabricate composites of MnO_2/graphite nanoplatelets (GNP) for potential applications as electrodes of supercapacitors [185]. Due to the homogeneous dispersion of MnO_2 nanorods on surfaces of the GNPs, the nanocomposites exhibited promising electrochemical performances, with optimized specific capacitance of 276.3 F·g^{-1}.

GNPs were prepared by using an ultrasonication exfoliation method. 1 g EG was dispersed in 400 mL distilled water, which ultrasonically treated for 10 hours. The GNPs suspensions, together with $MnSO_4·H_2O$, $(NH_4)_2S_2O_4$ and $(NH_4)_2SO_4$, with a molar ratio of $MnSO_4·H_2O$ and $(NH_4)_2S_2O_4$ to $(NH_4)_2SO_4$ of 1:1:2.5, were transferred into a three-necked flask, which were refluxed at 90°C for 12 hours. After reaction, the products were collected with filtration and thorough washing with distilled water, followed by drying at 60–80°C for 12 hours. MnO_2/GNP composites with contents of 5, 10, 20 and 40 wt% GNPs were obtained,

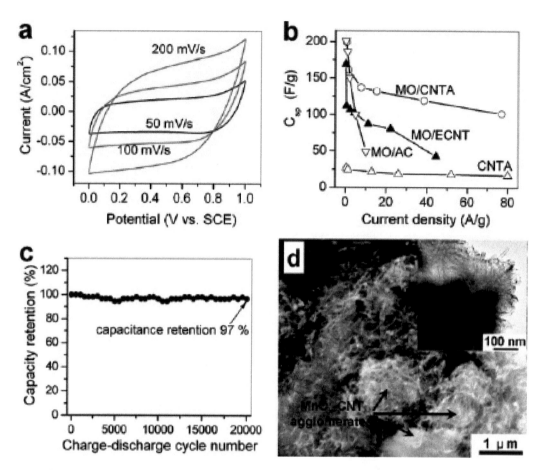

Fig. 4.99. Electrochemical properties of the manganese oxide/CNTA composite based electrode: (a) CV curves at 50–200 mV·s^{-1}, (b) specific capacitance of the manganese oxide/CNTA, manganese oxide/ECNT, manganese oxide/AC and original CNTA, as a function of discharge current density, and (c) charge–discharge cycle performance. (d) SEM and TEM images (inset) of the manganese oxide/ECNT composite. Reproduced with permission from [181], Copyright © 2008, American Chemical Society.

denoted as GNPs-5, GNPs-10, GNPs-20 and GNPs-40, respectively. For comparison, a sample with 20 wt% EG, i.e., MnO$_2$/EG composite, denoted as EG-20, was also prepared.

SEM results indicated that EG had a loose and porous structure consisting of graphite platelets. The loosely packed graphite nanosheets were exfoliated into individual nanoplatelets with a thickness of 30 nm and diameters of 3–10 μm with the aid of the strong ultrasonication. After hydrothermal reaction, MnO$_2$ nanorods were formed and distributed on surface of the EG. In comparison, GNPs with thin graphite nanoplatelets offered more active sites for the attachment of the MnO$_2$ nanoparticles during the hydrothermal reaction process. As a result, MnO$_2$ nanorods were uniformly distributed on surfaces of the GNPs. The MnO$_2$ nanorods consisted of two phases, i.e., α-MnO$_2$ and γ-MnO$_2$.

Figure 4.100 shows a schematic diagram used to describe the formation mechanism of the MnO$_2$/GNPs. Firstly, Mn^{2+} ions in the precursor solution were adsorbed on surface of the GNPs, which were then oxidized *in-situ* by (NH$_4$)$_2$SO$_4$, thus leading to the formation of amorphous MnO$_2$ during the hydrothermal reaction. The amorphous MnO$_2$ was prone to develop into layer structured δ-MnO$_2$ through a condensation reaction due to the relatively high concentration and high surface energy of the nanoparticles [193]. Eventually, the lamellar structures MnO$_2$ were evolved into nanorods, due to the presence of suitable amount of NH$_4^+$ on surface of the GNPs.

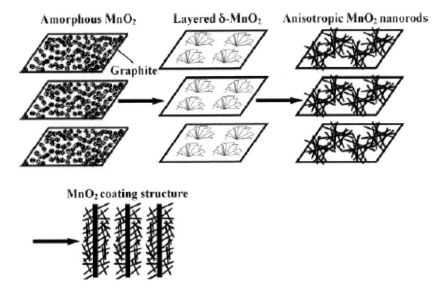

Fig. 4.100. Schematic diagram illustrating synthesis of the MnO$_2$/GNPs composites. Reproduced with permission from [185], Copyright © 2008, Elsevier.

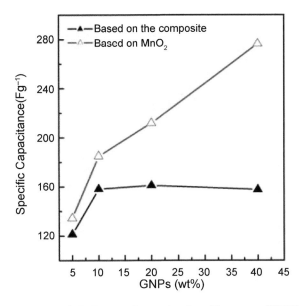

Fig. 4.101. Specific capacitances of the MnO$_2$/GNP composites as a function of the content of GNP. Reproduced with permission from [185], Copyright © 2008, Elsevier.

Specific capacitances, in terms of the mass of MnO$_2$, at current densities of 2, 10 and 20 mA·cm^{-2} were 179.9, 110.2 and 79.2 F·g^{-1} for EG-20, and 211.8, 149.9 and 104.0 F·g^{-1} for GNPs-20, respectively. In contrast, both GNPs and EG had specific capacitances of < 2 F·g^{-1}. Therefore, electrochemical effect of the composites was attributed mainly to the Faradaic redox transitions of MnO$_2$, while the contribution of the EG and GNPs could be neglected.

Figure 4.101 shows specific capacitance of the MnO$_2$/GNP composites as a function of the content of GNPs [185]. The specific capacitance based on the mass of MnO$_2$ was increased from 134.5 to 276.3 F·g^{-1},

262 *Nanomaterials for Supercapacitors*

while that based on the mass of composite was increased from 121 to 158 $F \cdot g^{-1}$, as the content of GNPs was increased from 5 to 10 wt%. However, further increase in the content of GNPs exhibited almost no effect on the magnitude of specific capacitance. The presence of the GNPs in the composites could greatly enhance electrical conductivity of the electrodes. Meanwhile, the higher the content of the GNPs, the more were the active sites that the composites could have. Therefore, an increase in the content of the GNPs led to an increase in specific capacitance at low concentration levels. However, an increase in the GNPs corresponded to a decrease in the loading level of the active material MnO_2. As a result, the increase in specific capacitance due to the GNPs was cancelled out by the decrease due to the reduced content of MnO_2.

MnO₂–other nanocarbons

Besides CNTs and graphene as discussed above, there have also been other nanocarbon materials, such as carbon aerogels or nanofoams, which can be incorporated with manganese oxide to achieve high performance electrodes of supercapacitors [194–203]. It can be attributed to their advantageous properties, including high specific surface areas of up to 1000 $m^2 \cdot g^{-1}$, high meso- and macroporosities, relatively strong mechanical integrity and high electrical conductivity of up to 100 $S \cdot cm^{-1}$, which have made them promising hosts for loadings of MnO_2 nanoparticles to form a new group of nanocomposites of carbon nanofoams and MnO_2 for electrochemical applications. Carbon aerogels are usually prepared by using microemulsion-templated sol–gel polymerization method, involving pyrolysis of Resorcinol-Formaldehyde (RF) gels.

For example, MnO_2–mesoporous carbon nanocomposites were prepared by using a solution method, which showed high value of redox pseudocapacitance at high scan rates of up to 100 mV s^{-1} [197]. In such nanocomposites, MnO_2 nanodomains with a thickness of about 1 nm and phase composition of birnessite were uniformly coated throughout the mesoporous carbon structure. The mesoporous carbon was synthesized thorough self-assembly of block copolymer and phenolic resin at acidic conditions [204]. 16.5 g resorcinol and 16.5 g F127 ($EO_{106}PO_{70}EO_{106}$) were dissolved in the mixture of 67.5 mL ethanol and 67.5 mL HCl solution (3.0 M). Then, 19.5 g formaldehyde (37 wt% in H_2O) was added with stirring at room temperature. During the polymerization of resorcinol and formaldehyde, a phase separation was induced, due to the *in-situ* self-assembly of the phenolic resin and F127.

The polymer-riched gel was collected and then dissolved in the mixture of 18 g THF and 12 g ethanol. The mixture was cast onto a substrate and dried at room temperature, followed by curing at 80°C for 24 hours. The samples were finally carbonized in N_2 at 400°C for 2 hours, followed by further heat treatment at 850°C for 3 hours. Thirty mg mesoporous carbon was mixed with 10 mL aqueous solution of $KMnO_4$ with concentrations in the range of 0.001–0.1 M. The suspensions were then filtered and thoroughly washed with DI water, so that MnO_2–C nanocomposite powders were obtained after calcination at 200°C for 6 hours in air.

Figure 4.102 shows SEM images of the mesoporous carbon and the MnO_2–C nanocomposites with different loading levels [197]. Mesopores of the nanostructured carbon could be clearly observed in the pure carbon sample and the sample with 2 wt% MnO_2, as illustrated in Fig. 4.102(a, b). As the loading level was increased to 16 wt%, the visibility of the porous structure was largely reduced, while short MnO_2 ribbon-like structures were present, as seen in Fig. 4.102(c). Further increase in the content of MnO_2 to 30 wt%, the MnO_2 ribbon-like structures were clearly demonstrated, which were uniformly distributed throughout the porous structure, as revealed in Fig. 4.102(d).

Figure 4.103 shows XRD patterns of the mesoporous carbon and the MnO_2–C nanocomposites with 21 wt% and 30 wt% MnO_2 [197]. The strong peaks at 23° and 43° were attributed to graphitization of the nanocarbon materials, while the relatively broad peaks implied that the samples exhibited a certain degree of graphitic characteristics that were derived from the aromatic carbon precursors. With the presence of MnO_2, such peaks disappeared, instead, several additional weak peaks could be observed at 36.8° and 65.7°, corresponding to the (006) and (119) planes of birnessite [205]. The MnO_2 nanocrystals in the samples with 21 wt% and 30 wt% MnO_2 had crystal sizes of 3.8 and 3.5 nm, respectively.

Oxide Based Supercapacitors: I-Manganese Oxides 263

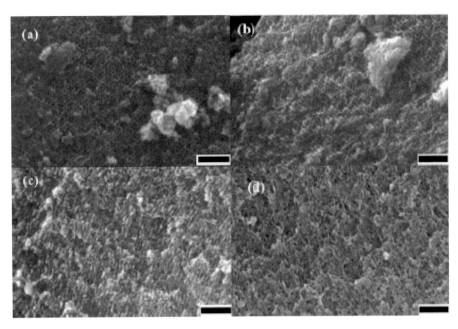

Fig. 4.102. SEM images of (a) mesoporous carbon and the samples modified with (b) 2 wt% MnO_2, (c) 16 wt% MnO_2 and (d) 30 wt% MnO_2. Black scale bar = 100 nm. Reproduced with permission from [197], Copyright © 2010, The Royal Society of Chemistry.

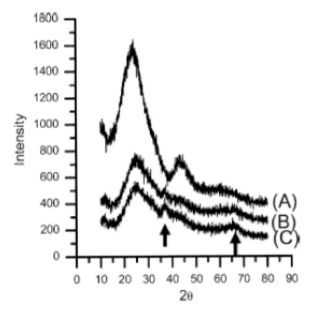

Fig. 4.103. XRD patterns of (A) mesoporous carbon and those modified with (B) 21 wt% MnO_2 and (C) 30 wt% MnO_2, with the presence of birnessite phase. Reproduced with permission from [197], Copyright © 2010, The Royal Society of Chemistry.

CV behaviours of the MnO$_2$–C nanocomposites were characterized in 1 M Na$_2$SO$_4$ over potentials between 0.4 and −0.4 V. The samples exhibited a relatively flat and rectangular profile, suggesting their EDLC behaviour, together with a certain level of redox pseudocapacitance contributed by MnO$_2$. At 2 mV·s^{-1}, the 2 wt% MnO$_2$ sample had a specific capacitance that was higher than that of the pure mesoporous carbon by about 40%. At higher loading levels of 16 wt% and 30 wt%, the flat profile was still observed, suggesting that electrical conductivity of the samples was still sufficiently high at the low scan rate, although the content of MnO$_2$ was quite high. Even at a scan rate of as high as 100 mV·s^{-1}, the flat profiles were still well retained, implying that the nanocomposites had both high ion transport and electrical conductivity. Figure 4.104 shows gravimetric capacitances at 2 and 100 mV·s^{-1}, as a function of MnO$_2$ mass and total composite mass (MnO$_2$ + carbon).

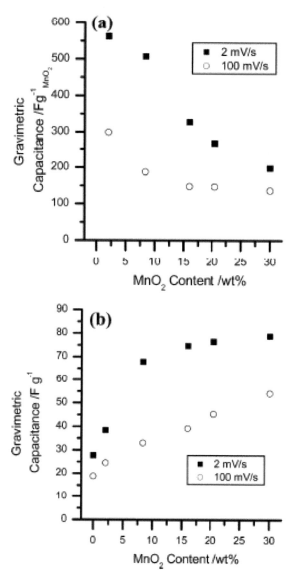

Fig. 4.104. Specific capacitances of the MnO$_2$–C nanocomposites at 2 mV·s^{-1} and 100 mV·s^{-1} in 1 M Na$_2$SO$_4$ based on (a) MnO$_2$ mass and (b) total mass. Reproduced with permission from [197], Copyright © 2010, The Royal Society of Chemistry.

It was found that the sample with 2 wt% MnO_2 had a specific capacitance C_{MnO2} of 560 $F \cdot g_{MnO2}^{-1}$. The specific capacitance was decreased as the loading level of MnO_2 was increased. At 2 mV s^{-1}, the C_{MnO2} was decreased in a linear way up to 21 wt% MnO_2 and kept to be about 200 $F \cdot g_{MnO2}^{-1}$ as the content of MnO_2 was increased from 20 wt% to 30 wt%. At 100 mV·s^{-1}, the saturation of C_{MnO2} was at 16 wt% with a C_{MnO2} of 137 $F \cdot g_{MnO2}^{-1}$ for the sample with 30 wt% MnO_2. As shown in Fig. 4.104(b), the C_g value was increased abruptly at the low mass loading regime and then saturated at 2 mV s^{-1}. The value of C_g of the 30 wt% sample was 79 $F \cdot g^{-1}$, which was nearly three times that of the pure mesoporous carbon. A similar increasing trend was observed at 100 mV·s^{-1}, but with a more linear relationship to the loading level of MnO_2. For the 30 wt% MnO_2–C sample, there is a 32% drop in the value of C_g was only decreased by 32%, while the scan rate was increased by 50 times (from 2 to 100 mV s^{-1}).

Ordered mesoporous carbons formed another group of nanostructured materials that could offer suitable electrolyte transport routes so as to achieve high performance electrodes of supercapacitors. Ordered mesoporous carbon materials are usually derived from carbon precursors inside silica or aluminosilicate mesoporous templates, followed by carbonization and removal of the templates [205–207]. The use of ordered mesoporous carbons is based on the fact that electrolyte ion diffusion within the electrodes is a rate-controlling step. Therefore, in order to enhance the rate capacity of electrochemical supercapacitors, one of the effective approaches is to optimize the transport paths of electrolyte ions, while electron transport is not compromised.

Silica spheres with 3D assemblies were used as a hard template to prepare porous carbon materials with large mesopores of about 100 nm and large surface areas of 900 $m^2 \cdot g^{-1}$, onto which birnessite MnO_2 nanoparticles were deposited by using a chemical co-precipitation method [205]. With increasing content of MnO_2, specific surface area of the nanocomposites was decreased monotonically, whereas specific capacitance was increased initially, reached a maximum value and then decreased at the loading level of 10 wt%. As the loading levels of MnO_2 were less than 10 wt%, all the MnO_2 nanoparticles were in contact with the surface of the carbon materials, thus establishing an effective electro-active material/substrate solid–solid interface. Therefore, the MnO_2 nanoparticles all participated in the electrochemical reaction. However, as the loading level was further increased, there would be MnO_2 nanoparticles that had not contribution to capacitive reaction and thus specific capacitance was decreased.

Silica spheres were synthesized by using the Stöber method, through hydrolysis of tetraethoxysilane (TEOS) as silica precursor [208]. TEOS was dissolved in dehydrated ethyl alcohol (EtOH), with an appropriate [TEOS]/[NH$_4$OH] ratio, while solution (28%) was used as the catalyst. To obtain silica spheres with diameters in the range of 150–200 nm, 24 g TEOS was dissolved in 720 ml EtOH solution, with addition of 4 g H_2O and 48 g NH_4OH, corresponding to a [TEOS]/[NH$_4$OH] = 1/6. The silica spheres were assembled into ordered 3D configuration, as the solvent was slowly evaporated. The 3D assembles were heated at 800°C for 6 hours to enhance the contact among the spheres, followed by immersing in sulfuric acid solution with pH = 3.5 for equilibration.

Silica 3D-assemble powder with a quantity of 1.2 g was thoroughly impregnated with 1 g furfuryl alcohol (FA) with a SiO_2/FA molar ratio of 2/1, followed by polymerisation at 80°C for 1 day. After the polymerization step, the sample was calcined in N_2 at 800°C for 16 hours, from which the hard template was removed from the silica spheres/carbon composite by etching in a 3 M NaOH solution for one week, in order to achieve interconnected porosity. Then, the carbon sample was thoroughly washed with distilled water and ethyl alcohol. To deposit birnessite MnO_2, the porous carbon 3D assembles were immersed in 0.5 M solution of $MnSO_4$ for 3 hours. After filtration, the powder was soaked for 5 minutes in $KMnO_4$ solutions with concentrations in the range of 5.10^{-4}–5.10^{-2} M. Therefore, the loading level of MnO_2 could be well controlled accordingly. A thermal treatment at 200°C for 6 hours was then conducted to obtain the final MnO_2/C composite powders with the birnessite phase of MnO_2.

Figure 4.105 shows SEM images of samples at different synthetic steps, from SiO_2 sphere assembles to the MnO_2/C nanocomposites [205]. As illustrated in Fig. 4.105(a), the silica spheres were assembled into 3D structure with hexagonal packing. It was found that slow solvent evaporation and thermal treated at 800°C for 6 hours were very critical to obtain the 3D assemblies. Otherwise, the silica spheres would be fused and the 3D structure collapsed. After NaOH etching for a short while, the silica spheres embedded in the carbon matrix became visible, due to the partial dissolution of the silica spheres, as seen

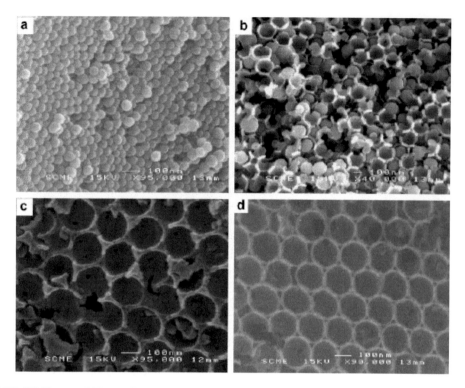

Fig. 4.105. SEM images of (a) assembled silica spheres prepared by using the Stöber method, (b) SiO$_2$/C composite after partial etching, (c) mesoporous carbon after complete removal of the template and (d) final MnO$_2$/C composite. Reproduced with permission from [205], Copyright © 2008, Elsevier.

in Fig. 4.105(b). After the silica spheres were entirely removed, fully open porous structure with high interconnectivity was developed, as demonstrated in Fig. 4.105(c). Although the MnO$_2$ particles could be clearly observed in the SEM images, their presence was confirmed by TGA results. The weight ratios of MnO$_2$ to C could be up to 100.

An *in-situ* reduction method was employed to prepare MnO$_2$/mesoporous carbon (MnC) composite, in which the MnO$_2$ nanoparticles were introduced into the mesoporous carbon wall of CMK-3 by using the redox reaction between permanganate ions and carbons [206]. The content of MnO$_2$ could be well controlled by using different concentrations of the KMnO$_4$ aqueous solution. MnC samples with different contents MnO$_2$ were derived from KMnO$_4$ aqueous solution with concentrations of 0.005, 0.01 and 0.05 M for 5 minutes, denoted as 0.005MnC-5, 0.01MnC-5 and 0.05MnC-5, corresponding to 9.4, 13 and 26 wt%, respectively. More importantly, an increase in the content of MnO$_2$ had almost no effect on pore size of the nanocomposites, because all the MnO$_2$ nanoparticles were attached on the walls of the pores. Specific capacitances of 200 and 600 F·g^{-1} were achieved for the MnC composite and MnO$_2$, respectively.

Figure 4.106 shows TEM images of 0.005MnC-5 and 0.05MnC-5 [206]. The ordered structure of CMK-3, an exactly negative replica of SBA-15 with a hexagonal arrangement of cylindrical mesoporous tubes, was well retained, although various levels of MnO$_2$ had been loaded into the porous structure. All the 0.005MnC-5, 0.01MnC-5 and 0.05MnC-5 possessed similar pore sizes of 3.5–4 nm in the mesopore range. The MnO$_2$ nanoparticles were uniformly distributed inside the CMK-3 porous structure, without the presence of significant aggregation. According to the data of pore size and BET surface area measurement results, no MnO$_2$ nanoparticles were found to block the pore channels, implying that they were all attached to the walls of CMK-3.

CVs of CMK-3 and the MnC composite electrodes with different contents of MnO$_2$ were measured in 2 M KCl aqueous solution over potentials in the range of 0–1.0 V at a scan rate of 5 mV·s^{-1}. The non-

Oxide Based Supercapacitors: I-Manganese Oxides 267

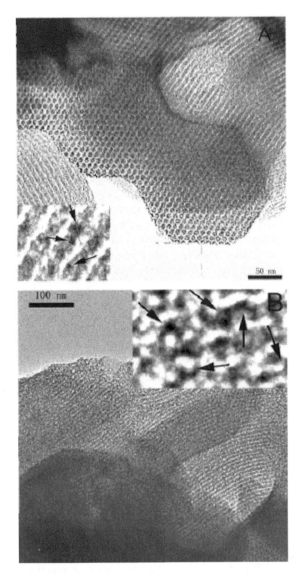

Fig. 4.106. TEM images of (A) 0.005MnC-5 and (B) 0.05MnC-5 samples, with the insets showing the images at high magnification. Reproduced with permission from [206], Copyright © 2006, American Chemical Society.

rectangles CVs suggested that the Ohmic resistance to the motion of the electrolyte inside the pores of the nanocarbon materials had influenced the double-layer formation mechanism. A potential difference between the mouth and bottom of the pores could be induced, thus leading to a delayed current response. As the content of MnO_2 was increased from 0 to 26 wt%, the specific current was increased gradually, due to the contribution of the pseudocapacitive MnO_2. The specific capacitance was increased almost linearly to 220 F/g at 26 wt%.

By subtracting the contribution of the mesoporous carbon CMK-3, the MnO_2 exhibited pretty high capacitances. For instance, for the samples with low levels of 9.4 and 13 wt%, capacitances based the MnO_2 were 605 and 628 $F·g^{-1}$, respectively. It has been recognized that only the surface has contribution to pseudocapacitive effect of MnO_2. In this work, the special mesoporous carbon structure ensured a large surface area to host the MnO_2 nanoparticles, at the same time, a high electrical conductivity was achieved by the nanocomposite due to the presence of the carbon component, both of which were beneficial to

268 *Nanomaterials for Supercapacitors*

electrochemical reaction of the electrodes. The MnO_2 based specific capacitance of the 0.05MnC-5 sample was slightly decreased to 547 $F\cdot g^{-1}$, which was readily attributed to the larger size of the MnO_2 nanoparticles as well as agglomeration.

Concluding Remarks

Manganese oxides have been acknowledged to be the most promising candidates as electrodes of supercapacitors. Different MnO_2 powders, synthesize by using various methods, such as chemical co-precipitation, hydrothermal or solvo, sol–gel and solution combustion process, have been extensively studied as electrodes of electrochemical supercapacitors in mild aqueous electrolytes. For the amorphous and crystalline MnO_2 compounds, a wide range of specific capacitance value has been reported in the open literature. Various factors, including porosity, morphology, defect chemistry, crystal structure and residual water content, could influence their charge storage processes. However, there is a lack of standard on structure and morphology of MnO_2 with optimized electrochemical performances. Because pure MnO_2 has a relatively low electrical conductivity to maintain high rate charge–discharge reactions, it is necessary to incorporate conductive components, such as conductive polymers and various nanocarbons.

Despite the significant progress, the potential of manganese oxide-based electrochemical supercapacitors has not been fully utilized. Innovative manufacturing processes are still necessary to modify the chemical and structural properties of manganese oxide materials, so as to create more electrochemically active sites and smoothen the transportation of both electrons and electrolyte cations. The basic requirements to achieve this include appropriate porosity, high specific surface area and high electrical conductivity. On the other hand, although the chemical and structural stability and flexibility of manganese oxides have been improved by incorporating with polymers to form composite materials, systematic information is not available regarding to the optimization of the chemistry and microstructure of the composites.

Besides the issues of materials, the engineering of electrodes is also an important consideration, which has not been well established in the open literature. Electrode materials with desired 3D architectures are crucial to the development of electrochemical supercapacitors with sufficiently high energy/power densities, to meet the requirement of practical applications. Moreover, the selection of counter electrodes, electrolytes (e.g., divalent cation-containing solutions, hydrogel polymers and ionic liquids), membrane separators, current collectors and packaging techniques, should also be systematically studied.

Various fundamental issues still remain to be clarified, such as characterization and understanding of electron transfer and ionic transport during the electrochemical interface reaction processes, especially inside the composite electrodes. It is strongly believed that, in order to fully achieve the potential of manganese oxide-based electrode materials, optimization of both the synthetic parameters and material characteristics will be the future research direction of supercapacitors.

Keywords: Supercapacitor, Pseudocapacitor, Pseudocapacitive reaction, Galvanostatic charge, Electrochemical Impedance Spectroscopy (EIS), Cyclic Voltammograms (CV), Cyclic voltammetry, Voltammetric charge, Cyclibility, Hydrothermal reaction, Solvo reaction, Sol-gel, Chemical precipitation, Chemical coprecipitation, Nanorod, Nanorods, Nanoflower, Nanoflowers, Nanowhisker, Nanowhiskers, Nanoneedle, Nanoflakes, Nanosheet, Manganese oxide, MnO_x, Mn_3O_4, Mn_2O_3, MnO_2, α-MnO_2, γ-MnO_2, λ-MnO_2, MnO_2-C, MnO_2-nanocarbon, MnO_2-carbon nanotubes, MnO_2-CNTs, MnO_2-MWCNTs, MnO_2-polyaniline, MnO_2-polypyrrole, MnO_2-polythiophene, MnO_2-poly(o-phenylenediamine), Mn-Fe mixed oxides, $MnFe_2O_4$, Mn-Co mixed oxides, $(Mn+Co)O_x\cdot nH_2O$, Mn-Ni mixed oxides, $Ru_nMn_{1-n}O_x$, MnO_x-RuO_2

References

[1] Lee HY, Goodenough JB. Supercapacitor behavior with KCl electrolyte. Journal of Solid State Chemistry. 1999; 144: 220–3.

[2] Lee HY, Manivannan V, Goodenough JB. Electrochemical capacitors with KCl electrolyte. Comptes Rendus De L Academie Des Sciences Serie Ii Fascicule C-Chimie. 1999; 2: 565–77.

[3] Pang SC, Anderson MA, Chapman TW. Novel electrode materials for thin-film ultracapacitors: Comparison of electrochemical properties of sol-gel-derived and electrodeposited manganese dioxide. Journal of the Electrochemical Society. 2000; 147: 444–50.

[4] Toupin M, Brousse T, Belanger D. Charge storage mechanism of MnO_2 electrode used in aqueous electrochemical capacitor. Chemistry of Materials. 2004; 16: 3184–90.

[5] Wei WF, Cui XW, Chen WX, Ivey DG. Manganese oxide-based materials as electrochemical supercapacitor electrodes. Chemical Society Reviews. 2011; 40: 1697–721.

[6] Brousse T, Toupin M, Dugas R, Athouel L, Crosnier O, Belanger D. Crystalline MnO_2 as possible alternatives to amorphous compounds in electrochemical supercapacitors. Journal of the Electrochemical Society. 2006; 153: A2171–A80.

[7] Desai BD, Fernandes JB, Dalal VNK. Manganese-dioxide—A review of a battery chemical, 2. Solid-state and electrochemiccal properties of manganese dioxides. Journal of Power Sources. 1985; 16: 1–43.

[8] Fernandes JB, Desai BD, Dalal VNK. Manganese-dioxide—A review of a battery chemical, 1. Chemical synthesis and X-ray diffraction studies of manganese dioxides. Journal of Power Sources. 1985; 15: 209–37.

[9] Preisler E. Electrodeposited manganese-dioxide with preferred crystal growth. Journal of Applied Electrochemistry. 1976; 6: 301–10.

[10] Pang SC, Anderson MA. Novel electrode materials for electrochemical capacitors: Part II. Material characterization of sol-gel-derived and electrodeposited manganese dioxide thin films. Journal of Materials Research. 2000; 15: 2096–106.

[11] Broughton JN, Brett MJ. Electrochemical capacitance in manganese thin films with chevron microstructure. Electrochemical and Solid State Letters. 2002; 5: A279–A82.

[12] Toupin M, Brousse T, Belanger D. Influence of microstucture on the charge storage properties of chemically synthesized manganese dioxide. Chemistry of Materials. 2002; 14: 3946–52.

[13] Yang XH, Wang YG, Xiong HM, Xia YY. Interfacial synthesis of porous MnO_2 and its application in electrochemical capacitor. Electrochimica Acta. 2007; 53: 752–7.

[14] Ghodbane O, Pascal JL, Favier F. Microstructural effects on charge-storage properties in MnO_2-based electrochemical supercapacitors. ACS Applied Materials & Interfaces. 2009; 1: 1130–9.

[15] Chen YS, Hu CC. Capacitive characteristics of binary manganese-nickel oxides prepared by anodic deposition. Electrochemical and Solid State Letters. 2003; 6: A210–A3.

[16] Nakayama M, Tanaka A, Konishi S, Ogura K. Effects of heat-treatment on the spectroscopic and electrochemical properties of a mixed manganese/vanadium oxide film prepared by electrodeposition. Journal of Materials Research. 2004; 19: 1509–15.

[17] Prasad KR, Miura N. Electrochemically synthesized MnO_2-based mixed oxides for high performance redox supercapacitors. Electrochemistry Communications. 2004; 6: 1004–8.

[18] Nakayama M, Tanaka A, Sato Y, Tonosaki T, Ogura K. Electrodeposition of manganese and molybdenum mixed oxide thin films and their charge storage properties. Langmuir. 2005; 21: 5907–13.

[19] Kim H, Popov BN. Synthesis and characterization of MnO_2-based mixed oxides as supercapacitors. Journal of the Electrochemical Society. 2003; 150: D56–D62.

[20] Li Y, Xie HQ. Mechanochemical-synthesized Al-doped manganese dioxides for electrochemical supercapacitors. Ionics. 2010; 16: 21–5.

[21] Fischer AE, Pettigrew KA, Rolison DR, Stroud RM, Long JW. Incorporation of homogeneous, nanoscale MnO_2 within ultraporous carbon structures via self-limiting electroless deposition: Implications for electrochemical capacitors. Nano Letters. 2007; 7: 281–6.

[22] Wang GX, Zhang BL, Yu ZL, Qu MZ. Manganese oxide/MWNTs composite electrodes for supercapacitors. Solid State Ionics. 2005; 176: 1169–74.

[23] Fan Z, Chen JH, Wang MY, Cui KZ, Zhou HH, Kuang W. Preparation and characterization of manganese oxide/CNT composites as supercapacitive materials. Diamond and Related Materials. 2006; 15: 1478–83.

[24] Raymundo-Pinero E, Khomenko V, Frackowiak E, Beguin F. Performance of manganese oxide/CNTs composites as electrode materials for electrochemical capacitors. Journal of the Electrochemical Society. 2005; 152: A229–A35.

[25] Subramanian V, Zhu HW, Wei BQ. Synthesis and electrochemical characterizations of amorphous manganese oxide and single walled carbon nanotube composites as supercapacitor electrode materials. Electrochemistry Communications. 2006; 8: 827–32.

[26] Wu YT, Hu CC. Effects of electrochemical activation and multiwall carbon nanotubes on the capacitive characteristics of thick MnO_2 deposits. Journal of the Electrochemical Society. 2004; 151: A2060–A6.

[27] Lee CY, Tsai HM, Chuang HJ, Li SY, Lin P, Tseng TY. Characteristics and electrochemical performance of supercapacitors with manganese oxide-carbon nanotube nanocomposite electrodes. Journal of the Electrochemical Society. 2005; 152: A716–A20.

[28] Long JW, Rhodes CP, Young AL, Rolison DR. Ultrathin, protective coatings of poly(o-phenylenediamine) as electrochemical proton gates: Making mesoporous MnO_2 nanoarchitectures stable in acid electrolytes. Nano Letters. 2003; 3: 1155–61.

270 *Nanomaterials for Supercapacitors*

[29] McEvoy TM, Long JW, Smith TJ, Stevenson KJ. Nanoscale conductivity mapping of hybrid nanoarchitectures: Ultrathin poly(o-phenylenediamine) on mesoporous manganese oxide ambigels. Langmuir. 2006; 22: 4462–6.

[30] Prasad KR, Miura N. Polyaniline-MnO_2 composite electrode for high energy density electrochemical capacitor. Electrochemical and Solid State Letters. 2004; 7: A425–A8.

[31] Rios EC, Rosario AV, Mello RMQ, Micaroni L. Poly(3-methylthiophene)/MnO_2 composite electrodes as electrochemical capacitors. Journal of Power Sources. 2007; 163: 1137–42.

[32] Sivakkumar SR, Ko JM, Kim DY, Kim BC, Wallace GG. Performance evaluation of CNT/polypyrrole/MnO_2 composite electrodes for electrochemical capacitors. Electrochimica Acta. 2007; 52: 7377–85.

[33] Liu R, Lee SB. MnO_2/Poly(3,4-ethylenedioxythiophene) coaxial nanowires by one-step coelectrodeposition for electrochemical energy storage. Journal of the American Chemical Society. 2008; 130: 2942–3.

[34] Miller JR, Simon P. Materials science—electrochemical capacitors for energy management. Science. 2008; 321: 651–2.

[35] Simon P, Gogotsi Y. Materials for electrochemical capacitors. Nature Materials. 2008; 7: 845–54.

[36] Frackowiak E, Beguin F. Electrochemical storage of energy in carbon nanotubes and nanostructured carbons. Carbon. 2002; 40: 1775–87.

[37] Jeong YU, Manthiram A. Nanocrystalline manganese oxides for electrochemical capacitors with neutral electrolytes. Journal of the Electrochemical Society. 2002; 149: A1419–A22.

[38] Ragupathy P, Vasan HN, Munichandraiah N. Synthesis and characterization of nano-MnO_2 for electrochemical supercapacitor studies. Journal of the Electrochemical Society. 2008; 155: A34–A40.

[39] Ragupathy P, Park DH, Campet G, Vasan HN, Hwang SJ, Choy JH et al. Remarkable capacity retention of nanostructured manganese oxide upon cycling as an electrode material for supercapacitor. Journal of Physical Chemistry C. 2009; 113: 6303–9.

[40] Xu CJ, Li BH, Du HD, Kang FY, Zeng YQ. Electrochemical properties of nanosized hydrous manganese dioxide synthesized by a self-reacting microemulsion method. Journal of Power Sources. 2008; 180: 664–70.

[41] Yuan CZ, Gao B, Su LH, Zhang XG. Interface synthesis of mesoporous MnO_2 and its electrochemical capacitive behaviors. Journal of Colloid and Interface Science. 2008; 322: 545–50.

[42] Devaraj S, Munichandraiah N. Electrochemical supercapacitor studies of nanostructured α-MnO_2 synthesized by microemulsion method and the effect of annealing. Journal of the Electrochemical Society. 2007; 154: A80–A8.

[43] Franger S, Bach S, Farcy J, Pereira-Ramos JP, Baffier N. Synthesis, structural and electrochemical characterizations of the sol-gel birnessite $MnO_{1.84} \cdot 0.6H_2O$. Journal of Power Sources. 2002; 109: 262–75.

[44] Thackeray MM, David WIF, Bruce PG, Goodenough JB. Lithium insertion into manganese spinels. Materials Research Bulletin. 1983; 18: 461–72.

[45] Tsuda M, Arai H, Nemoto Y, Sakurai Y. Electrode performance of sodium and lithium-type romanechite. Journal of the Electrochemical Society. 2003; 150: A659–A64.

[46] Shen XF, Ding YS, Liu J, Cai J, Laubernds K, Zerger RP et al. Control of nanometer-scale tunnel sizes of porous manganese oxide octahedral molecular sieve nanomaterials. Advanced Materials. 2005; 17: 805–9.

[47] Feng Q, Yanagisawa K, Yamasaki N. Hydrothermal soft chemical process for synthesis of manganese oxides with tunnel structures. Journal of Porous Materials. 1998; 5: 153–61.

[48] Ching S, Krukowska KS, Suib SL. A new synthetic route to todorokite-type manganese oxides. Inorganica Chimica Acta. 1999; 294: 123–32.

[49] Feng Q, Kanoh H, Miyai Y, Ooi K. Metal-ion extraction/insertion reactions with todorokite-type manganese oxide in the aqueous-phase. Chemistry of Materials. 1995; 7: 1722–7.

[50] Athouel L, Moser F, Dugas R, Crosnier O, Belanger D, Brousse T. Variation of the MnO_2 birnessite structure upon charge/discharge in an electrochemical supercapacitor electrode in aqueous Na_2SO_4 electrolyte. Journal of Physical Chemistry C. 2008; 112: 7270–7.

[51] Walanda DK, Lawrance GA, Donne SW. Hydrothermal MnO_2: Synthesis, structure, morphology and discharge performance. Journal of Power Sources. 2005; 139: 325–41.

[52] Devaraj S, Munichandraiah N. Effect of crystallographic structure of MnO_2 on its electrochemical capacitance properties. Journal of Physical Chemistry C. 2008; 112: 4406–17.

[53] Chabre Y, Pannetier J. Structural and electrochemical properties of the proton γ-MnO_2 system. Progress in Solid State Chemistry. 1995; 23: 1–130.

[54] Al-Sagheer FA, Zaki MI. Synthesis and surface characterization of todorokite-type microporous manganese oxides: implications for shape-selective oxidation catalysts. Microporous and Mesoporous Materials. 2004; 67: 43–52.

[55] Hunter JC. Preparation of a new cyrstal from of manganese-dioxide-λ-MnO2. Journal of Solid State Chemistry. 1981; 39: 142–7.

[56] Rziha T, Gies H, Rius J. RUB-7, a new synthetic manganese oxide structure type with a 2x4 tunnel. European Journal of Mineralogy. 1996; 8: 675–86.

[57] Xia GG, Tong W, Tolentino EN, Duan NG, Brock SL, Wang JY et al. Synthesis and characterization of nanofibrous sodium manganese oxide with a 2 x 4 tunnel structure. Chemistry of Materials. 2001; 13: 1585–92.

[58] Wang X, Li YD. Rational synthesis of α-MnO_2 single-crystal nanorods. Chemical Communications. 2002; 2002: 764–5.

[59] Subramanian V, Zhu HW, Vajtai R, Ajayan PM, Wei BQ. Hydrothermal synthesis and pseudocapacitance properties of MnO_2 nanostructures. Journal of Physical Chemistry B. 2005; 109: 20207–14.

[60] Subramanian V, Zhu HW, Wei BQ. Nanostructured MnO_2: Hydrothermal synthesis and electrochemical properties as a supercapacitor electrode material. Journal of Power Sources. 2006; 159: 361–4.

[61] Burke A. Ultracapacitors: why, how, and where is the technology. Journal of Power Sources. 2000; 91: 37–50.

[62] Hu CC, Tsou TW. Ideal capacitive behavior of hydrous manganese oxide prepared by anodic deposition. Electrochemistry Communications. 2002; 4: 105–9.

[63] Reddy RN, Reddy RG. Sol-gel MnO_2 as an electrode material for electrochemical capacitors. Journal of Power Sources. 2003; 124: 330–7.

[64] Bao SJ, He BL, Liang YY, Zhou WJ, Li HL. Synthesis and electrochemical characterization of amorphous MnO_2 for electrochemical capacitor. Materials Science and Engineering A-Structural Materials Properties Microstructure and Processing. 2005; 397: 305–9.

[65] Reddy RN, Reddy RG. Synthesis and electrochemical characterization of amorphous MnO_2 electrochemical capacitor electrode material. Journal of Power Sources. 2004; 132: 315–20.

[66] Xu MW, Kong LB, Zhou WJ, Li HL. Hydrothermal synthesis and pseudocapacitance properties of α-MnO_2 hollow spheres and hollow urchins. Journal of Physical Chemistry C. 2007; 111: 19141–7.

[67] Li BX, Rong GX, Xie Y, Huang LF, Feng CQ. Low-temperature synthesis of α-MnO_2 hollow urchins and their application in rechargeable Li^+ batteries. Inorganic Chemistry. 2006; 45: 6404–10.

[68] Bao SJ, Bao QL, Li CM, Chen TP, Sun CQ, Dong ZL et al. Synthesis and electrical transport of novel channel-structured β-$AgVO_3$. Small. 2007; 3: 1174–7.

[69] Tang N, Tian XK, Yang C, Pi ZB. Facile synthesis of α-MnO_2 nanostructures for supercapacitors. Materials Research Bulletin. 2009; 44: 2062–7.

[70] Yan J, Wei T, Cheng J, Fan ZJ, Zhang ML. Preparation and electrochemical properties of lamellar MnO_2 for supercapacitors. Materials Research Bulletin. 2010; 45: 210–5.

[71] Bao SJ, Bao QL, Li CM, Chen TP, Sun CQ, Dong ZL et al. Synthesis and electrical transport of novel channel-structured beta-$AgVO_3$. Small. 2007; 3: 1174–7.

[72] Armstrong AR, Bruce PG. Synthesis of layered $LiMnO_2$ as an electrode for rechargeable lithium batteries. Nature. 1996; 381: 499–500.

[73] Wang X, Li YD. Rational synthetic strategy. From layered structure to MnO_2 nanotubes. Chemistry Letters. 2004; 33: 48–9.

[74] Abou-El-Sherbini KS, Askar MH, Schollhorn R. Hydrated layered manganese dioxide Part I. Synthesis and characterization of some hydrated layered manganese dioxides from α-$NaMnO_2$. Solid State Ionics. 2002; 150: 407–15.

[75] Hu CC, Wu YT, Chang KH. Low-temperature hydrothermal synthesis of Mn_3O_4 and MnOOH single crystals: Determinant influence of oxidants. Chemistry of Materials. 2008; 20: 2890–4.

[76] Jiang H, Zhao T, Yan CY, Ma J, Li CZ. Hydrothermal synthesis of novel Mn_3O_4 nano-octahedrons with enhanced supercapacitors performances. Nanoscale. 2010; 2: 2195–8.

[77] Ye C, Lin ZM, Hui SZ. Electrochemical and capacitance properties of rod-shaped MnO_2 for supercapacitor. Journal of the Electrochemical Society. 2005; 152: A1272–A8.

[78] Qu QT, Zhang P, Wang B, Chen YH, Tian S, Wu YP et al. Electrochemical performance of MnO_2 nanorods in neutral aqueous electrolytes as a cathode for asymmetric supercapacitors. Journal of Physical Chemistry C. 2009; 113: 14020–7.

[79] Ni JP, Lu WC, Zhang LM, Yue BH, Shang XF, Lv Y. Low-temperature synthesis of monodisperse 3D manganese oxide nanoflowers and their pseudocapacitance properties. Journal of Physical Chemistry C. 2009; 113: 54–60.

[80] Beaudrouet E, La Salle ALG, Guyomard D. Nanostructured manganese dioxides: Synthesis and properties as supercapacitor electrode materials. Electrochimica Acta. 2009; 54: 1240–8.

[81] Zolfaghari A, Ataherian F, Ghaemi M, Gholami A. Capacitive behavior of nanostructured MnO_2 prepared by sonochemistry method. Electrochimica Acta. 2007; 52: 2806–14.

[82] Ghaemi M, Ataherian F, Zolfaghari A, Jafari SM. Charge storage mechanism of sonochemically prepared MnO_2 as supercapacitor electrode: Effects of physisorbed water and proton conduction. Electrochimica Acta. 2008; 53: 4607–14.

[83] Yu P, Zhang X, Chen Y, Ma YW. Solution-combustion synthesis of ε-MnO_2 for supercapacitors. Materials Letters. 2010; 64: 61–4.

[84] Wang XY, Wang XY, Huang WG, Sebastian PJ, Gamboa S. Sol-gel template synthesis of highly ordered MnO_2 nanowire arrays. Journal of Power Sources. 2005; 140: 211–5.

[85] Ma RH, Bando Y, Zhang LQ, Sasaki T. Layered MnO_2 nanobelts: Hydrothermal synthesis and electrochemical measurements. Advanced Materials. 2004; 16: 918–22.

[86] Zhao PT, Huang KX. Preparation and characterization of netted sphere-like CdS nanostructures. Crystal Growth & Design. 2008; 8: 717–22.

[87] Lopez CM, Choi KS. Electrochemical synthesis of dendritic zinc films composed of systematically varying motif crystals. Langmuir. 2006; 22: 10625–9.

[88] Chen CC, Tsay CY, Lin HS, Jheng WD, Lin CK. Effect of iron particle addition on the pseudocapacitive performance of sol-gel derived manganese oxides film. Materials Chemistry and Physics. 2012; 137: 503–10.

272 *Nanomaterials for Supercapacitors*

[89] Chen CC, Yang CY, Lin CK. Improved pseudo-capacitive performance of manganese oxide films synthesized by the facile sol-gel method with iron acetate addition. Ceramics International. 2013; 39: 7831–8.

[90] Chen CY, Wang SC, Tien YH, Tsai WT, Lin CK. Hybrid manganese oxide films for supercapacitor application prepared by sol-gel technique. Thin Solid Films. 2009; 518: 1557–60.

[91] Sarkar A, Satpati AK, Kumar V, Kumar S. Sol-gel synthesis of manganese oxide films and their predominant electrochemical properties. Electrochimica Acta. 2015; 167: 126–31.

[92] Nagarajan N, Humadi H, Zhitomirsky I. Cathodic electrodeposition of MnOx films for electrochemical supercapacitors. Electrochimica Acta. 2006; 51: 3039–45.

[93] Li J, Zhitomirsky I. Electrophoretic deposition of manganese oxide nanofibers. Materials Chemistry and Physics. 2008; 112: 525–30.

[94] Chen CY, Wang SC, Lin CY, Chen FS, Lin CK. Electrophoretically deposited manganese oxide coatings for supercapacitor application. Ceramics International. 2009; 35: 3469–74.

[95] Chin SF, Pang SC, Anderson MA. Material and electrochemical characterization of tetrapropylammonium manganese oxide thin films as novel electrode materials for electrochemical capacitors. Journal of the Electrochemical Society. 2002; 149: A379–A84.

[96] Long JW, Young AL, Rolison DR. Spectroelectrochemical characterization of nanostructured, mesoporous manganese oxide in aqueous electrolytes. Journal of the Electrochemical Society. 2003; 150: A1161–A5.

[97] Lin CK, Chuang KH, Lin CY, Tsay CY, Chen CY. Manganese oxide films prepared by sol-gel process for supercapacitor application. Surface & Coatings Technology. 2007; 202: 1272–6.

[98] Lin CC, Chen HW. Electrochemical characteristics of manganese oxide electrodes prepared by an immersion technique. Journal of Applied Electrochemistry. 2009; 39: 1877–81.

[99] Lin CC, Chen HW. Coating manganese oxide onto graphite electrodes by immersion for electrochemical capacitors. Electrochimica Acta. 2009; 54: 3073–7.

[100] Xu MW, Zhao DD, Bao SJ, Li HL. Mesoporous amorphous MnO_2 as electrode material for supercapacitor. Journal of Solid State Electrochemistry. 2007; 11: 1101–7.

[101] Kim KJ, Park YR. Sol-gel growth and structural and optical investigation of manganese-oxide thin films: structural transformation by Zn doping. Journal of Crystal Growth. 2004; 270: 162–7.

[102] Hu CC, Tsou TW. Capacitive and textural characteristics of hydrous manganese oxide prepared by anodic deposition. Electrochimica Acta. 2002; 47: 3523–32.

[103] Wu MS, Chiang PCJ. Fabrication of nanostructured manganese oxide electrodes for electrochemical capacitors. Electrochemical and Solid State Letters. 2004; 7: A123–A6.

[104] Broughton JN, Brett MJ. Variations in MnO_2 electrodeposition for electrochemical capacitors. Electrochimica Acta. 2005; 50: 4814–9.

[105] Zhou YK, Toupin M, Belanger D, Brousse T, Favier F. Electrochemical preparation and characterization of Birnessite-type layered manganese oxide films. Journal of Physics and Chemistry of Solids. 2006; 67: 1351–4.

[106] Kuo SL, Wu NL. Investigation of pseudocapacitive charge-storage reaction of $MnO_2.nH_2O$ supercapacitors in aqueous electrolytes. Journal of the Electrochemical Society. 2006; 153: A1317–A24.

[107] Chou SL, Cheng FY, Chen J. Electrodeposition synthesis and electrochemical properties of nanostructured γ-MnO_2 films. Journal of Power Sources. 2006; 162: 727–34.

[108] Shinomiya T, Gupta V, Miura N. Effects of electrochemical-deposition method and microstructure on the capacitive characteristics of nano-sized manganese oxide. Electrochimica Acta. 2006; 51: 4412–9.

[109] Wei WF, Cui XW, Chen WX, Ivey DG. Phase-controlled synthesis of MnO_2 nanocrystals by anodic electrodeposition: Implications for high-rate capability electrochemical supercapacitors. Journal of Physical Chemistry C. 2008; 112: 15075–83.

[110] Wei WF, Cui XW, Chen WX, Ivey DG. Electrochemical cyclability mechanism for MnO_2 electrodes utilized as electrochemical supercapacitors. Journal of Power Sources. 2009; 186: 543–50.

[111] Wei WF, Chen WX, Ivey DG. Defective rock-salt structure in anodically electrodeposited Mn-Co-O nanocrystals. Journal of Physical Chemistry C. 2007; 111: 10398–403.

[112] Babakhani B, Ivey DG. Anodic deposition of manganese oxide electrodes with rod-like structures for application as electrochemical capacitors. Journal of Power Sources. 2010; 195: 2110–7.

[113] Wei WF, Cui XW, Mao XH, Chen WX, Ivey DG. Morphology evolution in anodically electrodeposited manganese oxide nanostructures for electrochemical supercapacitor applications-Effect of supersaturation ratio. Electrochimica Acta. 2011; 56: 1619–28.

[114] Hu CC, Chang KH, Lin MC, Wu YT. Design and tailoring of the nanotubular arrayed architecture of hydrous RuO_2 for next generation supercapacitors. Nano Letters. 2006; 6: 2690–5.

[115] Xu CL, Bao SJ, Kong LB, Li H, Li HL. Highly ordered MnO_2 nanowire array thin films on Ti/Si substrate as an electrode for electrochemical capacitor. Journal of Solid State Chemistry. 2006; 179: 1351–5.

[116] Dong B, Xue T, Xu CL, Li HL. Electrodeposition of mesoporous manganese dioxide films from lyotropic liquid crystalline phases. Microporous and Mesoporous Materials. 2008; 112: 627–31.

Oxide Based Supercapacitors: I-Manganese Oxides   273

[117] Xu CL, Zhao YQ, Yang GW, Li FS, Li HL. Mesoporous nanowire array architecture of manganese dioxide for electrochemical capacitor applications. Chemical Communications. 2009: 7575–7.

[118] Xia H, Feng JK, Wang HL, Lai MO, Lu L. MnO_2 nanotube and nanowire arrays by electrochemical deposition for supercapacitors. Journal of Power Sources. 2010; 195: 4410–3.

[119] Chang JK, Hsu SH, Tsai WT, Sun IW. A novel electrochemical process to prepare a high-porosity manganese oxide electrode with promising pseudocapacitive performance. Journal of Power Sources. 2008; 177: 676–80.

[120] Pan SJ, Shih YJ, Chen JR, Chang JK, Tsai WT. Selective micro-etching of duplex stainless steel for preparing manganese oxide supercapacitor electrode. Journal of Power Sources. 2009; 187: 261–7.

[121] Xiao W, Xia H, Fuh JYH, Lu L. Electrochemical synthesis and supercapacitive properties of ε-MnO_2 with porous/nanoflaky hierarchical architectures. Journal of the Electrochemical Society. 2009; 156: A627–A33.

[122] Wei J, Nagarajan N, Zhitomirsky I. Manganese oxide films for electrochemical supercapacitors. Journal of Materials Processing Technology. 2007; 186: 356–61.

[123] Jacob GM, Zhitomirsky I. Microstructure and properties of manganese dioxide films prepared by electrodeposition. Applied Surface Science. 2008; 254: 6671–6.

[124] Zhitomirsky I, Cheong M, Wei J. The cathodic electrodeposition of manganese oxide films for electrochemical supercapacitors. JOM. 2007; 59: 66–9.

[125] Djurfors B, Broughton JN, Brett MJ, Ivey DG. Microstructural characterization of porous manganese thin films for electrochemical supercapacitor applications. Journal of Materials Science. 2003; 38: 4817–30.

[126] Broughton JN, Brett MJ. Investigation of thin sputtered Mn films for electrochemical capacitors. Electrochimica Acta. 2004; 49: 4439–46.

[127] Djurfors B, Broughton JN, Brett MJ, Ivey DG. Electrochemical oxidation of Mn/MnO films: formation of an electrochemical capacitor. Acta Materialia. 2005; 53: 957–65.

[128] Djurfors B, Broughton JN, Brett MJ, Ivey DG. Production of capacitive films from Mn thin films: Effects of current density and film thickness. Journal of Power Sources. 2006; 156: 741–7.

[129] Chang JK, Huang CH, Tsai WT, Deng MJ, Sun IW, Chen PY. Manganese films electrodeposited at different potentials and temperatures in ionic liquid and their application as electrode materials for supercapacitors. Electrochimica Acta. 2008; 53: 4447–53.

[130] Chang JK, Huang CH, Tsai WT, Deng MJ, Sun IW. Ideal pseudocapacitive performance of the Mn oxide anodized from the nanostructured and amorphous Mn thin film electrodeposited in $BMP-NTf_2$ ionic liquid. Journal of Power Sources. 2008; 179: 435–40.

[131] Chang JK, Huang CH, Lee MT, Tsai WT, Deng MJ, Sun IW. Physicochemical factors that affect the pseudocapacitance and cyclic stability of Mn oxide electrodes. Electrochimica Acta. 2009; 54: 3278–84.

[132] Yamagata M, Tachikawa N, Katayama Y, Miura T. Electrochemical behavior of several iron complexes in hydrophobic room-temperature ionic liquids. Electrochimica Acta. 2007; 52: 3317–22.

[133] Cheong M, Zhitomirsky I. Electrophoretic deposition of manganese oxide films. Surface Engineering. 2009; 25: 346–52.

[134] Zhang X, Yang WS. Electrophoretic deposition of a thick film of layered manganese oxide. Chemistry Letters. 2007; 36: 1228–9.

[135] Li J, Zhitomirsky I. Cathodic electrophoretic deposition of manganese dioxide films. Colloids and Surfaces A-Physicochemical and Engineering Aspects. 2009; 348: 248–53.

[136] Ching S, Petrovay DJ, Jorgensen ML, Suib SL. Sol-gel synthesis of layered birnessite-type manganese oxides. Inorganic Chemistry. 1997; 36: 883–90.

[137] Ma Y, Luo J, Suib SL. Syntheses of birnessites using alcohols as reducing reagents: Effects of synthesis parameters on the formation of birnessites. Chemistry of Materials. 1999; 11: 1972–9.

[138] Kuo SL, Wu NL. Electrochemical capacitor of $MnFe_2O_4$ with NaCl electrolyte. Electrochemical and Solid State Letters. 2005; 8: A495–A9.

[139] Kuo SL, Wu NL. Electrochemical characterization on $MnFe_2O_4$/carbon black composite aqueous supercapacitors. Journal of Power Sources. 2006; 162: 1437–43.

[140] Cai WH, Lai T, Dai WL, Ye JS. A facile approach to fabricate flexible all-solid-state supercapacitors based on $MnFe_2O_4$/graphene hybrids. Journal of Power Sources. 2014; 255: 170–8.

[141] Li B, Fu YS, Xia H, Wang X. High-performance asymmetric supercapacitors based on $MnFe_2O_4$/graphene nanocomposite as anode material. Materials Letters. 2014; 122: 193–6.

[142] Kotutha I, Swatsitang E, Meewassana W, Maensiri S. One-pot hydrothermal synthesis, characterization, and electrochemical properties of $rGO/MnFe_2O_4$ nanocomposites. Japanese Journal of Applied Physics. 2015; 54.

[143] Bai S, Shen XP, Zhong X, Liu Y, Zhu GX, Xu X et al. One-pot solvothermal preparation of magnetic reduced graphene oxide-ferrite hybrids for organic dye removal. Carbon. 2012; 50: 2337–46.

[144] Lee MT, Chang JK, Hsieh YT, Tsai WT. Annealed Mn-Fe binary oxides for supercapacitor applications. Journal of Power Sources. 2008; 185: 1550–6.

[145] Tsang C, Kim J, Manthiram A. Synthesis of manganese oxides by reduction of $KMnO_4$ with KBH_4 in aqueous solutions. Journal of Solid State Chemistry. 1998; 137: 28–32.

274 Nanomaterials for Supercapacitors

[146] Chuang PY, Hu CC. The electrochemical characteristics of binary manganese-cobalt oxides prepared by anodic deposition. Materials Chemistry and Physics. 2005; 92: 138–45.

[147] Chang JK, Hsieh WC, Tsai WT. Effects of the Co content in the material characteristics and supercapacitive performance of binary Mn-Co oxide electrodes. Journal of Alloys and Compounds. 2008; 461: 667–74.

[148] Luo JM, Gao B, Zhang XG. High capacitive performance of nanostructured Mn-Ni-Co oxide composites for supercapacitor. Materials Research Bulletin. 2008; 43: 1119–25.

[149] Wen JG, Ruan XY, Zhou ZT. Preparation and electrochemical performance of novel ruthenium-manganese oxide electrode materials for electrochemical capacitors. Journal of Physics and Chemistry of Solids. 2009; 70: 816–20.

[150] Hyun TS, Kang JE, Kim HG, Hong JM, Kim ID. Electrochemical properties of MnO_x-RuO_2 nanofiber mats synthesized by co-electrospinning. Electrochemical and Solid State Letters. 2009; 12: A225–A8.

[151] Machefaux E, Brousse T, Belanger D, Guyomard D. Supercapacitor behavior of new substituted manganese dioxides. Journal of Power Sources. 2007; 165: 651–5.

[152] Hsieh YC, Lee KT, Lin YP, Wu NL, Donne SW. Investigation on capacity fading of aqueous $MnO_2.nH_2O$ electrochemical capacitor. Journal of Power Sources. 2008; 177: 660–4.

[153] Chen L, Sun LJ, Luan F, Liang Y, Li Y, Liu XX. Synthesis and pseudocapacitive studies of composite films of polyaniline and manganese oxide nanoparticles. Journal of Power Sources. 2010; 195: 3742–7.

[154] Sun LJ, Liu XX. Electrodepositions and capacitive properties of hybrid films of polyaniline and manganese dioxide with fibrous morphologies. European Polymer Journal. 2008; 44: 219–24.

[155] Zhou ZH, Cai NC, Zhou YH. Capacitive of characteristics of manganese oxides and polyaniline composite thin film deposited on porous carbon. Materials Chemistry and Physics. 2005; 94: 371–5.

[156] Yuan CZ, Su LH, Gao B, Zhang XG. Enhanced electrochemical stability and charge storage of MnO_2/carbon nanotubes composite modified by polyaniline coating layer in acidic electrolytes. Electrochimica Acta. 2008; 53: 7039–47.

[157] Zhang X, Ji LY, Zhang SC, Yang WS. Synthesis of a novel polyaniline-intercalated layered manganese oxide nanocomposite as electrode material for electrochemical capacitor. Journal of Power Sources. 2007; 173: 1017–23.

[158] Kharade PM, Chavan SG, Salunkhe DJ, Joshi PB, Mane SM, Kulkarni SB. Synthesis and characterization of PANI/MnO_2 bi-layered electrode and its electrochemical supercapacitor properties. Materials Research Bulletin. 2014; 52: 37–41.

[159] Yang F, Xu MW, Bao SJ, Sun QQ. MnO_2-assisted fabrication of PANI/MWCNT composite and its application as a supercapacitor. RSC Advances. 2014; 4: 33569–73.

[160] Sharma RK, Rastogi AC, Desu SB. Manganese oxide embedded polypyrrole nanocomposites for electrochemical supercapacitor. Electrochimica Acta. 2008; 53: 7690–5.

[161] Zhang X, Yang WS, Ma YW. Synthesis of polypyrrole-intercalated layered manganese oxide nanocomposite by a delamination/reassembling method and Its electrochemical capacitance performance. Electrochemical and Solid State Letters. 2009; 12: A95–A8.

[162] Sharma RK, Karakoti A, Seal S, Zhai L. Multiwall carbon nanotube-poly(4-styrenesulfonic acid) supported polypyrrole/manganese oxide nano-composites for high performance electrochemical electrodes. Journal of Power Sources. 2010; 195: 1256–62.

[163] Grote F, Lei Y. A complete three-dimensionally nanostructured asymmetric supercapacitor with high operating voltage window based on PPy and MnO_2. Nano Energy. 2014; 10: 63–70.

[164] Li J, Que TL, Huang JB. Synthesis and characterization of a novel tube-in-tube nanostructured PPy/MnO_2/CNTs composite for supercapacitor. Materials Research Bulletin. 2013; 48: 747–51.

[165] Yuan LY, Wan CY, Zhao LL. Facial in-situ synthesis of MnO_2/PPy composite for supercapacitor. International Journal of Electrochemical Science. 2015; 10: 9456–65.

[166] Lu Q, Zhou YK. Synthesis of mesoporous polythiophene/MnO_2 nanocomposite and its enhanced pseudocapacitive properties. Journal of Power Sources. 2011; 196: 4088–94.

[167] Sharma RK, Zhai L. Multiwall carbon nanotube supported poly(3,4-ethylenedioxythiophene)/manganese oxide nano-composite electrode for super-capacitors. Electrochimica Acta. 2009; 54: 7148–55.

[168] Nijjer S, Thonstad J, Haarberg GM. Oxidation of manganese(II) and reduction of manganese dioxide in sulphuric acid. Electrochimica Acta. 2000; 46: 395–9.

[169] Wang MR, Zhang HH, Wang CY, Wang GX. Synthesis of MnO_2/poly-o-phenylenediamine composite and its application in supercapacitors. Electrochimica Acta. 2013; 106: 301–6.

[170] Chen Y, Liu CG, Liu C, Lu GQ, Cheng HM. Growth of single-crystal α-MnO_2 nanorods on multi-walled carbon nanotubes. Materials Research Bulletin. 2007; 42: 1935–41.

[171] Ma SB, Nam KW, Yoon WS, Yang XQ, Ahn KY, Oh KH et al. Electrochemical properties of manganese oxide coated onto carbon nanotubes for energy-storage applications. Journal of Power Sources. 2008; 178: 483–9.

[172] Yan J, Fan ZJ, Wei T, Cheng J, Shao B, Wang K et al. Carbon nanotube/MnO_2 composites synthesized by microwave-assisted method for supercapacitors with high power and energy densities. Journal of Power Sources. 2009; 194: 1202–7.

[173] Jiang RR, Huang T, Tang Y, Liu JL, Xue LG, Zhuang JH et al. Factors influencing MnO_2/multi-walled carbon nanotubes composite's electrochemical performance as supercapacitor electrode. Electrochimica Acta. 2009; 54: 7173–9.

Oxide Based Supercapacitors: I-Manganese Oxides 275

[174] An GM, Yu P, Xiao MJ, Liu ZM, Miao ZJ, Ding KL et al. Low-temperature synthesis of Mn_3O_4 nanoparticles loaded on multi-walled carbon nanotubes and their application in electrochemical capacitors. Nanotechnology. 2008; 19.

[175] Fan ZJ, Qie ZW, Wei T, Yan J, Wang SS. Preparation and characteristics of nanostructured MnO_2/MWCNTs using microwave irradiation method. Materials Letters. 2008; 62: 3345–8.

[176] Ma SB, Ahn KY, Lee ES, Oh KH, Kim KB. Synthesis and characterization of manganese dioxide spontaneously coated on carbon nanotubes. Carbon. 2007; 45: 375–82.

[177] Jin X, Zhou W, Zhang S, Chen GZ. Nanoscale microelectrochemical cells on carbon nanotubes. Small. 2007; 3: 1513–7.

[178] Cui XW, Hu FP, Wei WF, Chen WX. Dense and long carbon nanotube arrays decorated with Mn_3O_4 nanoparticles for electrodes of electrochemical supercapacitors. Carbon. 2011; 49: 1225–34.

[179] Li J, Zhitomirsky I. Electrophoretic deposition of manganese dioxide-carbon nanotube composites. Journal of Materials Processing Technology. 2009; 209: 3452–9.

[180] Wang YH, Zhitomirsky I. Electrophoretic deposition of manganese dioxide-multiwalled carbon nanotube composites for electrochemical supercapacitors. Langmuir. 2009; 25: 9684–9.

[181] Zhang H, Cao GP, Wang ZY, Yang YS, Shi ZJ, Gu ZN. Growth of manganese oxide nanoflowers on vertically-aligned carbon nanotube arrays for high-rate electrochemical capacitive energy storage. Nano Letters. 2008; 8: 2664–8.

[182] Fan Z, Chen JH, Zhang B, Sun F, Liu B, Kuang YF. Electrochemically induced deposition method to prepare gamma-MnO_2/multi-walled carbon nanotube composites as electrode material in supercapacitors. Materials Research Bulletin. 2008; 43: 2085–91.

[183] Wan C, Azumi K, Konno H. Effect of synthesis routes on the performance of hydrated Mn(IV) oxide-exfoliated graphite composites for electrochemical capacitors. Journal of Applied Electrochemistry. 2007; 37: 1055–61.

[184] Wan CY, Azumi K, Konno H. Hydrated Mn(IV) oxide-exfoliated graphite composites for electrochemical capacitor. Electrochimica Acta. 2007; 52: 3061–6.

[185] Yan J, Fan ZJ, Wei T, Qie ZW, Wang SS, Zhang ML. Preparation and electrochemical characteristics of manganese dioxide/graphite nanoplatelet composites. Materials Science and Engineering B-Advanced Functional Solid-State Materials. 2008; 151: 174–8.

[186] Gui DY, Chen W, Liu CL, Liu JH. Graphene-like membrane supported MnO_2 nanospheres for supercapacitor. Journal of Materials Science-Materials in Electronics. 2016; 27: 5121–7.

[187] Hao JN, Zhong YY, Liao YQ, Shu D, Kang ZX, Zou XP et al. Face-to-face self-assembly graphene/MnO_2 nanocomposites for supercapacitor applications using electrochemically exfoliated graphene. Electrochimica Acta. 2015; 167: 412–20.

[188] Li ZM, An YF, Hu ZG, An N, Zhang YD, Guo BS et al. Preparation of a two-dimensional flexible MnO_2/graphene thin film and its application in a supercapacitor. Journal of Materials Chemistry A. 2016; 4: 10618–26.

[189] Liu CL, Gui DY, Liu JH. Process dependent graphene-wrapped plate-like MnO_2 nanospheres for high performance supercapacitor. Chemical Physics Letters. 2014; 614: 123–8.

[190] Zhang JY, Yang XF, He YB, Bai YL, Kang LP, Xu H et al. δ-MnO_2/holey graphene hybrid fiber for all-solid-state supercapacitor. Journal of Materials Chemistry A. 2016; 4: 9088–96.

[191] Zhao Y, Meng YN, Wu HP, Wang Y, Wei ZX, Li XJ et al. *In situ* anchoring uniform MnO_2 nanosheets on three-dimensional macroporous graphene thin-films for supercapacitor electrodes. RSC Advances. 2015; 5: 90307–12.

[192] Zheng YH, Pann WX, Zhengn DY, Sun CX. Fabrication of functionalized graphene-based MnO_2 nanoflower through electrodeposition for high-performance supercapacitor electrodes. Journal of the Electrochemical Society. 2016; 163: D230–D8.

[193] Liu XM, Fu SY, Huang CJ. Synthesis, characterization and magnetic properties of β-MnO_2 nanorods. Powder Technology. 2005; 154: 120–4.

[194] Li J, Wang XY, Huang QH, Gamboa S, Sebastian PJ. A new type of MnO_2.xH_2O/CRF composite electrode for supercapacitors. Journal of Power Sources. 2006; 160: 1501–5.

[195] Fischer AE, Saunders MP, Pettigrew KA, Rolison DR, Long JW. Electroless deposition of nanoscale MnO_2 on ultraporous carbon nanoarchitectures: Correlation of evolving pore-solid structure and electrochemical performance. Journal of the Electrochemical Society. 2008; 155: A246–A52.

[196] Long JW, Sassin MB, Fischer AE, Rolison DR, Mansour AN, Johnson VS et al. Multifunctional MnO_2-carbon nanoarchitectures exhibit battery and capacitor characteristics in alkaline electrolytes. Journal of Physical Chemistry C. 2009; 113: 17595–8.

[197] Patel MN, Wang XQ, Wilson B, Ferrer DA, Dai S, Stevenson KJ et al. Hybrid MnO_2-disordered mesoporous carbon nanocomposites: synthesis and characterization as electrochemical pseudocapacitor electrodes. Journal of Materials Chemistry. 2010; 20: 390–8.

[198] Li GR, Feng ZP, Ou YN, Wu DC, Fu RW, Tong YX. Mesoporous MnO_2/carbon aerogel composites as promising electrode materials for high-performance supercapacitors. Langmuir. 2010; 26: 2209–13.

[199] Kiani MA, Khani H, Mohammadi N. MnO_2/ordered mesoporous carbon nanocomposite for electrochemical supercapacitor. Journal of Solid State Electrochemistry. 2014; 18: 1117–25.

[200] Patel MN, Wang XQ, Slanac DA, Ferrer DA, Dai S, Johnston KP et al. High pseudocapacitance of MnO_2 nanoparticles in graphitic disordered mesoporous carbon at high scan rates. Journal of Materials Chemistry. 2012; 22: 3160–9.

276 *Nanomaterials for Supercapacitors*

[201] Wang GX, Xu HF, Lu L, Zhao H. One-step synthesis of mesoporous MnO_2/carbon sphere composites for asymmetric electrochemical capacitors. Journal of Materials Chemistry A. 2015; 3: 1127–32.

[202] Wen ZQ, Li M, Zhu SJ, Wang T. Novel mesoporous carbon-carbonaceous materials nanostructures decorated with MnO_2 nanosheets for supercapacitors. International Journal of Electrochemical Science. 2016; 11: 1810–20.

[203] Zhao JC, Tang BH, Sun L, Zheng J, Cao J, Xu JL. Effect of loading amount on capacitance performances of MnO_2/ordered mesoporous carbon composites. Materials Technology. 2012; 27: 328–32.

[204] Liang CD, Dai S. Synthesis of mesoporous carbon materials via enhanced hydrogen-bonding interaction. Journal of the American Chemical Society. 2006; 128: 5316–7.

[205] Lei Y, Fournier C, Pascal JL, Favier F. Mesoporous carbon-manganese oxide composite as negative electrode material for supercapacitors. Microporous and Mesoporous Materials. 2008; 110: 167–76.

[206] Dong XP, Shen WH, Gu JL, Xiong LM, Zhu YF, Li Z et al. MnO_2-embedded-in-mesoporous-carbon-wall structure for use as electrochemical capacitors. Journal of Physical Chemistry B. 2006; 110: 6015–9.

[207] Zhang LL, Wei TX, Wang WJ, Zhao XS. Manganese oxide-carbon composite as supercapacitor electrode materials. Microporous and Mesoporous Materials. 2009; 123: 260–7.

[208] Stober W, Fink A, Bohn E. Controlled growth of monodisperse silica spheres in the micron size range. Journal of Colloid and Interface Science. 1968; 26: 62–9.

Oxide Based Supercapacitors
II-Other Oxides

Ling Bing Kong,[1,] Wenxiu Que,[2] Lang Liu,[3] Freddy Yin Chiang Boey,[1,a] Zhichuan J. Xu,[1,b] Kun Zhou,[4] Sean Li,[5] Tianshu Zhang[6] and Chuanhu Wang[7]*

Overview

As discussed in the last chapter, oxides exhibit strong pseudocapacitive effect, thus making them promising candidates as electrodes of supercapacitors. Besides manganese oxides, various other oxides and complex oxides, such as V_2O_5 [1–3], Fe_2O_3 [4, 5], Fe_3O_4 [6–8], Co_3O_4 [9], NiO [10], CuO [11, 12], ZnO [13, 14], WO_3 [15], MoO_3 [16], RuO_2 [17–20], $NiCo_2O_4$ [21–23] and ferrites [24–26], have been developed in various forms, such as pure oxides, nanohybrids and nanocomposites, with outstanding electrochemical performances as electrodes of supercapacitors. From the morphology point of view, they can be presented as nanoparticles, nanowires, nanorods, nanotubes, nanosheets, nanoneedles, nanoflakes and nanoflowers, which can be developed by using different methods, including hydrothermal, solvothermal, chemical precipitation, chemical bath deposition, sol-gel, electrodeposition and microwave assisted synthetic routes. They have also been incorporated with other components, such as CNTs, graphenes and conductive

---

[1] School of Materials Science and Engineering, Nanyang Technological University, 50 Nanyang Avenue, Singapore 639798.
[a] Email: mycboey@ntu.edu.sg
[b] Email: xuzc@ntu.edu.sg
[2] Electronic Materials Research Laboratory and International Center for Dielectric Research, Xi'an Jiaotong University, Xi'an 710049, Shaanxi Province, People's Republic of China.
 Email: wxque@xjtu.edu.cn
[3] School of Chemistry and Chemical Engineering, Xinjiang University, Urumqi 830046, Xinjiang, People's Republic of China.
 Email: llyhs1973@sina.com
[4] School of Mechanical and Aerospace Engineering, Nanyang Technological University, 50 Nanyang Avenue, Singapore 639798.
 Email: kzhou@ntu.edu.sg
[5] School of Materials Science and Engineering, The University of New South Wales, Australia.
 Email: sean.li@unsw.edu.au
[6] Anhui Target Advanced Ceramics Technology Co. Ltd., Hefei, Anhui, People's Republic of China.
 Email: 13335516617@163.com
[7] Department of Material and Chemical Engineering, Bengbu Univresity, Bengbu, Peoples's Republic of China.
 Email: bbxywch@126.com
* Corresponding author: elbkong@ntu.edu.sg

278   *Nanomaterials for Supercapacitors*

polymers, to form 0D–3D nanostructured materials. However, due to the space limitation, this chapter only covers Co_3O_4 [9] and NiO [10], as well as their corresponding complex oxide $NiCo_2O_4$ [21–23], as representatives.

Cobalt Oxides (Co_3O_4)

Cobalt-based compounds, especially cobalt oxides (Co_3O_4), together with their composites, have been extensively and intensively studied as pseudocapacitors in recent years. Significant progress have been achieved, with superior performances, including high specific capacitances of 3000 $F \cdot g^{-1}$, high areal capacitances of up to 25 $F \cdot cm^{-2}$, promising cycling stability up to 10,000 cycles, and so on [9, 27–29]. Due to the limitation of space, the focus will only be on Co_3O_4 based materials, which are classified into two groups: (i) power based electrodes and (ii) film based electrodes.

Powder based electrodes

Chemical precipitation

Chemical precipitation is one of the most widely used methods which can be used to synthesize Co_3O_4 powders with ultrafine structures [30–39]. There are various process parameters that can be utilized to achieve high performance Co_3O_4 based electrodes, such as composition, microstructure, morphology and so on [40–45]. Selected examples are presented and discussed as follows.

It was demonstrated that the morphologies of cobalt oxalates could be well controlled by adjusting the content, combination and feeding order of the reactants, which thus led to Co_3O_4 powders with the morphologies of the precursors after calcination [33]. In this case, Co_3O_4 nanostructures, including nanorods, nanowires and Layered Parallel Folding (LPF) structures, were obtained, with BET specific surface areas of 45.1, 35.6 and 75.9 $m^2 \cdot g^{-1}$, respectively. Their average pore diameters were 17.28, 19.37 and 8.58 nm, respectively. Correspondingly, the samples exhibited specific capacitances of 128, 103 and 203 $F \cdot g^{-1}$ for nanorods, nanowires and LPF nanostructures, respectively. The higher capacitances demonstrated by the LPF nanostructures can be attributed to their higher surface area, smaller pores and smaller particle sizes which shorten the diffusion length of OH^- ions. Due to their influence on the surface area and porosity of electrode materials, morphologies thus could be utilized to tailor the charge storage performances.

Heat treatment temperature also had a significant effect on textural properties of the synthesized materials. As the calcination temperature to convert $Co(OH)_2$ microflowers into Co_3O_4 porous microflowers was increased from 300 to 500°C, the crystal sizes was increased and the pores were enlarged. For instance, BET specific surface areas of the samples thermally treated at 300, 400 and 500°C were 77.33, 32.02 and 12 $m^2 \cdot g^{-1}$, respectively. The decrease in specific surface areas of was ascribed to aggregation of the nanoparticles. As a consequence, the samples exhibited specific capacitances of 160, 88 and 71 $F \cdot g^{-1}$, respectively.

The precursors, i.e., cobalt oxalate nanostructures, were synthesized by using a water-controlled precipitation process [33]. Two routes, namely, (i) water/stock solution route and (ii) stock solution/ water route, were employed for the one-step synthesis of the 1D cobalt oxalate nanostructures. The stock solutions were prepared by dissolving oxalic acid ($H_2C_2O_4$) and cobalt chloride hexahydrate ($CoCl_2 \cdot 6H_2O$) with an equal mole ratio in either *N,N*-dimethylacetamide (DMA) or dimethyl sulfoxide (DMSO), to form transparent blue solutions. In this case, water as the precipitation reagent.

To synthesize cobalt oxalate nanorods and nanowires, in the case of water/stock solution route, 1 mmol $H_2C_2O_4$ was dissolved in 15 mL of *N,N*-dimethylacetamide (DMA) to form a homogeneous solution with constant stirring. After the $H_2C_2O_4$ was completely dissolved, 1 mmol $CoCl_2$ was added into the reaction system, followed by dropwise addition of 10 mL deionized water. After reaction for 5 minutes, pink-colored precipitate was formed, which was collected through centrifugation, followed by thorough washing with ethanol and drying at 60°C in vacuum for 12 hours. As the stock solution/water route was adopted, the only difference was the feeding order of the two solutions, in which the stock solution was added dropwise into water with constant stirring.

The Layered Parallel Folding (LPF) nanostructures were obtained in a slightly different way. Two mmol H$_2$C$_2$O$_4$ was dissolved in 15 mL dimethyl sulfoxide (DMSO) to form a transparent solution. After that, 2 mmol CoCl$_2$ was introduced into the solution, as the H$_2$C$_2$O$_4$ was dissolved entirely. Then, 5 mL deionized water was added in a dropwise manner. The rest of the steps were similar to those for the nanorods and nanowires. To obtain mesoporous Co$_3$O$_4$ nanostructures, all the cobalt oxalate nanostructures were calcined at 400°C for 3 hours in air, at a relatively slow heating rate of 2°C min^{-1}. The Co$_3$O$_4$ nanostructures were denoted as samples C1 (nanorods), C2 (nanowires) and C3 (LPF nanostructures).

Representative SEM images of the cobalt oxalate nanorods obtained through the water/stock solution route (DMA as the solvent) are shown in Fig. 5.1(a, b). The nanorods had an average diameter of 50 nm and an average length of sub-micrometers. As the feeding order of the two solutions was exchanged, nanowires were formed, as illustrated in Fig. 5.1(c, d). They possessed an average diameter of about 100 nm and an average length of 10s of micrometers. In contrast, if DMSO was used as the solvent to form the stock solution, the synthesized cobalt oxalate nanostructures had a different morphology, as seen in Fig. 5.1(a, b). The sample consisted of a large number of layered parallel folding (LPF) structures, with an average length of 2 μm. The LPF nanostructures were composed of numerous curved nanosheets with a thickness of about 50 nm. If the feeding order of the two solutions was changed, irregular nanobelt-bundles were formed. The nanobelt-bundles had a diameter of about 500 nm, consisting of nanobelts with a thickness of about 50 nm with different lengths.

It has been well known that the growth of crystals with anisotropic structures is attributed to the significant difference in surface energy between the different crystallographic planes [46]. In this work, two C$_2$O$_4^{2-}$ ions were chelated with Co^{2+} to form a square planar configuration with the Co^{2+} to be at the center. Two H$_2$O molecules were coordinated at each side of the molecular plane, so that two molecular

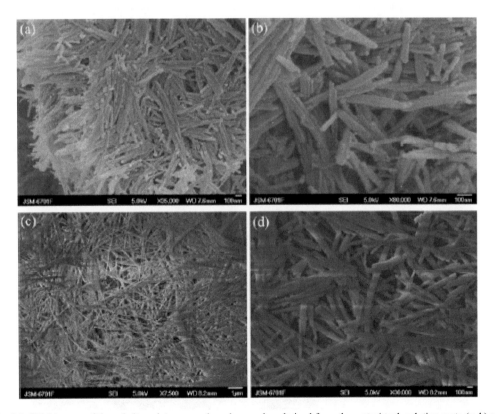

Fig. 5.1. SEM images of the cobalt oxalate nanorods and nanowires derived from the water/stock solution route (a, b) and stock solution/water route (DMA as the solvent) (c, d). Reproduced with permission from [33], Copyright © 2011, American Chemical Society.

planes could be linked by the H$_2$O molecule. Due to the hydrogen bonding and the π–π interaction within the adjacent molecules, the growth of the crystals in the direction parallel to the molecule planes was prevented, while that in the direction perpendicular to the planes was favored [47]. As a result, the square planes were stacked in a face-to-face manner, thus leading to the formation of columnar structures and then the presence of the cobalt oxalate nanorods [48].

Furthermore, the solvents also played an important role in determining morphologies of cobalt oxalate precursors. It was expected that nucleation and growth behaviors of the nanocrystals were closely related to the properties of the solvents. In addition, different solvents exhibit different solvation and coordination effects, so that the development of the cobalt oxalate seeds was different in different solvents. As DMA was used as the solvent, both the diameter and length of the cobalt oxalate nanorods were increased, once the stock solution was added into water, as shown in Fig. 5.2. With the solvent of DMSO, nanobelt-bundles were developed, as the stock solution was added into water.

In this case, water exhibited multiple functions during the reaction process. Firstly, it served as a precipitation reagent for the precipitation of cobalt oxalate, due to the low solubility of cobalt oxalate in water. Secondly, water molecules acted as a competitor with the organic solvent molecules to coordinate Co^{2+} ions. As a consequence, the concentration of [Co-solvent]$^{2+}$ was balanced, so as to control over the supersaturation concentration of Co^{2+} [49]. Thirdly, water molecules were involved in the formation of the cobalt oxalate precipitate, which contained two water molecules. As the stock solutions were added into water, the cobalt oxalates could be more easily precipitated due to the excessive reagent, as compared with the case in which water was added into the stock solutions. In other words, water molecules promoted the anisotropic growth of the cobalt oxalate nanostructures [50].

After calcination at 400°C for 3 hours, sample C1 contained Co$_3$O$_4$ nanorods, with an average diameter of 50 nm and an average length of 100s of nanometer. TEM image indicated that every individual Co$_3$O$_4$ nanorod consisted of a large number of nanoparticles, i.e., the Co$_3$O$_4$ nanorods possessed a polycrystalline nature. Selected Area Electron Diffraction (SAED) pattern demonstrated that the cobalt oxide was cubic Co$_3$O$_4$. Figure 5.3 shows SEM and TEM images of sample C2, confirming the presence of Co$_3$O$_4$ nanowires, with an average diameter of 100 nm in diameter and a length of 10s of micrometers, comparable with the dimensions of the precursor nanowires. These results implied that the morphologies of the cobalt oxalate precursors were well retained after they converted to oxides after the thermal calcination process.

Amorphous and mesoporous Co$_3$O$_4$ nanoparticles were synthesized by using the precipitation method, through the formation and disassociation of cobalt–citrate complex reaction [36]. The precipitated precursors were Co$_3$O$_4$/Co(OH)$_2$ nanosheets. In reaction, the stable cobalt–citrate complex was formed, because the Co nanoparticles were oxidized, due to the presence of sodium citrate and oxygen, while the cobalt–citrate complex was disassociated to Co(OH)$_2$ in an alkaline environment. The granular Co$_3$O$_4$ nanoparticles possessed a high specific surface area and mesoporous microstructure, which was attributed to the topotactic transformation during the solid-state oxidation process. Because of the mesoporous microstructure, the

Fig. 5.2. SEM images of the cobalt oxalate nanostructures derived from the water/stock solution route (DMSO as the solvent). Reproduced with permission from [33], Copyright © 2011, American Chemical Society.

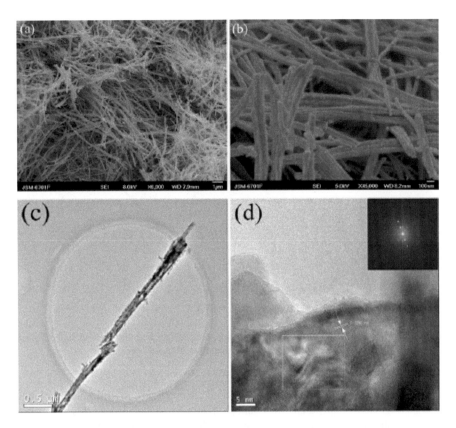

Fig. 5.3. Morphological and structural characterization of sample C2: (a, b) SEM images, (c) TEM image and (d) HRTEM image. The inset in panel (d) corresponds to its FFT pattern taken from the selected area marked with a square. The fringe spacing of 0.286 nm corresponds to the (220) of cubic Co_3O_4. Reproduced with permission from [33], Copyright © 2011, American Chemical Society.

Co_3O_4 nanoparticles prepared in this way demonstrated promising electrochemical performances, with a stable discharging capacity in terms of cycle lifetime and a high specific capacitance of 427 $F·g^{-1}$.

Figure 5.4 shows a schematic diagram illustrating synthesis process of the mesoporous Co_3O_4 nanoparticles [36]. Sixty mg $Co(NO_3)_2·6H_2O$ and 180 mg sodium citrate (Na-Cit) were dissolved in 30 ml distilled water to form aqueous solution, into which 5 ml ice-cold $NaBH_4$ with a concentration of 0.1 M was quickly added under vigorous stirring, which led to the formation of a black solution. After continuous stirring for 2 hours at room temperature, the Co nanoparticles completely oxidized to form the stable cobalt–citrate, which was monitored by observing the color change of the solution from black to violet. After that, a solution of 3 mg SDS and 60 mg NaOH dissolved in 5 ml distilled water was added into the reaction system. The reaction mixture was further stirred for 5 minutes, following by aging at 30°C for 36 hours. Then, a blue solid was precipitated, due to the formation of $Co_3O_4/Co(OH)_2$, which was collected through high speed centrifugation, followed by thorough washing with ethanol and distilled water and drying at 80°C for 4 hours. Finally, black powder of Co_3O_4 was obtained after calcination at 250°C for 3 hours in air.

Co_3O_4 had a mixed spinel phase, as confirmed by XRD measurement results. SEM observation results indicated that the $Co_3O_4/Co(OH)_2$ was present as loosely packed nanostructures with irregular and curly morphologies. The nanostructures contained nanosheets with an average thickness of 10 nm. Representative TEM images of precursor sample are shown in Fig. 5.5(a, b) [36], confirming that the $Co_3O_4/Co(OH)_2$ nanostructures consisted of nanosheets with a curly layer morphology. There were monodispersed nanoparticles with a size of about 40 nm. It was found that the self-assembled nanosheets were

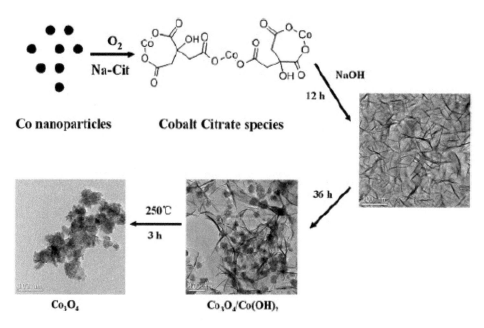

Fig. 5.4. Formation process of the mesoporous Co_3O_4 nanoparticles derived from the cobalt-citrate complex. Reproduced with permission from [36], Copyright © 2012, Elsevier.

formed during the early 12 hours, without the presence of nanoparticles. After reaction for longer than 12 hours, the nanosheets were enlarged and thus nanoparticles were developed. In this case, the surfactant SDS acted as a template to facilitate the formation of the loosely packed nanostructures.

After calcination, the loosely packed $Co_3O_4/Co(OH)_2$ nanostructures were converted to pure Co_3O_4 nanoparticles with a mesoporous microstructure, as demonstrated in Fig. 5.5(c, d) [36]. According to the widened XRD peaks, the Co_3O_4 nanoparticles had an average crystal size of as small as 8.9 nm, estimated with the Scherrer equation based on the diffraction peak of (311). TEM images revealed that the Co_3O_4 nanoparticles exhibited a granule morphology, with an irregular shape and mesoporous microstructure, as seen in Fig. 5.5(c, d). Selected-area electron diffraction (SAED) and lattice spacing displayed well-defined rings, demonstrating their polycrystalline nature, as observed in the inset of Fig. 5.5(d). The crystal structure of the Co_3O_4 observed with TEM was in a good agreement with the XRD results. In addition, the Co_3O_4 powder had a large surface area of 129 $m^2 \cdot g^{-1}$, with the pore size distribution centered at 3.7 nm and 4.7 nm.

A formation mechanism was proposed to explain the formation of the nanostructure, as schematically shown in Fig. 5.6 [36]. During the initial nucleation stage, a large number $Co(OH)_2$ nucleation centers were developed due to the disassociation of the cobalt–citrate complex. Then, the reaction processed to second stage, i.e., the growth of the nanosheet. Because the $Co(OH)_2$ crystals were of an anisotropic structure, the growing point of each crystal was located in certain specific directions, in which the cobalt–citrate complex was slowly disassociated to form the precursors. As a result, nanosheets were obtained. At the third stage, dissolution and growth of the nuclei took place. Thus, the nuclei of CoOOH was formed either in the external region or on flake boundaries of the dissolved $Co(OH)_2$ nanosheets. The presence of the intermediate CoOOH could be attributed to the fact that oxyhydroxides generally have lower surface energies than oxides, especially during the precipitation from aqueous solution [51]. Finally, transformation and growth stage was approached. At this stage, $Co(OH)_2$ was continuously converted into intermediate CoOOH, which could be reduced to Co_3O_4 at a low temperature, through the reaction with oxygen and hydroxyl ions in the alkaline media. The hybrid structure of the $Co_3O_4/Co(OH)_2$ nanoparticles was resulted from the solid-state reaction between the $Co(OH)_2$ nanosheets and the dissolved oxygen in the solution.

A simple and convenient method was reported to synthesize hollow Co_3O_4 nanostructures via a precursor without the use of surfactants or templates [38]. The precursor, $(Co_5(OH)_2(CH_3COO)_8 \cdot 2H_2O)$,

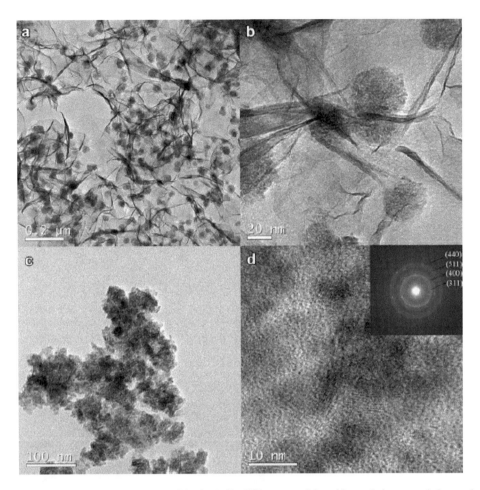

Fig. 5.5. (a, b) TEM characterization results of the Co$_3$O$_4$/Co(OH)$_2$ nanoparticles with a curly layer morphology and small nanoscale particles. (c, d) TEM and HRTEM images of the granular Co$_3$O$_4$ nanoparticles with a mesoporous microstructure, with the inset showing the SAED pattern of the Co$_3$O$_4$ nanoparticles. Reproduced with permission from [36], Copyright © 2012, Elsevier.

was recrystalized from the solution of Co(CH$_3$COO)$_2$·4H$_2$O in ethanol at a relatively low temperature. The precursor consisted of 1D prisms, with dimensions of 3 μm, 300 nm and 300 nm in length, width and height, respectively. Co$_3$O$_4$ nanoparticles, with a hollow box morphology, were simply obtained by calcining the precursor. The formation of the hollow box Co$_3$O$_4$ was ascribed to the Kirkendall effect [52]. The hollow Co$_3$O$_4$ boxes demonstrated promising electrochemical performances, with a high specific capacitance of 278 F·g^{-1} and superior cycle stability.

To synthesize the prism precursor, 0.8 g cobalt acetate tetrahydrate, Co(CH$_3$COO)$_2$·4H$_2$O, was dissolved in 500 mL cold ethanol, which was kept at −5°C for several days. During this time duration, precipitation took place. Then, the precipitate was collected and thoroughly washed with ethanol, by using a centrifugation–redispersion process. After that, the powder was dried at 40°C for 4 hours. To obtain Co$_3$O$_4$ powder, the precursor was calcined at 300°C for 10 minutes in air at a slow heating rate of 1°C·min^{-1}.

Figure 5.7(a) shows a representative SEM image of the precursor, indicating the presence of prism-like morphology with a dimension as mentioned earlier [38]. A typical TEM image of the precursor is shown in Fig. 5.7(b), further confirming the formation of the 1D prism particles. The particles were of a solid microstructure. The phase composition of the precursor was investigated by XRD. Phase composition of Co$_5$(OH)$_2$(CH$_3$COO)$_8$·2H$_2$O of the precursor was confirmed by the XRD analysis result. During this

284 Nanomaterials for Supercapacitors

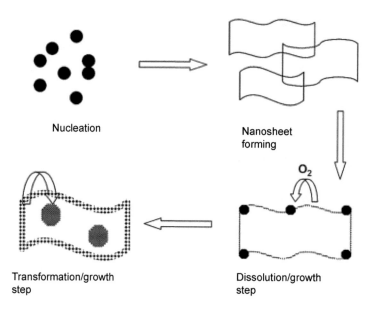

Fig. 5.6. Schematic diagram illustrating formation mechanism of the Co_3O_4/$Co(OH)_2$ precursor nanoparticles. Reproduced with permission from [36], Copyright © 2012, Elsevier.

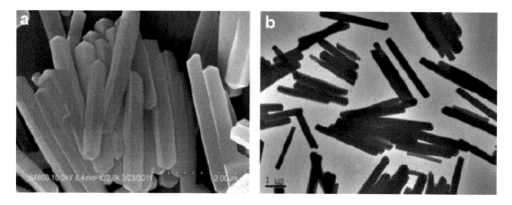

Fig. 5.7. (a) SEM and (b) TEM images of the precursor prisms. Reproduced with permission from [38], Copyright © 2013, Elsevier.

precipitation process, after the $Co(CH_3COO)_2 \cdot 4H_2O$ was dissolved in ethanol, the small amount of crystallization water trigger slow hydrolysis of $Co(CH_3COO)_2$, so as to facilitate the re-crystallization of $Co_5(OH)_2(CH_3COO)_8 \cdot 2H_2O$.

Thermal analysis results revealed that the cobalt precursor was converted to Co_3O_4 at a temperature of < 300°C [53]. Figure 5.8 shows XRD pattern of the product obtained by annealing the precursor at 300°C, confirming the cubic spinel structure of the converted oxide [38]. Figure 5.9(a) shows SEM image of the Co_3O_4 powder after calcination. The 1D prism morphology of the precursor was well retained, while the 1D crystals became hollow boxes. The boxes demonstrated a rough surface, implying that they consisted of Co_3O_4 nanoparticles. Figure 5.9(b) shows a representative TEM image of the hollow box-like structure, revealing that the box had uniform length, width and height, consisting of Co_3O_4 nanoparticles. Therefore, polycrystalline SAED pattern was observed, as illustrated by the top-right inset in Fig. 5.9(b).

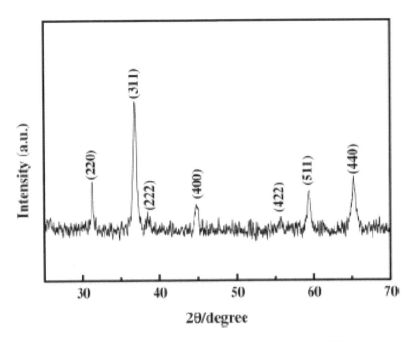

Fig. 5.8. XRD pattern of the Co$_3$O$_4$ powder derived from the precursor after calcination at 300°C. Reproduced with permission from [38], Copyright © 2013, Elsevier.

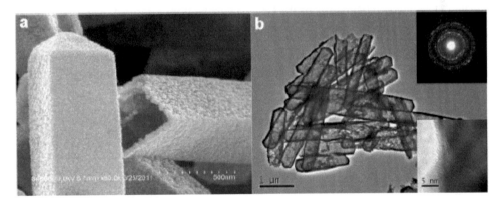

Fig. 5.9. (a) SEM and (b) TEM images of the Co$_3$O$_4$ boxes derived from the precursor prisms (insets: HRTEM image and SAED pattern). Reproduced with permission from [38], Copyright © 2013, Elsevier.

The crystalline nature of the Co$_3$O$_4$ boxes was further confirmed by high-resolution HRTEM image, shown as the bottom-right inset in Fig. 5.9(b).

Cyclic Voltammetry (CV) curves of the Co$_3$O$_4$ nanostructure were characterized in 3% KOH electrolyte, within the voltage window of 0–0.5 V, at a scanning rate of 5 mV·s^{-1} [38]. The CV curves were quite symmetrical, with two pairs of redox peaks, corresponding to the redox processes of Co$_3$O$_4$/CoOOH/CoO$_2$, due to the electrochemical pseudocapacitors from the reversible faradaic redox reactions within the electroactive materials [54, 55]. The Co$_3$O$_4$ box powder had a specific capacitance of 278 F·g^{-1}, which was much higher than that of the commercial Co$_3$O$_4$ powder (77 F·g^{-1}). The high electrochemical performance of the Co$_3$O$_4$ box powder could be readily attributed to the hollow structural morphology. The high porosity structure of the Co$_3$O$_4$ boxes provided both shortened ionic and electronic transport distances and thus enhanced kinetic performance of the electrode, ensuring the potential of the electrode materials for high power supercapacitor applications.

Co$_3$O$_4$ nanotube bundles have been fabricated by using a precipitation method, combined with Anodic Aluminum Oxide (AAO) as a hard template [32]. According to BET measurement result, the Co$_3$O$_4$ nanotubes had a specific surface area 218 m^2·g^{-1} and pore sizes of 38–110 nm. Electrode based on the Co$_3$O$_4$ nanotubes exhibited specific capacitances of 574, 551, 538 and 484 F·g^{-1} at 0.1, 0.2, 0.5 and 1 A·g^{-1}, respectively. The electrode also displayed an excellent cycling stability with 95% retention of the capacitance after charge-discharge for 1000 cycles.

The Co$_3$O$_4$ nanotubes were obtained by using porous AAO membranes as the template, which had a pore diameter of 200 nm and a thickness of 60 μm. The AAO template membranes were immersed in 2 mol·L^{-1} CoCl$_2$ solution for 4 hours, so that the CoCl$_2$ solution was thoroughly sucked into the pores of the templates due to the capillary force. After that, the CoCl$_2$ solution filled AAO templates were placed in 1 mol·L^{-1} NH$_4$OH solution for 1 hour, which led to the precipitation of Co(OH)$_2$ inside the pores of the templates. The samples were then thoroughly washed with distilled water, followed by drying in air at 60°C for 5 hours. The dried samples were calcined at 500°C for 2 hours in the air, in order to convert Co(OH)$_2$ to Co$_3$O$_4$. The AAO templates were removed by etching the calcined samples in 3 mol·L^{-1} NaOH solution for 5 hours, resulting in Co$_3$O$_4$ nanotube bundles.

Figure 5.10 shows XRD pattern of the Co$_3$O$_4$ nanotubes, confirming their cubic spinel structure. Figure 5.11 shows representative SEM and TEM images of the Co$_3$O$_4$ nanotubes. Obviously, the pores of the AAO templates were filled with tubular shaped materials, as seen in Fig. 5.11(a–c). The Co$_3$O$_4$ nanotubes had an average diameter of 300 nm and an average thickness of 50 nm, as illustrated in Fig. 5.11(d), which were closely related to the pore dimension of the AAO template, reaction parameters and properties of the precursor solutions. Pore sizes of the Co$_3$O$_4$ nanotubes were in the range of 38–110 nm, implying the presence of both mesopores and macropores. BET specific surface area of the Co$_3$O$_4$ nanotubes was 218 m^2·g^{-1}.

Cyclic voltammetric (CV) curves of the Co$_3$O$_4$ nanotube were measured over a potential range between −0.10 and 0.50 V (vs. SCE). Figure 5.12 shows the CV curves measured in 6 mol·L^{-1} KOH at scan rates of 5, 10, 20 and 50 mV·s^{-1}. The CV curves exhibited a pair of broad redox reaction peaks, due to the pseudocapacitive behavior of the Co$_3$O$_4$ nanotube based electrodes. In addition, the peak current was increased, as the scan rate was increased from 5 to 50 mV·s^{-1}, implying that fast charge–discharge

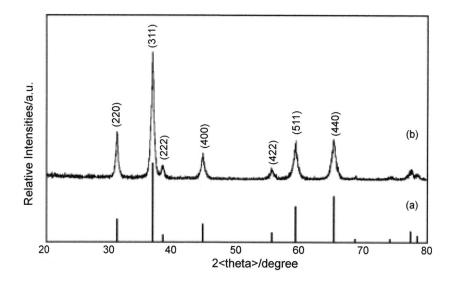

Fig. 5.10. XRD patterns of standard spectrum of Co$_3$O$_4$ (a) and the Co$_3$O$_4$ nanotubes (b). Reproduced with permission from [32], Copyright © 2010, Elsevier.

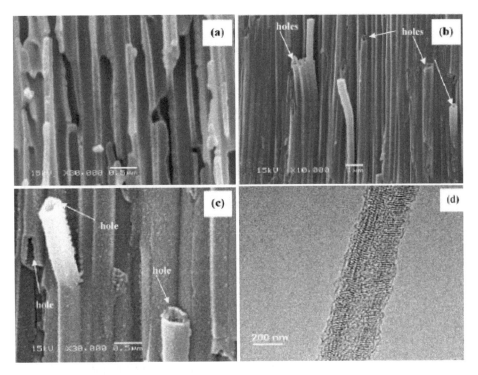

Fig. 5.11. SEM images of vacant AAO template (a), Co$_3$O$_4$ nanotubes embedded in the AAO template (b, c) and TEM image of an individual Co$_3$O$_4$ nanotube (d). Reproduced with permission from [32], Copyright © 2010, Elsevier.

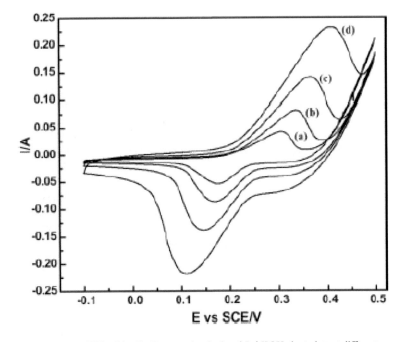

Fig. 5.12. Cyclic Voltammogram (CV) of the Co$_3$O$_4$ nanotubes in 6 mol·L^{-1} KOH electrolyte at different scan rates: (a) 5, (b) 10, (c) 20 and (d) 50 mV·s^{-1}. Reproduced with permission from [32], Copyright © 2010, Elsevier.

288 *Nanomaterials for Supercapacitors*

response was highly reversible [56]. Potential mechanism of the reaction was expressed as $Co_3O_4 + OH^- + H_2O = 3CoOOH + e^-$ [57]. It has been recognized that as high concentration alkali ions were adsorbed on the surface of electrodes, the electrolyte starvation near the electrode surface was reduced, which resulted in decrease in internal resistance of the electrode. As a result, the pseudocapacitive performances would be enhanced [58].

Various mesoporous Co_3O_4 nanoparticles were prepared by using the precipitation method, with $Co(NO_3)_2 \cdot 6H_2O$ as precursor and templates, such as mesoporous silicas, KIT-6 and SBA-15 [31]. It was found that calcination temperature had almost no effect on textural properties of the Co_3O_4 nanoparticles, while BET surface areas of the samples could be well controlled by adjusting the parameters of the KIT-6 templates. However, specific capacitances of the samples were decreased with increasing calcination temperature, due to the variation in the BET surface area. The large sizes pores and highly ordered mesopores ensured a smooth ion transportation, whereas specific capacitances of the Co_3O_4 nanoparticles were not obviously influenced by their mesoporous characteristics. The sample with the highest BET surface area of 184.8 $m^2 \cdot g^{-1}$ had a specific capacitance of 370 $F \cdot g^{-1}$.

To synthesize 3D cubic $Ia3d$ KIT-6 mesoporous silica templates, 4.0 g Pluronic P123 ($EO_{20}PO_{70}EO_{20}$) was dissolved in the mixture of 144 g distilled water and 7.9 g 35 wt% HCl solution at 35°C [59]. After constant stirring to be completely dissolved, 4.0 g BuOH was added, with further stirring for 1 hour. Then, 8.6 g tetraethoxysilane (TEOS) was added. The mixture was vigorously stirred at 35°C for 24 hours, followed by hydrothermal treatment at 100°C for 24 hours. The mesoporous product was collected through filtering without washing and the dried at 100°C for 24 hours. To remove the surfactant, the samples were washed gently with ethanol/HCl, followed by calcination at 550°C in air for 5 hours. The sample was denoted as KIT-6-100, with 100 to represent the hydrothermal temperature. The hydrothermal treatment was conducted in a temperature range of 40–135°C. 2D hexagonal SBA-15 mesoporous silica templates were prepared, combined with hydrothermal treatment 100°C, leading to sample SBA-15−100 [60].

Mesoporous Co_3O_4 samples were obtained by using the calcined 3D $Ia3d$ cubic KIT-6 and 2D hexagonal SBA-15 as templates [31]. For instance, 0.3 g KIT-6-100 was dispersed in 10 mL solution formed with 0.6 g $Co(NO_3)_2 \cdot 6H_2O$ in ethanol, with constant stirring for 12 hours at room temperature. Then, the mixture was heated at 40°C overnight to remove ethanol, resulting in a dried sample. After grinding for 10 minutes, the powder was calcined at 200°C in air for 5 hours. The silica templates were etched out by soaking in 5% HF aqueous solution at room temperature. Black Co_3O_4 powders were collected through centrifugation, followed by drying at 80°C, forming sample Co_3O_4-KIT-6-100-200. The calcination was carried out over a temperature range of 200–700°C. According to the pore volume of the corresponding KIT-6, for KIT-6-40, KIT-6-60, KIT-6-70 and KIT-6-80, 0.93, 0.56, 0.48 and 0.41 g KIT-6 templates were used, respectively. For templates KIT-6-100, KIT-6-120 and KIT-6-135, 0.3 g KIT-6 was used. Co_3O_4-SBA-15 was prepared in a similar way to that of Co_3O_4-KIT-6. Normal Co_3O_4 powder was synthesized by calcining $Co(NO_3)_2 \cdot 6H_2O$ at 500°C with the presence of any template.

Figure 5.13 shows wide-angle and small angle XRD patterns of Co_3O_4-KIT-6-100-200. The wide-angle pattern indicated the formation of cubic Co_3O_4 after calcination at 200°C, while the low-angle pattern confirmed the bulk structural ordering characteristics. Figure 5.14 shows SEM and TEM images of Co_3O_4-KIT-6-100-200, Co_3O_4-KIT-6-100-300 and Co_3O_4-KIT-6-100-500. As demonstrated in Fig. 5.14(a, b), the particle size was not influenced by the calcination temperature. The Co_3O_4 nanoparticles had diameters in the range of 50−200 nm, while the KIT-6 particles were at micrometer scale. All the particles exhibited an obvious mesoporous structure, as seen in Fig. 5.14(c, g, h). As revealed in Fig. 5.14(e), the Co_3O_4 nanoparticles possessed $Ia3d$ cubic mesostructure, following the mesostructure of the KIT-6 template. Figure 5.14(f) shows a high-resolution TEM image, demonstrating the crystalline nature of the walls of Co_3O_4-KIT-6-100-200. Representative SAED patterns are shown in Fig. 5.14(d, i), confirming the mesoporous nanoparticles possessed well crystallized walls.

As ethanol was removed, most of the nitrates were present on the outside surface of the silica template particles. Note that $Co(NO_3)_2 \cdot 6H_2O$ has a melting point of about 55°C and decomposes at about 74°C, before decomposition it was melted into a liquid phase, so as to fill in the mesopores of the KIT-6 template. Therefore, decomposition of the nitrate and crystallization of the Co_3O_4 both occurred inside the mesopores [61]. Finally, mesoporous Co_3O_4 nanoparticles were developed, once the silica template was removed.

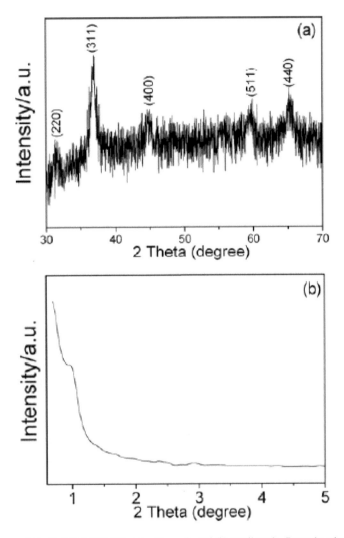

Fig. 5.13. XRD patterns of Co$_3$O$_4$-KIT-6-100-200: (a) wide-angle and (b) small-angle. Reproduced with permission from [31], Copyright © 2009, American Chemical Society.

Figure 5.15 shows N$_2$ adsorption–desorption isotherms and BJH pore size distribution profiles of the samples, which were derived with KIT-6-100 as the template and calcined at different temperatures [31]. The samples exhibited a typical sorption isotherm with a broad capillary condensation range from $p/p_0 = 0.5$ to $p/p_0 = 1$, implying the presence of a large textural porosity. The narrow pore size distribution at about 5 nm was attributed to the mesopores, which were formed after the pore walls of the SiO$_2$ template were removed. The size of the mesopores was slightly decreased with increasing calcination temperature, due to the gradually increased shrinkage of SiO$_2$ template. Another broad pore distribution was in the range of 15−50 nm, owing to the stacking of the nanoparticles. However, according to BET surface area and pore volume, textural parameters of mesoporous Co$_3$O$_4$ were not significantly affected by the calcination temperature. This implies that the pore structures of the samples prepared with hard-templates are stable at temperatures of up to 700°C.

Figure 5.16(a) shows CV curves of the Co$_3$O$_4$-KIT-6-100-200 based electrode [31]. The shape of the CV varied, as the scan rate was increased, which suggested that the measured capacitances were attributed to the redox reactions. Figure 5.16(b) shows galvanostatic charge–discharge curves, of the

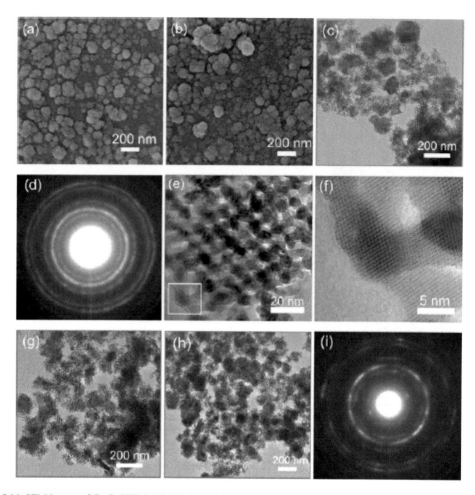

Fig. 5.14. SEM images of Co$_3$O$_4$-KIT-6-100-200 (a) and Co$_3$O$_4$-KIT-6-100-500 (b). TEM images of Co$_3$O$_4$-KIT-6-100-200 (c, e, f), Co$_3$O$_4$-KIT-6-100-300 (g) and Co$_3$O$_4$-KIT-6-100-500 (h). SAED images of Co$_3$O$_4$-KIT-6-100-200 (d) and Co$_3$O$_4$-KIT-6-100-500 (i). The square area in panel (e) corresponds to panel (f). Reproduced with permission from [31], Copyright © 2009, American Chemical Society.

Co$_3$O$_4$-KIT-6 samples synthesized at different temperatures, together with normal Co$_3$O$_4$, over 0–0.45 V, at the current of 10 mA after 50 cycles. The specific capacitance was slightly decreased with increasing calcination temperature, which was readily attributed to the decreased reactive activity of the samples. In addition, only about 1% loss in specific capacitance was observed after 300 cycles, implying a highly stable electrochemical activity of the electrode materials. Moreover, specific capacitance of the Co$_3$O$_4$-KIT-6-100-500 was larger than that of normal Co$_3$O$_4$ by almost eight times.

Mesoporous structures of the Co$_3$O$_4$ nanoparticles could also be regulated by using KIT-6 templates synthesized at different hydrothermal treatment temperatures. It is found that Co$_3$O$_4$-KIT-6-40-200 possessed the highest mesoporosity, which had a relatively broad pore size distribution. Comparatively, Co$_3$O$_4$-KIT-6-100-200, Co$_3$O$_4$-KIT-6-120-200, and Co$_3$O$_4$-KIT-6-135-200 exhibited a narrower pore size distribution, due to the presence of more ordered mesopores. BET surface areas of the mesoporous Co$_3$O$_4$ nanoparticles were increased with decreasing pore volume of the corresponding KIT-6 templates. This suggested that the BET surface area could be well controlled through the utilization of the templates, thus leading to electrodes with controllable electrochemical performances.

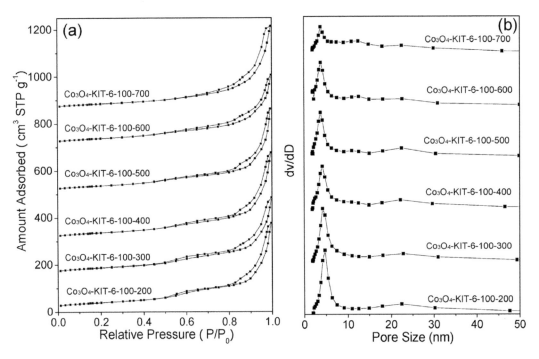

Fig. 5.15. N$_2$ adsorption–desorption isotherms (a) and BJH pore size distributions from adsorption branch (b) for the mesoporous Co$_3$O$_4$ samples prepared at different calcination temperatures with KIT-6-100 as the template. Reproduced with permission from [31], Copyright © 2009, American Chemical Society.

Hydrothermal/solvothermal method

Hydrothermal and solvothermal methods are meant to synthesize inorganic materials from aqueous and organic solutions, respectively, at both high vapor pressures and temperatures [62–77]. Hydrothermal/solvothermal methods allow in obtaining crystalline items, thus effectively avoiding the post-annealing/calcination step, which is usually required by precipitation and sol-gel methods. Furthermore, there are a plenty of parameters, such as starting materials, solvent, additives, minerals, precursor concentration, reaction temperature/time, which can be used to modify crystallization, morphology and dimension of the final products. Also, they have advantages, such as high reaction yield, low costs of facility and raw materials and high energy efficiency. Both powder or thin-film based electrodes have been developed by using hydrothermal or solvothermal methods.

A facile hydrothermal treatment induced homogeneous precipitation process was developed to obtain ultralayered Co$_3$O$_4$ structures with high porosity [70]. The Co$_3$O$_4$ structures were constructed with well-arranged micrometer dimensioned rectangular 2D flakes, which possessed high specific surface area, high porosity and relatively narrow pore size distribution. The ultralayered Co$_3$O$_4$ structures exhibited promising electrochemical performances, which could be readily attributed to the highly reversible redox reactions. According to the charge–discharge behavior, it was found that the materials demonstrated a specific capacitance of 548 F·g^{-1}, at the current density of 8 A·g^{-1}, which was retained by 66% at 32 A·g^{-1}. Electrode based on the ultralayered Co$_3$O$_4$ structures showed a highly stable specific capacitance retention capability, which was maintained to be 98.5% after 2000 charge–discharge cycles, at a current density of as high as 16 A·g^{-1}. The outstanding cyclic, structural and electrochemical stabilities at high current rates, along with nearly 100% Coulombic efficiency and very low ESR value, made the ultralayered Co$_3$O$_4$ structures to be a promising candidate to fabricate high-performance supercapacitors.

The synthesis of the ultralayered Co$_3$O$_4$ structures, Co(NO$_3$)$_2$·6H$_2$O, Triton X-100 and urea were used as the starting materials. 100 mL aqueous cobalt salt (20 mmol) solution was added dropwise to 100 mL

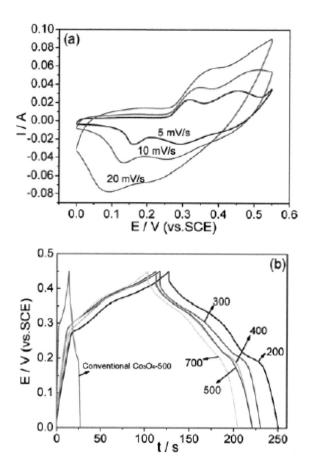

Fig. 5.16. (a) CV curves of Co$_3$O$_4$-KIT-6-100-200 at different scan rates. (b) Charge−discharge curves at 10 mA of the mesoporous Co$_3$O$_4$ samples obtained at different calcination temperatures with KIT-6-100 as the template. The Arabic numerals within panel (b) are the calcination temperatures of the samples. Reproduced with permission from [31], Copyright © 2009, American Chemical Society.

Triton X-100 (10 mmol) aqueous solution, followed by constant stirring for 1 hour. After that, 40 mmol urea was added, followed by continuous stirring for another 3 hours. The resultant solution was then hydrothermally treated for 24 hours at 120°C. After the hydrothermal reaction, a light pink solid precipitate was formed, which was collected through centrifugation at 3000 rpm, followed by thorough washing with water, ethanol–water mixture and pure ethanol successively. The sample was dried at 60°C in vacuum and then was calcined at 300°C in air at a heating rate of 1°C·min^{-1}.

Figure 5.17(a) shows wide-angel XRD patterns of the uncalcined and calcined samples [70]. It was observed that the precipitate was monoclinic cobalt hydroxide carbonate, i.e., (Co$_2$(OH)$_2$CO$_3$). In aqueous solution, urea decomposed and then was hydrolyzed at high temperatures, so that OH$^-$ and CO$_3^{2-}$ were released, which trigger the precipitation of Co^{2+} ions in the solution, thus forming the cobalt hydroxide carbonate precipitate. The hydrothermal treatment facilitated a better control over the structure and promoted the crystallization process. After calcination, nanostructured Co$_3$O$_4$ phase was obtained, as shown in Fig. 5.17(a).

Figure 5.17(b) shows N$_2$ adsorption–desorption isotherm of the Co$_3$O$_4$ powder measured at 77 K, demonstrating a rapid uptake in the p/p_0 region of 0–0.04, with characteristics of the 'type IV' hysteresis in the mesoporous region. It was suggested that the Co$_3$O$_4$ sample had mesopores as the majority pore, with a minor fraction of micropores, as illustrated by the insets in Fig. 5.17(b). The narrow and single pore size

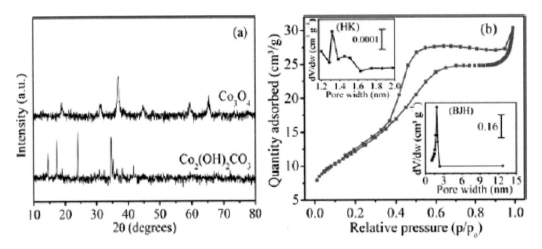

Fig. 5.17. (a) XRD patterns of uncalcined precursor (Co$_2$(OH)$_2$CO$_3$) and the product (Co$_3$O$_4$) after calcination at 300°C. (b) N$_2$ adsorption–desorption isotherm of the Co$_3$O$_4$ sample, with the two insets showing Horvath–Kawazoe (HK) and Barrett–Joyner–Halenda (BJH) pore size distributions of Co$_3$O$_4$, respectively. Reproduced with permission from [70], Copyright © 2011, American Chemical Society.

distribution at about 1.85 nm was similar to those of the nanomaterials synthesized with templates. The sample possessed BET specific surface area and pore volume to be 97 m$^2 \cdot$g^{-1} and 0.2 cm$^3 \cdot$g^{-1}, respectively. These microstructural properties made the ultralayered Co$_3$O$_4$ structures especially suitable as electrodes of supercapacitor [78].

Figure 5.18 shows SEM images of the cobalt–hydroxide–carbonate precipitate [70]. The sample possessed a layered structure formed with highly oriented 2D microsheets and a relatively low porosity. Figure 5.19 shows SEM images the Co$_3$O$_4$ sample derived from the precipitate after calcination. Moreover, higher magnification image revealed that the sample possessed a long-range structure, with very uniform micrometer length and interlayer spacing, as seen in Fig. 5.18(b). It was found that the layered structure of the precipitate was well retained in the oxide sample, suggesting a high feasibility of the method. Due to the formation and release of CO$_2$ and H$_2$O molecules during the decomposition process, additional porosity was formed in the Co$_3$O$_4$ product.

The formation of Co$_3$O$_4$ architectures could be well controlled by optimizing various experimental parameters, including concentration of metal salt solution, urea, reaction temperature and reaction time, due to their effects on the hydrolysis rate, nucleation and crystal growth behaviors [79]. Crystal growth of nanomaterials dependent on various factors, such as selective adsorption of solvents on specific crystallographic planes, application of inorganic additives and templating effect of surfactants [80]. Among these, through selective adsorption on certain crystal planes, their growth is thus retarded or inhibited, thus resulting in materials with special morphologies. Triton X-100 is a neutral surfactant, which acted as the template to develop the layered structure during the hydrothermal treatment of the sample. In this case, it had two functions, i.e., (i) selective adsorption and (ii) prevention of agglomeration of the 2D sheets. The 2D morphology formation mechanism with the presence of Triton X-100 could be related to the kinetic growth of layered structures, dependent on various factors, such as van der Waals forces, hydrophobic interactions, crystal face attraction, electrostatic and dipolar fields, hydrogen bonding, intrinsic crystal contraction and Ostwald ripening [81]. Accordingly, a possible mechanism has been proposed to describe the evolution of the layered Co$_3$O$_4$ structures, as schematically shown in Fig. 5.20 [70].

Mesoporous Co$_3$O$_4$ nanostructures, hexagonal nanosheets and microspheres, were prepared through calcination of β-Co(OH)$_2$ nanoparticles, which were obtained by using self-assembly method from nanosheets [65]. The β-Co(OH)$_2$ nanostructures could be controlled by tuning the volume ratio of water to ethanolamine (EA). BET specific surface area and micropore area of the Co$_3$O$_4$ nanosheets were 25.12 and 2.80 m$^2 \cdot$g^{-1}, while those of the Co$_3$O$_4$ microspheres were 21.53 and 1.29 m$^2 \cdot$g^{-1}, respectively. Average

294 *Nanomaterials for Supercapacitors*

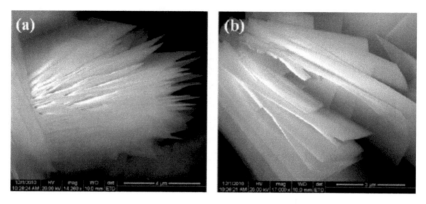

Fig. 5.18. (a) Low and (b) high magnification SEM images of the precursor $Co_2(OH)_2CO_3$. Reproduced with permission from [70], Copyright © 2011, American Chemical Society.

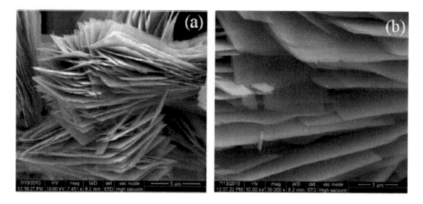

Fig. 5.19. (a) Low and (b) high magnification SEM images of the product Co_3O_4. Reproduced with permission from [70], Copyright © 2011, American Chemical Society.

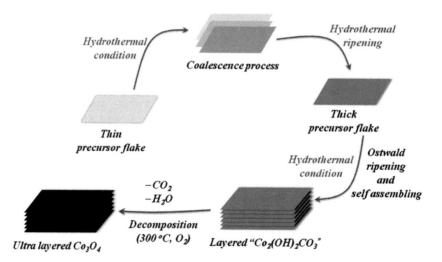

Fig. 5.20. Schematic diagram showing possible formation mechanism of the ultralayered Co_3O_4. Reproduced with permission from [70], Copyright © 2011, American Chemical Society.

pore diameters of the Co$_3$O$_4$ nanosheets and microspheres were 26.7 and 23.5 nm, respectively. The Co$_3$O$_4$ nanosheets exhibited specific capacitances of 92, 91, 90 and 85 F·g^{-1}, at current densities of 5, 10, 15 and 20 mA·cm^{-2}.

To synthesize β-Co(OH)$_2$ nanostructures, 2 mmol Co(Ac)$_2$·4H$_2$O was dissolved in distilled water to form a homogeneous solution. After that, a certain amount of EA was added with constant stirring for 10 minutes. Then, the mixture was subjected to hydrothermal reaction at 120–200°C for 20 hours. The products were collected and thoroughly washed with distilled water and absolute alcohol, followed by vacuum-drying. To obtain mesoporous Co$_3$O$_4$ nanostructures, the β-Co(OH)$_2$ nanostructured precursors were calcined at 400°C. Phase compositions of the samples before and after calcination were confirmed by using XRD results.

It was found that the volume ratio of water and EA was a determining factor for the formation of the β-Co(OH)$_2$ nanostructures with different morphologies. Figure 5.21 shows representative SEM images of the β-Co(OH)$_2$ nanostructure samples synthesized under different conditions [65]. As shown in Fig. 5.21(a, b), the sample prepared with the H$_2$O/EA volume ratio to be 20:20 at 160°C for 20 hours demonstrated monodisperse flower-like nanostructures, which exhibited a relatively narrow diameter

Fig. 5.21. SEM images of the β-Co(OH)$_2$ powders prepared under different conditions: (a, b) H$_2$O/EA (20:20 v/v) at 160°C for 20 hours, (c, d) H$_2$O/EA (30:10 v/v) at 120°C for 20 hours and (e, f) H$_2$O/EA (30:10 v/v) at 200°C for 20 hours. Reproduced with permission from [65], Copyright © 2009, Wiley-VCH Verlag GmbH & Co. KGaA, Weinheim.

distribution in the range of 4–5 μm. High-magnification SEM images indicated that the nanospheres consisted of nanosheets, as seen in Fig. 5.21(b).

Meanwhile, the hydrothermal reaction temperature also exerted a strong effect on morphology of the nanoparticles. For instance, with fixed volume ratio of H_2O/EA to 30:10, the hydrothermal reaction was conducted at 120°C or 200°C, shapes of the nanoparticles were largely different from that of the sample obtained at 160°C. As shown in Fig. 5.21(c, d), after reaction at 120°C for 20 hours, the sample contained copper-coin-like nanosheets, with diameters in the range of 4–5 nm. In comparison, as the reaction temperature was increased to 200°C, β-Co (OH)$_2$ microspheres were developed. Obviously, the β-Co (OH)$_2$ microspheres were assembled from nanosheets, as revealed in Fig. 5.21(e, f).

Figure 5.22 shows representative SEM images of the Co_3O_4 nanosheets and microspheres derived from the precursors after calcination at 160°C [65]. It was observed that both the Co_3O_4 original nanosheets and microspheres of the precursor were well retained after calcination. As seen in Fig. 5.22(c), a hexagonal sheet with a width of about 7 μm was present. High-magnification SEM image revealed that a large number of cracks and pores were observed on surface of the nanosheets, as illustrated in Fig. 5.22(d). The porous characteristics were confirmed by high-magnification TEM image, with an average pore diameter of about 20 nm. HRTEM image presented fringe spacing of about 0.46 nm, corresponding to the spacing value of the (111) plane of cubic Co_3O_4, consistent with XRD results.

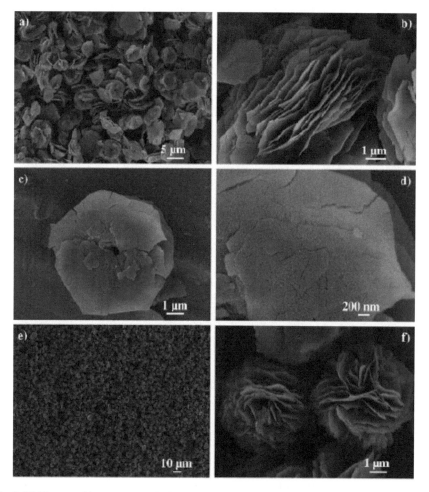

Fig. 5.22. (a–d) SEM images of the mesoporous Co_3O_4 nanosheets, (e, f) SEM images of the mesoporous Co_3O_4 microspheres. Reproduced with permission from [65], Copyright © 2009, Wiley-VCH Verlag GmbH & Co. KGaA, Weinheim.

A hydrothermal method was reported to synthesize nanoporous Co_3O_4 nanorods, consisting of textured aggregations of nanocrystals, which exhibited a specific capacitance of 280 $F \cdot g^{-1}$ [66]. To synthesize the Co_3O_4 nanorods, 1.2 g $CoCl_2$ and 0.06 g $CO(NH_2)_2$ (urea) were dissolved in 20 mL distilled water to form their respective homogeneous solutions. The urea solution was then added dropwise to the $CoCl_2$ solution with constant stirring. The mixture was then hydrothermally treated at 105°C for 6 hours, leading to a pink precipitate, which was collected through centrifugation, thorough washing with distilled water and ethanol and drying in vacuum. The dried precursor was calcined at 300°C for 3 hours to obtain black Co_3O_4 powders.

Figure 5.23 shows XRD pattern of the Co_3O_4 nanorods, confirming the presence of cubic spinel phase [66]. Figure 5.24(a) shows low magnification bright-field TEM image of the Co_3O_4 nanorods, which possessed lengths at the scale of micrometers. Figure 5.24(b) shows high magnification TEM image of a single Co_3O_4 nanorod, indicating its diameter to be about 400 nm. The Co_3O_4 nanorod was highly polycrystalline, with a nanoporous structure. The formation of the nanoporous Co_3O_4 nanorods involved two steps. Firstly, intermediate products, i.e., $Co^{II}(OH)_a(CO_3)_bCl_{(2-a-2b)} \cdot nH_2O$ nanorods, were developed during the hydrothermal reaction process. Secondly, the precursors were converted to porous Co_3O_4 nanorods through a post-calcination process.

The crystal structure of $Co^{II}(OH)_a(CO_3)_bCl_{(2-a-2b)} \cdot nH_2O$ contained Co–OH layers and counteranions in between the Co–OH layers, which were converted into Co_3O_4 nanorods, through dehydration and pyrolysis of the counteranions, accompanied by the generation of gases, e.g., CO_2 and HCl [82]. The nanopores were originated from spaces of the –OH and counterions. Selected Area Electron Diffraction (SAED) pattern is shown as the inset in Fig. 5.24(b), confirming the formation of cubic Co_3O_4 crystal. Figure 5.24(c) shows a high resolution TEM image of part of an individual Co_3O_4 nanorod, demonstrating a shape resembling elongated ellipses, with sizes of 20–30 nm. There were pores with sizes at nanometer scale.

The SAED shown in Fig. 5.24(c) was on the region along the [001] zone axis, resulting in a spot SAED pattern, which is shown as the inset in Fig. 5.24(c), revealing that the Co_3O_4 nanocrystals were grown along one direction and thus aggregated into the porous structure. Figure 5.24(d) shows a HRTEM image of part of an individual Co_3O_4 nanocrystal, with lattice resolution. An interplanar spacing of 0.285 nm was identified, corresponding to the (220) crystal planes of cubic Co_3O_4. The Co_3O_4 nanorods possessed a BET specific surface area of 232 $m^2 \cdot g^{-1}$.

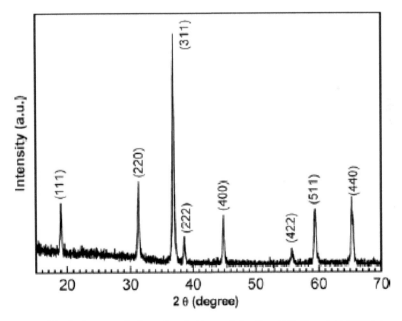

Fig. 5.23. XRD pattern of the Co_3O_4 nanorods. Reproduced with permission from [66], Copyright © 2009, American Chemical Society.

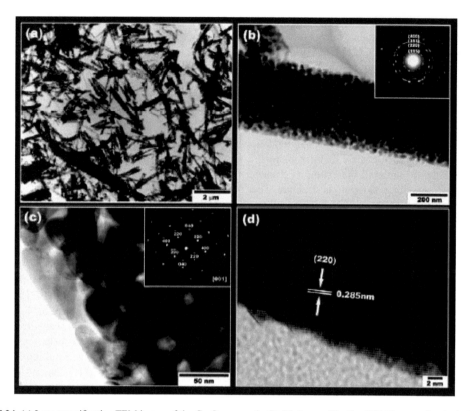

Fig. 5.24. (a) Low magnification TEM image of the Co_3O_4 nanorods. (b) High magnification TEM image of a single Co_3O_4 nanorod, showing the porous structure, with the inset show the corresponding electron diffraction pattern. (c) HRTEM image of part of a single Co_3O_4 nanorod, demonstrating crystal size and pore size of the sample, with inset showing the selected area electron diffraction pattern obtained by converge beam. (d) Lattice resolved HRTEM image of a single Co_3O_4 nanocrystal. Reproduced with permission from [66], Copyright © 2009, American Chemical Society.

A facile hydrothermal method has been reported to synthesize porous Co_3O_4 nanowires with large aspect ratio, derived from precursors—long $Co(CO_3)_{0.5}(OH) \cdot 0.11H_2O$ nanowires [72]. It was found that the amount of urea added in the reaction system played an important role in determining morphology of the precursor cobalt-carbonate-hydroxide, i.e., the concentration of urea influenced both uniformity and overall structure of the precursors. The calcination had almost no effect on morphology of the precursors when they were converted to Co_3O_4 nanowires. The derived Co_3O_4 nanostructures exhibited high performances, when used as electrodes of supercapacitors, with a specific capacitance of 240 $F \cdot g^{-1}$ after 2000 charge/discharge cycles, which corresponded to 98% retention of the initial capacitance.

To synthesize the precursor $Co(CO_3)_{0.5}(OH) \cdot 0.11H_2O$ nanowires, 0.56 g $CoSO_4 \cdot 7H_2O$ was dissolved in 40 mL mixture, consisting of 7 mL glycerol and 33 mL deionized water, which led to a transparent solution after stirring for about 10 minutes. After that, urea with amount of 0.05–0.2 g was added into the solution, following by stirring for 30 minutes. Then, the solution was hydrothermally treated at 170°C for 24 hours. After the reaction was finished, the precipitate was collected and thoroughly washed with deionized water and ethanol through centrifugation, followed by drying at 60°C. The dried $Co(CO_3)_{0.5}(OH) \cdot 0.11H_2O$ precursors were calcined at 400°C for 2 hours at a heating rate of $0.5°C \cdot min^{-1}$ obtain Co_3O_4.

Figure 5.25 shows representative SEM and TEM images of the $Co(CO_3)_{0.5}(OH) \cdot 0.11H_2O$ nanowires [72]. As revealed in Fig. 5.25(a), the nanowires were uniformly distributed and highly dispersed, with lengths of 10s of micrometres. High magnification image indicated that the nanowires possessed a uniform width in the axial direction, with an average diameter of several 10s of nanometers, as seen in Fig. 5.25(b). The nanowires were highly flexible, so that they could be easily bended, due to their pretty large aspect ratio.

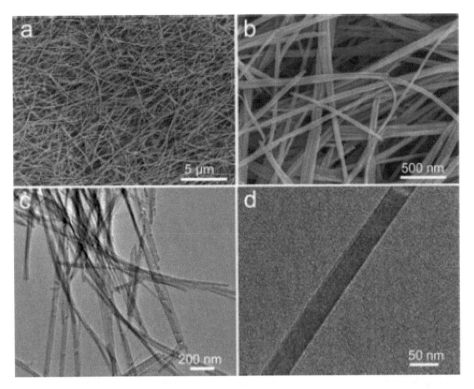

Fig. 5.25. SEM (a, b) and TEM (c, d) images of long Co(CO$_3$)$_{0.5}$(OH)·0.11H$_2$O precursor nanowires obtained with 0.1 g urea used during the hydrothermal reaction. Reproduced with permission from [72], Copyright © 2012, Royal Society of Chemistry.

The well-developed 1D microstructures, with a smooth surface, were also confirmed by the TEM results, as demonstrated in Fig. 5.25(c, d). During the chemical reactions under the hydrothermal conditions, both OH$^-$ and CO$_3^{2-}$ ions were slowly generated, due to the hydrolysis of urea, which facilitated the growth of the product in such a way that it was favorable in one specific direction, i.e., the longitudinal axis, so that nanowires were obtained.

XRD characterization results confirmed the crystal structure of orthorhombic Co(CO$_3$)$_{0.5}$(OH)·0.11H$_2$O, which was not only phase pure but also well-crystallized. The conversion of the precursor to Co$_3$O$_4$ was also evidenced by the XRD pattern. According to the results of N$_2$ adsorption–desorption isotherm, the Co$_3$O$_4$ nanowire samples exhibited a surface area of about 88 m^2·g^{-1} and a total pore volume of 0.62 cm^3·g^{-1}.

The effect of the concentration on morphology of the precursor samples was clarified by SEM results. For instance, if only 0.05 g urea was used, the hydrothermal product was not well dispersed and bundles were present, although a wire-like morphology was still observed. At the same time, the 1D nanostructures were much less uniform as compared with the sample discussed above. As the amount of urea was increased to 0.2 g, product with sphere-like morphology with a diameter of about 5 μm was developed. The microspheres were constructed with fine wire-like subunits pointing outward radially. In this case, the carbonate ions generated during the hydrolysis not only acted as a precipitation agent to facilitate anisotropic growth of the nanowires, but also served as coordinating agents to form more complicated hierarchical structure at high concentrations. Meanwhile, a control experiment confirmed the positive effect of glycerol in the formation of the nanowires.

Figure 5.26 shows SEM and TEM images of calcined product [72]. The obtained Co$_3$O$_4$ possessed a well-dispersed wire-like structure, as illustrated in Fig. 5.26(a), which suggested that morphology of the precursor was well retained after the calcination step. As observed in Fig. 5.26(b), high magnification image indicated that broken wires were occasionally present. The observations were confirmed by the

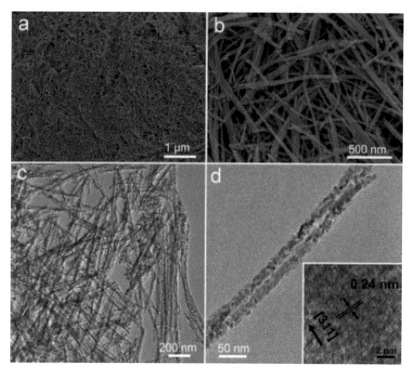

Fig. 5.26. SEM (a, b) and TEM (c, d) images of the Co$_3$O$_4$ nanowires obtained after calcination, with the inset in (d) showing a high-resolution TEM image of the sample. Reproduced with permission from [72], Copyright © 2012, Royal Society of Chemistry.

TEM analysis results, as seen in Fig. 5.26(c). As demonstrated in Fig. 5.26(d), twin nanowires with high porosity but consisting of densely packed nanoparticles were observed. Visible lattice fringes could be identified, with an inter-planar distance of ~ 0.24 nm, corresponding to the (311) plane of Co$_3$O$_4$, as depicted as the inset in Fig. 5.26(d).

An interesting method was reported to synthesize urchin-like Co$_3$O$_4$ microspherical hierarchical superstructures that were assembled from 1D Co$_3$O$_4$ nanowires, derived from precipitates of hydroxides, through calcination process [71]. Electrode based on the urchin-like Co$_3$O$_4$ superstructures exhibited promising electrochemical performances, with a specific capacitance (SC) of 614 F·g^{-1}, at a current density of 1 A·g^{-1}, which retained to be 536 F·g^{-1} at a high current density of 4 A·g^{-1} in 3 M KOH aqueous solution. Such special urchin-like hierarchical superstructures allowed for the electrolyte ions and electrons to transport the electroactive Co$_3$O$_4$ nanowires at high rates, so as to ensure sufficient Faradaic reactions for high power density energy storage. In addition, the specific capacitance was retained by 77%, after 5,000 continuous charge–discharge cycles, at a current density of 4 A·g^{-1}, confirming their excellent high rate electrochemical stability.

Commercially available Lysine and CoCl$_2$·6H$_2$O were used as the starting materials. All aqueous solutions were prepared by using high-purity water (18 MΩ cm resistance). Two g lysine and 1.3658 g CoCl$_2$·6H$_2$O were dissolved in 60 mL water, to form solution for hydrothermal reaction, which was carried out at 100°C for 24 hours. After the reaction was finished, a pink color product was obtained, which was collected through filtering, followed by thorough washing with water and drying at 100°C in vacuum. The dried sample was calcined at 250°C for 3 hours to obtain a black Co$_3$O$_4$ superstructure powder. The formation of Co$_3$O$_4$ cubic phase was confirmed by XRD results.

Figure 5.27 shows SEM images of the Co$_3$O$_4$ powder [71]. As seen in Fig. 5.27(a), the Co$_3$O$_4$ powder contained microspherical particles, with diameters of 1–2 μm. High magnification SEM image indicated that surfaces of the spherical structures were attached with a large number of nanowires, resulting in an

Fig. 5.27. SEM images of the as-synthesized Co_3O_4 sample. Reproduced with permission from [71], Copyright © 2011, Royal Society of Chemistry.

urchin-like appearance, as illustrated in Fig. 5.27(b). The formation of the ordered urchin-like superstructure was attributed to the synergistic effect of the lysine and the Cl⁻ ions, during the hydrothermal reaction. The urchin-like morphology of the precursor was well retained after calcination at 250°C for 3 hours in order to obtain Co_3O_4. Additionally, the Co_3O_4 nanowires led to abundant 'V-type' porous channels, in the radial direction, so that the electrolyte ions could easily penetrate the superstructures to contact the electroactive surface of the Co_3O_4 building blocks. As a result, energy storage capability was readily achieved. A schematic diagram of the special microstructure of the Co_3O_4 superstructures is shown in Fig. 5.28.

Figure 5.29 shows HRTEM images of the urchin-like Co_3O_4 microspherical hierarchical superstructures, in order to more clearly demonstrate their microstructures [71]. It was found that almost all the nanowires were oriented and assembled along the radial direction, pointing from the center to the surface of microspherical superstructures, as seen from Fig. 5.29(a–c). As a result, the superstructures possessed an urchin-like morphology, as demonstrated in Fig. 5.28(a). The individual Co_3O_4 nanowires consisted of numerous Co_3O_4 nanoparticles, with sizes in the range of 10–50 nm, as revealed in Fig. 5.29(b, c). The superstructures were so stable that they even could be damaged and broken into Co_3O_4 nanowires or nanoparticles when subject to long-time ultrasonication, implying that the special superstructures were well assembled from the Co_3O_4 nanosized building blocks. Figure 5.29(d) clearly shows lattice image of the Co_3O_4 nanoparticles, confirming well-crystallized cubic Co_3O_4 phase in the microspherical superstructures. The high crystallinity was also verified by the SAED pattern, as shows as the inset in Fig. 5.29(d).

3D hierarchical Co_3O_4 twin-spheres, with an urchin-like structure, which could be produced at a large scale and exhibited promising electrochemical performances, were synthesized, by using a solvothermal method [74]. The 3D nanoarchitectures were derived from a precursor, cobalt carbonate hydroxide hydrate, with a chemical composition of $Co(CO_3)_{0.5}(OH) \cdot 0.11H_2O$ and a 3D hierarchical twin-spherical structure. The twin-spheres were assembled from 1D Co_3O_4 nanochains with a granular structure, products of oxidation and decomposition of single-crystalline nanoneedles in the precursor. Moreover, according to the effect of reaction time on morphology of the precursor, a multistep-splitting growth mechanism was proposed to describe the formation of the 3D hierarchical nanostructures. Electrodes based on the materials possessed specific capacitances of 781, 754, 700, 670 and 611 F·g⁻¹ at current densities of 0.5, 1, 2, 4 and 8 A·g⁻¹, respectively. A high reversibility with an efficiency of 97.8% after cycling 1000 times, at a current density of 4 A·g⁻¹, was achieved. The high electrochemical performances of the electrodes were readily ascribed to the special 3D hierarchical structure of the Co_3O_4 twin-spheres.

For solvothermal reaction, 1.455 g $Co(NO_3)_2 \cdot 6H_2O$ (5.00 mmol) was dissolved in the mixture of 20 mL deionized water and 20 mL ethylene glycol ($C_2H_6O_2$), with the aid of constant stirring, until a transparent solution was obtained. After that, 0.005 g glucose (0.028 mmol) was added into the solution, with constant stirring for 30 minutes, followed by solvothermal treatment at 160°C for 16 hours. The precipitates were collected through centrifugation, followed by thorough washing with deionized water and ethanol and drying at 70°C for 12 hours. The dried powders were calcined at 300°C for 2 hours, thus leading to black powders of Co_3O_4.

302 *Nanomaterials for Supercapacitors*

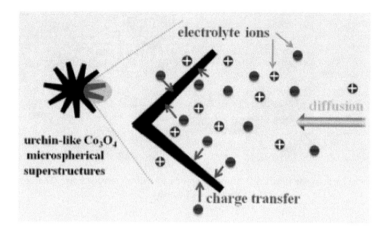

Fig. 5.28. Schematic diagram showing optimized ion diffusion path in the urchin-like Co_3O_4 microspherical hierarchical superstructures. Reproduced with permission from [71], Copyright © 2011, Royal Society of Chemistry.

Fig. 5.29. Bright-field (a, b and c) TEM, HR-lattice (d) TEM images and corresponding SAED (the inset in (d)) of the urchin-like Co_3O_4 microspherical superstructures. Reproduced with permission from [71], Copyright © 2011, Royal Society of Chemistry.

As shown in Fig. 5.30(a), XRD pattern revealed that the red precursor powder was orthorhombic cobalt carbonate hydroxide hydrate, $Co(CO_3)_{0.5}(OH)\cdot0.11H_2O$. SEM and TEM images of the precursor are shown in Fig. 5.30(b) and (c), respectively. The 3D hierarchical twin-spheres with an urchin-like structure had diameters of 10–15 μm. Precursor contained nanoneedles with a smooth surface and a plate-like shape, while their widths were in the range of 5–150 nm. HRTEM image of the segments of sample demonstrated fringes perpendicular to the axis of the nanoneedle, confirming single-crystalline characteristics of the

Oxide Based Supercapacitors: II-Other Oxides 303

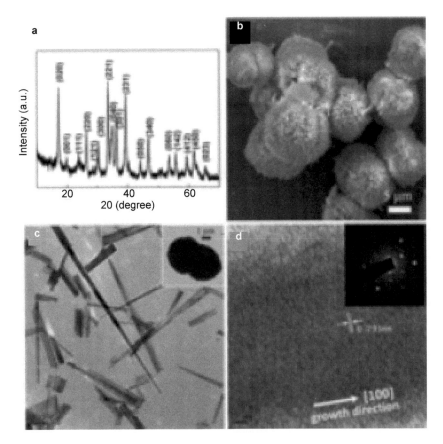

Fig. 5.30. (a) XRD pattern of cobalt carbonate hydroxide hydrate. (b) SEM image of the precursor. (c) TEM image of the nanoneedles disassembled from the precursor caused by ultrasound treatment, with the inset showing a low-magnification TEM image of one twin-sphere of the precursor. (d) HRTEM image and SAED pattern (inset) from one of the precursor nanoneedles. Reproduced with permission from [74], Copyright © 2012, WILEY-VCH Verlag GmbH & Co. KGaA, Weinheim.

nanoneedles in the precursor. The lattice spacing of 0.293 nm was the interlayer spacing of the (300) planes of Co$(CO_3)_{0.5}$(OH)·0.11H$_2$O, as observed in Fig. 5.30(d). It was suggested that the 1D nanoneedles were grown along the [100] direction. Single-crystalline nature of the precursor was also evidenced by the SAED, shown as the insert in Fig. 5.30(d).

Figure 5.30 shows SEM images and XRD pattern of the Co$_3$O$_4$ powder. The inset XRD pattern in Fig. 5.31(a) confirmed single phase cubic Co$_3$O$_4$ of the final powder. Low magnification SEM image revealed that the 3D hierarchical structure of the precursor was well retained after thermal annealing. High-magnification SEM images of the sample indicated that the hierarchical Co$_3$O$_4$ twin-spheres consisted of 1D nanochains with rough surfaces, which grew radically to form urchin-like structures from their centers, as illustrated in Fig. 5.31(b–d).

Figure 5.32(a) shows a TEM image of two Co$_3$O$_4$ twin-spheres at low magnification, while an enlarged TEM image of one sphere is depicted in Fig. 5.32(b), confirming that the 3D hierarchical nanostructures consisted of 1D nanochains formed with nanoparticles having diameters of 5–30 nm. Figure 5.32(c) shows HRTEM image of the sample. The d-spacing of 0.46 nm was ascribed to the (111) planes of cubic Co$_3$O$_4$, with the lattices being oriented randomly in different directions. Therefore, the nanoparticles had a high crystallinity, while the nanochains were polycrystalline. Figure 5.32(d) shows SAED pattern of the twin-spheres, further confirming their polycrystalline nature. The Co$_3$O$_4$ powder contained pores with diameters centered at 6.30 nm, pore volume of 0.19 cm^3·g^{-1} and a BET surface area of 22.99 m^2·g^{-1}.

Fig. 5.31. (a) Low-magnification SEM image of the 3D hierarchical Co_3O_4 twin-spheres after annealing at 300°C for 2 hours, with the inset showing the corresponding XRD pattern of the sample. (b-d) High-magnification SEM images of the 3D hierarchical Co_3O_4 twin-spheres. Reproduced with permission from [74], Copyright © 2012, WILEY-VCH Verlag GmbH & Co. KGaA, Weinheim.

More recently, a one-step low temperature hydrothermal method was developed to synthesize Co_3O_4 microspheres, used as electrodes of supercapacitors [77]. By using the easy oxidation characteristics of cobalt complex ammonia, the Co_3O_4 microspheres could be synthesized at a hydrothermal temperature of as low as 100°C. Without the use of any surfactant, the Co_3O_4 microspheres, with an average diameter of 500 nm and smooth surface, were obtained by using nitrate. It was found that both morphology and size distribution of the Co_3O_4 microspheres were influenced by the hydrothermal temperature and the concentration of nitrate. The Co_3O_4 microspheres possessed promising electrochemical performances, with specific capacitances of 850, 780, 700 and 630 F·g^{-1}, at current densities of 1, 2, 4 and 8 A·g^{-1}, respectively. After 1000 charge-discharge cycles, the Co_3O_4 microsphere based electrodes demonstrated high stability, with an efficiency of 90.8% at the current density of 2 A·g^{-1}.

To synthesize the precursor powder, 10 mM $Co(NO_3)_2$ and 10 mM $NaNO_3$ were dissolved in 20.0 mL deionized water to form homogeneous solution, into which 10 mL $NH_3·H_2O$ was added dropwise, with constant stirring for 30 minutes. After that, 2 mL H_2O_2 was added into the mixture, which was then hydrothermally treated at 100°C for 12 hours. After the reaction was finished, the precipitate was collected and thoroughly washed with deionized water, followed by drying in air at an elevated temperature. Finally, a black powder of Co_3O_4 microspheres was obtained.

As the hydrothermal reaction was conducted at 90°C, both the oxidation of the precursor $Co(NH_3)_6^{2+}$ by H_2O_2 and the driving force of Co_3O_4 recrystallization were largely weakened. Therefore, only intermediate products without regular morphology were formed, as revealed in Fig. 5.33(a). With increasing hydrothermal temperature, the recrystallization process of the products was significantly promoted. For instance, as the temperature was slightly increased to 95°C, more regularly spherical products were obtained, although there were still the intermediate products, as illustrated in Fig. 5.33(b). However, too high temperature would have a negative effect on the formation of Co_3O_4 microspheres. As demonstrated in Fig. 5.33(c), if

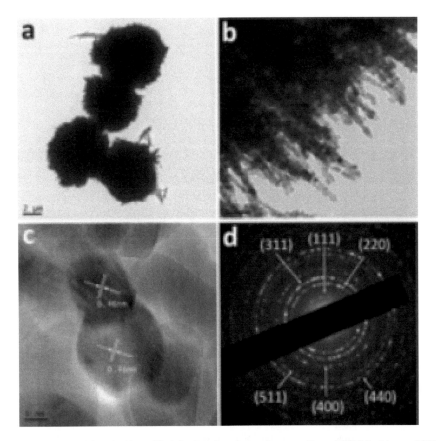

Fig. 5.32. (a, b) TEM images of the 3D hierarchical Co$_3$O$_4$ twin-spheres after annealing at 300°C for 2 hours. HRTEM image (c) and SAED pattern (d) of the Co$_3$O$_4$ nanochains consisting of Co$_3$O$_4$ nanoparticles. Reproduced with permission from [74], Copyright © 2012, WILEY-VCH Verlag GmbH & Co. KGaA, Weinheim.

the temperature was increased to 105°C, the Co$_3$O$_4$ microspheres became slightly irregular, as compared that synthesized at 100°C. Furthermore, the microspheres tended to aggregate, as observed in Fig. 5.33(d), for the sample reacted at 110°C. This observation could be attributed to the over-growth of the nanocrystals at the high temperatures. Therefore, 100°C was the optimal hydrothermal reaction temperature, so as to obtain perfect Co$_3$O$_4$ microspheres.

Due the presence with high concentrations, nitrate should have played a significant role in controlling growth kinetics of the crystals [83]. Also, owing to the long c-axis and charge balance requirement, the nitrate could be inserted into the Co–O bond, so as to wrap up the Co$_3$O$_4$ nanoparticles, thus leading to the formation of microspheres. The synergistic effect of the driving force to reduce surface energy and the presence of nitrate, the Co$_3$O$_4$ nanoparticles tended to aggregate to result in microspheres. To obtain perfect Co$_3$O$_4$ microspheres, 0.625 M was the optimal concentration of NO$_3^-$. As shown in Fig. 5.34(a), at a low concentration of NO$_3^-$ (e.g., 0.5 M), surface of the newly created Co$_3$O$_4$ microspheres could not be fully covered by the NO$_3^-$ ions. As a result, during the growth process, the particles would agglomerate. As the concentration of NO$_3^-$ was slightly increased to 0.575 M, Co$_3$O$_4$ microspheres with a diameter of 500 nm were present, although other smaller spheres were also observed, as demonstrated in Fig. 5.34(b). However, excessive NO$_3^-$ would have a negative effect on the growth behavior of the Co$_3$O$_4$ microspheres. As the concentration of NO$_3^-$ was further increased to 0.7 M, neither the morphology control nor the size limitation was weakened, due to the high concentration of the nitrate ions, as illustrated in Fig. 5.34(c). As seen in Fig. 5.34(d), at an even higher concentration of 0.75 M, inhomogeneous NO$_3^-$ covering had seriously damaged the uniform size distribution of the Co$_3$O$_4$ microspheres.

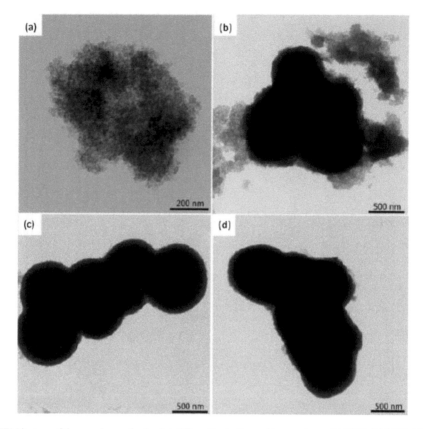

Fig. 5.33. TEM images of the samples synthesized at different hydrothermal temperatures: (a) 90°C, (b) 95°C, (c) 105°C and (d) 110°C. Reproduced with permission from [77], Copyright © 2015, Elsevier.

Sol–gel method

The sol-gel method has also been used to synthesize Co_3O_4 nanopowders for supercapacitor applications [84, 85]. For example, cobalt oxide aerogels were synthesized by using an epoxy-mediated sol–gel procedure, followed by supercritical CO_2 drying and calcination [84]. It was found that grain size of the Co_3O_4 powders was increased from 8 to 12 nm, as the calcination temperature was raised from 200 to 400°C. The sample calcined at 200°C had the highest specific surface area of 235 $m^2 \cdot g^{-1}$. The specific surface area was gradually decreased with increasing calcination temperature. All the aerogel samples possessed type IV isotherms and type H3 hysteresis loops, typical characteristics of mesoporous microstructure. The Co_3O_4 aerogels calcined at 200, 300 and 400°C exhibited specific capacitances of 623, 239 and 174 $F \cdot g^{-1}$, respectively.

The cobalt oxide aerogels were synthesized by using an epoxide addition process, combined with supercritical CO_2 drying. $CoCl_2 \cdot 6H_2O$ and $Co(NO_3)_2 \cdot 6H_2O$ were used as the starting materials. However, gels were only obtained with nitrate. $Co(NO_3)_2 \cdot 6H_2O$ (0.466 g or 1.6 mmol) was dissolved in 3.5 mL methanol, after which propylene oxide (0.93 g or 16 mmol) was added to form a clear solution. After stirring for 10 minutes, the solution was kept at room temperature for gelation, while gel was formed in about 12 hours. The wet gel was thoroughly washed with ethanol to remove the methanol and byproduct, followed by supercritical CO_2 drying, in order to obtain dried aerogel. The dried aerogels were cobalt hydroxide, which were converted to cobalt oxide after calcination at temperatures of $\geq$ 200°C for 5 hours.

Five different polar protic solvents, i.e., water, methanol, ethanol, isopropyl alcohol and acetone, were attempted for the two cobalt salts, i.e., nitrate and chloride. Both cobalt salts had high solubility in

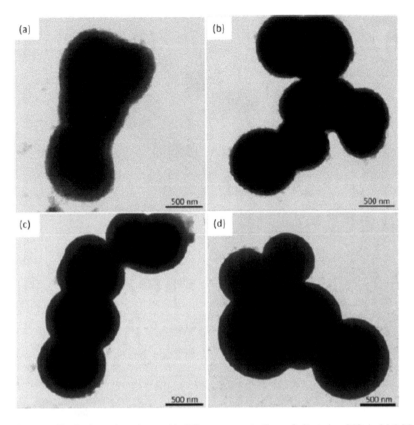

Fig. 5.34. TEM images of hydrothermal products with different concentrations of nitrate ion (NO$_3^-$): (a) 0.5 M, (b) 0.575, (c) 0.7 and (d) 0.75 M. Reproduced with permission from [77], Copyright © 2015, Elsevier.

water and methanol. However, the strong neucleophilicity of water attacked the protonated epoxide ring to generate undesired protons, which reduced pH value of the solution [86]. Also, ring-opening occurred, due to the attacks of chloride ions. Although no protons were produced, pH value of the solution was increased rapidly, as shown in Fig. 5.35(a). This took place due to the consumption of the protonated epoxide and the enhanced proton scavenging of epoxide from the hydrated metal ions, thus leading to the precipitation of nanoparticles. The precipitates of the chloride were unstable gel, while those of the nitrate were very stable gel, as seen in the inset of Fig. 5.35(a). For the other three solvents, i.e., ethanol, isopropyl alcohol and acetone, precipitates were formed, due to the relatively low solubility the precursors in them.

To obtain cobalt oxide powder with high porosity, containing both micropores and mesopores, MCM-41 was used as the template. In this case, 0.2 g MCM-41 powder was dispersed in 2 mL 0.8 M Co(NO$_3$)$_2$·6H$_2$O/EtOH solution, with constant stirring at room temperature for 1 hour. EtOH was then removed by heating the solution at 80°C for 24 hours. After EtOH was completely removed, the sample was calcined at 200°C for 5 hours, which led to the porous cobalt oxide. To remove the MCM-41 template, the powder was immersed in 1 M NaOH for 1 day, followed by thorough rinsing with DI water and drying at 60°C for 1 day.

Figure 5.35(b) shows XRD patterns of the aerogels calcined at 200, 300 and 400°C, denoted as samples A-200, A-300 and A-400, respectively. XRD pattern of the porous cobalt oxide with micropore calcined at 200°C (i.e., PCO-micro) is also included. The as-prepared aerogel was crystalline Co(OH)$_2$. After calcination, all samples were Co$_3$O$_4$. This implied that 200°C was sufficiently high to facilitate the conversion of the aerogel to cobalt oxide. Crystal size of the Co$_3$O$_4$ powders was slightly increased from 8 to 12 nm, as the calcination temperature was raised from 200 to 400°C.

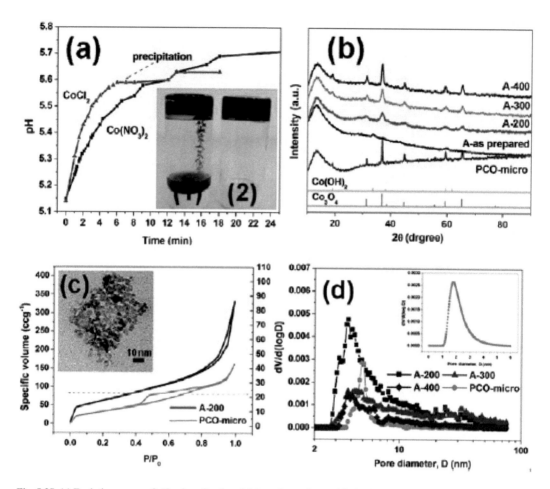

Fig. 5.35. (a) Evolution curves of pH value after the addition of propylene oxide for (1) CoCl$_2$ and (2) Co (NO$_3$)$_2$, with inset showing the falling of precipitates in the case of chloride. (b) XRD patterns of the cobalt oxide aerogel and samples (including A-200, A-300, A-400 and PCO-micro). The broad diffraction peak centered at the 2θ = 13° was contributed to the sample holder. (c) N$_2$ adsorption/desorption isotherms of samples A-200 and PCO-micro, measured at 77 K. The inset shows a representative TEM image of sample A-200. (d) Pore size distributions of samples A-200 (solid square), A-300 (solid triangle), A-400 (solid diamond) and PCO-micro (solid circle). The inset shows the micropore distribution of sample PCO-micro. Reproduced with permission from [84], Copyright © 2009, American Chemical Society.

Figure 5.35(c) shows N$_2$ adsorption/desorption isotherms of samples A-200 and PCO-micro. All the aerogel samples possessed type IV isotherms and type H3 hysteresis loops, which are typical characteristics of mesoporous materials. Because sample PCO-micro contained both micropores and mesopores, the types of isotherm and loop could not be clearly identified. Half of the specific surface area of the sample PCO-micro was due to the micropores. The presence of mesopores in the sample PCO-micro was evidenced by the hysteresis loop in the N$_2$ adsorption/desorption isotherm.

BET specific surface area results indicated that the value of the cobalt oxide was increased from 103 to 235 m^2·g^{-1} after calcination at 200°C, due to the removal of the physisorbed water molecules and the solvent to open the pores. However, further increase in calcination temperature resulted in a decrease in BET specific surface areas, which was attributed to both the densification of the aerogel and the growth of cobalt oxide grains. Pore size distributions of the three aerogel samples are shown in Fig. 5.35(d), with a similar single-modal distribution centered at about 3.8 nm. A representative TEM image of the

200°C-calcined aerogel sample is depicted as the inset in Fig. 5.35(c). As observed in Fig. 5.35(d), the sample PCO-micro possessed both micropores and mesopores, centered at about 4.8 nm and 1.7 nm, respectively.

A sol–gel route that was coupled with a freeze-drying method has been employed to synthesize interconnected macroporous and mesoporous Co_3O_4 nanocrystals, with promising electrochemical performances, when used as electrodes of supercapacitors [85]. The cryogels obtained in this way had several advantages: (i) high porosity caused by the low surface tension during the drying process, (ii) formation of macroporous structure due to the ice-templated effect and (iii) low degree of agglomeration. The Co_3O_4 nanocrystal powder derived from the template of triblock polymer P123 at pH 3 had a BET surface area of 82.8 $m^2 \cdot g^{-1}$. Electrodes based on the Co_3O_4 nanocrystal powder exhibited a specific capacitance of 742 $F \cdot g^{-1}$ over a potential window of 0.5 V. After 2000 cycles of charge-discharge process, the specific capacitance was retained by 86.2%.

Commercially available $Co(NO_3)_2 \cdot 6H_2O$, citric acid, monohydrate ($C_5H_8O_7 \cdot H_2O$) and triblock polymer P123 were used as the starting materials. Three groups of samples were prepared, i.e., Co–P123–PH3, Co–P123, and Co–CA complexes. To synthesize sample Co–P123–PH3, 210 mg citric acid monohydrate (1 mmol) and 232 mg P123 (0.04 mmol) were dissolved in 30 mL of deionized water, with constant stirring for 30 minutes. After that, 1 mmol $Co(NO_3)_2 \cdot 6H_2O$ was added into the solution, while its pH value was adjusted to 3.33, with 30% ammonia solution. The final solution was heated at 70°C in order to obtain gels with a dark red color. The gels were frozen in liquid N_2 and freeze-dried at 0.2 mbar for 1 day. The as-prepared cryogels were calcined to 300°C and for 5 hours.

Sample Co–P123 was prepared in a similar way, where pH value of the solution was not adjusted, after the $Co(NO_3)_2 \cdot 6H_2O$ was added, which was 2.43. The sample Co–CA complex was obtained by mixing 1 mmol citric acid and 1 mmol $Co(NO_3)_2 \cdot 6H_2O$ in 30 mL DI water. The solution was heated at 70°C to form a red gel, without the addition of P123 and pH value adjustment. The gels were then treated in the same way as those of other samples.

The reaction was a modified Pechini-type sol–gel process, which involved one-pot formation of metal complex and *in-situ* polymerization of organic components, as shown in schematically in Fig. 5.36.

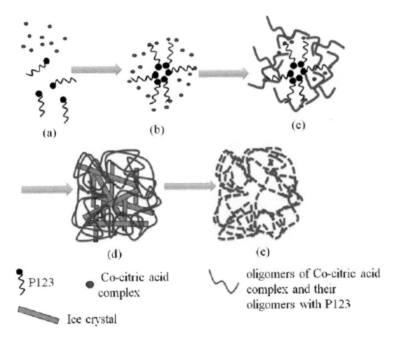

Fig. 5.36. (a) Precursors of P123 and the Co^{2+}–citric acid complex, (b) micelle formation of P123, (c) polymerization of the precursors and gel formation at 70°C, (d) ice crystal formation during freeze with liquid N_2 and (e) hierarchical pore system formation after evaporation of water. Reproduced with permission from [85], Copyright © 2012, American Chemical Society.

Initially, citric acid coordinated with the Co^{2+} ion to form a complex. Due to the evaporation of the solvent and the slow polyesterification process, i.e., citric acid with P123, during heating at 70°C, a transition from solution to sol took place. As the water content was continuously decreased, gel was formed due to the polymerization reaction. After freezing in liquid N$_2$, interconnected ice crystals were formed within the gel matrix. In vacuum, the ice started to unfreeze, thus leading to slow cross-linking within the gel. Eventually, the ice was entirely removed, so that a highly porous cryogel was obtained.

Figure 5.37(a) shows XRD pattern of the Co$_3$O$_4$ nanocrystal powder after calcination at 300°C, confirming the formation of single phase Co$_3$O$_4$. Figure 5.37(b) shows SEM image of powder, indicating

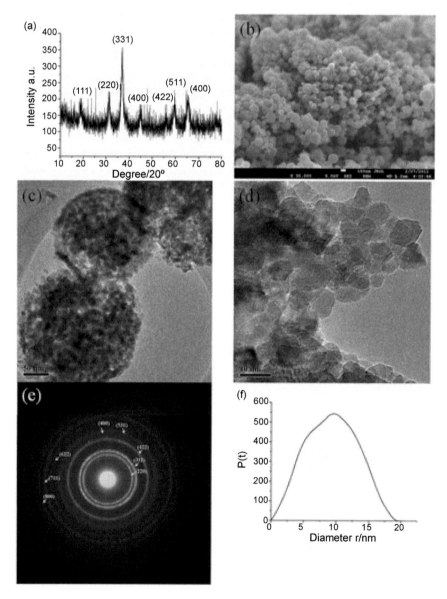

Fig. 5.37. (a) XRD pattern of the Co$_3$O$_4$ cryogel calcined at 300°C (sample Co–P123–PH3), (b) SEM image of the calcined Co$_3$O$_4$ cryogel (sample Co–P123–PH3), (c, d) TEM image of the Co$_3$O$_4$ cryogel at different magnifications (sample Co–P123–PH3), (e) SAED of sample Co–P123–PH3 and (f) distribution of Co$_3$O$_4$ particle size obtained from SAXS (sample Co–P123–PH3). Reproduced with permission from [85], Copyright © 2012, American Chemical Society.

that the macroporous structure of Co_3O_4 consisted of interconnected spheres, with an average diameter of 100 nm. Microstructure of the Co_3O_4 powder was also confirmed by the low-magnification TEM image, as demonstrated in Fig. 5.37(c). It was observed that the spheres consisted of numerous small crystals, which resulted in the mesoporous structure within the macroporous Co_3O_4 particles. High-magnification TEM image revealed that Co_3O_4 nanocrystals had an average diameter of 10 nm, as illustrated in Fig. 5.37(d). The size of the Co_3O_4 nanocrystals was further confirmed by X-ray small-angle scattering (SAXS) analysis result, as demonstrated in Fig. 5.37(f). Polycrystalline nature of the calcined cryogel was supported by the SAED pattern, as seen in Fig. 5.37(e).

It was found that the presence of P123 into the sol–gel system played a significant in determining the microstructure of the calcined cryogel. The micelles of P123 formed in the sol acted as a template, resulting in the formation of the uniform nanoparticles of sample Co–P123 [87]. Without the presence of P123 (i.e., sample Co–CA complex), an irregular macroporous structure was obtained, due to the absence of the sphere-like macroporous network. The morphology development was also influenced by pH value of the reaction systems. At a relatively low pH value of 1.03, the cryogel had a dense monolith like structure. As the pH value was increased to 3.33, the hierarchical sphere structure disappeared, while disordered structures were obtained instead.

Besides the template effect of P123, the ice crystal also served as a template during the freeze-drying process. The ice crystallized rapidly at $-196°C$ in liquid N_2, so as to form solid scaffolds to further promote the gelation process. In vacuum, the sublimation of ice resulted in cryogel that possessed hollow networks, which was responsible for the presence of both the mesoporous and macroporous characteristics [88]. The effect of pH value on pore diameter of the samples was attributed to the fact that the morphology of the micelles was varied with the pH value, due to the change in packing parameter [89]. The average pore diameter was decreased from 20 nm of Co–P123 to 10 nm of Co–P123–PH3. At the same time, BET surface areas of Co–CA, Co–P123, and Co–P123–PH3 were 52.76, 62.86, and 82.8 $m^2 g^{-1}$, respectively.

Thin film based electrodes

Chemical solution deposition

Chemical Solution Deposition (CSD) has been widely used to deposit various thin films on different substrates for a wide range of applications [58, 90, 91]. During CSD processes, a relatively slow chemical reaction takes place in solutions, which leads to the formation of a solid product on a given substrate.

Co_3O_4 nanowire arrays have deposited on Ni foam [58] and the Ti substrate [90] by using an ammonia-evaporation method [91]. The open space related to the loosely packed nanowires ensured a smooth diffusion of the electrolyte ions into the interior of the electrode, thus leading to largely decreased internal resistance and thus enhanced electrochemical performances. The sample on Ni foam, with a high mass loading of 16 $mg·cm^{-2}$, achieved high specific capacitance of 746 $F·g^{-1}$ in 6 M KOH [58]. However, the capacitance was decreased by about 14% after 500 charge-discharge cycles. When KOH electrolyte solutions with higher concentrations were used, the specific capacitance increased, which was accompanied by more distinguished redox peaks, because of the decreased internal resistance and the electrolyte starvation near surface of the electrode materials.

To synthesize Co_3O_4 nanowire arrays on Ni foams, 12 mmol $Co(NO_3)_2$ and 6 mmol NH_4NO_3 were dissolved in a solution of 15 mL ammonia (with concentration of 30 wt%) mixed in 35 mL H_2O. After stirring for 10 minutes, the solution was heated at 90°C for 2 hours. Then, a piece of Ni foam, with a dimension of $10 \times 10 \times 1.1$ mm^3, was thoroughly cleaned with acetone, etched with 6.0 M HCl for 5 minutes, rinsed with water and soaked in 0.1 mM $NiCl_2$ solution for 4 hours, followed by rinsing thoroughly with water. After that, it was immersed in the Co $(NO_3)_2$ solution for 14 hours at 90°C, where nanowires were grown on the Ni foam. The Ni foam with nanowires was then washed with water, dried at room temperature for 6 hours and then calcined at 300°C for 2 hours in air. In this case, the electrode with a size of 1 cm^2 contained 16 mg Co_3O_4 nanowires.

Figure 5.38 shows representative SEM and TEM images of the Co_3O_4 nanowire arrays and an individual nanowire. The Ni foam was densely covered by the Co_3O_4 nanowires that possessed relatively uniform sizes, with an average diameter of 250 nm and a length of up to 15 μm, as demonstrated in Fig. 5.38(A). TEM images revealed that the nanowire was developed through the packing of nanoplatelets in a nearly layer-by-layer manner, as observed in Fig. 5.38(B). In addition, as observed in Fig. 5.38(A), the Co_3O_4 nanowires were grown randomly in different directions, which could be attributed to the high concentration of $Co(NO_3)_2$ and NH_4NO_3, as well as the long growth time.

Co_3O_4 nanowires grown on Ti substrate were further coated with Ag nanoparticles, in order to enhance the electrochemical performances [90]. It was found that specific capacitance of the Co_3O_4 nanowires–Ag based electrode was 1006 $F \cdot g^{-1}$, as compared with 922 $F \cdot g^{-1}$ for the electrode based only on Co_3O_4 nanowires. In addition, the coating of Ag nanoparticles also improved the rate capability from 54.2 to 95%, as the capacitances at 10 $A \cdot g^{-1}$ and 2 $A \cdot g^{-1}$ were compared. The electrodes exhibited excellent cycling life by retaining > 95% of the initial capacitance after 5000 cycles.

To synthesize the Co_3O_4 nanowire arrays on Ti substrates, 15 mmol of $Co(NO_3)_2$ was dissolved in 15 mL DI water to form Co^{2+} solution [92]. After that, 60 mL 25% ammonia solution was added dropwise to the Co^{2+} solution for 5 minutes with strong magnetic stirring that was continued for 10 minutes. At the same time, Ti substrate was cleaned on one side and then placed in a Petri dish containing the Co^{2+} and ammonia mixture, with the cleaned side down-facing above the bottom of the dish by 2–3 mm. The reaction was carried out at 90°C for 14 hours to obtain Co_3O_4 nanowire arrays.

Ag nanoparticles were coated on surfaces of the Co_3O_4 nanowire arrays by using the conventional silver-mirror reaction method. 0.01 mol $AgNO_3$ was dissolved in 50 mL DI water at room temperature, into which 50 mL 0.2 M NaOH solution, thus resulting in a brown precipitate. Then, concentrated (10 M) ammonia solution was added dropwise into the mixed solution under vigorous stirring, until the brown

Fig. 5.38. SEM image of the Co_3O_4 nanowire arrays deposited on nickel foam skeleton (A) and TEM image of a single nanowire detached from the nickel foam (B). Reproduced with permission from [58], Copyright © 2010, Elsevier.

precipitate just started to be dissolved. Immediately, 50 mL 0.4 M glucose solution was introduced into the mixture with constant stirring. The reaction system was heated in water bath at 90°C. Ag coating was conducted by repeating the following procedure six times: (i) immersion of the Co_3O_4 nanowire arrays into the $AgNO_3$–glucose solution for 5 minutes, (ii) thorough washing with DI water, ethanol and acetone, and (iii) drying in a vacuum at 60°C for 2 hours.

Representative SEM images of the samples are shown in Fig. 5.39(a, b), which indicated that Co_3O_4 nanowires had diameters in the range of 400–900 nm and lengths of 10–5 μm, appearing as arrays on the Ti substrate. TEM images revealed that the Co_3O_4 nanowire possessed quasi-single-crystal characteristics, as seen in Fig. 5.39(d), whereas they exhibited a tubular structure with a mesoporous cavity going through the entire center of the nanowire, as illustrated in Fig. 5.39(e). Figure 5.39(c) shows SEM image of the sample coated with Ag nanoparticles. As demonstrated in Fig. 5.39(e), Ag nanoparticles on surface of the Co_3O_4 nanowire could be observed occasionally. EDX spectrum and the corresponding Co Kα1 and Ag Lα1 elemental mappings suggested that the content of Ag in the Co_3O_4–Ag composite nanowires was about 3 wt%. In addition, the Ag nanoparticles were uniformly attached on surfaces of the Co_3O_4 nanowires.

A porous Co_3O_4 thin film consisting of spherical coarse particles on an ITO coated glass substrate, which was used as electrode of supercapacitors [93]. The Co_3O_4 thin film electrode displayed a maximum specific capacitance of 227 $F·g^{-1}$ at 0.2 $A·g^{-1}$, which was decreased to 152 $F·g^{-1}$ (67%), as the current density was raised to 1.4 $A·g^{-1}$. The specific capacitance value was decayed by about 34% after 500 cycles, which remained to be stable up to 1000 cycles.

Before film deposition, ITO coated glass substrates were thoroughly cleaned with acetone, ethanol and DI water, with the aid of ultrasonication. To deposit Co_3O_4 thin film on the cleaned ITO glass substrate, a deposition solution was prepared by mixing 25 ml 1 M $CoSO_4$, 5 ml 25% $NH_3·H_2O$, 15 ml 0.25 M $K_2S_2O_8$ and 5 ml deionized water. The deposition was conducted for 2 hours at room temperature. After that, the sample was thoroughly washed with deionized water, followed by drying at room temperature. Then, the thin film electrode was obtained by calcining the dried sample at 200°C for 1 hour in air. Formation of cubic Co_3O_4 phase was confirmed by XRD pattern.

Fig. 5.39. (a) Cross-sectional SEM image, (b, c) plan-view SEM and (d, e) TEM images of the Co_3O_4 nanowire arrays before (a, b, d) and after (c, e) coating with Ag nanoparticles. Electron diffraction pattern of the Co_3O_4 nanowire before Ag coating is shown as the inset in panel (d). Reproduced with permission from [90], Copyright © 2010, Springer.

314 *Nanomaterials for Supercapacitors*

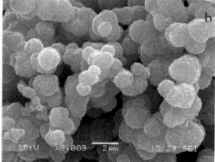

Fig. 5.40. SEM images of Co$_3$O$_4$ thin film: (a) low magnification and (b) high magnification. The inset in (a) is photograph of the Co$_3$O$_4$ thin film deposited on ITO coated glass substrate. Reproduced with permission from [93], Copyright © 2011, Elsevier.

Figure 5.40 SEM images of the Co$_3$O$_4$ thin film at low and high magnifications. A photograph of the Co$_3$O$_4$ thin film is shown as the inset in Fig. 5.40(a), which indicated that the Co$_3$O$_4$ thin film was black in color. According to the SEM images, it was found that the Co$_3$O$_4$ thin film contained spherical-like particles, with a size of < 2 μm and pores among the particles. Furthermore, high magnification SEM image revealed that the spherical-like particles were rough and had agglomeration. The presence of pores in between the particles and rough surface of the spherical-like particles could be beneficial to electrochemical performances of the Co$_3$O$_4$ thin film based electrode.

Electrodeposition

Electrodeposition is a versatile strategy that has been used to deposit cobalt oxide thin films as electrodes of supercapacitors [94–98]. For example, an electrodeposition method has been reported to synthesize cobalt oxide thin films, which were used as either positive or negative electrodes of supercapacitors, with promising electrochemical performances [94]. The as-deposited thin film was Co (OH)$_2$. As the Co (OH)$_2$ films were calcined at temperatures in the range of 100–350°C, the specific capacitance was gradually decreased. In addition, the films calcined at temperatures of ≤ 300°C, CVs showed exhibited a redox reaction besides the Electric Double Layer (EDL) behavior in 3 wt% KOH electrolyte. It was attributed to the fact that the films calcined at sufficiently low temperatures allowed H$^+$ ions to diffuse into them. However, after calcination at high temperatures, more OH$^-$ ions would be adsorbed on surface of the films. Electrochemical characterization results indicated that, the Co$_3$O$_4$ film processed at optimized conditions displayed specific capacitances of 76 and 164 F·g^{-1}, as negative and positive electrodes, respectively.

The Co (OH)$_2$ thin films were coated on Au foils that were masked with a sealant (Silicone™) in such a way that an area of 1.0 cm^2 was exposed to the deposition solution. The deposition was conducted at room temperature in a bath containing 0.175 M Co (NO$_3$)$_2$ and 0.075 M NaNO$_3$ in a solvent with 1:1 volume ratio of DI water and ethanol. After the application of a cathodic current density of 1.0 mA·cm^{-2} for 8 min, Co (OH)$_2$ films with an areal density of 360 μg·cm^{-2} could be obtained.

Co$_3$O$_4$ thin films have deposited on ITO glass substrates, without or with surfactants, cetyltrimethylammonium bromide (CTAB) and Sodium Dodecyl Sulfate (SDS) [95]. Without the presence of surfactant, the films had a dendrite like morphology. In comparison, if surfactants were used, morphologies of the films would be totally different. For instance, with the presence of 1% CTAB, the deposited films would display a sphere-like morphology, with pore sizes in the range of 3–4 nm and wall thicknesses of 3–4 nm. Specific capacitances of the Co$_3$O$_4$–CTAB and Co$_3$O$_4$–SDS films were 491 and 373 F·g^{-1}, which were higher than that (i.e., 255 F·g^{-1}) of the film without surfactant.

To deposit the Co$_3$O$_4$ thin films, 2.81 g CoSO$_4$·7H$_2$O was dissolved in 50 ml DI water to form the electrolyte solution. In case of using surfactants, 0.5 g CTAB or SDS (1 wt%) was added into the solution, so that the films would have a mesoporous structure. Electrodeposition was conducted potentiostatically at voltages of 1.0–1.6 V at 70°C. Before electrodeposition, the ITO coated glass substrates were thoroughly cleaned with the aid of ultrasonification for 15 minutes in a mixed solvent consisting of isopropanol and acetone with a volume ratio of 1:1. The counter and reference electrodes were Pt and Ag/AgCl electrodes, respectively. Co_3O_4 films without the use of a surfactant were deposited in a similar way. After deposition, the samples were thoroughly cleaned in a mixed solvent of ethanol and DI water, in order to completely remove the surfactant, followed by drying in air.

A cathodic electrodeposition process was employed to deposit Co_3O_4 films with vertically aligned interconnected nanowalls on Ni foams, for supercapacitor applications [96]. Due to the nanowalled structures, there were plenty of open spaces, with sizes of 30–300 nm, offering smooth paths for the diffusion of electrolyte ions. The porous Co_3O_4 films exhibited a specific capacitance of 325 and 230 $F \cdot g^{-1}$ at 2 $A \cdot g^{-1}$. Comparatively, the dense film had a specific capacitance of 230 $F \cdot g^{-1}$ at 2 $A \cdot g^{-1}$. Also, the Co_3O_4 nanowall film displayed lower internal resistance and smaller polarization, evidenced by its higher discharge plateau and lower charge plateau, as compared with the dense film. As the current density was increased 2 $A \cdot g^{-1}$ to 40 $A \cdot g^{-1}$, specific capacitance of the nanowall film was retained by 76.0%, whereas that of the dense film was maintained by 72.6%. Additionally, the porous film demonstrated higher cycling stability than the dense counterpart.

The nanowall Co_3O_4 film was deposited by using a cathodic electrodeposition, with a standard three-electrode configuration at room temperature. Ni foams with a dimension of 2.5×2.5 cm were used as the substrate and working electrode, while Saturated Calomel Electrode (SCE) and a Pt foil were employed as reference electrode and counter-electrode, respectively. Before electrodeposition, the Ni foam substrates were thoroughly cleaned with ethanol and acetone with the aid of ultrasonication for 15 minutes. The electrodeposition was carried out in electrolyte solution containing 0.9 M Co $(NO_3)_2$ and 0.075 M $NaNO_3$, at a constant cathodic current of 1 $mA \cdot cm^{-2}$ for 10 minutes. After the washing and drying processes, the deposited samples were thermally treated at 200°C for 1 hour in flowing Ar. For the purpose of comparison, dense Co_3O_4 films were also prepared in a similar way, with electrolyte solution consisting of 0.25 M cobalt sulfate, 0.25 M sodium sulfate and 0.2 M sodium acetate. The electrodeposition was conducted at a constant anodic current of 0.25 $mA \cdot cm^{-2}$ for 1 hour. After deposition, the sample was calcined at 200°C in Ar for 1 hour. The two samples had a thickness of 1.2 μm.

XRD patterns indicated that the as-deposited films were of single phase α-Co $(OH)_2$. After calcination, the films were converted to Co_3O_4 The Co_3O_4 films consisted of interconnected nanoflakes with an average thickness of 20 nm, as shown in Fig. 5.41(a, b). As seen from the cross-sectional SEM image in Fig. 5.41(a), the Co_3O_4 nanoflakes were vertically grown on the substrate, thus resulting in a porous nanowall structure with a large porosity. The nanoflakes had very small roughness. The porous nanowall structure of Co_3O_4 was retained from the porous nanowall α-Co(OH)$_2$ precursor film, as illustrated in Fig. 5.41(d). Due its layered structure, Co(OH)$_2$ always preferentially grows along the layered plane, so as to develop 2D nanoplates structure [99]. In comparison, the dense Co_3O_4 film displayed a uniform surface morphology, without the presence of pores, as demonstrated in Fig. 5.42(a, b). BET specific surface areas of the porous nanowall Co_3O_4 and dense Co_3O_4 films were 185 and 28 $m^2 \cdot g^{-1}$, respectively.

Figure 5.43 shows electrochemical stabilities of the two films, as a function of charge/discharge cycling, at a current density of 2 $A \cdot g^{-1}$. After 2000 cycles, the porous nanowall Co_3O_4 film exhibited a specific capacitance of 383 $F \cdot g^{-1}$, whereas the dense film had a value of 229 $F \cdot g^{-1}$, indicating the higher capacity retention of the porous nanowall Co_3O_4 film. Moreover, the porous nanowall Co_3O_4 film was more flexible in buffering volume variation during the electrochemical reactions. For instance, after 2000 cycles, the porous nanowall Co_3O_4 film showed almost no change in terms of the porous structure. However, many cracks were formed in the dense film.

A hierarchically porous Co_3O_4 film with interconnected porous nanowalls was prepared by using electrodeposition method, combined with the use of a liquid crystalline template [97]. The nanoflakes possessed a continuous net-like structure with fine pores of 5–25 nm in diameter. Meanwhile, the nanoflakes contained mesoporous walls with mesopores of 2–3 nm, due to the packing of nanoparticles. The hierarchical porous Co_3O_4 film showed a specific capacitance of 443 $F \cdot g^{-1}$, which was higher than that of the Co_3O_4 film prepared without the liquid crystalline template by 37%. In addition, the capacitance was retained by 94.3% after 3000 cycles.

Fig. 5.41. SEM and TEM images of the porous nanowall Co$_3$O$_4$ film (a–c) and Co(OH)$_2$ film (d). The upper-right inset in (a) corresponds to the side view of film. Reproduced with permission from [96], Copyright © 2011, Elsevier.

The hierarchically porous Co$_3$O$_4$ films with mesoporous walls were deposited by using electrodeposition method with a two-electrode glass cell at 80°C, where the Ni foam substrate with a size of 2 × 3 cm^2 also served as the working electrode, while a Pt foil acted as the counter-electrode. The electrolyte consisted of 50 wt% nonionic surfactant polyoxyethylene cetyl ether (Brij 56, C$_{16}$EO$_{10}$) and a 50 wt% aqueous solution containing 1.0 M Co(NO$_3$)$_2$ and 0.1 M NaNO$_3$ as the supporting electrolyte. The surfactant Brij 56 was heated to about 80°C, i.e., slightly above the melting point of the surfactant. Then, the aqueous solution containing 1.0 M Co(NO$_3$)$_2$ and 0.1 M NaNO$_3$ was added dropwise to form a homogeneous solution. The electrodeposition was conducted at a constant cathodic current of 2.0 mA·cm^{-2} for 5 minutes. After that, the samples were rinsed in in ethanol, 2-propanol and high purity water for 2 days, in order to completely remove the surfactant, followed by drying and calcination at 250°C for 1 hour in Ar. The final samples had a thickness of about 950 nm, with a loading weight of the hierarchically porous Co$_3$O$_4$ to be about 0.9 mg·cm^{-2}.

Figure 5.44(a) shows XRD pattern of film after calcination. In addition to the three strong peaks from the Ni foam substrate, all the rest peaks could be indexed to Co$_3$O$_4$ phase, suggesting that crystalline Co$_3$O$_4$ was formed after the calcination process. As observed in Fig. 5.44(b), the Co$_3$O$_4$ film possessed a hierarchically porous structure, which was assembled with interconnected Co$_3$O$_4$ nanoflakes, with thicknesses of 15–20 nm. Also, the Co$_3$O$_4$ nanoflakes were grown vertically on the substrate, so that a net-like structure with pores of 30–300 nm in diameter, as demonstrated as the inset in Fig. 5.44(b). Moreover, the Co$_3$O$_4$ nanoflakes exhibited a fine porous structure with pore diameters in the range of 5–25 nm, as illustrated in Fig. 5.44(c).

Figure 5.45 shows representative TEM and HRTEM images of the samples. As seen in Fig. 5.45 (a, b), the TEM image clearly indicated that the nanoflake possessed a continuous porous net-like structure, with finer pores having sizes in the range 5–25 nm, consisting of Co$_3$O$_4$ nanoparticles with sizes of 10–50 nm. Furthermore, the Co$_3$O$_4$ nanoparticles in the nanoflakes had mesoporous walls with mesopores

Oxide Based Supercapacitors: II-Other Oxides 317

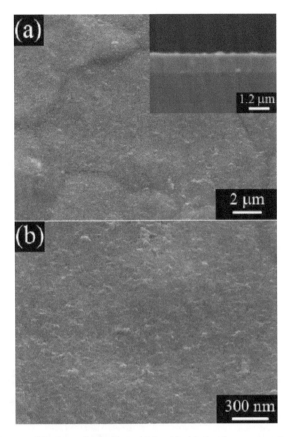

Fig. 5.42. SEM and TEM images of the dense Co$_3$O$_4$ film, with inset in (a) showing cross-sectional image. Reproduced with permission from [96], Copyright © 2011, Elsevier.

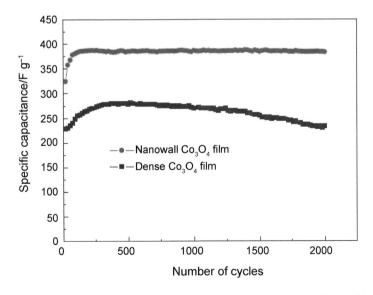

Fig. 5.43. Cycling performances of the porous and dense Co$_3$O$_4$ film electrodes at a current density of 2 A·g^{-1}. Reproduced with permission from [96], Copyright © 2011, Elsevier.

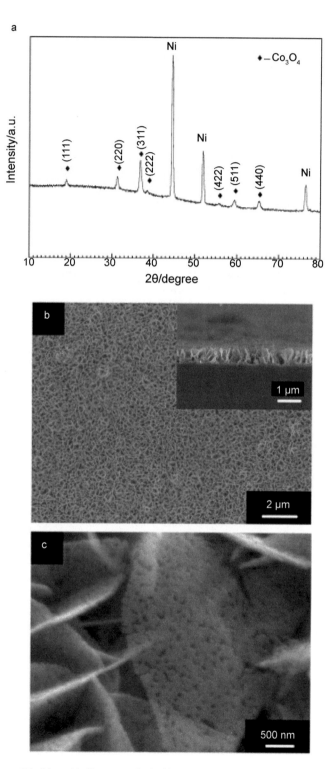

Fig. 5.44. (a) XRD pattern of the hierarchically porous Co$_3$O$_4$ film grown on Ni foam. (b) SEM image of the hierarchically porous Co$_3$O$_4$ film (with the inset showing cross-sectional view of the film) and (c) magnified SEM image of the Co$_3$O$_4$ nanoflakes. Reproduced with permission from [97], Copyright © 2011, Elsevier.

Oxide Based Supercapacitors: II-Other Oxides 319

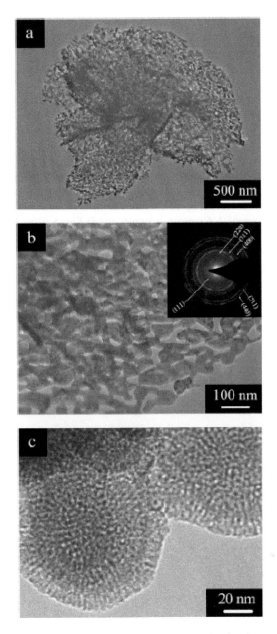

Fig. 5.45. TEM images of (a, b) the individual Co_3O_4 nanoflake (with the inset showing the corresponding SAED pattern) and (c) HRTEM image of the Co_3O_4 nanoflake. Reproduced with permission from [97], Copyright © 2011, Elsevier.

of 2–3 nm, as illustrated in Fig. 5.45(c). Also, in the Selected Area Electronic Diffraction (SAED) pattern of the nanoflakes, all the diffraction rings could be indexed to the spinel phase Co_3O_4, implying that the hierarchically porous Co_3O_4 film with mesoporous walls was highly polycrystalline, which was in a good agreement with the XRD result.

Co_3O_4 monolayer hollow sphere arrays have been prepared by using polystyrene (PS) spheres as a template [98]. The arrays demonstrated a skeleton that was present with interconnected hollow spheres of 600 nm in diameter, while the interstices between the hollow spheres were filled with Co_3O_4 nanoflakes. The Co_3O_4 monolayer based electrode had a specific capacitance of 358 F·g^{-1} at a current density of

2 A·g⁻¹, which was moreover retained by 85%, as the current density was increased to 40 A·g⁻¹. During the cycling testing, the specific capacitance was increased up to 500 cycles and remained to be constant thereafter until 4000 cycles, implying an excellent stability.

As demonstrated in Fig. 5.46(a), a close-packed monolayer array was formed with the PS spheres, with a long-range ordering in both a parallel and perpendicular direction along the Ni foil substrate. Once the PS spheres were removed through heat treatment, Co_3O_4 monolayer hollow-sphere array could be obtained. The final product was present as a skeleton of array, consisting of interconnected hollow spheres with a diameter of 600 nm, whereas the interstices in between the hollow spheres were made of Co_3O_4 nanoflakes with a thickness of about 20 nm, as illustrated in Fig. 5.46(b).

Figure 5.47 shows representative TEM and HRTEM images of the samples, from which a insight view of the microstructure of the individual hollow spheres and nanoflakes was revealed. It was interesting to notice that both the hollow spheres and the nanoflakes had mesoporous walls with finer pores with sizes in the range of 2–5 nm, as observed in Fig. 5.47(a, b). The mesopores were formed due to the pacing of Co_3O_4 nanoparticles, with sizes of 2–6 nm, as demonstrated in Fig. 5.47(c, d). Polycrystalline nature of the array was confirmed by SAED pattern and XRD pattern, which was spinel phase Co_3O_4. In addition, the lattice spacing of 2.44 Å was well consistent with the (311) interplanar distance of spinel Co_3O_4, as shown in Fig. 5.47(c).

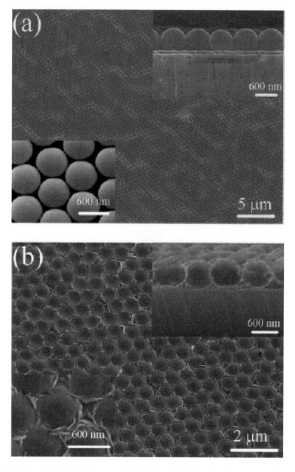

Fig. 5.46. Representative SEM images of (a) the monolayer PS sphere template (with insets showing the corresponding magnified top view and side view) and (b) the Co_3O_4 monolayer hollow-sphere array (with insets showing the corresponding magnified top view and side view). Reproduced with permission from [98], Copyright © 2011, Royal Society of Chemistry.

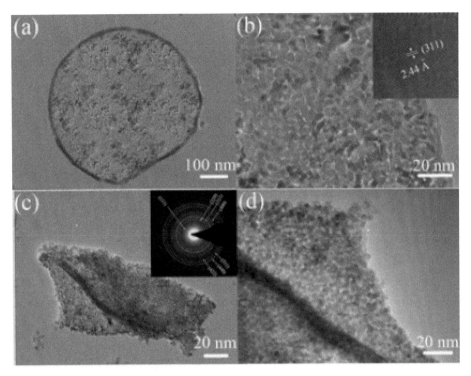

Fig. 5.47. TEM images of (a, b) an individual Co$_3$O$_4$ hollow sphere (with the inset demonstrating the corresponding HRTEM image) and (c, d) an individual Co$_3$O$_4$ nanoflake (with the inset giving the corresponding SAED pattern). Reproduced with permission from [98], Copyright © 2011, Royal Society of Chemistry.

Spray deposition

Spray pyrolysis is a solution deposition technique by spraying the solution (aerosol) directly on a heated substrate surface [100, 101]. For example, a solution spray pyrolysis technique was used to deposit uniform and adherent Co$_3$O$_4$ thin films on glass substrates from aqueous cobalt chloride solution [100]. Electrode based on the Co$_3$O$_4$ exhibited outstanding electrochemical performance in aqueous KOH electrolyte. The XRD pattern revealed that the thin film was resulted from the oriented growth of cubic Co$_3$O$_4$. According to electrical resistivity measurement results, it was found that the Co$_3$O$_4$ films had a semiconducting characteristics, with room temperature electrical resistivity to 1.5×10^3 $\Omega \cdot$cm. The supercapacitor assembled with the spray-deposited Co$_3$O$_4$ film demonstrated a specific capacitance of 74 F·g^{-1} or 32 mF·cm^{-2}.

A similar solution precursor plasma deposition route was reported to fabricate nanostructured and porous Co$_3$O$_4$ thin film on a stainless steel sheet [101]. Due to the effect of the high temperatures of the plasma plume, thermochemical conversion of cobalt acetate into Co$_3$O$_4$ agglomerates or fused particles was highly promoted. The Co$_3$O$_4$ thin film based electrode exhibited a specific capacitance of 250 F·g^{-1} at a current density of 0.2 A·g^{-1}, was decreased to 150 F·g^{-1}, as the current density was raised to 4.5 A·g^{-1}. In addition, the electrode experienced a loss in capacitance by 27.8% after 1000 cycles.

The precursor solution for the spray deposition was obtained by dissolving 0.2 mole cobalt acetate tetrahydrate in 1000 mL deionized water, with aid of vigorous stirring. The Co$_3$O$_4$ films were deposited on a stainless steel sheet with a thickness of 25 μm, which also acted as the current collector. The precursor solution was fed into the plasma plume by using an axial liquid atomizer. Figure 4.48 shows a schematic diagram of solution precursor plasma deposition set-up, which was employed to deposit the Co$_3$O$_4$ thin films for electrodes of supercapacitors. Fine droplets were formed, as the precursor solution was atomized by the compressed air in the atomizer, which were incorporated into the plasma plume. During that process,

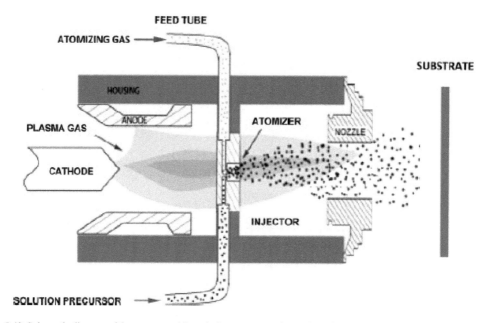

Fig. 5.48. Schematic diagram of the set-up used for solution precursor plasma deposition of the Co_3O_4 thin films. Reproduced with permission from [101], Copyright © 2012, Elsevier.

high temperatures of the plasma plume facilitated the thermo-chemical conversion of the precursor solution droplets into Co_3O_4 nanoparticles directly, which were then coated onto the stainless steel substrate.

The XRD pattern confirmed that polycrystalline cubic phase Co_3O_4 was present in the films. In the plasma deposition process, because the thermo-chemical conversion of precursor solution and the deposition of the nanoparticles on the substrate occurred in air, there is plenty of oxygen that could aid in the formation of Co_3O_4. The extent of thermo-chemical conversion of the precursor solution droplets was influenced by several parameters, including temperature of the plasma plume, size/mass of the atomized droplets, retention time of the droplets inside the plasma plume, and/or velocity of the droplet injection, which were actually interdependent. The smaller the droplets, the fast the thermo-chemical conversion would be, whereas the size of the droplets could be controlled by adjusting the pressure of the compressed air and the feed rate of the precursor solution. However, the pressure of the compressed air could be controlled along with the velocity of the plasma jet to vary the retention time. As a result, the three parameters were crucial to maximize the conversion of the precursor solution before deposition of the nanoparticles on the substrate.

Figure 5.49 shows SEM image of the Co_3O_4 thin film. During the deposition process, the thermo-chemically converted particles in the plasma plume could melt to a certain degree, thus leading to the formation of clusters, due to the high temperatures of the plasma plume. At the same time, the films had a certain level of porosity, as seen in Fig. 5.49, which was caused by the loose packing of the Co_3O_4 nanoparticles. The porosity of the Co_3O_4 thin films ensured the easy access of the electrolyte ions into the electrodes.

Figure 5.50 shows representative TEM images of the Co_3O_4 films. As demonstrated in Fig. 5.50(a), the high magnification TEM image indicated the presence of agglomerated Co_3O_4 nanoparticles, which could be readily related to the thermo-chemical conversion of the precursor solution. The Co_3O_4 nanoparticles had sizes in the range of 10–50 nm. The Co_3O_4 nanoparticles were loosely accumulated to create a large surface area, thus shortening the mass and charge diffusion distances for rapid electrochemical reactions. Figure 5.50(b) shows interplanar spacing between the adjacent two (111) planes of the Co_3O_4 crystal structure. The cubic phase of Co_3O_4 was also confirmed by the diffraction pattern shown in the inset of Fig. 5.50(a) and the HRTEM image illustrated in Fig. 5.50(b).

Fig. 5.49. SEM images of the Co$_3$O$_4$ thin films deposited by using the solution precursor plasma deposition technique, with the inset showing zoom-in view. Reproduced with permission from [101], Copyright © 2012, Elsevier.

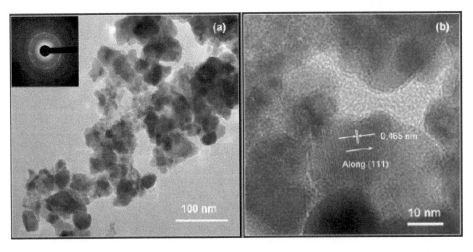

Fig. 5.50. TEM images of the Co$_3$O$_4$ thin films: (a) Co$_3$O$_4$ nanoparticles with the corresponding diffraction pattern and (b) high resolution lattice image of the Co$_3$O$_4$ nanoparticles. Reproduced with permission from [101], Copyright © 2012, Elsevier.

Hydrothermal/solvothermal processing

The hydrothermal or solvothermal processing technique has been also widely used to synthesize Co$_3$O$_4$ thin films on different substrates [102–109]. For instance, Co$_3$O$_4$ nanoflower structures consisting of nanosheets were deposited on Ni foam by using a solvothermal synthesis process, followed by post-annealing [102]. The Co$_3$O$_4$ nanosheets covered the entire surface of the Ni foam, which thus enabled efficient electron transport. The sample post-annealed at 250°C had specific capacitances of 1937, 1518, 1377 and 1309 F·g^{-1}, at the current densities of 0.2, 1, 2 and 3 A·g^{-1}, respectively, within a narrow potential window of 0–0.34 V. The Co$_3$O$_4$ thin film based electrode exhibited a loss in capacitance by 21.8% after 1000 cycles, which was attributed to the dissolution of the active material in the high concentration KOH electrolyte (6 M).

324 *Nanomaterials for Supercapacitors*

To prepare the reaction solutions, 0.5–2 g Co (NO$_3$)$_2$·6H$_2$O, 1 g hexadecyl trimethyl ammonium bromide (CTAB) and 0.5–4 ml water were dissolved in 23.5–20 mL absolute methanol. Solvothermal reaction was conducted with a 30 ml Teflon-lined stainless steel autoclave. A piece of Ni foam with a specific surface area of 420 g·m^{-2} was thoroughly cleaned, dried and then immersed in the growing solution, followed by reaction for 24 hours at 150–180°C. After reactions, the Ni foam deposited with Co$_3$O$_4$ nanostructures was thoroughly washed with H$_2$O and ethanol, followed by drying in vacuum at 120°C for 8 hours, with samples to be denoted as sample A-as-prepared. In a separate experiment, 1 g CTAB as added to the above solution, while all other parameters were the same, with the final products to be denoted as sample B-as-prepared.

The A-as-prepared samples were then calcined at 250°C for 4 hours, 250°C for 14 hours and 300°C for 4 hours, with the final products to be denoted as samples A-250-4, A-250-14 and A-300-4, respectively. The B-as-prepared sample was calcined at 300°C for 4 hours, with the final product to be denoted as sample B-300-4. The deposit weight of Co$_3$O$_4$ was accurately measured by weighing the Ni foam before and after the solvothermal reaction.

XRD results indicated that the as-obtained samples were β-Co(OH)$_2$ and cobalt carbonate nitrate hydroxide hydrate (Co(OH)$_{1.81}$(NO$_3$)$_{0.11}$(CO$_3$)$_{0.04}$·0.6H$_2$O). After calcination, only spinel Co$_3$O$_4$ was observed.

Figure 5.51 shows representative SEM images of the A-as-prepared samples. The individual nanoflower was assembled with numerous nanosheets having jagged edges. The nanoflowers had diameters in the range of 2–5 μm, while the nanosheets exhibited thicknesses of 60–110 nm, widths of 0.2–2.5 μm and lengths of 0.2–2.5 μm. The nanosheets also demonstrated a very smooth face and well-developed facet. It was observed that the entire surface of Ni foam was covered by the nanosheets, which made the films suitable as electrodes of supercapacitors. In this case, the surfactant CTAB could serve as the adsorbing species, so as to change the habits of crystallization and trigger the growth of the nanoflowers. As shown in Fig. 5.51(c, d), the flower-like morphology of precursor was well retained after the thermal calcination,

Fig. 5.51. Representative SEM images of sample A-as-prepared: (a) low and (b) high magnifications. Representative SEM images of the sample A-250-4: (c) low and (d) high magnifications. Reproduced with permission from [102], Copyright © 2011, Elsevier.

thus leading to the formation of Co_3O_4. At the same time, almost no change in diameters and sizes was observed before and after the transformation from the precursor to crystalline Co_3O_4.

Representative TEM images of the sample A-250-4 are presented in Fig. 5.52(a, b). As seen in Fig. 5.52(a), the petals were porous, due to the removal of water molecules at high temperatures. In addition, the sample A-250-4 had pores with sizes in the range of 0.7–12 nm. SAED pattern as the inset in Fig. 5.52(a) indicated that the nanoflowers had a polycrystalline structure. The corresponding HRTEM image shown in Fig. 5.52(b) confirmed the high crystallinity of the calcined product, with a distinct lattice spacing of 0.27 nm, corresponding to the (110) plane of Co_3O_4.

TEM images of the A-250-14, A-300-4 samples are shown in Fig. 5.52(c, d). The insets in Fig. 5.52(c, d) show the corresponding SEM images of A-250-14 and A-300-4 samples. Obviously, the flower morphology of precursor was almost entirely retained after the thermal calcination. Both samples of A-250-14 and A-300-4 were highly porous in microstructure. As the post-calcination duration was increased from 4 hours to 14 hours at 250°C, the pore sizes were kept almost unchanged. If the post-calcination temperature was raised from 250 to 300°C, the pore sizes were increased obviously, in the range of 2–60 nm.

Co_3O_4 nanoparticles have been deposited on Ni sheet by using a hydrothermal method without the requirement of post-annealing step [103]. It was found that the pre-etching step of the Ni sheet governed the phase of cobalt compounds during the hydrothermal reaction. β-Co(OH)$_2$ phase was formed on the HCl-etched Ni sheet, Co_3O_4 was obtained on the HNO_3-etched one. Co_3O_4 nanoparticles with sizes in the range of 50–150 nm and mesopores formed due to the loose packing of the nanoparticles. The presence of the mesopores would promote the electrolyte ions to penetrate into the bulk of the electrode, so as to maintain a larger electroactive surface for Faradaic reactions. A specific capacitance of 928 F·g^{-1} was achieved at

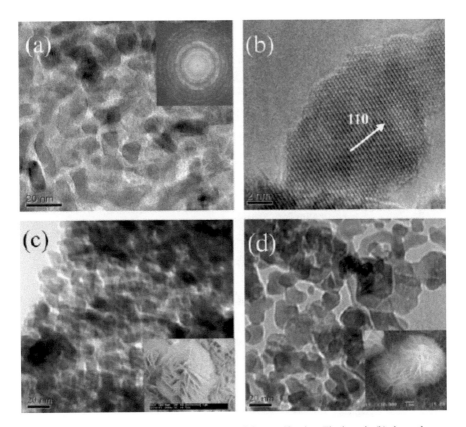

Fig. 5.52. HRTEM images of sample A-250-4: (a) low and (b) high magnification. The inset in (b) shows the corresponding SAED pattern of sample A-250-4. TEM of sample A-250-14 (c) and A-300-4 (d). The insets in (c) and (d) show the corresponding SEM images of samples A-250-14 and A-300-4. Reproduced with permission from [102], Copyright © 2011, Elsevier.

a current density of 1.2 A·g⁻¹. As the current density was increased to as high as 12 A·g⁻¹, 84% retention in specific capacitance was achieved, demonstrating high rate capability of the Co₃O₄ based electrode.

Representative SEM images are shown in Fig. 5.53(a, b), indicating the formation of the Co₃O₄ nanoparticles. The nanoparticles had a spherical morphology. The thickness of the Co₃O₄ nanoparticle layer was about 160 nm. Figure 5.53(c) shows TEM image of the Co₃O₄ nanoparticles, which were spherical in shape and had an average size of about 120 nm. SAED pattern revealed that the Co₃O₄ nanoparticles are polycrystalline in nature, as illustrated by the inset in Fig. 5.53(c). EDAX data indicated that the presence of Co and O elements, consistent with the phase of Co₃O₄, as observed in Fig. 5.53(d). The element composition of the Co₃O₄ nanoparticles was also confirmed by XPS characterization results.

Co₃O₄ nanowire arrays have been deposited on various seed-coated substrates, including Ni foil, Ni foam and Si wafer [104]. Interconnected Co(OH)₂ nanowalls were first developed, which were then converted to Co₃O₄ after thermal annealing. The nanowires possessed roughened walls and a hollow center, with highly porous structures. It was the high porosity that enabled effective penetration of electrolyte ions to the bulk of the electrode materials. At the same time, roughened walls and the hollow center led to a high surface-to-bulk ratio for the electrolyte ions to transport and diffuse. The hollow Co₃O₄ nanowire array based electrode showed specific capacitances of 295 and 599 F·g⁻¹ before and after 1000 charge-discharge cycles. A capacitance retention of 73% was observed as the current density was increased from 2 to 40 A·g⁻¹. For a Co₃O₄ mass loading of 15 mg·cm⁻², the corresponding areal capacitance was as high as 9 F·cm⁻². After 7500 cycles, the capacitance was retained by 91 and 82% at the current densities of 2 and 10 A·g⁻¹.

To deposit Co₃O₄ nanowire arrays, clean Ni foil, Ni foam and silicon wafer, with a dimension of 2.5 cm × 3 cm, were used as the substrates. To trigger the nucleation of hollow Co₃O₄ nanowire arrays, a

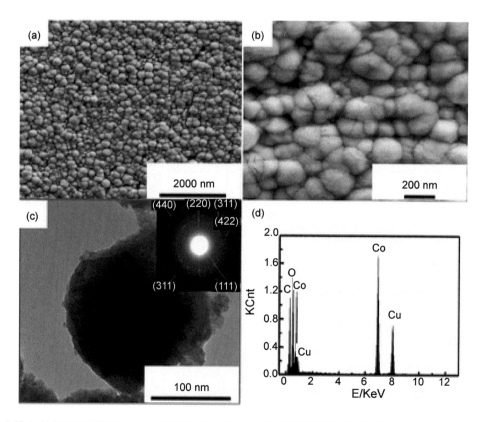

Fig. 5.53. (a, b) SEM, (c) TEM images and SAED pattern (inset in c) and (d) EDAX of the Co₃O₄ nanoparticles grown on Ni sheet. Reproduced with permission from [102], Copyright © 2011, Royal Society of Chemistry.

5 nm-thick Co$_3$O$_4$ thin film was first deposited as the seed layer on the substrates by using the spin coating method, with a 5 mM ethanol solution of cobalt acetate, followed by thermal annealing at 350°C for 0.5 hours in flowing Ar. Typically, 30 mmol Co(NO$_3$)$_2$ and 3 mmol NaNO$_3$ were dissolved in 60 mL 28 wt% ammonia solution and 90 mL H$_2$O to form a homogeneous solution, which was consistently stirred for 0.5 hour in air, until the color was changed from the original pink to black. After that, the solution was subjected to hydrothermal treatment together with the substrates that were fixed with a Teflon clamp with the seed layer side facing downwards, kept about 5 mm away from the bottom of the reaction vehicle. The top side of the substrates was covered with a polytetrafluoroethylene tape to prevent solution contamination. The hydrothermal reactions were carried out 120°C for 12 hours in an oxygen environment. After reaction, the samples were thoroughly washed with DI water, followed by drying and thermal annealing at 250°C for 1 hour in Ar.

Finally, Co$_3$O$_4$ nanowires were grown vertically onto the substrates, thus resulting in aligned and oriented nanowire arrays, as shown in Fig. 5.54. It was found that the substrates had almost no effect on the characteristics of the Co$_3$O$_4$ nanowires arrays. Regardless the flatness or roughness of the substrates, the Co$_3$O$_4$ nanowire arrays could be well grown out. The Co$_3$O$_4$ nanowires were 200 nm in diameter and 15 μm in length. In fact, the length of the nanowires could be maintained by simply controlling the growth

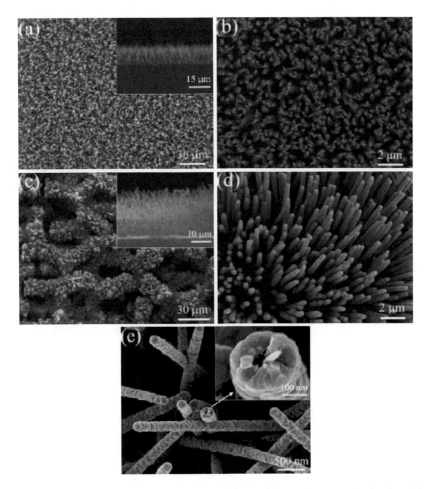

Fig. 5.54. SEM images of the self-supported Co$_3$O$_4$ nanowire arrays on different substrates: (a, b) Si wafer and (c, d) Ni foam substrate. (e) Magnified SEM image of the Co$_3$O$_4$ nanowires. The upper-right insets in (a), (c) and (e) corresponded to cross-sectional views of the Co$_3$O$_4$ nanowire arrays and magnified hollow center structure of the Co$_3$O$_4$ nanowires, respectively. Reproduced with permission from [104], Copyright © 2011, Royal Society of Chemistry.

time. Enlarged SEM image indicated that the individual nanowire had a rough surface, demonstrating a ring-like architecture with a hollow interior, as revealed in Fig. 5.54(e).

The same research group also successfully prepared Co_3O_4 nanowire arrays by the hydrothermal method with an annealing step. A similar method was developed by combining NH_4F as a surface-directing agent, smooth and very long (25 mm) single-crystalline cobalt carbonate hydroxide ($Co_2(OH)_2(CO_3)_2$) nanowires have been grown on Ni foam, which were then converted to Co_3O_4 nanowire arrays by using thermal annealing [106]. Similarly, after thermal annealing in Ar, rough Co_3O_4 nanowires were obtained, which were assembled by interconnected nanoparticles with high mesoporosity, due to the release of CO_2 and H_2O during the thermal annealing of the precursors. The Co_3O_4 nanowire array based electrodes exhibited specific capacitances of 754 and 610 $F \cdot g^{-1}$, at the current densities of 2 and 40 $A \cdot g^{-1}$, respectively, indicating an outstanding high rate stability. In addition, the electrode also demonstrated a pretty high cycling stability up to 4000 charge-discharge cycles.

To synthesize the cobalt carbonate hydroxide nanowires, 10 mmol $Co(NO_3)_2$, 20 mmol NH_4F and 50 mmol $CO(NH_2)_2$ were dissolved in 70 mL distilled water to form a solution, which was then subject to hydrothermal reaction, together with a piece of clean Ni foam substrate with a dimension of 3 × 7 cm². The Ni foam was placed inside the reactor in a similar way to that discussed above. The hydrothermal reaction was conducted at 120°C for 9 hours. After the reaction was finished, the precipitate was collected and thoroughly washed with distilled water. The precursor was finally thermally annealed at 350°C for 2 hours in Ar, which was converted to Co_3O_4, with a load weight of about 4.5 $mg \cdot cm^{-2}$.

Figure 5.55 shows SEM images of the $Co_2(OH)_2(CO_3)_2$ precursor and the Co_3O_4 films. Obviously, the $Co_2(OH)_2(CO_3)_2$ precursor film had 1D nanowire arrays architecture, as seen in Fig. 5.55(a, b). The skeletons of the Ni foam were uniformly covered by the nanowires, which were arranged in a dense and vertical way on the substrate. The morphology of the precursor was well retained after the thermal annealing. The Co_3O_4 nanowires possessed a sharp tip, with an average diameter of 70 nm, lengths of up to 25 μm, as observed in Fig. 5.55(c, d).

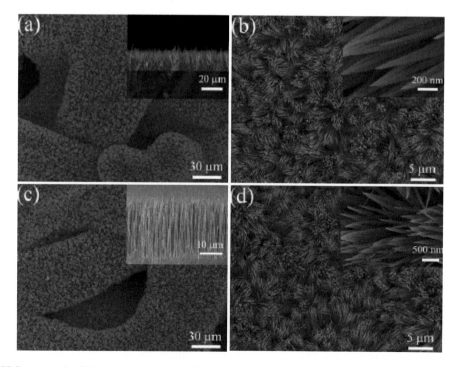

Fig. 5.55. Representative SEM images of (a, b) $Co_2(OH)_2(CO_3)_2$ precursor film and (c, d) Co_3O_4 nanowire array grown on Ni foam, with side view of the arrays and magnified top view presented as the insets. Reproduced with permission from [106], Copyright © 2012, Royal Society of Chemistry.

Figure 5.56 shows TEM images of the nanowires. The $Co_2(OH)_2(CO_3)_2$ nanowire possessed a smooth texture and single crystalline characteristics, as demonstrated in Fig. 5.56(a, b). The SAED pattern indicated that the $Co_2(OH)_2(CO_3)_2$ nanowires were grown in the direction along [010]. TEM images revealed that the average diameter of an individual Co_3O_4 nanowire was 70 nm, as illustrated in Fig. 5.56(c, d). The Co_3O_4 nanowire was formed due to the close packing of numerous nanoparticles. The SAED pattern of a specific Co_3O_4 nanowire revealed that the nanowires were of a single crystalline nature. It was found that the single crystalline Co_3O_4 nanowires were grown along the [110] direction. HRTEM image of the side of a Co_3O_4 nanowire revealed that a lattice spacing of about 2.42 Å nm was clearly identified, which corresponded to the (311) planes of Co_3O_4, confirming the crystalline nature of the nanowire.

A versatile method has been reported to synthesize hierarchical Co_3O_4 nanosheet @ nanowire arrays (NSWA) on Ni foam, as schematically shown in Fig. 5.57 [107]. The development of the hierarchical structures consisted of two steps: (i) formation of nanosheet arrays on Ni foam and (ii) growth of nanowires on surface of the nanosheets. It was found that the presence of NH_4F was a crucial requirement during the hydrothermal synthesis, without which the Co_3O_4 NSWA could only be grown with a mass loading of as low as 1.5 mg·cm^{-2}, lower that of the sample deposited when using NH_4F by about five times. Also, the use of NH_4F significantly improved the adhesion between the Co_3O_4 NSWA and the substrate.

Electrodes based on the Co_3O_4 NSWA exhibited specific capacitances of 715 and 491 F·g^{-1}, corresponding to 5.44 and 3.73 F·cm^{-2}, at current densities of 5 and 30 mA·cm^{-2}, respectively. In contrast, electrodes based on the single-component Co_3O_4 nanosheet arrays of the nanowire arrays had much lower specific capacitance and poorer rate stability. The superior capacitance and rate capability of the Co_3O_4 NSWA based electrode could be readily ascribed to the effect of the 3D architecture of the hierarchical structures. In other words, factors, including high electroactive surface area, hierarchical porosity and open space, enhanced electrolyte–electrode contact area and good adhesion between current-collector and active materials, all contributed to the high electrochemical performances.

To fabricate the Co_3O_4 NSWAs, nanosheet arrays (NSAs) and nanowire arrays (NWAs), 0.58 g Co $(NO_3)_2$·$6H_2O$ (2 mmol), 0.30 g NH_4F (8 mmol) and 0.6 g urea (10 mmol) were dissolved in 36 mL distilled

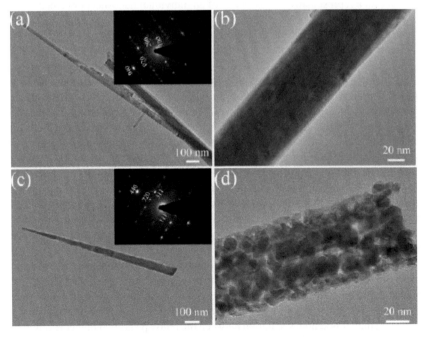

Fig. 5.56. Representative TEM images of (a, b) the $Co_2(OH)_2(CO_3)_2$ nanowire and (c, d) Co_3O_4 nanowire, with the insets in (a) and (c) showing the corresponding SAED patterns. Reproduced with permission from [106], Copyright © 2012, Royal Society of Chemistry.

Fig. 5.57. Schematic diagram demonstrating formation process of the 3D ordered nano-array structures. Reproduced with permission from [107], Copyright © 2012, Royal Society of Chemistry.

water, with constant stirring. Ni foam with a dimension of 3 cm × 2 cm was cleaned with concentrated HCl solution (37 wt%) with the aid of ultrasonication for 5 minutes, so as to remove the NiO layer, followed by thorough washing with DI water and absolute ethanol. The aqueous solution, together with the Ni foam, was subject to hydrothermal reaction at 100°C for 6 hours, 100°C for 9 hours, 120°C for 9 hours, in order to get Co_3O_4 NSAs, Co_3O_4 NSWAs and Co_3O_4 NWAs, respectively. After reactions, all the samples were thoroughly washed with distilled water and ethanol, under a strong ultrasonication, followed by drying at 80°C for 6 hours and thermal annealing at 250°C for 3 hours.

Representative SEM images of the Co_3O_4 NSWAs are shown in Fig. 5.58(A–C) [107]. The samples possessed an urchin-like structure, with multidirectional tertiary nanowires that were grown on the secondary nanosheets, leading to the formation of a highly dense film on the Ni foam. The nanosheets were vertically aligned on the substrate, with structure having a total size in the range of 6–8 μm. The nanowires exhibited an average diameter and an average length of 50–100 nm and 0.5–1 μm, respectively. This special structure had a relatively high surface area because of the hierarchical and free-standing nanowires, together with high morphology stability because of the large size of the nanosheets, which would be beneficial to achieve high electrochemical performances, owing to the enhanced access of the active ions into the electrode. Figure 5.58(D) shows a representative TEM image of an individual nanosheet branched with nanowires. It was observed that the Co_3O_4 nanowires were single-crystals, which were grown along the direction of [220], as demonstrated by HRTEM images and Fast Fourier Transform (FFT) patterns as seen in Fig. 5.58(E). Figure 5.58(F) shows of the Co_3O_4 NSWAs, indicating the highly crystalline phase.

Representative SEM images of the samples synthesized for different time durations are shown in Fig. 5.59(A–D), in order to illustrate the morphological and structural evolution from nanosheet array to nanosheet@nanowire array [107]. After reaction for 6 hours, the formation of the nanosheet array was started, with a relatively high density. One more hour later, i.e., after reaction for 7 hours, nanoparticles with a size of 10 nm started to nucleate on surfaces of the nanosheets. The nanoparticles possessed a thorn-like morphology, due to the anisotropic growth of the structure, as demonstrated by the inset in Fig. 5.59(B). As the reaction time was further increased to 8 hours, the thorn-like nanoparticles were elongated, thus forming rudiments of nanowires. At the same time, the length was still < 500 nm, i.e., shorter than the size of the final sample. Finally, after reaction for 9 hours, the assembling of the hierarchical nanostructures from the multidirectional nanowires was finished. The XRD pattern revealed that the as-synthesized samples were β–$Co(OH)_2$, which was transferred to Co_3O_4 after thermal annealing. Therefore, it suggested that length of the Co_3O_4 sub-nanowires could be maintained by controlling the reaction time, as schematically described in Fig. 5.59(E).

A hierarchical porous Co_3O_4 film, consisting of monolayer hollow sphere arrays and porous netlike nanoflakes, was fabricated by using monolayer polystyrene spheres as the template [108]. The individual nanoflakes were assembled with nanoparticles having sizes in the range of 5–15 nm, which thus resulted in a continuous porous structure, with pores of sizes in the range of 2–5 nm. Such hierarchical porous Co_3O_4 films demonstrated smaller potential separation between the redox peaks and higher reaction reversibility, as compared with those obtained without the use of the template. In addition, the hierarchical porous Co_3O_4 films exhibited a lower value of polarization during the charge–discharge processes. The hierarchical porous Co_3O_4 films and the Co_3O_4 nanoflakes displayed specific capacitances of 352 and 325 F·g^{-1}, respectively,

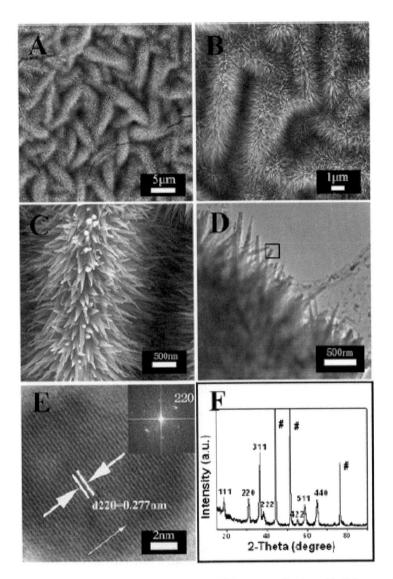

Fig. 5.58. (A, B, C) Representative SEM images of the NSWAs at different magnifications. (D, E) Representative TEM and HRTEM images of the NSWAs, with the inset showing the corresponding FFT pattern. (F) XRD pattern of the Co$_3$O$_4$ NSWAs. Reproduced with permission from [107], Copyright © 2012, Royal Society of Chemistry.

at a current density of 2 A·g^{-1}. Meanwhile, the specific capacitances were retained by 82.7 and 66.8%, respectively, as the discharge current density was increased from 2 to 40 A·g^{-1}. Moreover, the hierarchical porous Co$_3$O$_4$ films illustrated a higher cycling stability.

Figure 5.60 shows a schematic diagram demonstrating formation process of the hierarchically porous Co$_3$O$_4$ films [108]. Initially, a certain amount of polystyrene (PS) suspension was cast on the surface of Ni foil substrate. After holding stationary for 1 minute to ensure the well dispersion of the suspension, the substrate was gently immersed into DI water. A monolayer of PS spheres was formed to on the water surface on the surface of the Ni foil substrate, as the suspension was contacted with the water surface. After that, several drops of 2% dodecylsodiumsulfate solution were added to the water to vary its surface tension. In this case, the PS sphere monolayer that was suspended on the surface of the water was pushed aside due to the variation in the surface tension. To ensure that no additional PS spheres were attached onto

332 *Nanomaterials for Supercapacitors*

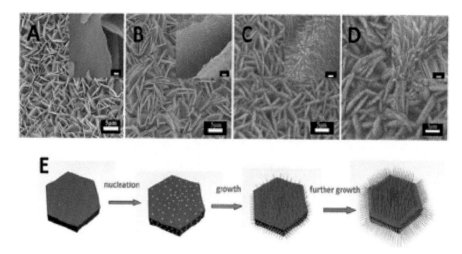

Fig. 5.59. SEM images of the samples after reaction for different time durations: (A) 6 hours, (B) 7 hours, (C) 8 hours and (D) 9 hours. The insets are the corresponding magnified SEM images with scale bars = 200 nm. (E) Scheme diagram illustrating formation process of the Co_3O_4 hierarchically structure. Reproduced with permission from [107], Copyright © 2012, Royal Society of Chemistry.

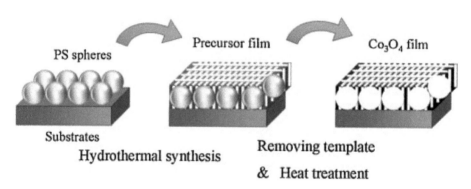

Fig. 5.60. Schematic diagram illustrating formation process of the hierarchically porous Co_3O_4 films. Reproduced with permission from [108], Copyright © 2012, Elsevier.

the monolayer during the pushing process, the substrate was lifted up through the clear area, which was then followed by heating the samples at 105°C for 2 hours, so that the PS sphere monolayer was strongly adhered on the Ni foil substrate.

To conduct the hydrothermal reaction, 5 mmol $Co(CH_3COO)_2$ and 15 mmol hexamethylenetetramine ($C_6H_{12}N_4$) were dissolved in 50 mL distilled water. The solution, together with the Ni foil substrate, was carried out at 100°C for 2 hours. After the hydrothermal reaction was finished, the samples were thoroughly washed with distilled water to remove the residual nanoparticle debris. Then, the samples were soaked in toluene for 24 hours to remove the PS sphere template, followed by drying at 85°C and then thermal annealing at 250°C for 1.5 hours in flowing Ar. The load weight of the Co_3O_4 film was about 0.8 mg·cm^{-2}.

A representative SEM image of the self-assembled PS sphere monolayer is shown in Fig. 5.61(a) [108]. The PS spheres were closely packed as a monolayer perpendicularly to the Ni foil substrate. After the PS sphere template was removed, highly porous Co_3O_4 films were thus obtained. Figure 5.61(b) shows a cross-sectional SEM image of the porous Co_3O_4 film, suggesting that it consisted of two types of structures. The substructure of the Co_3O_4 film contained a monolayer of closely packed hollow spheres,

Oxide Based Supercapacitors: II-Other Oxides 333

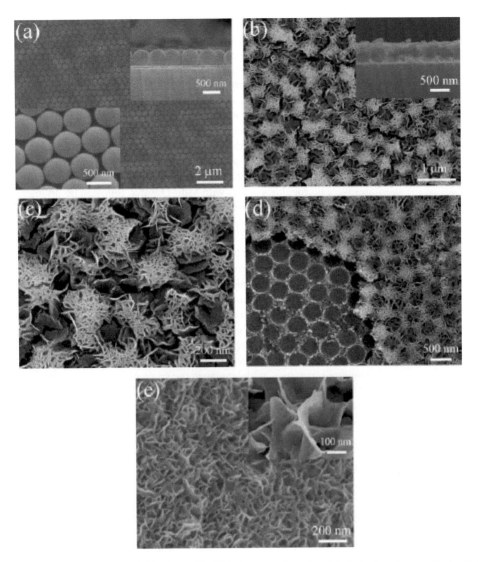

Fig. 5.61. SEM images of (a) top and side views of the PS sphere monolayer template (side view shown as the inset). (b, c) Top and side views of the hierarchically porous Co$_3$O$_4$ film (side view shown as the inset). (d) Representative SEM image of the film detached from the substrate. (e) SEM image of the Co$_3$O$_4$ film obtained without the use of the PS sphere template (fine structure shown as inset). Reproduced with permission from [108], Copyright © 2012, Elsevier.

with a diameter of 500 nm, whereas the superstructure was formed with free-standing Co$_3$O$_4$ nanoflakes, with a thickness of about 20 nm, as seen in Fig. 5.61(c). As a result, a porous net-like structure, with pore diameters in the range of 30–300 nm, was formed by the interconnected nanoflakes. The hierarchically porous structure could be detached from the substrate by using a blade, in the form of free-standing film, as observed in Fig. 5.61(d). In comparison, without the use of the template, the resultant Co$_3$O$_4$ film exhibited a randomly porous structure assembled with flakes that had a thickness of about 20 nm, as illustrated in Fig. 5.61(e). Microstructure of the individual nanoflake was examined by using TEM, which indicated that the individual flake possessed continuous pores with sizes in the range of 2–5 nm, which were achieved due to the packing of the nanoparticles with sizes of 5–15 nm. Crystalline nature of the Co$_3$O$_4$ film was confirmed by SAED and XRD patterns.

Co$_3$O$_4$ micro/nanostructures, with various morphologies, such as lath-like, necklace-like and net-like morphologies, have been grown on Ni foam, by using a hydrothermal synthesis method [110]. The different morphologies could be achieved by simply tuning the reaction temperature. It was found that the Co$_3$O$_4$ net-like structures exhibited the highest electrochemical performance, with a specific capacitance of as high as 1090 F·g^{-1}, with a mass loading of 1.4 mg·cm^{-2}.

The fabrication process involved two steps: (i) cobalt hydroxide carbonate precursor nanostructures were grown on the Ni foam substrates at 50–90°C from a solution consisting of cobalt precursor and urea and (ii) conversion of the hydroxide carbonate nanostructures into Co$_3$O$_4$ nanostructures through thermal calcination. The morphology of the samples was almost the same before and after the calcination process. To prepare the reaction solutions, CoCl$_2$·6H$_2$O and Co (NO$_3$)$_2$·6H$_2$O were dissolved in Milli-Q water. The solutions were then subject to hydrothermal reaction at different temperatures for 1–24 hours. After reaction, the samples on the substrates were thoroughly washed with Milli-Q water, followed by drying naturally in air. Finally, the dried samples were thermally annealed at temperatures of 170–300°C for 4 hours. The samples grown at 90°C from the Co(NO$_3$)$_2$·6H$_2$O solution and those grown at 90°C, 70°C and 50°C from the CoCl$_2$·6H$_2$O solution were denoted as S90N, S90Cl, S70Cl and S50Cl, respectively.

XRD pattern indicated that the cobalt chloride hydroxide carbonate had a chemical formula of Co (CO$_3$)$_{0.35}$Cl$_{0.20}$(OH)$_{1.10}$·1.74H$_2$O, which was readily converted to cubic Co$_3$O$_4$ after calcination. SEM image revealed that the calcined sample S90Cl exhibited a lath-like shape, with the laths that were assembled with interconnected Co$_3$O$_4$ nanoparticles, with sizes in the range of 18–28 nm and a mean diameter of 23 nm. The presence of porosity within the lath was confirmed by the TEM images. The formation of the porosity was closely related to the presence of water and CO2 molecules, due to the decomposition of the precursor.

Representative SEM images of sample S90N are shown in Fig. 5.62(a, b), indicating the flower-like Co$_3$O$_4$ structures which were derived from the Co (NO$_3$)$_2$·6H$_2$O solution at 90°C. The individual flower-like structure consisted of numerous nanorods, with lengths in the range of 2–4 μm. A high porosity was present due to the packing of the nanorods. The difference in morphology of Co$_3$O$_4$ nanostructures grown at 90°C from the CoCl$_2$·6H$_2$O and Co(NO$_3$)$_2$·6H$_2$O solutions could be ascribed to the unique space structure of Cl$^-$ and NO$_3^-$ ions in the solutions [41].

In case of the CoCl$_2$·6H$_2$O solution, the reaction temperature was shown to be an important factor to determine the morphology and dimensions of the Co$_3$O$_4$ nanostructures. At 70°C, the nanorod arrays exhibited a necklace-like morphology, as observed in Fig. 5.62(c, d). In this case, the nanorods were similar to those obtained at 90°C, with nanoparticle diameters of 18–28 nm. However, the sample S70Cl contained nanorods with an average diameter of < 50 nm, which was more convenient for the electrolyte ions to access the Co$_3$O$_4$. Representative SEM images of the sample S50Cl are shown in Fig. 5.62(e, f). This sample had the highest electrochemical performance. It had a net-like morphology, with a microstructure that was much finer than those of the samples obtained at higher temperatures.

Nickel Oxides

Another group of important oxides, which have been used as electrodes of supercapacitors and have been widely studied, are nickel oxides (NiO) [10, 111, 112]. NiO could exhibit multiple oxidation states, which is favorable to ensure a fast redox reaction, thus leading to potentially high specific capacitance. However, NiO has a low electrical conductivity [113], which has become a serious issue for supercapacitor applications. As a result, significant efforts have been made to improve the electrical conductivity of NiO, by forming nanocomposites to synergistically utilize the advantages of both the high electrochemical performance NiO and other high conductivity components.

Nanostructured NiO powders can be readily derived from Ni (OH)$_2$ through thermal decomposition [114, 115]. Various NiO nanostructures, such as nano/microspheres [116, 117], nanoflowers [118, 119], nanosheets [120] and nanofibers [121], have been synthesized by using different synthetic methods. Methods, including surfactant template [122], hard template [123, 124], sol-gel [125] and anodization [126], have been developed to fabricate porous NiO nanostructures. More recently, hierarchical porous NiO nanostructures with multiple sub-microstructures have been reported in the open literature [127–132]. As mentioned earlier, an appropriate porous nanostructure, with a sufficiently large surface area and high

Fig. 5.62. (a, b) SEM images of the sample S90N with flower-like morphology. (c, d) Necklace-like Co$_3$O$_4$ nanostructures of sample S70Cl. (e, f) Net-like Co$_3$O$_4$ nanostructures of the sample S50Cl. Reproduced with permission from [110], Copyright © 2011, American Chemical Society.

electrical conductivity, is an important factor to enhance electrochemical performance of an electrode [78, 133]. Strategies that have been employed to fabricate and synthesize NiO nanostructures used as the electrode of supercapacitors will be selectively described in a more detailed way next.

Hydrothermal/solvothermal synthesis

As discussed before, electrode materials with 3D porous nanostructures are promising candidates for supercapacitor applications, due to their relatively large surface area, smooth pathways for the electrolyte ions to access and high mechanical strength to accommodate volume variation during the charge-discharge cycling. Strategically, 3D nanostructured electrodes can be fabricated either by using metal foam as templates or by assembling the active materials based on certain synthetic routes. Usually, metallic foams, such as Ni foams, are used as the template to grow 3D NiO nanostructures, because of their highly porous microstructures and electrical conductivity.

A simple method was reported to develop NiO nanopowders, with a flake-like morphology, on Ni substrate, as electrodes of supercapacitor, by using a hydrothermal synthetic technique [134]. With H$_2$

as the reducing agent, composite was fabricated, which exhibited an optimal specific capacitance of 760 F·g⁻¹. The value was much higher than that of pure NiO based electrode (480 F·g⁻¹), which was attributed to the increased electrical conductivity of the composite as compared with the pure NiO.

To prepare NiO, 8 g Ni(NO$_3$)$_2$·6H$_2$O was dissolved in 200 mL mixed solvent of NMP and H$_2$O with a volume ratio of 39:1, under constant stirring at 190°C for 2 hours. After that, the suspension was subject to hydrothermal reaction at 190°C for 6 hours. The product was then collected through filtering and washed thoroughly with distilled water and ethanol, followed by drying at 60°C for 10 hours. To obtain NiO, the precursor powder was calcined at 250°C for 2 hours. To develop NiO/Ni composites, the NiO nanopowder was thermally treated at 190°C in H$_2$ for 20 minutes, 40 minutes and 60 minutes, thus leading samples denoted as NiO/Ni20, NiO/Ni40 and NiO/Ni60, respectively. The presence of both NiO and Ni was confirmed by XRD pattern.

Figure 5.63 shows representative SEM images of the NiO and NiO/Ni samples [134]. As seen in Fig. 5.63(a), pure NiO powder consisted of nanoflakelets with a thickness of about 10 nm, forming a hierarchical network-like structure. It was demonstrated that the NiO/Ni samples exhibited a gradual variation in morphology from flakelet to spherical, as reduction time was increased, as illustrated in Fig. 5.63(b–d). After reduction for 20 minutes, the sample NiO/Ni20 displayed an almost similar microstructure as the pure NiO, where nanoflakelets were dominant, as observed in Fig. 5.63(b). As the reduction time was increased to 40 minutes, NiO/Ni40 contained portions that became nanoparticles, although flakelet-like morphology was still a majority, as revealed in Fig. 5.63(c). In the sample NiO/Ni60, agglomerated larger particles became dominant, while almost no flakelet-like structures were observed, as demonstrated in Fig. 5.63(d). Therefore, reduction time was an important factor that could be used to control over morphology of the NiO/Ni composite powders. According to chemical titration results, the contents of Ni element in the samples were 67.63 (NiO), 68.48 (NiO/Ni20), 76.57 (NiO/Ni40) and 80.76% (NiO/Ni60), which was consistent with the XRD results.

Representative TEM images of the samples are shown in Fig. 5.64 [134]. Pure NiO powder contained flakelet-like items, with a thin hierarchical structure, as observed in Fig. 5.64(a). Darker areas of the TEM

Fig. 5.63. SEM images of the pure and composite samples: (a) NiO, (b) NiO/Ni20, (c) NiO/Ni40 and (d) NiO/Ni60. All the scale bars = 100 nm. Reproduced with permission from [134], Copyright © 2013, Royal Society of Chemistry.

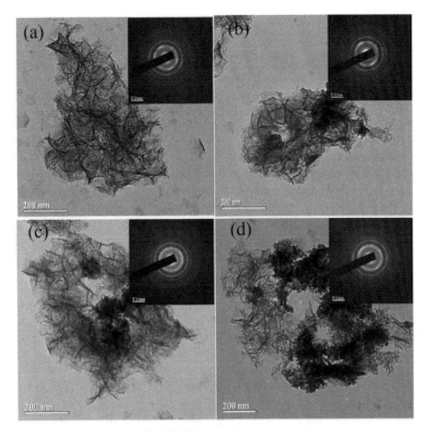

Fig. 5.64. TEM images of the samples: (a) NiO, (b) NiO/Ni20, (c) NiO/Ni40 and (d) NiO/Ni60. The insets were the corresponding SAED patterns. Reproduced with permission from [134], Copyright © 2013, Royal Society of Chemistry.

images revealed that the particles possessed sizes in the range of 15–30 nm, which were confirmed to be Ni particles according to the EDS results, as seen in Fig. 5.64(b–d). Obviously, the darker areas in the TEM images were increased gradually from NiO/Ni20 to NiO/Ni40 and NiO/Ni60, indicating an increase in the content of Ni. The SAED patterns (as insets), which were obtained in areas that had only nanoflakelets, were characterized with well-defined rings. They could be readily indexed to the crystal planes of NiO, confirming polycrystalline nature of the particles, which was thus consistent with the XRD results.

In the NiO/Ni composites, NiO nanoflakelets acted as active materials but had low electrical conductivity, whereas the Ni nanoparticles were highly conductive. As observed in the TEM images, the Ni nanoparticles were distributed in between the NiO nanoflakelets. Therefore, although the NiO nanoflakelets possessed a low electrical conductivity, the composites could be highly conductive, because the Ni particles provided a conduction path in the composites. Experimental data indicated that the conductivity of the NiO/Ni composites was higher than that of pure NiO nanoflakelets, while the conductivity of the NiO/Ni composites was increased with increasing content of Ni particles.

Lotus-root-like NiO nanosheets were fabricated by a hydrothermal method, but no electrochemical characterization was reported in the study [135]. During the experiment, the morphology of the NiO nanostructures was maintained by controlling the content of ammonia, while porous structures were created by using PVP. Ni(OH)$_2$ nanosheet precursors were prepared by using the chemical precipitation method and hydrothermal process. For example, to synthesize Ni(OH)$_2$ nanosheets, 1.426 g nickel nitrate hexahydrate and 0.5 g PVP were dissolved in 80 mL distilled water, with pH value that was adjusted to 8.30 with an ammonia solution. Then, the solution was subject to hydrothermal reaction at 150°C for 15 hours. After the reaction was finished, a greenish product was obtained, which was collected through centrifugation, followed by thorough washing with DI water and ethanol and then drying at 60°C in vacuum. After that, the hydroxide precursor was calcined at 400°C for 2 hours in air to be converted to lotus-root-like NiO nanosheets.

Phase composition and morphology of the NiO nanostructures were characterized by using XRD, FSEM and TEM, with representative results shown in Fig. 5.65 [135]. As demonstrated in Fig. 5.65(a), the NiO nanosheets had a morphology of lotus-root, with an average diameter of 400 nm. On the small-crystal 2D nanosheets, fine pores were uniformly distributed, with a mean diameter of 10 nm. Figure 5.65(b) shows a high-magnification SEM image, indicating that the small-crystal 2D nanosheets were assembled to NiO microspheres. The NiO microspheres possessed an average diameter of 2.5 μm, as illustrated in Fig. 5.65(c).

Figure 5.65(d) shows a representative TEM image of the NiO nanosheets, confirming the formation of the lotus-root-like structures, as demonstrated in the SEM images. Figure 5.65(e) shows TEM image of the NiO nanosheets from the microspheres. High magnification images indicated the individual particle consisted of numerous small intra-connected grains, with sizes of 4–5 nm, together with holes of 5–20 nm in size, as observed in Fig. 5.65(g, h). Crystalline structure of the nanosheets was demonstrated by the clearly observed d spacings in the HRTEM image, as shown in Fig. 5.65(i). The d spacings of 2.4 Å, 2.08 Å and 1.47 Å were attributed to the (111), (200) and (220) planes of NiO, respectively. Figure 5.65(f)

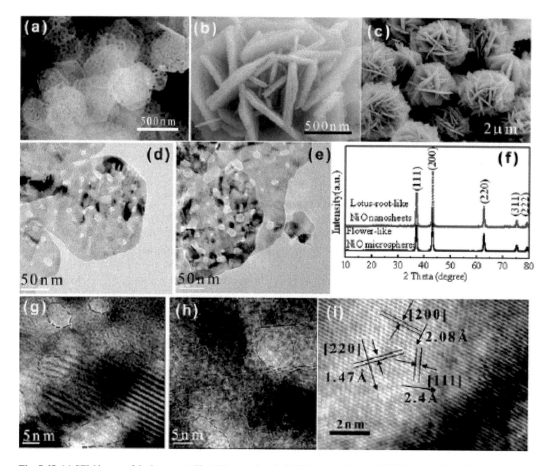

Fig. 5.65. (a) SEM image of the lotus-root-like NiO nanosheets. (b) High-magnification SEM image of the NiO microsphere. (c) SEM image of the flower-like NiO microspheres. (d) TEM image of the NiO nanosheets. (e) TEM image of the NiO microsphere. (f) XRD patterns of the NiO nanostructures. (g, h, i) HRTEM images. Reproduced with permission from [135], Copyright © 2011, Royal Society of Chemistry.

shows XRD patterns of the lotus-root-like NiO nanosheets and the NiO microspheres. Their diffraction peaks correspond to the (111), (200), (220), (311) and (222) planes, confirming their polycrystalline structure.

Mesoporous NiO nanowires have been synthesized by using a hydrothermal reaction combined with calcination [136]. The NiO nanowires exhibited a specific surface area of 85.18 $m^2 \cdot g^{-1}$ and a mean pore size of 12.5 nm. Electrode based on the NiO nanowires demonstrated a specific capacitance of 348 $F \cdot g^{-1}$. After the conversion from precursor Ni(OH)$_2$ to final NiO during the calcination process, the morphology was well retained. Mesoporous NiO nanowires were synthesized by the combination of a hydrothermal method and annealing. In a typical synthesis, 0.5 mmol of NiCl$_2 \cdot$6H$_2$O ($\geq$ 97% purity) was dissolved in 20 mL of distilled water. Then 0.03 g of K$_2$C$_2$O$_4$ (98% purity) and 8.0 mL of ethylene glycol (99% purity) were added under vigorous stirring. A clear transparent solution was formed gradually. The mixture was then transferred into a Teflon-lined stainless steel autoclave and heated to 220°C for 12 hours. After cooling to room temperature, the product was collected by centrifugation and washed thoroughly with distilled water and absolute ethanol several times, and then dried at 80°C for 12 hours in a vacuum oven. Finally, the as-prepared precursor materials were annealed at 400°C for 2 hours.

Figure 5.66 shows SEM images of the NiC$_2$O$_4 \cdot$2H$_2$O precursor nanowires and the NiO nanowires after calcination [136]. The NiC$_2$O$_4 \cdot$2H$_2$O precursor nanowires had lengths of up to 10s of micrometers, as observed in Fig. 5.66(a). After calcination, the morphology of the precursor was perfectly retained, without the occurrence of breaking, as demonstrated in Fig. 5.66(b). Figure 5.67 shows XRD patterns of the NiC$_2$O$_4 \cdot$2H$_2$O precursor and the NiO nanowires, confirming the phase compositions supposed to be.

Figure 5.68 TEM images of the precursor and the final NiO nanowires, to further reveal their microstructures [136]. As illustrated in Fig. 5.68(a), the precursor nanowires exhibited diameters in the

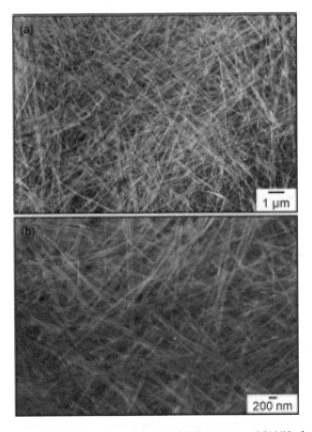

Fig. 5.66. SEM images of the nanowires: (a) as-synthesized NiC$_2$O$_4 \cdot$2H$_2$O precursor and (b) NiO after calcination. Reproduced with permission from [136], Copyright © 2012, WILEY-VCH Verlag GmbH & Co. KGaA, Weinheim.

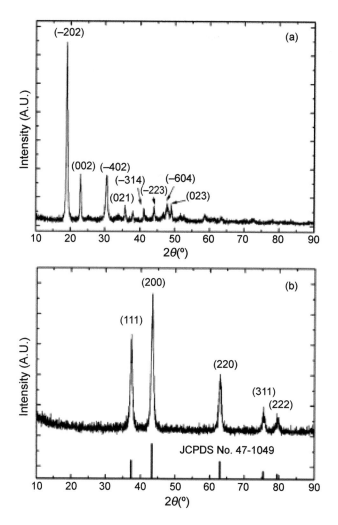

Fig. 5.67. XRD patterns of the nanowires: (a) $NiC_2O_4 \cdot 2H_2O$ precursor and (b) NiO nanowires after calcination. Reproduced with permission from [136], Copyright © 2012, WILEY-VCH Verlag GmbH & Co. KGaA, Weinheim.

range of 30–50 nm, with a very smooth surface. It was found that the precursor nanowires were unstable under the irradiation electron beam, due to the decomposition of $NiC_2O_4 \cdot 2H_2O$ to NiO. Crystal structure of the $NiC_2O_4 \cdot 2H_2O$ precursor nanowires was further confirmed by HRTEM image and SAED pattern. Figure 5.68(b) shows TEM image of an individual NiO nanowire. The diameters of NiO nanowires were in the range of 40–60 nm, which were slightly thicker than those of the precursor ones. The NiO nanowires were highly polycrystalline, comprising of numerous small nanoparticles. Figure 5.68(c) shows a high-magnification TEM image of an individual NiO nanowire, which consisted of NiO nanocrystals with sizes at the nanometer scale. In between the NiO nanoparticles, there were mesopores. The SAED pattern could be readily indexed to the cubic NiO, as shown as the inset in Fig. 5.68(c). As demonstrated in Fig. 5.68(d), the lattice-resolved HRTEM image of an individual NiO nanowire indicated an interplanar distance of 0.242 nm, corresponding to the (111) crystal plane of NiO.

The formation of NiO nanostructure from α-Ni(OH)$_2$ could be explained in terms of anisotropic Ostwald ripening [129]. During the early stage of reaction, the precursors would be dissolved into the solution, because of their high surface energy due to the high surface area. As the precipitating agent was introduced into the precursor solution, Ni^{2+} ions would be precipitated onto particles with the crystallographic facet

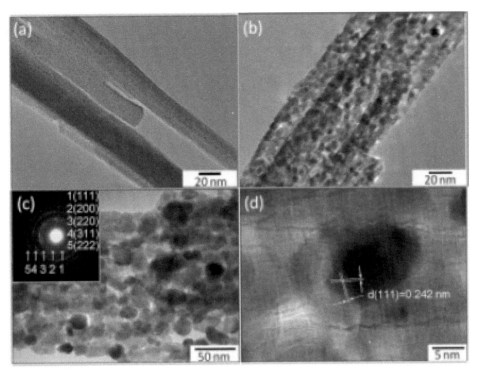

Fig. 5.68. (a) Low-magnification TEM image of the NiC$_2$O$_4$·2H$_2$O precursor nanowires. (b) TEM image of an individual NiO nanowire. (c) High magnification TEM image of an individual NiO nanowire, with the inset to be the corresponding SAED pattern. (d) Lattice-resolved HRTEM image of the NiO nanocrystals. Reproduced with permission from [136], Copyright © 2012, WILEY-VCH Verlag GmbH & Co. KGaA, Weinheim.

of high energy being present, thus resulting in the growth of large particles. At the same time, because the attachment and coalescence of the particles tended to take place having favorable crystallographic planes, oriented attachments would occurred, which led to the formation of large aggregated spherical particles. As the particles continued to grow, hierarchically porous morphologies were formed on the surface. Figure 5.69 shows a schematic diagram demonstrating formation mechanism.

It has been acknowledged that, during the decomposition of urea, which is one of the most widely used precursors in hydrothermal synthesis methods, CO$_3^{2-}$ and NH$_4^+$ ions are released, thus leading to the formation of HCO$_3^-$ and OH$^-$ ions, when the hydrolysis process takes place. For instance, as a relatively low level of urea was used, OH$^-$ ions were gradually released [137, 138]. However, if a relatively high amount of urea was employed, e.g., > 12 mmol [116], HCO$_3^-$ became the majority item.

A hydrothermal method was employed to synthesize NiO nanostructures, including nanobelts, nanospheres and nanoparticles, via the precursors of Ni(SO$_4$)$_{0.3}$(OH)$_{1.4}$ and Ni(HCO$_3$)$_2$, by using urea as the precipitating agent [116]. The NiO nanobelts, nanospheres and nanoparticles had BET surface areas of 44.2, 18.1 and 58.5 m^2·g^{-1}, respectively. Specific capacitances of the NiO nanospheres, nanobelts and nanoparticles were 45.2, 556.2 and 609.5 F·g^{-1}, respectively. Therefore, the observation implied that the larger surface area the NiO had, the higher the electrochemical performance would be achieved. Furthermore, the optimal specific capacitance of the NiO nanoparticles was retained by 62% after 3750 charge-discharge cycles.

To synthesize the Ni(SO$_4$)$_{0.3}$(OH)$_{1.4}$ and Ni(HCO$_3$)$_2$ nanostructure precursors, 2 mmol NiSO$_4$·(NH$_4$)$_2$SO$_4$·6H$_2$O and a certain amount of urea were dissolved in 30 mL DI water, to form a solution with the aid of sonication. The homogeneous solution was then subject to hydrothermal reaction at 180°C for 18 hours. After that, the samples were collected through centrifugation and thoroughly washed with

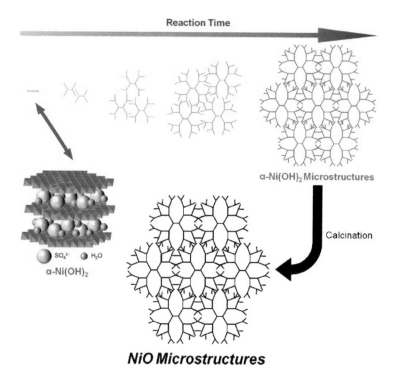

Fig. 5.69. Schematic diagram demonstrating of growth mechanism of α-Ni(OH)$_2$ with hierarchical microstructures. Reproduced with permission from [129], Copyright © 2011, Elsevier.

DI water and ethanol, followed by drying at 80°C for 12 hours. The solutions, with mole ratios of NiSO$_4$· (NH$_4$)$_2$SO$_4$·6H$_2$O to urea of 1:1, 1:6 and 1:15, resulted in samples denoted as P1, P2 and P3, respectively. NiO nanostructured powders with different morphologies were obtained by calcining the Ni (SO$_4$)$_{0.3}$(OH)$_{1.4}$ and Ni(HCO$_3$)$_2$ nanostructure precursors at 450°C for 2 hours in air.

XRD pattern of the sample P1 confirmed the phase composition of Ni (SO$_4$)$_{0.3}$(OH)$_{1.4}$. At the same time, SEM images revealed that the as-obtained sample P1 exhibited belt-like structures, with lengths at the micrometer level and an average width of about 100 nm. As the mole ratio of NiSO$_4$·(NH$_4$)$_2$SO$_4$·6H$_2$O/ urea was decreased to 1:6, XRD pattern of sample P2 indicated that the phase composition was changed to Ni (HCO$_3$)$_2$. Meanwhile, the Ni (HCO$_3$)$_2$ powder consisted of anocrystals, according to the broadened XRD diffraction peaks. When the mole ratio of NiSO$_4$· (NH$_4$)$_2$SO$_4$·6H$_2$O/urea was further decreased to 1:15, the phase composition was still Ni(HCO$_3$)$_2$, while the powder was composed of microspheres, which were assembled with nanoparticles.

Figure 5.70 shows representative SEM and TEM images of NiO nanosized powders derived from the Ni-based precursors after calcination. As seen in Fig. 5.70(a, b), NiO nanobelts were derived from the belt-like Ni (SO$_4$)$_{0.3}$(OH)$_{1.4}$ precursor. From the spherical Ni (HCO$_3$)$_2$ nanostructured precursor, NiO nanospheres consisting of small nanoparticles were obtained, as illustrated in Fig. 5.70(c, d). With sample P3 as the precursor, NiO nanoparticles with an average diameter of 30 nm were synthesized, as observed in Fig. 5.70(e, f). These results implied that NiO nanostructures with different morphologies could be readily developed by using different Ni-based nanostructure precursors whose morphologies could be simply controlled by adjusting the composition of the hydrothermal reaction solutions.

Porous NiO nanoslices, nanoplates and nanocolumns have been prepared by using a simple hydrothermal reaction method, which were used as electrodes of supercapacitors [139]. Electrochemical characterization results indicated the porous NiO nanocolumns had a specific capacitance of 390 F·g^{-1}, which was higher than the values of the other two nanostructures. The higher electrochemical performance

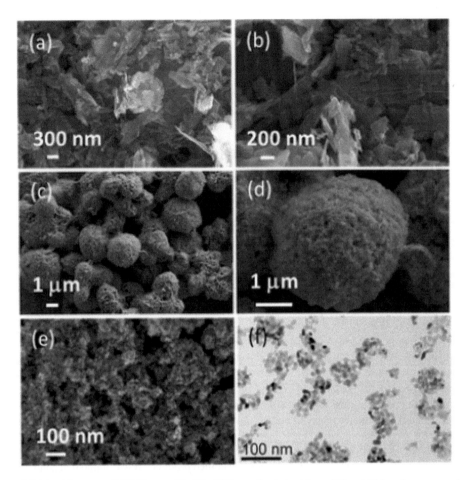

Fig. 5.70. SEM (a–e) images and TEM images (f) of the NiO samples derived from different Ni-based precursors: (a, b) P1, Ni (SO$_4$)$_{0.3}$(OH)$_{1.4}$ nanobelts, (c, d) Ni(HCO$_3$)$_2$ nanospheres and (e, f) Ni(HCO$_3$)$_2$ nanoparticles. Reproduced with permission from [116], Copyright © 2015, Royal Society of Chemistry.

of the nanocolumns was attributed to their surface area due to the stacking of the β-Ni(OH)$_2$ nano-slices/plates favors, which resulted in a large number of small pores in the final NiO nanocrystals after calcination.

To obtain Ni (OH)$_2$ precursors with different morphologies, 0.02 mmol NiCl$_2$·6H$_2$O was dissolved in 10 mL DI water to form solutions, with pH values adjusted to be 12, 13, or 14 by adding NaOH solution, followed by vigorous stirring for 1 hour. Then, the solutions were hydrothermally treated at 160°C for 8 hours. After the reaction was finished, the products were collected and thoroughly washed with distilled water and ethanol, combined with centrifugation. Porous NiO nanostructured powders were obtained by calcining the Ni (OH)$_2$ nanocrystals at 400°C for 2 hours.

Figure 5.71 shows SEM images of the Ni (OH)$_2$ nanostructured precursors with different morphologies, which were obtained at different synthetic conditions. In the crystal structure of Ni (OH)$_2$, every two Ni^{2+} ions in the adjacent (100) planes are linked through one OH$^-$ ion, in such a way that the two Ni^{2+} ions in two adjacent (001) planes are separated by two layers of OH$^-$ ions. This is the reason why the morphology of Ni (OH)$_2$ can be tailored by simply varying pH value of the hydrothermal reaction solutions. At pH = 14, the sample was present as hexagonal nanoslices, with an average diameter of 300 nm and thicknesses in the range of 5–10 nm, as demonstrated in Fig. 5.71(a, d). As the pH value was decreased to 12, Ni (OH)$_2$ nanoplates were obtained, which had thicknesses in the range of 20–50 nm, as illustrated in Fig. 5.71 (b, e). At pH = 13, Ni(OH)$_2$ nanoplates were formed, with their basal surfaces stacking to form columnar

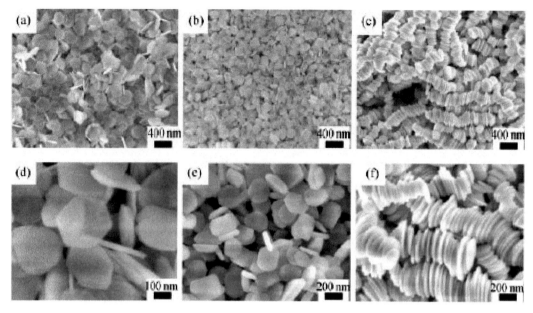

Fig. 5.71. Representative SEM images of the β-Ni (OH)$_2$ nanostructures: (a, d) nanoslices with diameters of 200–400 nm and thicknesses of 5–10 nm, (b, e) nanoplates with diameters of 100–300 nm and thicknesses of 20–50 nm and (c, f) nanocolumns with diameters of 200–400 nm and heights of 1–2 μm. Reproduced with permission from [139], Copyright © 2010, Springer.

structures, with diameters in the range of 200–400 nm and heights in the range of 1–2 μm, as observed in Fig. 5.71(c, f). HRTEM images revealed that Ni (OH)$_2$ nanoslices and nanoplates were of single crystal nature, while the nanocolumns were stacked with the nanoplates along the [001] direction. XRD results indicated that the precursors were β-Ni (OH)$_2$, whereas the calcined products were all single phase NiO.

Figure 5.72 shows SEM and TEM images of the calcined products. It was found that nanopores were formed in the NiO nanocrystals. Sizes of the pores were dependent on the nanostructural characteristics of the β-Ni (OH)$_2$ precursors. After calcining the β-Ni (OH)$_2$ nanoslices, the sample possessed large pores with sizes of 15–30 nm, as seen in Fig. 5.72(a, d). Besides the large pores with diameters in the range of 15–30 nm, the samples derived from the nanoplates and nanocolumns also had smaller pores, with average diameters of 10 nm and 2 nm, as revealed in Fig. 5.72(b, e) and Fig. 5.72(c, f), respectively. The presence of smaller pores in the nanocolumns could be attributed to their special structure, i.e., the stacking of the slices/plates through the overlapping of their basal surfaces. In this case, the reduced surface area slowed down the thermal decomposition of the precursor, thus leading to slower water loss and pore-generation. As a consequence, smaller pores were formed in the nanocolumns.

Also with urea, a hydrothermal synthetic route has been reported to prepare NiO precursor at various temperatures, which was then transferred to nanostructured NiO with a flake-like morphology after thermal annealing at a relatively low temperature [140]. The NiO nanoflakes had widths of 50–80 nm and an average thickness of 20 nm. Electrochemical capacitive characterization results indicated that the NiO nanostructures exhibited promising performances in 2 M KOH electrolyte solution. An outstanding cycle stability was observed, with the initial specific capacitance to be retained by 91.6% after 1000 charge–discharge cycles. An electrochemical impedance spectroscopy study indicated that the performance of the NiO electrode was determined by the mass transfer limitation in the active material, which displayed an internal resistance of 0.2 Ω. An optimal specific capacitance of 137.7 F·g^{-1} was obtained at the current density of 0.2 A·g^{-1}, within the potential window of 0–0.46 V. The high performance of the NiO nanoflakes was readily attributed to the high surface area of the materials, which ensured transport of electrolyte ions during the charge/discharge process.

Oxide Based Supercapacitors: II-Other Oxides 345

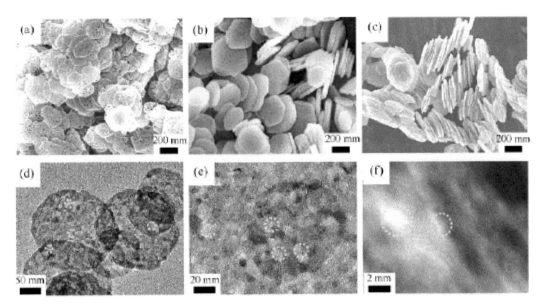

Fig. 5.72. SEM images of the porous NiO nanocrystals: (a) nanoslices, (b) nanoplates and (c) nanocolumns. Representative HRTEM images of the porous NiO nanocrystals: (d) nanoslices with pore sizes of 10–30 nm, (e) nanoplates with pore sizes of 10–25 nm and (f) nanocolumns with pore sizes of 2–5 nm. The dotted circles in (d)–(f) are used to highlight the pores in the NiO nanocrystals. Reproduced with permission from [139], Copyright © 2010, Springer.

To synthesize the precursors of NiO, Ni $(NO_3)_2$ and urea with a molar ratio of 1:4 were dissolved in water to form a solution with a concentration of 0.1 M, under vigorous stirring for 1 hour. The solution was then subjected to hydrothermal reaction at 140°C and 150°C for 6 hours. After reaction, the greenish Ni $(OH)_2$ precipitates were collected through filtration, followed by thorough washing with super-pure water and ethanol and drying at 105°C for 12 hours. The dried precursors were calcined at 300°C for 3 hours to obtain NiO nanostructures, with the phase confirmed by XRD patterns. The products obtained at the hydrothermal reaction temperatures of 140°C and 150°C were denoted as sample 1 and sample 2, respectively.

Figure 5.73 shows SEM images of the NiO nanostructures derived from the precursors hydrothermally reacted at the two temperatures [140]. As seen in Fig. 5.73(a), Sample 1 consisted of micaceous nanoflakes with sandwich and cauliflower morphologies, which could be ascribed to the presence of urea in the reaction solutions that acted as the hydrolysis-controlling agent. The nanostructured NiO had a BET surface area of 107.5 $m^2 \cdot g^{-1}$. Sample 2 also exhibited a nanoflake-like morphology, but with a denser packing state, as illustrated in Fig. 5.73(b). As a consequence, it had a lower specific surface area.

During the hydrothermal reactions, concentrations of the reagents also played a significant role in determining morphology of the hydroxide precursors and thus the morphology of the final metal oxides. For example, when using the hydrothermal method to synthesize NiO nanoplatelets on conductive substrate, it was found that the effects of reaction temperature, reaction time, type and concentration of the Ni^{2+} source (such as NO^{3-}, Cl^- and SO_4^{2-}) on morphology of the Ni $(OH)_2$ nanoarrays were very weak [141]. However, the concentrations of NH_3 and $K_2S_2O_8$ had demonstrated a significant effect on morphology evolution of the Ni $(OH)_2$ nanoarray precursors. At a relatively low concentration of NH_3, the growth of Ni $(OH)_2$ took place at a high rate, thus leading to a poor control over the morphology. In contrast, if the concentration of NH_3 was sufficiently high, short and thick nanoplatelets were obtained. It could be attributed to the fact a high concentration of NH_3 ensured a high surface coverage, so that stitching up of the surface with each other was facilitated to trigger the formation of a more compact architecture [142, 143]. Also, excessive NH_3 could hinder the growth of Ni $(OH)_2$ to a certain degree, because of the strong chelation effect of NH_3 on Ni^{2+} ions. The porous film of NiO nanoplatelet arrays showed a capacitance of up to 64 mF cm^{-2}

Fig. 5.73. SEM images of the NiO nanostructures: (a) sample 1 and (b) NiO sample 2. Reproduced with permission from [140], Copyright © 2009, Elsevier.

at a current density of 4 mA cm^{-2}. Also, the NiO film based electrode exhibited promising electrochemical performances, in terms of specific capacitance, recharge time and cyclic stability.

Similarly, the concentration of $K_2S_2O_8$ was also very crucial to the morphology development of the Ni(OH)$_2$ nanoplatelet arrays [141]. It was found that, without the presence of $K_2S_2O_8$, the Ni (OH)$_2$ crystals were grown randomly in all directions, thus resulting in a dense Ni (OH)$_2$ film. Once $K_2S_2O_8$ was present in the reaction solutions, the $S_2O_8^{2-}$ ions were selectively adsorbed onto the (001) surface of the crystals [144]. Therefore, growth of the (100) planes was limited, so that the Ni (OH)$_2$ crystals were stacked in the direction vertical to the surface of the substrate. With increasing concentration of $S_2O_8^{2-}$ ions, both the height of the array and the porosity of the film were increased.

To grow the single-crystalline Ni (OH)$_2$ platelet nanoarray, a solution was prepared by dissolving 1.16 g Ni(NO$_3$)$_2$·5H$_2$O and 0.2 g $K_2S_2O_8$ in 35 mL distilled water, into which 5 mL condensed aqueous ammonia (25–28%) was added, followed by stirring for 10 minutes. Then, the solution was subjected to hydrothermal reaction, together with a piece of Fluorine-doped Tin Oxide (FTO) coated glass substrate, at 150°C for 10 hours. After reaction, a green Ni (OH)$_2$ array film were deposited on the side of FTO. The film was then calcined at 400°C for 2 hours, to form NiO films dark or yellow-dark color depending on the thickness of the films. XRD patterns indicated that the as-deposited film was β-Ni(OH)$_2$, while the calcined sample was NiO.

A representative SEM image of the β-Ni(OH)$_2$ array is shown in Fig. 5.74(a). The film had a porous microstructure, consisting of interconnected uniform platelets, with a thickness of < 100 nm and a length of 1–2 μm. Cross-sectional view revealed that the platelets were aligned vertically to the substrate, with a height of 3–4 μm, as seen in Fig. 5.74(c). Figure 5.74(d) shows an enlarged cross-sectional SEM image, indicating that the nanoplatelets were constructed with multilayers of densely stacked thin nanosheets, owing to the Layered Double Hydroxide (LDH) characteristic of Ni (OH)$_2$ [145]. After calcination, almost no change in morphology was observed, as illustrated in Fig. 5.74(b). Thickness of the film could be maintained by controlling hydrothermal reaction time.

Figure 5.75 shows representative TEM and HRTEM images of the nanoarrays, confirming the microstructures observed by using SEM. SAED pattern inset in Fig. 5.75(a) further indicated that the Ni (OH)$_2$ nanoplatelets were of single-crystalline nature. Figure 5.75(b) shows HRTEM image of an individual Ni (OH)$_2$ nanoplatelet, with distinct lattice fringes further revealing its single-crystalline characteristics. According to the nature of the hexagonal structure, the planes perpendicular to the electron beam were (001), which implied that the nanoplatelets were formed due to the stacking of the thin nanosheets with exposed (001) planes along the [0001] direction.

A low-resolution TEM image of a NiO platelet is shown in Fig. 5.75(c). Although the nanoplatelet motif was well retained after calcination, pores were formed in the NiO sample. Single-crystal nature of the NiO was confirmed by the SAED pattern to have a cubic structure, as shown as the inset in Fig. 5.75(c). The SAED pattern revealed a hexagonal alignment of (220) spots, corresponding to the diffraction pattern

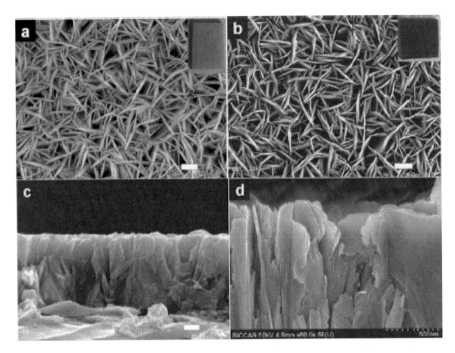

Fig. 5.74. SEM images of the array films: (a) Ni (OH)$_2$ and (b) NiO after calcination. The insets present the photographs of the arrays on the FTO substrate. (c, d) Cross-sectional SEM images of the nanoarrays at different magnifications. The scale bars in (a, b, c) stand for 1 μm. Reproduced with permission from [141], Copyright © 2011, Royal Society of Chemistry.

along the [111] zone axis of the NiO crystal [146]. Figure 5.75(d) shows an HRTEM image, demonstrating that the 2D lattice fringes with a spacing of 0.148 nm belonged to the (220) planes of the Face-Centered-Cubic (FCC) NiO, in good agreement with the SAED pattern [147]. Therefore, the NiO nanoplatelets were single crystals and the two dominant surfaces corresponded to the (111) planes.

To examine growth kinetics of the Ni (OH)$_2$ nanoarrays, various hydrothermal reaction parameters were studied, such as temperature (120–180°C), reaction time (2, 5, 10 and 24 hours), type Ni^{2+} precursor (NO$_3^-$, Cl$^-$ and SO$_4^{2-}$) and the solution concentration (0.1, 0.2 and 0.3 M). Typical SEM images of the samples are shown in Fig. 5.76(a, b). Figure 5.76(c) shows a schematic diagram describing formation mechanism of the Ni (OH)$_2$ nanoarrays.

When using hydrothermal template guided synthetic methods, reaction parameters, such as properties of the starting solutions, the amount of template, reaction temperature and time, could have significant effects on the morphologies of NiO powders and thus their electrochemical performances as the electrodes of supercapacitors. For example, as a solution consisting of NiCl$_2$·6H$_2$O, ammonia and glucose as the template was subject to hydrothermal reaction with normal conditions, e.g., at 140°C, the sample was comprised of carbonaceous polysaccharide microspheres [148]. Because there were a large number of hydroxyl and carbonyl groups, the surfaces became highly hydrophilic. As a result, the Ni^{2+} ions were easily adsorbed onto the carbonaceous polysaccharide microspheres, where they reacted with OH$^-$ ions released during the hydrolysis of ammonia, thus forming NiO hollow spheres consisting of nanoparticles. If the reaction temperature was too high, NiO microspheres would have smooth surface, so that the Ni^{2+} ions would react with more OH$^-$ ions, thus leading to Ni (OH)$_2$ spheres with thicker shells. Such NiO hollow spheres exhibited a very low specific capacitance of 51.7 F·g^{-1}, due mainly to high crystallinity of NiO items, as evidenced by the XRD patterns with very sharp diffraction peaks.

To synthesize the precursor of NiO hollow spheres, 0.5348 g NiCl$_2$·6H$_2$O and 0.2 mL ammonia were dissolved in 19.8 mL distilled water to form aqueous solution, into which 2.0 g glucose was added with constant stirring for 5 minutes. After that, the solution was subject to hydrothermal reaction with a 25 mL

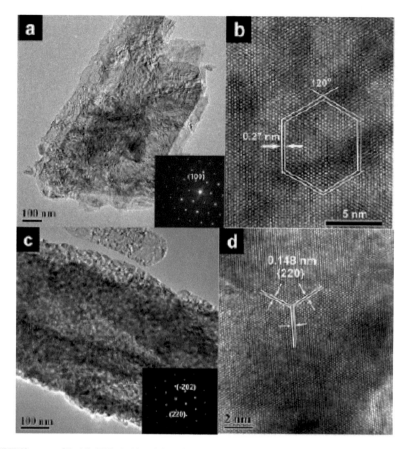

Fig. 5.75. HRTEM images of the Ni (OH)$_2$ (a, b) and the NiO nanoplatelets (c, d). The insets in (a) and (c) are the corresponding SAED patterns. Reproduced with permission from [141], Copyright © 2011, Royal Society of Chemistry.

Teflon-lined stainless-steel autoclave at 140°C for 12 hours. After the end of the reaction, the products were collected through filtration, followed by thorough washing with distilled water and absolute ethyl alcohol and drying at 60°C for 6 hours, thus leading to black precursor powders. NiO hollow spheres were obtained by calcining the precursor powders at 450°C for 2 hours. Hydrothermal reaction temperatures 80°C, 140°C and 180°C were studied to check the effect on morphology of the precursor and the final NiO samples.

Figure 5.77 shows SEM images of the Ni(OH)$_2$ and NiO samples prepared at different hydrothermal reaction temperatures [148]. At a low reaction temperature of 80°C, the powder contained only nanoparticles, as observed in Fig. 5.77(a). After calcination, the nanoparticles were aggregated into large particles, as seen in Fig. 5.77(b). As the reaction temperature was increased to 140°C, spherical Ni (OH)$_2$ particles with diameters of 4–5 μm were obtained, as illustrated in Fig. 5.77(c). Calcination of the spherical precursor led to NiO hollow spheres, with an average diameter of 2 μm, as demonstrated in Fig. 5.77(d). The hollow spheres had holes on the rough surface. If the reaction temperature was raised to 180°C, Ni (OH)$_2$ microspheres with smooth surface were developed, as revealed in Fig. 5.77(e). As a result, NiO microspheres with smooth surface were obtained after calcination, as depicted in Fig. 5.77(f), suggesting that the spherical morphology of the precursor could well survive the high temperature treatment.

A hydrothermal reaction method was used to synthesize NiO nanosheet nanostructures, with anionic surfactant Sodium Dodecyl Sulfate (SDS) as the template [149]. The NiO nanosheets prepared at 160°C exhibited the optimal specific capacitance of 989 F·g^{-1}, due to its porous microstructure to effectively facilitate the smooth transport of the electrolyte ions. To prepare precursors of the NiO nanosheets,

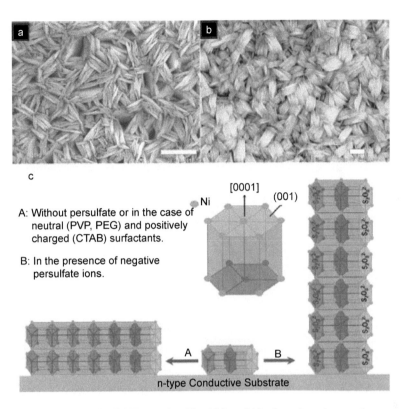

Fig. 5.76. (a) SEM image of the 10 h Ni(OH)$_2$ sample with addition of 10 mL condensed ammonia aqueous solution. (b) SEM image of the 10 h Ni(OH)$_2$ sample without the presence of potassium persulfate. (c) Possible growth mechanism of the Ni(OH)$_2$ (001) nanoarrays. The scale bars in (a) and (b) are 5 μm and 1 μm, respectively. Reproduced with permission from [141], Copyright © 2011, Royal Society of Chemistry.

0.1081 g SDS was dissolved in 50 mL double distilled water, while 0.4362 g Ni(NO$_3$)$_2$·6H$_2$O was dissolved in 15 mL water. The latter solution was added slowly to the SDS solution at room temperature with constant stirring, into which 0.1801 g urea dissolved in 10 mL water was added, followed by constant stirring for 2 hours. The solution was then used to conduct hydrothermal reaction at 120, 140, 160 and 180°C for 24 hours. After the reaction, green precipitates were collected through centrifugation, followed by thorough washing with ethanol and double distilled water. To obtain NiO nanostructure powders, the dried precursors were calcined at 300°C for 2 hours in air. The samples were denoted as N120, N140, N160 and N180, respectively.

SEM images confirmed the N120 and N140 samples consisted of hierarchical microspheres, which were assembled with 2D nanosheet arrays interwoven to form the 3D. The N180 sample possessed a plate-like morphology, due to the agglomeration of nanorods. Figure 5.78 shows a schematic diagram describing the possible formation mechanism of the NiO nanosheets. A sheet-like structure was developed first, owing to the presence of the anion surfactant, so that lamellar micelles were formed, which were further assembled to the microscaled architectures. Initially, the hydrothermal reaction to trigger the formation of Ni(OH)$_2$, due to the release of OH$^-$ ions during the hydrolysis of urea. At different reaction temperatures, both the nucleation and growth rates were different. At 120 and 140°C, the assembly of nanosheets was relatively slow, so that microspheres were obtained. With increasing reaction temperature, surface tension of the solution was gradually decreased, corresponding to weak electrostatic interaction. Therefore, during the reaction at 160°C, the aggregation was weakened and microspheres were formed with well-resolved nanosheets, because of the reduced surface tension. However, as the reaction temperature was increased to 180°C, both the nucleation and growth rates were too high, so that there was no sufficient time for the

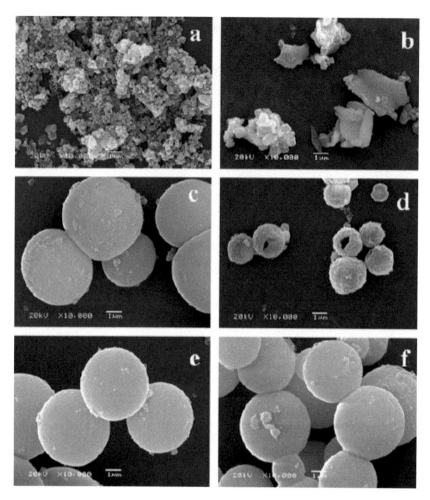

Fig. 5.77. SEM images of the Ni (OH)$_2$ (left-hand side) and NiO (right-hand side) samples hydrothermally treated at different reaction temperatures: (a, b) 80°C, (c, d) 140°C and (e, f) 180°C. Reproduced with permission from [148], Copyright © 2010, Elsevier.

anion surfactant and cation to assemble, thus leading to the formation of thick nanorod-assembled plates [150, 151].

In a separate study, surfactant cetyltrimethyl ammonium bromide (CTAB) was used as the template to synthesize nanoporous pine-cone structured NiO, with a high surface area of 265 m^2·g^{-1} and an average pore diameter of 4.19 nm, which exhibited an optimal specific capacitance of 337 F·g^{-1} [152]. To prepare the reaction solution, 20 mmol Ni (NO$_3$)$_2$·6H$_2$O and 10 mmol CTAB were dissolved in 100 mL triple distilled water, respectively, to form solutions. Then, the Ni (NO$_3$)$_2$·6H$_2$O solution was dropwise added into the CTAB solution, with constant stirring for 1 hour, so that a homogeneous green solution was obtained. After that, 40 mmol urea was added to the solution, with stirring for 3 hours to ensure desired homogeneity. The solution was then hydrothermally reacted at 120°C for 24 hours, which led to fluffy grass-green precipitate. The green precipitate was then collected through high speed centrifugation, followed by thorough washing with triple distilled water, then a mixture of absolute ethanol and water and finally absolute ethanol. The washed product was dried at 60°C for 24 hours in vacuum. To get NiO powders, the precursor was calcined at 300°C for 3 hours in flowing air.

Figure 5.79 shows representative SEM images of NiO sample. Low magnification SEM images indicated that NiO powder possessed a hierarchical porous spherical morphology, with a relatively uniform

Oxide Based Supercapacitors: II-Other Oxides 351

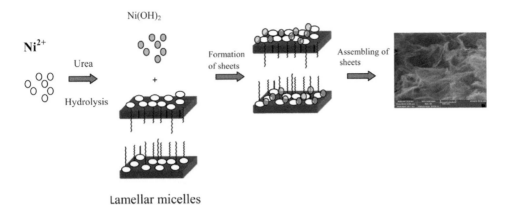

Fig. 5.78. Schematic diagram showing possible formation mechanism of the NiO nanosheet assembled nanostructures. Reproduced with permission from [149], Copyright © 2013, American Chemical Society.

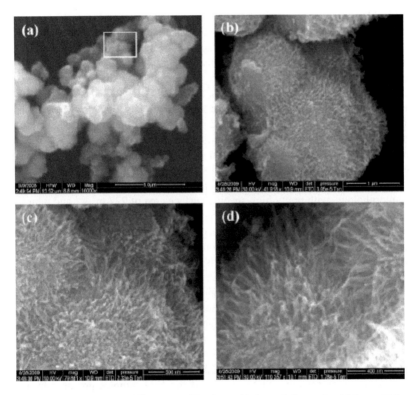

Fig. 5.79. SEM images of NiO powder at different magnifications with the scale bars to be: (a) 5 μm, (b) 1 μm, (c) 500 nm and (d) 400 nm. Reproduced with permission from [152], Copyright © 2010, Elsevier.

size distribution. High magnification images revealed the pine-cone fibrous structures were grown on surface of the porous NiO microspheres. Representative TEM images of the NiO powder are shown in Fig. 5.80, confirming that the NiO microspheres had aggregates of nanofibers on their surface. The porous sphere-like structures possessed an average diameter of about 2 μm, as seen in Fig. 5.80(a), while the surface nanofibers had an average diameter of about 15 nm, as illustrated in Fig. 5.80(b). The formation of the

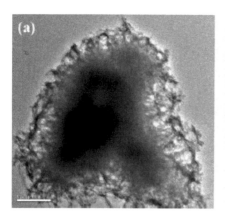 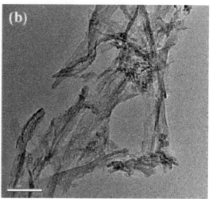

Fig. 5.80. TEM images of the NiO powder showing the porous structures made of nanofibers at different magnifications: (a) 0.2 μm and (b) 50 nm. Reproduced with permission from [152], Copyright © 2010, Elsevier.

porous hierarchical NiO powder was attributed to the removal of water and surfactant residues during the calcination process. It was found that this method is feasible to develop uniform porous microstructured NiO powder, consisting of microspheres whose surface was characterized by nano-fiber assembled pine-cone items, with high productivity and reproducibility.

It has been demonstrated that surface area, pore size and pore volume of NiO nanostructures could be controlled by using different types of surfactants, such as CTAB, SDS and Triton X-100, which thus further influenced electrochemical properties of the active materials. For instance, NiO nanostructures with a spherical morphology have been synthesized by using these organic surfactants as the templates and urea as the hydrolysis agent [153]. It was found that nucleation behavior of Ni (OH)$_2$ was strongly dependent on factors like adsorption mechanism and nature of the interaction between the surfactant and inorganic precursors, which was then influenced physical properties of the final product, such as surface area, pore volume, pore size, crystallite size, morphology and microstructure. The NiO-SDS displayed a nanoflake-like morphology, with pretty high specific surface area of 227 $m^2 \cdot g^{-1}$ and a high pore volume of 0.352 $m^3 \cdot g^{-1}$. Specific capacitances of the NiO powders were increased in the order of NiO-Triton < NiO-CTAB < NiO-SDS, with experimental values of 144, 239 and 411 $F \cdot g^{-1}$, respectively.

The precursor solutions for hydrothermal reactions were prepared by using 20 mmol Ni(NO$_3$)$_2 \cdot$4H$_2$O, 40 mmol urea and 10 mmol surfactants. They were dissolved in 200 mL triple-distilled water, with stirring for 2 hours at room temperature to obtain respective transparent solutions, which were then subject to hydrothermal reaction, with a 250 mL Teflon-lined stainless steel autoclave, at 120°C for 24 hours. After the reactions were finished, greenish precipitates were obtained, which were collected through centrifugation, followed by thorough washing with distilled water and ethanol. After being dried at 60°C for 24 hours, the precipitates were calcined at 300°C for 3 hours in air to obtain NiO powders. Surfactants, including Sodium Dodecyl Sulfate (SDS), cetyltrimethyl ammonium bromide (CTAB) and Triton X-100, were used, corresponding samples denoted as NiO-S, NiO-C and NiO-T, respectively.

Figure 5.81 shows representative SEM images of the NiO samples prepared by using different surfactants. Both NiO-C and NiO-T possessed a sphere-like morphology, while NiO-S had a porous nanoflake-like morphology. The presence of the flake-like morphology in the NiO-S sample was further confirmed by high magnification SEM and HRTEM images. Due to its unique morphology, the sample NiO-S demonstrated the highest electrochemical performances, due to the high accessibility to OH$^-$ ions into the pores of the active materials. Therefore, as stated above, the NiO-S based electrode exhibited higher specific capacitance than those based on NiO-C and NiO-T. The experimental results also revealed that the hydrolysis rate of nickel nitrate was well facilitated by using urea as the precipitating agent at 120°C, which was responsible for the formation of the stable porous nickel hydroxide skeleton. Furthermore, after calcination, the pore skeleton of the sample was well retained.

Oxide Based Supercapacitors: II-Other Oxides 353

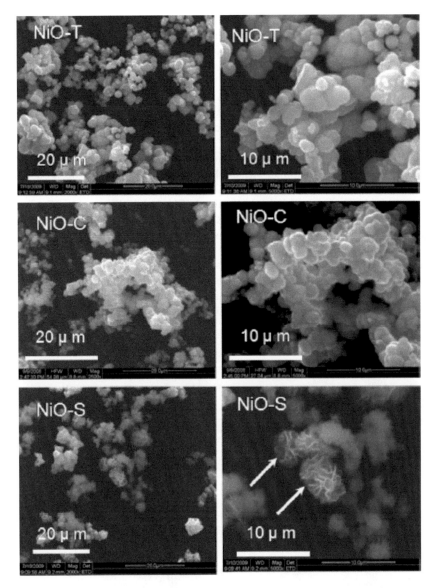

Fig. 5.81. SEM images of the NiO samples calcined at 300°C at different magnifications. Reproduced with permission from [153], Copyright © 2010, American Chemical Society.

In similar study, CTAB was also used as the template to synthesize porous NiO powder, which exhibited an optimal specific capacitance of 279 F·g^{-1} [154]. In this case, the effect of type of anions, including Cl$^-$, CH$_3$COO$^-$ and NO$_3^-$ ions, on morphology and electrochemical behavior of the final NiO powders, was systematically evaluated. It was found that different anions resulted in NiO samples with different hierarchical surface morphologies. Among them, the powder derived from the precursor of Cl$^-$ ions possessed a nanoflower-like morphology, while the one from the CH$_3$COO$^-$ ion based solution had a honeycomb type morphology. Comparatively, the powder obtained in the presence of NO$_3^-$ ions demonstrated the highest specific surface area, intermediate porosity and a novel pine-cone morphology. Accordingly, this sample showed the highest specific capacitance of 279 F g^{-1} at the scan rate of 5 mV s^{-1}, as mentioned above.

Ni(NO$_3$)$_2$·6H$_2$O, Ni(CH$_3$CO$_2$)$_2$·4H$_2$O and NiCl$_2$·6H$_2$O were used as the sources of Ni^{2+}, with different anions, leading to samples denoted as NiO-N, NiO-A and NiO-C, respectively. Cetyltrimethyl ammonium bromide (CTAB) and urea were used as the surfactant and precipitation agent, respectively. To prepare the reaction solutions, 20 mmol Ni salt was dissolved in 100 mL triple distilled water, which was then added dropwise to CTAB solution (10 mmol in 100 mL triple distilled water). After stirring for 1 hour, homogeneous green colored solutions were formed. Then, 40 mmols urea was added to the solutions, followed by constant stirring for 3 hours to ensure high homogeneity, characterized with a transparent green color. The solutions were subject to hydrothermal reaction in 250 mL Teflon lined stainless steel autoclave at 120°C for 24 hours. After reactions, fluffy grass green-like precipitates were obtained, which were collected through centrifugation at 3000 rpm. The samples were all thoroughly washed with triple distilled water, mixture of absolute ethanol and water and absolute ethanol, followed by drying at 60°C for 24 hours in vacuum. The dried samples finally calcined at 300°C for 3 hours in air to obtain NiO nanostructured powders. The samples before calcination were named as NiON-uc, NiOA-uc and NiOC-uc, respectively.

Figure 5.82 shows SEM images of the NiO samples from different precursors. Nevertheless, morphology of the precursor Ni(OH)$_2$ samples was well retained in the final NiO powders, after thermal calcination. Because porosity of the NiO samples was closely related to evaporation of water and decomposition of the anions in the Ni(OH)$_2$ precursors, the type of the anions should have an indirect effect on the porosity of NiO. Low magnification SEM images revealed that NiO samples all exhibited a porous sphere-like morphology. However, high magnification indicated that they possessed different surface profiles, as mentioned before. The difference in surface morphology among the NiO powders was confirmed by HRTEM characterization results.

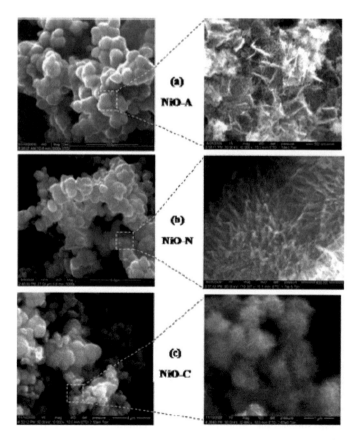

Fig. 5.82. SEM images of the samples NiO-N (a), NiO-A (b) and NiO-C (c), with pine-cone, honeycomb and flower-like morphologies, respectively. Reproduced with permission from [154], Copyright © 2011, Royal Society of Chemistry.

It is widely accepted that crystallographic morphology of a precipitate is determined by evolution of the crystallographic planes, which are closely related to their corresponding surface energies. It is also known that organic and inorganic additives adsorb onto certain crystallographic planes and modify the relative order of surface energies during the crystal growth process [155]. The preferential adsorption on specific surfaces would lower the surface energy, so that corresponding morphology was developed. Expectedly, different anions would be adsorbed on different crystallographic planes, so as to result in precipitating precursors with different morphologies.

The NiO-A sample had a honeycomb-like macroporous structure, consisting of hierarchically oriented hairy plate-like items, as seen in Fig. 5.82(a). The CH_3COO^- ions with a relatively large size resulted in sample with a highly porous honeycomb morphology. High magnification images indicated that the NiO-N sample displayed a unique pine-cone like surface morphology, comprising of a oriented and densely packed nanofiber structure, as illustrated in Fig. 5.82(b), which implied that the NiO-N particles were assembled with the NiO nanofibers. The NiO-C sample exhibited a flower-like morphology, made of oriented and self-assembled micropetals, as observed in Fig. 5.82(c).

Therefore, it could be temporally concluded that the structures of the precipitates were primarily aggregates of the nanosized particles with different morphologies, which were nucleated in the supersaturated solutions although the kinetic pathway of crystallization, with the presence of different anions. During the nucleation and crystallization process, various factors, such as van der Waals forces, hydrophobic interactions, crystal-face attraction, electrostatic and dipolar fields, and hydrogen bonds, could play their roles in determining the morphology of the precipitates [156].

On one hand, nanostructured NiO powders usually have relatively low electrical conductivity, due to the damage of the crystal structures [157]. On the other hand, nanostructures experiences more serious variation in volume during the charge-discharge cycling process. These two aspects have become the main issues for nanostructured NiO for electrode applications of supercapacitors. One of the strategies to address these problems is to construct hierarchical structures by transferring the nanosized items to 3D ones, so as to increase the connectivity of the materials and enhance the mechanical strength of the whole structures [128, 158, 159]. Meanwhile, 3D hierarchical structures have enlarged active surface areas and increased porosity, which will offer smoother transportation of the electrolyte ions.

For instance, 3D mesoporous NiO network-like hierarchical microspheres assembled with ultrathin nanowires have been developed by using the hydrothermal method, which exhibited an optimal specific capacitance of 555 $F \cdot g^{-1}$, together with outstanding electrochemical stability [128]. To prepare the reaction solution, 1 mmol $NiCl_2 \cdot H_2O$ was dissolved into 40 mL deionized water with the aid of magnetic stirring, into which a desired amount of urea was added, with further stirring for 5 minutes at room temperature. To conduct hydrothermal reaction, the solution was transferred into a reactor. The hydrothermal treatment was kept at 100°C for 20 hours, with a 60 mL Teflon-lined stainless steel autoclave. After reaction, the precipitates were collected and washed thoroughly with distilled water and ethanol, followed by vacuum drying. To obtain NiO hierarchical nanostructures, the dried precipitate precursors were calcined at 300°C for 2 hours.

It was found that nickel hydroxide hierarchical microspheres could be readily developed by simply controlling the molar ratio of $NiCl_2$ and urea. Representative SEM images of the α-Ni $(OH)_2$ nanostructures, derived from the solution with a molar ratio of $NiCl_2/CO(NH_2)_2 = 1:1$, are depicted in Fig. 5.83(a, b). The powder had uniform sphere-like morphology, with diameters of 3–4 μm. The network-like hierarchical microspheres consisted of ultrathin nanowires with diameters of 5–6 nm. XRD pattern confirmed the phase composition of α-Ni $(OH)_2$, as demonstrated in Fig. 5.83(c). The presence of the (110) reflection peak suggested the layered structural feature was attributed to the separated hydroxyl layers and the weak bonding interactions between intercalated ions and the hydroxyl layers. Figure 5.83(d) illustrates a schematic diagram of the α-Ni $(OH)_2$ phase, with the layered structure to be intercalated by water molecules. In different α-Ni $(OH)_2$ samples, the basal spacing of the NiO_6 octahedral slabs could be modulated by intercalating with different anions or molecules.

Morphology and dimensions of the α-Ni $(OH)_2$ precipitates could be controlled by varying the reaction conditions. For instance, by using different concentrations of urea, morphogenesis of the α-Ni $(OH)_2$ could be adjusted. With urea concentrations of 1.0, 2.0, 4.0 and 6.0 mmol, the corresponding morphologies of

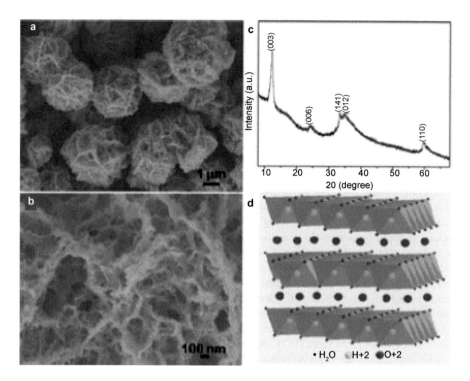

Fig. 5.83. (a–b) SEM images of the α-Ni(OH)$_2$ network-like microspheres at different magnifications. (c) Corresponding XRD pattern and (d) schematic diagram of the layered structure of α-Ni (OH)$_2$ phase intercalated by water molecules. Reproduced with permission from [128], Copyright © 2012, Royal Society of Chemistry.

the hydroxide precipitates were network-like spheres consisting of ultrathin self-assembled nanowires, flower-like spheres constructed with sheets, coexistence of spheres with sheets and nanorods self-assembly and urchin-like spheres with nanorods, respectively. Also, if NaOH or ethanol amine was used to replace urea, β-Ni (OH)$_2$ was obtained instead of α-Ni (OH)$_2$. In other words, the formation of α-Ni (OH)$_2$ was also closely related to the special properties and structure of urea.

XRD results confirmed that calcination at 300°C for 2 hours was sufficient for the α-Ni (OH)$_2$ precursors to be transferred to NiO. Representative SEM images of the NiO powders are demonstrated in Fig. 5.84(a–d). It was observed that morphologies of the α-Ni (OH)$_2$ precursors were well retained after the thermal calcination, with almost no change in both the shape and size. For example, the NiO hierarchical microspheres self-assembled with ultrathin nanowires were same as those of the precursors, as seen in Fig. 5.84(a). High-magnification SEM and TEM images further confirmed the structure of NiO, as illustrated in Fig. 5.84(b, c), i.e., the network-like hierarchical microspheres were composed of ultrathin nanowires with diameters of 2–5 nm. The SAED pattern as the inset in Fig. 5.84(c) supported the presence of NiO. In the HRTEM image in Fig. 5.84(d), a lattice spacing of 0.21 nm was identified, which was the (200) facet of FCC NiO, in good agreement with the XRD results.

During the topotactic conversion from hexagonal phase of the precursor to cubic phase of NiO, there was a fundamental relationship, i.e., the [0001]$_H$ was parallel to [111]$_C$ [160]. Schematic diagrams of α-Ni (OH)$_2$ along [0001]$_H$ and NiO along [111]$_C$ are shown in Fig. 5.84(e) and (f), respectively. In the crystal structure of α-Ni (OH)$_2$, Ni (OH)$_6$ octahedron are linked with one another by sharing the sides, thus leading to presence of the single layer configuration. Meanwhile, the NiO$_6$ octahedron has a rock salt-like structure, by sharing all the sides in the 3D structure, with an interconnecting format. Along the [111]$_C$ axis, the octahedrons were arranged in a side-by-side manner, which resulted in the 2D layer, as observed in Fig. 5.84(f), which was similar to the lamellar stacking way of α-Ni(OH)$_2$. Therefore, the octahedron-stacked symmetry provided with a platform for the topotactic conversion from the precursor to NiO.

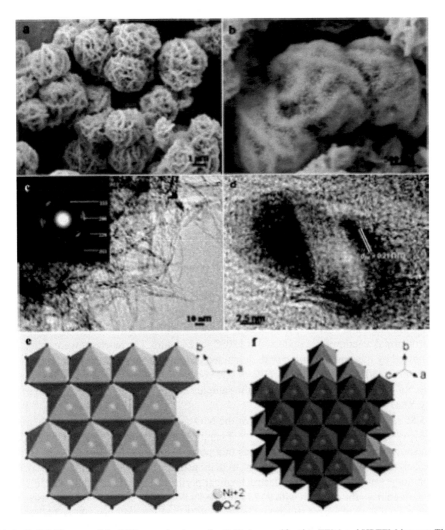

Fig. 5.84. (a, b) SEM images of the NiO nanostructures. (c, d) High-magnification TEM and HRTEM images. The inset is the corresponding SAED pattern obtained from panel (c). (e) Schematic diagrams of α-Ni (OH)$_2$ along [0001]$_H$ and (f) FCC-NiO along [111]$_C$, illustrating the structural relationship between the precursor and the final oxide product. Reproduced with permission from [128], Copyright © 2012, Royal Society of Chemistry.

NiO nanoflakes have been grown on flexible carbon cloth substrate by using a hydrothermal method, which was employed as the cathode and rGO on 3D Ni foam as the anode to assemble asymmetric supercapacitor that exhibited a high areal capacitance of 248 mF·cm^{-2}, together with promising capacitance retention of 95% after 3000 charge-discharge cycles [161]. To synthesize precursor Ni (OH)$_2$ nanoflakes, well-cleaned carbon cloth substrate was soaked in 20 mM Ni(NO$_3$)$_2$ ethanol solution, followed by heat treatment at 300°C for 10 minutes. This cycle of soaking and drying was repeated five times to develop seeded substrate, which was attached to a glass slide and then transferred into a Teflon-lined stainless steel autoclave for hydrothermal reaction. The reaction solution was obtained by dissolving 0.72 g Ni (NO$_3$)$_2$·6H$_2$O, 0.12 g ammonium persulfate and 3.12 ml ammonium hydroxide (28%) in 25 ml distilled water. The substrate was placed in the autoclave, with the seeded carbon cloth to face down. Hydrothermal reaction was conducted at 150°C for 6 hours. After reaction, the substrate was covered with a uniform layer of Ni (OH)$_2$, which was thoroughly washed with water and ethanol. After the drying process, the sample was calcined at 300°C for 1 hour, so that Ni (OH)$_2$ was decomposed to NiO.

XRD results indicated that the as-grown sample on carbon cloth was β-Ni (OH)$_2$, with high crystallinity and high purity, as shown in Fig. 5.85(a). After calcination, the β-Ni (OH)$_2$ was decomposed to NiO with cubic crystal structure. Figure 5.85(b) shows SEM image of the NiO film, revealing that vertically aligned NiO nanoflakes were uniformly grown on the carbon fibers of the carbon cloth substrate. The nanoflakes had an average thickness of 100 nm. Because the NiO nanoflakes were deposited on the conductive substrate, it was not necessary to used polymer binders when making electrodes of supercapacitors. Figure 5.85(c) depicts TEM image of the NiO nanoflakes, demonstrating that they were porous in microstructure that was attributed to the rapid dehydration of Ni (OH)$_2$ [162]. The porous structure of the NiO film offered more accessible surface area for the electrolyte ions and thus led to high electrochemical performances.

A hollow flowerlike NiO nanostructure has been synthesized by using a hydrothermal self-assembly process, with Pluronic P123 as the template [163]. The presence of Pluronic P123 retarded the growth of NiO grains, thus leading the formation of flowers assembled with nanoparticles. The flower-like NiO sample had a BET surface area of 96 m^2·g^{-1}, which was much higher than that of the NiO synthesized without the use of template (41 m^2·g^{-1}). As a result, electrode based on the flower-like NiO powder exhibited a specific capacitance of 619 F·g^{-1} that was much higher than that of pure NiO (347 F·g^{-1}). This observation implied that the flower-like NiO sample offered shorter diffusion path for the electrolyte ions into the electrode, so as to facilitate effective electrode/electrolyte contact at the interface. The Pluronic P123 micelles were preferably adsorbed onto the well-aligned Ni (OH)$_2$ surfaces, through either hydrogen bonding or loose coordination, which triggered the assembly of the Ni (OH)$_2$ nanoparticles into flowers.

The reaction solution was prepared by dissolving 0.5 g NiSO$_4$·6H$_2$O and 0.4 g P123 in 10 mL distilled water. Then, 4 mL ammonia was dropwise added into the solution at room temperature with the aid of stirring. Hydrothermal reaction was conducted at 180°C for 24 hours. The reactants were collected through filtration, followed by thorough washing with deionized water. After drying, the powder was calcined at 500°C for 2 hours, which was denoted as NiO/P. The control sample prepared without the use of P123 was named as NiO/WP. Phase compositions of the samples before and after the calcination process were confirmed by XRD results.

Figure 5.86 shows SEM and TEM images of the NiO samples prepared with and without the P123 template [163]. As shown in Fig. 5.86(a), the NiO/WP sample consisted of densely-packed nanosheets. TEM investigation shows that although the sample was of a porous structure, the porosity was relatively low, due to the densely-packed nanosheet morphology, as demonstrated by the inset in Fig. 5.86(a). Obviously, such a morphology was not desired for electrochemical performances of the materials. In contrast, the NiO sample derived from the solution with P123 exhibited a flowerlike structure that was assembled from nanosheets as the building block, as illustrated in Fig. 5.86(b). As a result, the calcined NiO powder had a relatively high porosity, as revealed in Fig. 5.86(d). Moreover, the templated NiO sample possessed a multilayered structure, with highly ordered texture and a hole at the center, as seen in Fig. 5.86(c), which was the reason why the NiO/P sample had a higher specific surface area the NiO/WP one.

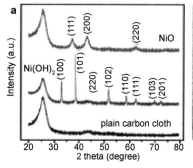  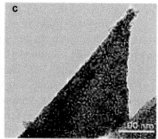

Fig. 5.85. (a) XRD patterns of the carbon cloth substrate, as-prepared Ni (OH)$_2$ and NiO films. (b) SEM image of the NiO nanoflakes grown on the carbon cloth substrate. (c) TEM image of an individual NiO nanoflake. Reproduced with permission from [161], Copyright © 2013, Royal Society of Chemistry.

Fig. 5.86. SEM and TEM images of the samples: (a) SEM and TEM (inset) images of the as-prepared NiO/WP, (b) SEM images of the as-prepared NiO/P at low and high magnifications (inset), (c) SEM and TEM (inset) images of the as-prepared NiO/P (hollow structure) and (d) TEM images of NiO/P at low and high magnifications (inset). Reproduced with permission from [163], Copyright © 2014, Royal Society of Chemistry.

According to the difference in morphology between the samples with and without the template, a possible growth mechanism was proposed, as schematically shown in Fig. 5.87 [163]. During the hydrothermal reactions, the Ni(OH)$_2$ nanoparticles with the P123 micelles were adsorbed preferably on surfaces of the well-aligned Ni(OH)$_2$ nanostructures [164]. As a consequence, with the presence of the P123 template, the Ni(OH)$_2$ nanoparticles were assembled into the flower-like structures, during the further growth and crystallization processes, at the appropriate hydrothermal reaction temperature. After calcination at 500°C, the flower-like microstructure was well retained in the NiO powders.

More recently, an urchin-like porous NiO nanostructure was prepared by using a simple hydrothermal method, which was used as an electrode of supercapacitors [165]. The urchin-like architectures were assembled with NiO nanorods, which consisted of irregular nanoparticles, with diameters in the range of 5–10 nm. The mesoporous NiO nanourchin based electrode demonstrated a specific capacitance of 540.5 F·g^{-1}, at a current density of 1 A·g^{-1}, together with outstanding cycling stability. The high electrochemical performance of the NiO nanostructure was ascribed to three factors. Firstly, it had a large specific surface area, so as to provide with more active sites, due mainly to the unique hierarchical architectures. Secondly, effective contact and diffusion between the electrode and electrolyte were ensured, because of the porous urchin-like structure. Lastly, the interconnected interspaces among the nanoparticles effectively shorten the ion diffusion path.

In the experiments, 2.377 g NiCl$_2$·6H$_2$O was dissolved in 20 mL deionized water to form a solution, into which 10 mL urea solution with a concentration of 1 M was added dropwise, with the aid of stirring for 30 minutes. Then, the mixture was hydrothermally treated at 110°C for 6 hours. The resultant precipitates were collected through filtration and thoroughly washed with deionized water and ethanol. The obtained powders were dried at 70°C for 12 hours first, followed by calcination at 400°C for 3 hours.

Figure 5.88(a) shows XRD pattern of the NiO nanostructured powder, confirmed the formation of cubic NiO, with high purity, as evidenced by the absence of diffraction peaks of other phases.

360 *Nanomaterials for Supercapacitors*

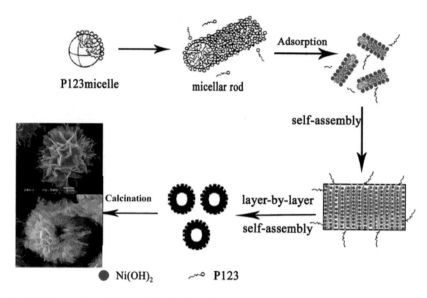

Fig. 5.87. Schematic diagram illustrating possible growth mechanism of the flower-like NiO nanostructures. Reproduced with permission from [163], Copyright © 2014, Royal Society of Chemistry.

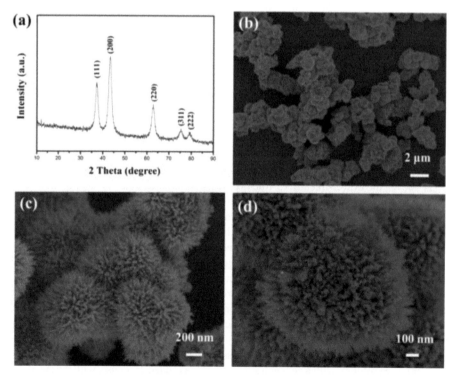

Fig. 5.88. (a) XRD pattern of the NiO nanourchins. (b–d) low-, medium- and high magnification SEM images of the NiO nanourchin architectures. Reproduced with permission from [163], Copyright © 2016, Elsevier.

Figure 5.88(b) shows a low-magnification SEM image of the sample, demonstrating the overall morphology of the as-prepared NiO powder. The NiO sample was composed of monodispersed microspheres, with an average diameter of about 1 μm. Medium and high magnification SEM images of the NiO powder are shown in Fig. 5.88(c, d), illustrating that the microspheres of the NiO sample displayed an urchin-like architecture, with the urchins being assembled with numerous nanorods. The nanorods were uniform and fine, with diameters in the range of 20–50 nm.

Solution and sol-gel deposition

Solution and sol-gel deposition techniques have also been used to prepare NiO nanostructures to serve as electrodes of supercapacitors. For instance, a porous NiO film was grown by using a chemical bath deposition method, with an aqueous solution of nickel nitrate [166]. The porous NiO sample had a rock salt structure, as evidenced by the XRD pattern. Electrochemical properties of the NiO film based electrode were systematically characterized in two electrolyte solutions, i.e., NaOH and KOH, with specific capacitances of 129.5 and 69.8 $F \cdot g^{-1}$, respectively. The electrode exhibited a high electrochemical stability up to 1500 charge–discharge cycles.

The NiO films were deposited on glass substrates that were coated with Indium doped Tin Oxide (ITO) as the conductive layer, with surface resistances of $25–27 \, \Omega \cdot cm^{-2}$. Before film deposition, the ITO substrates were thoroughly cleaned in acetone, methanol and deionized water, with the aid of ultrasonication. The reaction bath to deposit the porous NiO film consisted of 0.25 M nickel nitrate and 0.25 M urea, with a pH value to be 5. During the deposition experiments, the cleaned ITO glass substrates were placed vertically in the deposition bath. The deposition process was conducted at 80°C for 90 minutes. After the deposition was finished, the films were cleaned in deionized water, followed by drying in air. The dried films were finally calcined at 350°C for 2 hours in air to increase their structural quality and adherence to the substrates.

NiO monolayer hollow-sphere array composed of porous net-like NiO nanoflakes film, with a specific surface area 325 $m^2 \cdot g^{-1}$, was fabricated by using a chemical bath deposition, with polystyrene sphere template [167]. The porous NiO films had a specific capacitance of 311 $F \cdot g^{-1}$, together with a high capacitance retention, which was attributed to the porous structure to buffer the distortion in structure of the electrode related to the volume expansion during the cycling process.

To fabricate the monolayer polystyrene (PS) sphere template, 5 drops of the PS sphere suspension were dropped onto clean Ni foil substrate with a dimension of $2 \times 2 \, cm^2$. Uniform dispersion of the suspension was achieved by holding the substrate stationary for 1 minute. Then, the substrate was slowly immersed into deionized water. Once the suspension contacted the water surface, a monolayer of PS spheres was formed on the water surface and on the surface of the Ni foil substrate. The substrate was immersed to avoid overloading and non-uniform deposition. After that, a few drops of 2% dodecylsodiumsulfate solution were added to the water to decrease its surface tension. In this case, the monolayer of PS spheres suspending on the water surface could be pushed aside, due to the decrease in the surface tension of the water. Then, the substrate was lifted up through the clear area. It was necessary to make sure that no more PS spheres were deposited onto the monolayer. The samples were heated at 110°C for 5 minutes to adhere the monolayer onto the Ni foil substrate.

For chemical bath deposition, 80 ml 1 M nickel sulfate, 60 ml 0.25 M potassium persulfate and 20 ml aqueous ammonia (25–28%) were mixed in a 250 ml pyrex beaker at room temperature to obtain the reaction solution. The substrates with PS template layers were placed vertically in the solution at 25°C for 30 minutes to deposit the precursor film. To avoid any contamination, the back side of the substrate was tightly covered with polyimide tape. After deposition, the precursor films were thoroughly washed with deionized water. After taking off the tape masks, the samples were immersed in toluene for 24 hours to remove the PS sphere template. After that, the precursor films were calcined at 350°C in Ar for 1.5 hours. NiO films without the use of PS template were similarly prepared for comparison.

XRD results confirmed the cubic phase of the NiO films after calcination. PS spheres with three particle sizes, i.e., 1 μm, 600 nm and 200 nm were used. All the three types of PS spheres were well-organized into a long-range ordered hexagonal close-packed structure, with alignment perpendicular to the Ni foil substrate. As a result, hierarchically porous NiO films were readily obtained after the PS spheres were removed. For example, by using the templated of 1 μm PS spheres, the NiO film exhibited a typical hierarchically porous structure, comprising of two parts, as demonstrated in Fig. 5.89(a, b). The film was featured by a substructure consisting of monolayer close-packed hollow-spheres, with a diameter of 1 μm, i.e., the diameter of the PS spheres. At the same time, the superstructure was uniformly coated with a thin layer of free-standing NiO nanoflakes, with a thickness of about 30 nm, as illustrated in Fig. 5.89(b). Meanwhile, the NiO nanoflakes were distributed vertically to the monolayer of NiO hollow-sphere array, as seen in Fig. 5.89(a). The NiO nanoflakes were interconnected to one and another, leading to the presence of extended network structures, with pore diameters in the range of 30–300 nm.

The hierarchically porous structure could be detached from the substrate by using a blade, with the bottom side showing an ordered bowl-like array, as observed in Fig. 5.89(c), while the left part on the substrate was the bowl-like array layer, as revealed in Fig. 5.89(d). The hierarchical pore structure was also confirmed by TEM characterization results. Figure 5.89(e) shows TEM image of a piece of free-standing

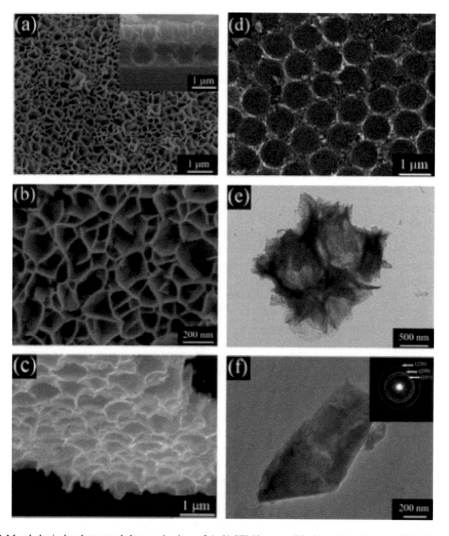

Fig. 5.89. Morphological and structural characterizations of: (a, b) SEM images of the hierarchically porous NiO film deposited with 1 μm PS spheres, with the inset showing a side view images. (c) SEM image of the film detached from the substrate. (d) SEM image of the residual pattern left on the substrate. (e, f) TEM images of the detached film sample, with the inset in (f) presenting a corresponding SAED pattern. Reproduced with permission from [167], Copyright © 2011, Royal Society of Chemistry.

porous NiO film, revealing a circle structure connected with nanoflakes, in a good agreement with the SEM images. Furthermore, the nanoflakes were characterized by a thin and flat feature, as demonstrated in Fig. 5.89(f). The diffraction rings in the SAED patterns belonged to the cubic phase of NiO, confirming polycrystalline nature of the film, as revealed by the XRD characterization results. At the same time, similar hierarchically porous structure could be obtained, when using the PS spheres with diameters of 600 and 200 nm, as illustrated in Fig. 5.90.

In comparison, the NiO thin film prepared without the presence of the PS sphere template possessed a highly porous net-like structure, consisting of numerous interconnected nanoflakes, with an average thickness of 30 nm. The nanoflakes were vertically grown on the substrate, resulting in the net-like structure and thus forming pores with sizes in the range of 30–300 nm. Also, the film deposited by using the eletrodeposition method had a dense microstructure. The flakes in the dense film displayed a rough surface, due to the packing of nanoparticles. These films were also crystalline in nature.

A sol-gel process has been reported to fabricate NiO nanostructures with different morphologies, including flower, slice and particle shapes [168]. The nanoflower-shaped NiO with a 3D network and the highest pore volume exhibited the highest electrochemical performances, with a specific capacitance of 480 $F \cdot g^{-1}$. The capacitance value was higher than that of the nanoparticles, even though the latter had a higher specific surface area of 233 $m^2 \cdot g^{-1}$. The observation was attributed to the fact that the NiO nanoflower had an appropriate average pore size and nanochannels, so as to ensure the contact between the electrode materials and electrolyte, thus offering a smooth transport of the electrolyte ions.

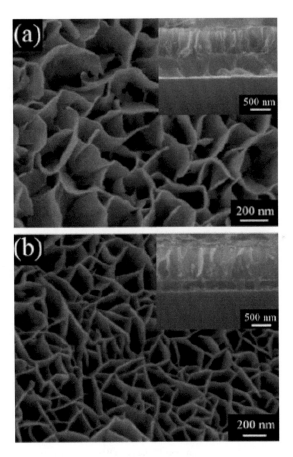

Fig. 5.90. SEM images of the hierarchically porous NiO film prepared with the PS spheres of different sizes: (a) 600 nm and (b) 200 nm. The insets in (a) and (b) were side views of the films. Reproduced with permission from [167], Copyright © 2011, Royal Society of Chemistry.

During the synthesis of the Ni (OH)$_2$ nanostructures, Ni(CH$_3$COO)$_2$·H$_2$O and Ni(NO$_3$)$_2$·6H$_2$O were used as precursors of Ni (OH)$_2$, while NH$_4$OH, LiOH and hexamethylene tetramine (HMTA) were employed to control the reaction conditions. To obtain Ni (OH)$_2$ flower nanostructure, 100 mmol Ni (NO$_3$)$_2$·6H$_2$O and 100 mmol HMTA were dissolved in deionized water with the aid of stirring. To synthesize Ni (OH)$_2$ nanoslice and nanoparticle, 100 mmol Ni(CH$_3$COO)$_2$·H$_2$O, 4 mL NH$_4$OH and 100 mmol LiOH were added into the Ni (CH$_3$COO)$_2$·H$_2$O solution, instead of HMTA as the acidity adjusting agents. Hydrothermal reaction for all the solutions was conducted at 100°C for 4 hours. After reactions, the precipitates were collected through centrifugation and thoroughly washed with distilled water and ethanol. The precursor samples were dried at 60°C for 8 hours in vacuum. To obtain nanostructured powders of NiO, the Ni (OH)$_2$ precursor samples were all calcined at 300°C for 2 hours. The conversion of Ni (OH)$_2$ into NiO was evidenced by the color change from light green to gray.

Figure 5.91 shows a schematic diagram demonstrating formation process of the Ni (OH)$_2$ precursors, with different morphologies derived from solutions at different conditions, together with their corresponding SEM images [167]. It was observed that three types of powders, with morphologies of nanoflower, nanoslice and nanoparticle, could be obtained by simply using different Ni-precursors combined with acidity of the reaction solutions. Flower-like Ni (OH)$_2$ nanostructure, with diameters in the range of about 3–5 μm, was formed by using HMTA solution, as shown in Fig. 5.91(a), nanoslice-like Ni (OH)$_2$, with diameters of 300–530 nm could be derived from the NH$_4$OH based solution, as seen in Fig. 5.91(b). With the presence of strongly basic LiOH, the Ni (OH)$_2$ nanostructure consisted of nanoparticles with an average diameter of 50 nm, as illustrated in Fig. 5.91(c).

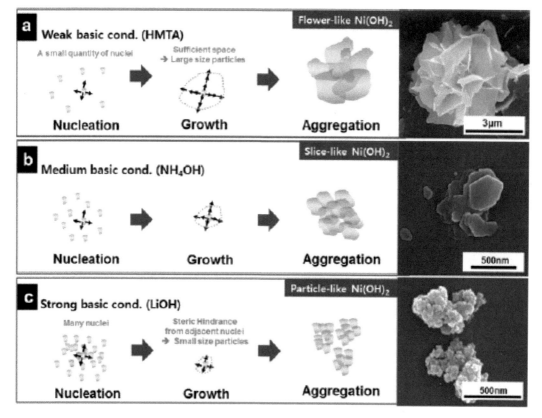

Fig. 5.91. Schematic diagram showing formation process of the Ni (OH)$_2$ precursors with different morphologies together with their corresponding SEM images: (a) nanoflower, (b) nanoslice and (c) nanoparticle. Reproduced with permission from [168], Copyright © 2013, American Chemical Society.

The formation of the Ni(OH)$_2$ nanostructures with different morphologies could be explained in terms of the difference in dissociation rate among the bases, because it determined the kinetics of the reaction between the OH$^-$ and the Ni^{2+} ions. The formation of Ni(OH)$_2$ starts from the nuclei, due to the reaction of the Ni^{2+} and OH$^-$ ions. After that, the nuclei aggregated or self-assembled into larger particles, with different crystal orientation and thus different morphology, driven by the force to minimize the surface energy [119]. The nanoslice was formed due to the stacking of the hexagonal shaped nanoplates in a medium alkaline solution. In comparison, a weak basic HMTA generated OH$^-$ ions at a relatively slow rate, so that the nuclei had sufficient time and space to grow into large particles [169]. This leads to the creation of larger size flower-shaped Ni(OH)$_2$ with an energetically favored state, which can be explained by Ostwald ripening [170]. With the presence of the strongest base of LiOH, the formation rate of OH$^-$ was fast, thus leading to a bust of Ni(OH)$_2$ nuclei. In this case, the chance of the nuclei to grow into large particles was significantly decreased, because the amount of the precursor ions in the solution was decreased and the steric or spherical hindrance due to the adjacent nuclei was increased, so that small particles were derived.

A sol-gel process was developed to synthesize NiO$_x$ xerogels, combined with a thermal treatment at ambient pressure [171]. NiO$_x$ xerogels thermally treated at 250°C exhibited an optimal specific capacitance of 696 F·g^{-1} at a current density of 2.0 mA·cm^{-2} in 7 M KOH solution. To prepare the Ni(OH)$_2$ gels, 30 g·L^{-1} NiSO$_4$·6H$_2$O, 13 g·L^{-1} Na$_3$C$_6$H$_5$O$_7$·2H$_2$O and 5 g·L^{-1} NaOOCCH$_3$ were dissolved in DI water to form a solution, with the aid of stirring at 80°C. Then, 3 M KOH aqueous solution was dropwise added into the solution, until the solution had a pH value of 12. The suspension was centrifuged and rinsed thoroughly with distilled water and ethanol, to obtain Ni(OH)$_2$ gels. The Ni(OH)$_2$ gels were dried at 80°C to form xerogels. The dried gels were calcined at temperatures in the range of 110–450°C for 1 hour to obtain NiO$_x$ xerogels.

Figure 5.92 shows XRD patterns of the samples calcined at different temperatures, demonstrating the effect of the calcination temperature on the crystalline behavior of the NiO$_x$ xerogels. Below 200°C, all the diffraction peaks belonged to β-Ni(OH)$_2$. As the temperature was increased from 250 to 280°C, diffraction peaks of NiO started to appear. Above 280°C, all the peaks corresponded to NiO. At the same time, intensity of the peaks was heightened with increasing calcination temperature, suggesting a rise in the crystallinity of the samples.

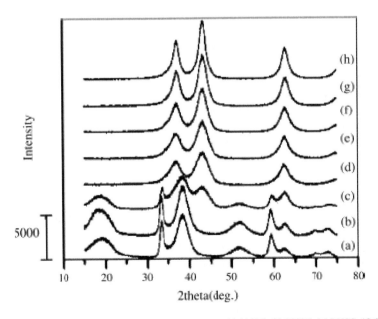

Fig. 5.92. XRD of the NiO$_x$ xerogels calcined at different temperatures: (a) 110°C, (b) 200°C, (c) 250°C, (d) 280°C, (e) 300°C, (f) 350°C, (g) 400°C and (h) 450°C. Reproduced with permission from [171], Copyright © 2006, Elsevier.

366 *Nanomaterials for Supercapacitors*

Chemical precipitation

Mesoporous Ni (OH)$_2$ was prepared, with Sodium Dodecyl Sulfate (SCS) as the template and urea as the hydrolysis-controlling agent, by using a chemical precipitation method [172]. Mesoporous NiO powder, with a centralized pore-size distribution, was obtained by calcining the Ni (OH)$_2$ precursor at different temperatures. The NiO sample calcined at 250°C had a BET specific surface area of 477.7 m$^2 \cdot$g^{-1}, corresponding to a specific capacitance of 124 F$\cdot$g^{-1}. To synthesize the Ni (OH)$_2$ precursor, NiCl$_2 \cdot$6H$_2$O, SDS and urea dissolved in distilled water, with the aid of stirring for 2 hours. The solution was then heated at 80°C for 6 hours, leading to Ni (OH)$_2$ precipitate, which was collected through centrifugation, followed by drying at 100°C overnight. After that, the sample was thoroughly washed with ethanol and then dried in air. Calcination of the Ni (OH)$_2$ precursor powder at different temperatures resulted in NiO powders.

A simple precipitation method was reported to prepare nanoporous NiO powders with polyhedron morphologies, high specific surface areas and narrow pore distributions, from oxalate precursors [173]. Experimental parameters, including heating rate, calcination temperature and particle size of the oxalate precursors, could be used to control pore structures of the final products. The mesoporous NiO powder obtained at the optimal conditions had a specific surface area of 179 m$^2 \cdot$g^{-1}, with a narrow pore distribution at about 1.0 nm and 6.0 nm, respectively. A high specific capacitance of 165 F$\cdot$g^{-1} was achieved by using the mesoporous NiO as the electrode of supercapacitor.

To prepare NiC$_2$O$_4 \cdot$2H$_2$O, NiCl$_2 \cdot$6H$_2$O and sodium dioctylsulfosuccinate (AOT), with a ratio of NiCl$_2$:oxalic acid:AOT of 1:0.6:0.2, were dissolved in water at 80°C in an oil bath with stirring for 2 hours. After that, oxalic acid solution, with a small amount of phosphoric acid, was slowly dropped into the solution. Cyan colored precipitates were formed during the reaction process. The precipitated products were quickly cooled down with ice water, collected through centrifugation, and then thoroughly washed with deionized water and ethanol, followed by drying overnight. The nanoporous NiO powders were obtained by thermally decomposing the NiC$_2$O$_4 \cdot$2H$_2$O at temperatures from 300°C to 400°C for 1 hour.

Representative SEM images of the NiC$_2$O$_4 \cdot$2H$_2$O and NiO after thermal decomposition are shown in Fig. 5.93(a, b). The NiC$_2$O$_4 \cdot$2H$_2$O was highly crystallized, with the particles to have polyhedron morphologies, i.e., cubic and/or rhombus-like. After thermal decomposition of the precursor, the morphologies were well retained in the NiO powders, as seen in Fig. 5.93(b). According to the SEM images of the precursor and the decomposition product, there was almost no shrinkage in dimension, while the weight loss was about 55%, which implied that the NiO particles were highly porous. TEM images of the NiO powders, which were calcined at 300°C and 340°C, are illustrated in Fig. 5.93(c, d), confirming their porous microstructures. The pores were randomly distributed over the bulk of the NiO polyhedron particles. Both pore and particle sizes were increased slightly, as the calcination temperature was increased from 300°C to 340°C.

A two-step method was developed to prepare self-supported hexagonal NiO nanoplatelet arrays on Ni foam, through the precipitation of hydroxides combined with post-thermal calcination [174]. The nanoplatelets consisted of multilayered ultrathin mesoporous NiO nanosheets, supported by Ni foam. Electrode based on the mesoporous NiO nanoplatelet arrays facilitated smooth ion and electron transport in the materials, large electroactive surface area and outstanding structural stability, thus leading to promising electrochemical performance, with specific capacitances of up to 1124 F$\cdot$g^{-1} and high cycling stability up to 5000 charge-discharge cycles.

Before the synthesis experiment, Ni foams were cleaned with HCl (6 M) solution, followed by thorough washing with DI water and absolute ethanol. Then, 2 g block polymer P123, i.e., $(C_2H_4O)_{20}(C_3H_6O)_{70}(C_2H_4O)_{20}$ or EO$_{20}$PO$_{70}$EO$_{20}$, with MW = 5800, was dissolved in 40 mL milli-Q water, under magnetic stirring for 20 minutes, in a three-necked flask, into which 1.84 g Ni (NO$_3$)$_2 \cdot$6H$_2$O was added, so as to form a green solution. After stirring for 1 hour, a piece of Ni foam placed inside a glass tube with two ends to be opened was soaked into the solution. After that, 20 mL ammonia solution with a concentration of 25 wt% was added, followed by refluxing at 120°C for 8 hours. The Ni foam was coated with a layer of green precursors, which was washed with the aid of ultrasonication for 10 minutes, followed by drying at 80°C for 6 hours. To obtain NiO, the precursors were calcined at 300°C for 1 hour.

Oxide Based Supercapacitors: II-Other Oxides   367

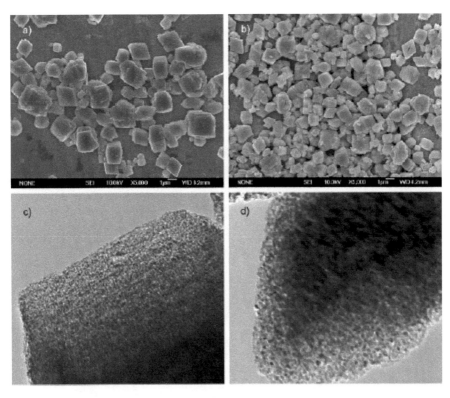

Fig. 5.93. SEM images of the NiC$_2$O$_4$·2H$_2$O (a) and NiO calcined at 300°C (b). TEM images of the nanoporous NiO calcined at 300°C (c) and 340°C (d). Reproduced with permission from [173], Copyright © 2008, WILEY-VCH Verlag GmbH & Co. KGaA, Weinheim.

Figure 5.94(a) shows SEM images of the NiO nanoplatelet arrays deposited on Ni foams, which revealed that the 3D grid structure and hierarchical porosity of the Ni foam could be clearly observed. Figure 5.94(b) depicts a high-magnification SEM image of the rectangle region of the inset in Fig. 5.94(a). It was found that the NiO layer was present as a uniform nanoplatelet array. As illustrated in Fig. 5.94(c, d), the NiO nanoplatelets exhibited a hexagonal morphology, with an average thickness of < 50 nm and lengths in the range of 200–500 nm. The NiO nanoplatelets were uniformly distributed on the Ni foam, which had high density growth while being separated in an appropriate manner, so as to result in macropores that ensured the fast transport of electrolyte ions. Moreover, the strong adhesion of the NiO nanoplatelet arrays on the Ni foam substrate offered mechanical stability of the materials.

A reflux method has been developed to prepare porous NiO powders, with NaBH$_4$-EG and NaOH-EG as the alkaline precipitants [175]. The type of the precipitants has a significant influence on morphology and structure of the NiO powders, which further had an effect on their specific capacitances when they were used as the electrodes of supercapacitors. Comparatively, NaBH$_4$-EG was more versatile in controlling the morphology of NiO than NaOH-EG. It was found that the NiO powder prepared with NaBH$_4$-EG as the precipitant possessed spherical spongy morphology, with a high specific surface area, while that obtained by using NaOH-EG displayed an irregular structure with agglomeration, thus leading to specific capacitances of 930 F·g^{-1} and 510 F·g^{-1}, respectively, at a current density of 15 A·g^{-1}. Furthermore, the NiO powder made with NaBH$_4$-EG exhibited a very specific capacitance of 1396 F·g^{-1}, after 1000 charge-discharge cycles, at a current density of 4 A·g^{-1}. The NaBH$_4$-EG-NiO based electrode prepared by the former NiO demonstrated lower charge transfer resistance and ion diffusion resistance, which were responsible for the higher electrochemical performance.

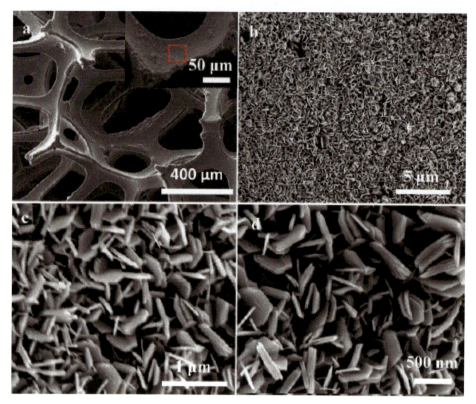

Fig. 5.94. SEM images of the mesoporous NiO nanoplatelet arrays grown on Ni foam substrate at different magnifications. Reproduced with permission from [174], Copyright © 2012, Elsevier.

To synthesize the precursors, 1.2 g NaBH$_4$ was dissolved in 25 mL ethylene glycol (EG), until the bubbles disappeared, which led to NaBH$_4$-EG solution [175]. Meanwhile, 600 mg Ni (NO$_3$)$_2$·6H$_2$O was dissolved in 100 mL EG and heated to 190°C, into which the NaBH$_4$-EG solution was added. The mixture was kept at 190°C with continuous stirring for 1 hour. The products were thoroughly washed with distilled water and ethanol, followed by drying at 100°C. To get NiO, the precursors were heated at 250°C for 2 hours. At the same time, NiO was also prepared by using NaOH to replace NaBH$_4$. The two NiO samples were denoted as NiO–NaBH$_4$ and NiO–NaOH. The NaBH$_4$-EG solution was dried in vacuum at 170°C to obtain dry powder, denoted as dry NaBH$_4$-EG.

Representative SEM and TEM images of the NiO powders are shown in Fig. 5.95 [175]. As shown in Fig. 5.95(a, b), the NiO–NaBH$_4$ sample consisted of highly homogenous spongy-shaped spheres, with sizes in the range of 200–300 nm. The individual sphere was assembled with a large number of rippled-shaped thin nanoflakelets interconnected to one and another. In comparison, the NiO–NaOH powder contained irregular blocks with serious agglomeration, with sizes to be much larger than that of the NiO–NaBH$_4$ one, as demonstrated in Fig. 5.95(c, d). Also, the NiO–NaBH$_4$ sample was more porous than the NiO–NaOH one. The highly porous structure of the NiO–NaBH$_4$ powder was further confirmed by TEM images. The irregular structure of the NiO–NaOH was formed with randomly aggregated particles, with an average size of 5 nm.

The presence of the alkaline precipitants was crucial to the reaction process. Without the use of the alkaline precipitants, almost no precipitates could be formed. At the same time, the difference in morphology between the two NiO powders was closely related to the different alkalinities of the two precipitants. Because NaOH is a very strong alkaline, the NaOH-EG contained a high concentration of hydroxide ion, which led to the quick formation of precursor particles with a size of about 5 nm. The small particles were

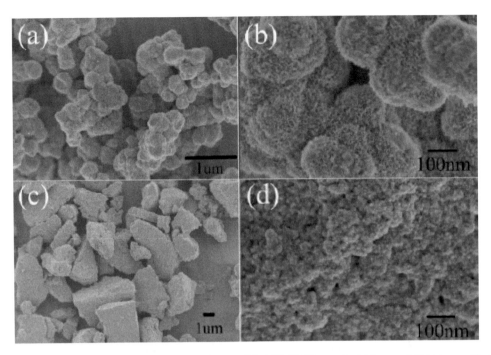

Fig. 5.95. (a) Low- and (b) high-magnification SEM images of the NiO–NaBH$_4$ sample. (c) Low- and (d) high-magnification SEM images of the NiO–NaOH powder. Reproduced with permission from [175], Copyright © 2013, Elsevier.

then assembled to irregular blocks, due to the rapid mass transport and directional fusion of the particles. In contrast, NaBH$_4$ is a well-known reducing agent and is a mild alkaline, in the form of complex. In this case, the precursor particles were formed slowly and uniformly, thus leading to uniformly distributed spheres.

Precursor samples with reactions at different time intervals were collected, in order to clarify the formation process of the porous structure. It was found that, after reaction for 4 minutes, nanoflakelets started to interconnect one another randomly, which could be attributed to the macromolecular precipitant. Because the NaBH$_4$ precipitant was mild, with a pH value of about 9.6, the formation rate of nickel alkoxide was not so fast. At the same time, the macromolecule EG acted as a structure-directing agent, which triggered the formation of the flakelet structure. As the reaction time was increased to 10 minutes, the morphology of the precursor remained almost unchanged, but the nanoflakelets became more pronounced. After reaction for 30 minutes, spongy-structured spheres were formed through the assembling of the nanoflakelets. Prolonged reaction led to more regular spherical spongy morphology of the samples. Figure 5.96 shows a schematic diagram describing the formation mechanism of the NiO nanostructures. The reaction process involved three stages: (i) formation and random agglomeration of nanoflakelets, (ii) self-assembly of nanoflakelets into spherical structure and (iii) growth of uniform spheres.

NiO hollow spheres were formed through the assembly of nanosheets, which were derived from Ni$_2$CO$_3$(OH)$_2$ nanosheets that were synthesized on sulfonated polystyrene hollow spheres by using a low temperature solution method [158]. Due to their hollow interior and hierarchical structure, the NiO nanosheet hollow spheres exhibited a high specific capacitance of 415 F·g^{-1}, with a retention of capacitance by 91% after 1000 charge-discharge cycles. The high electrochemical performances of the NiO nanosheet hollow spheres could be attributed to the accommodation capability to the volume distortion caused by the charge-discharge cycling.

To synthesize the Ni$_2$CO$_3$(OH)$_2$ nanosheets, 0.2 g sPSHSs was dispersed in a solution consisting of 20 mL DI water and 20 mL ethanol, with the aid of sonication for 10 minutes. After that, Ni(NO$_3$)$_2$·6H$_2$O and 1.2 g urea were added into the solution, with the aid of stirring for 5 minutes. The reaction solutions were then heated at 80°C for 12 hours. The precipitates were collected through centrifugation at 8000

370 *Nanomaterials for Supercapacitors*

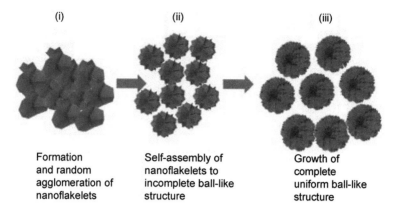

Fig. 5.96. Schematic diagram demonstrating formation mechanism of the spherical spongy structures. Reproduced with permission from [175], Copyright © 2013, Elsevier.

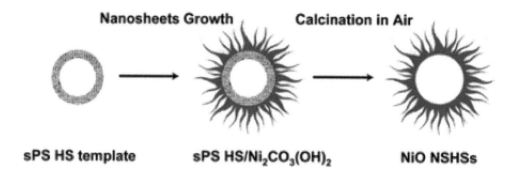

Fig. 5.97. Schematic diagram describing formation mechanism of the NiO nanosheet hollow spheres. Reproduced with permission from [158], Copyright © 2011, Royal Society of Chemistry.

rpm for 6 minutes and thoroughly washed with DI water and ethanol. The powders were dried at 80°C for 12 hours. To obtain NiO HSs, the sPSHS@Ni$_2$CO$_3$(OH)$_2$ precursor powders were calcined at 450°C for 2 hours. Morphology and size of the NiO hollow spheres were closely related to the amount of Ni (NO$_3$)$_2$·6H$_2$O used in the precursor synthesis.

Figure 5.95 shows a schematic diagram demonstrating formation mechanism of the NiO nanosheet hollow spheres (NSHSs). During the precursor synthesis, Ni$_2$(CO$_3$)$_2$(OH)$_2$ nanosheets were assembled on the sPSHSs template, because the surface of the PSHSs was functionalized with sulfonate groups (–SO$_3$H) [176]. The NiO NSHSs were obtained after the sPSHS@Ni$_2$CO$_3$(OH)$_2$ composite spheres were calcined, without the loss of the sphere morphology. Since hollow PS templates instead of solid PS templates were used, the amount of gas produced during the template removal through calcination was significantly reduced. As a result, the collapse of the hollow structures was effectively prevented. This was the reason why well-preserved NiO hierarchical nanosheet hollow spheres could be developed through the calcination, which was favorable to electrochemical performances of the materials when they were used as electrodes of supercapacitors.

A self-assembly liquid-crystalline phase-templating method was reported for the preparation of highly dispersed NiO nanoparticles embedded in lignin-derived mesoporous carbon frameworks with Pluronic F127 [177]. Both the glutaraldehyde and formaldehyde were employed as the cross-linking agents. It was found that the glutaraldehyde led to a specific capacitance of 880.2 F·g^{-1}, which was slightly higher that of the sample from the formaldehyde (860.4 F·g^{-1}), which was attributed to the relatively lower surface area for the latter. Due to the difference in molecular structure between glutaraldehyde and formaldehyde, the glutaraldehyde has a long carbon chain structure, which resulted in a laxer assembly with the liquid crystalline phase, thus leading to a larger surface area. The nanostructures derived from the laxer assembly has at least three advantages: (i) an improved conductivity of NiO due to the coating of the carbon layer, (ii) the outer mesoporous carbon shells to offer a conducting pathway for electron transfer and prevent the inner nanoparticles from aggregation and pulverization and (iii) the mesoporous morphology of carbon shell to buffer the volume variation during the charge-discharge cycling.

The synthesis of the precursor was started by dissolving 0.5 g $Ni(NO_3)_2 \cdot 6H_2O$ and 0.5 g Pluronic F127 in 10.0 g ethanol first, followed by the addition of 1.0 g sodium lignosulfonate. After that, 2.0 g formalin solution (containing 0.67 g formaldehyde) and 1.0 g KOH were slowly introduced sequentially, with the aid of stirring. The mixture sol was cast into Petri dishes, so that the ethanol and water were evaporated. The samples were then heated at 80°C for 8 hours to trigger the polymerization. After that, the powders were collected from the dishes and then heated in N_2 at 600°C for 2 hours for carbonization to obtain NiO@MPCs. The sample was denoted as F-NiO@C-1, with the first alphabet to represent the CLA of formaldehyde, G to stand for the CLA of glutaraldehyde and the last number to mean the weight ratio of Ni/lignin in the starting solution. During the experiments, the masses of the lignin and Pluronic P123 were kept to be constant, while the amount of $Ni(NO_3)_2 \cdot 6H_2O$ was varied in the range of 0.5–2.5 g to control the content of NiO.

Low magnification TEM images of the samples are shown in Fig. 5.98(a–c), demonstrating that the monodispersed nanoparticles with an average diameter of 20 nm were uniformly distributed inside the mesoporous carbon matrix. The carbon mesostructure well confined the NiO nanoparticles, so that they were not aggregated, thus preventing phase separation between the NiO and the carbon framework during the heating process. The carbon matrix exhibited a disordered mesoporous feature, which was confirmed by high magnification TEM image, as shown in Fig. 5.98(d). With increasing content of NiO, the size of the nanoparticles was increased.

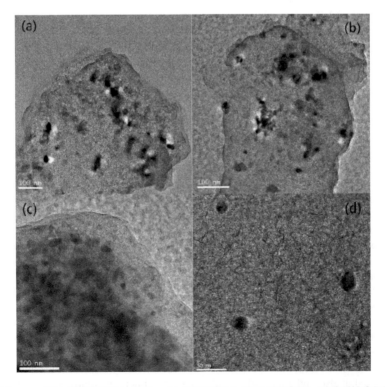

Fig. 5.98. TEM images of the NiO@MPCs samples: (a) G-NiO@C-1, (b) F-NiO@C-1, (c) G-NiO@C-3 and (d) higher magnification of G-NiO@C-3. Reproduced with permission from [177], Copyright © 2013, Royal Society of Chemistry.

372 *Nanomaterials for Supercapacitors*

Electrochemical deposition

A mesoporous 1D core-shell structure was fabricated by using Chemical Vapor Deposition (CVD), followed by electrochemical deposition of hexagonal lyotropic $Ni(OH)_2$, with a lyotropic liquid crystalline template [178]. To avoid the low conductivity problem of NiO, Ti and ITO (Indium Titanium Oxide) were used as the current collector and substrates to deposit NiO thin films. Due to the higher conductivity of the ITO than Ti substrate, It was found that NiO/ITO sample had a specific capacitance of 1025 $F \cdot g^{-1}$, which was high than that of the NiO/Ti one (416 $F \cdot g^{-1}$), due to the higher electrical conductivity of ITO than Ti.

To synthesize indium–tin oxide nanowires, ITO NWs, a carbothermal reduction method was used, with a horizontal double-tube system. Graphite was employed as the reducing agent, while the titanium substrate was coated with a thin layer of Au as catalyste. In_2O_3, SnO_2 and graphite powders, with a weight ratio of 0.3750:0.0415:0.0830, were thoroughly mixed. After that, the mixture was placed at the sealed end of the slender quartz tube. The Au-coated titanium substrates were placed at the downstream side for the deposition of the products. Before reaction the tube was purged with Ar at a flow of 200 sccm for 10 minutes. The reaction was conducted at 1000°C for 2 hours, in Ar flow at a rate of 50 sccm. During the reaction to grow ITO NWs, the working pressure was kept in the range of 5–6 mbar.

To electrically deposit mesoporous NiO, the Lyotropic Liquid Crystalline (LLC) template was obtained by thoroughly mixing 50 wt% Brij56 (polyoxyethylene(10) cetyl ether, $C_{16}[EO]_{10}$) and 50 wt% Ni^{2+} electrolyte consisting of 1.8 M $Ni(NO_3)_2$ and 0.075 M $NaNO_3$ as the supporting electrolyte. A standard three-electrode cell was assembled, with the ITO NWs/Ti as the working electrode, a piece of platinum as the counter electrode and a standard Ag/AgCl electrode. Potentiostatic electrodeposition for H_1–e (hexagonal lyotropic) Ni $(OH)_2$ was conducted at 40°C at a voltage of −0.7 V (vs. Ag/AgCl). After the electrodeposition experiment, the surfactant on the electrode was removed by soaking in ethanol, with was repeated for four times every 2 hours, followed by immersing in 2-propanol for 2 hours. Finally, the electrode was thoroughly cleaned with DI water and then calcined at 270°C for 2 hours. NiO on pure titanium substrate was deposited similarly for comparison. Figure 5.99 shows a schematic diagram demonstrating preparation process of the NiO/ITO NWs electrode.

SEM images indicated that the as-grown ITO NWs on the titanium substrate had lengths in the range of 30–50 μm. High magnification SEM image revealed that the ITO NWs were separated from one and another. The growth of nanowires was started from the hexagonal (111) crystal face, whereas the nanowires were single crystal in nature. Figure 5.100 shows representative SEM and TEM images of the mesoporous NiO/ITO NWs heterostructure. As observed in Fig. 5.100(a–c), the NiO nanoflakes were present in a densely interlaced manner. A uniform layer of NiO with thicknesses in the range of 50–100 nm was coated on surface of the ITO NWs. The NiO in the coating layer was composed of ultrathin nanosheets with a certain degree of aggregation. Although no significant difference was observed in morphology between NiO and Ni $(OH)_2$, mesopores were present in the NiO matrix, with cylindrical pores that had diameters of 2–3 nm, according to high magnification TEM image. The formation of the pores was due to the removal of the surfactant with alcohol and the calcination of the precursor. In addition, the release of H_2O molecules generated due to the decomposition of Ni $(OH)_2$ at the calcination temperature could also produce pores. Average interpore distance calculated from the low-angle XRD pattern was about 6.8 nm. Due to the presence of the mesoporous structure, the NiO sample had a high specific surface area, which was beneficial to the pseudocapacitive reaction, thus leading to high specific capacitance of the materials.

Core-shell nanowires and nanowire arrays have formed a special group of materials that have been used as the electrodes of supercapacitors with unique properties [179–182]. For instance, a novel method has been employed to prepare coaxial nickel oxide/nickel (NiO/Ni) nanowire (NW) arrays for electrodes of pseudocapacitors with promising electrochemical performances [179]. The arrays were developed by using a direct electrochemical deposition process to fill Ni into the nanopores of alumina template, followed by oxidation of the outer shell into NiO through oxygen plasma annealing. The core/shell structured 1D NiO arrays stood vertically, with the inner Ni core to act as conductor and the support of the whole structure. Due to the large effective surface area, the arrays possessed not only high electrical conductivity due to the inner Ni pillars but also high electrochemical behavior owing to the redox-active NiO shells.

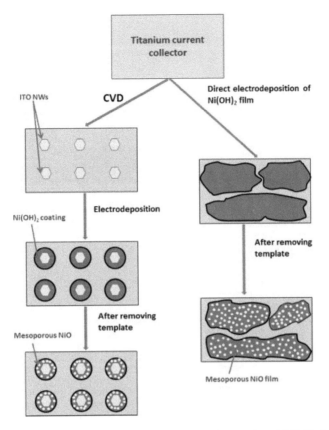

Fig. 5.99. Schematic diagram illustrating formation process of the NiO/ITO NWs and NiO/Ti electrodes, by using the chemical vapor deposition and potentiostatic electrodeposition techniques (top view). Reproduced with permission from [178], Copyright © 2013, Elsevier.

It was found that the morphology, crystal structure and shell thickness of NiO could be well controlled by adjusting the annealing conditions, such as input power and plasma annealing time. The coaxial NiO/Ni NW arrays, with the optimized thickness of the NiO shell, exhibited a capacitance of 0.36 F·cm^{-2}. In addition, the array based electrode demonstrated fast charge–discharge kinetics and high long-term cycling stability, with almost no degradation in capacitance after charge-discharge cycles of up to 2000, at current density of 5 mA·cm^{-2}. Such electrochemical performances could be ascribed to the unique vertically aligned core/shell nanoarchitectures, which ensured a smooth transport of electron and ion during the charge–discharge process.

Whatman alumina membranes, with nanopore diameters in the range of 250–300 nm, length of 60 μm and density of 1×10^9 cm^{-2}, were used as the templates. A layer of Ni with a thickness of 400 nm was coated on one side of the template, which was used as the working electrode. A Cu wire was connected to the Ni layer with a silver paste. A low stress nickel sulfamate bath solution was prepared from nickel sulfamate, boric acid, nickel bromide and wetting agent ANKOR (R) F, with concentrations of 0.37 mol·dm^{-3}, 0.64 mol·dm^{-3}, 0.18 mol·dm^{-3} and 10 ml·L^{-1}, respectively. The solution was adjusted to have a pH value of 3.8, by using 1 mol·dm^{-3} sulfonic acid. The deposition was carried out at the constant temperature of 50°C, while it was stirred at a slow rate of 100 rpm. A two electrodes cell was used for the deposition, which consisted of Ni pellets in Ti-basket as the anode and Ni/porous alumina template as the cathode (working electrode). A 50 mA·cm^{-2} (as the ratio of the total template area) constant current was applied to the working electrode and the deposition rate was found to be 0.6 μm·min^{-1}.

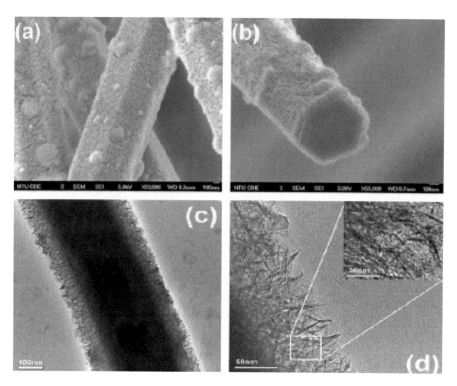

Fig. 5.100. (a) SEM image of the NiO coated on surface of the ITO nanowires. (b) SEM image of tip of the NiO/ITO NWs heterostructure. (c) Low magnification TEM images of NiO/ITO NWs heterostructure, showing uniform distribution of NiO nanoflakes and (d) high magnification TEM image of the NiO coating, demonstrating its mesoporous structure. Reproduced with permission from [178], Copyright © 2013, Elsevier.

Vertically standing Ni NW arrays were released by etching out the template with 1 mol·dm^{-3} KOH solution for 1 hour, followed by thorough washing with DI water and drying in air. A thin film of Ni with thicknesses of 3–4 μm was also used to coat the back side of the seed layer, where the Ni nanostructures were grown inside the pores, so as to prevent the NWs from collapsing as the template was etched out. The obtained Ni NW arrays were treated with plasma for different time durations, at input powers in the range of 25–50 W, in O$_2$ flow at a constant rate of 30 cm^3·min^{-1}. The Ni NW arrays present as a free-standing film were attached onto a Si substrate with adhesive tape at the edge, to cover the back side, so that only the surface of Ni NWs was oxidized to NiO. In this case, only a small portion of the seed layer was exposed to plasma, at the front side of the seed layer (NWs side), due to the high density of the NWs, as mentioned above. After that, the plasma treated films were peeled off from the Si substrate for further characterization.

Figure 5.101 shows a schematic diagram demonstrating formation process of the NiO/Ni NW arrays, through the direct electrochemical deposition of metallic Ni inside the nanopores of alumina membrane template, combined with oxygen plasma oxidation. SEM images indicated that the as-deposited and plasma treated NWs had different surface morphologies, with surface roughness to be increased upon plasma treatment, because of the formation of NiO. As a result, Ni NWs were coated with a thin layer NiO, illustrating the efficiency of oxygen plasma in oxidizing Ni.

Figure 5.102 shows SEM and TEM images of the vertically aligned Ni NWs with different plasma treating times. SEM images taken at the angle of 45° of the samples with 120 seconds, 300 seconds and 600 seconds are in Fig. 5.102(a–c). After annealing for 120 seconds, thickness of the oxide layer could not be clearly identified by using SEM. The oxide was so thin that it was only visible in HRTEM Bright Filed (BF) image, as demonstrated as the inset in Fig. 5.102(d). With increasing annealing time, the thickness of the oxide layer was gradually increased, as revealed in Fig. 5.102(d–f). Therefore, at the fixed input

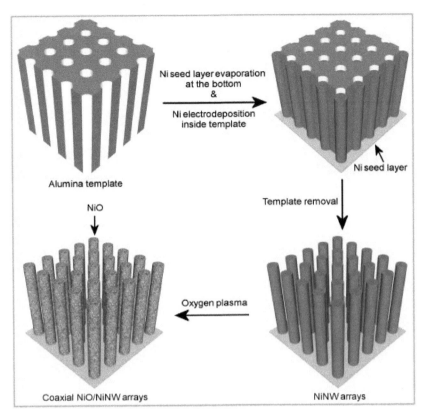

Fig. 5.101. Schematic diagram indicating the steps involved in the fabrication of the coaxial NW arrays, by using the simple electrodeposition into nanoporous alumina template, followed by subsequent conversion of the outer shell into oxide with oxygen plasma annealing after the template was removed. Reproduced with permission from [179], Copyright © 2012, Elsevier.

power of 50 W, the annealing time could be used to control the thickness of the outer shell NiO layer. As illustrated in Fig. 5.102(b, c), after annealing for 300 seconds and 600 seconds, the surface roughness of the oxide layers was increased. Meanwhile, prolonged annealing led to the formation of agglomerated bundles on the surface of the Ni NWs, which resulted in a decrease in total electroactive surface area available for the electrochemical reactions.

Electrophoretic deposition

A NiO nanostructure was derived from Ni(OH)$_2$ that was directed by the electrophoretic deposition (EPD) method [183]. After annealing at 300°C for 3 hours, the Ni(OH)$_2$ nanoplatelet was converted to NiO. It was found that the Ni(OH)$_2$ nanoplatelets were only formed in isopropyl alcohol suspension with iodine and water as additives. In that suspension, H$^+$ ions were formed due to the reaction between iodine and water, which were then adsorbed on the nickel hydroxide, thus leading to positively charged particles. In addition, positively charged nanoplatelets with a high zeta potential was favorable for dispersion of the items and EPD efficiency. Electrode was directly obtained by EPD which had high uniformity than that deposited by using dip coating method. Although the EPD electrode had a relatively low specific capacitance of 112 F·g^{-1} in 0.5 M KOH, at the scan rate of 500 mV·s^{-1}, this value was higher that of the dip-coated NiO electrode. Moreover, the EPD electrode exhibited outstanding cycling stability, with almost no degradation in the specific capacitance after 5000 charge-discharge cycles.

Ni(OH)$_2$ powder was first prepared by using chemical precipitation method. EPD bath was prepared by dispersing 50 mL isopropyl alcohol (IPA), 0.05 g iodine, 1.5 g water and 0.5 g Ni(OH)$_2$ powder,

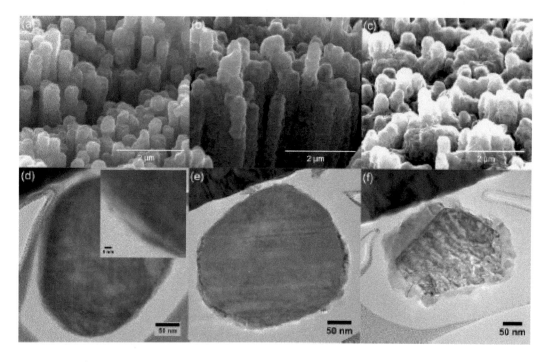

Fig. 5.102. 45°-view SEM images of the NiO/Ni NW arrays after oxygen plasma annealing at 50 W for different time durations: (a) 120 seconds, (b) 300 seconds and (c) 600 seconds. TEM-bright field (TEM-BF) images of the NiO/Ni NWs annealed for different time durations: (d) 120 seconds, (e) 300 seconds and (f) 600 seconds. Reproduced with permission from [179], Copyright © 2012, Elsevier.

with the aid of ultrasonication for 1 hour. The EPD experiment was conducted at a voltage of 10 V for 30 seconds, with a piece of Stainless Steel (SS) as the working (negative) electrode and platinum as the counter (positive) electrode, as a distance of 1 cm. The back side of the SS substrate was covered with a piece of insulating polymer to avoid possible deposition. The EPD sample was dried and then calcined at 300°C for 3 hours in air to form NiO electrode.

For dip-coating deposition, Ni(OH)$_2$ colloid was obtained by ultrasonically dispersing 0.5 g Ni(OH)$_2$ in 50 mL IPA. Then, polished SS foil substrate, with one side to be covered with insulating polymer that was soaked in the Ni(OH)$_2$ suspension and withdrawn at a controlled speed, which was repeated several times, in order to obtain desired thickness. After the dip-coating step, the film was thoroughly washed with DI water, followed by drying and calcination at 300°C for 3 hours in air to obtain NiO electrode.

After calcination at the optimal temperature of 300°C for 3 hours [184], the NiO powder had BET specific surface area and average pore diameter of 170 m$^2 \cdot$g^{-1} and 6 nm, respectively. Figure 5.103 shows XRD patterns of the NiO films prepared by using the two methods described above, together with that of the powder sample. The diffraction peaks could be well indexed to cubic NiO, with lattice parameter of $a = 4.1771$ Å and space group $Fm3m$ (225). The EPD and dip-coating NiO electrodes exhibited stronger diffraction peaks than the NiO powder, indicating their higher crystalline nature.

Microwave-assisted deposition

The microwave-assisted method has a distinct advantage of a rapid process, which has been used to prepare NiO nanostructures with various morphologies, demonstrating its versatile capabilities [119, 185–187]. A gas/liquid interfacial microwave-assisted process was reported to synthesize high surface area flower-like NiO nanostructures, with a high specific capacitance of 585 F·g^{-1} [119]. Ethylene Glycol (EG) was used as the solvent, due to its strong polarity and coordination capability with metal ions to form metal alkoxide.

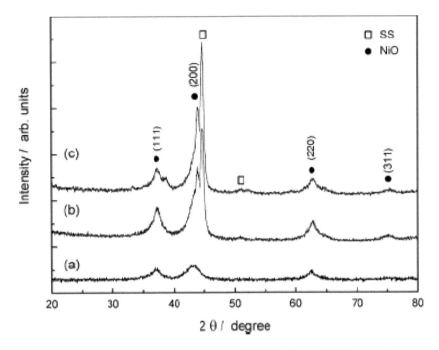

Fig. 5.103. XRD patterns of the samples after thermal annealing at 300°C for 3 hours: (a) NiO powder from precipitation, (b) dip-coating film and (c) EPD film. Reproduced with permission from [183], Copyright © 2009, Elsevier.

Ammonia was introduced into the reactor of EG solution of Ni (Ac)$_2$. At the gas/liquid interface, ammonia reacted with acetylacetone ions, thus triggering the reaction. The presence of the ammonia vapor would increase the pressure inside the reactor, which could be an additional factor that was responsible for the formation of the flower-like NiO hollow nanosphere precursors.

To obtain the precursor, 0.5 mmol Ni(Ac)$_2$ was dissolved in 5 mL EG in a beaker, which was then put into a 70 mL Teflon-lined autoclave that was filled with 6 mL ammonia solution. The autoclave was sealed and heated to 170°C for 3 minutes, by using microwave irradiation and then kept at the temperature for 30 minutes. After reaction, green precipitates of NiO precursors were collected through centrifugation, followed by thorough washing with ethanol. The precursors were finally calcined at 300°C for 4 hours to convert to NiO.

Figure 5.104(a) shows SEM image of the sample after microwave heating at 170°C for 30 minutes. The sample consisted of uniform flower-like spheres, an average diameter of 300 nm. Figure 5.104(b) shows a high-magnification SEM image of the sample, revealing detailed morphology of the flower-like nanostructures. It was found that the individual sphere was assembled with a large number of twisted nanosheets, which were connected to one another to form 3D flower-like architectures. The hollow structure of the spheres was demonstrated by the broken ones, as shown as the inset in Fig. 5.104(b). Figure 5.104(c) shows a TEM image of some spheres, confirming their hollow structural characteristics. After calcination, the morphology of the precursor was well retained, after it was converted to NiO.

Mesoporous NiO with morphologies of nanoslices (NSs) and nanoplatelets (NPLs) has been synthesized, by using a microwave-assisted process [185]. The NiO-NPLs exhibited a specific capacitance of 1200 F·g^{-1}, which was higher as compared with that of the NiO-NSs (600 F·g^{-1}). The NiO-NPLs possessed mesopores with a preferred thickness with controlled size and near-perfect hexagonal shape, which ensured smooth diffusion of OH$^-$ ions to the active sites of the platelets, resulting in the high specific capacitance. It was also found that the microwave-assisted synthesis led to materials with enhanced charge storage and stability, as compared with those prepared by using other methods, such as hydrothermal reaction.

378 *Nanomaterials for Supercapacitors*

Fig. 5.104. (a) SEM image of the NiO precursor. (b) High magnification SEM image of the NiO precursor, with the inset showing a broken nanosphere, scale bar = 100 nm. (c) TEM image of the NiO precursor. Reproduced with permission from [119], Copyright © 2011, Royal Society of Chemistry.

To obtain hexagonal NiO nanoplatelets, 2.0 g anhydrous nickel chloride and 0.5 g sodium oleate (SO) were dissolved in 50 mL distilled water, with pH value to be adjusted to 10, by drop wise adding ammonia solution (28%). The solution was loaded into a 75 mL Teflon-lined microwave reactor and subject to microwave radiation at the maximum power of 600 W at 160°C for 1 hour. After reaction, the precipitate was collected, followed by washing and drying at 45°C. Finally, the sample was calcined at 300°C for 3 hours in O_2 flow. NiO nanoslices were prepared in a similar way, by dissolving 2.5 g anhydrous nickel chloride in 50 mL distilled water.

The formation of the mesoporous NiO nanostructures experienced three steps, i.e., (i) nucleation, (ii) growth and (iii) dehydroxylation or decomposition. The crystals were grown through the aggregation of nickel hydroxides precursor. Due to the slow and continuous reaction under homogeneous precipitation conditions, the nascent nickel hydroxide nuclei could be assembled along a specific crystal direction. The driving force for the crystals to grow was to decrease the surface energy, so as to develop the slice- or platelet-like nanostructures. The SO acted as a directing agent, so that the hydroxide nuclei were arranged to form smooth hexagonal morphology, i.e., mesoporous platelets. Because of the mild thermal calcination, the mesoporous network of precursor nanostructure was well retained. Figure 5.105 shows representative SEM images of the samples, illustrating that the rectangular slices had average diameters in the range of 300–350 nm and an average thickness of about 40 nm, while the hexagonal platelets possessed diameters of 150–200 nm and a thickness of about 10 nm.

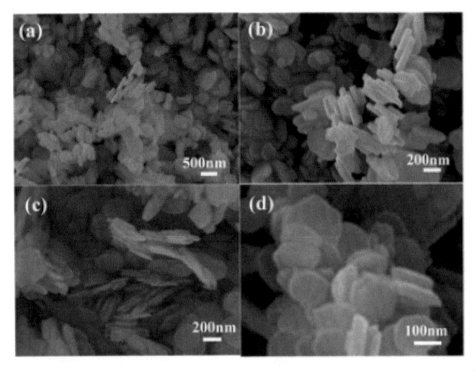

Fig. 5.105. SEM images of the NiO nanostructures: (a, b) nanoslices and (c, d) nanoplatelets. Reproduced with permission from [185], Copyright © 2013, Royal Society of Chemistry.

Other deposition techniques

An ammonia-evaporation method was reported to prepare hierarchically porous NiO film on Ni foam substrate, with promising electrochemical properties [188]. The film was assembled with NiO triangular prisms and porous NiO nanoflakes. It was demonstrated that, in 1 M KOH, the hierarchically porous NiO film had specific capacitances of 232, 229, 213 and 200 F·g^{-1}, at current densities of 2, 4, 10 and 20 A·g^{-1}, respectively. The specific capacitance was retained by 87% as the current density was increased from 2 A·g^{-1} to 20 A·g^{-1}. After 4000 charge-discharge cycles, the porous NiO film still possessed a specific capacitance of 348 F·g^{-1}.

To the prepare the precursor, 30 ml aqueous ammonia with concentrations in the range of 25–28% was added to 30 ml solution of 0.1 M Ni(NO$_3$)$_2$·H$_2$O, under constant stirring. Ni foam with a dimension of 2 × 3 cm was pressed to the thin plate by applying a pressure of 10 MPa and then cleaned in ethanol for 10 minutes with the aid of ultrasonication. One side of the Ni foam was covered with a polytetrafluoroethylene tape to avoid solution contamination and then immersed into the precursor solution, followed by heating at 90°C for 5 hours. After that, a green film was grown on the Ni foam substrate, which was thoroughly washed with DI water and then dried in air. Finally, the film was calcined at 300°C for 2 hours in flowing Ar.

XRD patterns revealed that the green film was well-crystallized β-Ni(OH)$_2$. After calcination at 300°C, the β-Ni(OH)$_2$ precursor was converted to NiO. Figure 5.106 shows SEM images of the NiO film after calcination. The Ni foam substrate was fully covered by the NiO film, while the film consisted of homogeneous small triangular prisms, as seen in Fig. 5.106(a). The triangular prisms had a dimension at the micron scale. As demonstrated in Fig. 5.106(b), the individual prism was comprised of two parts, i.e., (i) regular stacking of triangular platelets and (ii) randomly porous NiO nanoflakes. The NiO nanoflakes were grown vertically on the triangular platelets, thus resulting in a porous surface. The unique morphological characteristics of the NiO film was also confirmed by TEM images, with the nanoflakes having a thickness of about 100 nm, which consisted of nanoparticles with diameters in the range of 5–10 nm.

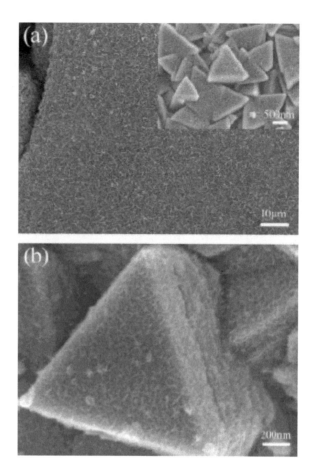

Fig. 5.106. Low (a) and high (b) magnification SEM images of the NiO film after calcination at 300°C for 2 hours. Reproduced with permission from [188], Copyright © 2012, Elsevier.

Nickel Cobalt Complex Oxide (NiCo$_2$O$_4$)

Brief overview

Spinel NiCo$_2$O$_4$ is a ternary complex metal oxide, which has been extensively studied [189, 190], due to its applications in fields of magnetic [191–193], electrocatalytic water splitting [23, 194–197] and lithium ion batteries [198–202]. More recently, NiCo$_2$O$_4$ has also become a hot material to be promising candidates as electrodes of supercapacitors [21]. As compared with the two single component oxides, NiO and Co$_3$O$_4$, NiCo$_2$O$_4$ has exhibited a much richer redox chemistry, owing to the combination of the advantages from both Ni^{2+} and Co^{3+}/Co^{2+} [21].

NiO has simple rock salt structure with octahedral Ni^{2+} and O^{2-} sites, while Co$_3$O$_4$ exhibits normal spinel structure, with Co^{2+} ions at the tetrahedral sites and Co^{3+} ions at the octahedral sites, which are formed through the close packing of O^{2-} ions [203, 204]. The spinel structure is a face-centered cubic close packing of O^{2-} ions, with one half of the octahedral sites and one eighth of the tetrahedral sites to be taken by the cations. Because of the requirement of charge balance, one third of the cations have an oxidation state of 2+, while two thirds are at 3+ oxidation state. Experimental data indicated that the nickel cobaltite has a general formula of $(Co^{2+}_{1-x}Co^{3+}_{x})_{tet}[Co^{3+}Ni^{2+}_{x}Ni^{3+}_{1-x}]_{oct}O_4$ ($0 < x < 1$). Therefore, Ni^{2+}/Ni^{3+} ions occupy the octahedral sites, while Co^{2+} ions occupy the tetrahedral sites and Co^{3+} ions are distributed at both sites [205]. Because there are mixed valences for the two cations, the spinel cobaltite NiCo$_2$O$_4$ has higher electrical conductivity than both nickel and cobalt oxides [206]. As a result, $NiCo_2O_4$ as electrode materials of supercapacitors possessed higher electrochemical performances, due to its higher electrical conductivity, as compared with those based on the single cation oxides [207].

The performance of electrode materials has a close relation to behaviors of the electro-active species, the electron transportation, electrolytes and reactants [208, 209]. The properties of a given electrode material is mainly determined by its microstructure, which includes morphology, size and size distribution, macro/mesopores and nano-architecture [210–213]. $NiCo_2O_4$ as an electrode of supercapacitors has been presented in two forms: (i) powder based electrodes and (ii) conducting substrate supported electrodes. Examples with more details will be present and discussed below.

$NiCo_2O_4$ nanostructures

Various methods, such as sol–gel, hydrothermal and precipitation, have been employed to develop nanostructures, with outstanding electrochemical performance as electrodes of supercapacitors [214–227]. Generally, sol–gel and precipitation processes led to porous and homogeneous 0D nanostructures or nanoparticles, while hydrothermal reaction resulted in nanostructures, with well-defined controllable 1D, 2D and 3D architectures by adjusting the reaction parameters. Comparatively, 3D nanostructures are advantageous over 1D and 2D ones, because (i) they offer continuous electron transport channels for electrolyte ions, (ii) have higher electrical conductivity and (iii) possess higher structural mechanical stability to withstand the volume expansion during the charge-discharge cycles. Selected examples for each method will be presented and discussed next.

Sol–gel process

A sol–gel synthetic method was employed to prepare cubic $NiCo_2O_4$ crystals, with citric acid as the chelating ligand and H_2O-N,N-dimethylformamide (H_2O-DMF) as the solvent [228]. The crystals had both submicron-sized coral like morphology and nanosized particles. The submicron-sized $NiCo_2O_4$ exhibited an optimal specific capacitance of $217\ F \cdot g^{-1}$, with capacity retention of 96.3% after 600 charge–discharge cycles. Moreover, the materials experienced only 7.4% capacity degradation, as the current density was increased from 1 to $10\ mA \cdot cm^{-2}$, demonstrating the electrode had an outstanding rate capability.

Experimental parameters, which were varied to optimize the performances of the final products, included initial concentration of the reactants, evaporation time (8 hours and 16 hours), type of solvent (H_2O, H_2O-DMF) and calcination temperature (400°C, 500°C and 600°C). In a typical synthesis process, $8 \times 10^{-3}\ M\ Co(Ac)_2 \cdot 4H_2O$, $4 \times 10^{-3}\ M\ Ni(Ac)_2 \cdot 4H_2O$ and $2.4 \times 10^{-2}\ M$ citric acid were dissolved in H_2O at constant stirring, to form a solution with a light pink color. The solution was heated at 120°C for 16 hours, which resulted in precursor (I). It was pulverized and calcined at 400°C for 3 hours in air. During the heat treatment at 120°C, the light pink colored solution was changed gradually to solid because of the polymerization. For comparison, precursor (II) was prepared by using the concentration of $Co(Ac)_2 \cdot 4H_2O$, $Ni(Ac)_2 \cdot 4H_2O$ and citric acid, while the drying time was 8 hours. Precursor (III) was obtained by using a mixed solvent of H_2O-DMF, with a v/v ratio of 1:1 and drying time of 16 hours. Calcination was conducted at 400°C for 3 hours in air. The calcination temperature was selected according to the DSC-TGA and XRD results.

Figure 5.107(a) shows a representative SEM image of sample (I) calcined at 400°C for 3 hours, revealing a morphology that was like coral reefs, with a stable network consisting of irregular branches with a dimension of about 1 μm. The high magnification image indicated that the sample contained micron-sized $NiCo_2O_4$, with a porous structure, where the window-pores had a size of about 200 nm, as seen in Fig. 5.107(b). BET specific surface area of the porous structure was $41.5\ m^2 \cdot g^{-1}$. The formation of the reef-like structure involved three steps [229]. Initially, primary $NiCo_2O_4$ nanoparticles with sizes of 10–20 nm were developed. After that, the primary $NiCo_2O_4$ particles grew to irregular secondary particles, with sizes in the range of 200–300 nm. Eventually, branch structures were evolved from the secondary $NiCo_2O_4$ particles.

SEM images of sample (III) are shown in Fig. 5.107(c, d). In this case, submicron-sized particles were obtained, as illustrated in Fig. 5.107(d). Nanosized spheres with a relatively large size distribution were

382 *Nanomaterials for Supercapacitors*

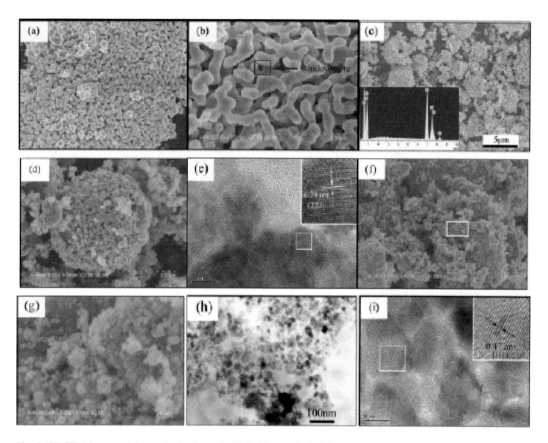

Fig. 5.107. SEM images of the as-obtained sample (I) (a, b), sample (III) (c, d) and sample (II) (f, g). TEM image (h) and HRTEM image (i) of sample (II). Inset in (c) is EDS analysis result of sample (III). (e) HRTEM image of sample (III). Reproduced with permission from [228], Copyright © 2011, Elsevier.

observed in the sample. The presence of the nanosized spheres could be closely related to the effect of the mixed solvent of H_2O-DMF. Figure 5.107(e) shows HRTEM image of sample (III), together with lattice fringes of the selected area. As shown in the inset, its interfringe distance was 0.29 nm, corresponding to the (220) planes of $NiCo_2O_4$. The EDS analysis result confirmed that sample (III) contained O, Co and Ni, with an atomic ratio of Co:Ni to be about 2.03, as observed as the inset in Fig. 5.107(c). The submicron-sized $NiCo_2O_4$ particles had sizes in the range of 90–220 nm, with a mean size of 170 nm.

Figure 5.107(f) shows a low magnification SEM image of sample (II), indicating the presence of irregular shaped particles, which were agglomerates of $NiCo_2O_4$ nanoparticles with an average size of 13 nm, as demonstrated in Fig. 5.107(g). Figure 5.107(h) shows TEM image of sample (II), revealing that the nanoparticles were in the range of 11–15 nm, as further confirmed by the high magnification image in Fig. 5.107(g). Figure 5.107(i) shows HRTEM image of the nanoparticles, with lattice fringes of the selected area. The inset image indicated that the nanoparticle exhibited an interfringe distance of 0.47 nm, corresponding to the (111) planes of $NiCo_2O_4$. The crystalline nature of the $NiCo_2O_4$ nanoparticles were confirmed by the SAED pattern.

A sol-gel method, combined with rapid microwave-assisted hydrothermal treatment, was used to prepare $NiCo_2O_4$ nanocrystals from nickel–cobalt oxy-hydroxides $(Ni/2Co)(OH)_2$ [230]. The Ni–Co oxy-hydroxide had an optimal specific capacitance of 215 F·g^{-1}, at the current density of 5 mA·cm^{-2}, in 1 M NaOH electrolyte. The phase transformation could be further promoted by introducing GO as the microwave adsorbents, which thus led to $NiCo_2O_4$/rGO composite, with smaller crystal size. Accordingly, the NiCo$_2$O$_4$/rGO composite exhibited a total specific capacitance of 410 F·g^{-1}, corresponding to a specific capacitance of 730 F·g^{-1} for NiCo$_2$O$_4$, which was much higher than the pure NiCo$_2$O$_4$ sample. The enhanced electrochemical performance was attributed to the increased electrical conductivity of the composite, due to the presence of the highly conductive rGO nanosheets.

To synthesize the mixed Ni–Co hydroxide, NiCl$_2$·6H$_2$O and CoCl$_2$·6H$_2$O, with a molar ratio of 1:2, were dissolved in a mixed solution with equal volume of water and ethanol, with the aid of stirring for 30 minutes to form the organometallic intermediate phase. Then, 1 M NaOH solution was added into the solution to trigger the precipitation of the hydroxide, once the pH value was > 8. After that, the hydroxide precipitates were collected through centrifugation, followed by thorough washing with pure water, until the pH value was about 7. To obtain dry precipitates, some powder was kept in a vacuum oven at room temperature for 8 hours. Wet precipitates were directly dispersed in DI water, with the aid of ultrasonication for 20 minutes and then treated in a microwave reactor, at temperatures of 100, 150, 200 and 210°C for 15 minutes, thus forming Ni–Co oxy-hydroxide powders, which were denoted as NCOH100, NCOH150, NCOH200 and NCOH210, respectively. TEM images revealed that average sizes of the spherical NiCo$_2$O$_4$ particulates were 30.2, 43.3 and 57.4 nm for the samples treated at 150, 200 and 210°C, respectively.

NiCo$_2$O$_4$ aerogels were prepared by an epoxide-driven process, which exhibited a specific capacitance of as high as 1400 F·g^{-1} [231]. Such a superb electrochemical performance was ascribed to the high mesoporosity and well-connected networked structures of the NiCo$_2$O$_4$ aerogels, which offered significantly reduced mass-transfer resistance, during the electrolyte penetration and ion diffusion and promoted the electron hopping in between adjacent NiCo$_2$O$_4$ nanoparticles.

To prepare nickel cobaltite aerogels, a mixture of chlorides, CoCl$_2$·6H$_2$O and NiCl$_2$·6H$_2$O, with the molar ratio of 2:1, were dissolved in ethanol. After that, propylene oxide with the molar ratio of propylene oxide and metal ion to be 11:1, thus leading to gel at room temperature. The wet gel was washed thoroughly with ethanol and then dried through supercritical drying process. The dried aerogel was finally calcined at 200 and 300°C for 5 hours to convert the hydroxide precursor into NiCo$_2$O$_4$. In addition, cobalt oxide and nickel oxide aerogels were similarly prepared as a comparison, with Co(NO$_3$)$_2$·6H$_2$O and NiCl$_2$·6H$_2$O, as the starting materials, respectively. The formation of the corresponding oxides was confirmed by the XRD results.

Representative TEM images of the samples are shown in Fig. 5.108(a, b), clearly demonstrating the presence of the mesoporous microstructure. According to SAED patterns, with well-defined rings, it was suggested the samples were all of polycrystalline characteristics. Most of the rings could be attributed to the crystal planes of NiCo$_2$O$_4$, while only one or two rings belonged to Ni(OH)$_2$. Therefore, NiCo$_2$O$_4$ was the dominant phase in the samples. Obviously, the Ni–Co–O–200 sample had higher crystallinity than the

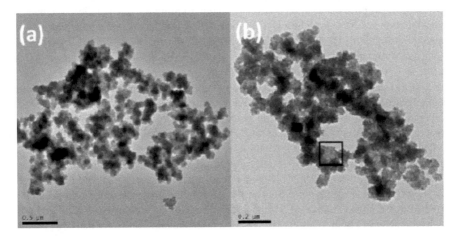

Fig. 5.108. TEM images of (a) sample Ni–Co–O and (b) sample Ni–Co–O–200. Reproduced with permission from [231], Copyright © 2009, WILEY-VCH Verlag GmbH & Co. KGaA, Weinheim.

384 *Nanomaterials for Supercapacitors*

Ni–Co–O one, implying that the crystallinity was significantly increased after the calcination at 200°C. Pore sizes derived from the N_2-adsorption/desorption isotherm revealed that the majority of the pores were in the range of 2–5 nm, which was favorable for supercapacitor applications [232, 233].

Hydrothermal reaction

The hydrothermal process has been extensively explored to synthesize $NiCo_2O_4$ nanomaterials with various morphologies for electrode applications of supercapacitors. For instance, porous hexagonal $NiCo_2O_4$ nanoplates, with an average diameter of 100 nm and thickness of 25 nm, were developed by using a hydrothermal method, with hexadecyl trimethyl ammonium bromide (CTAB) as the complexing agent [234]. The $NiCo_2O_4$ samples possessed BET surface areas in the range of 51.88–617.08 $m^2 \cdot g^{-1}$, with mesoporous structures. An optimal specific capacitance of 294 $F \cdot g^{-1}$ was achieved, at a current density of 1 $A \cdot g^{-1}$, with cyclic stability of 89.8% after 2200 charge-discharge cycles.

The hydrothermal reaction precursor was prepared by dissolving 0.476 g $CoCl_2 \cdot 6H_2O$ (4 mmol), 0.249 g $Ni(Ac)_2 \cdot 4H_2O$ (2 mmol) and 0.1 g hexadecyl trimethyl ammonium bromide (CTAB) in 40 mL DI water, with stirring for 1 hour, leading to the formation of a light pink colored solution. Then, 3.2 g NaOH was added to the solution at room temperature and stirred for 10 minutes, which was subject to hydrothermal reaction at 160°C for 20 hours. After the reaction was finished, the precipitate was collected and thoroughly washed with DI water and ethanol, followed by drying at 60°C for 8 hours. The precursor powders were calcined at 250, 300, 350 and 400°C in air for 2 hours, thus resulting in black powders, which were noted as C250, C300, C350 and C400. For comparison, precursor without the hydrothermal process was also prepared similarly and calcined at 350°C for 2 hours to form sample S350.

Figure 5.109(a) shows SEM image of the precursor, revealing its hexagonal nanoplate morphology. SEM images of the C300 sample are shown in Fig. 5.109(b, c), with a similar morphology to the precursor, which suggested that low temperature calcination had no negative effect on the morphology. However, the $NiCo_2O_4$ nanoplates contained a large number of pores, as demonstrated in Fig. 5.109(c). Furthermore, at the calcination temperature was increased to 350°C, the nanoplate morphology was still retained, as illustrated in Fig. 5.109(d, e) for the sample C350. In contrast, without the processing hydrothermal reaction, no regular hexagonal shape was observed in the sample S350, as revealed in Fig. 5.109(f). Therefore, the hydrothermal treatment played a significant role in promoting the crystallization of the precursors.

Urchin-like $NiCo_2O_4$ nanostructures, with a BET surface area of 99.3 $m^2 \cdot g^{-1}$, which were grown radially from the center, were prepared by using a hydrothermal method [235]. The $NiCo_2O_4$ urchins possessed specific capacitances of 1650 and 1348 $F \cdot g^{-1}$, at current densities of 1 and 15 $A \cdot g^{-1}$, respectively, with a retention of 81.7%, as current density was increased from 1 to 15 $A \cdot g^{-1}$. To prepare hydrothermal reaction solution, 0.948 g $CoCl_2 \cdot 6H_2O$, 0.474 g $NiCl_2 \cdot 6H_2O$ and 1.08 g urea were dissolved in 30 mL DI water, which had a transparent pink color, after stirring for about 10 minutes. The solution was then subject to hydrothermal reaction at 120°C for 6 hours. After the reaction was finished, the precipitate was collected and thoroughly washed with DI water, followed by vacuum drying at 80°C for 6 hours. To obtain the final $NiCo_2O_4$ nanostructures, the precipitate powder was calcined at 300°C for 3 hours, at a heating rate of 5°C min^{-1}.

Figure 5.110(a) shows XRD pattern of the $NiCo_2O_4$ sample, indicating the formation of face-centered cubic structured $NiCo_2O_4$. Figure 5.110(b) shows SEM image of the $NiCo_2O_4$ powder, revealing the presence of spherical urchin-like nanostructures, with an average diameter of about 5 μm. The individual $NiCo_2O_4$ urchins were assembled from numerous nanorods, which were radially grown from the center. Careful inspection demonstrated that the $NiCo_2O_4$ nanorods exhibited diameters in the range of 100–200 nm and an average length of 2 μm. Due to the hierarchical packing of the nanorods, the $NiCo_2O_4$ nanostructures possessed a high specific surface area. This unique nanostructure was responsible for the high specific capacitance, by offering easy access of the active materials to the interfaces during the redox process.

Figure 5.110(c) a representative TEM image of an individual urchin-like $NiCo_2O_4$ nanostructure, confirming its diameter of about 5 μm. It also evidenced that the nanorods were densely packed pointing from the center of the sphere in the radical direction. High magnification TEM image, as shown in Fig. 6.110(d), illustrated that the $NiCo_2O_4$ nanorods were highly porous in microstructure, with diameters

Oxide Based Supercapacitors: II-Other Oxides 385

Fig. 5.109. Representative SEM images of the samples: (a) precursor, (b, c) C300 sample at different magnifications, (d, e) C350 sample at different magnifications and (f) S350 sample. Reproduced with permission from [231], Copyright © 2013, Elsevier.

of 100–200 nm, in agreement with the SEM observation. Figure 5.110(e) shows HRTEM image of an individual NiCo$_2$O$_4$ nanorod, revealing the polycrystalline characteristic of the nanorods. The lattice fringes of 0.28 and 0.24 nm corresponded to the (220) and (311) planes of spinel structured NiCo$_2$O$_4$. Such a polycrystalline nature was further confirmed by the SAED pattern, as seen in Fig. 5.110(f).

The formation mechanism of the urchin-like NiCo$_2$O$_4$ nanostructures we studied by varying the pH value of the hydrothermal reaction solutions. Figure 5.111 shows SEM images of the samples derived from the solutions with different pH values. As the pH value was lower than 5, no precipitate was formed due to the acidic nature of the solution. If the pH value was slightly increased to 5.5, nanorods started to appear, while some nanorod bundles with small nanorods extruding from the two ends were also observed, as demonstrated in Fig. 5.111(a). As the pH value was further increased to 6.1, straw-like nanorod bundles were obtained, as illustrated in Fig. 5.111(b). Therefore, in this case, the bifurcation phenomenon occurred in the solutions with relatively low pH values, which was more and more pronounced, as the pH value was gradually increased. As a consequence, straw-like nanorod bundles were continuously developed, finally leading to the formation of urchin-like spheres from the solution with pH value of 6.8, as seen in

386 *Nanomaterials for Supercapacitors*

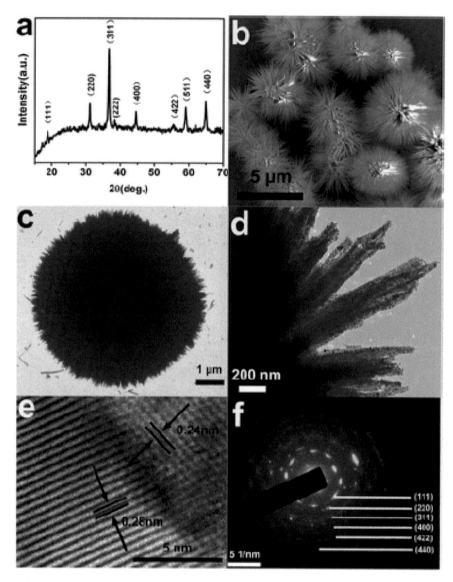

Fig. 5.110. Phase composition and microstructures of the urchin-like NiCo$_2$O$_4$ nanostructures: (a) XRD pattern, (b) SEM image, (c, d) TEM images, (e) HRTEM image and (f) SAED pattern. Reproduced with permission from [235], Copyright © 2012, Royal Society of Chemistry.

Fig. 5.111(c). With this observation, a transformation mechanism of 'nanorod-to-straw-bundles-to-urchin spheres' could be used to explain the formation of the urchin-like NiCo$_2$O$_4$ nanostructures, as shown schematically in Fig. 5.111(d–f).

Thermal decomposition

A simple scalable thermal decomposition was reported to prepare spinel NiCo$_2$O$_4$ nanocrystals [236]. CTAB (C$_{16}$H$_{33}$) N (CH$_3$)$_3$Br was used as surfactant that could significantly enhance the homogeneity and porosity of the spinel NiCo$_2$O$_4$. The porosity and high specific surface area of NiCo$_2$O$_4$ ensured the outstanding electrochemical performances, due to the presence of smooth paths for the electrolyte to penetrate the

Oxide Based Supercapacitors: II-Other Oxides 387

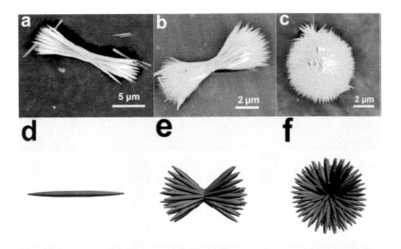

Fig. 5.111. (a–c) SEM images of the NiCo$_2$O$_4$ nanostructures after hydrothermal reaction with pH values of 5.5, 6.1 and 6.8, respectively. (d–f) Schematic illustrations of the growth mechanism of the urchin-like NiCo$_2$O$_4$ nanostructures developing from nanorods to straw bundles and finally urchin-like spheres. Reproduced with permission from [235], Copyright © 2012, Royal Society of Chemistry.

electrode and reach the inner active sites. BET specific surface area, pore volume and average pore diameter of the NiCo$_2$O$_4$ powder were 124 m^2·g^{-1}, 0.203 cm^3·g^{-1} and 3.484 nm, respectively. As a result, a high specific capacitance of 764 F·g^{-1} was achieved at the current density of 2 mV·s^{-1}, with long term stability up to 5600 cycles by losing about only 10% and negligible loss after 10,000 cycles.

Ni (AcO)$_2$·4H$_2$O and Co(AcO)$_2$·4H$_2$O were dissolved in methanol to form the precursor solution. CTAB was added into the solution at different concentrations as the surfactant. After stirring for 1 hour, the mixed solutions were cast into a glass disk and then heated at 300°C. The concentrations of CTAB were 0, 0.5, 2.5, 5 and 10 mM, resulting in samples denoted as NCO-0, NCO-05, NCO-25, NCO-50 and NCO-100, respectively.

A mechanism was proposed to explain the interactions between the precursors and the CTAB surfactant, as well as the formation of the porous NiCo$_2$O$_4$ nanoparticles, as demonstrated in Fig. 5.112 [236]. Firstly, cationic surfactant CTAB and Ni–Co precursors were dissolved in methanol to form a homogeneous precursor solution at step (A). As the solvent was evaporated, the concentration of CTAB was increased gradually, so that micelles containing Ni and Co precursors started to appear, because of the affinity of Ni–Co ions to the hydrophilic side of CTAB, which was denoted as step (B). After the removal of the surfactant, highly porous NiCo$_2$O$_4$ structure was obtained, which was step (C). The concentration of CTAB played a significant role in determining the morphology and size of the final NiCo$_2$O$_4$ nanostructure, which implied that the porous architecture of NiCo$_2$O$_4$ could be well controlled by optimizing the surfactant concentration in the precursor solutions.

3D NiCo$_2$O$_4$ micro-spheres with radial chain-like nanowires were prepared with urea as the oxidizing agent in a mixed solvent of water and ethanol, which exhibited specific capacitances of 1284 and 986 F·g^{-1} at current densities of 2 and 20 A·g^{-1}, respectively, corresponding to a capacity retention of 77%, with respect to the increase in current density [237]. Furthermore, the NiCo$_2$O$_4$ microspheres had capacitance of only 2.5% after 3000 charge-discharge cycles. The promising electrochemical performances of the NiCo$_2$O$_4$ micro-spheres was ascribed to the high conductivity of chain-like nanowires, by offering an effective path for electrons to transport, favorable electron collection behavior, short electron transport and electrolyte ion diffusion paths.

To prepare the starting solution, 1 mmol Ni (NO$_3$)$_2$·6H$_2$O, 2 mmol Co(NO$_3$)$_2$·6H$_2$O and 4 mmol urea were dissolved in a mixed solution consisting of 10 mL ethanol and 40 mL water. The solution was further stirred for 30 minutes and then hydrothermally treated at 100°C for 6 hours. After the reaction was finished,

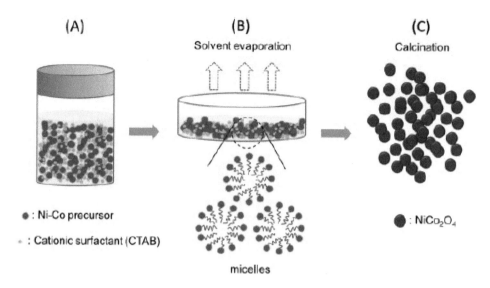

Fig. 5.112. Schematic diagram showing formation process of the NiCo$_2$O$_4$ nanocrystals through the thermal decomposition process with the presence of CTAB: (A) Ni–Co acetates and CTAB dissolved in methanol, (B) the formation of CTAB micelles containing Ni–Co precursors during methanol evaporation and (C) formation of NiCo$_2$O$_4$ nanostructures due to thermal decomposition. Reproduced with permission from [236], Copyright © 2013, Elsevier.

the precipitate was collected and thoroughly washed with DI water and ethanol through centrifugation, followed by drying at 60°C. The precursor microspheres were finally calcined at 300°C for 2 hours to form chain-like NiCo$_2$O$_4$ nanowires.

Figure 5.113(a) shows SEM image of the precursor 3D micro-spheres, with an average diameter of about 4 μm. The micro-spheres were uniformly assembled from radial ultrafine nanowires, with a dandelion-like morphology. The inset in Fig. 5.113(a) revealed that numerous straight and smooth nanowires were observed in the microspheres. The nanowires possessed lengths of up to several micrometers and diameters of less than several nanometers. SEM images the NiCo$_2$O$_4$ samples after calcination at 300°C are shown in Fig. 5.113(b, c), suggesting that the 3D microsphere morphology was well retained. Similarly, a large number of nanowires were present on surface of the microspheres. However, the nanowires became coarser than those in the as-obtained precursor sample before calcination, which was ascribed to the crystallization of NiCo$_2$O$_4$. High magnification SEM image indicated that the NiCo$_2$O$_4$ nanowires experienced shrinkage after the calcination and exhibited a 1D chain-like morphology, composing of nanoparticles, as demonstrated in Fig. 5.113(c). Figure 5.112(d) shows XRD pattern of the 3D microspheres after calcination, confirming the formation of NiCo$_2$O$_4$.

Figure 5.114(a) shows TEM image of an individual calcined microsphere with diameter of about 4 μm. Similar to that observed in the SEM image, the sphere surface consisted of ultrafine nanowires. High magnification TEM image of the sphere confirmed that the nanowires were constructed with nanoparticles that were aligned in 1D configuration, diameters in the range of 5–15 nm, as observed in Fig. 5.114(b). The chain-line structure of the nanoparticles was further evidenced by HRTEM image, as seen in Fig. 5.114(c). No amorphous layers were present on surface of the nanoparticles, which implied that the NiCo$_2$O$_4$ phase was completely crystallized.

In the top-right inset in Fig. 5.114(c), the d-spacings of 0.24 nm and 0.24 nm corresponded to the (220) and ($2\bar{2}0$) planes of NiCo$_2$O$_4$ lattice fringes. The corresponding Fast Fourier Transformation (FFT) pattern could be indexed to spinel structure, as shown in the bottom-left inset of Fig. 5.114(c). The pattern belonged to the [001] zone of spinel structure, which implied that the nanoparticle surface terminated with the (001) plane. Meanwhile, the two lattice fringes with a d-spacing of 0.47 nm were the (111) and ($\bar{1}11$) lattice spacing of NiCo$_2$O$_4$. The FFT pattern corresponded to the [01$\bar{1}$] zone of spinel structure, i.e., the

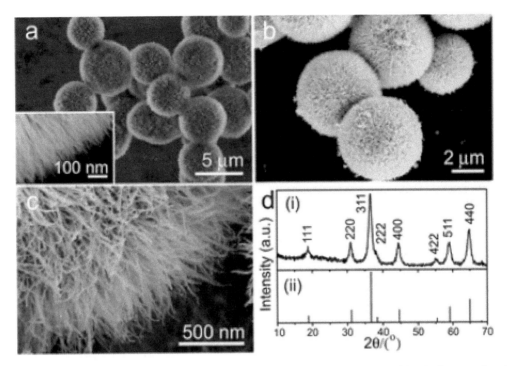

Fig. 5.113. (a) SEM image of the precursor microspheres, with the inset showing SEM image of the ultrafine nanowires. (b, c) Low- and high-magnification SEM images of the NiCo$_2$O$_4$ microspheres. (d) XRD pattern of the 3D NiCo$_2$O$_4$ microspheres after calcination. Reproduced with permission from [237], Copyright © 2013, Royal Society of Chemistry.

nanoparticle top surface was terminated with the (01$\bar{1}$) plane. In addition, the atom configuration at the transition region between every two adjacent nanocrystals had a twin crystal structure that was indicated with an arrow, so that nanocrystals were linked to form the chain-like morphology. The EDX spectrum of the chain-like nanowire reveals the presence of Co, Ni and O, as depicted in Fig. 5.114(d).

Chemical precipitation

A polymer-assisted precipitation method was reported to obtain porous NiCo$_2$O$_4$ nanostructures, with ethylenediaminetetraacetic acid (EDTA) and polyethyleneimine (PEI) as the complexing and structure directing agents, respectively [238]. The NiCo$_2$O$_4$ nanostructures having a hierarchical porous network exhibited a specific capacitance of 587 F·g^{-1} at the current density of 2 A·g^{-1}, which was only slightly degraded to 518 F·g^{-1}, as the current density was increased to 16 A·g^{-1}.

The precursor solution was obtained by dissolving 2 mM Co(NO$_3$)$_2$·6H$_2$O and 1 mM Ni(NO$_3$)$_2$·6H$_2$O in 30 mL DI water, into which 1 g EDTA and PEI were added, followed by vigorous stirring. The mixture was first heated to evaporate the excess water and then calcined at 500°C for 1 hour, which led to the formation of a black loose foam-like powder. At the same time, NiCo$_2$O$_4$ samples, denoted as NiCo$_2$O$_4$–EDTA and NiCo$_2$O$_4$–PEI, were also prepared similarly, without the presence of PEI and EDTA, respectively.

The fact that the NiCo$_2$O$_4$ powder possessed an intimately interconnected network-like structure was evidenced by SEM images. Such a structure resulted in the presence of inter-particle pores. Representative TEM images are shown in Fig. 5.115(a–c), which further confirmed formation of the hierarchical meso- and/or macroporous network-like NiCo$_2$O$_4$ framework, which had a high degree of pore connectivity. The NiCo$_2$O$_4$ nanosized building blocks had sizes in the range of 20–30 nm, as observed in Fig. 5.115(c). The SAED pattern, as the inset in Fig. 5.115(c), confirmed polycrystalline characteristics of the NiCo$_2$O$_4$ powder. Figure 5.115(d) shows HRTEM image of the NiCo$_2$O$_4$ powder, revealing that the nanoparticles could have

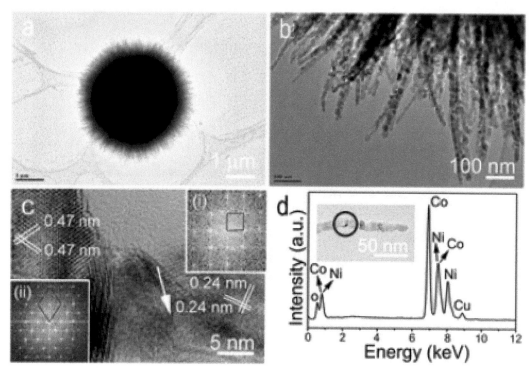

Fig. 5.114. (a) TEM image of an individual NiCo$_2$O$_4$ microsphere. (b) High magnification TEM image revealing the chain-like morphology of the NiCo$_2$O$_4$ nanowires. (c) HRTEM images confirming the chain-like NiCo$_2$O$_4$ nanowires, with (i) top-right inset showing the FFT pattern along the [001] zone axis of the NiCo$_2$O$_4$ crystal and (ii) the bottom-left inset confirming the FFT pattern along the [01$\bar{1}$] zone axis of the NiCo$_2$O$_4$ crystal. (d) EDX spectrum taken from a selected area of the chain-like nanowire. Reproduced with permission from [237], Copyright © 2013, Royal Society of Chemistry.

different orientations overlapping with one and another, so as to form the interconnected network-like architecture. The hierarchical porous network-like framework of the precursor was well retained, which was attributed to the intimated interconnection of the NiCo$_2$O$_4$ nanoparticles.

Mesoporous hollow NiCo$_2$O$_4$ sub-microspheres, consisting of ultrathin nanosheets, were developed by using silica spheres as the hard template [239]. The NiCo$_2$O$_4$ sub-microspheres exhibited a specific capacitance of 678 F·g^{-1} at the current density of 1 A·g^{-1}, while it was degraded by only 13% after 3500 charge-discharge cycles at a current density of 10 A·g^{-1}. Silica spheres were dispersed in 30 mL solution of ethanol (15 mL), DI water (15 mL) and hydroxypropyl cellulose (HPC, 0.05 g), followed by 30 minutes ultrasonication. After that, 0.5 mM Ni(NO$_3$)$_2$·6H$_2$O, 1 mM Co(NO$_3$)$_2$·6H$_2$O and 4 mM urea were added into the mixture, which was then refluxed at 80°C for 3 hours with constant stirring. After the refluxing was finished, the product was collected, followed by washing, drying and calcination at 300°C for 1 hour. The dried powder was redispersed in 100 mL 3 M NaOH solution, which were continuously stirred for 24 hours. The sample was then collected through centrifugation and thoroughly washed with DI water and ethanol. Finally, the dried sample was calcined at 300°C for 0.5 hour, at a heating rate of 1°C min^{-1}. NiCo$_2$O$_4$ samples were also prepared with solutions having different precursor ion concentrations. For sample NiCo$_2$O$_4$-1, 0.25 mM Ni(NO$_3$)$_2$·6H$_2$O, 0.5 mM Co(NO$_3$)$_2$·6H$_2$O and 2 mM of urea were used. For sample NiCo$_2$O$_4$-2, 0.75 mM Ni(NO$_3$)$_2$·6H$_2$O, 1.5 mM Co(NO$_3$)$_2$·6H$_2$O and 6 mM urea were used.

Figure 5.116 shows SEM images of the NiCo$_2$O$_4$ hollow sub-microspheres [239]. The sample consisted of uniform mono-dispersed sub-microspheres with an average diameter of about 650 nm, as illustrated in Fig. 5.116(a, b), close to that of the silica templates with sizes in the range of 600–700 nm. No NiCo$_2$O$_4$ nanoparticles were observed, which implied all the NiCo$_2$O$_4$ nanoparticles were assembled onto surface of the silica spheres. At the same time, the spherical morphology was well retained during the removal of

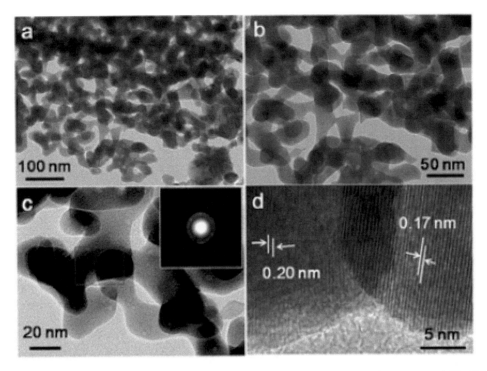

Fig. 5.115. TEM (a–c), HRTEM (d) images and SAED pattern (the inset in c) of the hierarchical porous network-like NiCo$_2$O$_4$ nanostructure. Reproduced with permission from [238], Copyright © 2013, Royal Society of Chemistry.

the silica template in NaOH. In addition, hollow sphere morphology was clearly demonstrated by using those with cracked shells, as demonstrated by the inset in Fig. 5.116(a). Furthermore, the NiCo$_2$O$_4$ sub-microspheres exhibited a rough surface, where evident 'wrinkles' with a thickness of about 4 nm could be clearly observed, as seen in Fig. 5.116(c, d). Such a surface characteristic offered more electroactive sites for electrochemical energy storage.

Figure 5.117 shows a schematic diagram describing formation process of the NiCo$_2$O$_4$ hollow sub-microspheres [239]. During refluxing, the Co–Ni precursor was uniformly deposited on the surface of the silica spheres, which was then transferred to NiCo$_2$O$_4$ after calcination and removal of the NaOH template. It was found that, the morphology of the derived NiCo$_2$O$_4$ samples was closely related to the concentrations of Ni^{2+}, Co^{2+} and urea. For example, under the condition of sample NiCo$_2$O$_4$-1, hollow architecture could be formed, instead, they all collapsed. This was simply because the number of the NiCo$_2$O$_4$ nanoparticles was not sufficient to cover the whole surface of the silica sphere template, due to the relatively low concentrations of the precursor ions. As a consequence, after the temperature was removed, the NiCo$_2$O$_4$ shell was too thin to ensure the integrity of the hollow spherical architecture. In comparison, in the sample NiCo$_2$O$_4$-2, well-developed hollow sub-microspheres were obtained, with thicker shells in the range of 110–116 nm.

Nickel cobalt oxide with a composition of Ni$_{0.3}$Co$_{2.7}$O$_4$ was derived from the nickel cobalt oxalate hydrate (Ni$_{0.1}$Co$_{0.9}$C$_2$O$_4$·nH$_2$O) precursor through calcination [240]. The Ni$_{0.3}$Co$_{2.7}$O$_4$ based electrode exhibited a specific capacitance of as high as 1931 F·g^{-1} at a scan rate of 5 mV·s^{-1}, with a cyclic stability by only 0.6% degradation after charge-discharge cycles of up to 2000. Moreover, the Ni$_{0.3}$Co$_{2.7}$O$_4$ based electrode possessed power densities in the range of 141–1409 W·kg^{-1}, with minimized sacrifice of energy densities of 27.1–22.8 Wh·kg^{-1}. The promising electrochemical performance was readily attributed to the appropriate composition and the special hierarchical mesoporous structures, which facilitated electrolyte ions to penetrate the electrode materials and approach surface for electrochemical reactions.

For precursor synthesis, 1 mL 0.1 M Ni(NO$_3$)$_2$ aqueous solution, 9 mL 0.1 M CoCl$_2$ aqueous solution and 10 mL ethanol were mixed for 30 minutes with the aid of stirring at room temperature, into which the

392 *Nanomaterials for Supercapacitors*

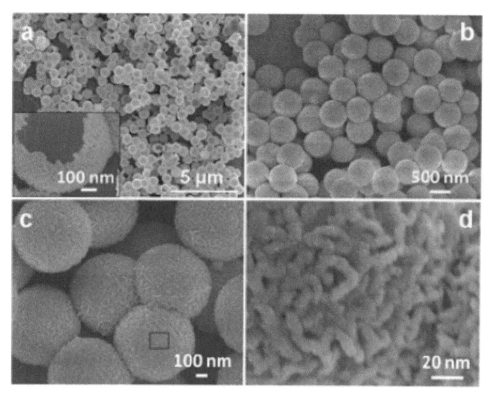

Fig. 5.116. (a–d) SEM images of the hollow NiCo$_2$O$_4$ sub-microspheres at different magnifications. The inset in panel (a) shows a magnified image. Reproduced with permission from [239], Copyright © 2013, Royal Society of Chemistry.

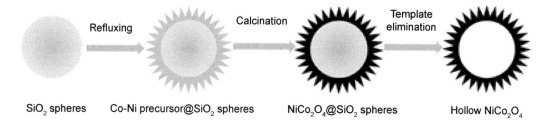

Fig. 5.117. Schematic diagram describing formation process of the mesoporous hollow NiCo$_2$O$_4$ sub-microspheres. Reproduced with permission from [239], Copyright © 2013, Royal Society of Chemistry.

10 mL 0.1 M Na$_2$C$_2$O$_4$ solution was quickly added. A pink precipitate was formed, which was collected after reaction for 1 hour, thus resulting in precursor Ni$_{0.1}$Co$_{0.9}$C$_2$O$_4$·nH$_2$O. The precursor was calcined at 400°C for 10 minutes, at a heating rate of 1°C min^{-1}, lead to sample denoted as P1. After calcination at 450 and 550°C for 10 minutes, the corresponding samples were named as P2 and P3, respectively.

Phase composition of the precursor was confirmed to be Ni$_{0.1}$Co$_{0.9}$C$_2$O$_4$·nH$_2$O by using XRD pattern. SEM images revealed that the precursor contained spherical particles, with an average size of about 4 µm. The individual particle assembled from rod-like subunits, with an average diameter of about 50 nm and a length of about 2 µm, pointing outward radially and thus leading to a dandelion-like morphology. In addition, some nanorods were agglomerated to form bundles with relatively sizes. Figure 5.118 shows

Fig. 5.118. SEM images of the mesoporous $Ni_{0.3}Co_{2.7}O_4$ hierarchical architectures: (a, b) P1, (c, d) P2 and (e, f) P3. Reproduced with permission from [240], Copyright © 2013, Royal Society of Chemistry.

SEM images of the $Ni_{0.3}Co_{2.7}O_4$ hierarchical structures, which suggested that morphology of the precursor was well retained after calcination.

$NiCo_2O_4$ with conductive substrate

Similarly, the electrochemical performance of $NiCo_2O_4$ nanostructures could be significantly improved, by using conductive substrates to support thin layers of $NiCo_2O_4$, because of the greatly enhanced electrical conductivity of the whole structures. In this respect, the substrates are usually highly porous, with high specific surface area, which include Ni foam, Ti foil, Carbon Fiber Paper (CFP), flexible carbon fabric and so on. Therefore, electrodes obtained in this way are called hybrid electrodes, where the substrates also act as current collectors, while the interconnected pores provide a continuous channel for electrolyte ion to diffuse and penetrate the electrode materials. $NiCo_2O_4$ coatings on different conducting substrates have been fabricated by using various deposition methods, such as solution method (e.g., Chemical Bath Deposition or CBD), hydrothermal reaction and electrodeposition [113, 241–252].

Solution processes

NiCo$_2$O$_4$ nanoneedles were directly deposited on to Ni foam and Ti foil, by using a solution process, with water and ethanol as mixed solvent and urea as the complexing agent [253]. The samples possessed an areal (interfacial) capacitance of 3.12 F·cm^{-2} at the current density of 1.11 mA·cm^{-2}. The high electrochemical performance of the electrode was attributed to the fact that the NiCo$_2$O$_4$ nanoneedles had a high specific surface area of 136.3 m^2·g^{-1}, which provided easy access for the electrolyte ions. In addition, the direct growth of the active materials on the substrates offered high mechanical stability and high electrical conductivity.

To prepare the deposition solutions, 2 mmol Ni(NO$_3$)$_2$·6H$_2$O and 4 mmol Co(NO$_3$)$_2$·6H$_2$O were dissolved in a mixed solution consisting of 40 mL ethanol and 40 mL H$_2$O, into which 24 mmol urea was then added. The solution was stored in a 100 mL bottle. The Ni foam and Ti foil were cleaned with 3 M HCl solution for 15 minutes to remove surface oxide layer. To deposit NiCo$_2$O$_4$ layers, the two substrates were immersed in solution at 85°C for 8 hours. After that, the samples were cleaned by ultrasonication to remove the loosely attached products on the surface, followed by drying at 60°C. To convert the precursor to NiCo$_2$O$_4$ nanowires, the samples were calcined at 250°C for 90 minutes.

Figure 5.119(A) shows SEM image of the Ni–Co precursor deposited on Ni foam substrate, revealing 1D nanostructures with a needle-like morphology, which were uniformly distributed on the substrate as an array structure [253]. The array structure could be grown on the substrate over a large area, as observed in the corresponding inset. After calcination, NiCo$_2$O$_4$–Ni had almost no change in surface profile as compared with bare Ni foam. The presence of NiCo$_2$O$_4$ phase was confirmed by XRD results. The uniform array structure and morphology of the NiCo$_2$O$_4$ nanoneedles were not affected by the calcination process, as illustrated in Fig. 5.119(B, C). Figure 5.119(D) shows SEM image of the NiCo$_2$O$_4$ nanoneedle arrays at a different angle, confirming the needle-like morphology. It was found that, at the initial stage, nanoparticles

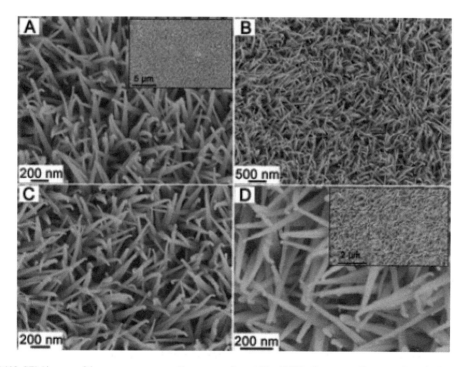

Fig. 5.119. SEM images of the precursor nanoneedle arrays and crystallized NiCo$_2$O$_4$ nanoneedle arrays deposited on Ni foam: (A) precursor nanoneedle arrays with the inset showing a large-area view image, (B, C) top-view and (D) tilted view of the crystallized NiCo$_2$O$_4$ nanoneedle arrays, with the inset in (D) showing a large-area view image. Reproduced with permission from [253], Copyright © 2012, Royal Society of Chemistry.

were uniformly deposited first, while nanorods were formed later and nanoneedles were developed finally after reaction for 8 hours.

Figure 5.120(A) shows XRD patterns of bare Ti foil and the sample NiCo$_2$O$_4$–Ti, confirming the formation of cubic phase NiCo$_2$O$_4$ [253]. SEM images of the sample NiCo$_2$O$_4$–Ti are shown in Fig. 5.120(B–D), indicating the presence of highly uniform nanoneedle arrays on the flat Ti foil substrate. TEM images revealed that the nanoneedles had sharp tips, with an average length of about 1.5 μm and diameters decreasing about 100 nm at the bottom to several nanometers at the tips. The formation of the needle-like morphology was attributed to the gradual depletion of the precursor growth process that proceeded. According to the HRTEM lattice image and corresponding FFT analysis results, the growth of the single-crystalline nanoneedles was along the <110> direction, i.e., the (110) planes were grown during the crystallization.

A simple solution method was developed to deposit NiCo$_2$O$_4$ nanosheets various conducting substrates, including Ni foam, Ti foil, stainless-steel foil and flexible graphite paper [254]. The NiCo$_2$O$_4$ nanosheets possessed a BET surface area of 112.6 m^2·g^{-1} and pores having sizes in the mesoporous regime of 2–5 nm, exhibiting an areal capacitance of 3.51 F·cm^{-2} at the current density of 1.8 mA·cm^{-2}. The solution preparation and growth process were similar to the previous one. Briefly, 1 mmol Ni(NO$_3$)$_2$·6H$_2$O and 2 mmol of Co(NO$_3$)$_2$·6H$_2$O were dissolved into the mixed solution of 40 mL H$_2$O and 20 mL ethanol, into which 6 mmol hexamethylene-tetramine was added. The solution was then transferred to a capped bottle (80 mL in capacity). Cleaned conductive substrates were immersed in the solution and heated at 90°C for 10 hours. After reaction, the samples were all dried at 60°C and then calcined at 320°C for 2 hours to obtain NiCo$_2$O$_4$ mesoporous nanosheets.

Representative SEM images of the Ni-Co precursor nanosheets on Ni foam are shown in Fig. 5.121 (A–C) [254]. The nanosheets were grown on the Ni foam in a uniform manner. As demonstrated in

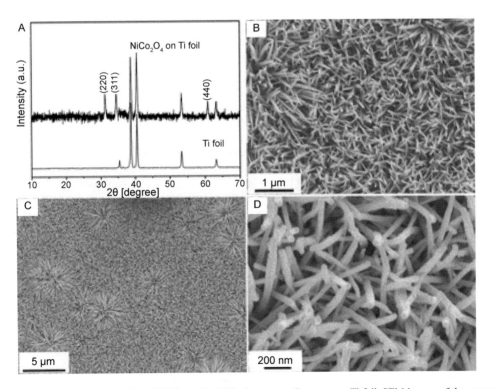

Fig. 5.120. (A) XRD patterns of bare Ti foil and the NiCo$_2$O$_4$ nanoneedle arrays on Ti foil. SEM images of the precursor nanoneedle arrays (B) and crystallized NiCo$_2$O$_4$ nanoneedle arrays on Ti foil (C, D). Reproduced with permission from [253], Copyright © 2012, Royal Society of Chemistry.

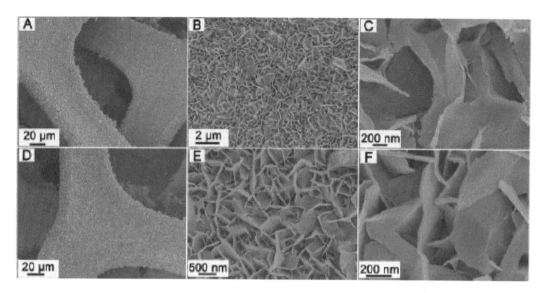

Fig. 5.121. Typical FESEM images at different magnifications: (A–C) Ni-Co precursor nanosheets on Ni foam and (D–F) $NiCo_2O_4$ nanosheets on Ni foam after annealing at 320°C for 2 hours in air at a heating rate of 2°C min^{-1}. Reproduced with permission from [254], Copyright © 2012, WILEY-VCH Verlag GmbH & Co. KGaA, Weinheim.

Fig. 5.121(B, C), the nanosheets were interconnected to form a network structure. After calcination, the Ni-Co precursor nanosheets were crystallized to $NiCo_2O_4$ nanosheets, without variation in the nanosheet morphology. As seen in Fig. 5.121(D), the overall profile was almost the same as that shown in Fig. 5.121(A). The formation of the $NiCo_2O_4$ phase was confirmed by XRD results. The network structure after calcination was well retained, as illustrated in Fig. 5.121(E, F). Besides Ni foam, other conductive substrates, including Ti foil, stainless-steel foil and flexible graphite paper, could also be used to deposit the mesoporous $NiCo_2O_4$ nanosheets.

Mesoporous characteristics of the $NiCo_2O_4$ nanosheets were confirmed by TEM images [254]. The nanosheets were continuous and had a smooth surface. Mesopores were uniformly distributed over the surface of the nanosheets. The formation of the mesopores could be attributed to the release of the gases generated due to the decomposition of the Ni-Co precursor. SAED pattern revealed that the nanosheets were of polycrystalline nature, which was in a good agreement with the XRD characterization results.

A facile scalable template-free method was developed to prepare 1D multilayered mesoporous $NiCo_2O_4$ nanowires (NWs), which assembled from quasi-single-crystalline $NiCo_2O_4$ nanosheets as the building blocks [255]. The mesoporous nanowire based electrode displayed a specific capacitance of 401 F·g^{-1} at the current density of 1 A g^{-1}, with a rate capability of 75% retention at 8 A g^{-1} and high cycling stability of 90% retention after 5000 charge-discharge cycles. The promising electrochemical performance was ascribed to the special multilayered mesoporous nanowire structures, with a high specific surface area of 97 m^2·g^{-1}, an average pore size of 6.3 nm and a mesoporous volume of 0.24 cm^3·g^{-1}.

Ni foams, with a dimension of 2 cm × 1 cm × 1 mm, were cleaned with acetone, etched with 6 M HCl for 20 minutes and then thoroughly washed with DI water and absolute ethanol. 10 mmol $Co(NO_3)_2·6H_2O$, 5 mmol $Ni(NO_3)_2·6H_2O$ and 5 mmol NH_4NO_3 were dissolved in 35 ml water to form the reactoin solution. After stirring for 10 minutes, 20 ml $NH_3·H_2O$ was added, followed by further stirring for about 1 hour. Gradually, a light blue gray precipitate was first observed, which then disappeared, thus leading to a dark brown solution with pH = 10.8. The solution was transferred into a three-necked bottle. A glass tube with two open ends, Cleaned Ni foams were inserted into the two open ends of a glass tube, which was immersed into the dark brown solution by hanging. The whole reaction system was refluxed in an oil bath at 110°C for 6 hours with continuous stirring. Finally, the Ni foams were covered with black coating, which were thoroughly washed with DI water, followed by ultrasonication in absolute ethanol for 1 hour. Powders were collected through centrifugation, followed by drying and calcination at 300°C for 2 hours, at a slow heating rate of 1°C min^{-1}. Samples were also collected from the bottom of the three-necked bottle reactor in a similar way.

Figure 5.122 shows representative SEM images of the nickel cobaltite NWs, which were detached from the NiCo$_2$O$_4$/Ni foam samples through ultrasonication. The NiCo$_2$O$_4$ NWs possessed a relatively high aspect ratio, with lengths of several micrometers and diameters in the range of 200–300 nm. Obviously, the NW morphology of the precursor was well retained after conversion to NiCo$_2$O$_4$. High magnification image indicated that bumps and holes were observed on surface of the NWs, due to the multilayered microstructure, i.e., the NWs were assembled from nanosheets, along the thickness direction, as illustrated in Fig. 5.122(b). Comparatively, Co$_3$O$_4$ and NiO components prepared in a similar way exhibited different morphologies. Co$_3$O$_4$ NW had a hollow structure and polycrystalline characteristics, whereas the NiO was also polycrystalline and demonstrated a mesoporous nanosheet morphology. Therefore, the NiCo$_2$O$_4$ NWs seemed to adopt the NW feature of Co$_3$O$_4$ and the nanosheet structure of NiO in a combining way.

It was found that the NiCo$_2$O$_4$ powder collected from the bottom of the three-neck bottle reactor, rather than from surface of the Ni foam, contained microspherical aggregation of nanoparticles. Therefore, the ammonia-evaporation-induced mechanism for Co$_3$O$_4$ NWs, as discussed previously, could not be used to explain the formation of the 1D NiCo$_2$O$_4$ NW structure [91, 92]. In fact, the Ni foam substrate should have played a certain role in triggering the formation of the 1D NW structure, although the underlying mechanism still reverses further clarification.

NiCo$_2$O$_4$ (nanowire)@MnO$_2$ (nanoflakes) core–shell heterostructures were deposited on Ni foams, which exhibited promising electrochemical performance as electrode of supercapacitors [256]. The integrated NiCo$_2$O$_4$@MnO$_2$ based electrode had an areal capacitance of 3.31 F·cm^{-2} at the current density of 2 mA·cm^{-2}, with an outstanding stability of 88%, at a relatively high current density of 10 mA·cm^{-2}, after 2000 charge-discharge cycles. In the hybrid structure, the ultrathin MnO$_2$ nanoflakes offered a fast reversible faradaic reaction and a short ion diffusion path, while ensuring structural stability of the core during the charge–discharge cycles. At the same time, the NiCo$_2$O$_4$ nanowires acted as both the backbone and electron 'superhighway' for charge storage and delivery. Also, the mesoporous microstructure resulted in a large interface for electrode-electrolyte contact.

Ni foams, with a dimension of 2 cm × 10 cm, in rectangular shape, were immersed in 3 M HCl solution for 15 minutes to etch out the surface oxide layer. 1.16 g Co(NO$_3$)$_2$·6H$_2$O, 0.58 g Ni(NO$_3$)$_2$·6H$_2$O and 1.44 g urea were dissolved in 160 mL mixed solution with ethanol and H$_2$O volume ratio of 1:1 at room temperature to prepare the reaction solution. The solution was transferred into a 250 mL bottle, into which the cleaned Ni foams were immersed into the solution. The reactor was sealed tightly and then heated at 90°C for 8 hours. The samples were collected with ultrasonication, followed by drying at 60°C in air. The Ni foams with the precursor of NiCo$_2$O$_4$ were hydrothermally treated in 1.6 mM KMnO$_4$ solution

Fig. 5.122. SEM images of the multilayered mesoporous NiCo$_2$O$_4$ NWs: (a) low- and (b) high-magnification. Reproduced with permission from [255], Copyright © 2012, Royal Society of Chemistry.

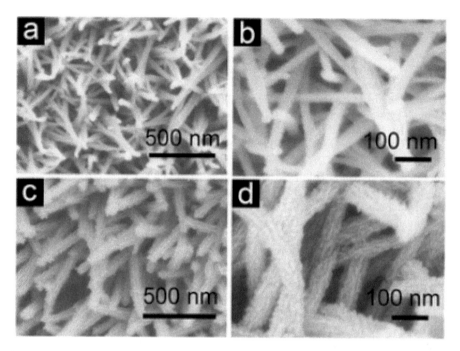

Fig. 5.123. SEM images of the NiCo$_2$O$_4$ (a, b) NW arrays and hierarchical NiCo$_2$O$_4$@MnO$_2$ core–shell heterostructured NW arrays (c, d) grown on Ni foam. Reproduced with permission from [256], Copyright © 2013, Royal Society of Chemistry.

at 160°C for 30 minutes. KMnO$_4$ solutions with high concentrations of 14 and 28 mM were also used to prepare NiCo$_2$O$_4$@MnO$_2$ core-shell NW arrays for the purpose of comparison. After the hydrothermal reaction, all samples were calcined at 350°C for 2 hours, thus leading hierarchical NiCo$_2$O$_4$@MnO$_2$ core-shell heterostructured NW arrays.

Representative top-view SEM images of the NiCo$_2$O$_4$ NW arrays that were grown on Ni foam are shown in Fig. 5.123(a, b), revealing that the NiCo$_2$O$_4$ NWs had a sharp tip and were uniformly distributed on the substrate, with sufficiently large inter-NW spacing. Therefore, the nanoarrays exhibited a highly open and porous structure. Representative SEM images of the NiCo$_2$O$_4$@MnO$_2$ core–shell NW arrays are depicted in Fig. 5.123(c, d). It was found that the MnO$_2$ nanoflakes were all attached onto the NiCo$_2$O$_4$ NWs, so that no MnO$_2$ nanoparticles were observed in between the NiCo$_2$O$_4$ NWs. Therefore, such core–shell NWs offered convenient diffusion channels for the electrolyte ions to access for electrochemical reactions, thus benefiting the energy storage capability of the materials.

Hydrothermal/solvothermal synthesis

A hydrothermal process was used to deposit NiCo$_2$O$_4$ nanoflake-nanowire hetero-structure on Ni foams [257]. The hetero-structure based electrode showed a specific capacitance of 891 F·g^{-1} at the current density of 1 A·g^{-1}, which was attributed to the special morphology and microstructure, with high specific surface areas of 57.26–99.96 m^2·g^{-1} and pore sizes of 30–120 nm within the macro-porous region. In addition, the NiCo$_2$O$_4$ based electrode exhibited an energy density of 26 Wh·kg^{-1} at a high power density of 11 kW·kg^{-1}, as well as an outstanding stability of 97.2%, at the current density of 2 A·g^{-1}, after 8000 charge-discharge cycles. The special hetero-structure ensured smooth ion and electron transport, due to the large surface area and high strain accommodating capability.

To develop the self-assembled porous NiCo$_2$O$_4$ arrays on Ni foam, 1 mmol Ni(NO$_3$)$_2$, 2 mmol Co(NO$_3$)$_2$, 6 mmol NH$_4$F and 15 mmol urea were dissolved in 70 mL DI water, with constant stirring for 30 minutes [257]. Ni foam substrate with dimension of 3.5 × 5.0 cm in size was pressed to be a thin plate, followed by ultrasonic cleaning in ethanol for 15 minutes. The top side of the Ni foam was covered with a polytetrafluoroethylene tape to prevent contamination. The solution, together with the Ni foam substrate were subjected to hydrothermal reaction at 120°C for different time durations. After the reaction was finished, the samples were thoroughly washed with distilled water. To convert cobalt–nickel hydroxide in to $NiCo_2O_4$, the samples were calcined at 350°C for 2 hours in Ar flow.

Figure 5.124 shows SEM images of the calcined samples, together with the precursor hydrothermally reacted for 8 hours [257]. It was demonstrated that the hetero-architecture morphology was well retained after the calcination. Growth process of the nanoflake–nanowire $NiCo_2O_4$ array was monitored by examining the morphologies of the samples collected at different growth stages in terms of reaction time. For example, after reaction for 3 hours, the Ni foam had been almost fully covered a layer of nanoflakes,

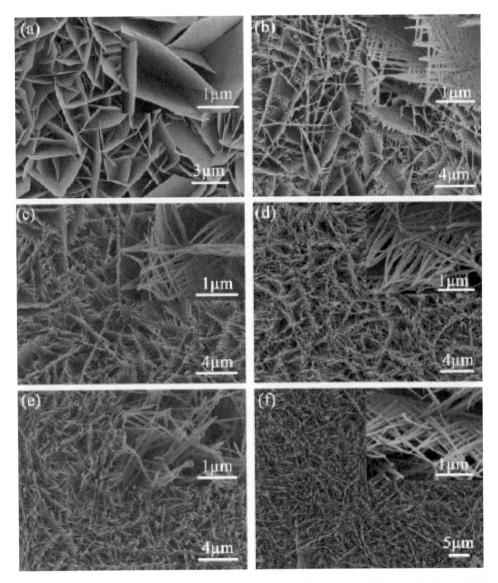

Fig. 5.124. SEM images of the products after hydrothermal reaction for different times: (a) 3 hours, (b) 5 hours, (c) 7 hours, (d) 8 hours, (e) 12 hours and (f) the precursor obtained after hydrothermal reaction for 8 hours. Reproduced with permission from [257], Copyright © 2013, Elsevier.

while no nanowire was observed at this stage, as seen in Fig. 5.124(a). As the reaction time was gradually increased, nanowires appeared at the edge of the nanoflakes, forming a 2D network-like structure, as illustrated in Fig. 5.124(b–e). Eventually, a hetero-structured array was developed. After reaction for 8 hours, nanowires grown at the edge of the nanoflakes became more pronounced. The NiCo$_2$O$_4$ nanowires possessed an average diameter of 100 nm and lengths of up to 2 μm, as depicted in Fig. 5.124(f). In addition, there were no nanowires in between the nanoflakes, they were preferentially grown from the edge. At the same time, the length of the nanowires could be readily controlled by controlling the growth time.

Figure 5.125 shows a schematic diagram describing possible growth mechanism of the special hetero-structures [257]. One of the main factors was the difference in solubility (K_{sp}) between the nickel-rich nanoflakes and nickel-poor nanowires. During the reaction process, mesocrystals with a common crystallographic orientation started to be present in the supersaturated solution. After a certain time, they combined to form particles at the planar interface. The nickel-rich particles were easily formed due to their relatively lower K_{sp}, as compared with their nickel-poor (i.e., cobalt-rich) counterparts. Therefore, the nickel-rich particles were dominant in the initial growth stage, which resulted in the formation of the nanoflakes. Since the Co^{2+}/Ni^{2+} ratio in the starting solution was 2:1, the presence of the nickel-rich phase implied that there would be extra Co^{2+}, once the reaction was started. Because of that, nickel-poor nanowires were developed. Meanwhile, the nanoflakes served as a backbone to support the nickel-poor nanowires. Because the surface energy at the edges was higher than that at the plain surface, the cobalt-rich particles were preferentially grown at the edge of the nanoflakes, thus leading to the self-assembled porous NiCo$_2$O$_4$ hetero-structure array.

Two types of NiCo$_2$O$_4$ nanoarchitectures, i.e., nanowires and nanosheets, have been grown on carbon cloth, which demonstrated specific capacitances of 245 F·g^{-1} and 123 F·g^{-1}, respectively, at the current density of 1 A·g^{-1} [258]. To prepare the hydrothermal reaction solution, 1 mmol Ni(NO$_3$)$_2$·6H$_2$O, 2 mmol Co(NO$_3$)$_2$·6H$_2$O, 2 mmol NH$_4$F and 6 mmol CO(NH$_2$)$_2$ were dissolved in 80 mL mixed solvent of ethanol and H$_2$O with a volume ratio of 1:1, with the aid of stirring. The solution, together with a piece of well-cleaned carbon cloth, was hydrothermally treated at 95°C for 10 hours. After the reaction was finished, the product was collected, thoroughly washed and vacuum-dried, followed by calcination at 350°C for 6 hours, which led to NiCo$_2$O$_4$ nanowires. Mass loading of the NiCo$_2$O$_4$ nanowires on the carbon cloth was 0.52 mg·cm^{-2}.

To obtain NiCo$_2$O$_4$ nanosheets, 1 mmol Ni(NO$_3$)$_2$·6H$_2$O, 2 mmol Co(NO$_3$)$_2$·6H$_2$O, and 1 g hexadecyltrimethyl ammonium bromide were dissolved in 80 mL mixed solvent of methanol and H$_2$O with a volume ratio of 5:1. In this case, the hydrothermal reaction was conducted at 180°C for 12 hours. All other processes are similar to those of NiCo$_2$O$_4$ nanowires. Finally, NiCo$_2$O$_4$ nanosheets deposited on the carbon cloth had a mass loading of 0.6 mg·cm^{-2}.

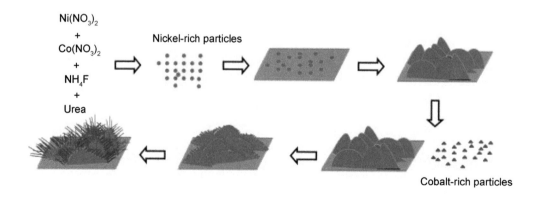

Fig. 5.125. Schematic diagram showing growth process of the self-assembled porous NiCo$_2$O$_4$ arrays. Reproduced with permission from [257], Copyright © 2013, Elsevier.

Figure 5.126 shows SEM images of the $NiCo_2O_4$ nanowires grown on carbon cloth [258]. The $NiCo_2O_4$ nanowires were uniformly grown on the carbon cloth, with a high density, in the form of 3D hierarchical structure, as illustrated in Fig. 5.126(a). Although an ordered woven structure covered the carbon cloth, it was still highly fleasible. High-magnification SEM images revealed more details the $NiCo_2O_4$ nanowires, as observed in Fig. 5.126(b–d). According to Fig. 5.126(b), the carbon fiber after growth of $NiCo_2O_4$ nanowires had a thickness of about 12 μm, as compared with 9 μm for the pristine fiber in the carbon cloth. It was demonstrated that the $NiCo_2O_4$ nanowires had a sharp tip and were grown vertically, as seen in Fig. 5.126(c, d). More importantly, the $NiCo_2O_4$ nanowires were so firmly attached to the carbon cloth that they were not easily detached under the strong ultrasonication. TEM images indicated that the nanowires had diameters in the range of 10–40 nm, with mesopores to be present on their surface.

Figure 5.127 shows SEM images of the $NiCo_2O_4$ nanosheets on carbon cloth at different magnifications [258]. As shown in Fig. 5.126(a), the woven carbon fibers were not altered in the carbon cloth after the growth of the $NiCo_2O_4$ nanosheets. All the individual carbon fibers were uniformly decorated with $NiCo_2O_4$ nanosheets, as revealed in Fig. 5.127(b). Similarly, the nanosheets were vertically grown on the carbon fibers, with a cross-linking configuration, as illustrated in Fig. 5.127(c, d). TEM images suggested that the $NiCo_2O_4$ nanosheets were assembled from nanoparticles with average size of 20 nm, thus resulting in the mesoporous characteristics.

Ultrathin $NiCo_2O_4$ nanosheets were grown on Carbon Fiber Paper (CFP) by using a solvothermal method, which exhibited a specific capacitance of as high as 999 $F \cdot g^{-1}$, at a high current density of 20 $A \cdot g^{-1}$ [259]. CFP pieces, with dimensions of 1×4 cm^2, were cleaned with 5 M HCl aqueous solution, absolute ethanol and DI water for 15 minutes each, with the aid of ultrasonication. Meanwhile, 0.5 mmol Co

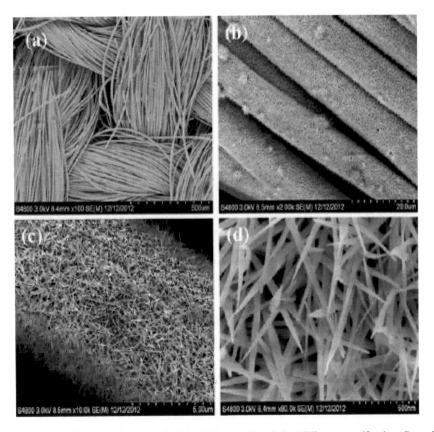

Fig. 5.126. (a–d) SEM images of the $NiCo_2O_4$ nanowires grown on carbon cloth at different magnifications. Reproduced with permission from [258], Copyright © 2013, American Chemical Society.

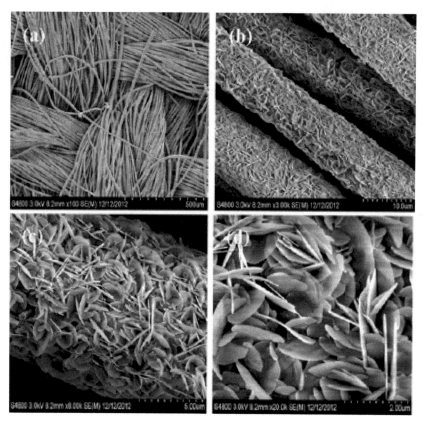

Fig. 5.127. (a–d) SEM images of the NiCo$_2$O$_4$ nanosheets grown on carbon cloth at different magnifications. Reproduced with permission from [258], Copyright © 2013, American Chemical Society.

(NO$_3$)$_2$·6H$_2$O, 0.25 mmol Ni(NO$_3$)$_2$·6H$_2$O, 1.5 mmol hexamethylenetetramine were dissolved into 12 mL of absolute methanol to form a transparent pink solution with constant stirring for solvothermal reaction. The solution together with a piece of the pre-treated CFP was subject to solvothermal treatment at 180°C for 12 hours. After the reaction, the samples were thoroughly cleaned with DI water and ethanol, followed by drying at 60°C for 2 hours. The samples were finally calcined at 350°C for 3 hours to convert the precursor hydroxide phase to NiCo$_2$O$_4$ mesoporous nanosheets on CFP, with a mass of about 0.8 mg·cm^{-2}. Figure 5.128 shows a schematic diagram describing fabrication process of the NiCo$_2$O$_4$ nanosheets on CFP.

Figure 5.129 shows SEM images of the Ni–Co precursor and the NiCo$_2$O$_4$ nanostructures on CFP substrate. The Ni–Co precursor nanosheets were uniformly grown on the carbon nanofibers, as seen in Fig. 5.129(a–c). High magnification images indicated that the nanosheets were interconnected with one and another, as demonstrated in Fig. 5.129(b, c), so as to form a wall-like structure and have high mechanical strength [28]. Furthermore, the NiCo$_2$O$_4$ nanosheets appeared as a hierarchical array, with wide spacing to be present in between the nanosheets. Such a microstructure ensured a smooth access of the electrolyte ions into the materials for electrochemical reactions, thus leading to high electrochemical performance. By using a relatively slow heating rate of 1°C·min^{-1}, the Ni–Co precursor nanosheets were transferred to crystalline NiCo$_2$O$_4$ without losing the morphological characteristics, as illustrated in Fig. 5.129(d–f).

Hierarchical NiCo$_2$O$_4$@NiCo$_2$O$_4$ core/shell nanoflake arrays on Ni foams have been fabricated by using a combination of hydrothermal and chemical bath deposition (CBD) methods [260]. The ultrathin NiCo$_2$O$_4$ nanoflakes were synthesized firstly by using the hydrothermal method as 'core', whereas the NiCo$_2$O$_4$ nanoflakes as 'shell' were prepared by using the CBD method. The NiCo$_2$O$_4$@NiCo$_2$O$_4$ nanoflake array based electrode had an areal capacitance of 1.55 F·cm^{-2} at the current density of 2 mA·cm^{-2} in 2 M

Oxide Based Supercapacitors: II-Other Oxides   403

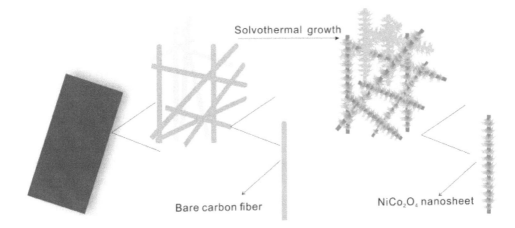

Fig. 5.128. Schematic illustration for the formation processes of the NiCo$_2$O$_4$ nanosheets on CFP. Reproduced with permission from [259], Copyright © 2014, Elsevier.

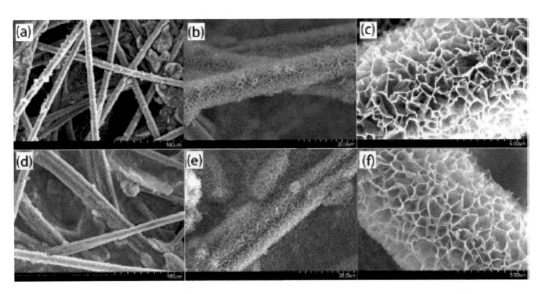

Fig. 5.129. SEM images of the samples at different magnifications: (a–c) Ni–Co precursor nanosheets on CFP and (d–f) final NiCo$_2$O$_4$ nanosheets on CFP after calcination at 350°C for 3 h. Reproduced with permission from [259], Copyright © 2014, Elsevier.

KOH electrolyte. In addition, the electrode possessed a capacity retention of 98.6% after 4000 charge-discharge cycles.

To develop NiCo$_2$O$_4$ nanoflake arrays on Ni foam, 1 mmol Ni (NO$_3$)$_2$, 2 mmol Co(NO$_3$)$_2$, 6 mmol NH$_4$F and 15 mmol urea were dissolved in 70 mL of DI water, with continous stirring for 30 minutes, which was then subjected to hydrothermal reaction at 120°C for 3 hours, together with a piece of Ni foam substrate with a dimension of 3.5 × 5.0 cm. After that, the autoclave was heated, and then cooled to room temperature. After the reaction, the samples were taken out from the reaction and thoroughly washed with distilled water. To convert the cobalt–nickel hydroxide precursor to NiCo$_2$O$_4$, the samples were then calcined at 350°C for 2 hours in Ar flow.

The NiCo$_2$O$_4$ nanoflake arrays on Ni foam were then further coated with NiCo$_2$O$_4$ through the CBD method. To do that, the NiCo$_2$O$_4$ nanoflake array samples were vertically placed in a 250 mL Pyrex beaker, into which CBD solution was filled. The CBD solution was prepared by mixing 20 mL aqueous ammonia (25–28%) with the mixture containing 50 mL 1 M nickel sulfate, 50 mL 2 M cobalt sulfate and 80 mL 0.25 M potassium peroxydisulfate. The CBD process was carried out for 8 minutes. After that, the precursor samples calcined similarly. Eventually, the NiCo$_2$O$_4$@NiCo$_2$O$_4$ core/shell nanoflake arrays had a total loading mass of 1.97 mg cm^{-2}, containing 1.34 and 0.63 mg core and the shell NiCo$_2$O$_4$, respectively. A schematic diagram to present the fabrication process of the NiCo$_2$O$_4$@NiCo$_2$O$_4$ core/shell nanoflake arrays is shown in Fig. 5.130 [260].

Figure 5.131(a) shows SEM image of the bare NiCo$_2$O$_4$ nanoflakes after the hydrothermal reaction, demonstrating quite a smooth surface. The formation of the nanoflake film was governed by heterogeneous nucleation and growth mechanism, because of the low interfacial nucleation energy on the Ni foam substrate. Figure 5.131(b) depicts SEM image of the core–shell nanoflake array after the CBD process was applied. Obviously, thickness of the nanoflakes was increased, while the surface was decorated with leaf-like ultrathin nanoflakes, thus leading to a highly porous core–shell architecture. The nanoflake shells were interconnected to form a network-like structure, whereas there were still areas to be exposed. The highly

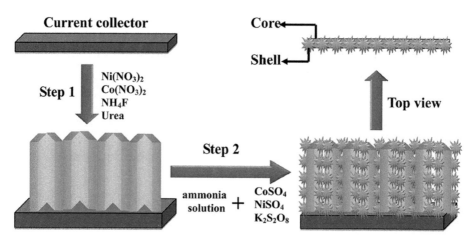

Fig. 5.130. Schematic diagram of the two-step synthesis of NiCo$_2$O$_4$@NiCo$_2$O$_4$ core/shell nanoflake arrays directly on Ni foam substrate. Reproduced with permission from [260], Copyright © 2013, American Chemical Society.

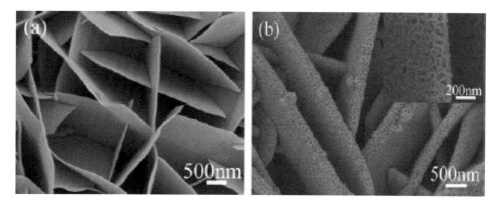

Fig. 5.131. SEM images of (a) the as-prepared NiCo$_2$O$_4$ nanoflake array and (b) the NiCo$_2$O$_4$@NiCo$_2$O$_4$ core/shell nanoflake array. Reproduced with permission from [260], Copyright © 2013, American Chemical Society.

porous structure would facilitate smooth transportation of electrolyte ions, during the charge–discharge reaction process. Therefore, the core/shell nanoflakes became highly accessible to the electrolyte ions, which was responsible for the presence of high specific capacitance. Also, because both the 'core' and the 'shell' were ultrathin, the electrolyte ions could readily penetrate the whole electrode, so that the active materials were fully utilized for the electrochemical charge storage process.

Figure 5.132 shows SEM images of the samples with different CBD growth times, revealing the morphology evolution of the $NiCo_2O_4$ shell. After deposition for 1.5 minutes, the nanoparticles began to appear on the surface of the core nanoflakes, as demonstrated in Fig. 5.132(a). As the deposition time was increased, the coverage of $NiCo_2O_4$ shell on the surface of the $NiCo_2O_4$ core was gradually enlarged. Therefore, after the core surface was entirely covered by the shell, the thickness of the $NiCo_2O_4$ shell could be controlled by controlling the CBD coating time.

$NiCo_2O_4$ nanowires have been directly grown on Carbon Fiber Paper (CFP), which were further coated with a 3D hybrid structure of cobalt and nickel double hydroxide, $Co_xNi_{1-x}DHs$ [261]. Figure 5.133 shows a schematic diagram describing the two-step synthesis process of the hybrid structures. The $NiCo_2O_4$ nanowires on the CFP served as a backbone to support the $Co_xNi_{1-x}DHs$ coatings, so that there were sufficiently large surface areas for the electrolyte ions to access. At the same time, the CFP also offered electrical connection to ensure a high conductivity of the electrode. Also, the $Co_xNi_{1-x}DHs$ were fully utilized. The 3D $Co_xNi_{1-x}DHs/NiCo_2O_4/CFP$ hybrid electrodes demonstrated areal capacitances of 0.61, 1.52, 2.17 and 1.88 $F \cdot cm^{-2}$, at the current density of 10 $mA \cdot cm^{-2}$, for the samples with $x = 1$, 0.67, 0.5 and 0.33, respectively. Besides, the hybrid with $Co_{0.5}Ni_{0.5}DHs$ exhibited an energy density of 58.4 $Wh \cdot kg^{-1}$, corresponding to a power density of 41.3 $kW \cdot kg^{-1}$, while the $Co_{0.33}Ni_{0.67}DHs$ hybrid possessed the highest cycling stability than other compositions.

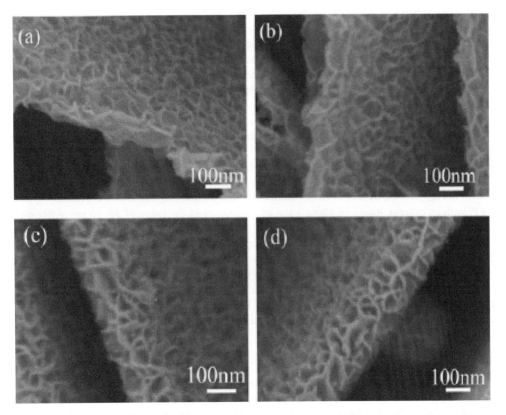

Fig. 5.132. SEM images of the $NiCo_2O_4@NiCo_2O_4$ core/shell nanoflake arrays after chemical bath deposition for various times: (a) 1.5, (b) 3, (c) 6 and (d) 8 minutes. Reproduced with permission from [260], Copyright © 2013, American Chemical Society.

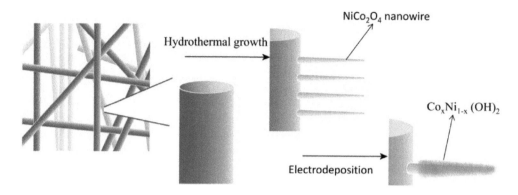

Fig. 5.133. Schematic diagram illustrating the processes to hydrothermally grow $NiCo_2O_4$ nanowires on CFP and electrodepositing the Co_xNi_{1-x}DHs on the $NiCo_2O_4$ nanowires. Reproduced with permission from [261], Copyright © 2013, American Chemical Society.

To hydrothermally develop self-supported $NiCo_2O_4$ nanowire arrays, 1.33 g Co $(NO_3)_2·6H_2O$, 0.66 g $Ni(NO_3)_2·6H_2O$, and 0.48 g urea were dissolved in 160 mL DI water to form the precursor solution. Together with two pieces of CFP, with a dimension of 3 × 4 cm², the precursor solution was subject to hydrothermal reaction at 120°C for 16 hours. After that, the samples were ultrasonically cleaned for 5 minutes in the DI water and rinsed with ethanol, followed by drying at 80°C. The dried samples were calcined at 300°C in air for 2 hours to obtain the $NiCo_2O_4$ nanowires on CFP. The $NiCo_2O_4$ nanowires on the CFP exhibited a mass loading of 0.3 mg·cm⁻².

To coat the self-supported $NiCo_2O_4$ nanowire arrays with Co_xNi_{1-x}DHs, $Co_xNi_{1-x}(OH)_2$ nanosheet shells were grown by using a simple cathodic electrodeposition method [261]. The electrodeposition was conducted with a standard three-electrode configuration at room temperature, where the self-supported $NiCo_2O_4$ nanowire arrays acted as the working electrode, Saturated Calomel Electrode (SCE) was used as the reference electrode and a Pt wire was employed as the counter-electrode. The electrolyte for electrodeposition was 70 ml metal ion solution, with a concentration of 0.1 M and Ni^{2+}/Co^{2+} concentration ratios to 0:1, 1:2, 1:1 and 2:1. The $Co_xNi_{1-x}(OH)_2$ nanosheet shells were formed after deposition for 7 minutes at the potential of −1.0 V. After the deposition was finished, the samples were thoroughly cleaned with DI water and ethanol and then dried in air. The Co $(OH)_2$ and $Co_xNi_{1-x}(OH)_2$ (x≠1) nanosheet shells had mass loadings of 0.8 and 0.7 mg·cm⁻², respectively.

Representative SEM images of the samples after the growth of the $NiCo_2O_4$ nanowire arrays on CFP are shown in Fig. 5.134(a, b). The $NiCo_2O_4$ nanowires with a pretty high density were radially grown on the CFP, with an average length of about 3 μm and diameters in the range of 30–80 nm. After electrodeposition, the $NiCo_2O_4$ nanowires were coated with Co_xNi_{1-x}DHs nanosheets, as illustrated in Fig. 5.134(e). The $NiCo_2O_4$ nanowires and the Co_xNi_{1-x}DHs coating could be readily identified by using TEM image, as revealed in Fig. 5.134(f). Also, the $NiCo_2O_4$ nanowires were composed of nanocrystallites with sizes of 10–20 nm and highly porous, with pores having diameters in the range of 2–4 nm, as illustrated in Fig. 5.134(c). SAED pattern and HRTEM image also indicated that the mesoporous $NiCo_2O_4$ nanowires were polycrystalline, as observed in Fig. 5.134(d). The Co_xNi_{1-x}DHs coatings were well deposited on surface of the $NiCo_2O_4$ nanowires, with a thickness at the nanometer scale.

Electrodeposition

A direct electrodeposition was employed to deposit $Ni_xCo_{2x}(OH)_{6x}$ into self-standing titanium nitride nanotube array (TiN-NTA), which resulted in a coaxial nanostructure, with promising electrochemical performances [262]. Electrode based on the $Ni_xCo_{2x}(OH)_{6x}$/TiN nanostructure had a specific capacitance of 2543 F·g⁻¹ at the current of 5 mV·s⁻¹, with a high rate performance of 660 F·g⁻¹ at a high current density

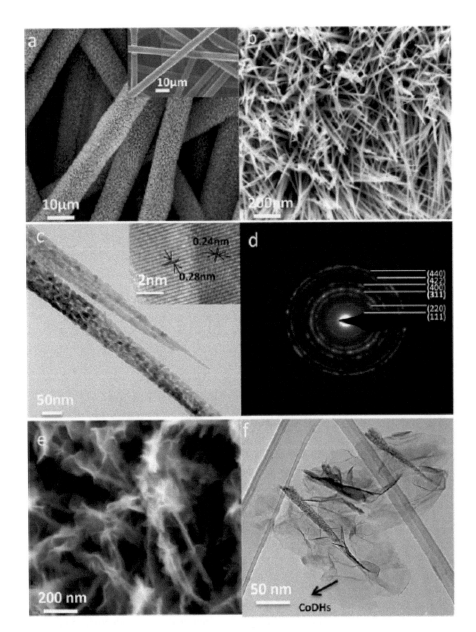

Fig. 5.134. (a) SEM image of the CFP before (inset) and after growth of the NiCo$_2$O$_4$ nanowires. (b) High-magnification SEM image of the NiCo$_2$O$_4$ nanowires on CFP. (c) TEM image and HRTEM image (inset) of the NiCo$_2$O$_4$ nanowires. (d) Diffraction pattern of an individual NiCo$_2$O$_4$ nanowire. (e) SEM image of a piece of CoDHs coating on the NiCo$_2$O$_4$ nanowire. (f) TEM image of the CoDHs/NiCo$_2$O$_4$ heterostructured nanowires. Reproduced with permission from [261], Copyright © 2013, American Chemical Society.

of 500 mV·s^{-1}. In addition, the electrode experienced a very low capacitance loss of about 6.25% after 5000 charge-discharge cycles, which was ascribed to the highly reversible redox reaction capabilities of the Ni$_x$Co$_{2x}$(OH)$_{6x}$/TiN electrode.

Figure 5.135(a) shows a schematic diagram to demonstrate preparation process of the Ni$_x$Co$_{2x}$(OH)$_{6x}$/TiN NTA, involving steps of anodization, calcination in ammonia and electrodeposition [262]. After calcination, the TiN NTA appeared as uniform nanotube arrays, with outer diameters of 100–120 nm, as

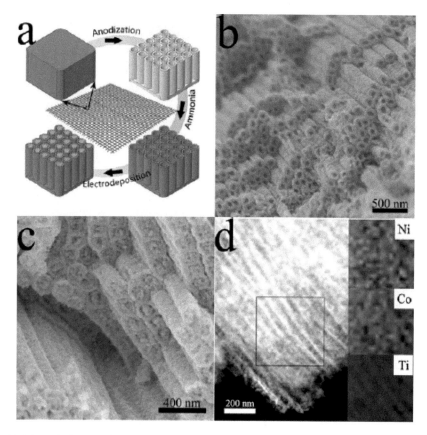

Fig. 5.135. (a) Schematic diagram of fabrication process of the Ni$_x$Co$_{2x}$(OH)$_{6x}$/TiN NTA. (b) SEM image of a pristine TiN NTA. (c) SEM image of a TiN NTA after deposition of Ni$_x$Co$_{2x}$(OH)$_{6x}$. (d) STEM image of the Ni$_x$Co$_{2x}$(OH)$_{6x}$/TiN NTA. Reproduced with permission from [262], Copyright © 2013, American Chemical Society.

observed in Fig. 5.135(b). Such one-dimensional arrays on the Ti mesh substrate acted as a promising current collector, due to its high electrical conductivity. Under optimal conditions, Ni–Co hydroxides were electrochemically deposited into the TiN nanotubes, which led to coaxial nanowire arrays, as revealed in Fig. 5.135(c, d). Elemental mapping profiles, as inset in Fig. 5.135(d), suggested that both Ni and Co were uniformly distributed inside the TiN NTA tubes.

Electrodes based on this nanostructure possessed various advantages, including (i) enhanced electrical conductivity due to the highly conductive TiN substrate, (ii) promoted ionic conductivity and boosted access of the electrolyte ions without the presence of any binder and (iii) high utilization of the active materials because of the large surface area of the NTA substrates. More importantly, the Ni–Co precursor hydroxides inside the NTA were highly porous and thus presented a nanosheet similar to rippled silk, which further enlarged the surface area of the active materials and thus improved the electrochemical performance by providing charge storage in the double layer.

Similarly, ultrathin NiCo$_2$O$_4$ nanosheets have been deposited into Ni foams by using an electrodeposition method, which displayed a high capacitance of 2010 F·g^{-1}, at the current density of 2 A·g^{-1} [263]. Such a high electrochemical performance could be readily attributed to the ultrathin and mesoporous characteristics of the nanosheets, so as to offer a very high surface area, enriched electrochemical reactive sites for redox reaction to store charges.

The electrodeposition process was conducted with the standard three-electrode configuration, with pre-cleaned Ni foam as the working electrode, a platinum plate as counter electrode and a saturated calomel reference electrode (SCE), at 10 ± 1°C. 4 mM Co(NO$_3$)$_2$·6H$_2$O and 2 mM Ni(NO$_3$)$_2$·6H$_2$O aqueous mixed electrolyte solution was used to deposit bimetallic (Ni, Co) hydroxide precursors. The electrodeposition potential was set to be –1.0 V (vs. SCE). After electrodeposition for 5 minutes, the Ni foam became green, which was cleaned with DI water and absolute ethanol with the aid of ultrasonication, followed by naturally drying in air. The dried samples were calcined at 300°C for 2 hours, as slow heating rate of 1°C min^{-1}, in order to obtain ultrathin mesoporous NiCo$_2$O$_4$ nanosheets.

Figure 5.136(a) shows a low magnification SEM image of the bimetallic (Ni, Co) hydroxide precursor that was deposited onto the Ni foam. The presence of the hydroxide precursor had almost no effect on the 3D grid structure with hierarchical macro-porosity of the Ni foam. Figure 5.136(b) shows a high-magnification top-view SEM image of the sample, revealing that the (Ni, Co) hydroxide precursor had a nanosheet microstructure, with a morphology similar to rippled silk, owing mainly to its ultrathin structure. After conversion to NiCo$_2$O$_4$ through calcination, the morphology was well retained, without damage caused by the calcination treatment. The nanosheets exhibited a lateral size of several hundred nanometers, which intercrossed one another, thus leading to highly porous nanostructures with large open spaces and electroactive surface sites for electrochemical reaction and charge storage.

The electrodeposition method has also been adopted to deposit NiCo$_2$O$_4$ nanosheets on flexible carbon fabric (CF) [264]. Electrode based on the NiCo$_2$O$_4$ nanosheets exhibited a specific capacitance of as high as 2658 F·g^{-1} at the current density of 2 A g^{-1}, together with outstanding rate capability. In addition, it demonstrated a relatively high cycling performance with 20% loss in capacitance after 3000 charge-discharge cycles, at a current density of 10 A·g^{-1}. The ultrathin interconnected NiCo$_2$O$_4$ nanosheets with a thickness of about 10 nm generated an inter-particle porous network, which created more electroactive sites and thus ensured a smooth transportation for the electrolyte ions into the active materials.

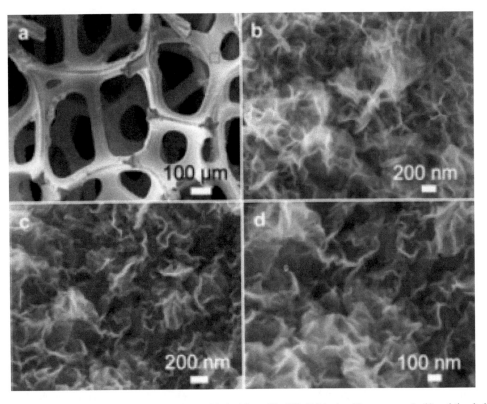

Fig. 5.136. SEM images of the Ni foam deposited with the bimetallic (Ni, Co) hydroxide precursor (a, b) and the derived NiCo$_2$O$_4$ ultrathin nanosheets/Ni foam (c, d) after calcination. Reproduced with permission from [263], Copyright © 2012, WILEY-VCH Verlag GmbH & Co. KGaA, Weinheim.

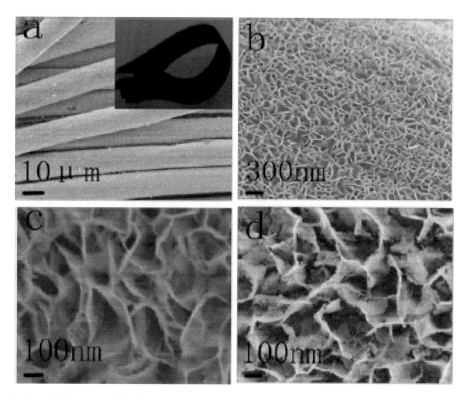

Fig. 5.137. (a–c) SEM images of NiCo$_2$O$_4$ precursor supported on CF. The inset in panel (a) is a photograph of the precursor supported on CF. (d) SEM image of the NiCo$_2$O$_4$/CF composites. Reproduced with permission from [263], Copyright © 2012, American Chemical Society.

The reaction solution was prepared by mixing 5 mmol Ni (NO$_3$)$_2$·6H$_2$O and 10 mmol Co (NO$_3$)$_2$·6H$_2$O in 20 mL distilled water, with the aid of stirring. The NiCo$_2$O$_4$ precursor was electrodeposited on carbon fabric with the reaction solution at room temperature. The washed CF was used as the working electrode, Hg/Hg$_2$Cl$_2$ was adopted as the reference electrode and a Pt foil was employed as the counter electrode. The electrodeposition was carried out at a potential of −1.0 V (vs. SCE). After electrodeposition for 20 minutes, the sample was cleaned with the aid of ultrasonication in ethanol and distilled water, followed by drying in air. To convert the precursor to NiCo$_2$O$_4$ nanosheet arrays, the sample was calcined at 300°C for 2 hours.

Figure 5.137 shows SEM images of the hydroxide precursor and the NiCo$_2$O$_4$/CF nanocomposite samples. As observed in Fig. 5.137(a, b), the hydroxide precursor was uniformly coated on the carbon fabric. The NiCo$_2$O$_4$/CF nanocomposite was still highly flexible, without losing any mechanical strength, as demonstrated in the inset in Fig. 5.137(a). Figure 5.137(c) indicated that the (Ni, Co) hydroxide precursor nanosheets were interconnected to one and another, so that highly ordered nanosheet arrays were developed. Furthermore, the (Ni, Co) hydroxide precursor nanosheets possessed an average thickness of about 10 nm. After calcination, the nanosheet arrays were well retained, without any damage in integrity, as illustrated in Fig. 5.137(d). TEM images revealed that the NiCo$_2$O$_4$ nanosheets consisted of NiCo$_2$O$_4$ nanoparticles, with an average diameter of 5 nm, together with numerous interparticle pores.

The ultrathin porous NiCo$_2$O$_4$ nanosheet arrays on flexible CF could be used as electrode of supercapacitors directly, without the requirement of any binder. The binder-free electrode has various advantages that were responsible for the high electrochemical performance. Firstly, the NiCo$_2$O$_4$ nanostructures were strongly attached on the CF, so that the electrode would have a high electrical conductivity. Secondly, the highly porous characteristics of the NiCo$_2$O$_4$ nanosheets offered more electroactive sites to allow the electrolyte ions to access and promote the electrochemical reaction. Thirdly, the interconnected nanosheets ensured mechanical stability of the electrode, so that it could withstand the structural variation during charge-discharge cycling.

Summary

By using NiO, Co_3O_4 and $NiCo_2O_4$ as the representatives, oxides other than manganese oxides, have been demonstrated to exhibit promising electrochemical performances as the electrodes of supercapacitors. Similarly, these oxides are also semiconductors, with electrical conductivities that are less than the requirement to achieve high specific capacitances. To increase their conductivities, they are usually incorporated either by forming hybrids or composites with conductive components, such as various nanocarbons, or directly deposited on conductive substrates, such as Ni foams, carbon fiber papers/cloths or metal plates. As a result, sufficiently high electrical conductivity could be readily achieved. At the same time, the electroactive sites for the electrolyte ions could be enriched.

To fully explore the potential of the oxide based electrode materials for the applications of supercapacitors, optimization of both the synthetic parameters and material properties should be taken in account in an equal manner. Also, it is necessary to address the issues of high specific capacitance and long term cycling stability at the same time, which could be the main focus of near future and long term research activities. In addition to synthetic methodologies, other issues, such as measurement techniques and electrode fabrication standards, also need to be taken into account. Moreover, engineering factors, including large-scale fabrication of electrodes, selection of electrolytes, preparation of membrane separators and device packaging technologies, should be established, especially for industrial applications, which remain to be open research topics.

Keywords: Supercapacitor, Pseudocapacitor, Pseudocapacitive effect, V_2O_5, Fe_2O_3, Fe_3O_4, Co_3O_4, NiO, NiO_x, CuO, ZnO, WO_3, MoO_3, RuO_2, $NiCo_2O_4$ and ferrites, Thin film, Chemical Solution Deposition (CSD), Electrodeposition, Spray deposition, Sol-gel deposition, Hydrothermal/solvothermal processing, Hybrids, Nanocomposites, Nanostructures

References

[1] Qu QT, Zhu YS, Gao XW, Wu YP. Core-shell structure of polypyrrole grown on V_2O_5 nanoribbon as high performance anode material for supercapacitors. Advanced Energy Materials. 2012; 2: 950–5.

[2] Foo CY, Sumboja A, Tan DJH, Wang J, Lee PS. Flexible and highly scalable V_2O_5-rGO electrodes in an organic electrolyte for supercapacitor devices. Advanced Energy Materials. 2014; 4.

[3] Rauda IE, Augustyn V, Saldarriaga-Lopez LC, Chen XY, Schelhas LT, Rubloff GW et al. Nanostructured pseudocapacitors based on atomic layer deposition of V_2O_5 onto conductive nanocrystal-based mesoporous ITO scaffolds. Advanced Functional Materials. 2014; 24: 6717–28.

[4] Xia H, Hong CY, Li B, Zhao B, Lin ZX, Zheng MB et al. Facile synthesis of hematite quantum-dot/functionalized graphene-sheet composites as advanced anode materials for asymmetric supercapacitors. Advanced Functional Materials. 2015; 25: 627–35.

[5] Sun S, Lang J, Wang R, Kong L, Li X, Yan X. Identifying pseudocapacitance of Fe_2O_3 in an ionic liquid and its application in asymmetric supercapacitors. Journal of Materials Chemistry A. 2014; 2: 14550–6.

[6] Sun HY, Chen B, Jiao XL, Jiang Z, Qin ZH, Chen DR. Solvothermal synthesis of tunable electroactive magnetite nanorods by controlling the side reaction. Journal of Physical Chemistry C. 2012; 116: 5476–81.

[7] Liu M, Sun J. *In situ* growth of monodisperse Fe_3O_4 nanoparticles on graphene as flexible paper for supercapacitor. Journal of Materials Chemistry A. 2014; 2: 12068–74.

[8] Xu S, You LJ, Zhang P, Zhang YT, Guo J, Wang CC. Fe_3O_4@coordination polymer microspheres with self-supported polyoxometalates in shells exhibiting high-performance supercapacitive energy storage. Chemical Communications. 2013; 49: 2427–9.

[9] Lee KK, Chin WS, Sow CH. Cobalt-based compounds and composites as electrode materials for high-performance electrochemical capacitors. Journal of Materials Chemistry A. 2014; 2: 17212–48.

[10] Sk MM, Yue CY, Ghosh K, Jena RK. Review on advances in porous nanostructured nickel oxides and their composite electrodes for high-performance supercapacitors. Journal of Power Sources. 2016; 308: 121–40.

[11] Deng MJ, Wang CC, Ho PJ, Lin CM, Chen JM, Lu KT. Facile electrochemical synthesis of 3D nano-architectured CuO electrodes for high-performance supercapacitors. Journal of Materials Chemistry A. 2014; 2: 12857–65.

412 *Nanomaterials for Supercapacitors*

[12] Xue CF, Lv YY, Zhang F, Wu LM, Zhao DY. Copper oxide activation of soft-templated mesoporous carbons and their electrochemical properties for capacitors. Journal of Materials Chemistry. 2012; 22: 1547–55.

[13] Zhang Z, Ren L, Han WJ, Meng LJ, Wei XL, Qi X et al. One-pot electrodeposition synthesis of ZnO/graphene composite and its use as binder-free electrode for supercapacitor. Ceramics International. 2015; 41: 4374–80.

[14] Lu T, Pan LK, Li HB, Zhu GA, Lv TA, Liu XJ et al. Microwave-assisted synthesis of graphene-ZnO nanocomposite for electrochemical supercapacitors. Journal of Alloys and Compounds. 2011; 509: 5488–92.

[15] Jo C, Hwang J, Song H, Dao AH, Kim YT, Lee SH et al. Block-copolymer-assisted one-pot synthesis of ordered mesoporous WO_{3-x}/carbon nanocomposites as high-rate-performance electrodes for pseudocapacitors. Advanced Functional Materials. 2013; 23: 3747–54.

[16] Ji HM, Liu XL, Liu ZJ, Yan B, Chen L, Xie YF et al. *In situ* preparation of sandwich MoO_3/C hybrid nanostructures for high-rate and ultralong-life supercapacitors. Advanced Functional Materials. 2015; 25: 1886–94.

[17] Chang KH, Hu CC. Hydrothermal synthesis of hydrous crystalline RuO2 nanoparticles for supercapacitors. Electrochemical and Solid State Letters. 2004; 7: A466–A9.

[18] Gujar TP, Shinde VR, Lokhande CD, Kim W-Y, Jung K-D, Joo O-S. Spray deposited amorphous RUO2 for an effective use in electrochemical supercapacitor. Electrochemistry Communications. 2007; 9: 504–10.

[19] Park JH, Ko JM, Park OO. Carbon nanotube/RuO2 nanocomposite electrodes for supercapacitors. Journal of the Electrochemical Society. 2003; 150: A864–A7.

[20] Zhou Z, Zhu Y, Wu Z, Lu F, Jing M, Ji X. Amorphous RuO2 coated on carbon spheres as excellent electrode materials for supercapacitors. Rsc Advances. 2014; 4: 6927–32.

[21] Dubal DP, Gomez-Romero P, Sankapal BR, Holze R. Nickel cobaltite as an emerging material for supercapacitors: An overview. Nano Energy. 2015; 11: 377–99.

[22] Wu Z, Zhu Y, Ji X. $NiCo_2O_4$-based materials for electrochemical supercapacitors. Journal of Materials Chemistry A. 2014; 2: 14759–72.

[23] Garg N, Basu M, Upadhyaya K, Shivaprasad SM, Ganguli AK. Controlling the aspect ratio and electrocatalytic properties of nickel cobaltite nanorods. RSC Advances. 2013; 3: 24328–36.

[24] Kumbhar VS, Jagadale AD, Shinde NM, Lokhande CD. Chemical synthesis of spinel cobalt ferrite ($CoFe_2O_4$) nano-flakes for supercapacitor application. Applied Surface Science. 2012; 259: 39–43.

[25] Wang R, Li Q, Cheng L, Li H, Wang B, Zhao XS et al. Electrochemical properties of manganese ferrite-based supercapacitors in aqueous electrolyte: The effect of ionic radius. Colloids and Surfaces a-Physicochemical and Engineering Aspects. 2014; 457: 94–9.

[26] Wang W, Hao Q, Lei W, Xia X, Wang X. Ternary nitrogen-doped graphene/nickel ferrite/polyaniline nanocomposites for high-performance supercapacitors. Journal of Power Sources. 2014; 269: 250–9.

[27] Kim SH, Kim YI, Park JH, Ko JM. Cobalt-manganese oxide/carbon-nanofiber composite electrodes for supercapacitors. International Journal of Electrochemical Science. 2009; 4: 1489–96.

[28] Wang Y-L, Zhao Y-Q, Xu C-L. May 3D nickel foam electrode be the promising choice for supercapacitors? Journal of Solid State Electrochemistry. 2012; 16: 829–34.

[29] Lu Z, Yang Q, Zhu W, Chang Z, Liu J, Sun X et al. Hierarchical Co_3O_4@Ni-Co-O supercapacitor electrodes with ultrahigh specific capacitance per area. Nano Research. 2012; 5: 369–78.

[30] Tronel F, Guerlou-Demourgues L, Menetrier M, Croguennec L, Goubault L, Bernard P et al. New spinel cobalt oxides, potential conductive additives for the positive electrode of Ni-MH batteries. Chemistry of Materials. 2006; 18: 5840–51.

[31] Zheng MB, Cao J, Liao ST, Liu JS, Chen HQ, Zhao Y et al. Preparation of mesoporous Co_3O_4 nanoparticles via solid-liquid route and effects of calcination temperature and textural parameters on their electrochemical capacitive behaviors. Journal of Physical Chemistry C. 2009; 113: 3887–94.

[32] Xu J, Gao L, Cao J, Wang W, Chen Z. Preparation and electrochemical capacitance of cobalt oxide (Co_3O_4) nanotubes as supercapacitor material. Electrochimica Acta. 2010; 56: 732–6.

[33] Wang DW, Wang QH, Wang TM. Morphology-controllable synthesis of cobalt oxalates and their conversion to mesoporous Co_3O_4 nanostructures for application in supercapacitors. Inorganic Chemistry. 2011; 50: 6482–92.

[34] Godillot G, Guerlou-Demourgues L, Taberna PL, Simon P, Delmas C. Original conductive nano-Co_3O_4 investigated as electrode material for hybrid supercapacitors. Electrochemical and Solid State Letters. 2011; 14: A139–A42.

[35] Chen X, Cheng JP, Shou QL, Liu F, Zhang XB. Effect of calcination temperature on the porous structure of cobalt oxide micro-flowers. CrystEngComm. 2012; 14: 1271–6.

[36] Wang RT, Kong LB, Lang JW, Wang XW, Fan SQ, Luo YC et al. Mesoporous Co_3O_4 materials obtained from cobalt-citrate complex and their high capacitance behavior. Journal of Power Sources. 2012; 217: 358–63.

[37] Gong LY, Liu XH, Su LH, Wang LQ. Synthesis and electrochemical capacitive behaviors of Co_3O_4 nanostructures from a novel biotemplating technique. Journal of Solid State Electrochemistry. 2012; 16: 297–304.

[38] Du W, Liu RM, Jiang YW, Lu QY, Fan YZ, Gao F. Facile synthesis of hollow Co_3O_4 boxes for high capacity supercapacitor. Journal of Power Sources. 2013; 227: 101–5.

[39] Wang Y, Lei Y, Li J, Gu L, Yuan H, Xiao D. Synthesis of 3D-nanonet hollow structured Co_3O_4 for high capacity supercapacitor. Acs Applied Materials & Interfaces. 2014; 6: 6739–47.

[40] Ishikawa T, Matijevic E. Formation of uniform particles of cobalt compounds and cobalt. Colloid and Polymer Science. 1991; 269: 179–86.

[41] Xu R, Zeng HC. Dimensional control of cobalt-hydroxide-carbonate nanorods and their thermal conversion to one-dimensional arrays of Co_3O_4 nanoparticles. Journal of Physical Chemistry B. 2003; 107: 12643–9.

[42] Wang WW, Zhu YJ. Microwave-assisted synthesis of cobalt oxalate nanorods and their thermal conversion to Co_3O_4 rods. Materials Research Bulletin. 2005; 40: 1929–35.

[43] Zhao ZG, Geng FX, Bai JB, Cheng HM. Facile and controlled synthesis of 3D nanorods-based urchinlike and nanosheets-based flowerlike cobalt basic salt nanostructures. Journal of Physical Chemistry C. 2007; 111: 3848–52.

[44] Meher SK, Rao GR. Effect of microwave on the nanowire morphology, optical, magnetic, and pseudocapacitance behavior of Co_3O_4. Journal of Physical Chemistry C. 2011; 115: 25543–56.

[45] Cheng JP, Chen X, Wu JS, Liu F, Zhang XB, Dravid VP. Porous cobalt oxides with tunable hierarchical morphologies for supercapacitor electrodes. CrystEngComm. 2012; 14: 6702–9.

[46] Jun YW, Choi JS, Cheon J. Shape control of semiconductor and metal oxide nanocrystals through nonhydrolytic colloidal routes. Angewandte Chemie-International Edition. 2006; 45: 3414–39.

[47] Yang L, Wang GZ, Tang CJ, Wang HQ, Zhang L. Synthesis and photoluminescence of corn-like ZnO nanostructures under solvothermal-assisted heat treatment. Chemical Physics Letters. 2005; 409: 337–41.

[48] Pang H, Lu QY, Lia YC, Gao F. Facile synthesis of nickel oxide nanotubes and their antibacterial, electrochemical and magnetic properties. Chemical Communications. 2009; 2009: 7542–4.

[49] Wang DW, Wang QH, Wang TM. Controlled growth of pyrite FeS_2 crystallites by a facile surfactant-assisted solvothermal method. CrystEngComm. 2010; 12: 755–61.

[50] Peng XG. Mechanisms for the shape-control and shape-evolution of colloidal semiconductor nanocrystals. Advanced Materials. 2003; 15: 459–63.

[51] Navrotsky A, Mazeina L, Majzlan J. Size-driven structural and thermodynamic complexity in iron oxides. Science. 2008; 319: 1635–8.

[52] Wang JX, Ma C, Choi YM, Su D, Zhu YM, Liu P et al. Kirkendall effect and lattice contraction in nanocatalysts: A new strategy to enhance sustainable activity. Journal of the American Chemical Society. 2011; 133: 13551–7.

[53] Xu ZP, Zeng HC. Thermal revolution of cobalt hydroxides: a comparative study of their various structural phases. Journal of Materials Chemistry. 1998; 8: 2499–506.

[54] Palmas S, Ferrara F, Vacca A, Mascia M, Polcaro AM. Behavior of cobalt oxide electrodes during oxidative processes in alkaline medium. Electrochimica Acta. 2007; 53: 400–6.

[55] Wang GL, Cao DX, Yin CL, Gao YY, Yin JL, Cheng L. Nickel foam supported-Co_3O_4 nanowire arrays for H_2O_2 electroreduction. Chemistry of Materials. 2009; 21: 5112–8.

[56] Lin C, Ritter JA, Popov BN. Characterization of sol-gel-derived cobalt oxide xerogels as electrochemical capacitors. Journal of the Electrochemical Society. 1998; 145: 4097–103.

[57] Barbero C, Planes GA, Miras MC. Redox coupled ion exchange in cobalt oxide films. Electrochemistry Communications. 2001; 3: 113–6.

[58] Gao YY, Chen SL, Cao DX, Wang GL, Yin JL. Electrochemical capacitance of Co_3O_4 nanowire arrays supported on nickel foam. Journal of Power Sources. 2010; 195: 1757–60.

[59] Kim TW, Kleitz F, Paul B, Ryoo R. MCM-48-like large mesoporous silicas with tailored pore structure: Facile synthesis domain in a ternary triblock copolymer-butanol-water system. Journal of the American Chemical Society. 2005; 127: 7601–10.

[60] Zhao DY, Feng JL, Huo QS, Melosh N, Fredrickson GH, Chmelka BF et al. Triblock copolymer syntheses of mesoporous silica with periodic 50 to 300 angstrom pores. Science. 1998; 279: 548–52.

[61] Yue W, Zhou WZ. Synthesis of porous single crystals of metal oxides via a solid-liquid route. Chemistry of Materials. 2007; 19: 2359–63.

[62] Wang X, Li YD. Solution-based synthetic strategies for 1-D nanostructures. Inorganic Chemistry. 2006; 45: 7522–34.

[63] Rao CNR, Agrawal VV, Biswas K, Gautam UK, Ghosh M, Govindaraj A et al. Soft chemical approaches to inorganic nanostructures. Pure and Applied Chemistry. 2006; 78: 1619–50.

[64] Mao Y, Park TJ, Zhang F, Zhou H, Wong SS. Environmentally friendly methodologies of nanostructure synthesis. Small. 2007; 3: 1122–39.

[65] Xiong SL, Yuan CZ, Zhang MF, Xi BJ, Qian YT. Controllable synthesis of mesoporous Co_3O_4 nanostructures with tunable morphology for application in supercapacitors. Chemistry-A European Journal. 2009; 15: 5320–6.

[66] Wang GX, Shen XP, Horvat J, Wang B, Liu H, Wexler D et al. Hydrothermal synthesis and optical, magnetic, and supercapacitance properties of nanoporous cobalt oxide nanorods. Journal of Physical Chemistry C. 2009; 113: 4357–61.

[67] Cui L, Li J, Zhang XG. Preparation and properties of Co_3O_4 nanorods as supercapacitor material. Journal of Applied Electrochemistry. 2009; 39: 1871–6.

[68] Zhu T, Chen JS, Lou XW. Shape-controlled synthesis of porous Co_3O_4 nanostructures for application in supercapacitors. Journal of Materials Chemistry. 2010; 20: 7015–20.

414 *Nanomaterials for Supercapacitors*

[69] Wang Y, Zhong Z, Chen Y, Ng CT, Lin J. Controllable synthesis of Co_3O_4 from nanosize to microsize with large-scale exposure of active crystal planes and their excellent rate capability in supercapacitors based on the crystal plane effect. Nano Research. 2011; 4: 695–704.

[70] Meher SK, Rao GR. Ultralayered Co_3O_4 for high-performance supercapacitor applications. Journal of Physical Chemistry C. 2011; 115: 15646–54.

[71] Hou LR, Yuan CZ, Yang L, Shen LF, Zhang F, Zhang XG. Urchin-like Co_3O_4 microspherical hierarchical superstructures constructed by one-dimension nanowires toward electrochemical capacitors. RSC Advances. 2011; 1: 1521–6.

[72] Wang B, Zhu T, Wu HB, Xu R, Chen JS, Lou XW. Porous Co_3O_4 nanowires derived from long $Co(CO_3)_{0.5}(OH).0.11 H_2O$ nanowires with improved supercapacitive properties. Nanoscale. 2012; 4: 2145–9.

[73] Xiao YH, Zhang AQ, Liu SJ, Zhao JH, Fang SM, Jia DZ et al. Free-standing and porous hierarchical nanoarchitectures constructed with cobalt cobaltite nanowalls for supercapacitors with high specific capacitances. Journal of Power Sources. 2012; 219: 140–6.

[74] Xiao YH, Liu SJ, Li F, Zhang AQ, Zhao JH, Fang SM et al. 3D hierarchical Co_3O_4 twin-spheres with an urchin-like structure: Large-scale synthesis, multistep-splitting growth, and electrochemical pseudocapacitors. Advanced Functional Materials. 2012; 22: 4052–9.

[75] Shi WD, Song SY, Zhang HJ. Hydrothermal synthetic strategies of inorganic semiconducting nanostructures. Chemical Society Reviews. 2013; 42: 5714–43.

[76] He G, Li J, Chen H, Shi J, Sun X, Chen S et al. Hydrothermal preparation of Co_3O_4@graphene nanocomposite for supercapacitor with enhanced capacitive performance. Materials Letters. 2012; 82: 61–3.

[77] Feng C, Zhang JF, Deng YD, Zhong C, Liu L, Hu WB. One-pot fabrication of Co_3O_4 microspheres via hydrothermal method at low temperature for high capacity supercapacitor. Materials Science and Engineering B-Advanced Functional Solid-State Materials. 2015; 199: 15–21.

[78] Simon P, Gogotsi Y. Materials for electrochemical capacitors. Nature Materials. 2008; 7: 845–54.

[79] Patzke GR, Zhou Y, Kontic R, Conrad F. Oxide nanomaterials: Synthetic developments, mechanistic studies, and technological innovations. Angewandte Chemie-International Edition. 2011; 50: 826–59.

[80] Whitesides GM, Boncheva M. Beyond molecules: Self-assembly of mesoscopic and macroscopic components. Proceedings of the National Academy of Sciences of the United States of America. 2002; 99: 4769–74.

[81] Colfen H, Antonietti M. Mesocrystals: Inorganic superstructures made by highly parallel crystallization and controlled alignment. Angewandte Chemie-International Edition. 2005; 44: 5576–91.

[82] Hosono E, Fujihara S, Honma I, Zhou HS. Fabrication of morphology and crystal structure controlled nanorod and nanosheet cobalt hydroxide based on the difference of oxygen-solubility between water and methanol, and conversion into Co_3O_4. Journal of Materials Chemistry. 2005; 15: 1938–45.

[83] Qian LW, Zai JT, Chen Z, Zhu J, Yuan YP, Qian XF. Control of the morphology and composition of yttrium fluoride via a salt-assisted hydrothermal method. CrystEngComm. 2010; 12: 199–206.

[84] Wei TY, Chen CH, Chang KH, Lu SY, Hu CC. Cobalt oxide aerogels of ideal supercapacitive properties prepared with an epoxide synthetic route. Chemistry of Materials. 2009; 21: 3228–33.

[85] Wang X, Sumboja A, Khoo E, Yan CY, Lee PS. Cryogel synthesis of hierarchical interconnected macro-/mesoporous Co_3O_4 with superb electrochemical energy storage. Journal of Physical Chemistry C. 2012; 116: 4930–5.

[86] Gash AE, Tillotson TM, Satcher JH, Poco JF, Hrubesh LW, Simpson RL. Use of epoxides in the sol-gel synthesis of porous iron(III) oxide monoliths from Fe(III) salts. Chemistry of Materials. 2001; 13: 999–1007.

[87] Kim HC, Park SM, Hinsberg WD. Block copolymer based nanostructures: Materials, processes, and applications to electronics. Chemical Reviews. 2010; 110: 146–77.

[88] Nishihara H, Mukai SR, Yamashita D, Tamon H. Ordered macroporous silica by ice templating. Chemistry of Materials. 2005; 17: 683–9.

[89] Wan Y, Zhao DY. On the controllable soft-templating approach to mesoporous silicates. Chemical Reviews. 2007; 107: 2821–60.

[90] Cheng H, Lu ZG, Deng JQ, Chung CY, Zhang KL, Li YY. A facile method to improve the high rate capability of Co_3O_4 nanowire array electrodes. Nano Research. 2010; 3: 895–901.

[91] Li YG, Tan B, Wu YY. Freestanding mesoporous quasi-single-crystalline Co_3O_4 nanowire arrays. Journal of the American Chemical Society. 2006; 128: 14258–9.

[92] Li YG, Tan B, Wu YY. Mesoporous CO_3O_4 nanowire arrays for lithium ion batteries with high capacity and rate capability. Nano Letters. 2008; 8: 265–70.

[93] Li YH, Huang KL, Yao ZF, Liu SQ, Qing XX. Co_3O_4 thin film prepared by a chemical bath deposition for electrochemical capacitors. Electrochimica Acta. 2011; 56: 2140–4.

[94] Srinivasan V, Weidner JW. Capacitance studies of cobalt oxide films formed via electrochemical precipitation. Journal of Power Sources. 2002; 108: 15–20.

[95] Lee JK, Kim GP, Kim KH, Song IK, Baeck SH. Fabrication of mesoporous cobalt oxide (Co_3O_4) film by electrochemical method for electrochemical capacitor. Journal of Nanoscience and Nanotechnology. 2010; 10: 3676–9.

[96] Wu JB, Lin Y, Xia XH, Xu JY, Shi QY. Pseudocapacitive properties of electrodeposited porous nanowall Co_3O_4 film. Electrochimica Acta. 2011; 56: 7163–70.

[97] Yuan YF, Xia XH, Wu JB, Huang XH, Pei YB, Yang JL et al. Hierarchically porous Co_3O_4 film with mesoporous walls prepared via liquid crystalline template for supercapacitor application. Electrochemistry Communications. 2011; 13: 1123–6.

[98] Xia XH, Tu JP, Wang XL, Gu CD, Zhao XB. Mesoporous Co_3O_4 monolayer hollow-sphere array as electrochemical pseudocapacitor material. Chemical Communications. 2011; 47: 5786–8.

[99] Xia XH, Tu JP, Xiang JY, Huang XH, Wang XL, Zhao XB. Hierarchical porous cobalt oxide array films prepared by electrodeposition through polystyrene sphere template and their applications for lithium ion batteries. Journal of Power Sources. 2010; 195: 2014–22.

[100] Shinde VR, Mahadik SB, Gujar TP, Lokhande CD. Supercapacitive cobalt oxide (Co_3O_4) thin films by spray pyrolysis. Applied Surface Science. 2006; 252: 7487–92.

[101] Tummala R, Guduru RK, Mohanty PS. Nanostructured Co_3O_4 electrodes for supercapacitor applications from plasma spray technique. Journal of Power Sources. 2012; 209: 44–51.

[102] Qing XX, Liu SQ, Huang KL, Lv KZ, Yang YP, Lu ZG et al. Facile synthesis of Co_3O_4 nanoflowers grown on Ni foam with superior electrochemical performance. Electrochimica Acta. 2011; 56: 4985–91.

[103] Yuan C, Yang L, Hou L, Shen L, Zhang F, Li D et al. Large-scale Co_3O_4 nanoparticles growing on nickel sheets via a one-step strategy and their ultra-highly reversible redox reaction toward supercapacitors. Journal of Materials Chemistry. 2011; 21: 18183–5.

[104] Xia X-h, Tu J-p, Mai Y-j, Wang X-l, Gu C-d, Zhao X-b. Self-supported hydrothermal synthesized hollow Co_3O_4 nanowire arrays with high supercapacitor capacitance. Journal of Materials Chemistry. 2011; 21: 9319–25.

[105] Zhang F, Yuan CZ, Lu XJ, Zhang LJ, Che Q, Zhang XG. Facile growth of mesoporous Co_3O_4 nanowire arrays on Ni foam for high performance electrochemical capacitors. Journal of Power Sources. 2012; 203: 250–6.

[106] Xia X-h, Tu J-p, Zhang Y-q, Mai Y-j, Wang X-l, Gu C-d et al. Freestanding Co_3O_4 nanowire array for high performance supercapacitors. Rsc Advances. 2012; 2: 1835–41.

[107] Yang Q, Lu ZY, Chang Z, Zhu W, Sun JQ, Liu JF et al. Hierarchical Co_3O_4 nanosheet@nanowire arrays with enhanced pseudocapacitive performance. RSC Advances. 2012; 2: 1663–8.

[108] Duan BR, Cao Q. Hierarchically porous Co_3O_4 film prepared by hydrothermal synthesis method based on colloidal crystal template for supercapacitor application. Electrochimica Acta. 2012; 64: 154–61.

[109] Yang Q, Lu ZY, Sun XM, Liu JF. Ultrathin Co_3O_4 nanosheet arrays with high supercapacitive performance. Scientific Reports. 2013; 3.

[110] Wang HT, Zhang L, Tan XH, Holt CMB, Zahiri B, Olsen BC et al. Supercapacitive properties of hydrothermally synthesized Co_3O_4 nanostructures. Journal of Physical Chemistry C. 2011; 115: 17599–605.

[111] Feng LD, Zhu YF, Ding HY, Ni CY. Recent progress in nickel based materials for high performance pseudocapacitor electrodes. Journal of Power Sources. 2014; 267: 430–44.

[112] Augustyn V, Simon P, Dunn B. Pseudocapacitive oxide materials for high-rate electrochemical energy storage. Energy & Environmental Science. 2014; 7: 1597–614.

[113] Wang L, Wang X, Xiao X, Xu F, Sun Y, Li Z. Reduced graphene oxide/nickel cobaltite nanoflake composites for high specific capacitance supercapacitors. Electrochimica Acta. 2013; 111: 937–45.

[114] Wan X, Yuan M, Tie SL, Lan S. Effects of catalyst characters on the photocatalytic activity and process of NiO nanoparticles in the degradation of methylene blue. Applied Surface Science. 2013; 277: 40–6.

[115] Wang Y, Zhu QS, Zhang HG. Fabrication of beta-Ni(OH)$_2$ and NiO hollow spheres by a facile template-free process. Chemical Communications. 2005: 5231–3.

[116] Cheng G, Yan YN, Chen R. From Ni-based nanoprecursors to NiO nanostructures: Morphology-controlled synthesis and structure-dependent electrochemical behavior. New Journal of Chemistry. 2015; 39: 676–82.

[117] Yuan C, Zhang X, Su L, Gao B, Shen L. Facile synthesis and self-assembly of hierarchical porous NiO nano/micro spherical superstructures for high performance supercapacitors. Journal of Materials Chemistry. 2009; 19: 5772–7.

[118] Lang J-W, Kong L-B, Wu W-J, Luo Y-C, Kang L. Facile approach to prepare loose-packed NiO nano-flakes materials for supercapacitors. Chemical Communications. 2008: 4213–5.

[119] Cao C-Y, Guo W, Cui Z-M, Song W-G, Cai W. Microwave-assisted gas/liquid interfacial synthesis of flowerlike NiO hollow nanosphere precursors and their application as supercapacitor electrodes. Journal of Materials Chemistry. 2011; 21: 3204–9.

[120] Zhu JX, Gui Z. From layered hydroxide compounds to labyrinth-like NiO and Co_3O_4 porous nanosheets. Materials Chemistry and Physics. 2009; 118: 243–8.

[121] Qiu YJ, Yu J, Zhou XS, Tan CL, Yin J. Synthesis of porous NiO and ZnO submicro- and nanofibers from electrospun polymer fiber templates. Nanoscale Research Letters. 2009; 4: 173–7.

[122] Brezesinski K, Wang J, Haetge J, Reitz C, Steinmueller SO, Tolbert SH et al. Pseudocapacitive contributions to charge storage in highly ordered mesoporous group V transition metal oxides with iso-oriented layered nanocrystalline domains. Journal of the American Chemical Society. 2010; 132: 6982–90.

[123] Zhang ZY, Zuo F, Feng PY. Hard template synthesis of crystalline mesoporous anatase TiO_2 for photocatalytic hydrogen evolution. Journal of Materials Chemistry. 2010; 20: 2206–12.

416　*Nanomaterials for Supercapacitors*

[124] Kresge CT, Leonowicz ME, Roth WJ, Vartuli JC, Beck JS. Ordered mesoporous molecular-sieves synthesized by a liquid-crystal template mechanism. Nature. 1992; 359: 710–2.

[125] Padmanathan N, Selladurai S. Electrochemical capacitance of porous NiO-CeO_2 binary oxide synthesized via sol-gel technique for supercapacitor. Ionics. 2014; 20: 409–20.

[126] Frey S, Keipert S, Chazalviel JN, Ozanam F, Carstensen J, Foll H. Electrochemical formation of porous silica: Toward an understanding of the mechanisms. Physica Status Solidi A-Applications and Materials Science. 2007; 204: 1250–4.

[127] Lang X, Hirata A, Fujita T, Chen M. Nanoporous metal/oxide hybrid electrodes for electrochemical supercapacitors. Nature Nanotechnology. 2011; 6: 232–6.

[128] Li XW, Xiong SL, Li JF, Bai J, Qian YT. Mesoporous NiO ultrathin nanowire networks topotactically transformed from α-$Ni(OH)_2$ hierarchical microspheres and their superior electrochemical capacitance properties and excellent capability for water treatment. Journal of Materials Chemistry. 2012; 22: 14276–83.

[129] Lee JW, Ahn T, Kim JH, Ko JM, Kim JD. Nanosheets based mesoporous NiO microspherical structures via facile and template-free method for high performance supercapacitors. Electrochimica Acta. 2011; 56: 4849–57.

[130] Ni X, Zhang Y, Tian D, Zheng H, Wang X. Synthesis and characterization of hierarchical NiO nanoflowers with porous structure. Journal of Crystal Growth. 2007; 306: 418–21.

[131] Zhu LP, Liao GH, Yang Y, Xiao HM, Wang JF, Fu SY. Self-assembled 3D flower-like hierarchical β-$Ni(OH)_2$ hollow architectures and their *in situ* thermal conversion to NiO. Nanoscale Research Letters. 2009; 4: 550–7.

[132] Yuan CZ, Xiong SL, Zhang XG, Shen LF, Zhang F, Gao B et al. Template-free synthesis of ordered mesoporous NiO/poly (sodium-4-styrene sulfonate) functionalized carbon nanotubes composite for electrochemical capacitors. Nano Research. 2009; 2: 722–32.

[133] Xia XH, Tu JP, Zhang YQ, Wang XL, Gu CD, Zhao XB et al. High-quality metal oxide core/shell nanowire arrays on conductive substrates for electrochemical energy storage. ACS Nano. 2012; 6: 5531–8.

[134] Liu MM, Chang J, Sun J, Gao L. A facile preparation of NiO/Ni composites as high-performance pseudocapacitor materials. RCS Advances. 2013; 3: 8003–8.

[135] Cui YF, Wang C, Wu SJ, Liu G, Zhang FF, Wang TM. Lotus-root-like NiO nanosheets and flower-like NiO microspheres: Synthesis and magnetic properties. CrystEngComm. 2011; 13: 4930–4.

[136] Su DW, Kim HS, Kim WS, Wang GX. Mesoporous nickel oxide nanowires: Hydrothermal synthesis, characterisation and applications for lithium-ion batteries and supercapacitors with superior performance. Chemistry-A European Journal. 2012; 18: 8224–9.

[137] Wen W, Wu JM, Lai LL, Ling GP, Cao MH. Hydrothermal synthesis of needle-like hyperbranched $Ni(SO4)_{0.3}(OH)_{1.4}$ bundles and their morphology-retentive decompositions to NiO for lithium storage. CrystEngComm. 2012; 14: 6565–72.

[138] Yang DG, Liu PC, Gao Y, Wu H, Cao Y, Xiao QZ et al. Synthesis, characterization, and electrochemical performances of core-shell $Ni(SO_4)_{0.3}(OH)_{1.4}$/C and NiO/C nanobelts. Journal of Materials Chemistry. 2012; 22: 7224–31.

[139] Zhang XJ, Shi WH, Zhu JX, Zhao WY, Ma J, Mhaisalkar S et al. Synthesis of porous NiO nanocrystals with controllable surface area and their application as supercapacitor electrodes. Nano Research. 2010; 3: 643–52.

[140] Zheng YZ, Ding HY, Zhang ML. Preparation and electrochemical properties of nickel oxide as a supercapacitor electrode material. Materials Research Bulletin. 2009; 44: 403–7.

[141] Li JT, Zhao W, Huang FQ, Manivannan A, Wu NQ. Single-crystalline $Ni(OH)_2$ and NiO nanoplatelet arrays as supercapacitor electrodes. Nanoscale. 2011; 3: 5103–9.

[142] Sarkar S, Pradhan M, Sinha AK, Basu M, Negishi Y, Pal T. An aminolytic approach toward hierarchical β-$Ni(OH)_2$ nanoporous architectures: A bimodal forum for photocatalytic and surface-enhanced Raman cattering activity. Inorganic Chemistry. 2010; 49: 8813–27.

[143] Zhou W, Yao M, Guo L, Li YM, Li JH, Yang SH. Hydrazine-linked convergent self-assembly of sophisticated concave polyhedrons of β-$Ni(OH)_2$ and NiO from nanoplate building blocks. Journal of the American Chemical Society. 2009; 131: 2959–64.

[144] Dong LH, Chu Y, Sun WD. Controllable synthesis of nickel hydroxide and porous nickel oxide nanostructures with different morphologies. Chemistry-A European Journal. 2008; 14: 5064–72.

[145] Qi YJ, Qi HY, Lu CJ, Yang Y, Zhao Y. Photoluminescence and magnetic properties of β-$Ni(OH)(2)$ nanoplates and NiO nanostructures. Journal of Materials Science-Materials in Electronics. 2009; 20: 479–83.

[146] Matsui K, Kyotani T, Tomita A. Hydrothermal synthesis of single-crystal $Ni(OH)_2$ nanorods in a carbon-coated anodic alumina film. Advanced Materials. 2002; 14: 1216–+.

[147] Hu JC, Zhu KK, Chen LF, Yang HJ, Li Z, Suchopar A et al. Preparation and surface activity of single-crystalline NiO(111) nanosheets with hexagonal holes: A semiconductor nanospanner. Advanced Materials. 2008; 20: 267–+.

[148] Wang L, Hao YJ, Zhao Y, Lai QY, Xu XY. Hydrothermal synthesis and electrochemical performance of NiO microspheres with different nanoscale building blocks. Journal of Solid State Chemistry. 2010; 183: 2576–81.

[149] Purushothaman KK, Babu IM, Sethuraman B, Muralidharan G. Nanosheet-assembled NiO microstructures for high-performance supercapacitors. Acs Applied Materials & Interfaces. 2013; 5: 10767–73.

[150] Kuang DB, Lei BX, Pan YP, Yu XY, Su CY. Fabrication of novel hierarchical β-Ni(OH)$_2$ and NiO microspheres via an easy hydrothermal process. Journal of Physical Chemistry C. 2009; 113: 5508–13.

[151] Liu JY, Guo Z, Meng FL, Jia Y, Luo T, Li MQ et al. Novel single-crystalline hierarchical structured ZnO nanorods fabricated via a wet-chemical route: Combined high gas sensing performance with enhanced optical properties. Crystal Growth & Design. 2009; 9: 1716–22.

[152] Meher SK, Justin P, Rao GR. Pine-cone morphology and pseudocapacitive behavior of nanoporous nickel oxide. Electrochimica Acta. 2010; 55: 8388–96.

[153] Justin P, Meher SK, Rao GR. Tuning of capacitance behavior of NiO using anionic, cationic, and nonionic surfactants by hydrothermal synthesis. Journal of Physical Chemistry C. 2010; 114: 5203–10.

[154] Meher SK, Justin P, Rao GR. Nanoscale morphology dependent pseudocapacitance of NiO: Influence of intercalating anions during synthesis. Nanoscale. 2011; 3: 683–92.

[155] Siegfried MJ, Choi KS. Electrochemical crystallization of cuprous oxide with systematic shape evolution. Advanced Materials. 2004; 16: 1743–6.

[156] Colfen H, Mann S. Higher-order organization by mesoscale self-assembly and transformation of hybrid nanostructures. Angewandte Chemie-International Edition. 2003; 42: 2350–65.

[157] Yeager MP, Su D, Marinkovic NS, Teng XW. Pseudocapacitive NiO fine nanoparticles for supercapacitor reactions. Journal of the Electrochemical Society. 2012; 159: A1598–A603.

[158] Ding SJ, Zhu T, Chen JS, Wang ZY, Yuan CL, Lou XW. Controlled synthesis of hierarchical NiO nanosheet hollow spheres with enhanced supercapacitive performance. Journal of Materials Chemistry. 2011; 21: 6602–6.

[159] Meher SK, Justin P, Rao GR. Microwave-mediated synthesis for improved morphology and pseudocapacitance performance of nickel oxide. ACS Applied Materials & Interfaces. 2011; 3: 2063–73.

[160] Tian L, Zou HL, Fu JX, Yang XF, Wang Y, Guo HL et al. Topotactic conversion route to mesoporous quasi-single-crystalline Co$_3$O$_4$ nanobelts with optimizable electrochemical performance. Advanced Functional Materials. 2010; 20: 617–23.

[161] Luan F, Wang GM, Ling YC, Lu XH, Wang HY, Tong YX et al. High energy density asymmetric supercapacitors with a nickel oxide nanoflake cathode and a 3D reduced graphene oxide anode. Nanoscale. 2013; 5: 7984–90.

[162] Dong XY, Wang L, Wang D, Li C, Jin J. Layer-by-layer engineered Co-Al hydroxide nanosheets/graphene multilayer films as flexible electrode for supercapacitor. Langmuir. 2012; 28: 293–8.

[163] Fan MQ, Ren B, Yu L, Liu Q, Wang J, Song DL et al. Facile growth of hollow porous NiO microspheres assembled from nanosheet building blocks and their high performance as a supercapacitor electrode. CrystEngComm. 2014; 16: 10389–94.

[164] Kleitz F, Blanchard J, Zibrowius B, Schuth F, Agren P, Linden M. Influence of cosurfactants on the properties of mesostructured materials. Langmuir. 2002; 18: 4963–71.

[165] Zhang YY, Wang JX, Wei HM, Hao JH, Mu JY, Cao P et al. Hydrothermal synthesis of hierarchical mesoporous NiO nanourchins and their supercapacitor application. Materials Letters. 2016; 162: 67–70.

[166] Inamdar AI, Kim Y, Pawar SM, Kim JH, Im H, Kim H. Chemically grown, porous, nickel oxide thin-film for electrochemical supercapacitors. Journal of Power Sources. 2011; 196: 2393–7.

[167] Xia XH, Tu JP, Wang XL, Gu CD, Zhao XB. Hierarchically porous NiO film grown by chemical bath deposition via a colloidal crystal template as an electrochemical pseudocapacitor material. Journal of Materials Chemistry. 2011; 21: 671–9.

[168] Kim S-I, Lee J-S, Ahn H-J, Song H-K, Jang J-H. Facile route to an efficient NiO supercapacitor with a three-dimensional nanonetwork morphology. Acs Applied Materials & Interfaces. 2013; 5: 1596–603.

[169] Ren Y, Gao LA. From three-dimensional flower-like α-Ni(OH)$_2$ nanostructures to hierarchical porous NiO nanoflowers: Microwave-assisted fabrication and supercapacitor properties. Journal of the American Ceramic Society. 2010; 93: 3560–4.

[170] Lou XW, Yuan CL, Rhoades E, Zhang Q, Archer LA. Encapsulation and Ostwald ripening of Au and Au-Cl complex nanostructures in silica shells. Advanced Functional Materials. 2006; 16: 1679–84.

[171] Cheng J, Cao GP, Yang YS. Characterization of sol-gel-derived NiO$_x$ xerogels as supercapacitors. Journal of Power Sources. 2006; 159: 734–41.

[172] Xing W, Li F, Yan ZF, Lu GQ. Synthesis and electrochemical properties of mesoporous nickel oxide. Journal of Power Sources. 2004; 134: 324–30.

[173] Yu CC, Zhang LX, Shi JL, Zhao JJ, Gao JH, Yan DS. A simple template-free strategy to synthesize nanoporous manganese and nickel oxides with narrow pore size distribution, and their electrochemical properties. Advanced Functional Materials. 2008; 18: 1544–54.

[174] Yuan CZ, Li JY, Hou LR, Yang L, Shen LF, Zhang XG. Facile growth of hexagonal NiO nanoplatelet arrays assembled by mesoporous nanosheets on Ni foam towards high-performance electrochemical capacitors. Electrochimica Acta. 2012; 78: 532–8.

[175] Liu MM, Chang J, Sun J, Gao L. Synthesis of porous NiO using NaBH$_4$ dissolved in ethylene glycol as precipitant for high-performance supercapacitor. Electrochimica Acta. 2013; 107: 9–15.

418 *Nanomaterials for Supercapacitors*

[176] Ding SJ, Zhang CL, Yang M, Qu XZ, Lu YF, Yang ZZ. Template synthesis of composite hollow spheres using sulfonated polystyrene hollow spheres. Polymer. 2006; 47: 8360–6.

[177] Chen F, Zhou WJ, Yao HF, Fan P, Yang JT, Fei ZD et al. Self-assembly of NiO nanoparticles in lignin-derived mesoporous carbons for supercapacitor applications. Green Chemistry. 2013; 15: 3057–63.

[178] Dam DT, Wang X, Lee JM. Mesoporous ITO/NiO with a core/shell structure for supercapacitors. Nano Energy. 2013; 2: 1303–13.

[179] Hasan M, Jamal M, Razeeb KM. Coaxial NiO/Ni nanowire arrays for high performance pseudocapacitor applications. Electrochimica Acta. 2012; 60: 193–200.

[180] Kim JY, Lee SH, Yan YF, Oh J, Zhu K. Controlled synthesis of aligned Ni-NiO core-shell nanowire arrays on glass substrates as a new supercapacitor electrode. RSC Advances. 2012; 2: 8281–5.

[181] Singh AK, Sarkar D, Khan GG, Mandal K. Unique hydrogenated Ni/NiO core/shell 1D nanoheterostructures with superior electrochemical performance as supercapacitors. Journal of Materials Chemistry A. 2013; 1: 12759–67.

[182] Kim JH, Kang SH, Zhu K, Kim JY, Neale NR, Frank AJ. Ni-NiO core-shell inverse opal electrodes for supercapacitors. Chemical Communications. 2011; 47: 5214–6.

[183] Wu MS, Huang CY, Lin KH. Electrophoretic deposition of nickel oxide electrode for high-rate electrochemical capacitors. Journal of Power Sources. 2009; 186: 557–64.

[184] Wu MS, Hsieh HH. Nickel oxide/hydroxide nanoplatelets synthesized by chemical precipitation for electrochemical capacitors. Electrochimica Acta. 2008; 53: 3427–35.

[185] Khairy M, El-Safty SA. Mesoporous NiO nanoarchitectures for electrochemical energy storage: influence of size, porosity, and morphology. RSC Advances. 2013; 3: 23801–9.

[186] Vijayakumar S, Nagamuthu S, Muralidharan G. Supercapacitor studies on NiO nanoflakes synthesized through a microwave route. ACS Applied Materials & Interfaces. 2013; 5: 2188–96.

[187] Xu LP, Ding YS, Chen CH, Zhao LL, Rimkus C, Joesten R et al. 3D flowerlike α-nickel hydroxide with enhanced electrochemical activity synthesized by microwave-assisted hydrothermal method. Chemistry of Materials. 2008; 20: 308–16.

[188] Zhang YQ, Xia XH, Tu JP, Mai YJ, Shi SJ, Wang XL et al. Self-assembled synthesis of hierarchically porous NiO film and its application for electrochemical capacitors. Journal of Power Sources. 2012; 199: 413–7.

[189] Cui B, Lin H, Li JB, Li X, Yang J, Tao J. Core-ring structured $NiCo_2O_4$ nanoplatelets: Synthesis, characterization, and electrocatalytic applications. Advanced Functional Materials. 2008; 18: 1440–7.

[190] Alcantara R, Jaraba M, Lavela P, Tirado JL. $NiCo_2O_4$ spinel: First report on a transition metal oxide for the negative electrode of sodium-ion batteries. Chemistry of Materials. 2002; 14: 2847–8.

[191] Liu ZQ, Xiao K, Xu QZ, Li N, Su YZ, Wang HJ et al. Fabrication of hierarchical flower-like super-structures consisting of porous $NiCo_2O_4$ nanosheets and their electrochemical and magnetic properties. RSC Advances. 2013; 3: 4372–80.

[192] Verma S, Joshi HM, Jagadale T, Chawla A, Chandra R, Ogale S. Nearly monodispersed multifunctional $NiCo_2O_4$ spinel nanoparticles: Magnetism, infrared transparency, and radiofrequency absorption. Journal of Physical Chemistry C. 2008; 112: 15106–12.

[193] Cabo M, Pellicer E, Rossinyol E, Estrader M, Lopez-Ortega A, Nogues J et al. Synthesis of compositionally graded nanocast $NiO/NiCo_2O_4/Co_3O_4$ mesoporous composites with tunable magnetic properties. Journal of Materials Chemistry. 2010; 20: 7021–8.

[194] Ding R, Qi L, Jia MJ, Wang HY. Facile synthesis of mesoporous spinel $NiCo_2O_4$ nanostructures as highly efficient electrocatalysts for urea electro-oxidation. Nanoscale. 2014; 6: 1369–76.

[195] Zhang GQ, Xia BY, Wang X, Lou XW. Strongly coupled $NiCo_2O_4$-rGO hybrid nanosheets as a methanol-tolerant electrocatalyst for the oxygen reduction reaction. Advanced Materials. 2014; 26: 2408–12.

[196] Prasad R, Sony, Singh P. Low temperature complete combustion of a lean mixture of LPG emissions over cobaltite catalysts. Catalysis Science & Technology. 2013; 3: 3223–33.

[197] Ding R, Qi L, Jia MJ, Wang HY. Sodium dodecyl sulfate-assisted hydrothermal synthesis of mesoporous nickel cobaltite nanoparticles with enhanced catalytic activity for methanol electrooxidation. Journal of Power Sources. 2014; 251: 287–95.

[198] Huang G, Zhang LL, Zhang FF, Wang LM. Metal-organic framework derived $Fe_2O_3@NiCo_2O_4$ porous nanocages as anode materials for Li-ion batteries. Nanoscale. 2014; 6: 5509–15.

[199] Li Y, Zou LL, Li J, Guo K, Dong XW, Li XW et al. Synthesis of ordered mesoporous $NiCo_2O_4$ via hard template and its application as bifunctional electrocatalyst for $Li-O_2$ batteries. Electrochimica Acta. 2014; 129: 14–20.

[200] Chen YJ, Zhuo M, Deng JW, Xu Z, Li QH, Wang TH. Reduced graphene oxide networks as an effective buffer matrix to improve the electrode performance of porous $NiCo_2O_4$ nanoplates for lithium-ion batteries. Journal of Materials Chemistry A. 2014; 2: 4449–56.

[201] Chen YJ, Zhu J, Qu BH, Lu BA, Xu Z. Graphene improving lithium-ion battery performance by construction of $NiCo_2O_4$/graphene hybrid nanosheet arrays. Nano Energy. 2014; 3: 88–94.

[202] Sun B, Zhang JQ, Munroe P, Ahn HJ, Wang G. Hierarchical $NiCo_2O_4$ nanorods as an efficient cathode catalyst for rechargeable non-aqueous $Li-O_2$ batteries. Electrochemistry Communications. 2013; 31: 88–91.

[203] Suchow L. Detailed, simple crystal-field consideration of normal spinel structure of Co_3O_4. Journal of Chemical Education. 1976; 53: 560–.

[204] Xiao XL, Liu XF, Zhao H, Chen DF, Liu FZ, Xiang JH et al. Facile shape control of Co_3O_4 and the effect of the crystal plane on electrochemical performance. Advanced Materials. 2012; 24: 5762–6.

[205] Marco JF, Gancedo JR, Gracia M, Gautier JL, Rios EI, Palmer HM et al. Cation distribution and magnetic structure of the ferrimagnetic spinel $NiCo_2O_4$. Journal of Materials Chemistry. 2001; 11: 3087–93.

[206] Hu LF, Wu LM, Liao MY, Hu XH, Fang XS. Electrical transport properties of large, individual $NiCo_2O_4$ nanoplates. Advanced Functional Materials. 2012; 22: 998–1004.

[207] Li YG, Hasin P, Wu YY. $Ni_xCo_{3-x}O_4$ nanowire arrays for electrocatalytic oxygen evolution. Advanced Materials. 2010; 22: 1926–9.

[208] Rolison DR. Catalytic nanoarchitectures—The importance of nothing and the unimportance of periodicity. Science. 2003; 299: 1698–701.

[209] Long JW, Dunn B, Rolison DR, White HS. Three-dimensional battery architectures. Chemical Reviews. 2004; 104: 4463–92.

[210] Zhang QF, Uchaker E, Candelaria SL, Cao GZ. Nanomaterials for energy conversion and storage. Chemical Society Reviews. 2013; 42: 3127–71.

[211] Walcarius A. Mesoporous materials and electrochemistry. Chemical Society Reviews. 2013; 42: 4098–140.

[212] Orilall MC, Wiesner U. Block copolymer based composition and morphology control in nanostructured hybrid materials for energy conversion and storage: solar cells, batteries, and fuel cells. Chemical Society Reviews. 2011; 40: 520–35.

[213] Gao MR, Xu YF, Jiang J, Yu SH. Nanostructured metal chalcogenides: Synthesis, modification, and applications in energy conversion and storage devices. Chemical Society Reviews. 2013; 42: 2986–3017.

[214] Wang X, Liu WS, Lu XH, Lee PS. Dodecyl sulfate-induced fast faradic process in nickel cobalt oxide-reduced graphite oxide composite material and its application for asymmetric supercapacitor device. Journal of Materials Chemistry. 2012; 22: 23114–9.

[215] Xiao JW, Yang SH. Sequential crystallization of sea urchin-like bimetallic (Ni, Co) carbonate hydroxide and its morphology conserved conversion to porous $NiCo_2O_4$ spinel for pseudocapacitors. RSC Advances. 2011; 1: 588–95.

[216] Lu Q, Chen YP, Li WF, Chen JGG, Xiao JQ, Jiao F. Ordered mesoporous nickel cobaltite spinel with ultra-high supercapacitance. Journal of Materials Chemistry A. 2013; 1: 2331–6.

[217] Zhu W, Lu ZY, Zhang GX, Lei XD, Chang Z, Liu JF et al. Hierarchical $Ni_{0.25}Co_{0.75}(OH)_2$ nanoarrays for a high-performance supercapacitor electrode prepared by an *in situ* conversion process. Journal of Materials Chemistry A. 2013; 1: 8327–31.

[218] Yu L, Wu HB, Wu T, Yuan CZ. Morphology-controlled fabrication of hierarchical mesoporous $NiCo_2O_4$ micro-/nanostructures and their intriguing application in electrochemical capacitors. RSC Advances. 2013; 3: 23709–14.

[219] Zhou WW, Kong DZ, Jia XT, Ding CY, Cheng CW, Wen GW. $NiCo_2O_4$ nanosheet supported hierarchical core-shell arrays for high-performance supercapacitors. Journal of Materials Chemistry A. 2014; 2: 6310–5.

[220] Sun YJ, Xiao XP, Ni PJ, Shi Y, Dai HC, Hu JT et al. DNA-templated synthesis of nickel cobaltite oxide nanoflake for high-performance electrochemical capacitors. Electrochimica Acta. 2014; 121: 270–7.

[221] An CH, Wang YJ, Huang YA, Xu YA, Xu CC, Jiao LF et al. Novel three-dimensional $NiCo_2O_4$ hierarchitectures: solvothermal synthesis and electrochemical properties. CrystEngComm. 2014; 16: 385–92.

[222] Kuang M, Zhang W, Guo XL, Yu L, Zhang YX. Template-free and large-scale synthesis of hierarchical dandelion-like $NiCo_2O_4$ microspheres for high-performance supercapacitors. Ceramics International. 2014; 40: 10005–11.

[223] Ding R, Qi L, Jia MJ, Wang HY. Hydrothermal and soft-templating synthesis of mesoporous $NiCo_2O_4$ nanomaterials for high-performance electrochemical capacitors. Journal of Applied Electrochemistry. 2013; 43: 903–10.

[224] Cai D, Liu B, Wang D, Wang L, Liu Y, Li H et al. Construction of unique $NiCo_2O_4$ nanowire@$CoMoO_4$ nanoplate core/shell arrays on Ni foam for high areal capacitance supercapacitors. Journal of Materials Chemistry A. 2014; 2: 4954–60.

[225] Lei Y, Li J, Wang Y, Gu L, Chang Y, Yuan H et al. Rapid microwave-assisted green synthesis of 3D hierarchical flower-shaped $NiCo_2O_4$ microsphere for high-performance supercapacitor. Acs Applied Materials & Interfaces. 2014; 6: 1773–80.

[226] Qian ZY, Peng T, Wang J, Qu LT. Construction of hybrid supercapacitor-batteries with dual-scale shelled architecture. ChemSusChem. 2014; 7: 1881–7.

[227] Ding R, Qi L, Wang HY. Scalable electrodeposition of cost-effective microsized $NiCo_2O_4$ electrode materials for practical applications in electrochemical capacitors. ECS Electrochemistry Letters. 2012; 1: A43–A6.

[228] Wu YQ, Chen XY, Ji PT, Zhou QQ. Sol-gel approach for controllable synthesis and electrochemical properties of $NiCo_2O_4$ crystals as electrode materials for application in supercapacitors. Electrochimica Acta. 2011; 56: 7517–22.

[229] Kim H, Kim Y, Joo JB, Ko JW, Yi J. Preparation of coral-like porous gold for metal ion detection. Microporous and Mesoporous Materials. 2009; 122: 283–7.

[230] Hu CC, Hsu CT, Chang KH, Hsu HY. Microwave-assisted hydrothermal annealing of binary Ni-Co oxy-hydroxides for asymmetric supercapacitors. Journal of Power Sources. 2013; 238: 180–9.

420 *Nanomaterials for Supercapacitors*

[231] Wei T-Y, Chen C-H, Chien H-C, Lu S-Y, Hu C-C. A cost-effective supercapacitor material of ultrahigh specific capacitances: Spinel nickel cobaltite aerogels from an epoxide-driven sol-gel process. Advanced Materials. 2010; 22: 347–+.

[232] Futaba DN, Hata K, Yamada T, Hiraoka T, Hayamizu Y, Kakudate Y et al. Shape-engineerable and highly densely packed single-walled carbon nanotubes and their application as super-capacitor electrodes. Nature Materials. 2006; 5: 987–94.

[233] Zhou HS, Li DL, Hibino M, Honma I. A self-ordered, crystalline-glass, mesoporous nanocomposite for use as a lithium-based storage device with both high power and high energy densities. Angewandte Chemie-International Edition. 2005; 44: 797–802.

[234] Pu J, Wang J, Jin X, Cui F, Sheng E, Wang Z. Porous hexagonal $NiCo_2O_4$ nanoplates as electrode materials for supercapacitors. Electrochimica Acta. 2013; 106: 226–34.

[235] Wang QF, Liu B, Wang XF, Ran SH, Wang LM, Chen D et al. Morphology evolution of urchin-like $NiCo_2O_4$ nanostructures and their applications as psuedocapacitors and photoelectrochemical cells. Journal of Materials Chemistry. 2012; 22: 21647–53.

[236] Hsu CT, Hu CC. Synthesis and characterization of mesoporous spinel $NiCo_2O_4$ using surfactant-assembled dispersion for asymmetric supercapacitors. Journal of Power Sources. 2013; 242: 662–71.

[237] Zou RJ, Xu KB, Wang T, He GJ, Liu Q, Liu XJ et al. Chain-like $NiCo_2O_4$ nanowires with different exposed reactive planes for high-performance supercapacitors. Journal of Materials Chemistry A. 2013; 1: 8560–6.

[238] Yuan CZ, Li JY, Hou LR, Lin JD, Zhang XG, Xiong SL. Polymer-assisted synthesis of a 3D hierarchical porous network-like spinel $NiCo_2O_4$ framework towards high-performance electrochemical capacitors. Journal of Materials Chemistry A. 2013; 1: 11145–51.

[239] Yuan CZ, Li JY, Hou LR, Lin JD, Pang G, Zhang LH et al. Template-engaged synthesis of uniform mesoporous hollow $NiCo_2O_4$ sub-microspheres towards high-performance electrochemical capacitors. RSC Advances. 2013; 3: 18573–8.

[240] Wu HB, Pang H, Lou XW. Facile synthesis of mesoporous Ni0.3Co2.7O4 hierarchical structures for high-performance supercapacitors. Energy & Environmental Science. 2013; 6: 3619–26.

[241] Chang J, Sun J, Xu C, Xu H, Gao L. Template-free approach to synthesize hierarchical porous nickel cobalt oxides for supercapacitors. Nanoscale. 2012; 4: 6786–91.

[242] Lu XH, Huang X, Xie SL, Zhai T, Wang CS, Zhang P et al. Controllable synthesis of porous nickel-cobalt oxide nanosheets for supercapacitors. Journal of Materials Chemistry. 2012; 22: 13357–64.

[243] Padmanathan N, Selladurai S. Solvothermal synthesis of mesoporous $NiCo_2O_4$ spinel oxide nanostructure for high-performance electrochemical capacitor electrode. Ionics. 2013; 19: 1535–44.

[244] Salunkhe RR, Jang K, Yu H, Yu S, Ganesh T, Han SH et al. Chemical synthesis and electrochemical analysis of nickel cobaltite nanostructures for supercapacitor applications. Journal of Alloys and Compounds. 2011; 509: 6677–82.

[245] Hsu HY, Chang KH, Salunkhe RR, Hsu CT, Hu CC. Synthesis and characterization of mesoporous Ni-Co oxy-hydroxides for pseudocapacitor application. Electrochimica Acta. 2013; 94: 104–12.

[246] Shakir I, Sarfraz M, Rana UA, Nadeem M, Al-Shaikha MA. Synthesis of hierarchical porous spinel nickel cobaltite nanoflakes for high performance electrochemical energy storage supercapacitors. Rsc Advances. 2013; 3: 21386–9.

[247] Padmanathan N, Selladurai S. Controlled growth of spinel $NiCo_2O_4$ nanostructures on carbon cloth as a superior electrode for supercapacitors. Rsc Advances. 2014; 4: 8341–9.

[248] Deng FZ, Yu L, Sun M, Lin T, Cheng G, Lan B et al. Controllable growth of hierarchical $NiCo_2O_4$ nanowires and nanosheets on carbon fiber paper and their morphology-dependent pseudocapacitive performances. Electrochimica Acta. 2014; 133: 382–90.

[249] Wei Y, Chen S, Su D, Sun B, Zhu J, Wang G. 3D mesoporous hybrid $NiCo_2O_4$@graphene nanoarchitectures as electrode materials for supercapacitors with enhanced performances. Journal of Materials Chemistry A. 2014; 2: 8103–9.

[250] Huang L, Chen DC, Ding Y, Wang ZL, Zeng ZZ, Liu ML. Hybrid composite $Ni(OH)_2$@$NiCo_2O_4$ grown on carbon fiber paper for high-performance supercapacitors. ACS Applied Materials & Interfaces. 2013; 5: 11159–62.

[251] Li G, Li W, Xu K, Zou R, Chen Z, Hu J. Sponge-like $NiCo_2O_4$/MnO_2 ultrathin nanoflakes for supercapacitor with high-rate performance and ultra-long cycle life. Journal of Materials Chemistry A. 2014; 2: 7738–41.

[252] Luo Y, Zhang H, Guo D, Ma J, Li Q, Chen L et al. Porous $NiCo_2O_4$-reduced graphene oxide (rGO) composite with superior capacitance retention for supercapacitors. Electrochimica Acta. 2014; 132: 332–7.

[253] Zhang GQ, Wu HB, Hoster HE, Chan-Park MB, Lou XW. Single-crystalline NiCo2O4 nanoneedle arrays grown on conductive substrates as binder-free electrodes for high-performance supercapacitors. Energy & Environmental Science. 2012; 5: 9453–6.

[254] Zhang GQ, Lou XW. General solution growth of mesoporous $NiCo_2O_4$ nanosheets on various conductive substrates as high-performance electrodes for supercapacitors. Advanced Materials. 2013; 25: 976–9.

[255] Yuan CZ, Li JY, Hou LR, Yang L, Shen LF, Zhang XG. Facile template-free synthesis of ultralayered mesoporous nickel cobaltite nanowires towards high-performance electrochemical capacitors. Journal of Materials Chemistry. 2012; 22: 16084–90.

[256] Yu L, Zhang GQ, Yuan CZ, Lou XW. Hierarchical $NiCo_2O_4$@MnO_2 core-shell heterostructured nanowire arrays on Ni foam as high-performance supercapacitor electrodes. Chemical Communications. 2013; 49: 137–9.

[257] Liu XY, Zhang YQ, Xia XH, Shi SJ, Lu Y, Wang XL et al. Self-assembled porous $NiCo_2O_4$ hetero-structure array for electrochemical capacitor. Journal of Power Sources. 2013; 239: 157–63.

[258] Wang HW, Wang XF. Growing nickel cobaltite nanowires and nanosheets on carbon cloth with different pseudocapacitive performance. ACS Applied Materials & Interfaces. 2013; 5: 6255–60.

[259] Deng FZ, Yu L, Cheng G, Lin T, Sun M, Ye F et al. Synthesis of ultrathin mesoporous $NiCo_2O_4$ nanosheets on carbon fiber paper as integrated high-performance electrodes for supercapacitors. Journal of Power Sources. 2014; 251: 202–7.

[260] Liu XY, Shi SJ, Xiong QQ, Li L, Zhang YJ, Tang H et al. Hierarchical $NiCo_2O_4$@$NiCo_2O_4$ core/shell nanoflake arrays as high-performance supercapacitor materials. ACS Applied Materials & Interfaces. 2013; 5: 8790–5.

[261] Huang L, Chen DC, Ding Y, Feng S, Wang ZL, Liu ML. Nickel-cobalt hydroxide nanosheets coated on $NiCo_2O_4$ nanowires grown on carbon fiber paper for high-performance pseudocapacitors. Nano Letters. 2013; 13: 3135–9.

[262] Shang CQ, Dong SM, Wang S, Xiao DD, Han PX, Wang XG et al. Coaxial $Ni_xCo_{2x}(OH)_{6x}$/TiN nanotube arrays as supercapacitor electrodes. ACS Nano. 2013; 7: 5430–6.

[263] Yuan C, Li J, Hou L, Zhang X, Shen L, Lou XW. Ultrathin mesoporous $NiCo_2O_4$ nanosheets supported on Ni foam as advanced electrodes for supercapacitors. Advanced Functional Materials. 2012; 22: 4592–7.

[264 Du J, Zhou G, Zhang HM, Cheng C, Ma JM, Wei WF et al. Ultrathin porous $NiCo_2O_4$ nanosheet arrays on flexible carbon fabric for high-performance supercapacitors. ACS Applied Materials & Interfaces. 2013; 5: 7405–9.

6

Flexible Supercapacitors

Ramaraju Bendi, Vipin Kumar and *Pooi See Lee**

Brief Introduction

As mentioned before, the supercapacitor is a high power density energy storage device with a particularly high reversibility and long life cycle [1–4]. With a great potential in the consumer electronics and high power applications, such as in communications devices, hybrid electrical vehicles (start-stop application), aviation and power industries, the development of supercapacitor devices has gained significant interest in recent years, due to their capability to produce large power and energy density, as compared with the conventional capacitors [5, 6]. Based on the type of electrode materials and working principles, supercapacitors can be categorized into three main classes, including Electrochemical Double-Layer Capacitors (EDLCs), pseudocapacitors and hybrid capacitors, as demonstrated in Fig. 6.1. In the EDLCs, the energy is stored via reversible adsorption and desorption of ions at the electrode/electrolyte interfaces, while there is no net charge transfer across the interfaces. Due to the difference in electrochemical potential between the electrodes, the open circuit potential of the cell is varied, which responds to the passage of the charges in the external circuit. The energy stored in this process is purely electrical, in which there is no contribution from the electrode chemical reactions or faradaic reactions. Carbon based supercapacitors are typical EDLCs [2, 7–12]. In comparison, pseudocapacitors are based on Faradaic reactions, i.e., oxidation-reduction reactions [2, 13, 14], with only a slight contribution of EDLCs by < 5%. Almost all the transition metal oxides, nitrides, carbides and conducting polymers can be used as electrode materials of pseudocapacitors. The energy storage mechanism of hybrid capacitors is the combination of the Faradaic and non-Faradaic reaction, such as the supercapacitors based on porous carbons and metal oxides.

A schematic illustration of typical conventional supercapacitor is illustrated in Fig. 6.2, consisting of two electrodes, an electrolyte and a separator. In terms of storage mechanism, pure carbon-based supercapacitors are dependent on the Specific Surface Area (SSA) of the electrode materials and typically follow EDLCs. EDLCs have high cycling stabilities, due to the highly reversible charging and discharging processes. In this case, charges are accumulated at the electrode surfaces, without the presence of chemical reactions. Positive and negative ions from the electrolyte solution across the separator to enter the active sites/pores of the electrode materials. Because of the applied voltage, the opposite charges will gather between the electrolyte and the electrode interface. Because the recombination of the ions is precluded

---

School of Materials Science and Engineering, Nanyang Technological University, 50 Nanyang Avenue, Singapore 639798.
* Corresponding author: pslee@ntu.edu.sg

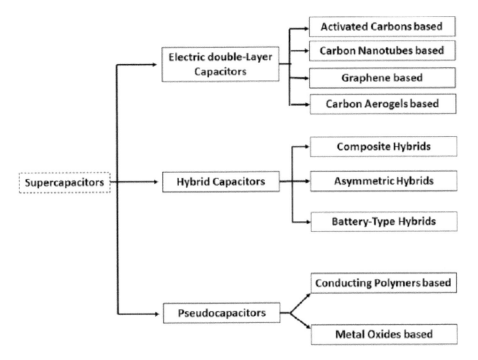

Fig. 6.1. Classification of supercapacitors.

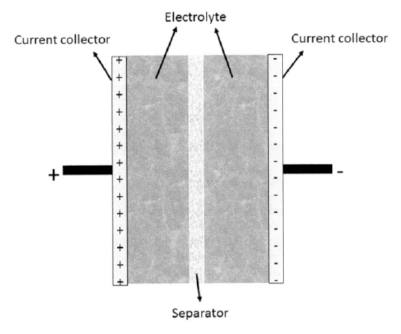

Fig. 6.2. Schematic diagram of conventional SC.

424 *Nanomaterials for Supercapacitors*

by careful engineering of the electrodes, a double-layer of charge is produced. Therefore, EDLCs store energy with the electrochemical double-layer of charge. Previously, high surface area, high stability and low cost porous carbons were widely used as the electrode material for an electrochemical capacitor. More recently, various carbon materials, such as Activated Carbons (ACs), carbon aerogels, carbon nanotubes and graphene, have been employed as electrodes of EDLCs. Due to the fact that not all the specific surface area (SSA) is in contact with electrolyte during the electrochemical reaction, the gravimetric capacitance of carbon based materials does not always increase with increasing SSA, thus resulting a low gravimetric energy density [15]. As a result, pseudocapacitors have been developed, with Conductive Polymers (CP), such as polyaniline (PANI) or polypyrrole (PPy) and Metal Oxides (MOs) as the electrode materials [16]. Compared with EDLCs, pseudocapacitors exhibited higher specific capacitance and energy density, because of the Faradaic processes. However, the cycling stability of pseudocapacitors is usually lower than that of EDLCs.

To overcome the disadvantages of the two types of supercapacitors, hybrid supercapacitors are proposed [2]. According to electrode configuration, there are three types of hybrid supercapacitors, i.e., asymmetric, composite and battery-type hybrid supercapacitors. Asymmetric hybrid supercapacitors are comprised of EDLC and pseudocapacitor electrodes [17–19], which have higher electrochemical cycling stability than the individual pseudocapacitors. Also, they can accomplish higher energy density and power density than EDLCs. Composite hybrids are made with composite materials, by integrating carbon materials with pseudocapacitive materials [20]. In this case, the porous carbon materials can act as a backbone with high surface area to improve the contact between the conducting polymer materials or metal oxides and the electrolyte. At the same time, the conducting polymer materials or metal oxides can further enhance the capacitance due to the faradaic reactions. Battery-type hybrids are completely different from above the two types of hybrid supercapacitors, because the electrode is based on battery [21–23]. Batteries have higher energy densities than supercapacitors, whereas supercapacitors possess much higher power density. Therefore, battery-type hybrid supercapacitors have a potential to fill the gap in the Ragone plot, in terms of energy density and power density.

Flexible Supercapacitors

For some special applications, thinner, cheaper and lighter supercapacitors have to be used. However, the currently developed supercapacitors are usually too heavy and bulky, due to the utilization of various unwanted components, like binders, conducting additives and current collectors. Therefore, it is urgent to design eco-friendly, lightweight, low-cost supercapacitors as flexible energy storage devices [24–26]. Various flexible supercapacitors have been reported in the open literature [25–30]. Recent progress in the development of flexible supercapacitors will be reviewed comprehensively here.

Carbon/graphene based flexible supercapacitors

Flexible supercapacitors could be used as flexible/wearable power sources for the future generation all-in-one portable and wearable electronics. Flexible electrodes are primarily based on carbon materials [25]. Carbon materials are mostly used as electrode materials in flexible EDLC-SCs, due to their large surface areas and robust mechanical properties. As stated in previous chapters, carbon materials have a variety of geometric shapes, such as zero-Dimensional (0D) fullerene or carbon particles, one Dimensional (1D) carbon nanotubes (CNTs) or Carbon Fibers (CFs) and two-Dimensional (2D) graphene or graphite sheets. Among them, 1D and 2D carbon materials have been widely used in flexible electrodes, because they can readily form highly conductive and flexible carbon networks, with outstanding electrochemical performances, as described in Fig. 6.3 [31].

Carbon fiber (CF) based flexible supercapacitors

Carbon networks, consisting of carbon fabric, cloth, film, coating, paper or textiles, are important in the development of flexible electrodes. These carbon architectures are generally made of single 1D and/or 2D carbon components, through aggregation via hydrogen bonds or van der Waals forces [32–34]. Various methods have been used to produce carbon networks from 1D or 2D carbon particles, such as weaving, Chemical Vapor Deposition (CVD), printing, filtration, evaporation and dipping-drying, schematically shown in Fig. 6.3 [31]. Activated Carbon (AC), with large specific surface area, relatively low cost, chemical stability and wide availability, is the most promising electrode material of EDLCs [35]. The best-known network used for flexible electrodes is Carbon Fabric (CF), which can be fabricated by carbon fibers through a commercial weaving method (Fig. 6.3, (F1)). Carbon fabrics manufactured with a loom commonly have three main weave styles, i.e., plain, stain and twill weaves, as illustrated in Fig. 6.4 [31].

Carbon Fiber Papers (CFPs) have been a promising candidate to be used as current collector or conductive substrate, due to their special properties, including high flexibility, conductivity, porosity and corrosive resistance. However, CFPs constructed with sp^2 carbon (90–95 wt%) and adhesive agent of PTFE (5–10 wt%) can only be used as a current collector, because they cannot store charges, due to the hydrophobic nature [36, 37]. Surface modification and functionalization of CFPs have been conducted

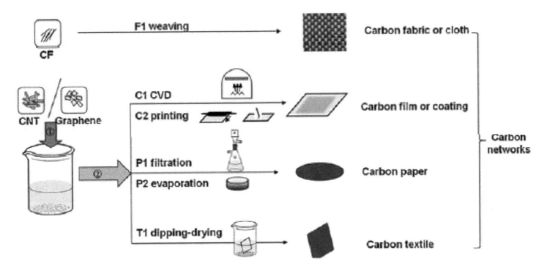

Fig. 6.3. Fabrication of carbon networks with 1D Carbon Fibers (CF), 1D carbon nanotubes (CNTs) and 2D graphene with a variety of methods. With carbon fibers as the raw material, carbon fabric or cloth is fabricated by weaving (F1). With CNTs or graphene as the raw material, carbon films or coatings are fabricated by using CVD (C1) or printing technology (C2), carbon paper is prepared by using a filtration method (P1) or an evaporation process (P2) and carbon textile is prepared by using dipping-drying method (T1). Reproduced with permission from [31], Copyright © 2013, Elsevier.

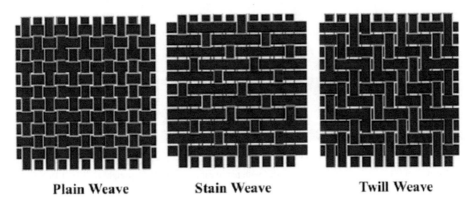

Fig. 6.4. Three weave styles of carbon fabrics. Reproduced with permission from [31], Copyright © 2013, Elsevier.

through various acid treatments. Interestingly, it is found that the functionalized CFPs (f-CFPs), with various oxygenated functional groups, showed pseudocapacitance, related to a surface redox reaction at the solid-liquid interface [38]. Representative FE-SEM images of CFP and f-CFP are shown in Fig. 6.5(a, b). In this case, the carbon fibers of both bare CFP and f-CFP had a diameter of 8 μm. Generally, they were coated with PTFE (polytetrafluoroethylene), acting as an adhesive material in the preparation process of CFPs. Static contact angle measurements with a water droplet on the surfaces of CFP and f-CFP for the wettability behavior are shown as inset images in Fig. 6.5(a, b). The contact angles for the CFP and f-CFP were 130° and 65°, respectively, which indicated that the hydrophobic nature of the bare CFP was modified to hydrophilic nature through the oxygenated functionalization. The optimal areal and volumetric capacitances of the f-CFP, produced with the acid treatment at 3:1 v/v concentration of H_2SO_4:HNO at 60°C for 1 hour, were 115.1 mF·cm^{-2} (ca.16.8 F·g^{-1}) and 4.6 F·cm^{-3} at an areal current of 1.15 mA cm^{-2}, as observed in Fig. 6.5(c).

In addition, fibers can be mixed to obtain hybrid fabrics. Woven fabric exhibited promising mechanical strength and stiffness, as well as excellent flexibility. However, carbonaceous fibers cannot meet the requirements to achieve high energy density and specific capacitance. Therefore, they are practically incorporated with pseudocapacitive materials, i.e., metal oxides, like RuO_2 [39], NiO [40] and MnO_2 [41, 42] or conducting polymers [43–46]. Excellent capacitive energy storage properties of pseudocapacitive components have been developed with CFs that have high electrical conductivity and mechanical flexibility [47, 48]. Growth of pseudocapacitive materials on well conductive carbon substrates can not only facilitate the diffusion of electrolyte ions but also can improve the transport of electrons, thus enhancing the electrochemical properties [49–54]. Metal oxide or conducting polymers on CFs core-shell fibers could be fabricated by using electrodeposition processes, as schematically demonstrated in Fig. 6.6.

The surface modified/unmodified CFs were used as working electrode, platinum and SCE were used as counter and reference electrodes, respectively. The coating precursor was used as the electrodeposition solution. Then, metal oxide or conducting polymers nanostructures were electrodeposited on the surface of the CFs, resulting in a core shell configuration [54]. For instance, Zhang et al. deposited MnO_2 on CFs and they found that both the volumetric capacitance and specific capacitance were higher as the MnO_2 layers were sufficiently thin, because of the shorter diffusion path for the electrolyte ions [54]. SEM images of pristine CFs indicated that there were notches with a diameter of about 7 μm on their surface, as seen in Fig. 6.7(a, b). After electrodeposition, MnO_2 nanostructures were uniformly distributed on the surface of CFs, without the presence of serious aggregations, as illustrated in Fig. 6.7(c, d). Magnified SEM images further manifested that the MnO_2 appeared as nanosheets that were uniformly and perpendicularly grown on the surface of the CFs. Therefore, a porous structure was developed, as seen in Fig. 6.7(e). The single MnO_2/CFs fiber electrode exhibited a maximum specific volumetric capacitance of 58.7 F cm^{-3} at 0.1 A·g^{-1}, which was nearly 28 times larger than that of the bare CFs (2.1 F·cm^{-3}). Figure 6.7(f) shows that volumetric capacitance of the MnO_2/CFs electrodes was increased with the deposition time of MnO_2.

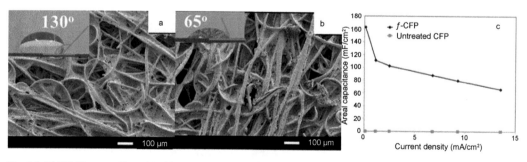

Fig. 6.5. (a) SEM images of bare CFP, (b) SEM images of f-CFP. (c) Areal capacitance of bare CFP and f-CFP at the current densities of 0.12–13.5 mA·cm^{-2}. Reproduced with permission from [38], Copyright © 2015, Elsevier.

Flexible Supercapacitors 427

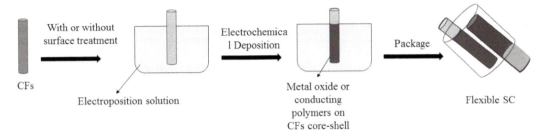

Fig. 6.6. Schematic illustration for the fabrication of metal oxide or conducting polymer/CFs-based flexible solid-state fiber-like supercapacitors.

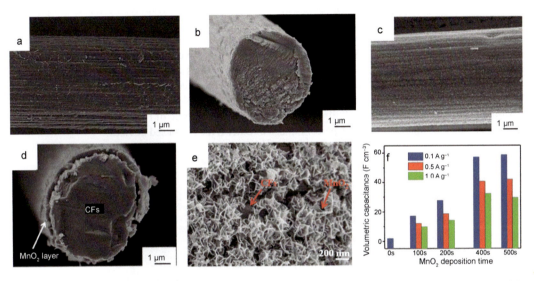

Fig. 6.7. (a) SEM images of pristine carbon fibers. (b) Cross-sectional image of the pristine carbon fibers. (c) SEM images of the MnO$_2$/CFs. (d) Cross-sectional image of the MnO$_2$/CFs. (e) High magnification SEM images of the MnO$_2$/CFs. (f) Calculated specific volumetric capacitances of the MnO$_2$/CFs samples at different current densities. Reproduced with permission from [54], Copyright © 2016, Elsevier.

CNTs based flexible supercapacitors

Carbon nanotubes (CNTs) have attracted large interest in recent years, due to their unique properties, including low mass density with high surface area, special internal structure, high electrical conductivity and remarkable chemical stability [55–60]. CNTs are suitable for flexible electrode, owing to their high thermal and electronic conductivity, excellent mechanical properties, thermal and chemical stability, and large surface area [61–63]. CNTs, especially SWCNTs, with an open and accessible mesopores network, have intrinsically elaborate properties as an electrode material. The mesopores are interconnected to one and another, resulting a continuous charge distribution. As a result, almost all the surface area can be utilized for charge storage. Also, the mesoporous network is more feasible for the electrolyte ions to access, thus leading to a low Equivalent Series Resistance (ESR). In order to further reduce the ESR of CNT materials, several new fabrication techniques have been developed for the preparation process. For example, CNTs can be developed directly onto the current collectors or subjected to thermal treatment. Chen et al. [64]

developed a simplified process to fabricate electrode, in which CNTs were directly grown onto graphite foil. As a consequence, ESR was effectively reduced and the electrochemical performance was significantly enhanced. A maximum specific capacitance of 115.7 F·g^{-1} was achieved in aqueous electrolyte of 1.0 M H$_2$SO$_4$ at a potential scan of 100 mV·s^{-1}.

Generally, the specific surface area of CNTs is lower than that of Activated Carbon (AC). Therefore, many efforts have been made to increase the surface area, so as to enhance the capacitance of CNTs based supercapacitors. To do this, CNTs were activated by with activating agents like KOH [65]. It was showed that KOH activated CNTs exhibited enhanced specific capacitance, which was linearly increased with increasing specific surface area, because the KOH activated CNT possessed pores that were sufficiently large for ion transfer. However, rate performance of the KOH activated CNTs was lower than that of raw CNTs. In this respect, it is necessary to further optimize the KOH activation process to achieve high rate performance.

An approach has been reported to develop all-solid-state flexible supercapacitors with CNTs, which were deposited on office paper, while an ionic-liquid-based silica gel was employed as an electrolyte [66]. The all-solid-state flexible supercapacitors demonstrated excellent electrochemical performance and high mechanical stability. The specific capacitance was 135 F·g^{-1} at a current density of 227 A·g^{-1}. The maximum energy and power density were 41 Wh·kg^{-1} and 164 kW·kg^{-1}, respectively. Kaempgen et al. [41] developed printable thin film flexible supercapacitors on a plastic substrate. In this study, the electrodes were fabricated with sprayed networks of SWCNTs, which were also used as charge collectors. As shown in Fig. 6.8, the SWCNTs were arranged as an entangled random network on a flexible polyethylene terephthalate (PET) substrate. This type SWCNT network film printed on a flexible substrate of PET showed high power and energy densities, comparable with the performances of other SWCNT-based devices.

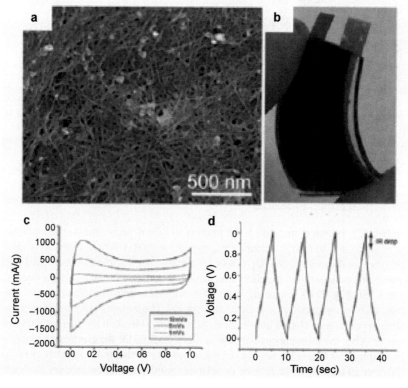

Fig. 6.8. (a) SEM image of the SWCNT networks. (b) Thin film supercapacitor based on sprayed SWCNT films on PET as electrodes and a gel electrolyte as both electrolyte and separator. Reproduced with permission from [41], Copyright © 2009, American Chemical Society.

Apart from PET, textiles such as polyester or cotton are also impressive substrates for the deposition of carbon based materials like CNTs. Cui et al., used extremely simple 'dipping and drying' process to deposit SWCNT ink on conductive textile substrates [33]. The electrical conductivity is depended on the thickness of these conductive textiles films. By decreasing the thickness of the cotton sheets from $\sim$ 2 mm to $\sim$ 80 μm under mechanical pressing it was observed that the conductivity was increased from $\sim$ 5 to $\sim$ 125 $S \cdot s^{-1}$. The conductive textiles have strong binding of SWCNTs with outstanding mechanical properties, flexibility, and stretchability, making them an ideal candidate for flexible and stretchable energy storage and wearable electronics applications. Conductive textiles based supercapacitors shows a large areal capacitance (0.48 $F \cdot cm^{-2}$) and high specific energy (20 $Wh \cdot kg^{-1}$) at a specific power of 10 $kW \cdot kg^{-1}$. A thinner flexible solid-state SCs developed with a thickness of 1.3 mm by using poly (vinyl alcohol) (PVA)/phosphoric acid electrolyte [67].

Prepared SC can be bent above 90° without showing any damage. Thus the prepared supercapacitor exhibits the specific energy of 48.86 $kW \cdot kg^{-1}$ and specific capacitance of 115.83 $F \cdot g^{-1}$. Carbon films are usually fabricated by a CVD method and printing process. A buckled single-walled carbon nanotube (SWCNT) film has been prepared with a simple CVD method on an elastomeric polydimethylsiloxane (PDMS) substrate followed by the relaxation of the pre strained substrate [68]. The electrochemical performance of the supercapacitor assembled by SWCNTs macrofilms is not influenced by the change in the bending degree as shown in experiments performed at different levels of pre-strain load on the substrate [32, 69, 70]. An approach has been developed by using 1-butyl-3-methylimidazolium bis(trifluoromethanesulfonyl)imide (BMITFSI) ionic liquid, which could be used to uniformly disperse SWCNTs as chemically stable dopants in a vinylidene fluoride-hexafluoropropylene copolymer matrix to form a composite film [71]. In this case, the content of SWCNT was increased to 20 wt%, without loosening the mechanical flexibility or softness of the copolymer. The SWCNT composite film was further coated with dimethylsiloxane–based rubber.

Carbon materials, especially CNTs, have a long cycle life and excellent mechanical properties. However, the charge storage capability of pure CNTs is still limited. To enhance the performance of the CNT-based supercapacitors, various active materials, such as MnO_2 [72–75], graphene nanosheets [76–79], $NiCo_2O_4$ [80] and conductive polymers [81, 82], have been incorporated in order to achieve high capacities. For example, MnO_2 nanowires were electrodeposited onto CNT paper, leading to MnO_2 nanowire/CNT composite paper (MNCCP), which exhibited excellent electrochemical performances together with quite high cyclability [83].

Figure 6.9 shows SEM images and Energy Dispersive Spectroscopy (EDS) patterns of the MNCCP. As shown in Fig. 6.9(a), a cross-sectional SEM image indicated that three layers were present, labeled as layer 1 to layer 3. Layer 1 was the top layer of the electrode, with a thickness of 1 μm, consisting of MnO_2 nanowires. In the layer 1, the MnO_2 nanowires with an average diameter of 30 nm were cross-linked with one and another, as shown in Fig. 6.9(b). In Fig. 6.9(c), the EDS pattern confirmed that layer 1 contained only Mn and O. Layer 2 had a thickness of 2 μm thickness (Fig. 6.9(a)), which was the transition layer between the MnO_2 nanowire layer (layer 1) and the CNT layer (layer 3). The EDS analysis result confirmed that the layer 2 contains elements included Mn, O and C, as seen in Fig. 6.9(c). This might happen due to the Mn^{2+} ions that could penetrate the CNT paper, during the electrochemical deposition of MnO_2 nanowires onto the CNT paper. Due to the high density of the CNT paper, we observed the 3rd layer of pure CNT layer, which acted as a flexible substrate and maintained good mechanical properties of the pristine CNT paper. The inset picture in Fig. 6.9(a) proved that the MNCCP based electrode was free-standing and highly flexible.

Solid-state supercapacitors of CNT-pseudocapacitive MoS_2 (molybdenum disulfide) nanosheets composite electrodes shows high stretchability ($\sim$ 240%) with a high capacitance of $\sim$ 13.16 F/cm with good cycling stability (98% cycling retention observed even after 10,000 charge-discharge cycles) [84]. This report demonstrated that by using chemical vapor deposition, CNT array was developed vertically on to the silicon wafer substrate which is precoated with an iron catalyst. After deposition of the CNT array, manually pressed with glass rod against an elastic polydimethylsiloxane (PDMS) polymer film to get horizontally aligned CNTs with PDMS surface (CNTs/PDMS). Then, MoS_2 was coated on the CNTs/PDMS by the simple drop coating method. In this process stretchable and flexible symmetric supercapacitors

430 *Nanomaterials for Supercapacitors*

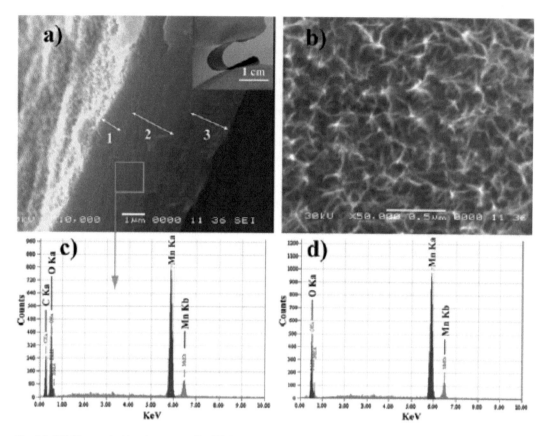

Fig. 6.9. SEM images (a, b) and corresponding EDS patterns (c, d) of the MNCCP with cross-sectional view (a, c) and top view (b, d). The inset of (a) is a typical photograph of the MNCCP electrode held between two fingers. Reproduced with permission from [83], Copyright © 2008, Elsevier.

were fabricated by using the CNT/MoS$_2$ hybrids electrodes and aqueous solution of polyvinyl alcohol and phosphoric acid gel electrolyte was sandwiched between these electrodes. In this SCs there is no need any additional current collector because CNT/MoS$_2$ hybrid electrode can act as a current collector.

The CNTs were deposited on silicon wafer with pre-coated Fe/Al$_2$O$_3$ used as a substrate. Vertically aligned CNT arrays were developed on the pre-coated silicon wafer with ethylene as the carbon source at 760°C under a mixture of H$_2$ and Ar carrier gas. The as-prepared CNT arrays showed highly aligned structure, with an average height of 240 μm. Then, the CNT arrays were transferred to an elastic substrate of pre-prepared PDMS by pressing on the CNT arrays with a glass rod. A horizontally aligned compact CNT films were affiliated with the PDMS substrate was obtained after separating the silicon wafer [84]. Representative SEM images of the CNTs on PDMS substrate are shown in Fig. 6.10(b, c). The horizontal arrays are suitable to be used for stretchable electronics, because of the movement of CNTs position in the aligned films has a minor effect on the electrochemical property of the electrode during the stretching process.

After that, aligned MoS$_2$/CNT composite films were fabricated by directly dropping MoS$_2$ solution onto the horizontally aligned CNT/PDMS films. By varying the amount of the MoS$_2$ solution, the content of MoS$_2$ on the CNT/PDMS substrate could be well controlled. Figure 6.10(d) shows SEM image demonstrating morphology of the MoS$_2$ nanosheets, which possessed a flower-like structure, due to the nanosheet aggregation. TEM images of the MoS$_2$ nanosheets are presented in the Fig. 6.10(e), revealing 2D structure and typical polycrystalline structure (inset of Fig. 6.10(e)). Figure 6.10(f, g) show TEM images of the composite of CNT/MoS$_2$, in which CNTs and MoS$_2$ nanosheets could be obviously observed [84].

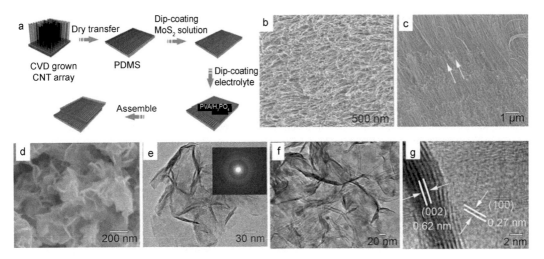

Fig. 6.10. (a) Schematic of fabrication process for the flexible and stretchable supercapacitors. (b, c) Top-view SEM images of the as-transferred CNTs on PDMS substrate before (b) and after (c) solvent treatment. The white arrows indicate the direction of pressing force and the aligned direction of CNTs. (d) SEM image of the as-synthesized MoS_2. (e) TEM image of the MoS_2 nanosheets, with inset to be electron diffraction pattern. (f, g) TEM images of the CNTs/MoS_2 composite after dispersed in dimethylformamide, at different magnifications. Reproduced with permission from [84], Copyright © 2016, John Wiley and Sons.

Flexible asymmetric supercapacitors (ASCs) was assembled with a hybrid thin-film electrodes based on SWCNTs with metal-oxide nanowire [85]. The main advantage of this hybrid nanostructured films is the presence of the uniform layered structure, which ensured high mechanical flexibility, mesoporous surface morphology and processing simplicity. In this ASCs, the MnO_2 nanowire/SWCNT hybrid films served as the positive electrode, while In_2O_3 nanowire/SWCNT films worked as the negative electrode. In this system, charges can be stored not only *via* EDLC related to SWCNTs but also through a reversible faradic process due to the metal-oxide nanowires. In addition, static electrochemical behavior was characterized throughout the charging/discharging cycles within a 2.0 V potential window. The hybrid metal-oxide nanowires/SWCNTs based ASCs exhibited a superior electrochemical performance, with a specific capacitance of 184 $F \cdot g^{-1}$ with 90% of columbic efficiency and exhibited high energy density and power density of 25.5 $Wh \cdot kg^{-1}$ and 50.3 $kW \cdot kg^{-1}$, respectively [85].

Graphene based flexible supercapacitors

Graphene is an allotrope of carbon with the form of 2D material, which consists of a single layer of carbon atoms bonded in a hexagonal lattice and possesses intriguing mechanical and electrical properties. It is the basic building unit of all the graphitic carbons, such as fullerenes, CNTs and graphite [86]. With charge carrier mobility of 200,000 $cm^2 \cdot V^{-1} s^{-1}$ and specific surface area of ~ 2,630 $m^2 \cdot g^{-1}$, graphene has been highly attractive for various electrochemical applications [87, 88]. Graphene could be simply produced from graphite by using micromechanical cleavage [89]. This approach allowed easy production of high quality graphene crystallites but has very low yield. Large scale production of graphene has been realized through exfoliation of graphite in the liquid phase by using chemical conversion or surfactant/solvent stabilization [90–93]. Due to its unique characteristics, graphene has been a potential candidate for applications in inflexible and stretchable electrode of supercapacitors [94, 95]. Theoretically, graphene based EDLC is able to reach a capacitance up to ~ 550 $F \cdot g^{-1}$, which is the highest value of intrinsic capacitance among all carbon-based electrodes [96].

Ultrathin free-standing graphene papers have attracted much attention, because of their unique properties, such as lightweight and flexibility, which are the most essential characteristics for flexible

432 *Nanomaterials for Supercapacitors*

electronic devices [97, 98]. In recent years, various solution processing methods have been developed to fabricate graphene-based films, such as Langmuir-Blodgett [99], spin-coating [100], Langmuir-Blodgett [99], interfacial self-assembly [101], vacuum filtration [102] and layer-by-layer deposition [103].

Wallace et al. first reported flexible graphene films as electrode materials, which were fabricated by using a simple filtration assembly method [104]. However, during the fabrication process, the bare graphene sheets aggregated and restacked, owing to the van der Waals forces and interlunar π-π interactions. This agglomeration could reduce surface area of the graphene films, thus decreasing the transport behavior of electrolyte ions, which resulted in a drop in electrochemical performance. Later, a number of strategies were employed to the increased surface area of graphene films, including addition of spacers, so as to prevent aggregation [88, 105, 106] and crumpling of the graphene sheets [107]. Template-assisted growth is another strategy [108].

Flexible graphene paper was employed as a high-performance electrode material for supercapacitors, which was fabricated by using the simple vacuum filtration method, with Carbon Black (CB) nanoparticles as spacers [105]. By inserting carbon black nanoparticles, the restacking of individual graphene layers was prevented, resulting in an open structure that was favorable for electrochemical performance for charge storage and ion diffusion. Electrode based on the pillared graphene paper exhibited an excellent specific capacitance of ~ 138 $F \cdot g^{-1}$ at a scan rate of 10 $mV \cdot s^{-1}$ and only ~ 4% degradation was observed after 2000 charge-discharge cycles at a current density of 10 $A \cdot g^{-1}$ in aqueous electrolyte. However, in this case, the carbon black served only as a conductive additive, thus having only inconsiderable contribution to the capacitance of the whole device.

As compared with the intercalation method, solution based strategies could be used to fabricate graphene layers with less agglomeration. Also, these methods have several advantages, such as effectiveness, processing, simplicity and cheap raw materials. As shown in Fig. 6.11(a), functionalized reduced Graphene Oxide (f-rGO) thin films were used as electrode, with which solid-state flexible SCs were assembled that contained solvent-cast Nafion electrolyte membrane films as an electrolyte and separator [109]. In this study, the f-rGOs were prepared through chemical reduction of GOs and further functionalized with Nafion polymer through the mutual hydrophobic and hydrophilic interactions among both amphiphilic features of GOs and Nafion [109]. Figure 6.11(b) shows SEM images, demonstrating surface morphology of the f-rGO films that were wrinkled and curved. Figure 6.11(c) shows cross-sectional views, in which highly packed and inter-locked sheet-like arrangement of the isolated nanosheets was observed. Favorable interfacial contact between the Nafion membrane and f-rGO films was achieved, when the Nafion electrolyte solution was infiltrated into the pores of f-rGO films. The integration of Nafion, an amphiphilic molecule, not only prevented the restacking of graphene sheets, but also developed interfacial wettability between the electrodes and electrolyte. As demonstrated in Fig. 6.11(d), the f-rGO based SCs showed excellent rate capability (90% retention at 30 $A \cdot g^{-1}$), along with a high specific capacitance of 118.5 $F \cdot g^{-1}$ at 1 $A \cdot g^{-1}$.

In general, due to the electroactive surface, there is transport resistance of electrolyte ions and the pores within the graphene films have a wide size distribution, the reversible capacitance of pure graphene is limited for EDLC energy storage. Therefore, pseudocapacitive materials with much larger capacitances have been incorporated to develop SCs with high energy densities, including conductive polymers [46, 110–113], transition metal oxides and hydroxides [40, 114–127]. Usually, these materials suffer from particle aggregation, structural degradation, poor electrical conductivities and inherent rigidity, thus leading to poor cycling stability, marginal flexibility and low power density. Because graphene possesses a light-weight, large surface area, flexibility and high electrical conductivity, it can be used as a scaffold to host these pseudocapacitive materials. As a result, it is expected that high performance flexible supercapacitors can be achieved by combining the high capacitance pseudocapacitive materials with graphene.

As stated in previous chapters, transition metal oxides, especially MnO_2, have gained attention due to high theoretical specific capacitance, low-cost and environmental friendly nature [128]. GO (Graphene Oxide) and MnO_2 could be integrated through the oxygenated functional groups of GO that behave as anchor sites for the growth of MnO_2. For instance, a template free synthesis method was reported to fabricate flexible reduced Graphene Oxide (rGO)/MnO_2 paper electrode, which exhibited a high areal capacitance of 897 $mF \cdot cm^{-2}$ [128]. The rGO/MnO_2 paper was fabricated by using vacuum filtration of GO/MnO_2 suspension, which was prepared by dispersing MnO_2 nanoparticles in GO solution. Figure 6.12(a) shows

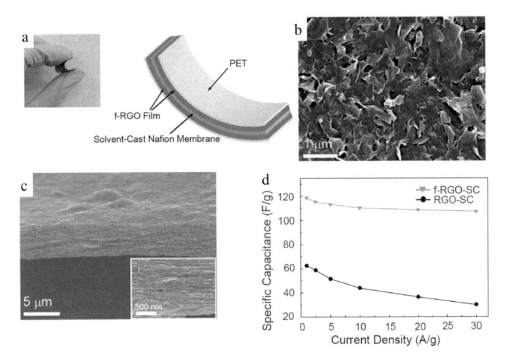

Fig. 6.11. (a) Photograph and schematic diagram of the f-rGO based SCs. (b) Surface SEM of the f-rGO films. (c) Cross-sectional SEM images of the f-rGO films (inset is high-magnification SEM image). (d) Specific capacitances of the rGO-SC and f-rGO-SC at current densities of 1, 2.5, 5, 10 20, and 30 A·g^{-1}. Reproduced with permission from [109], Copyright © 2011, American Chemical Society.

cross-sectional SEM image of the rGO/MnO$_2$ composite paper, clearly illustrating the well-separated and continuously aligned rGO layers.

The inset image in Fig. 6.12(a) shows that the rGO sheets have an open layered structure. Figure 6.12(b) indicates that the nano-spherical MnO$_2$ particles are sandwiched within the rGO layers. This hybrid rGO/MnO$_2$ paper, with macroscopic layered feature of interconnecting rGO sheets, behaved as an excellent mechanical support. Figure 6.12(c) demonstrates that the rGO/MnO$_2$ paper electrode had a high flexibility against bending, folding and twisting. Figure 6.12(d) shows areal capacitances at various applied currents. The areal capacitance of the hybrid rGO/MnO$_2$ paper is about 802 mF cm^{-2} (217 F·g^{-1}) at 100 mA·g^{-1}. When the applied current rises to 500 mA·g^{-1}, the areal capacitance of the rGO/MnO$_2$ paper decreases to 454 mF/cm^{-2}. Furthermore, the areal capacitance can be as high as 897 mF/cm^{-2} (243 F·g^{-1}), as the applied current was further decreased to 50 mA·g^{-1}.

Vanadium pentoxide-reduced Graphene Oxide (VGO) based free standing electrodes with sheet resistivity of 29.1 Ω cm have been reported for flexible SC applications [129]. Figure 6.13(a) shows the top view SEM image of the rGO sheet, revealing a crumpled and wrinkled surface, due to the defects from the breaking of graphite's sp^2 bonds. Figure 6.13(b) shows cross-sectional SEM image of the VGO, which exhibited an excellent exfoliation with well-defined rGO layers. Also, the rGO layers in the VGO had an average spacing of 30 μm, with the largest inter-layer spacing to be 230 μm. High magnification top view image of the VGO paper revealed that V$_2$O$_5$ nanoparticle were sandwiched within and between the rGO sheets, confirming the successful dispersion of the V$_2$O$_5$ nanoparticles inside the rGO paper, as observed in Fig. 6.13(c).

In the asymmetric devices, free-standing rGO was used as the cathode, while VGO was employed as the anode (inset of Fig. 6.13(d)). Figure 6.13(d) shows CV curves of the device in bent and flat states, revealing ideal symmetric CV curves with negligible deviation at both bent and flat states, which demonstrated its high flexibility due to the strong mechanical strength of the nanocomposite materials. At current density

434 Nanomaterials for Supercapacitors

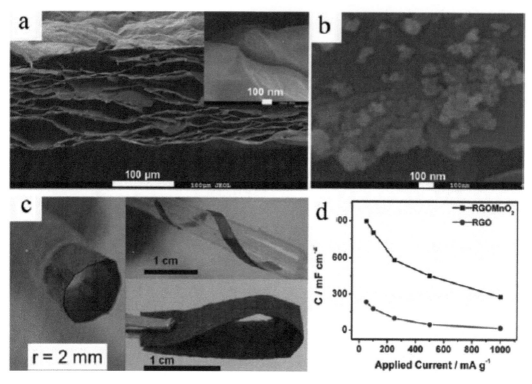

Fig. 6.12. (a) Cross-sectional SEM of the rGO/MnO$_2$ paper at low and high magnifications, with the inset showing the edge of the rGO/MnO$_2$ paper from the top view. (b) SEM Image of the rGO/MnO$_2$ paper showing the presence of MnO$_2$ nanoparticles. (c) Photographs of the rGO/MnO$_2$ paper. (d) Areal capacitance of the rGO/MnO$_2$ and rGO papers at different applied currents. Reproduced with permission from [128], Copyright © 2013, John Wiley and Sons.

of 0.1 A g^{-1}, the bent device was able to achieve an areal capacitance of 50.6 mF cm^{-2}, which was an almost similar performance in the flat state (52.5 mF cm^{-2}). The rate capability of the VGO electrode at different current densities was shown in Fig. 6.13(e). The maximum areal capacitance for the bent device was ~ 108.7 mF·cm^{-2} at 0.01 A g^{-1}. Figure 6.13(f) shows areal power and energy density of the device in the bent and flat states at various current densities. In the flat state, power density as high as 625 W·kg^{-1} (4.17 mW·cm^{-2}) can be achieved at 0.5 A g^{-1}, with an energy density of 1.22 Wh·kg^{-1} (8.1 μWh·cm^{-2}).

Among the various polymeric pseudocapacitive materials, polyaniline (PANI) has been widely studied for flexible SC applications, due to its unique conducting mechanism, excellent environmental stability and unusual dedoping/doping chemistry [130]. PANI nanowires have exhibited high pseudocapacitance and high specific surface area, while the storage performance can be improved with an optimized ion-diffusion pathway, as compared with their bulk counterparts. By using a facile template method, a freestanding 3D porous graphene film was developed [112]. The 3D porous graphene possessed porous interconnected structure with excellent flexibility. This porous structure ensured the access of electrolyte ions to the graphene film's internal surface, which led to high charge-discharge rate performance. By combining the PANI nanowire with 3D porous graphene, a hierarchical composite film of 3D porous graphene film /PANI was developed. Both the 3D porous graphene and 3D porous graphene/PANI showed excellent rate performance, whereas the 3D porous graphene/PANI film exhibited higher performance than the 3D porous graphene film, because of the pseudocapacitive contribution of PANI.

To produce flexibility and interconnected porous 3D porous graphene film, CaCO$_3$ particles were prepared in GO dispersion to facilitate the formation of the porous structure. In this case, CaCO$_3$ is used only as a sacrificial template, which was removed by washing with dilute acidic solution. Figure 6.14(a) illustrates a schematic diagram to show preparation process of the 3D porous graphene film and 3D porous

Flexible Supercapacitors 435

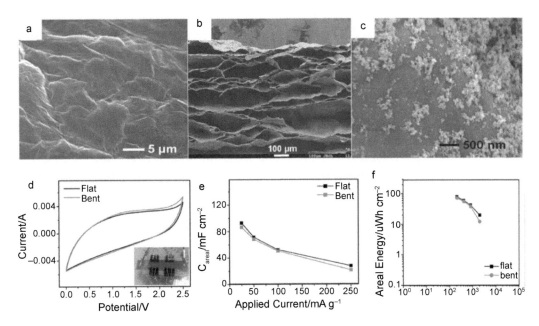

Fig. 6.13. (a) Top view SEM image of the free-standing rGO electrode. (b) Cross-sectional SEM of the VGO paper. (c) High magnification top view image of the VGO paper showing the presence of the V_2O_5 nanoparticles. (d) Cyclic voltammograms of the asymmetric device in flat and bent states at 25 mV s^{-1}, with the inset showing the device to be tested in the flat state. (e) Areal capacitance of the asymmetric device in flat and bent states at different applied currents. (f) Areal energy density of the asymmetric device in flat and bent states at different current densities. Reproduced with permission from [129], Copyright © 2014, John Wiley and Sons.

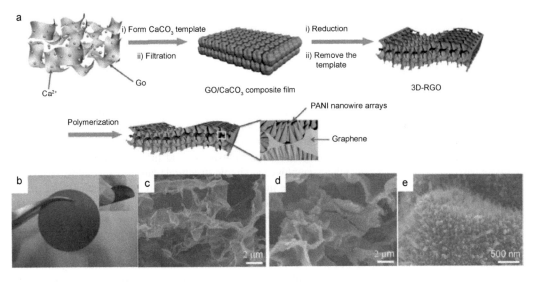

Fig. 6.14. (a) Schematic diagram of preparation of the 3D porous graphene film and 3D porous graphene/PANI films. (b) Photographs, with the inserted image showing the flexibility of the film. (c) Cross-sectional SEM images of the 3D porous graphene film at different magnifications; (d, e) Cross-sectional SEM images of the hierarchical composite film (3D porous graphene film/PANI films) at different magnifications. Reproduced with permission from [112], Copyright © 2013, John Wiley and Sons.

436   *Nanomaterials for Supercapacitors*

graphene/PANI papers. Firstly, $CaCl_2$ was dispersed in GO suspension. Then, CO_2 was bubbled through the mixture, $CaCl_2$ was turned to $CaCO_3$ particles, which were *in-situ* wrapped inside the GO sheets. Composite films of GO and $CaCO_3$ were obtained by using the simple vacuum filtration method. During the reduction of GO with hydrazine vapor, $CaCO_3$ was removed by washing with dilute acid solution. As a result, bendable 3D porous graphene skeletons were produced. With the skeletons as substrate, PANI nanowire arrays were developed through dilute polymerization on the outer and inner surface of the 3D porous graphene papers, which resulted in 3D porous graphene/PANI composite papers.

Figure 6.14(b) shows photographs of the free-standing dark-gray color 3D porous graphene film, with. The film has excellent flexibility and can be easily bent as shown in the inserted picture of Fig. 6.14(b). SEM images of the free-standing 3D porous graphene film is shown in Fig. 6.14(c). It shows the graphene sheets crumpled with some wrinkles on their surfaces, which form several micrometers sized pores. These pores are interconnected with each other, which is facilitated with other functional materials. As the PANI nanowire arrays were developed on the graphene sheet of 3D porous graphene, the 3D porous graphene's interconnected channels were available for diffusion of aniline monomers, and a low concentration of aniline monomers in the polymerization process guaranteed a well-controlled nucleation growth condition for the PANI nanowires. Hence, vertically aligned nanowire arrays were generated on the graphene surface of 3D porous graphene film (Fig. 6.14(d, e)). PANI nanoarrays developed in both the outer and the interior surface of graphene in the freestanding film and it showed a hierarchical structure while maintaining interconnected 3D pores, which facilitated easy ion diffusion during the charge discharge cycles.

Cellulose based flexible supercapacitors

Recently significant research has aimed at improving eco-friendly supercapacitors by introducing various bio-polymeric materials, such as proteins or polysaccharides. In addition to the sustainability and recyclability of such novel energy storage devices, these polymers also provide stable cycling performance, lightweight nature and flexibility, which are of tremendous importance for applications related to wearable electronics. Among the different sustainable natural polymers, cellulose, which is the most abundant biopolymer on earth and the major component in paper, is particularly promising because of its environmental friendliness, high mechanical performance, flexibility, low-cost, versatility, and tailorable surface functionalities. Consequently, research on electrically active cellulose-based supercapacitors has been increased since 2010. Cellulose, which is biodegradable, hydrophilic, odorless and water insoluble, derives from the condensation of D-glucose units through $\beta(1 \rightarrow 4)$ glycosidic bonds and is organized in fibrous arrangements. Cellulose fibers demonstrated high aspect ratios, high porosity, high surface area, excellent mechanical properties, extraordinary in flexibility and the ability to bind to other conductive materials, enabling extensive application inflexible energy-storage devices [131, 132]. Although the interest in designing electrically active cellulose-based devices for energy storage applications was started in the new century, research in the field has noticeably increased since 2012 [24].

Flexible supercapacitor devices require bendable, lightweight and high-performance electrodes. Cellulose based materials have been widely employed as simple structural platforms for depositing active conducting materials [133–135]. In this case, the most important point is their mechanical properties, including mechanical strength and/or flexibility when being used as the electrode of supercapacitors, even though other interesting properties, such as chemical stability, transparency and lightweight can also be indirectly incurred. The combination of CNTs with cellulose paper [25, 136] was well suitable for flexible energy storage devices and deliver a significant improvement for various nanocarbon based soft electrodes [137–140]. In spite of this, it is necessary to conduct optimization in the selection and functionalization of the various carbon materials, so as to develop cellulose-carbon composite materials with desired electrochemical performances.

Graphene-cellulose paper membrane is a flexible electrode, which exhibited low electrical resistance, large capacitance and high strength for polymer supercapacitors (Poly-SCs) [141]. The graphene-cellulose paper material is the combination of a filter paper and graphene nanosheets, in which the graphene nanosheets are attached to the cellulose fibers with the superabundant functional groups on the cellulose fibers in the filter paper and allow a strong interactive sites to bind graphene nanosheets. As a result, cellulose fibers are covered with the graphene nanosheets to form a conductive interwoven network with distributed macroporous texture. This graphene-cellulose paper membrane merged as macroporous texture with high strength of cellulose and the electroactivity and electrical conductivity of the graphene nanosheets. The adsorption property of cellulose fibers in the graphene-cellulose composite electrode is so strong that they could absorb the electrolyte to facilitate ion transportation [141]. Apart from these features, graphene-cellulose composite electrodes also demonstrated high electrochemical cyclic stability, with a pretty good geometric area capacitance of 81 $mF \cdot cm^{-2}$ and excellent rate capability. In addition, the graphene-cellulose composite membranes have excellent mechanical flexibility due to the cellulose fibers. For example, they could endure over 1000 repeated bending tests with only 6% increase in electrical resistivity [142].

The graphene-cellulose composite material was prepared by using a simple filtration process, in which the graphene suspension was filtered through a filter paper, as schematically shown in Fig. 6.15(a) [142]. During this filtration process, the graphene suspension diffused throughout the filter paper, while the graphene nanosheets were deposited on the cellulose fibers in the filter paper through electrostatic interactions. As the processing time was increased, the surfaces of the cellulose fibers and voids in between the fibers in the filter paper were gradually covered with the graphene nanosheets. Finally, the graphene nanosheets in the suspension were totally coated onto the cellulose fibers in the filter paper after repeating the filtration process, until colorless filtrate was observed. The filter paper was turned over after each filtration to obtain a homogeneous graphene-cellulose paper membrane. Eventually, as shown in the Fig. 6.15(a), the white color filter paper turned into a black color graphene-cellulose composite membrane with excellent flexibility. Filter paper with maximum pore sizes ($> 10 \mu m$) can allow the penetration of graphene nanosheets ($< 2 \mu m$). The functional groups present on the cellulose fibers strongly hold with graphene nanosheets, while the abundant porosity of the filter paper ($\sim 30\%$) ensures that it is able to hold a large number of graphene nanosheets. As seen in Fig. 6.15(c–e), it is evident that, as the content of graphene nanosheets was gradually increased, they were first anchored on the surface of the cellulose fibers and then filled the voids in between the filter paper.

Representative SEM and TEM are shown in Fig. 6.15(f) and 6.15(g), which confirmed that the cellulose fibers in the filter paper are chemically bond with the graphene nanosheets, which was perhaps ascribed to the electrostatic interaction between the negatively charged graphene nanosheets and the functional groups on the cellulose fibers. Moreover, the cross-sectional images of the graphene-cellulose paper composite film and the pure filter paper confirm that graphene nanosheets uniformly penetrate the filter paper. From Fig. 6.15(h) and 6.15(i), it was also found that the graphene nanosheets are not only attached on the paper surface but also deposited in the voids in between the cellulose fibers. This structure endowed the graphene-cellulose paper composite film with good mechanical flexibility and excellent electrochemical performance, with excellent specific capacitance, high power performance and very good cyclic stability. Figure 6.15(j) shows electrochemical capacitance retentions of the graphene-cellulose paper composite and G-paper electrodes. Obviously, the graphene-cellulose paper composite electrode has higher rate capability than the bare G-paper. The maximum values for the gravimetric capacitance of the graphene-cellulose paper composite and G-paper electrodes were identical, which were about 120 $F \cdot g^{-1}$ based on the mass of GNSs at 1 $mV \cdot s^{-1}$. This is reasonable due to the weak influence of ion transport at a relatively low rate. As binder-free and freestanding electrodes, graphene-cellulose paper electrode exhibited a capacitance per geometric area of 81 $mF \cdot cm^{-2}$ [142].

Significant efforts have been made to design paper based flexible electro active conducting polymer composite supercapacitor electrodes with improved capacitances. For example, Poly(ethylenedioxythiphene) (PEDOT) deposited cellulose based electrodes have exhibited a specific volumetric capacitance of 145 $F \cdot cm^{-3}$ at 0.4 $mA \cdot cm^{-3}$, as normalized with respect to the active material volume, but was only $F \cdot cm^{-3}$ when normalized with respect to the volume of the entire electrode [143]. Flexible PANI/Au/paper has shown a volumetric capacitance of 800 $F \cdot cm^{-3}$ at 1 $mA \cdot cm^{-2}$ based on volume of the active PANI layer [144]. For practical applications, the volumetric capacitances, based on the full electrode volume, as well as the rate capability of the electrodes still need to be significantly improved. PPy-nano cellulose fiber (PPy-NCFs) electrodes can be compressed to yield volumetric capacitances exceeding 236 $F \cdot cm^{-3}$ (or 192 $F \cdot g^{-1}$) based on the full electrode volume at 1 $F \cdot cm^{-3}$ [145]. Wang et al. used surface modified nanocellulose fibers (NCFs) to obtain flexible PPy-NCF electrodes with high gravimetric and volumetric capacitances [146].

438 *Nanomaterials for Supercapacitors*

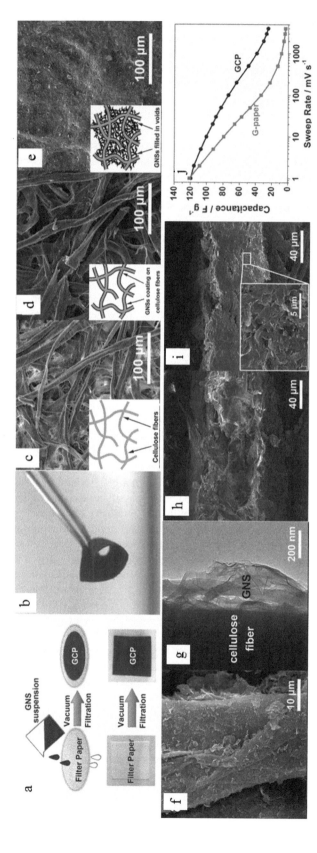

Fig. 6.15. (a) Schematic of the fabrication process of graphene-cellulose paper membrane. (b) Photograph of a Graphene-cellulose paper membrane demonstrating its flexibility. (c) SEM images of the 0 wt% pristine filter paper the filter paper (Inset Illustration of the structural evolution). (d) SEM images of 2.3 wt% filter paper the filter paper (Inset Illustration of the structural evolution). (e) SEM images of the 7.5 wt% filter paper the filter paper (Inset Illustration of the structural evolution). (f) SEM images of a cellulose fiber in a graphene-cellulose paper membrane showing graphene nanosheets anchored on the fiber surface; (g) TEM images of a cellulose fiber in a graphene-cellulose paper membrane showing graphene nanosheets anchored on the fiber surface. (h) SEM cross-section images of the pristine filter paper. (i) SEM cross-section images of the graphene-cellulose paper membrane (insert illustrating the location of graphene nanosheets from the surface to the bulk of the filter paper). Reproduced with permission from [142], Copyright © 2011, John Wiley and Sons.

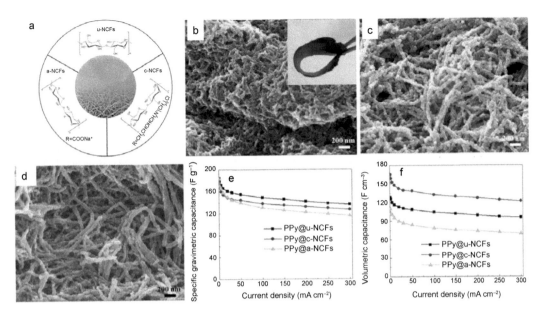

Fig. 6.16. (a) Photograph of NCFs paper and molecular structure of NCFs. (b) SEM image for PPy@c-NCFs. (c) SEM image for PPy@u-NCFs. (d) SEM image for PPy@a-NCFs. (e) Gravimetric capacitances of the samples at different charging/discharging current densities. (f) Volumetric capacitances at different charging/discharging current densities. Reproduced with permission from [146], Copyright © 2015, American Chemical Society.

Two well-known NCF modification processes were employed to introduce carboxylate and quaternary amine groups, resulting in fibers with anionic (a-NCFs) and cationic (c-NCFs) surface charges (at neutral pH and low ionic strength), respectively, as shown in Fig. 6.16(a). In these composites, the PPy@c-NCFs demonstrated excellent mechanical flexibility and could be bent several times (inset of Fig. 6.16(b)), without the formation of cracks in the material. The PPy@u-NCFs and PPy@a-NCFs composites were found to be less flexible and behaved as rather rigid, freestanding paper sheets. Micrographs of the three composites are shown in Fig. 6.16(b–d), revealing the differences at the micro and nanoscale.

Although all three composites displayed fibrous and porous structures. PPy@u-NCFs and PPy@a-NCFs appeared to have a loosely interconnected fibrous structure, while PPy@c-NCFs had more densely packed microstructure comparatively. The fact that the PPy@c-NCFs were significantly denser than the other two samples was confirmed through porosity determinations. By using symmetric supercapacitor cell configuration, with 2.0 M NaCl as the electrolyte, the charge storage properties of the different PPy-NCFs composites were studied. Figure 6.16(e) shows the plot of specific gravimetric electrode capacitance, Cg, which was obtained from galvanostatic charge/discharge curves, as a function of the current density. As seen from Fig. 6.16, the three composites exhibited practically the same gravimetric capacitances at a current density of 1 m A·cm^{-2}, which is ~ 180 F·g^{-1}. Also, at 300 m A·cm^{-2}, the Cg values of the different devices were rather close to the highest value of PPy@u-NCFs (136 F·g^{-1}) and the lowest value of PPy@a-NCFs (116 F·g^{-1}). Therefore, it was demonstrated that all the three composites exhibited promising rate capabilities. On the other hand, volumetric capacitances (Cv) of the composites were found to differ substantially, as illustrated in Fig. 6.16(f). The composite PPy@c-NCFs, with the highest bulk density, exhibited the highest Cv value, 173 F cm^{-3} at a current density of 1 mA cm^{-2} and 122 F cm^{-3} at 300 mA cm^{-2} (or 33 A g^{-1}) [146].

Conducting polymer-based flexible supercapacitors

Because of the rapid absorption and desorption of ions at the interface of electrode and electrolyte, EDLCs shows high power capability. However, the electrochemical capacitance is limited due to the deficiency of charge transfer across the interface. In contrast, the pseudocapacitor materials exhibit 10–100 times higher electrochemical capacitance than that of the EDLCs because of the involvement of redox (Faradaic) reactions. As discussed earlier, conducting polymer-based pseudocapacitors, with high conductivity, high-redox active capacitance, and high intrinsic flexibility have led to the high performance flexible supercapacitor applications [147–149]. In the recent development of conducting polymer-based flexible energy storage applications, various approaches have been adopted to hire either conducting polymers alone [150–153] or composites of conducting polymers with other materials like carbon materials or metal oxides or sulfides, etc., as the electrode materials [154].

In Fig. 6.17 the electrochemical capacitance of the EDLCs and pseudocapacitor systems based on various active materials and their composites have been summarized [150–154]. In general, carbon based electrodes have typical specific capacitance of ~ 300 F·g^{-1}, while conducting polymers and MOx can achieve levels of up to 200 F·g^{-1}. Apart from these categories, composite materials made by mixing various above mentioned materials can exhibit very huge electrochemical capacitance [154]. Several hybrid systems have been reported with promising electrochemical performances. However, electrodes based on these composite materials demonstrated performances, not only depending on the properties of individual components but also on their morphology and the interfacial characteristics.

For flexible supercapacitors, inherent flexible polymeric nature of conducting polymers are the most promising electrode materials, as compared with other active electrode materials. Meanwhile, electrochemical polymerization (electro polymerization) and wet chemical polymerization processes are the most commonly used approaches for the design of conducting polymer-based supercapacitors. However, the design of conducting polymer-based flexible supercapacitors with slurry-based wet chemical polymerization method is still limited in terms of practical device construction. Conventionally, the chemically polymerized slurry-based conducting polymers have to be combined with various binders/additives to fabricate electrodes. Such electrodes showed relatively poor performances, because they suffered from sluggish ion transportation throughout the redox reaction, due to the high interfacial and the inherent resistance of the binder materials.

In order to overcome these limitations, direct growth process has been adopted, in which the conducting polymers are directly integrated on flexible current collectors, similar to those used for the fabrication of fuel cells [155, 156]. For instance, Wang et al. used this technique to directly grow CNTs on Carbon Cloth (CC) electrodes, which exhibited enhanced electron transport [157]. With this concept, Horng et

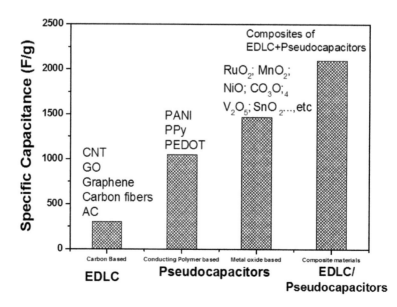

Fig. 6.17. Specific capacitances of different materials. Reproduced with permission from [154], Copyright © 2014, John Wiley and Sons.

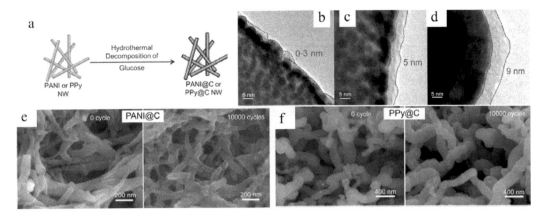

Fig. 6.18. (a) Schematic illustration of the carbonaceous coating procedure using glucose as carbon precursor. (b-d) TEM images of the PPy@C samples with different carbonaceous shell thicknesses, which was controlled through reaction time for hydrothermal deposition of glucose: (b) 1, (c) 2 and (d) 3 hours. (e) SEM images of the PANI@C-2h electrodes before and following the cycling test for 10,000 cycles. (f) SEM images of the PPy@C-2h electrodes before and after 10,000 cycles. Reproduced with permission from [159], Copyright © 2014, American Chemical Society.

al. [158] used an electro-polymerization method to grow PANI nanowires on CC. Therefore, by using suitable flexible substrates, conducting polymers can be directly grown to form composite electrode without the presence of binders, thus resulting in high electrochemical capacitance, due to the lower interfacial resistance and facile electron transport, which has become a promising technique towards the realization of the flexible supercapacitors.

Amarnath et al. demonstrated a two-step method to deposit conductive PANI electrodes with various morphologies, like nanospheres and nanorods, on ITO (Indium-Tin Oxide) substrate [150]. Lie et al. reported a facile strategy used to deposit a thin carbonaceous shell onto conducting polymers to enhance their electrochemical cycling stability of polymers nanowires (like PPy and PANI) based electrodes of pseudocapacitors [159]. The carbonaceous layer of coating with controllable thicknesses was deposited onto PPy and PANI nanowires by using hydrothermal reaction. In this process, glucose was used as the carbon precursor (Fig. 6.18(a)). In this electrode, it is very challenging to identify the interface between the two layers of carbonaceous shell and polymer core, because both the two materials are amorphous in nature and contain elements with similar molecular weight [159]. Therefore, a thin gold layer (average diameter of ~ 2 nm) was first sprayed on bare PPy or PANI nanowires, prior to the hydrothermal reactions. As shown in Fig. 6.18(b–d), gold particles were clearly present between the polymer core and the carbonaceous shell. TEM images revealed that the carbonaceous shell thickness could be controlled through the hydrothermal reaction time (Fig. 6.18(b–d)). Significantly, carbonaceous PPy and PANI electrodes exhibited exceptional capacitance retention of 85 and 95% after 10,000 cycles, as compared with the bare PPy and PANI electrodes [159]. SEM images also confirmed that the carbonaceous PPy and PANI electrodes had no obvious structural change after being tested for 10,000 cycles (Fig. 6.18(e, f)).

Concluding Remarks

Research and development in the past several years have witnessed the progress in the development of various flexible supercapacitors. Numerous attempts have been directed toward carbon, conducting polymer, metal oxide and cellulose based supercapacitors, as a new platform for flexible energy storage devices. New thoughts in the quest of intellectual designs for more efficient flexible devices are required. However, it is obvious that there are various issues that remain to be overcome to achieve desirable device performances, like energy density and electrochemical cycle stability. Also, it is still necessary to further increase specific power and energy density. In this respect, eco-friendly ionic liquid electrolyte system could be a promising

442 *Nanomaterials for Supercapacitors*

candidate, due to the larger working potential window. Additional efforts are also needed to promote large-scale commercial applications. A mature and better understanding must be established in the near future, in order to elucidate the factors to enhance the electrochemical performances of freestanding and flexible materials, so as to create a platform for flexible energy storage devices, including flexible supercapacitors.

Keywords: Flexible supercapacitor, Flexible electronics, Wearable electronics, Flexibility, Stretchability, Carbon fibers, Carbon cloth, Carbon Fiber Papers (CFPs), Nanocellulose fiber (NCFs), CNTs, SWCNTs, MWCNTs, Graphene, Plain weave, Strain weave, Twill weave, Conducting polymer, Cellulose, Polydimethylsiloxane (PDMS)

References

[1]   Zhang LL, Zhao XS. Carbon-based materials as supercapacitor electrodes. Chemical Society Reviews. 2009; 38: 2520–31.

[2]   Zhang Y, Feng H, Wu X, Wang L, Zhang A, Xia T et al. Progress of electrochemical capacitor electrode materials: A review. International Journal of Hydrogen Energy. 2009; 34: 4889–99.

[3]   Simon P, Gogotsi Y. Materials for electrochemical capacitors. Nat Mater. 2008; 7: 845–54.

[4]   Winter M, Brodd RJ. What are batteries, fuel cells, and supercapacitors? Chemical Reviews. 2004; 104: 4245–70.

[5]   Hall PJ, Bain EJ. Energy-storage technologies and electricity generation. Energy Policy. 2008; 36: 4352–5.

[6]   Bose S, Kuila T, Mishra AK, Rajasekar R, Kim NH, Lee JH. Carbon-based nanostructured materials and their composites as supercapacitor electrodes. Journal of Materials Chemistry. 2012; 22: 767–84.

[7]   Frackowiak E, Abbas Q, Beguin F. Carbon/carbon supercapacitors. Journal of Energy Chemistry. 2013; 22: 226–40.

[8]   Nasibi M, Golozar MA, Rashed G. Nanoporous carbon black particles as an electrode material for electrochemical double layer capacitors. Materials Letters. 2013; 91: 323–5.

[9]   Yang L, Hou LR, Zhang YW, Yuan CZ. Facile synthesis of mesoporous carbon nanofibres towards high-performance electrochemical capacitors. Materials Letters. 2013; 97: 97–9.

[10] Zheng JP, Goonetilleke PC, Pettit CM, Roy D. Probing the electrochemical double layer of an ionic liquid using voltammetry and impedance spectroscopy: A comparative study of carbon nanotube and glassy carbon electrodes in [EMIM]+[EtSO4]−. Talanta. 2010; 81: 1045–55.

[11] Qu D. Studies of the activated carbons used in double-layer supercapacitors. Journal of Power Sources. 2002; 109: 403–11.

[12] Shen H, Liu E, Xiang X, Huang Z, Tian Y, Wu Y et al. A novel activated carbon for supercapacitors. Materials Research Bulletin. 2012; 47: 662–6.

[13] Jiang J, Kucernak A. Electrochemical supercapacitor material based on manganese oxide: preparation and characterization. Electrochimica Acta. 2002; 47: 2381–6.

[14] Lang X, Hirata A, Fujita T, Chen M. Nanoporous metal/oxide hybrid electrodes for electrochemical supercapacitors. Nat Nano. 2011; 6: 232–6.

[15] Talpone JI, Puleston PF, More JJ, Griñó R, Cendoya MG. Experimental platform for development and Evaluation of hybrid generation systems based on fuel cells. International Journal of Hydrogen Energy. 2012; 37: 10346–53.

[16] Rudge A, Davey J, Raistrick I, Gottesfeld S, Ferraris JP. Conducting polymers as active materials in electrochemical capacitors. Journal of Power Sources. 1994; 47: 89–107.

[17] Arbizzani C, Mastragostino M, Soavi F. New trends in electrochemical supercapacitors. Journal of Power Sources. 2001; 100: 164–70.

[18] Mastragostino M, Arbizzani C, Soavi F. Conducting polymers as electrode materials in supercapacitors. Solid State Ionics. 2002; 148: 493–8.

[19] Laforgue A, Simon P, Fauvarque JF, Mastragostino M, Soavi F, Sarrau JF et al. Activated carbon/conducting polymer hybrid supercapacitors. Journal of the Electrochemical Society. 2003; 150: A645–A51.

[20] Peng C, Zhang S, Jewell D, Chen GZ. Carbon nanotube and conducting polymer composites for supercapacitors. Progress in Natural Science. 2008; 18: 777–88.

[21] Pell WG, Conway BE. Peculiarities and requirements of asymmetric capacitor devices based on combination of capacitor and battery-type electrodes. Journal of Power Sources. 2004; 136: 334–45.

[22] Li H, Cheng L, Xia Y. A hybrid electrochemical supercapacitor based on a 5 V Li-ion battery cathode and active carbon. Electrochemical and Solid-State Letters. 2005; 8: A433–A6.

[23] Wang X, Zheng JP. The optimal energy density of electrochemical capacitors using two different electrodes. Journal of the Electrochemical Society. 2004; 151: A1683–A9.

[24] Nyholm L, Nyström G, Mihranyan A, Strømme M. Toward flexible polymer and paper-based energy storage devices. Advanced Materials. 2011; 23: 3751–69.

[25] Pushparaj VL, Shaijumon MM, Kumar A, Murugesan S, Ci L, Vajtai R et al. Flexible energy storage devices based on nanocomposite paper. Proceedings of the National Academy of Sciences. 2007; 104: 13574–7.

[26] Wei D, Wakeham SJ, Ng TW, Thwaites MJ, Brown H, Beecher P. Transparent, flexible and solid-state supercapacitors based on room temperature ionic liquid gel. Electrochemistry Communications. 2009; 11: 2285–7.

[27] Jost K, Perez CR, McDonough JK, Presser V, Heon M, Dion G et al. Carbon coated textiles for flexible energy storage. Energy & Environmental Science. 2011; 4: 5060–7.

[28] Cherenack K, Zysset C, Kinkeldei T, Münzenrieder N, Tröster G. Woven electronic fibers with sensing and display functions for smart textiles. Advanced Materials. 2010; 22: 5178–82.

[29] Le VT, Kim H, Ghosh A, Kim J, Chang J, Vu QA et al. Coaxial fiber supercapacitor using all-carbon material electrodes. ACS Nano. 2013; 7: 5940–7.

[30] Fu Y, Cai X, Wu H, Lv Z, Hou S, Peng M et al. Fiber supercapacitors utilizing pen ink for flexible/wearable energy storage. Advanced Materials. 2012; 24: 5713–8.

[31] Shi S, Xu C, Yang C, Li J, Du H, Li B et al. Flexible supercapacitors. Particuology. 2013; 11: 371–7.

[32] Hu L, Cui Y. Energy and environmental nanotechnology in conductive paper and textiles. Energy & Environmental Science. 2012; 5: 6423–35.

[33] Hu L, Pasta M, Mantia FL, Cui L, Jeong S, Deshazer HD et al. Stretchable, porous, and conductive energy textiles. Nano Letters. 2010; 10: 708–14.

[34] Hu L, Wu H, La Mantia F, Yang Y, Cui Y. Thin, flexible secondary li-ion paper batteries. ACS Nano. 2010; 4: 5843–8.

[35] Sevilla M, Mokaya R. Energy storage applications of activated carbons: supercapacitors and hydrogen storage. Energy & Environmental Science. 2014; 7: 1250–80.

[36] Sawangphruk M, Krittayavathananon A, Chinwipas N. Ultraporous palladium on flexible graphene-coated carbon fiber paper as high-performance electro-catalysts for the electro-oxidation of ethanol. Journal of Materials Chemistry A. 2013; 1: 1030–4.

[37] Sawangphruk M, Srimuk P, Chiochan P, Krittayavathananon A, Luanwuthi S, Limtrakul J. High-performance supercapacitor of manganese oxide/reduced graphene oxide nanocomposite coated on flexible carbon fiber paper. Carbon. 2013; 60: 109–16.

[38] Suktha P, Chiochan P, Iamprasertkun P, Wutthiprom J, Phattharasupakun N, Suksomboon M et al. High-performance supercapacitor of functionalized carbon fiber paper with high surface ionic and bulk electronic conductivity: Effect of Organic Functional Groups. Electrochimica Acta. 2015; 176: 504–13.

[39] Miller JM, Dunn B, Tran TD, Pekala RW. Deposition of ruthenium nanoparticles on carbon aerogels for high energy density supercapacitor electrodes. Journal of the Electrochemical Society. 1997; 144: L309–L11.

[40] Yuan C, Zhang X, Su L, Gao B, Shen L. Facile synthesis and self-assembly of hierarchical porous NiO nano/micro spherical superstructures for high performance supercapacitors. Journal of Materials Chemistry. 2009; 19: 5772–7.

[41] Kaempgen M, Chan CK, Ma J, Cui Y, Gruner G. Printable thin film supercapacitors using single-walled carbon nanotubes. Nano Letters. 2009; 9: 1872–6.

[42] Yu G, Hu L, Vosgueritchian M, Wang H, Xie X, McDonough JR et al. Solution-processed graphene/MnO2 nanostructured textiles for high-performance electrochemical capacitors. Nano Letters. 2011; 11: 2905–11.

[43] Meng C, Liu C, Chen L, Hu C, Fan S. Highly flexible and all-solid-state paperlike polymer supercapacitors. Nano Letters. 2010; 10: 4025–31.

[44] Frackowiak E, Khomenko V, Jurewicz K, Lota K, Beguin F. Supercapacitors based on conducting polymers/nanotubes composites. Journal of Power Sources. 2006; 153: 413–8.

[45] Zhang K, Zhang LL, Zhao XS, Wu J. Graphene/polyaniline nanofiber composites as supercapacitor electrodes. Chemistry of Materials. 2010; 22: 1392–401.

[46] Wu Q, Xu Y, Yao Z, Liu A, Shi G. Supercapacitors based on flexible graphene/polyaniline nanofiber composite films. ACS Nano. 2010; 4: 1963–70.

[47] Masarapu C, Wang L-P, Li X, Wei B. Tailoring electrode/electrolyte interfacial properties in flexible supercapacitors by applying pressure. Advanced Energy Materials. 2012; 2: 546–52.

[48] Leela Mohana Reddy A, Estaline Amitha F, Jafri I, Ramaprabhu S. Asymmetric flexible supercapacitor stack. Nanoscale Research Letters. 2008; 3: 145.

[49] He Y-B, Li G-R, Wang Z-L, Su C-Y, Tong Y-X. Single-crystal ZnO nanorod/amorphous and nanoporous metal oxide shell composites: Controllable electrochemical synthesis and enhanced supercapacitor performances. Energy & Environmental Science. 2011; 4: 1288–92.

[50] Ghaemi M, Ataherian F, Zolfaghari A, Jafari SM. Charge storage mechanism of sonochemically prepared MnO2 as supercapacitor electrode: Effects of physisorbed water and proton conduction. Electrochimica Acta. 2008; 53: 4607–14.

[51] Dong S, Chen X, Gu L, Zhou X, Li L, Liu Z et al. One dimensional MnO2/titanium nitride nanotube coaxial arrays for high performance electrochemical capacitive energy storage. Energy & Environmental Science. 2011; 4: 3502–8.

[52] Bao L, Zang J, Li X. Flexible Zn2SnO4/MnO2 core/shell nanocable–carbon microfiber hybrid composites for high-performance supercapacitor electrodes. Nano Letters. 2011; 11: 1215–20.

[53] Yu M, Zhai T, Lu X, Chen X, Xie S, Li W et al. Manganese dioxide nanorod arrays on carbon fabric for flexible solid-state supercapacitors. Journal of Power Sources. 2013; 239: 64–71.

444 *Nanomaterials for Supercapacitors*

[54] Zhang J, Zhao X, Huang Z, Xu T, Zhang Q. High-performance all-solid-state flexible supercapacitors based on manganese dioxide/carbon fibers. Carbon. 2016; 107: 844–51.

[55] Iijima S. Helical microtubules of graphitic carbon. Nature. 1991; 354: 56–8.

[56] Saito R, Dresselhaus G, Dresselhaus MS. Electronic structure of double-layer graphene tubules. Journal of Applied Physics. 1993; 73: 494–500.

[57] Dresselhaus MS, Dresselhaus G, Saito R. C60-related tubules. Solid State Communications. 1992; 84: 201–5.

[58] Issi JP, Langer L, Heremans J, Olk CH. Electronic properties of carbon nanotubes: Experimental results. Carbon. 1995; 33: 941–8.

[59] Ebbesen TW, Lezec HJ, Hiura H, Bennett JW, Ghaemi HF, Thio T. Electrical conductivity of individual carbon nanotubes. Nature. 1996; 382: 54–6.

[60] Frackowiak E, Jurewicz K, Delpeux S, Béguin F. Nanotubular materials for supercapacitors. Journal of Power Sources. 2001; 97-98: 822–5.

[61] Schnoor TIW, Smith G, Eder D, Koziol KKK, Tim Burstein G, Windle AH et al. The production of aligned MWCNT/polypyrrole composite films. Carbon. 2013; 60: 229–35.

[62] Hughes M, Chen GZ, Shaffer MSP, Fray DJ, Windle AH. Electrochemical capacitance of a nanoporous composite of carbon nanotubes and polypyrrole. Chemistry of Materials. 2002; 14: 1610–3.

[63] Tsarfati Y, Strauss V, Kuhri S, Krieg E, Weissman H, Shimoni E et al. Dispersing perylene diimide/SWCNT hybrids: Structural insights at the molecular level and fabricating advanced materials. Journal of the American Chemical Society. 2015; 137: 7429–40.

[64] Chen JH, Li WZ, Wang DZ, Yang SX, Wen JG, Ren ZF. Electrochemical characterization of carbon nanotubes as electrode in electrochemical double-layer capacitors. Carbon. 2002; 40: 1193–7.

[65] Xu B, Wu F, Su Y, Cao G, Chen S, Zhou Z et al. Competitive effect of KOH activation on the electrochemical performances of carbon nanotubes for EDLC: Balance between porosity and conductivity. Electrochimica Acta. 2008; 53: 7730–5.

[66] Yu Jin K, Haegeun C, Chi-Hwan H, Woong K. All-solid-state flexible supercapacitors based on papers coated with carbon nanotubes and ionic-liquid-based gel electrolytes. Nanotechnology. 2012; 23: 065401.

[67] Hu S, Rajamani R, Yu X. Flexible solid-state paper based carbon nanotube supercapacitor. Applied Physics Letters. 2012; 100: 104103.

[68] Yu C, Masarapu C, Rong J, Wei B, Jiang H. Stretchable supercapacitors based on buckled single-walled carbon-nanotube macrofilms. Advanced Materials. 2009; 21: 4793–7.

[69] Chen P, Chen H, Qiu J, Zhou C. Inkjet printing of single-walled carbon nanotube/ruo2 nanowire supercapacitors on cloth fabrics and flexible substrates. Nano Research. 2010; 3: 594–603.

[70] Grande L, Chundi VT, Wei D, Bower C, Andrew P, Ryhänen T. Graphene for energy harvesting/storage devices and printed electronics. Particuology. 2012; 10: 1–8.

[71] Sekitani T, Noguchi Y, Hata K, Fukushima T, Aida T, Someya T. A rubberlike stretchable active matrix using elastic conductors. Science. 2008; 321: 1468.

[72] Tao J, Liu N, Ma W, Ding L, Li L, Su J et al. Solid-state high performance flexible supercapacitors based on polypyrrole-mno(2)-carbon fiber hybrid structure. Scientific Reports. 2013; 3: 2286.

[73] Gu T, Wei B. Fast and stable redox reactions of MnO2/CNT hybrid electrodes for dynamically stretchable pseudocapacitors. Nanoscale. 2015; 7: 11626–32.

[74] Yun TG, Hwang Bi, Kim D, Hyun S, Han SM. Polypyrrole–MnO2-coated textile-based flexible-stretchable supercapacitor with high electrochemical and mechanical reliability. ACS Applied Materials & Interfaces. 2015; 7: 9228–34.

[75] Liu Y, Shi K, Zhitomirsky I. Azopolymer triggered electrophoretic deposition of MnO2-carbon nanotube composites and polypyrrole coated carbon nanotubes for supercapacitors. Journal of Materials Chemistry A. 2015; 3: 16486–94.

[76] Sun R, Chen H, Li Q, Song Q, Zhang X. Spontaneous assembly of strong and conductive graphene/polypyrrole hybrid aerogels for energy storage. Nanoscale. 2014; 6: 12912–20.

[77] Kashani H, Chen L, Ito Y, Han J, Hirata A, Chen M. Bicontinuous nanotubular graphene–polypyrrole hybrid for high performance flexible supercapacitors. Nano Energy. 2016; 19: 391–400.

[78] Liu L, Bian X-M, Tang J, Xu H, Hou Z-L, Song W-L. Exceptional electrical and thermal transport properties in tunable all-graphene papers. RSC Advances. 2015; 5: 75239–47.

[79] Zhao Y, Liu J, Hu Y, Cheng H, Hu C, Jiang C et al. Highly compression-tolerant supercapacitor based on polypyrrole-mediated graphene foam electrodes. Advanced Materials. 2013; 25: 591–5.

[80] Xu S, Yang D, Zhang F, Liu J, Guo A, Hou F. Fabrication of NiCo2O4 and carbon nanotube nanocomposite films as a high-performance flexible electrode of supercapacitors. RSC Advances. 2015; 5: 74032–9.

[81] Liang B, Qin Z, Zhao J, Zhang Y, Zhou Z, Lu Y. Controlled synthesis, core-shell structures and electrochemical properties of polyaniline/polypyrrole composite nanofibers. Journal of Materials Chemistry A. 2014; 2: 2129–35.

[82] Niu Z, Luan P, Shao Q, Dong H, Li J, Chen J et al. A "skeleton/skin" strategy for preparing ultrathin free-standing single-walled carbon nanotube/polyaniline films for high performance supercapacitor electrodes. Energy & Environmental Science. 2012; 5: 8726–33.

[83] Chou S-L, Wang J-Z, Chew S-Y, Liu H-K, Dou S-X. Electrodeposition of MnO2 nanowires on carbon nanotube paper as free-standing, flexible electrode for supercapacitors. Electrochemistry Communications. 2008; 10: 1724–7.

[84]   Lv T, Yao Y, Li N, Chen T. Highly Stretchable supercapacitors based on aligned carbon nanotube/molybdenum disulfide composites. Angewandte Chemie International Edition. 2016; 55: 9191–5.

[85]   Chen P-C, Shen G, Shi Y, Chen H, Zhou C. Preparation and characterization of flexible asymmetric supercapacitors based on transition-metal-oxide nanowire/single-walled carbon nanotube hybrid thin-film electrodes. Acs Nano. 2010; 4: 4403–11.

[86]   Ban S, Malek K, Huang C, Liu Z. A molecular model for carbon black primary particles with internal nanoporosity. Carbon. 2011; 49: 3362–70.

[87]   Huang L, Li C, Shi G. High-performance and flexible electrochemical capacitors based on graphene/polymer composite films. Journal of Materials Chemistry A. 2014; 2: 968–74.

[88]   Qiu L, Yang X, Gou X, Yang W, Ma Z-F, Wallace GG et al. Dispersing carbon nanotubes with graphene oxide in water and synergistic effects between graphene derivatives. Chemistry—A European Journal. 2010; 16: 10653–8.

[89]   Zhou M, Wang Y, Zhai Y, Zhai J, Ren W, Wang F et al. Controlled synthesis of large-area and patterned electrochemically reduced graphene oxide films. Chemistry—A European Journal. 2009; 15: 6116–20.

[90]   Yu P, Lin Y, Xiang L, Su L, Zhang J, Mao L. Molecular films of water-miscible ionic liquids formed on glassy carbon electrodes: Characterization and electrochemical applications. Langmuir: The ACS Journal of Surfaces and Colloids. 2005; 21: 9000–6.

[91]   Kosmulski M, Osteryoung RA, Ciszkowska M. Diffusion coefficients of ferrocene in composite materials containing ambient temperature ionic liquids. Journal of the Electrochemical Society. 2000; 147: 1454–8.

[92]   Barisci JN, Wallace GG, MacFarlane DR, Baughman RH. Investigation of ionic liquids as electrolytes for carbon nanotube electrodes. Electrochemistry Communications. 2004; 6: 22–7.

[93]   Miller JR, Outlaw RA, Holloway BC. Graphene double-layer capacitor with ac line-filtering performance. Science. 2010; 329: 1637–9.

[94]   Lu W, Qu L, Henry K, Dai L. High performance electrochemical capacitors from aligned carbon nanotube electrodes and ionic liquid electrolytes. Journal of Power Sources. 2009; 189: 1270–7.

[95]   Wu ZS, Parvez K, Feng X, Müllen K. Graphene-based in-plane micro-supercapacitors with high power and energy densities. Nature Communications. 2013; 4: 2487.

[96]   Stoller MD, Park S, Zhu Y, An J, Ruoff RS. Graphene-based ultracapacitors. Nano Letters. 2008; 8: 3498–502.

[97]   Zhang LL, Zhou R, Zhao XS. Graphene-based materials as supercapacitor electrodes. Journal of Materials Chemistry. 2010; 20: 5983–92.

[98]   Choi H-J, Jung S-M, Chang DW, Dai L, Baek J-B. Graphene for energy conversion and storage in fuel cells and supercapacitors. Nano Energy. 2012; 1: 534–51.

[99]   Li X, Zhang G, Bai X, Sun X, Wang X, Wang E et al. Highly conducting graphene sheets and Langmuir-Blodgett films. Nat Nano. 2008; 3: 538–42.

[100]   Becerril HA, Mao J, Liu Z, Stoltenberg RM, Bao Z, Chen Y. Evaluation of solution-processed reduced graphene oxide films as transparent conductors. ACS nano. 2008; 2: 463–70.

[101]   Gan S, Zhong L, Wu T, Han D, Zhang J, Ulstrup J et al. Spontaneous and fast growth of large-area graphene nanofilms facilitated by oil/water interfaces. Advanced Materials. 2012; 24: 3958–64.

[102]   Li X, Zhang G, Bai X, Sun X, Wang X, Wang E et al. Highly conducting graphene sheets and Lagnmuir-Blodgett films. Nat Nanotech. 2008; 3: 538–42.

[103]   Güneş F, Shin H-J, Biswas C, Han GH, Kim ES, Chae SJ et al. Layer-by-layer doping of few-layer graphene film. ACS Nano. 2010; 4: 4595–600.

[104]   Wang C, Li D, Too CO, Wallace GG. Electrochemical properties of graphene paper electrodes used in lithium batteries. Chemistry of Materials. 2009; 21: 2604–6.

[105]   Wang G, Sun X, Lu F, Sun H, Yu M, Jiang W et al. Flexible pillared graphene-paper electrodes for high-performance electrochemical supercapacitors. Small. 2012; 8: 452–9.

[106]   Lei Z, Christov N, Zhao XS. Intercalation of mesoporous carbon spheres between reduced graphene oxide sheets for preparing high-rate supercapacitor electrodes. Energy & Environmental Science. 2011; 4: 1866–73.

[107]   Liu C, Yu Z, Neff D, Zhamu A, Jang BZ. Graphene-based supercapacitor with an ultrahigh energy density. Nano Letters. 2010; 10: 4863–8.

[108]   Vickery JL, Patil AJ, Mann S. Fabrication of graphene–polymer nanocomposites with higher-order three-dimensional architectures. Advanced Materials. 2009; 21: 2180–4.

[109]   Choi BG, Hong J, Hong WH, Hammond PT, Park H. Facilitated ion transport in all-solid-state flexible supercapacitors. ACS Nano. 2011; 5: 7205–13.

[110]   Li Y, Zhang Q, Zhao X, Yu P, Wu L, Chen D. Enhanced electrochemical performance of polyaniline/sulfonated polyhedral oligosilsesquioxane nanocomposites with porous and ordered hierarchical nanostructure. Journal of Materials Chemistry. 2012; 22: 1884–92.

[111]   Sujith K, Asha AM, Anjali P, Sivakumar N, Subramanian KRV, Nair SV et al. Fabrication of highly porous conducting PANI-C composite fiber mats via electrospinning. Materials Letters. 2012; 67: 376–8.

[112]   Meng Y, Wang K, Zhang Y, Wei Z. Hierarchical porous graphene/polyaniline composite film with superior rate performance for flexible supercapacitors. Advanced Materials. 2013; 25: 6985–90.

446 *Nanomaterials for Supercapacitors*

[113] D'Arcy JM, El-Kady MF, Khine PP, Zhang L, Lee SH, Davis NR et al. Vapor-phase polymerization of nanofibrillar poly(3,4-ethylenedioxythiophene) for supercapacitors. ACS Nano. 2014; 8: 1500–10.

[114] Zhang J, Jiang J, Li H, Zhao XS. A high-performance asymmetric supercapacitor fabricated with graphene-based electrodes. Energy & Environmental Science. 2011; 4: 4009–15.

[115] Bi R-R, Wu X-L, Cao F-F, Jiang L-Y, Guo Y-G, Wan L-J. Highly dispersed RuO2 nanoparticles on carbon nanotubes: Facile synthesis and enhanced supercapacitance performance. The Journal of Physical Chemistry C. 2010; 114: 2448–51.

[116] Hu C-C, Chang K-H, Lin M-C, Wu Y-T. Design and tailoring of the nanotubular arrayed architecture of hydrous ruo2 for next generation supercapacitors. Nano Letters. 2006; 6: 2690–5.

[117] Choi BG, Chang S-J, Kang H-W, Park CP, Kim HJ, Hong WH et al. High performance of a solid-state flexible asymmetric supercapacitor based on graphene films. Nanoscale. 2012; 4: 4983–8.

[118] Mu J, Chen B, Guo Z, Zhang M, Zhang Z, Zhang P et al. Highly dispersed Fe3O4 nanosheets on one-dimensional carbon nanofibers: Synthesis, formation mechanism, and electrochemical performance as supercapacitor electrode materials. Nanoscale. 2011; 3: 5034–40.

[119] Pendashteh A, Mousavi MF, Rahmanifar MS. Fabrication of anchored copper oxide nanoparticles on graphene oxide nanosheets via an electrostatic coprecipitation and its application as supercapacitor. Electrochimica Acta. 2013; 88: 347–57.

[120] Lu Q, Lattanzi MW, Chen Y, Kou X, Li W, Fan X et al. Supercapacitor electrodes with high-energy and power densities prepared from monolithic NiO/Ni nanocomposites. Angewandte Chemie. 2011; 123: 6979–82.

[121] Xia X, Tu J, Mai Y, Chen R, Wang X, Gu C et al. Graphene sheet/porous NiO hybrid film for supercapacitor applications. Chemistry—A European Journal. 2011; 17: 10898–905.

[122] Wei W, Cui X, Chen W, Ivey DG. Manganese oxide-based materials as electrochemical supercapacitor electrodes. Chemical Society Reviews. 2011; 40: 1697–721.

[123] Cheng Q, Tang J, Ma J, Zhang H, Shinya N, Qin L-C. Graphene and nanostructured MnO2 composite electrodes for supercapacitors. Carbon. 2011; 49: 2917–25.

[124] Chen S, Zhu J, Wu X, Han Q, Wang X. Graphene oxide–MnO2 nanocomposites for supercapacitors. ACS Nano. 2010; 4: 2822–30.

[125] Wu Z-S, Ren W, Wang D-W, Li F, Liu B, Cheng H-M. High-energy MnO2 nanowire/graphene and graphene asymmetric electrochemical capacitors. ACS Nano. 2010; 4: 5835–42.

[126] Liu J, Jiang J, Cheng C, Li H, Zhang J, Gong H et al. Co3O4 nanowire@MnO2 ultrathin nanosheet core/shell arrays: A new class of high-performance pseudocapacitive materials. Advanced Materials. 2011; 23: 2076–81.

[127] Yuan C, Yang L, Hou L, Shen L, Zhang X, Lou XW. Growth of ultrathin mesoporous Co3O4 nanosheet arrays on Ni foam for high-performance electrochemical capacitors. Energy & Environmental Science. 2012; 5: 7883–7.

[128] Sumboja A, Foo CY, Wang X, Lee PS. Large areal mass, flexible and free-standing reduced graphene oxide/manganese dioxide paper for asymmetric supercapacitor device. Advanced Materials. 2013; 25: 2809–15.

[129] Foo CY, Sumboja A, Tan DJH, Wang J, Lee PS. Flexible and highly scalable V2O5-rGO electrodes in an organic electrolyte for supercapacitor devices. Advanced Energy Materials. 2014; 4: 1400236–n/a.

[130] Miao Y-E, Fan W, Chen D, Liu T. High-performance supercapacitors based on hollow polyaniline nanofibers by electrospinning. ACS Applied Materials & Interfaces. 2013; 5: 4423–8.

[131] Paakko M, Vapaavuori J, Silvennoinen R, Kosonen H, Ankerfors M, Lindstrom T et al. Long and entangled native cellulose I nanofibers allow flexible aerogels and hierarchically porous templates for functionalities. Soft Matter. 2008; 4: 2492–9.

[132] Hu L, Zheng G, Yao J, Liu N, Weil B, Eskilsson M et al. Transparent and conductive paper from nanocellulose fibers. Energy & Environmental Science. 2013; 6: 513–8.

[133] Liu L, Niu Z, Zhang L, Zhou W, Chen X, Xie S. Nanostructured graphene composite papers for highly flexible and foldable supercapacitors. Advanced Materials. 2014; 26: 4855–62.

[134] Lu X, Yu M, Wang G, Tong Y, Li Y. Flexible solid-state supercapacitors: design, fabrication and applications. Energy & Environmental Science. 2014; 7: 2160–81.

[135] Fei H, Yang C, Bao H, Wang G. Flexible all-solid-state supercapacitors based on graphene/carbon black nanoparticle film electrodes and cross-linked poly(vinyl alcohol)–H2SO4 porous gel electrolytes. Journal of Power Sources. 2014; 266: 488–95.

[136] Hu L, Choi JW, Yang Y, Jeong S, La Mantia F, Cui L-F et al. Highly conductive paper for energy-storage devices. Proceedings of the National Academy of Sciences. 2009; 106: 21490–4.

[137] Futaba DN, Hata K, Yamada T, Hiraoka T, Hayamizu Y, Kakudate Y et al. Shape-engineerable and highly densely packed single-walled carbon nanotubes and their application as super-capacitor electrodes. Nat Mater. 2006; 5: 987–94.

[138] Kaempgen M, Ma J, Gruner G, Wee G, Mhaisalkar SG. Bifunctional carbon nanotube networks for supercapacitors. Applied Physics Letters. 2007; 90: 264104.

[139] Ci L, Manikoth SM, Li X, Vajtai R, Ajayan PM. Ultrathick freestanding aligned carbon nanotube films. Advanced Materials. 2007; 19: 3300–3.

[140] Ding W, Pengcheng S, Changhong L, Wei W, Shoushan F. Highly oriented carbon nanotube papers made of aligned carbon nanotubes. Nanotechnology. 2008; 19: 075609.

[141] Wang D-W, Li F, Liu M, Lu GQ, Cheng H-M. 3D Aperiodic hierarchical porous graphitic carbon material for high-rate electrochemical capacitive energy storage. Angewandte Chemie International Edition. 2008; 47: 373–6.

[142] Weng Z, Su Y, Wang D-W, Li F, Du J, Cheng H-M. Graphene–cellulose paper flexible supercapacitors. Advanced Energy Materials. 2011; 1: 917–22.

[143] Anothumakkool B, Soni R, Bhange SN, Kurungot S. Novel scalable synthesis of highly conducting and robust PEDOT paper for a high performance flexible solid supercapacitor. Energy & Environmental Science. 2015; 8: 1339–47.

[144] Yuan L, Xiao X, Ding T, Zhong J, Zhang X, Shen Y et al. Paper-based supercapacitors for self-powered nanosystems. Angewandte Chemie. 2012; 124: 5018–22.

[145] Wang Z, Tammela P, Zhang P, Stromme M, Nyholm L. High areal and volumetric capacity sustainable all-polymer paper-based supercapacitors. Journal of Materials Chemistry A. 2014; 2: 16761–9.

[146] Wang Z, Carlsson DO, Tammela P, Hua K, Zhang P, Nyholm L et al. Surface modified nanocellulose fibers yield conducting polymer-based flexible supercapacitors with enhanced capacitances. ACS Nano. 2015; 9: 7563–71.

[147] Snook GA, Kao P, Best AS. Conducting-polymer-based supercapacitor devices and electrodes. Journal of Power Sources. 2011; 196: 1–12.

[148] Mike JF, Lutkenhaus JL. Recent advances in conjugated polymer energy storage. Journal of Polymer Science Part B: Polymer Physics. 2013; 51: 468–80.

[149] Ramya R, Sivasubramanian R, Sangaranarayanan MV. Conducting polymers-based electrochemical supercapacitors—Progress and prospects. Electrochimica Acta. 2013; 101: 109–29.

[150] Amarnath CA, Chang JH, Kim D, Mane RS, Han S-H, Sohn D. Electrochemical supercapacitor application of electroless surface polymerization of polyaniline nanostructures. Materials Chemistry and Physics. 2009; 113: 14–7.

[151] Zhang H, Zhao Q, Zhou S, Liu N, Wang X, Li J et al. Aqueous dispersed conducting polyaniline nanofibers: Promising high specific capacity electrode materials for supercapacitor. Journal of Power Sources. 2011; 196: 10484–9.

[152] Park JH, Park OO. Hybrid electrochemical capacitors based on polyaniline and activated carbon electrodes. Journal of Power Sources. 2002; 111: 185–90.

[153] Prasad KR, Munichandraiah N. Fabrication and evaluation of 450 F electrochemical redox supercapacitors using inexpensive and high-performance, polyaniline coated, stainless-steel electrodes. Journal of Power Sources. 2002; 112: 443–51.

[154] Shown I, Ganguly A, Chen L-C, Chen K-H. Conducting polymer-based flexible supercapacitor. Energy Science & Engineering. 2015; 3: 2–26.

[155] Sun C-L, Chen L-C, Su M-C, Hong L-S, Chyan O, Hsu C-Y et al. Ultrafine platinum nanoparticles uniformly dispersed on arrayed CNx nanotubes with high electrochemical activity. Chemistry of Materials. 2005; 17: 3749–53.

[156] Chen LC, Wen CY, Liang CH, Hong WK, Chen KJ, Cheng HC et al. Controlling steps during early stages of the aligned growth of carbon nanotubes using microwave plasma enhanced chemical vapor deposition. Advanced Functional Materials. 2002; 12: 687–92.

[157] Wang CH, Du HY, Tsai YT, Chen CP, Huang CJ, Chen LC et al. High performance of low electrocatalysts loading on CNT directly grown on carbon cloth for DMFC. Journal of Power Sources. 2007; 171: 55–62.

[158] Horng Y-Y, Lu Y-C, Hsu Y-K, Chen C-C, Chen L-C, Chen K-H. Flexible supercapacitor based on polyaniline nanowires/carbon cloth with both high gravimetric and area-normalized capacitance. Journal of Power Sources. 2010; 195: 4418–22.

[159] Liu T, Finn L, Yu M, Wang H, Zhai T, Lu X et al. Polyaniline and polypyrrole pseudocapacitor electrodes with excellent cycling stability. Nano Letters. 2014; 14: 2522–7.

Microsupercapacitors

Ling Bing Kong,[1,] Wenxiu Que,[2] Lang Liu,[3] Freddy Yin Chiang Boey,[1,a] Zhichuan J. Xu,[1,b] Kun Zhou,[4] Sean Li,[5] Tianshu Zhang[6] and Chuanhu Wang[7]*

Brief Introduction

The demand of micropower sources and small-scale energy storage devices has been tremendously increased in recent years, because of the rapid development of miniaturized and portable electronic devices [1, 2]. Micropower devices and systems are especially important in the development of implantable biosensors, remote and mobile environmental sensors, nanorobotics, microelectromechanical systems (MEMS) and wearable electronics [3]. Examples include Wireless Sensor Networks (WSNs) [4], microbatteries [5], nanogenerators [6–8].

As discussed in previous chapters, supercapacitors are electrochemical energy storage devices with high power densities, as well as rapid charge and discharge capabilities. Different from the conventional supercapacitors, microsupercapacitors should have total device footprint area at the scale of centimeter or even millimeter ranges. On the one hand, the principles and design considerations of the convention macroscale supercapacitors could be used as a reference in fabrication and development of microsupercapacitors. On the other hand, microsupercapacitors have their own special requirements.

---

[1] School of Materials Science and Engineering, Nanyang Technological University, 50 Nanyang Avenue, Singapore 639798.
[a] Email: mycboey@ntu.edu.sg
[b] Email: xuzc@ntu.edu.sg
[2] Electronic Materials Research Laboratory and International Center for Dielectric Research, Xi'an Jiaotong University, Xi'an 710049, Shaanxi Province, People's Republic of China.
  Email: wxque@xjtu.edu.cn
[3] School of Chemistry and Chemical Engineering, Xinjiang University, Urumqi 830046, Xinjiang, People's Republic of China.
  Email: llyhs1973@sina.com
[4] School of Mechanical and Aerospace Engineering, Nanyang Technological University, 50 Nanyang Avenue, Singapore 639798.
  Email: kzhou@ntu.edu.sg
[5] School of Materials Science and Engineering, The University of New South Wales, Australia.
  Email: sean.li@unsw.edu.au
[6] Anhui Target Advanced Ceramics Technology Co. Ltd., Hefei, Anhui, People's Republic of China.
  Email: 13335516617@163.com
[7] Department of Material and Chemical Engineering, Bengbu Univresity, Bengbu, Peoples's Republic of China.
  Email: bbxywch@126.com
* Corresponding author: elbkong@ntu.edu.sg

For example, microsupercapacitors are generally made of either thin film electrodes with a thickness of ≤ 10 μm or arrays of microelectrodes with micrometer sizes. In the case of microelectrodes, they can be either two-Dimensional (2D) or three-Dimensional (3D) architectures, consisting of nanoscale or microscale building blocks. As a result, special fabrication techniques, such as thin film deposition, silicon process technology and MEMS, should be employed to develop microsupercapacitors. This chapter aims to discuss the recent progress in design and fabrication of microsupercapacitors.

Thin Film Microsupercapacitors

Microsupercapacitors based on thin film electrodes can be classified into two groups: (i) stacking configuration and (ii) in-plane interdigital configuration. No matter which configuration is adopted, one of the key requirements is the quality of the thin film electrodes, which should have desirable electrochemical performances and mechanical stability for device integration. Various nanocarbons and pseudocapacitive materials have been employed to deposit thin film electrodes for the fabrication of microsupercapacitors. Various methods have been developed to fabricate thin film electrodes, which will be described and discussed as follows.

Sputtering has been a widely used physical deposition technique that can be used to fabricate a wide range of thin films for various applications. More recently, it has also been employed to deposit thin film electrodes for microsupercapacitors. However, thin films prepared by using sputtering techniques are generally of internal porosity, so that they are not favorable for ion transport, which thus limits their potential for high performance supercapacitor electrodes.

As discussed in previous chapters, nanocarbon materials, such as carbon nanotubes (CNTs), graphene and Onion Like Carbon (OLC), have relatively high electrical conductivity, as well as high porosity, are suitable for electrodes of supercapacitors [9]. Due to their high electrical conductivity, these materials ensured fast transport of electrons. Meanwhile, owing to their planar or exohedral structures, their surface exhibited high accessibility to ions, thus leading to significantly high ion transport efficiency. However, the sputtering technique is not suitable to deposit nanocarbon based thin films. Instead, various chemical methods, such as Chemical Vapor Deposition (CVD) [10, 11], Layer-by-Layer (LbL) deposition [12, 13], electrophoretic deposition [14, 15], ink-jet printing [15–17], electrostatic spray deposition [18, 19], and so on, have been used to fabricate carbon thin films for microsupercapacitor applications.

Sputtering

Cobalt oxide films (Co_3O_4) were deposited by using a sputtering technique, with different sputtering gas-ratios of $O_2/(Ar+O_2)$, to evaluate the effects on microstructural properties and thus electrochemical performances [20]. The Co_3O_4 thin films were used to assemble all solid-state thin-film supercapacitors (TFSCs), with the Co_3O_4 as electrodes and an amorphous LiPON thin-film as electrolyte. The $Co_3O_4/$LiPON/Co_3O_4 TFSCs exhibited a bulk-type supercapacitor behavior. It was found that electrochemical performance of the TFSCs could be optimized through the sputtering gas-ratio.

The Co_3O_4 films were deposited on Pt/Ti/Si substrates by using a DC reactive sputtering method, with Co metal target at room temperature. The Pt films on top of the Ti adhesive layers were used as the current-collectors. The lower Co_3O_4 electrode, with a thickness of 300 nm, was deposited on the Pt current-collector at a DC power of 100 W, with oxygen gas flow ratios [$O_2/(Ar+O_2)$] of 0.1, 0.2 and 0.3. A LiPON electrolyte film, with a thickness of 1.3 μm, was grown with a Li_3PO_4 target, at an RF power of 300 W in pure N_2 at a flow of 50 sccm, which resulted in LiPON film with a composition of $Li_{2.94}PO_{2.37}L_{0.75}$. After deposition of the LiPON film, an upper Co_3O_4 film was deposited under that the same growth conditions of the underlying electrode film. Figure 7.1 shows a schematic diagram of the $Co_3O_4/$LiPON/Co_3O_4 TFSCs.

Charge–discharge characteristics of the $Co_3O_4/$LiPON/Co_3O_4 TFSCs were measured at a current of 50 μA·cm^{-2}, over 0–2 V. Figure 7.2 shows the capacitances of the TFSCs as a function of cycle number. It was found that the higher the oxygen gas flow ratio, the higher the capacitance was. The values of capacitance per volume of the TFSCs after one cycle were 5×10^{-3}, 8.2×10^{-3} and 2.5×10^{-2} F·cm^{-2}·μm, corresponding to the flow ratios of 0.1, 0.2 and 0.3, respectively. A degradation in capacitance with

450 *Nanomaterials for Supercapacitors*

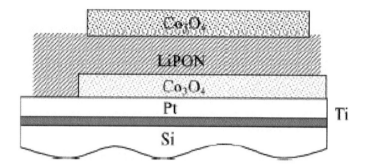

Fig. 7.1. Schematic diagram of the TFSCs based on Co$_3$O$_4$ thin film electrode deposited by using a DC sputtering method. Reproduced with permission from [20], Copyright © 2001, Elsevier.

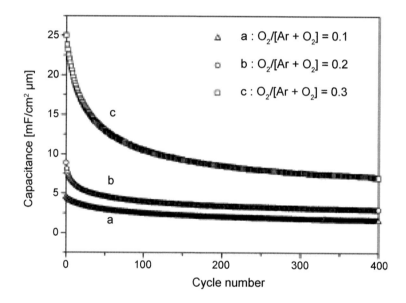

Fig. 7.2. Charge–discharge capacitances of the Co$_3$O$_4$/LiPON/Co$_3$O$_4$ TFSCs as function of cycle number. Reproduced with permission from [20], Copyright © 2001, Elsevier.

increasing cycle number was observed in all the TFSCs. After 400 cycles, the charge–discharge efficiency was decreased by more than 60%.

XRD patterns of the Co$_3$O$_4$ electrode films were employed to check their phase composition [20]. The sample deposited with a flow ratio of 0.1 exhibited typical diffraction peaks of cubic Co$_3$O$_4$, with a preferred orientation of (311). The Co$_3$O$_4$ film was comprised of fine grains, as evidenced by the broadened XRD peaks. The 0.2 sample had a similar diffraction pattern to that the 0.1 one. After deposition with a flow ratio of 0.3, the intensity of the peak (311) was largely decreased, which suggested a high gas flow ratio led to the formation of amorphous films. This was because the gas flow ratio influenced the sputtering mechanism and thus the microstructural characteristics of the films. At low oxygen flow ratios (≤ 0.2), Co atoms sputtered from the target reacted with oxygen at the surface of the substrate, so as to form Co$_3$O$_4$ phase with a preferred orientation. In comparison, if the oxygen flow ratio was sufficiently high, Co-oxide layer could be formed at the surface of the target. As a result, amorphous Co$_3$O$_4$ was formed instead.

The charge–discharge behaviors of the Co$_3$O$_4$/LiPON/Co$_3$O$_4$ TFSCs shown in Fig. 7.2 could be attributed to the pseudocapacitance mechanism, due to the redox processes involving the reduction and oxidation between Co_3O_4 and $LiCo_3O_4$. Figure 7.3 shows a possible pseudocapacitance mechanism that governed the charge–discharge behaviors of the Co_3O_4/LiPON/Co_3O_4 TFSCs [20]. As illustrated in Fig. 7.3(a), Li^+ ions and electrons were incorporated into the lower Co_3O_4 electrode, during the charge process. During the discharge process, the Li^+ ions were de-intercalated into the LiPON electrolyte layer, accompanied by the release of electrons, as demonstrated in Fig. 7.3(b).

Initially, the Li^+ ions were transported to the negative electrode, whereas then the redox reaction took place on the Co_3O_4 TFSCs involving Li^+ ion insertion between the positive and the negative electrode, alternately. It is found that capacity degradation was ascribed to the formation of crystallites and interfacial phases. In this study, it was readily expected that microstructure of the Co_3O_4 thin film electrodes was closely related to the gas flow ratio, as demonstrated by the XRD results mentioned above. Because an amorphous structure is favorable to high cycling stability of Li^+ intercalation materials [21], the TFSCs based on the Co_3O_4 thin film electrodes deposited at a high gas flow ratio of 0.3 exhibited the highest electrochemical performance.

Thin films of manganese oxide have also been deposited by using the sputtering technique to be used as electrodes of microsupercapacitors. The manganese oxide thin films could be obtained by either

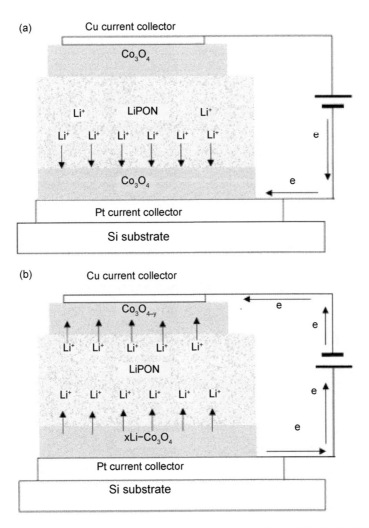

Fig. 7.3. Possible pseudocapacitance mechanism governing charge-discharge behavior of the Co_3O_4/LiPON/Co_3O_4 TFSCs: (a) charge process and (b) discharge process. Reproduced with permission from [20], Copyright © 2001, Elsevier.

depositing Mn films followed by electrochemical oxidation [22] or directly sputtering oxide targets [23]. Mn films with thicknesses in the range of 20–200 nm were deposited on Pt/Si substrates in Ar, which were then transferred to manganese oxide through electrochemical oxidation [22]. The manganese films possessed a porous dendritic microstructure structure, with promising pseudocapacitive behavior. An optimal specific capacitance (C_s) of 700 F·g^{-1} at a current density of 160 µA·cm^{-2}.

Figure 7.4 shows SEM images of the Mn thin films with different thicknesses after electrochemical oxidation. The convoluted dendritic structure became coarser as the film thickness was increased up to 100 nm, which usually did not vary from 100 nm to 200 nm. At the same time, surface area was increased significantly with increasing thickness of the films. Figure 7.5 shows areal capacitances as a function of current and sweep rate, for the four films with thicknesses of 20, 50, 100 and 200 nm. All the samples showed a decreasing trend in capacitance as the rate of current injection was increased, no matter whether current or voltage sweep methods were used. Figure 7.6 shows areal capacitances of the films measured at 160 µA·cm^{-2} as a function of thickness of the unoxidized films. The increase in capacitance was almost linear with the initial deposition thickness. The areal capacitance of the blank platinum films was about 100 µF·cm^{-2}, corresponding to double layer capacitance that was much smaller than the pseudocapacitance of the manganese films.

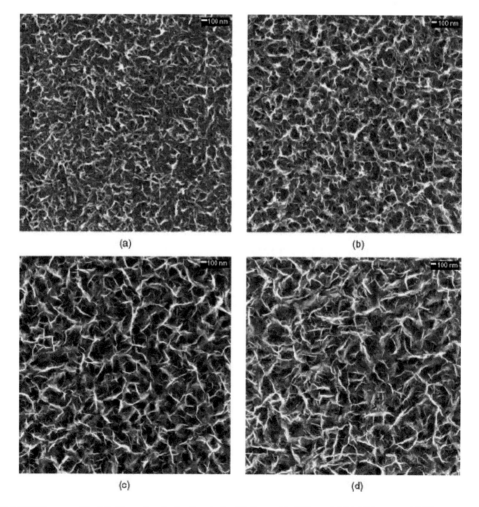

Fig. 7.4. SEM images of the Mn films after electrochemical oxidation, with different initial thicknesses: (a) 20 nm, (b) 50 nm, (c) 100 nm and (d) 200 nm. Reproduced with permission from [22], Copyright © 2004, Elsevier.

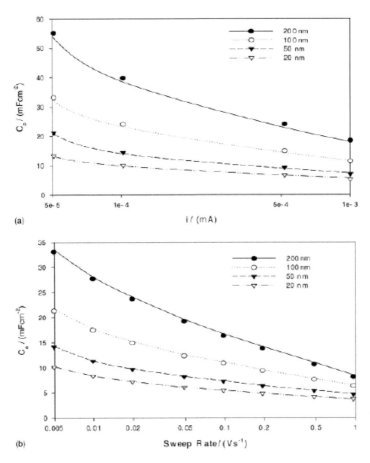

Fig. 7.5. Areal capacitance (C_a) as function of current (CP) and sweep rates (CV). Reproduced with permission from [22], Copyright © 2004, Elsevier.

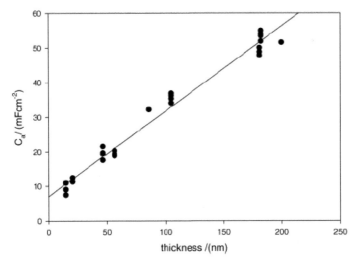

Fig. 7.6. Areal capacitance (C_a) as a function of initial thickness of the films. Reproduced with permission from [22], Copyright © 2004, Elsevier.

Chemical Vapor Deposition (CVD)

An in-plane fabrication approach was used to fabricate ultrathin supercapacitors, with electrodes based on pristine graphene and multilayer reduced graphene oxide [10]. The in-plane configuration was to utilize the surface of the individual graphene layer for energy storage. The thinnest devices could be fabricated with the 1–2 graphene layers, with specific capacities of up to 80 $\mu F \cdot cm^{-2}$. By using multilayered reduced graphene oxide (rGO), as specific capacitance of as high as 394 $\mu F \cdot cm^{-2}$ could be achieved.

Figure 7.7(a) shows schematic diagram of the in-plane device as supercapacitors. The 2D in-plane structure made full use of the advantages of the atomic layered and flat graphene nanosheets, making them typical 2D devices. In this case, the interaction of the electrolyte ions with the carbon nanosheets were significantly enhanced, as seen in Fig. 7.7(a). Therefore, the high surface area of the graphene layers were entirely utilized. Moreover, such a 2D structure could also effectively use the special electrochemical properties of edges of the graphene nanosheets along with the basal planes of graphene [24, 25]. Therefore, it was expected that usage efficiency of the electrochemical surface area of the graphene nanosheets could be maximized, due to the 2D in-plane design.

The Graphene (G) was synthesized by using a chemical vapor deposition method (CVD) [26], while multilayer graphene films, i.e., reduced Multilayer Graphene Oxide or rMGO, were obtained by chemically reducing the Graphene Oxide (GO) films prepared with a Layer-by-Layer (LbL) assembly method. The G and rMGO films were deposited on Cu foil and quartz substrates, respectively. In the conventional stacking geometry of supercapacitors, the graphene nanosheets were randomly oriented with reference to the current collectors. In contrast, the in-plane configuration allowed the graphene conductive planar sheets to be isolated, so as to form two electrodes, with a gap at the micrometer scale. Gold layers were deposited on external edges of the two electrodes as current collectors. A polymer-gel (PVA-H_3PO_4)

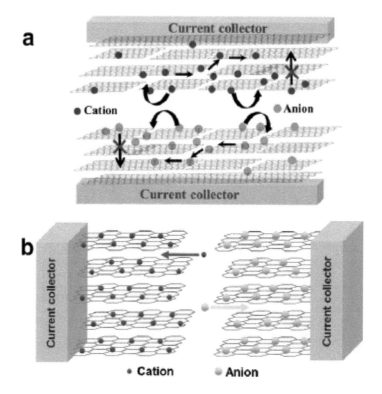

Fig. 7.7. (a) Schematic diagram of the stacking geometry of the thin film based supercapacitor. (b) Schematic diagram of the working principle of the in-plane supercapacitor used to evaluate the performance of the graphene electrodes. Reproduced with permission from [10], Copyright © 2011, American Chemical Society.

electrolyte filled the gap, as demonstrated in Fig. 7.8(a). In fact, the polymer-gel electrolyte acted as both the ionic electrolyte and the separator. Figure 7.8(b) shows a photograph of the 2D devices, which were compact, ultrathin, flexible and optically transparent.

The 2D in-plane structured rMGO films were deposited on quartz substrates by using LbL assembly method, followed by chemical reduction with hydrazine. Figure 7.8(d) shows a photograph of a representative LbL GO and rGO films on poly(ethyleneimine) (PEI). As illustrated in Fig. 7.8(e), the flat-layered structure contained no agglomeration, with a typical film thickness of 10 nm, which corresponded to about 21 layers of graphene nanosheets with an interlayer spacing of about 0.5 nm. Graphene films (G) were grown on Cu foils at 950°C with hexane as the carbon source in Ar/H$_2$. The G film exhibited single-layer characteristics, with 2–3 layers of graphene nanosheets as minority, as seen in Fig. 7.8(c). Electrochemical devices were fabricated on quartz substrates with the G films transferred from Cu substrates.

Vertically oriented graphene nanosheets were grown by using CVD method directly on metal current collectors, which were used to fabricate supercapacitors with ultra-high power performance [27]. Due to the special design, both the electronic and ionic resistances were minimized, so that the supercapacitors exhibited RC time-constants of < 200 μs. Moreover, because edge planes of the graphene nanosheets were largely exposed to the electrolyte, the charge storage capacity of the materials was significantly increased, as compared with those based on only basal plane surfaces. As a result, electrodes made with the vertically oriented graphene nanosheets were expected to have high-frequency EDLCs.

The vertically oriented graphene nanosheets were deposited directly on heated nickel substrates by using an RF plasma enhanced CVD method [28]. Before deposition, the substrates were plasma etched for 10 minutes in 40 vol% Ar + 60 vol% H$_2$, at a total pressure of 7 Pa. After that, Ni substrate was heated to 1000°C in H$_2$, followed by introduction of CH$_4$ and 1000 W plasma that was generated by using 40 vol% CH$_4$ and 60 vol% H$_2$ at a total pressure of 11 Pa. Figure 7.9 shows a schematic diagram of the growth process. In the beginning, graphitic islands were formed on the Ni substrate, known as Volmer-

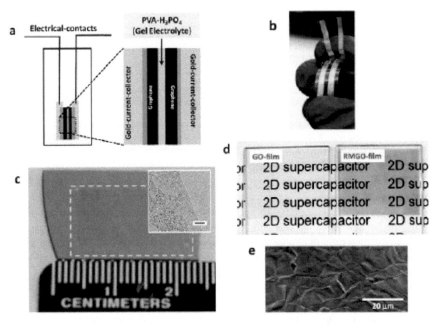

Fig. 7.8. (a) Schematic diagram of the 2D in-plane supercapacitors. (b) A prototype flexible 2D in-plane supercapacitor based on rMGO films. (c) Photograph of the large area single-layer graphene transferred onto a SiO$_2$ substrate from Cu foils, with the inset showing TEM image of the single-layer graphene. (d) A real-time photograph of the multilayer graphene oxide (GO) film obtained by using LbL (left) method and the rMGO film (right) after chemical reduction with hydrazine. (e) SEM image of inner layers of the rMGO film after cleaving the top surface using scotch tape. Reproduced with permission from [10], Copyright © 2011, American Chemical Society.

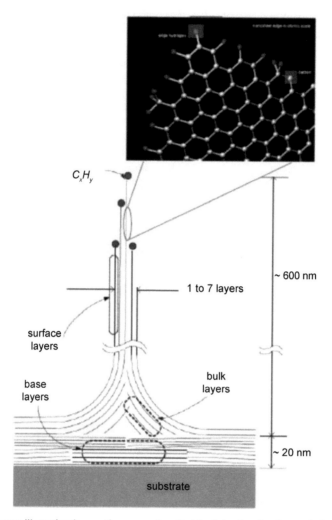

Fig. 7.9. Schematic diagram illustrating the growth process where the islands impinged so that vertical structures were formed, leading to the vertically oriented graphitic nanosheets with a height of about 600 nm and cross-sectional thickness of < 1 nm, corresponding to an average thickness of about three layers. Reproduced with permission from [27], Copyright © 2011, Elsevier.

Weber planar growth, followed by 2D growth, before they were impinged with other islands. The basal layers had thicknesses in the range of 10–15 nm. As the newly formed grain boundaries pushed the sp^2 bonds upwards, the hexagonal lattice was continuously grown, because there were free carbon atoms in the plasma. Due to the high growth temperature and the weak van der Waals bonding to surface of the 2D sheets, the formation of multilayer structures related to accumulation was effectively prevented. After the growth was conducted for 20 minutes at a substrate temperature of 700°C, a layer of vertically oriented graphitic nanosheets with a height of about 600 nm and a cross-section thickness of < 1 nm was developed. The vertical walls averagely consisted of three layers.

Figure 7.10(A) shows plane view SEM images of samples, demonstrating an irregular surface morphology that was caused by the defects generated because of the stress and hydrogen incorporation [27]. The structures were vertically grown on the metal substrate, so that there was a close contact between each other. The structure had a specific surface area of ~ 1100 m^2·g^{-1}, only slightly lower than half of the theoretical maximum value (2630 m^2·g^{-1}). Detailed views of edges of the vertically oriented nanosheet are shown in Fig. 7.10(B, C). It was found that the edge planes were largely exposed. Figure 7.11 shows SEM

Microsupercapacitors 457

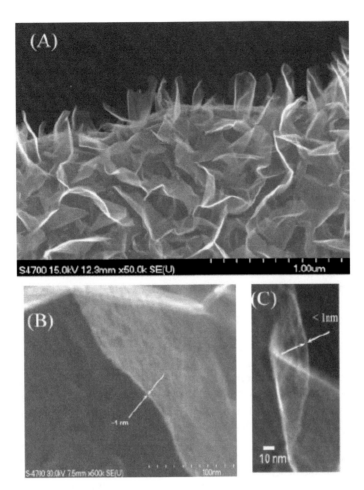

Fig. 7.10. (A) Shallow-angle view SEM images of the coated Ni electrode. (B, C) High magnification SEM images showing details of the vertically oriented graphene nanosheet. Reproduced with permission from [27], Copyright © 2011, Elsevier.

Fig. 7.11. Plane and cross-sectional SEM images of the vertical oriented graphene grown structures on a silicon substrate with hydrogen and acetylene gases as the carbon source, respectively. Reproduced with permission from [27], Copyright © 2011, Elsevier.

images of the vertically oriented graphene that was grown by using acetylene as the carbon source. As seen in Fig. 7.11(A), the vertically upright graphene nanosheets had a width of about 50 nm. The sample exhibited an open structure with direct surface accessibility, as demonstrated in Fig. 7.11(B).

Layer-by-layer (LbL) deposition

The LbL deposition technique has been widely used to prepare thin film electrodes of supercapacitors [12, 13, 29–31]. LbL derived thin films usually have a pretty dense microstructure, but are fairly permeable to ions, thus allowing to fabricate electrodes with energy density. For example, MWCNT films were deposited by using a LbL method, with negatively charged carboxylic acid functionalized MWCNTs (MWCNTs–COOH) and positively charged amine functionalized MWCNTs (MWCNTs–NH_2), for electrochemical supercapacitor applications [12]. Figure 7.12 shows a schematic diagram to describe fabrication process of the LbL thin films. The electrodes showed an average capacitance of 132 ± 8 F cm^{-3} (159 ± 10 F·g^{-1}) in 1 M H_2SO_4 electrolyte, which was much higher than the performances of VA-CNT electrodes.

Commercial MWCNT powder, with 95% purity, length of 1–5 μm and outer diameter 15 ± 5 nm, was used as the starting materials. The MWCNT powder was refluxed in concentrated H_2SO_4/HNO_3 (3/1 v/v, 96 and 70%, respectively) at 70°C, resulting in carboxylic acid functionalized MWCNTs, i.e., MWCNT-COOH, followed by thorough washing with deionized Milli-Q water (18 MΩ·cm) with nylon membrane filter with pore diameter of 0.2 μm. After drying, the carboxylated MWCNT sample was chlorinated by refluxing in $SOCl_2$ at 70°C for 12 hours. The $SOCl_2$ was then completely removed through evaporation, which was then used to prepare amine functionalized MWCNTs, MWNT-NH_2, through the reaction with $NH_2(CH_2)_2NH_2$ in dehydrated toluene at 70°C for 24 hours. The sample was thoroughly washed with ethanol and deionized water to obtain MWCNT-NH_2 powder after drying at 50°C in vacuum for 24 hours.

Both the MWCNT-COOH and MWCNT-NH_2 powders were dispersed in Milli-Q water with the aid of strong ultrasonication to obtain stable dispersions. The suspensions were dialyzed against Milli-Q water for a sufficiently long time to remove all the by-product and residual items during the powder preparation. After that, concentrations (0.5 mg/ml) and pHs of solutions of the suspensions were adjusted for thin film deposition through LbL assembly on various substrates. The LbL deposition was started by first immersing the substrates in the MWCNT-NH_2 suspension for 30 minutes, followed by rinsing the samples in three baths of Milli-Q water each for 2, 1 and 1 minutes. Then, the similar steps were applied with the MWCNT-COOH. After this cycle, a bilayer of MWCNT-NH_2 and MWCNT-COOH, i.e., MWCNT-NH_2/

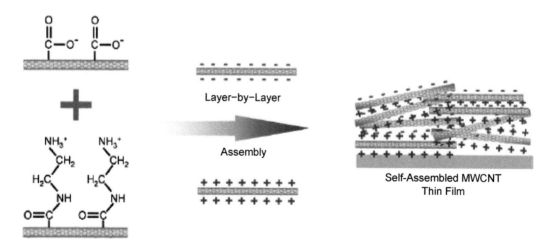

Fig. 7.12. Layer-by-layer assembled MWCNT thin films with positively and negatively charged MWCNT suspensions. Reproduced with permission from [12], Copyright © 2009, American Chemical Society.

MWCNT-COOH, was formed. By repeating this cycle, MWCNT films with designed thicknesses could be readily obtained.

Figure 7.13 shows surface SEM images of the cast film using the raw MWCNT suspension after sonication for 2 hours and the 20 bilayers MWCNT thin film assembled at pH values of 2.5 (+)/2.5 (−) [12]. In contrast to cast films from the raw MWCNTs, the LbL assembled MWCNT thin films exhibited a network structure with high porosity. In addition, the cast MWCNT film contained tube aggregation and aligned bundles, which was attributed to van der Waals interactions and capillary forces among the CNTs. As a result, such films demonstrated poor mechanical strength, uncontrollable thickness and dense packing microstructure, resulting in relatively low surface area. Comparatively, in the LbL assembled MWCNT films, the CNTs were randomly oriented without the presence of agglomeration, so that uniform porosity was developed, making them more suitable as electrodes of supercapacitors.

These differences come from the electrostatic repulsion between nanotubes that act against van der Waals interactions that yield close packed aggregates, as well as the electrostatic cross-linking between positively and negatively charge MWCNTs during the LbL process that lead to randomly oriented, kinetically driven CNT arrangements in the film. A cross-sectional view of a MWCNT thin film (Fig. 6.14a) demonstrates the conformal and uniform nature of the coating generated with LBL of MWCNTs on silicon wafer, suggesting their application to substrates without any geometric constraint. A torn film shown in the Fig. 6.14a insert reveals that most of the MWCNTs in the film are not parallel with the substrate, but form an interpenetrated structure with random orientations of the nanotubes. An angled cross-sectional view was created by cutting films on a slant (Fig. 6.14b); it more clearly shows the internal structure of the film, proving that a highly interpenetrated and porous structure is created. Since the MWCNTs have intrinsically high electrical conductivity and higher surface area, these porous network structures can work as fast electronic and ionic conducting channels, providing the basis for design of the ideal matrix structure for energy conversion as well as energy storage devices.

Figure 7.15 shows Cyclic Voltammogram (CV) curves of heat-treated MWCNT thin films, with a dimension of 0.7 cm × 2 cm, deposited on ITO-coated glass assembled at pH 2.5 (+)/3.5 (−), with different numbers of bilayers, were characterized in 1.0 M H_2SO_4 solution. CV curves the samples with different thicknesses exhibited similar rectangular shapes, due to the capacitive behavior of carbon materials. Integrated surface charge, according to the adsorbed and desorbed ions on the MWCNT thin film electrode, showed a linear relationship with thickness of the thin film, as seen in Fig. 7.15(b). An average capacitance of 132 ± 8 F·cm^{-3} or 159 ± 10 F·g^{-1} was achieved.

Similarly, Chemically Reduced Graphene (CRG) sheets were separated with MWCNTs by the LbL assembly method, for electrochemical micro-capacitor applications [13]. Layered structure, with submicron thin films of amine-functionalized MWCNTs (MWCNT-NH$_2$) and CRG, possessing high packing densities of up to 70%, were cross-linked with amide bonds. Electrodes based on the hybrid thin films exhibited a volumetric capacitance of 160 F cm^{-3} in an acidic electrolyte of 0.5 M H_2SO_4. It was found that the electrodes had a much lower capacitance in a neutral electrolyte of 1 M KCl, which confirmed that the high capacitances of the hybrid thin film electrodes in the acidic electrolyte was ascribed to the redox reactions between protons and surface oxygen-containing groups on the carbon materials. These hybrid thin films utilized the advantages of the high specific surface area of graphene and pseudo-capacitive effect of the functional groups on the surface of graphene.

To functionalize MWCNTs with amine groups, the pristine MWCNT was first treated to form MWCNT-COOH, with carboxylic acid groups on surface. Then, amine-substitution reaction introduce amine groups, thus leading to MWCNT-NH$_2$, as discussed above [12]. The MWCNT-NH$_2$ powder was dispersed in DI water to obtain suspension with a concentration of 0.5 mg·mL^{-1}, with the aid of sonication and constant stirring. After being dialyzed for 2 days, stable MWCNT-NH$_2$ suspension was formed, which could be used for LbL assembly. GO suspension with a concentration of 0.18 mg·mL^{-1} was obtained from a concentrated GO solution through sonication, centrifugation and dilution processed. pH values of the MWCNT-NH$_2$ and GO suspensions were adjusted to 2.5 and 3.5, respectively. Substrates, including ITO-coated glass, quartz and Si/SiO$_2$, were used to deposit the hybrid thin films. Before deposition, all the substrates were air-plasma treated for 5 minutes. During the LbL process, the substrates were dipped in (i) (+) MWCNT-NH$_2$ suspension, (ii) 3 times of washing aqueous solutions with pH = 2.5, (iii) (−) GO

460 *Nanomaterials for Supercapacitors*

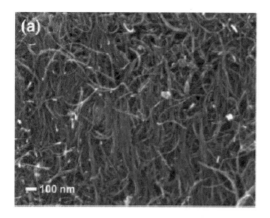 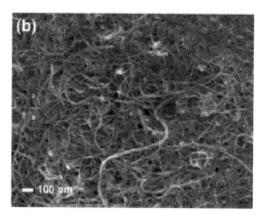

Fig. 7.13. Comparison of the topology of MWNTs using scanning electron microscopy (SEM) images. (a) Deposition of raw MWNT solution on silicon wafer after 2 hours sonication. (b) LBL assembled MWNT thin film at (pH 2.5 (+)/2.5 (−))$_{20}$. Reproduced with permission from [12], Copyright © 2009, American Chemical Society.

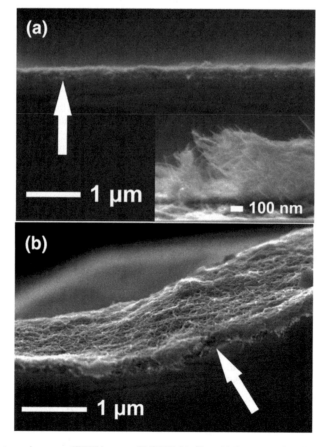

Fig. 7.14. Scanning electron microscopy (SEM) images of MWNT thin films. (a) A cross-section view of 15 bilayers of MWNT film (pH 2.5 (+)/3.5 (−)), and (b) a titled cross-sectional view of a 20 bilayer MWNT film (pH 2.5 (+)/4.5 (−)). White arrows indicate the interface between silicone substrate and MWNT thin films. Reproduced with permission from [12], Copyright © 2009, American Chemical Society.

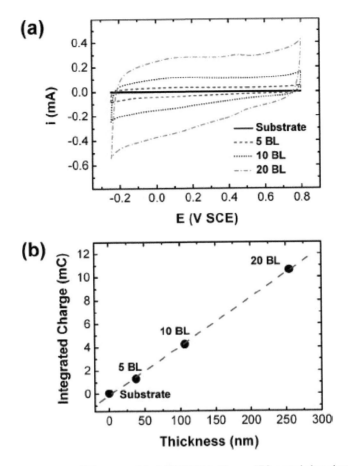

Fig. 7.15. (a) Cyclic Voltammograms (CV) curves of the MWCNT thin films on ITO-coated glass electrode (after the second heat-treatment) in 1.0 M H$_2$SO$_4$ at room temperature, measured at a scan rate of 50 mV·s^{-1}. (b) Integrated charge obtained from CV curves as a function of thickness of the films. Reproduced with permission from [12], Copyright © 2009, American Chemical Society.

suspension and (iv) 3 times of washing aqueous solutions with pH = 3.5. After this cycle from process (i) to (iv), 1 bilayer of MWCNT/GO was obtained. The dipping times of MWCNT-NH$_2$/GO suspension and three aqueous washing solutions were 30, 2, 1 and 1 minutes, respectively. The cycle was repeated to reach a desired number of bilayers then the film substrates were dried in air.

The LbL assembled MWCNT/GO thin films were reduced in hydrazine vapor at 120°C for 24 hours to form MWCNT/CRG hybrids. Figure 7.16 shows representative SEM and TEM images of the MWCNT/CRG samples. High-resolution image indicated that MWCNTs and CRG sheets were uniformly arranged in the MWCNT/CRG films, from the views of both surface and cross-section, as shown in Fig. 7.16(a, b). The CRG sheets were embedded in the MWCNT matrix. Moreover, it was revealed in TEM images that the MWCNT/CRG hybrid films possessed a random distribution of MWCNTs and CRG sheets, as demonstrated in Fig. 7.16(c, d).

Uniform ultrathin films with multilayer assemblies of polyaniline (PANi) and reduced Graphene Oxide (rGO) were prepared by using the LbL method, for the applications as electrodes of electrochemical supercapacitors [29]. The films possessed high electrical conductivity and large surface areas, making them suitable electrochemical capacitor applications. The electrochemical properties of the Multilayer Film (MF-)based electrodes, such as sheet resistance, volumetric capacitance, and charge/discharge ratio, could be optimized, through morphological modification and ways to reduce the GO to rGO in the multilayer

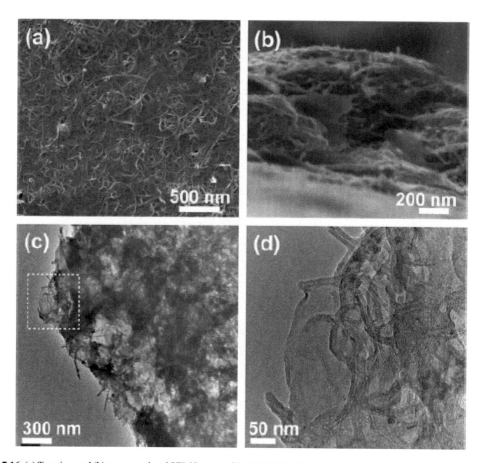

Fig. 7.16. (a) Top view and (b) cross-sectional SEM images of the MWCNT/CRG film with 20 bilayers. (c) Low-magnification TEM image and (d) magnified image of the boxed region in (c) of a piece from the 20 bilayer MWCNT/CRG thin films. Reproduced with permission from [13], Copyright © 2011, Elsevier.

films. Morphology of the GO monolayer in the multilayer films could be controlled through adjusting the concentrations of both PANi and GO. Among various reduction agents, it was found that HI was the most efficient one to reduce GO to rGO, whereas the damage to the virgin state of the acid-doped PANi was minimized. Electrochemical performances of the MF based electrodes could be further optimized. The MF-electrode composed of 15 PANi/rGO bilayers possessed a volumetric capacitance of 584 F·cm^{-3} at 3.0 A·cm^{-3}. Although this value decreased exponentially as the current density increased, approaching a value of 170 F·cm^{-3} at 100 A·cm^{-3}.

To prepare PANi solution, 200 mg emeraldine base form of PANi, with MW = 50000, was dissolved in 10 mL dimethylacetamide (DMAc), under constant stirring overnight, followed by sonicating for 10 hours, so that the PANi was completely dissolved, which led to a solution with a concentration of 20 mg·mL^{-1}. This solution was diluted with H$_2$O 10 times, with pH value to be adjusted to 3.0 with 3 N HCl solution. The pH of the solution was then quickly adjusted to 2.6 with 1 N HCl solution, in order to obtain hydrochlorated PANi. Before deposition, the PANi solution was filtered through a membrane with pore size of 0.2 μm and diluted with H$_2$O to different concentrations. In order to minimize the transformation of emeraldine salt into the emeraldine base, i.e., from partially doped state to undoped state, pH values of the washing and GO solutions were adjusted to be 2.6 and 3.5, respectively.

All the substrates were thoroughly cleaned with the aid of sonicating first in a piranha solution, with H$_2$SO$_4$/H$_2$O$_2$ = 7/3, and then in a mixture of H$_2$O/H$_2$O$_2$/NH$_3$ (5:1:1) at 80°C for 1 hour. After that, they were thoroughly rinsed with H_2O. The ITO glass substrates were cleaned with acetone, ethanol and H_2O successively, followed by drying with gentle stream of N_2, which were then immersed in a freshly prepared piranha solution at room temperature for 10 seconds. LbL assembly was carried out by dip-coating the freshly cleaned substrates. They were first immersed into the solution of positively charged PANi, with a concentration of 2 mM and pH = 2.6, for 15 minutes, followed by rinsing in H_2O at pH = 2.6 for 1 minute each in the three washing steps. After that, the the PANi coated substrates were immersed in the negatively charged GO solution, with a concentration of 0.5 mg·mL^{-1} and pH = 3.5, for 15 minutes. The LbL cycle was repeated to obtain PANi/GO bilayers with desired thicknesses.

Because the range of pH of the PANi solution was 2.6–2.7, 7% the amino groups of PANi were protonated to yield an ammonium group [32]. Figure 7.17(a) shows characteristic absorbance of the multilayer films at 310 and 880 nm, with the deposition of up to 15 alternative PANi and GO bilayers. The deposition of the PANi/GO bilayers on Si substrate was measured to have a thickness of 2.39 ± 0.08 nm, as illustrated in Fig. 7.17(b), which was confirmed by the thickness of about 2.34 nm, according to cross-sectional SEM image of the multilayer film consisting of 50 PANi/GO bilayers, as demonstrated in Fig. 7.18(a). The PANi and GO monolayers possessed thicknesses of 1.06 and 1.32 nm, respectively. As shown in Fig. 7.18(b), as the GO was reduced to RGO with HI, the multilayer films turned darker in color. For the LbL-assembled films on ITO-coated glass slides, as the number of PANi/RGO bilayers was increased, the samples were darkened more and more obviously. In addition, the chemical reduction process had almost no damage to the PANi/GO multilayer films, as revealed by the AFM images shown in Fig. 7.18(c). Only slight increase in rms roughness was observed, as the PANi/GO bilayers were HI-reduced to PANi/RGO films.

Multilayer films of Co–Al layered double hydroxide nanosheets (Co–Al LDH-NS) and Graphene Oxide (GO) were fabricated through LbL assembly, as shown schematically in Fig. 7.19, for flexible supercapacitor applications [30]. The Co–Al LDH-NS/GO multilayer films possessed a specific capacitance of 880 F·g^{-1} and an area capacitance of 70 F·cm^{-2}, at the scan rate of 5 mV·s^{-1}. The film also demonstrated outstanding cycle stability over 2000 cycles. After the films thermally treated at 200°C in H_2, both the specific capacitance and area capacitance were significantly increased to 1204 F·g^{-1} and 90 F·cm^{-2}, which was attributed to partial reduction of the GO nanosheets.

To synthesize and exfoliate Co–Al LDH-NS, cobalt chloride, aluminum chloride and urea were dissolved in 1 L Milli-Q water, so that their concentrations were 10, 5 and 35 mM, respectively [33]. After that, the solutions were mixed and refluxed at a high temperature, with continuous magnetic stirring for 2 days. Co–Al LDH-NS powder with a pink color was collected and thoroughly washed with water, following by air-drying at room temperature.

1.0 g Co–Al LDH was dispersed in 1 L NaCl aqueous solution with 1 M NaCl and 3.3 mM HCl. After purging with N_2 gas, the solution was stirred for 12 hours at room temperature, leading to NaCl–HCl exchanged Co–Al LDH, which was filtered and thoroughly washed with water and anhydrous ethanol. The sample was then dispersed in formamide with a concentration of 1.0 g·L^{-1}, followed by vigorous stirring for 2 days. As a result, a pink translucent suspension was obtained, which was centrifuged at 2000 rpm for 10 minutes to remove the unexfoliated LDH, resulting in a pink, transparent solution of Co–Al LDH-NS.

Co–Al LDH-NS/GO multilayer films were LbL assembled on thoroughly cleaned ITO substrates. The substrates were immersed into a PDDA aqueous solution with a concentration of 1.0 mg·mL^{-1} for 30 minutes to obtain a cationic surface prior to LBL assembly. For one cycle deposition, PDDA-coated ITO was first immersed into the Co–Al LDH-NS suspension with a concentration of 1.0 mg·mL^{-1} for 20 minutes, followed by washing with water and drying N_2 blow. Then, the substrates were immersed into the GO suspension of the same concentration for 20 minutes, followed by washing with water and drying N_2 blow. The cycle was repeated to reach the desired number of bilayers of Co–Al LDH-NS and GO films, i.e., 10, 20, 30, and 40 bilayers. The Co–Al LDH-NS/GO multilayer films could also be deposited on PET substrates as flexible electrodes.

According to AFM characterization results, the Co–Al LDH-NS and GO nanosheets had average thicknesses of 1.0 nm and < 1.2 nm, respectively. Figure 7.20(A) shows photographs of the bare ITO substrate and Co–Al LDH-NS/GO films with different numbers of bilayers deposited on the ITO substrates. With increasing number of the bilayers, the color of the Co–Al LDH-NS/GO films became darker and

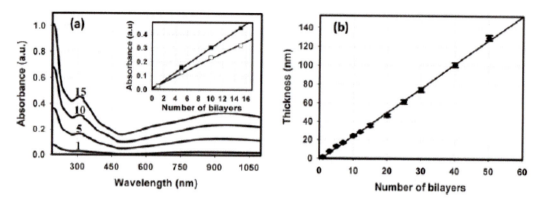

Fig. 7.17. (a) UV/visible absorption spectra of the multilayer films with different numbers of PANi/GO bilayers deposited on a fused silica substrate, with the inset showing plots of the absorbance curves, at 310 (■) and 880 nm (□), as a function of number of the bilayers. (b) Plot of film thickness as a function of number of the bilayers. Reproduced with permission from [29], Copyright © 2012, American Chemical Society.

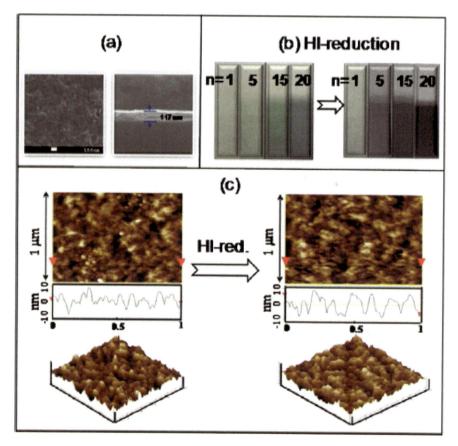

Fig. 7.18. (a) Surface and cross-sectional SEM images of the multilayer film with 50 PANi/GO bilayers on silicon substrate. (b) Representative photographs of the multilayer films composed of PANi/GO and PANi/RGO bilayers after HI-reduction, LbL-assembled on ITO glass slides, with n indicating the number of the bilayers. (c) 2D and 3D AFM images of the multilayer film with 10 PANi/GO and PANi/RGO bilayers (after HI-reduction) on silicon substrate, with rms roughnesses for the multilayer films with 10 PANi/GO and 10 PANi/RGO bilayers to be 3.75 ± 0.34 and 4.37 ± 0.42 nm, respectively. Reproduced with permission from [29], Copyright © 2012, American Chemical Society.

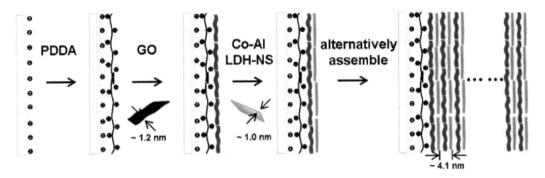

Fig. 7.19. Schematic diagram demonstrating LbL assembly process to fabricate the multilayer films of positively charged Co–Al LDH-NS and negatively charged GO nanosheets. Reproduced with permission from [30], Copyright © 2012, American Chemical Society.

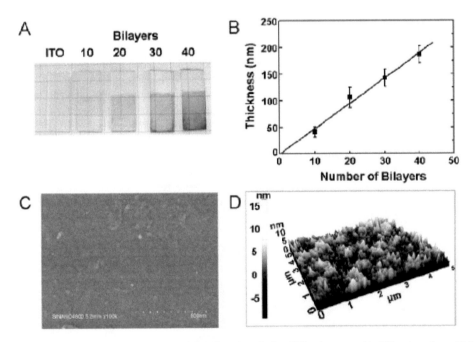

Fig. 7.20. (A) Photographs of the Co–Al LDH-NS/GO films deposited on ITO substrates with different numbers of bilayers. (B) Thickness of the Co–Al LDH-NS/GO films as a function of the number of bilayers. (C) SEM image and (D) 3D-AFM image of the 40-bilayer film. Reproduced with permission from [30], Copyright © 2012, American Chemical Society.

darker. Figure 7.20(B) shows thickness of the Co–Al LDH-NS/GO films as a function of the number of bilayers, which exhibited a linear relation, suggesting the uniform deposition of the individual layers. The average thickness of one bilayer was about 4.0 nm. Representative SEM and AFM images of the Co–Al LDH-NS/GO films are shown in Fig. 7.20(C) and (D), respectively, demonstrating their homogeneity.

Electrical deposition

Electrical deposition methods include electrochemical and electrophoretic depositions. These versatile deposition methods have been widely used to fabricate electrodes of the supercapacitor. During electrical depositions, charged species are driven by an external electric field to migrate towards an electrode, which acts as both the substrate and electrode. Due to the simple requirement on facilities and large-scale production capability, they are relatively cost-effective, as compared with other deposition techniques. Thin films of various nanocarbon materials, including activated carbon [34], CNTs [35], and graphene [36], as well as transition metal oxides [37, 38] and conducting polymers [39], have been fabricated by using electrochemical and electrophoretic deposition methods, for the application of electrodes of thin film supercapacitors.

Microsupercapacitors were developed with a several-micrometer-thick layer of nanostructured Onion-Like Carbon (OLC) with diameters of 6–7 nm, which was deposited by using electrophoretic deposition method [34]. The microsupercapacitors exhibited powers per volume that were comparable to electrolytic capacitors, but with capacitances that were four orders of magnitude higher and energies per volume that were an order of magnitude higher. At discharge rates of up to 200 V·s^{-1}, their performance was three orders of magnitude higher than the conventional supercapacitors. The performance could be further improved, by integrating the nanoparticles into a microdevice with a high surface-to-volume ratio, by using organic binders and polymer separators, because the ions could easily access the active materials.

OLCs with a quasi-spherical morphology are composed of concentric graphitic shells, which can be cost-effectively synthesized by annealing detonation nanodiamond powders at a large scale. OLCs can achieve specific surface area of up to 500 m^2·g^{-1}, which is lower than that of activated carbons. However their surface can be fully accessed by the ions of electrolytes, due to the absence of porous network inside the particles, as shown in Fig. 7.21(a). As a result, OLCs exhibited gravimetric capacitance of about one-third of activated carbons [40]. In this respect, it was expected that carbon onions can be used fabricate thin film electrodes for microsupercapacitors, because of the accessible external surface.

OLC nanoparticles were obtained by annealing nanodiamond powder at 1,800°C, with the morphology as shown in Fig. 7.21(b). The OCL nanoparticles were used to form colloidal suspensions, for electrophoretic deposition (EPD) onto silicon wafers with patterned interdigital gold current collectors. By using this way, the electrodes could be prepared without the use of organic binder, as illustrated in Fig. 7.21(c). No short circuit between the electrodes occurred, as seen in Fig. 7.21(d). The microsupercapacitor was constructed with 16 interdigital electrodes with thicknesses of 7 μm, as demonstrated in Fig. 7.21(e).

Cyclic Voltammograms (CVs) of the microdevices were tested at scan rates in the range of 1–200 V·s^{-1}. Reproducible and stable capacitive behavior up to 10,000 cycles could be achieved over a

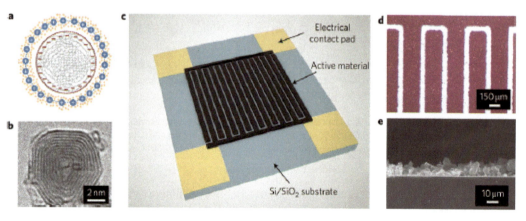

Fig. 7.21. (a) Cross-section of a charged zero-dimensional OLC (gray) capacitor, consisting of two layers of charges (blue and pink) forming the inner and outer spheres, respectively. (b) TEM image of a carbon onion synthesized at 1,800°C. Lattice spacing between the bent graphitic layers in the onions is close to 0.35 nm. (c) Schematic diagram of the microdevice with an area of 25 mm^2. Two gold current collectors made of 16 interdigital fingers were deposited by evaporation on an oxidized silicon substrate and patterned with the conventional photolithography/etching process. Carbon onions (active material) were then deposited by electrophoretic deposition onto the gold current collectors. (d) Photograph of the interdigital fingers with 100-μm spacing. (e) Scanning electron microscope image of the cross-section of the carbon onion electrode. A volumetric power density of 1 kW·cm^{-3} was obtained with a deposited layer thickness in the micrometre range, not the nanometre range. Reproduced with permission from [34], Copyright © 2010, Macmillan Publishers Limited.

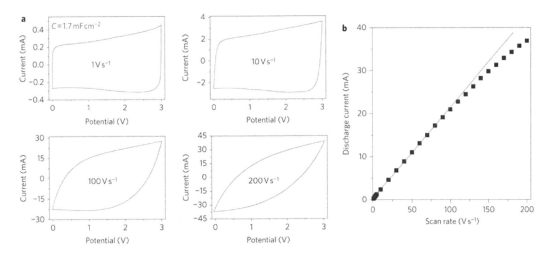

Fig. 7.22. (a) CVs measured at different scan rates in 1 M Et$_4$NBF$_4$/anhydrous propylene carbonate on a 16-interdigital electrochemical microcapacitor with a 7-μm-thick OLC layer. A typical rectangular shape, as expected for double-layer capacitive materials, is observed at an ultrahigh scan rate over a 3 V potential window. (b) Evolution of the discharge current as a function of scan rate. A linear dependence is obtained up to at least 100 V·s^{-1} in the capacitive region, indicating an ultrahigh power ability for the microdevices. Reproduced with permission from [34], Copyright © 2010, Macmillan Publishers Limited.

potential window of 3 V in 1 M solution of tetraethylammonium tetrafluoroborate in propylene carbonate, with a linear dependence of the discharge current on the scan rate and low resistance at scan rates of up to 100 V·s^{-1}, as shown in Fig. 7.22. It was found that power extracted from the microdevice could be simply increased by increasing the density of the interdigital electrodes. The microsupercapacitor possessed a specific capacitance of 0.9 mF·cm^{-2} at 100 V·s^{-1}.

Spray deposition

Spray deposition is a versatile technique to deposit thin films of carbon nanomaterials for microsupercapacitor applications. For example, thin films of MWCNTs were prepared by using an Electrostatic Spray Deposition (ESD) method [18]. It was found that areal capacitance of the MWCNTs thin film electrodes was increased with increasing deposition time and thickness of the films. A capacitance of about 80 mF cm^{-2} was achieved when a 28 μm thick film was used as the electrode. Due to their high conductivity and large pore size, the MWCNTs film electrodes retained about 85% of the initial capacitance at a scan rate of as high as 500 mV s^{-1}.

Two commercial MWNTs, i.e., (i) 'CNT-A' having a diameter of about 5 nm, lengths of 10–20 μm and BET surface area of 400 m^2·g^{-1} and (ii) 'CNT-B' having diameters of 10–20 nm, lengths of 10–50 μm and BET surface area of 200 m^2·g^{-1}, were used in this study. The CNT films were fabricated with two steps. Firstly, CNTs were dispersed in aqueous media. Hydrophilic functional groups were introduced onto surface of the MWCNTs, by treating them with nitric acid at 80°C for 4 hours, followed by washing with distilled H$_2$O, filtering and drying. The functionalized CNTs were sonicated in distilled H$_2$O for 1 hour, to form suspensions with concentrations of 0.04 and 0.12 wt% CNTs, which were then mixed with ethanol. Secondly, the CNT suspensions were used to conduct electrostatic spraying to deposit CNT films, with the ESD set-up being shown schematically in Fig. 7.23. The precursor MWCNT suspensions were pumped at flow rates in the range of 1–10 mL·h^{-1} into a stainless nozzle above the substrate by 2–10 cm, while the substrate was kept at temperatures of 80–200°C. The voltages in the range of 6–20 kV were applied between the spraying nozzle and the substrate.

Figure 7.24 shows representative SEM images of the CNT (CNT-A) films, indicating the presence of mesopores with spacing between the individual CNTs to several 10s of nanometers, as observed in the inset of (a). Because of the well-entangled and interconnected porous structures, the CNT films were

468 *Nanomaterials for Supercapacitors*

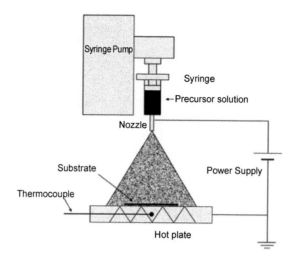

Fig. 7.23. Schematic diagram of the electrostatic spray deposition (ESD) system used to deposit the MWCNT films. Reproduced with permission from [18], Copyright © 2006, Elsevier.

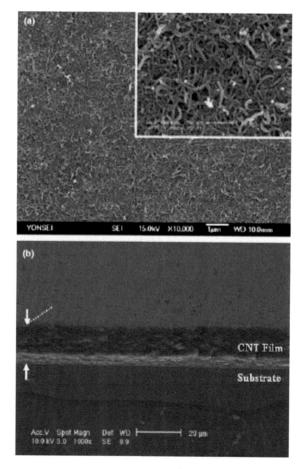

Fig. 7.24. SEM images of the CNT film electrode: (a) plane and (b) cross-sectional views. Reproduced with permission from [18], Copyright © 2006, Elsevier.

strongly adhered to the substrate. As illustrated in Fig. 7.24(b), the CNT films had a uniform thickness and flat surface, which was well attributed to the advantages of the ESD deposition method. In this case, tiny aerosol CNT droplets were formed and sprayed onto the substrate. Cyclic voltammograms of the CNT (CNT-A) film based electrode is shown in Fig. 7.25(a), within a potential window from −0.2 to 0.8 V$_{SCE}$. The curves revealed that the typical electric double layer capacitive behavior of the carbon based CNTs, with a pair of broad redox peaks.

Thin-films based on graphene nanoplatelets (GNPs) were prepared by using the Electrostatic Spray Deposition (ESD) technique, used as electrodes of microsupercapacitors [19]. The technique led to binder-free electrodes, while the materials possessed a fine pore structure, resulting in promising electrochemical performances for supercapacitor applications. For instance, cyclic voltammetry curves of the 1-μm-thick electrodes exhibited a nearly perfect rectangular shape, even at a scan rate of as high as 20 V·s^{-1}. At the scan rate of 5 V·s^{-1}, thin-film based electrodes had specific power and energy of 75.46 kW·kg^{-1} and 2.93 W·h·kg^{-1}, respectively. In addition, initial specific capacitance of the electrodes at low scan rates could be retained by about 53% at a high scan rate of 20 V·s^{-1}. It was found that rate capability of

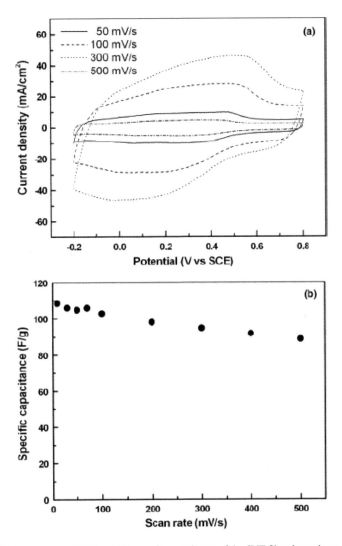

Fig. 7.25. (a) Cyclic Voltammograms (CVs) and (b) specific capacitance of the CNT film electrode as a function of potential scan rate. Reproduced with permission from [18], Copyright © 2006, Elsevier.

470 *Nanomaterials for Supercapacitors*

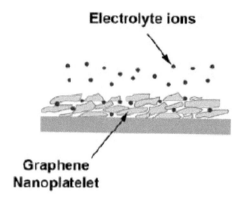

Fig. 7.26. Schematic diagram showing formation and microstructure of the thin GNP films with open pore structure which would allow for easy accessibility of the electrolyte ions. Reproduced with permission from [19], Copyright © 2012, Springer.

Fig. 7.27. SEM image of the deposited GNP film with a thickness of 6 μm: (a) cross-sectional view and (b) surface view. (c) TEM image of the commercially available GNPs. Reproduced with permission from [19], Copyright © 2012, Springer.

electrodes was influenced by thickness of the thin films. However, as thickness was increased to 6 μm, the initial capacitance still could be retained by about 30% at the scan rate of 20 V·s^{-1}.

Commercially available GNPs, with specific surface areas of 600–750 m^2·g^{-1} and average platelet diameter of <2 μm, were used to deposit the thin films [41]. To prepare the suspension for film deposition, 6 mg GNPs was added to 20 ml 1,2-propanediol in an ice bath, with the aid of ultrasonication, thus leading to a suspension with GNP concentration of 0.3 mg·ml^{-1}. The suspension was then fed to a stainless needle through a syringe pump at a feeding rate of 0.5 or 3 ml·h^{-1}, where voltages of 4–5 kV were applied between the needle and the stainless steel substrate, which was placed on a hotplate and heated at 250°C. GNP thin films with thicknesses of 1–6 μm were obtained by controlling the deposition time. Figure 7.26 shows a schematic diagram describing formation and microstructure of the GNP thin films.

Representative SEM images of the GNP thin films are shown in Fig. 7.27(a, b) [19]. It was observed that the GNPs interacted with one and another, so that an open pore structure was formed, which ensured smooth diffusion of electrolyte during the electrochemical reaction process, as illustrated in Fig. 7.27(b). Figure 7.27(c) shows a typical TEM image of the GNPs, which were corrugated and scrolled, because 2D sheet structure are thermodynamically stable in the bending state [42]. Electrodes based on the GNP thin film exhibited a moderate specific capacitance with outstanding high-power capability.

Ink-jet printing

Ink-jet printing is a very versatile technique that has been widely used to fabricate various thin films, with a wide range of applications, including in solar cells, thin film transistors and sensors [43]. More recently, this method has also applied to the fabrication of thin film electrodes of microsupercapacitors. For example, as a new method, ink-jet printing has been used to prepare graphene electrodes (IPGEs) for electrodes of supercapacitors [15]. Suspension of GO nanosheets dispersed in water were ink-jet printed on Ti foils, followed by thermal reduction at 200°C in N_2. Electrode based on the films had specific capacitances in the range of 48–132 $F \cdot g^{-1}$, at potential scan rates of 0.5–0.01 $V \cdot s^{-1}$, measured with 1 M H_2SO_4 as the electrolyte.

A commercially available GO suspension was used as the starting material, with a concentration of 2 $g \cdot mL^{-1}$ in water. The GO suspension was sonicated for 15 minutes, filtered with a 450 nm Millex syringe filter and then loaded into a print-head cartridge. Flexible Ti foils, with a thickness of 100 μm, were used as both the substrate and current collector. The substrates were thoroughly cleaned with acetone and deionized water to ensure a strong adhesion of the deposited films. The ink-jet printed GO films were reduced in N_2 flow at 200°C for 12 hours.

The GO ink was sufficiently stable, due to the presence of hydrophilic functional groups on surface of the GO nanosheets [44]. Figure 7.28(a) shows photograph of a bottle of the GO ink. The GO ink exhibited a viscosity and surface tension of 1.06 $mPa \cdot s$ and 68 $mN \cdot m^{-1}$, respectively, at room temperature, which were very close to those of deionized water, i.e., 0.99 $mPa \cdot s$ and 72 $mN \cdot m^{-1}$. Spherical ink droplets were generated with the GO ink at a velocity of about 7.5 $m \cdot s^{-1}$, as demonstrated in Fig. 7.28(b). By manipulating the firing voltage of the piezoelectric nozzles as a function of time, it was able to control the generation of the droplets.

For 10 pL droplet, a disk-shaped GO dot with a diameter of about 50 μm was formed on the Ti foil substrate after spreading and evaporation of the ink. As shown in Fig. 7.28(c), the circular GO dot was derived from 20 printing passes in 20 minutes. The GO thickness was sufficient for microscopy characterization, while the GO thickness with a minimum spatial resolution of 50 μm could be controlled through drop-to-drop placement and alignment. Continuous GO thin films with a lateral dimension 1 cm × 1 cm on the Ti foil substrate could be readily deposited, by allowing the adjacent droplets to be overlapped at a spacing of 15 μm between their centers, as demonstrated in Fig. 7.28(d). For effective electrochemical measurements, the printing step was repeated 100 times to deposit GO films with sufficient thickness. Resistance of the as-printed GO films on Kapton after thermal reduction was not higher than 1 MΩ. At the same time, the GO film color was varied from light brown to black after thermal reduction.

Island features with dimensions of 1–2 mm were present on surface of the IPGE, as seen in Fig. 7.28(d). SEM characterization results revealed that graphene was absent in the boundary areas. The islands were formed after the inkjet printing step, so that they were not related to the thermal reduction step. It was found that, if more hydrophilic substrates were used, less islands would be developed, which implied that their formation would be to be dependent on the hydrophobicity of the substrate. Inside the individual island, graphene nanosheets were densely stacked, so that a continuous network over average distances of 20–30 μm were formed, as revealed by the SEM image shown in Fig. 7.28(e). High magnification image of Fig. 7.28(f) indicated that the graphene nanosheets were more wrinkled and stacked less uniformly at the boundaries, as compared with the areas within the boundaries.

A similar microsupercapacitor was reported in a more detailed way [17]. As shown in Fig. 7.29(a), the evaporated current collectors had a lateral dimension 2 × 2 cm, with a lead, whereas the graphene was printed on collectors with a square area of 1.2 × 1.2 cm. Commercially available Kapton, with and without an FEP coating, was used as the substrate. The current collectors were used, i.e., (i) 50 nm Ti/300 nm Pt films on FEP-coated Kapton made with electron-beam evaporation and (ii) 20 nm Ti/230 nm Au films on bare Kapton. Ink-jet printing of GO film and the thermal reduction process were similar to those discussed above.

Supercapacitors were assembled with the reduced electrodes, as illustrated in Fig. 7.29. Firstly, the three sides of the supercapacitor with the printed electrodes face to face were thermally sealed, at 375°C for 60 seconds. If single-side FEP-coated Kapton (200FN011, 51-μm Kapton with 51-μm FEP) was used, the thermal healing step could be conducted directly. Otherwise, if plain Kapton (HN200, 51-μm Kapton) was employed, an intervening frame of double side FEP-coated Kapton (300FN929, 51-μm Kapton with 12.7-μm FEP on both sides) was required, as demonstrated in Fig. 7.29(b). The electrode separator could be either 10-μm polypropylene filter paper or Celgard 3501 separator material, which was inserted in

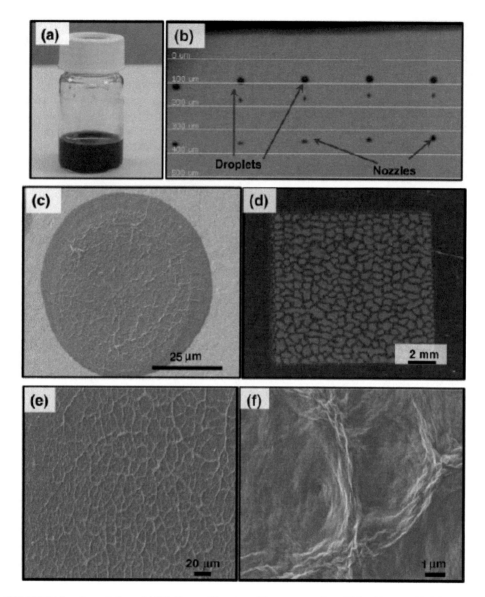

Fig. 7.28. IPGE ink and morphology: (a) GO dispersed in water with a concentration of 0.2 wt% as a stable ink, (b) spherical ink droplets generated with the piezoelectric nozzles, (c) SEM image of a circular GO dot printed on the Ti foil substrate after 20 printing passes at a spatial resolution of about 50 μm. (d–f) SEM images of the IPGE printed on the Ti foil substrate used for electrochemical characterization. Reproduced with permission from [15], Copyright © 2011, Elsevier.

between the electrodes. Then, 40 μl 0.5 M K_2SO_4 or 1 M H_2SO_4 electrolyte was injected, followed by the last seal to minimize the amount of trapped air in the device.

The ionic liquid electrolyte, 1-butyl-3-methylimidazolium tetrafluoroborate ($BMIM \cdot BF_4$), could not wet the reduced electrodes. To address this problem, the $BMIM \cdot BF_4$ was deposited on the unreduced GO electrodes, followed by thermal annealing at 250°C for 30 minutes. This is because both the $BMIM \cdot BF_4$ and GO are hydrophilic. The thermal annealing would decrease the viscosity of $BMIM \cdot BF_4$, so that it's wetting behavior was enhanced. At the same time, the presence of the ionic liquid during the reduction could hinder the shrinkage of the pores in between the graphene sheets, so that the ionic liquid ions would not be excluded.

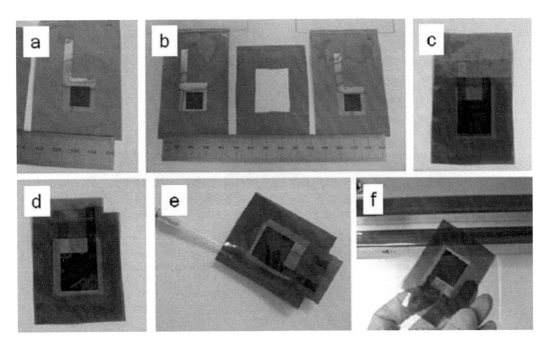

Fig. 7.29. Flexible supercapacitor assembly process: (a) graphene oxide printed on a metal current collector on Kapton and reduced at 200°C, (b) two electrodes were assembled with an intervening frame of FEP-coated Kapton, (c) electrodes were heat sealed face to face on three sides, (d) the polypropylene separator was placed between the electrodes, (e) the electrolyte was added and (f) the final seal was made. Reproduced with permission from [17], Copyright © 2014, Elsevier.

A printable aqueous gel electrolyte, together with an organic liquid electrolyte, was employed to fabricate microsupercapacitors, with energy and power densities of 6 kW·kg^{-1} for both electrolytes and 23 and 70 kW·kg^{-1} for aqueous gel electrolyte and organic electrolyte, respectively [16]. The flexible thin film supercapacitors were assembled with the printable electrolytes on plastic substrates. Meanwhile, the active electrodes were prepared with SWCNTs that were sprayed to form network structures, which acted as both electrodes and current collectors.

Commercially available SWCNTs powder was dispersed in water at concentrations of 1–2 mg·mL^{-1}, with the aid of ultrasonication. The stable suspension was then sprayed onto polyethylene–therephthalate (PET) substrates that were placed on a hot plate, with an air brush pistol. An entangled random network of SWCNTs was thus deposited on the PET, due to the evaporation of water during the spraying, as shown in Fig. 7.30(a). The SWCNT-coated PET substrates were directly used as thin film electrodes. Such SWCNT films had sheet resistances of 40–50 Ω, with an optical transmittance of about 12% and a thickness of about 0.6 μm. To make gel electrolyte, 1 g poly (vinyl alcohol) (PVA) powder, 10 mL water were mixed and 0.8 g concentrated phosphoric acid were mixed. The electrolyte was solidified, as the excessive water was evaporated. The SWCNT networks and the gel electrolyte were sandwiched to form the microsupercapacitor, as seen in Fig. 7.30(b). Liquid electrolytes at the same time, 1 M solutions of H_3PO_4, H_2SO_4 and NaCl in water, and 1 M $LiPF_6$ at a weight ratio of 1:1 in ethylene carbonate/diethylcarbonate (EC:DEC), were also prepared and evaluated as a comparison.

Cyclic voltammetry performances of the SWCNT network based electrodes are shown in Fig. 7.31 (a, b). It was found that the $LiPF_6$/EC: DEC electrolyte was stable up to a potential of as high as 3 V, implying the availability of high energy densities. Representative galvanostatic charge/discharge performances of the microsupercapacitors, with the aqueous polymer electrolyte and organic electrolytes, are shown in Fig. 7.31(c, d), measured at a current density of 1 mA·cm^{-2} (i.e., 30 mA·mg^{-1}). The charge/discharge curves demonstrated promising capacitive behavior.

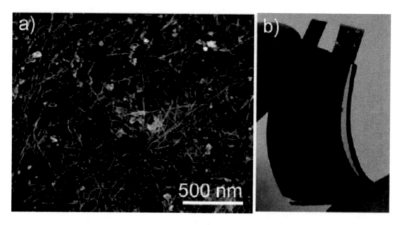

Fig. 7.30. (a) Scanning electron microscopy image of as-deposited SWCNT networks. (b) Thin film supercapacitor using sprayed SWCNT films on PET as electrodes and a PVA/H$_3$PO$_4$ based polymer electrolyte as both electrolyte and separator. Reproduced with permission from [16], Copyright © 2009, American Chemical Society.

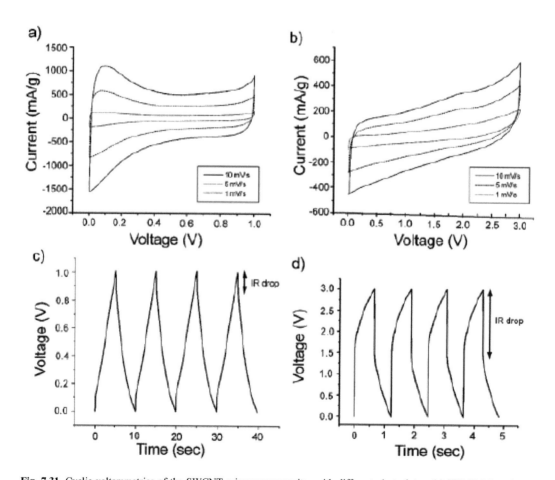

Fig. 7.31. Cyclic voltammetries of the SWCNT microsupercapacitor with different electrolytes: (a) PVA/H$_3$PO$_4$ polymer electrolyte and (b) LiPF$_6$/EC: DEC. Galvanostatic charge/discharge curves measured at a current density 1 mA·cm^{-2} or 30 mA·mg^{-1} for the thin film SWCNT supercapacitor with different electrolytes: (c) PVA/H$_3$PO$_4$ polymer electrolyte and (d) LiPF$_6$/EC:DEC. Reproduced with permission from [16], Copyright © 2009, American Chemical Society.

Other methods

The electrodes of microsupercapacitors should have sufficient mechanical strength to ensure the device fabrication processes. In this regard, several microporous carbon films have been demonstrated to possess these characteristics. For instance, films based on monolithic Carbide Derived Carbon (CDC) were prepared by chlorinating thin films of carbides, such as SiC and TiC, could be potentially used for such a purpose [45].

TiC thin films were deposited by using reactive DC magnetron sputtering with titanium target and carbon source of acetylene (C_2H_2) [46, 47]. The deposition was conducted with a power of 200 W at an overall pressure of 30 mTorr [47]. The feeding gas was a mixture of Ar and C_2H_2 at flow rates of 40 and 2.5 sccm, respectively. As the substrate temperature was about 700°C, textured TiC thin films with (111) orientation were produced, at a deposition rate of about 25 nm·min^{-1}. Substrates, including Si wafer (resistivity 1–100 Ohm cm with 200 nm wet thermal oxide), glassy carbon (one side diamond polished, Ra < 10 nm), Al_2O_3 (0001) oriented epi-polished and highly ordered pyrolytic graphite (HOPG), were used to deposit the TiC thin films. After deposition, the TiC thin films were etched by Cl_2 gas to form CDC films, at 250–500°C, through the reaction of $TiC + 2Cl_2 \rightarrow TiCl_4 + C$.

Electrochemical cells were assembled with stainless steel tube fittings and rods [47]. Two stainless steel rods with a diameter of 15 mm were place together, while the carbon film was deposited on a glassy carbon substrate and activated carbon electrode were in-between, separated by a PTFE separator, as shown schematically in Fig. 7.32. 1.5 M tetraethylammonium tetrafluoroborate ($TEABF_4$) in acetonitrile was used as the electrolyte. Activated carbon electrode of > 20 mg with 5 wt% PTFE was used as the counter electrode. The sample treated at 400°C exhibited an optimal volumetric capacitance of 180 F·cm^{-3} at a scan rate of 20 mV s^{-1}.

Figure 7.33 shows SEM images of the CDC films deposited on different substrates [47]. To avoid cracking and delamination of the films, thickness of initial TiC and chlorination temperature could be optimized, because the removal of Ti atoms generated tensile stress in the films during the chlorination process, which was increased with increasing thickness of the films. Although graphite and HOPG substrates were softer and possessed a rougher surface, as compared with Al_2O_3, oxidized Si and polished glassy carbon, films on different substrates showed no significant difference in cracking. Porous carbon films with a homogeneous microstructure could be deposited on Si substrates, with thicknesses of up to 3 μm through chlorination at a temperature of < 400°C.

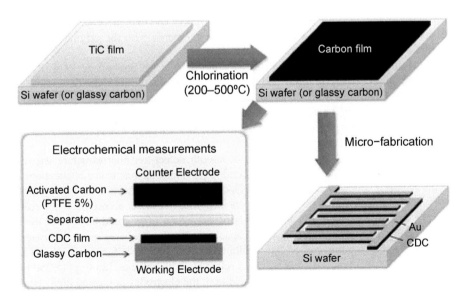

Fig. 7.32. Schematic diagram of the sample preparation and electrochemical characterization. Reproduced with permission from [47], Copyright © 2011, Royal Society of Chemistry.

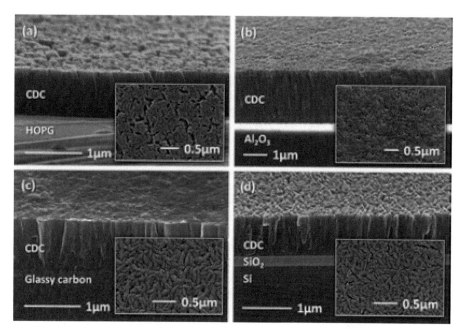

Fig. 7.33. SEM images of the CDC films on different substrates: (a) highly ordered pyrolytic graphite (HOPG) treated at 500°C, (b) on Al$_2$O$_3$ produced at 400°C, (c) on glassy carbon produced at 300°C and (d) oxidized Si wafer produced at 300°C. Reproduced with permission from [47], Copyright © 2011, Royal Society of Chemistry.

Since the TiC films deposited by using the reactive magnetron sputtering method exhibited a columnar structure, cracks were formed in lateral direction in the CDC film during the dry etching process if the temperature was too high. It was found that grain boundary cracks with sizes in the range of 100 nm were present, but they were only formed as the chlorination temperature was ≥ 400°C, as demonstrated in Fig. 7.33(a). Both density and size of the cracks were heightned with increasing chlorination temperature and thickness of the films. As illustrated in Fig. 7.33(c), the TiC film deposited on glassy carbon possessed a uniform thickness of about 1 μm and smooth surface, while CDC films produced at temperatures of 250–350°C had no almost no change in morphology as compared with the original TiC films, with thicknesses of 0.8–1 μm, very close to that of the TiC films.

As the TiC thin films were deposited on Si substrate without the buffer layer of SiO$_2$, N$_2$ reacted with both titanium atoms in the TiC films and the Si of the substrate at the same time, thus leading to delamination of the carbon film from the substrate. Therefore, a layer of thermal oxide (SiO$_2$) of about 200 nm was necessary to avoid the damage of the Si substrate and thus delamination of the carbon films, as revealed in Fig. 7.33(d). The protection of the oxide layer was effective at temperatures of up to 500°C for time duration of up to 30 minutes.

Activated Carbon Films (ACFs), with continuous, smooth, defect-free microstructure, were deposited on silicon substrates, followed by lithographic patterning for microsupercapacitor applications [48]. By controlling the synthesis conditions, the porous ACFs could be strongly attached to the original substrate or separated and thus transferred to new either dense or porous substrate, according to the requirement of applications. Also, both surface area and porosity of the carbon films could be tailored by adjusting the activation conditions. Electrical double-layer capacitor microdevices were assembled with the thin ACF. Electrodes based on the thin carbon film exhibited a pretty high specific capacitance of 510 F·g^{-1} (> 390 F·cm^{-3}) at a slow scan rate of 1 mV·s^{-1} and a relatively high value of 325 F·g^{-1} (> 250 F·cm^{-3}) at a very high current density of 45,000 mA·g^{-1}. In addition, the electrodes possessed high stability after 10,000 galvanostatic charge–discharge cycles.

Sucrose, DI water and H_2SO_4, with a mass ratio of 1:0.5:0.1, were mixed to form a solution. Si wafers with a diameter of 10 cm and SiO_2 layer of 1 μm were pre-etched in piranha solution in order to hydroxylate the surface to be extremely hydrophilic. After that, the solution was spin-coated on the Si substrates. Thickness of the sucrose solution coating was controlled by repeating the coating. The sucrose solution coatings were then carbonized and annealed in vacuum at 700°C for 2 hours. After carbonization and annealing, the Si wafer with carbon coatings were heated to 900°C in Ar flow, followed by activation using pure CO_2 gas at a flow rate of 500 mL·min^{-1} for time durations of 15 minutes, 1 hour and 2 hours, with carbon film samples to be denoted as ACF-15min, ACF-1h, and ACF-2h, respectively.

Before and after physical activation of the sucrose-derived porous carbon films on the wafers were continuous and very uniform, with no microcracks present, as shown in Fig. 7.34. As the activation time was increased from 15 minutes to 2 hours, thickness of the activated carbon films was slightly decreased from 1.8 μm to 1.5 μm. The decrease in thickness of the films was attributed to two factors, (i) compaction of the carbon film and (ii) surface oxidation. This is one of the issues of physical activation procedures, because of the preferential oxidation of carbon at the surface, especially when high concentration of an oxidant is used.

The ACFs could be peeled off from the substrate after prolonged soaking in proper solvents, e.g., acetone. The films demonstrated strong mechanical integrity. For example, 1–2 μm thick films could be bent to a large degree without failure. They could also withstand cutting or tearing and could be transferred and redeposited onto various types of substrates, including tubular, as seen in Fig. 7.35(a), and porous planar structures, such as alumina membrane, as revealed in Fig. 7.35(b–d). If the ACFs were annealed in Ar at 1100°C, they were strong attached to the SiO_2/Si substrates, so that they could be subjected to

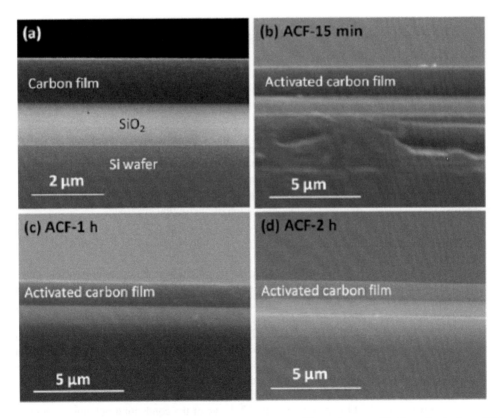

Fig. 7.34. Cross-sectional SEM images of the carbon films deposited on silicon substrates with after carbonization at 700°C (a) without physical activation and with physical activation in CO_2 at 900°C for (b) 15 minutes, (c) 1 hour and (d) 2 hours. Reproduced with permission from [48], Copyright © 2013, American Chemical Society.

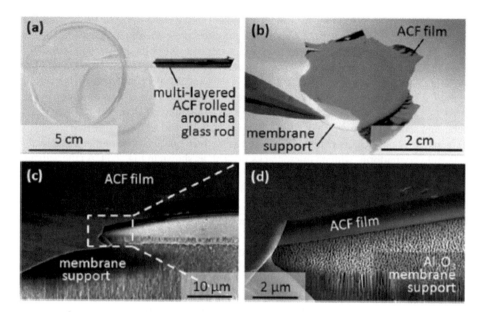

Fig. 7.35. ACF redeposited on various substrates. Photographs of the ACF on (a) a glass rod and (b) an alumina membrane. (c, d) SEM images of the cleaved membrane based on ACF deposited onto anodic alumina. Reproduced with permission from [48], Copyright © 2013, American Chemical Society.

lithographical patterning, which was attributed to the fact covalent bonds between Si and C were formed due to the sufficiently high temperatures [49].

On-Chip Microsupercapacitors

In-plane configuration

For in-plane configuration microsupercapacitors, interdigital electrodes are usually used as current collectors, which can be fabricated in two ways, i.e., (i) below active materials and (ii) on top of active materials. In the first case, interdigital current collectors are first deposited on a substrate, onto which active electrode materials are coated in the form of thin films. In the second configuration, thin film electrodes are first grown on a substrate, followed by the fabrication of interdigital electrodes.

Conducting polymers

Initially, on-chip microsupercapacitors with in-plane configuration were based on conducting polymers deposited by using the electrodeposition method. Conducting polymers used for such microsupercapacitors included polypyrrole (PPy) and poly (3-phenylthiophene) (PPT), which were coated on interdigital electrodes on Si/SiO$_2$ substrates [50, 51]. For instance, all-solid-state electrochemical microsupercapacitors have been fabricated by using photolithography, combined with electrochemical polymerization and solution casting [51]. Au interdigital microelectrode arrays were fabricated by using ultraviolet photolithography and a wet-etching method. Conducting polymers, polypyrrole (PPy), were potentiostatically deposited onto the microelectrodes. Microsupercapacitors were constructed with 50 parallel-connected pairs of the Au microelectrodes. The width of the microelectrodes and the distance between them are each 50 μm. Cell capacitance was controlled by the total synthesis charge of the conducting polymers, where a cell potential of 0.6 V was obtained. Figure 7.36 shows a schematic diagram illustrating fabrication procedure of the all-solid-state microsupercapacitors.

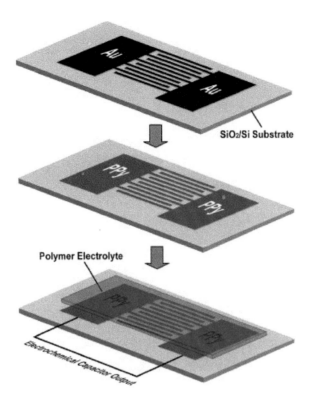

Fig. 7.36. Fabrication procedure of the all-solid-state microsupercapacitors. Reproduced with permission from [51], Copyright © 2004, Elsevier.

PPy was electrochemically synthesized on the gold microelectrode arrays. A three-electrode set-up was used, with the gold microelectrode array as the working electrode and a platinum plate as the counter electrode. Either Ag|AgCl (saturated KCl) or Ag|0.1 M AgNO$_3$(Ag|Ag$^+$) was used as the reference electrode, depending on the electrolyte medium. PPy was electrosynthesized on the Au microelectrode arrays at 0.65 V vs. Ag|AgCl in a 0.1 M pyrrole aqueous solution, with 0.1 M H$_3$PO$_4$ (85%) as the supporting electrolyte at 25°C or at 0.55 V vs. Ag/Ag$^+$ in 0.1 M pyrrole–propylene carbonate solution with 0.2 M lithium triflate (LiCF$_3$SO$_3$) as the supporting electrolyte at 25°C, until desired charges were reached. 2 vol% water was added into the PC solution to promote the electrosynthesis of PPy. Two polymer electrolytes, (i) aqueous-based PVA–H$_3$PO$_4$–H$_2$O gel and (ii) non-aqueous-based PAN/LiCF$_3$SO$_3$-EC/PC gel, were used for electrochemical characterization of the microsupercapacitors.

Figure 7.37 shows SEM images of the all-solid-state microsupercapacitors. A cathode–anode pair of the microsupercapacitor was formed with two adjacent electrodes. Capacities and powers of the microsupercapacitors could be readily designed by adjusting configuration of the PPy microelectrodes, such as the distance between fingers and the total number of microelectrodes. After the PPy films were deposited on the Au current-collectors, a PVA-based polymer electrolyte was coated on the PPy microelectrode arrays. It was found that the contact between the conducting polymer and the electrolyte was quite intimate. Small voids were observed at edges of the PPy electrodes, due mainly to the shrinkage of PVA caused by the coating for SEM experiments, as seen in Fig. 7.37(a, b). In addition, because the electric field was strengthened at the edges of the electrode, electrosynthesis of PPys was enhanced, as compared with the flat areas. As a result, the films were thicker at the edge area position, but still uniform in other areas, as illustrated in Fig. 7.37(c).

Figure 7.38(a) shows cyclic voltammograms of the all-solid microsupercapacitors with PPy electrodes and PVA polymer electrolytes. Four charges were used to deposit the PPy films. It was observed that as

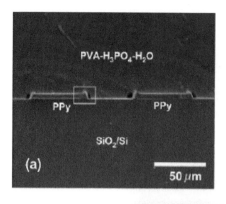
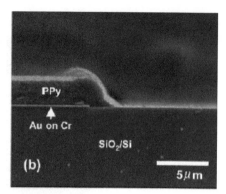
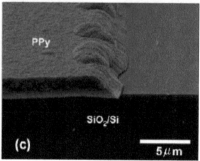

Fig. 7.37. SEM images of the all-solid-state microcapacitors composed of PPy electrodes and PVA–H$_3$PO$_4$ electrolytes: (a, b) cross-sectional views and (c) slightly tilted view with the PVA film being peeled off. Reproduced with permission from [51], Copyright © 2004, Elsevier.

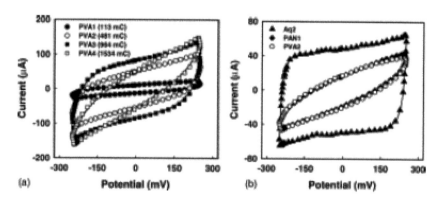

Fig. 7.38. Cyclic voltammograms of the microcapacitors: (a) effect of synthesis charge of PPy (50 mV·s^{-1}) and (b) comparison of microcapacitors with three different electrolytes (10 mV·s^{-1}). Reproduced with permission from [51], Copyright © 2004, Elsevier.

the synthesis charge of PPy was increased, the current output was increased, if the charge level was below 1000 mC. Once the charge was further increased to over 1000 mC, a resistor-like behavior was observed in the CV, while the current output was decayed quickly. Such a non-ideal behavior has been attributed to the low ionic conductivity in bulk of the relatively thick PPy films.

Figure 7.38(b) shows CVs of three microsupercapacitors. The two microsupercapacitors with polymer ionic conductors as electrolytes (PAN1 and PVA2 cells) exhibited quite similar shapes and charging currents, whereas CV curve of the one operated with aqueous H_3PO_4 electrolyte (Aq2 cell) was rectangular. The difference in electrochemical behavior between the liquid and polymer electrolyte cells was directly related to their difference in cell resistance, which was determined by two factors, i.e., (i) ionic conductivity of the electrolyte and (ii) wettability of the electrolyte within the electrodes [51].

Nanocarbons

Microsupercapacitors based on nanostructured carbon materials, including activated carbon, CDC, OLC, CNTs and graphene, have been well demonstrated [34, 46, 52]. A carbon-based microsupercapacitor was developed by using the ink-jet printing method on Si substrate, with potential applications as energy storage devices in self-powered modules [53]. The ink was prepared by mixing activated carbon powder with PTFE binder in ethylene glycol, which was stabilized with a surfactant. The carbon ink was coated on patterned gold current collectors on Si substrates. Figure 7.39 shows a schematic diagram of the microsupercapacitors, with two gold current collectors made of 20 interdigital fingers that were deposited by using an evaporation method and patterned by using the conventional photolithography/etching process. To ensure high coating homogeneity, the substrate was heated at 140°C. The microsupercapacitors exhibited promising electrochemical performances, with typical capacitive behavior over a wide potential range of 2.5 V and an optimal capacitance of 2.1 mF·cm^{-2}.

Figure 7.40(a) shows a photograph of a microsupercapacitor with dimensions of 20 fingers, 40 μm wide (w), 400 μm long (L) and 40 μm of interspace (i). Gold microwires were bonded from the microdevice to the package for electrochemical characterization, as shown in Fig. 7.40(b). The activated carbon was homogeneously deposited on the microsupercapacitor, with a well-defined pattern and no short circuit occurred between the interdigital electrodes. Thickness of the activated carbon film was 1–2 μm, depending on the micro-device samples.

A sequential CVD process was employed to fabricate microsupercapacitors with electrodes based on 3D hybrid of graphene and CNT (G/CNTCs) [54]. Cr/Ni interdigital electrodes were first deposited on Si/SiO$_2$ substrates and patterned, onto which Few-Layer Graphene (FLG) was coated as current collectors. Then, catalyst particles were deposited for the growth of CNTs. Finally, dense CNT layers were grown on the graphene current collectors. With the presence of the graphene layer, electrical contact of the CNT electrodes was significantly enhanced. The microsupercapacitors exhibited outstanding frequency behavior, with an impedance phase angle of as large as −81.5° at 120 Hz, which was very close to the phase angle of −90°, required for electrolytic capacitors used in filtering applications. The devices had specific capacitances of 2.16 mF·cm^{-2} and 3.93 mF·cm^{-2}, in aqueous electrolytes and ionic liquid electrolytes,

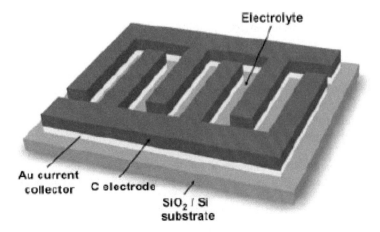

Fig. 7.39. Schematic drawing of the interdigital microsupercapacitor. Reproduced with permission from [53], Copyright © 2010, Elsevier.

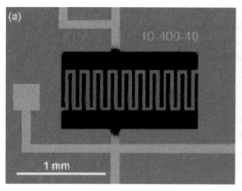 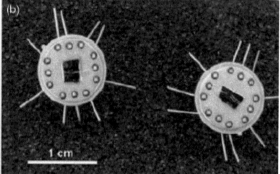

Fig. 7.40. Photographs of (a) a microsupercapacitor with 20 fingers, 40 μm wide, 400 μm long and interspaced by 40 μm and (b) the chips at lower magnification. Reproduced with permission from [53], Copyright © 2010, Elsevier.

respectively. In addition, the microsupercapacitors could be reversibly charged and discharged at a rate of as high as 500 V s^{-1}.

Figure 7.41(a) shows a schematic diagram of the patterned Cr/Ni interdigital electrodes, together with the 3D G/CNTCs hybrid (inset) [54]. The catalysts of Fe/Al$_2$O$_3$ were deposited on top of the CNTCs-graphene-Ni pillars. To fabricate the microsupercapacitors, interdigital-finger geometries on Si/SiO$_2$ substrates were first patterned by using the conventional photolithography, onto which Cr (10 nm) adhesion layer and Ni (450 nm) graphene growth catalyst layer were coated. Then, few-layer graphene (FLG) was grown on the patterned Ni electrodes by using a CVD [55]. After that, catalyst particles of Fe/Al$_2$O$_3$ were patterned and deposited on the FLG layer. Finally, CNTCs were grown on the FLG layer by using CVD.

Figure 7.41(b) shows a SEM image of the G/CNTCs-MC. It was observed that the CNTCs were split into CNT pitches of 1–2 μm in diameter, with exposed tip-ends to be capped by the catalyst particles of Fe/Al$_2$O$_3$, i.e., following the Odako growth behavior [56]. Expectedly, these individual pitches would facilitate an efficient access of electrolyte into the active CNTs, so as to achieve high specific capacitance and desired frequency response. The graphene was used as a buffer layer to block the Fe catalyst particles to react with the Ni electrodes. Representative TEM images of the CNTs are shown in Fig. 7.41(c–e), revealing that there were single-, double- and few-walled CNTs, with diameters ranging from 4 to 8 nm. Due to the densely packed CNTs with small diameters, the sample could have ultrahigh surface area [56]. By adjusting the growth time, height of the CNTCs could be well controlled. After growth for 1, 2.5 and 5 minutes, the CNTCs had heights of 10, 15 and 20 μm, respectively, with SEM images as illustrated in Fig. 7.41(f–h). The SEM images also indicated that growth direction of the CNTCs was vertical to the surface of the graphene layer, which ensured electrical conduction between the active material and the current collectors.

Electrochemical performances of microsupercapacitors can be further improved by using high surface area graphene as the electrodes. Due to the open planar structure, graphene offered minimized diffusion barriers for the transport of electrolyte ions. An ultra-high power handling the microsupercapacitor has been reported, with micro-patterned interdigital electrodes based on composites of reduced Graphene Oxide (rGO) and carbon nanotube (CNT) [57]. The binder-free microelectrodes were fabricated by using the Electrostatic Spray Deposition (ESD) method, combined with the photolithography lift-off technique. During the ESD deposition process, GO nanosheets had been reduced to rGO ones, so that no thermal or chemical reduction was required. More importantly, the in-plane interdigital design of the microelectrodes allowed highly efficient accessibility of the electrolyte ions into stacked rGO nanosheets via the electro-activation process.

By forming hybrid with CNTs, the restacking issue of rGO nanosheets could be readily addressed, which thus led to improved energy and power densities of the microsupercapacitors. The microdevices exhibited a specific capacitance of 6.1 mF·cm^{-2} at a scan rate of 0.01 V·s^{-1}. Moreover, a pretty high

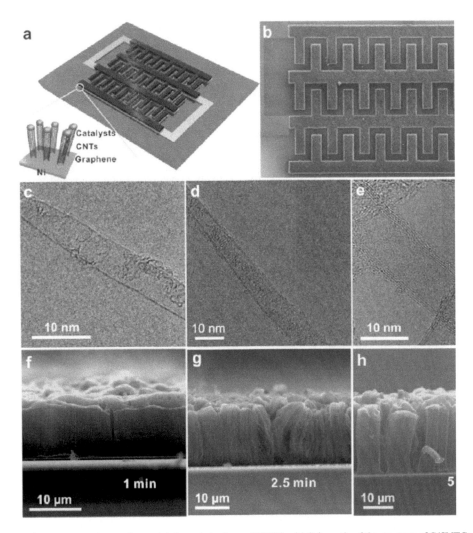

Fig. 7.41. Microsupercapacitors and material characterizations of CNTCs. (a) Schematic of the structure of G/CNTCs-MCs, with inset showing an enlarged scheme of the Ni-G-CNTCs pillar structure. (b) SEM image of a G/CNTCs-MC. (c–e) TEM images of individual single-, double- and few-wall CNTs. (f–h) Cross-sectional SEM images of the CNTCs after growth for 1, 2.5 and 5 minutes. Reproduced with permission from [54], Copyright © 2011, American Chemical Society.

specific capacitance of 2.8 mF·cm^{-2} was achieved, even at the scan rate of as high as 50 V·s^{-1}. The microsupercapacitors had a high-frequency response, with resistive-capacitive time constants to be as low as 4.8 ms, which was attributed to the introduction of CNTs, electrolyte-accessible and binder-free microelectrodes, together with the interdigitated in-plane configuration. Therefore, such interdigitated rGO–CNT composite electrodes would be promising candidates for on-chip energy storage application with high power capability.

The active materials were integrated on interdigital Ti/Au microelectrodes to fabricate the microsupercapacitors. The microelectrodes were fabricated by using a removable microfabricated photoresist mask, which covered the contact pads and the space between the microelectrodes. The microsupercapacitors had 20 in-plane interdigital microelectrodes, 10 positive and 10 negative ones. An individual microelectrode had a width of 100 μm and a length of 2500 μm, with a distance between the adjacent microelectrodes to be 50 μm. The layers of rGO and CNTs were derived from the solutions containing 100% GO and 100% CNTs, respectively, with samples of rGO-CNT-9-1 and rGO-CNT-8-2

484 *Nanomaterials for Supercapacitors*

to mean the GO:CNT weight ratios of 9:1 and 8:2, respectively. Thicknesses of all the electrodes were fixed to be 6 μm.

Figure 7.42 shows SEM image of the rGO microelectrodes, revealing that stacked layers of the graphene nanosheets had micron-sized wrinkles, due to the bending of the GO nanosheets during the deposition process. Local folding and non-uniform stacking were observed in the rGO layers, as seen in Fig. 7.42(b). The stacking of the graphene nanosheets resulted in extended irregular porous structures, which could serve as diffusion channels, so as to promote transportation of the electrolyte ions inside the active materials. Representative SEM images of the rGO–CNT hybrid electrodes are shown in Fig. 7.42(c–f). It was observed that CNTs were uniformly present in between the rGO nanosheets, with much less degree of rGO stacking.

In this case, CNTs acted as a nano-spacer to effectively prevent the aggregation and restacking of the graphene nanosheet, which ensured highly accessible surface area for the electrolyte ions. Also, the binder-free deposition of ESD indirectly increased the relative content of the active materials, thus leading to a positive effect on electrochemical performance of the microsupercapacitors. In addition, the narrow spacing between the microelectrodes would have shortened the mean diffusion pathway of the electrolyte ions. Furthermore, due to the small size of the electrodes and side-by-side in-plane configuration, electrolyte ions could easily diffuse in between the rGO nanosheets throughout the entire thickness of the electrodes.

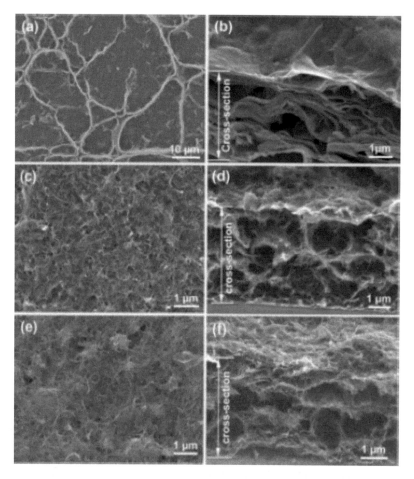

Fig. 7.42. Representative SEM images of the electrodes: (a) top view, (b) tilted-35° view, (c, d) rGO-CNT-9-1 microelectrodes and (e, f) rGO-CNT-8-2 microelectrodes. Reproduced with permission from [57], Copyright © 2012, WILEY-VCH Verlag GmbH & Co. KGaA, Weinheim.

It is expected that electrochemical performances of the microsupercapacitors can be further improved, by optimizing the electrode compositions and structural configuration.

It has been found that Graphene Quantum Dots (GQDs) could be used as electrode materials of microsupercapacitors [58]. Symmetric microsupercapacitors, GQDs//GQDs, were fabricated by using a simple electro-deposition approach. The GQDs microsupercapacitor demonstrated a superior rate capability at scan rates of up to 1000 V s^{-1}, outstanding power response with very short relaxation time constants, $\tau_0 = 103.6$ μs in aqueous electrolyte and $\tau_0 = 53.8$ μs in ionic liquid electrolyte, and excellent cycle stability. Furthermore, by employing MnO_2 nanoneedles as the positive electrode and GQDs as the negative electrode in aqueous electrolyte, asymmetric microsupercapacitor, GQDs//MnO_2, was assembled, which exhibited promising electrochemical performances.

GQDs were derived from GO power by using a one-step solvothermal method [59]. 540 mg GO powder was dispersed in 40 mL DMF, with the aid of ultrasonication for 30 minutes, which was then hydrothermally treated at 200°C for 8 hours. After the reaction was finished, the product was collected by filtering with a 0.22 μm microporous membrane, leading to a brown filter solution, which was suspension of GQDs in DMF. Dry powder of GQDs was obtained by heating the brown filter solution at 80°C to evaporate the solvent.

To fabricate GQDs//GQDs symmetric microsupercapacitors, GQDs were electrodeposited on pre-formed Au interdigital electrodes, with 50 mL suspension in DMF containing 3.0 mg GQDs and 6.0 mg $Mg(NO_3)_2 \cdot 6H_2O$, at a constant voltage of 80 V for 30 minutes. During the electrodeposition, two Au electrodes were connected each other with a Cu wire as the cathode, while a platinum sheet was used as the anode. After deposition, the interdigital Au electrodes coated with GQDs were washed thoroughly with ultrapure water, followed by drying in air.

To prepare asymmetric microsupercapacitors, GQDs//MnO_2, the layer of GQDs was deposited onto one side of the interdigital Au electrodes by using the same deposition method. MnO_2 layer was electrochemically deposited onto the other side of the Au electrodes, with an electrolyte solution containing 0.02 M $Mn(NO_3)_2$ and 0.1 M $NaNO_3$, at a constant current density of 1 mA·cm^{-2}, within a potential window from -1.2 V to 1.2 V for 5 minutes. A platinum sheet was used as the counter electrode, while an Ag/AgCl electrode was employed as the reference electrode. After deposition, the electrodes were washed thoroughly with ultrapure water, followed by drying in air.

The microsupercapacitors had 32 in-plane interdigital Au microelectrodes, 16 positive and 16 negative microelectrodes. The individual microelectrode had a width of 230 μm, a length of 10 mm and a finger distance of 200 μm. Figure 7.43(a) shows SEM image of the planar interdigital microsupercapacitor, suggesting that the GQDs had been uniformly coated on the Au electrodes. The deposited GQDs aggregated to form nanoparticles, as illustrated in Fig. 7.43(b, c). Figure 7.43(d) shows cross-sectional SEM image of the Au microelectrode, revealing that the layer of GQDs had a thickness of 312 nm. Meanwhile, the GQDs layer had an intimate contact with the Au-electrode.

The GQDs//GQDs symmetric microsupercapacitor exhibited a superior rate capability with the scan rate up to 1000 V s^{-1}, an excellent power response with a small RC time constant of 103.6 μs, a high area specific capacitor of 468.1 μF·cm^{-2}, and an excellent cycle stability in 0.5 M Na_2SO_4 aqueous electrolyte solution. However, if 2 M EMIMBF$_4$/AN electrolyte solution was used, the GQDs//GQDs symmetric microsupercapacitor would have a larger voltage window of 2.7 V, a smaller RC time constant of 53.8 μs and a much higher energy density comparatively. The GQDs//MnO_2 asymmetric microsupercapacitor possessed a much higher specific capacitance of 1107.4 μF·cm^{-2}, corresponding to an energy density of 0.154 μWh·cm^{-2}.

An innovative method was reported to directly reduce and pattern hydrated GO to fabricate microsupercapacitors by using a laser beam [60]. The microsupercapacitors consisted of patterned parts that were reduced from GO to rGO and served as electrodes and unreduced hydrated GO between the electrodes that exhibited high ionic conductivity and high electrical resistivity to act as both separator and electrolyte. Figure 7.44 shows both in-plane and conventional sandwich supercapacitor designs, with different patterns and shapes.

The laser-patterned microsupercapacitors, with a composition structure of rGO–GO–rGO, demonstrated a promising electrochemical effect, without the presence of external electrolytes. Since the exact mass of

486 *Nanomaterials for Supercapacitors*

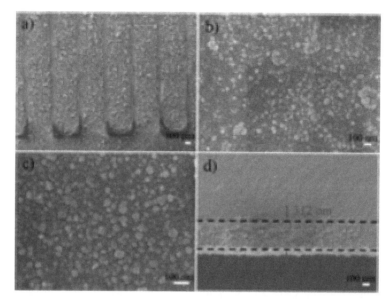

Fig. 7.43. SEM images of the interdigital electrodes with the coating of GQDs: (a–c) top-view images of two Au electrodes coated with GQDs at different magnifications and (d) cross-sectional image of an Au electrode finger coated with GQDs. Reproduced with permission from [58], Copyright © 2013, WILEY-VCH Verlag GmbH & Co. KGaA, Weinheim.

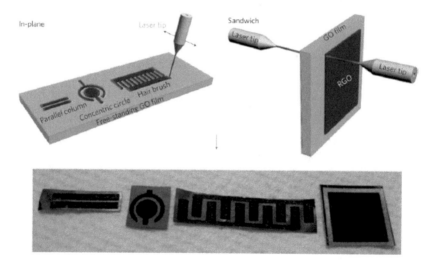

Fig. 7.44. Schematics diagram of the CO_2 laser-patterning of free-standing hydrated GO films to fabricate rGO–GO–rGO microsupercapacitors with in-plane and sandwich configurations. Reproduced with permission from [60], Copyright © 2011, Macmillan Publishers Limited.

the laser-reduced active electrode material could not be measured, only capacitance values in the area and volume density units were used to characterize electrochemical properties of the microsupercapacitors. The capacitance values were dependent on the geometry of devices, because the ionic mobilities and transport distances were anisotropic and different for different configurations. The in-plane supercapacitor structure with a circular geometry exhibited the highest capacitance of 0.51 $mF \cdot cm^{-2}$, as compared with those of other ones, e.g., the sandwich structure. Control experiment confirmed no capacitance was measured, without the formation of rGO.

The active electrode material of rGO, derived from the hydrated GO with laser beam, was of a porous microstructure, because various gases were produced due to the decomposition of the functional groups and evaporation of water molecules caused by the localized laser heating, as seen in Fig. 7.45(b). Both the resistivity of rGO and reduction depth of the GO layers were dependent on the power of the laser, with resistivity that could be reduced by as large as four orders of magnitude [61]. The rGO derived in this way had a long-range ordered structure, which ensured a smooth diffusion of electrolyte ions inside the electrodes [62]. In practical devices, the randomly arranged rGO flakes resulted in high resistance to ion transport, due to their contact resistance. Cyclic stability of the microsupercapacitors was characterized. It was found that, after 10,000 cycles, the capacitance of the in-plane circular device was dropped by about 30%, while the sandwich devices had a drop by about 35% drop.

Fabrication of graphene thin film electrodes with laser irradiation has also been reported by using a commercial light scribe DVD burner [63]. With this special device fabrication technique, microsupercapacitors with various design considerations have been demonstrated [64]. For the device with 16 interdigital electrodes, a stack capacitance of 3.05 $F \cdot cm^{-3}$, corresponding to an areal capacitance of 2.32 $mF \cdot cm^{-2}$, was achieved at a current density of 16.8 $mA \cdot cm^{-3}$. This value could be retained by 60%, i.e., 1.35 $mF \cdot cm^{-2}$, if the device was operated at a current density of as high as 18.4 $A \cdot cm^{-3}$.

Figure 7.46 shows schematic diagram of the fabrication process, by using the unusual photothermal effect GO that absorbs high-intensity light to convert into rGO, also known as Laser-Scribed Graphene or LSG [65, 66]. Due to the precision of the laser, GO film can be patterned to form desired circuits of rGO, by using computer-aided pattern design [67]. Representative interdigitated graphene electrodes on the disk are shown in Fig. 7.46(a). Because of its almost insulative behavior, GO acted as an effective separator to isolate the positive and negative rGO interdigitated electrodes, so as to form planar microsupercapacitors, as demonstrated in Fig. 7.46(b). As compared with the conventional microfabrication methods, no masks, expensive materials, post-processing and clean room operations are required by using the direct 'writing' method. More importantly, it has a potential for large scale production. As illustrated in Fig. 7.46(d), 112 microsupercapacitors could be generated on a piece of GO layer that was coated on a DVD disk. The lateral spatial resolution could reach about 20 μm, so that this direct 'writing' technique allows for the fabrication of microsupercapacitors with high-resolutions.

When GO was converted to rGO, a color change was accompanied, i.e., from golden-brown to black, which could be used as a sign to monitor the reduction process. Figure 7.47(a) shows a photograph of the LSG microsupercapacitors (LSG-MSC), with 16 interdigitated microelectrodes (8 positive and

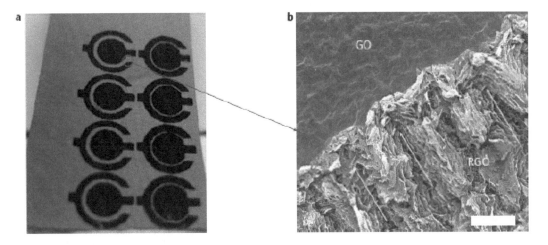

Fig. 7.45. (a) Photograph of an array of concentric circular patterns fabricated on a free-standing hydrated GO film. (b) SEM image of the interface between GO and rGO (scale bar = 100 μm), with the arrows indicating a long-range pseudo-ordered structure generated by the laser-beam heating. Reproduced with permission from [60], Copyright © 2011, Macmillan Publishers Limited.

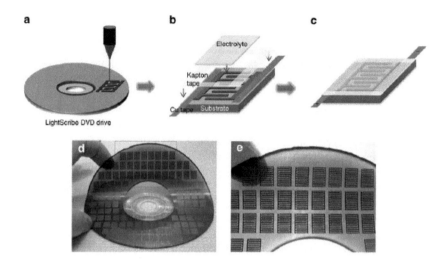

Fig. 7.46. (a–c) Schematic diagram showing fabrication process of the LSG microsupercapacitors. A GO film deposited on a PET substrate was placed on a DVD media disc, which was inserted into a LightScribe DVD drive and a computer-designed circuit was etched onto the film. The laser inside the drive reduced the golden-brown GO into black LSG at precise locations to produce interdigitated graphene circuits (a). Copper tape was applied along the edges to improve the electrical contacts, while the interdigitated area was defined with polyimide (Kapton) tape (b). An electrolyte overcoat was applied to produce a planar microsupercapacitor (c). (d, e) The technique is of potential to directly write micro-devices with high areal density. The micro-devices were highly flexible and thus could be produced on various substrates. Reproduced with permission from [64], Copyright © 2013, Macmillan Publishers Limited.

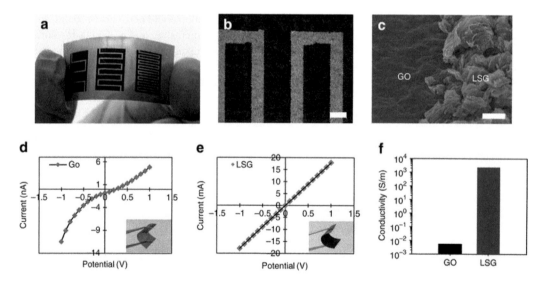

Fig. 7.47. (a) Photograph of the laser-scribed microsupercapacitors with 4 (LSG-MSC$_4$), 8 (LSG-MSC$_8$) and 16 interdigital electrodes (LSG-MSC$_{16}$). (b) Optical microscope image of the LSG-MSC$_{16}$ with interdigitated finger spacing of 150 μm. The dark area corresponds to LSG and the light area is GO. Scale bar = 200 μm. (c) Tilted-view (45°) SEM image of the interface, indicating the direct reduction and expansion of the GO film with the laser beam. Scale bar = 10 μm. (d, e) I–V curves of GO and LSG, respectively. LSG has a current enhanced by about 6 orders of magnitude, confirming the reduction of nearly insulating GO to conducting LSG. (f) Electrical conductivity values of GO and LSG. Reproduced with permission from [64], Copyright © 2013, Macmillan Publishers Limited.

8 negative electrodes). The well-defined pattern had no short circuits between the microelectrodes. Figure 7.47(c) indicated that the film was expanded, after the treatment of laser, which ensured the accessibility of the electrolyte ions to the surface of the electrode. In the microsupercapacitors, the graphene layer had a thickness of 7.6 μm. I–V curves of GO and LSG are shown in Fig. 7.47(d) and (e), respectively. The GO film had a nonlinear and slightly asymmetric behavior, with differential conductivities in the range of 8.07×10^{-4}–$5.42 \times 10^{-3}\,S \cdot m^{-1}$, which was dependent on the applied gate voltage. After the GO was laser reduced to rGO, I–V curve of the LSG sample became linear. At the same time, conductivity of the film was largely increased to $2.35 \times 10^3\,S \cdot m^{-1}$, as seen in Fig. 7.47(f). Due its high electrical conductivity and high surface area of > 1500, $m^2 \cdot g^{-1}$, the laser derived LSG could be used as both active materials and current collector.

Oxides and their hybrids

In order to further increase energy density, pseudocapacitive materials, including metal oxides and conducting polymers, have been included to construct of microsupercapacitors, with either symmetric or asymmetric configurations. Examples are presented and discussed as follows. For instance, microsupercapacitors have been fabricated, with films of manganese oxide nanofiber arrays, combined with micro-fluidic etching [68]. Fabrication of the microsupercapacitors involved two main steps. Manganese oxide nanoparticles were synthesized on a glass substrate by using the electrospinning method first and then the nanoparticles were incorporated into the microsupercapacitor structure on the solid polymer electrolyte film with the aid of contact transfer and microfluidic etching. With H_3PO_4–PVA gel as the electrolyte, the microsupercapacitors exhibited a specific capacitance of 338 F·g^{-1}, at a low discharge rate of 0.5 mA·cm^{-2}. However, at high discharge rate, the microdevices showed relatively high resistance, which was attributed to the poor contact between the electrode materials and the current collectors.

It has been acknowledged that ordering characteristics of electrospun fibers have a beneficial effect on the electrochemical performance of energy devices, e.g., electrospun fiber-based films possess surface with uniform spacing that could facilitate smooth transportation of electrolyte ions and electrons [69–71]. Figure 7.48(a) shows a schematic diagram demonstrating the fabrication procedure of the nanofiber-based MnO_2 thin film, by using the electrospinning method, with a mixed solution of poly(vinylpyrrolidone) (PVP) and manganese acetate. Aligned nanofibers were deposited on the collectors with micro-gold-arrays between two parallel electrodes. Figure 7.48(b) shows SEM image of four parallel gold microarrays covered with cross-aligned nanofibers. The electrospun fibers on the electrodes were highly oriented, while those out of the electrode region were randomly distributed, as revealed in Fig. 7.48(c).

The samples with electrospun nanofibers were calcined in Ar at temperatures of 300–750°C to form MnO_2 nanoparticles. The optimal calcination temperature for efficient crystallization of MnO_2 was 450°C, with a carbon content of 5.1%. A purification step was followed by immersing the samples in a sulfuric acid solution. XRD results indicated that the oxide phase was α-MnO_2. SEM results suggested that the MnO_2 nanofibers had a diameter of about 100 nm, consisting of MnO_2 nanoparticles with an average diameter of < 10 nm.

Figure 7.49 shows a diagram demonstrating fabricating procedure of the MnO_2-based all-solid-state microsupercapacitors [68]. After the MnO_2 nanofibers were formed, they were transferred onto solid polymer electrolyte film, as seen in the step (a) in Fig. 7.49. The solid electrolyte film, containing polyvinyl alcohol (PVA) and phosphoric acid solution, was semi-dried at 35°C for 6 hours. After that, the semi-dried electrolyte film and the layer of MnO_2 nanofibers were glued together at a pressure of about 0.1 MPa for 5 minutes. In the second step (Fig. 7.49(b)), indium tin oxide (ITO) film was coated to act as the current collector. Then, MnO_2/ITO interdigital fingers were developed by using microfluidic etching, as shown in Fig. 7.49(c). Therefore, all-solid-state microsupercapacitors were fabricated.

As shown in Fig. 7.50(a), the microsupercapacitors were highly transparent and flexible [68]. It was found that conductance of the MnO_2/ITO electrodes was varied by only 5% upon bending, as demonstrated by the right image in Fig. 7.50(a). The maximum loss in conductance was < 15%, after repeated bending

490 *Nanomaterials for Supercapacitors*

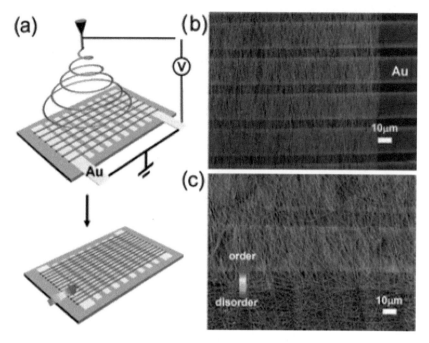

Fig. 7.48. (a) Schematic diagram illustrating the electrospinning set-up to deposit the aligned nanofiber films onto micro-gold-electrode collectors, which were micro-gold-arrays between two electrodes on the microscope slide glass. (b) SEM image of the aligned nanofibers deposited on the surface of the micro-gold-electrodes based collectors. (c) SEM image showing the orientation of nanofibers at edges of the gold electrodes. Reproduced with permission from [68], Copyright © 2011, Royal Society of Chemistry.

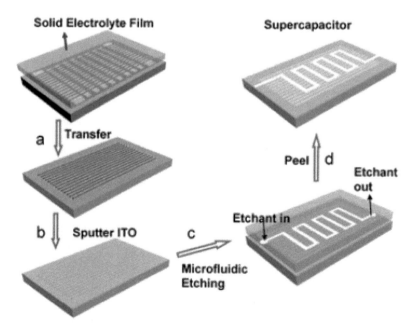

Fig. 7.49. Schematic diagram showing the fabrication of microsupercapacitors by using the microfluidic etching. Reproduced with permission from [68], Copyright © 2011, Royal Society of Chemistry.

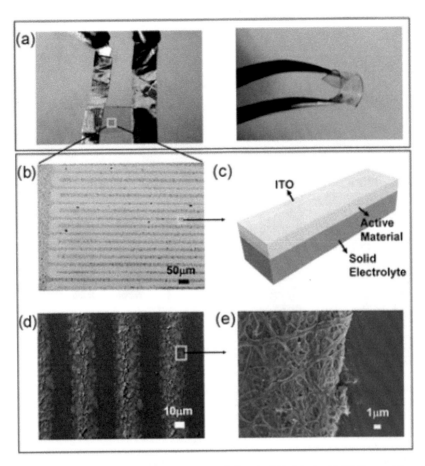

Fig. 7.50. (a) Optical images of microsupercapacitor on a paper (left) and a folded device (right). Two metal conductive tapes were connected to the electrodes of the capacitor separately for characterization of electrochemical performances. (b) Optical image of the MnO_2/ITO interdigital fingers on a solid electrolyte film. (c) Schematic diagram showing the material arrangement on one single interdigital finger. (d) SEM image of the MnO_2/ITO interdigital fingers and (e) a magnified image at a tilted angle of 20° with respect to the plane. Reproduced with permission from [68], Copyright © 2011, Royal Society of Chemistry.

many times. The interdigital fingers on the electrolyte film had a thickness of about 20 μm column features, as observed in Fig. 7.50(b, d), which was determined by the width of the microfluidic channel. Each finger had three layers, i.e., solid electrolyte film, MnO_2 active layer and ITO current collector layer from bottom to top, as illustrated in Fig. 7.50(c). Figure 7.50(e) shows a magnified SEM image of the MnO_2/ITO layer. It was observed that the MnO_2/ITO layer had a thickness of about 1 μm, which was electrospun for 20 minutes. Meanwhile, the MnO_2 layer displayed a smooth surface with very few cavities, which was beneficial to the electrochemical performance of the microsupercapacitors. In addition, besides flexibility and relatively high transparency, the use of the solid electrolyte film effectively avoided the problem of harmful liquid leakage in the conventional thin-film microsupercapacitors.

A microsupercapacitor based on MnO_2 on 3 μm wide microelectrodes was fabricated by using an electrodeposition method, which exhibited enhanced electrochemical performances [72]. Electrode pads with several pairs were first deposited and patterned on Si wafers by using photolithography (PL). After that, the wafers were cut into small pieces with desired dimensions, in which the electrode pads were used as markers for electron Beam Lithography (eBL). Then, micro-electrodes were developed on the chip, after the eBL patterning and electrode deposition. Representative SEM images of the micro-electrodes are shown in Fig. 7.51(a–c), indicating that they had a uniform shape and well defined configuration, together

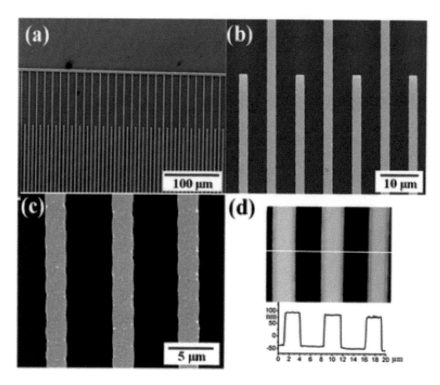

Fig. 7.51. (a–c) SEM images of the micro-electrode arrays on SiO$_2$/Si substrate at different magnifications. (d) AFM topographic image and cross-sectional characteristics of the micro-electrodes. Reproduced with permission from [72], Copyright © 2013, Royal Society of Chemistry.

with a flat electrode surface. Figure 7.51(d) shows an AFM topographical image and cross-sectional profile of the micro-electrodes. The individual finger had a width of 3.07 μm, while the average spacing between two adjacent electrodes was 4.03 μm. The total area of the micro electrode arrays was 1.87 × 10^{-2} cm^{-2}.

For the electrochemical deposition of MnO$_2$, 20 mM Mn(CH$_3$COO)$_2$ and 0.1 M NaNO$_3$ solutions were used as the manganese source and electrolyte, respectively. A low anodic current density of +0.5 mA·cm^{-2} was applied to the micro-electrode patterns for different time durations. The samples with different quantities of electron (Q), 0.3 C·cm^{-2}, 0.6 C·cm^{-2} and 0.9 C·cm^{-2}, were denoted as M-0.3C, M-0.6C and M-0.9C, respectively. The three samples exhibited a similar morphology. Figure 7.52 shows morphology characterization results of sample MC-0.3C, as an example. As seen in Fig. 7.52(a, b), manganese oxide was only deposited within the electrode area, while the adjacent electrodes were not connected, so that there was no short circuit problem between the cathode and anode, during the electrochemical characterization. High magnification SE image indicated that manganese oxide clusters were present on the surface of the electrode, appearing as nanosheets, as illustrated in Fig. 7.52(d).

A microsupercapacitor was constructed by using MnO$_2$ nanoparticles that were deposited on vertically aligned carbon nanotubes (VA-CNT) [73]. Hybrid electrodes based on MnO$_2$ and CNT coated on interdigital stack layers of Fe–Al/SiO$_2$ and Fe–Al/Au/Ti/SiO$_2$, which were pre-patterned by using photolithography and thin-film deposition technique. Electrochemical performances of the microsupercapacitors were characterized in 0.1 M Na$_2$SO$_4$ electrolyte solution. The asymmetric device based Fe–Al/SiO$_2$ exhibited a specific power 0.96 kW·kg^{-1}, corresponding to a specific energy 10.3 Wh·kg^{-1}, whereas those of the Fe–Al/Au/Ti/SiO$_2$ one were 1.16 kW·kg^{-1} and 5.71 Wh·kg^{-1}, respectively.

Figure 7.53(a) shows a top-view SEM image of the patterned comb-like CNT film [73]. The individual CNT finger had a thickness of 34, a width of 40 μm and a length of 2.7 mm. The fingers had a lateral distance of 20 μm. Every pattern consisted of 30 fingers, with a lateral dimension of 4 mm × 3 mm. A

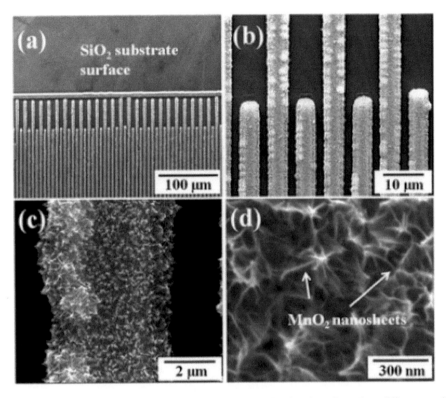

Fig. 7.52. SEM images of the manganese oxide (sample M-0.3C) deposited on the micro-electrodes at different magnifications. Reproduced with permission from [72], Copyright © 2013, Royal Society of Chemistry.

schematic diagram of the pattern with dimensional parameters is shown in Fig. 7.53(c). Figure 7.53(b) shows a tilted-view SEM image, demonstrating the CNT fingers grown on the substrate, with more detailed microstructure revealed in Fig. 7.53(d, e). From the sidewall image, interwoven nanotubes could be clearly observed, with diameters ranging from 10 and 20 nm and a number density of about 10^9 cm^{-2}. Figure 7.53(f) shows SEM image of sample coated with MnO_2 particles, with sizes to be increased from about 2 μm to 6 μm, in the direction from bottom to top of the film.

The Au layer in the structure of Fe–Al/Au/Ti/SiO$_2$ increased conductivity of the current collectors. Figure 7.53(g) shows a top-view SEM image of the patterned CNT film deposited on Fe–Al/Au/Ti/SiO$_2$. As seen in the inset of the figure, the film contained randomly oriented CNTs, with an average height of 1.4 μm. It is worth mentioning that thicker CNT film should be avoided, otherwise, randomly oriented CNTs could be grown into the spacing area, thus leading to a possible short circuit problem. Figure 7.53(h) shows a tilted-view SEM image of the CNT film on Fe–Al/Au/Ti/SiO$_2$ coated with MnO_2 particles. XRD results indicated that manganese oxide was a rutile phase of β-MnO_2.

Figure 7.54(a) shows log–log plots of specific power versus specific energy the CNT_CNT, MnO_2_MnO_2 and MnO_2 (+)_CNT devices. Following the common features of electrochemical capacitors, the energy density is decreased with increasing power density or discharge current. Due to the slow decrease rate in specific energy at low discharge current than at high current, the Ragone plot usually has a typical hook shape. As the CNT electrode was more responsive but less capacitive, the CNT_CNT device had high power density, but low energy density. In contrast, the MnO_2_MnO_2 device exhibited low power density, but high energy density, since the MnO_2 electrode had a pseudocapacitive behavior, with a sluggish characteristic. Therefore, with a positive electrode of MnO_2, the MnO_2(+)_CNT device could achieve a pretty high energy density, at only a relatively less expend of power density.

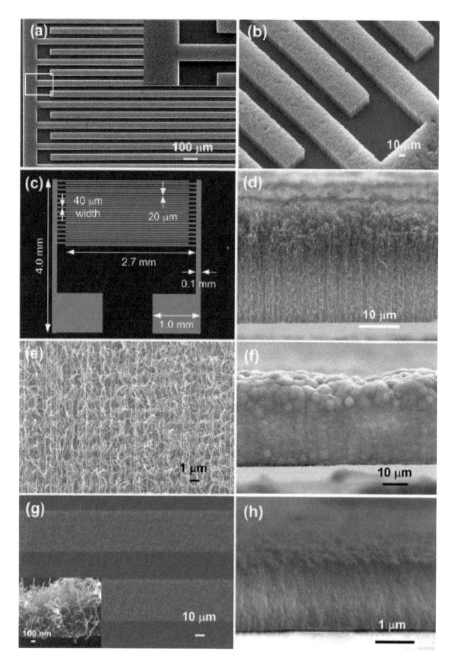

Fig. 7.53. (a–f) Morphology and characteristics of the patterned CNT and MnO$_2$–CNT films on Fe–Al/SiO$_2$: (a) top-view SEM image of the CNT interdigital film, with the inset to be an enlarged image, (b) 30° perspective view of the CNT film with vertical sidewall, (c) schematic comb-like pattern with linear dimension parameters, (d) cross-sectional image of the CNT film showing two sections of different alignment, (e) SEM image showing alignment detail of nanotubes in the lower section, (f) cross-sectional image of the 35 μm MnO$_2$–CNT film. (g, h) Morphology of the CNT and MnO$_2$–CNT films on Fe–Al/Au/Ti/SiO$_2$: (g) top-view image of the patterned CNT film with the inset showing the cross-sectional image of the CNT film and (h) cross-sectional image (10° tilted) of the MnO$_2$–CNT film. Reproduced with permission from [73], Copyright © 2011, Elsevier.

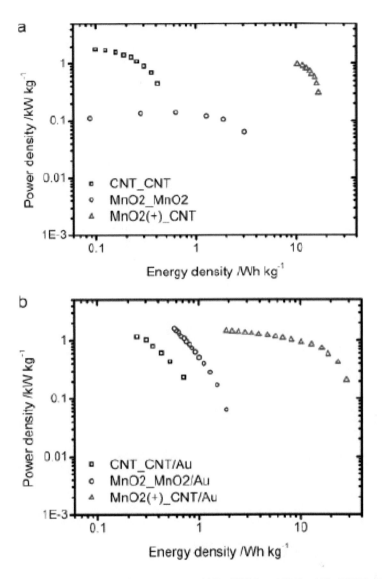

Fig. 7.54. Ragone plots of the three cells on different substrates: (a) Fe–Al/SiO$_2$ and (b) Fe–Al/Au/Ti/SiO$_2$. Reproduced with permission from [73], Copyright © 2011, Elsevier.

Other materials

Other materials, such as nitrides and sulfides, have also been explored to fabricate microsupercapacitors during the last years. For instance, a microsupercapacitor based on Vanadium Nitride (VN) as a negative electrode and NiO as a positive electrode has been reported in the open literature [74]. The VN thin film based electrode was deposited by using a magnetron sputtering method, while the NiO electrode was coated by using an electrodeposition techique. An optimal specific capacitance of 1.24 mF·cm^{-2} at a discharge current density of 2.7 mA cm^{-2} was achieved in 1 M KOH electrolyte.

An ammonia-assisted route was developed to exfoliate VS$_2$ flakes into ultrathin VS$_2$ nanosheets with less than five S–V–S single layers, which could be used as thin film electrode to construct in-plane supercapacitors [75]. An optimal specific capacitance of 4760 µF·cm^{-2} was achieved by using a 150 nm in-plane microsupercapacitor. The microdevice exhibited almost no decay after 1000 charge-discharge cycles. To synthesize VS$_2$, 3 mmol Na$_3$VO$_4$·12H$_2$O and 15 mmol thioacetamide (TAA) were dissolved in 40 mL distilled water. The solution was subject to hydrothermal reaction at 160°C for 24 hours. After the reaction was finished, black precipitate was collected through centrifugation, followed by thorough washing with distilled water. The wet precipitate was immediately exfoliated, in order to avoid the decomposition of the precursor VS$_2$·NH$_3$ to VS$_2$ flakes which were more difficult to be exfoliated.

For exfoliation, 20 mg VS$_2$·NH$_3$ was dispersed in 30 mL water, into which Ar gas was blown to eliminate the dissolved oxygen, which could oxidize V(IV) to V(V). After that, the VS$_2$·NH$_3$ suspension was ultrasonicated for 3 hours in iced water, followed by filtration with a medium-speed qualitative paper filter to remove the unexfoliated flakes, leading to a translucent solution of VS$_2$ nanosheets. With vacuum filtration by using a cellulose membrane with pore size of 0.22 µm, uniform films of VS$_2$ nanosheets thus would be obtained. XRD results indicated that VS$_2$·NH$_3$ was quickly converted to VS$_2$ in only 10 minutes. At the same time, single-phase VS$_2$ was obtained, without any damage in the S–V–S layers.

To construct the VS$_2$ nanosheet-based in-plane supercapacitor, a series of blades were stacked side by side accurately as the mold, so that VS$_2$ ribbons with micrometer-scale gaps were obtained. Au current collectors were deposited on the VS$_2$ ribbons, by using a specifically shaped mask. Then, BMIMBF$_4$–PVA electrolyte was coated onto the ribbon-shaped electrodes, so that an in-plane microsupercapacitor was fabricated.

Figure 7.55(A) shows SEM image of the precursor VS$_2$·NH$_3$, demonstrating a typical layered structure [75]. It was observed that thickness of the VS$_2$·NH$_3$ flakes was about 110 nm, as revealed by the inset in Fig. 7.55(A). The tapping mode AFM image of the VS$_2$ nanosheets is shown in Fig. 7.55(B), indicating a

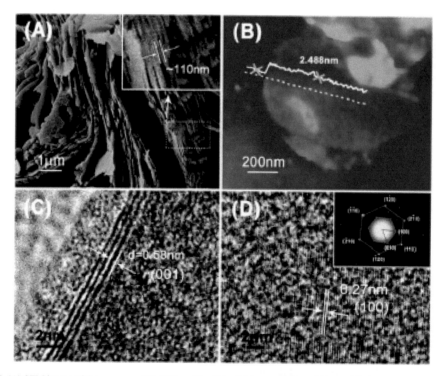

Fig. 7.55. (A) SEM image of the precursor VS$_2$·NH$_3$, with a flake thickness of about 110 nm. (B) Tapping mode AFM image of the exfoliated VS$_2$ nanosheets, showing a thickness of only 3 nm, confirming the successful exfoliation of the precursor. (C, D) HRTEM image and SAED pattern of the VS$_2$ nanosheets, with facets of (001) and (100) to be well indexed. Also, microscopic c orientation could be observed according to the SAED pattern with hexagonal symmetry shown in the inset of (D). Reproduced with permission from [75], Copyright © 2011, American Chemical Society.

thickness of 3.0 nm, which suggested the successful exfoliation of the precursor. Because the c parameter of VS_2 is 5.73 Å, the thickness of 2.488 nm corresponded to 4–5 layers of S–V–S. As illustrated in Fig. 7.55(C), four–five dark and bright patterns could be distinguished, confirming the AFM results. In addition, the interlayer spacing was measured to be about 0.58 nm, in a good agreement with the c parameter of VS_2 (0.573 nm). Also, most of the VS_2 nanosheets were c oriented, as evidenced by well-defined hexagonal symmetric SAED patterns, as shown by the inset in 7.55(D).

After filtration, gray to black homogeneous thin films were formed on the membrane, with thicknesses that could be controlled by adjusting the volume of the suspension, as shown in Fig. 7.56(A) [75]. The VS_2 thin films could be easily transferred onto other substrates, by etching away the membrane. Due to the microscopic self-orientation of the quasi-2D VS_2 nanosheets, the thin films obtained in this way also had a c orientation, which was confirmed by the XRD pattern, as shown in Fig. 7.56(B). Figure 7.56(C) shows planar resistivity of the thin film as a function of temperature. Obviously, with increasing temperature, the electric resistivity was monotonically increased, confirming metallic behavior of the VS_2 nanosheets. Upon being transferred onto flexible substrates, the thin VS_2 films exhibited a high flexibility. As observed in Fig. 7.56(D), the conductivity was almost unchanged after 200 bending cycles.

2D MoS_2 film-based microsupercapacitors, with large-scale fabrication and high capacitance, have been developed, by using simple and low-cost spray painting method [76]. The MoS_2 nanosheets were deposited on Si/SiO_2 chip, followed by laser patterning. The microsupercapacitors consisted of 10 interdigitated electrodes (five electrodes per polarity), with a length of 4.5 mm, finger width of 820 μm, spacing between two adjacent electrodes of 200 μm and thickness of 0.45 μm. The MoS_2-based microsupercapacitor displayed an optimal area capacitance of 8 mF·cm^{-2}, corresponding to a volumetric capacitance of 178 F·cm^{-3}, along with outstanding cyclic stability, making it suitable for flexible electronic applications.

h-MoS2 nanosheets were synthesized by using the hydrothermal method. To prepare hydrothermal precursor solution, 10 mmol MoO_3, 25 mmol potassium thiocyanate and 64 mg Sodium Dodecyl Sulfate

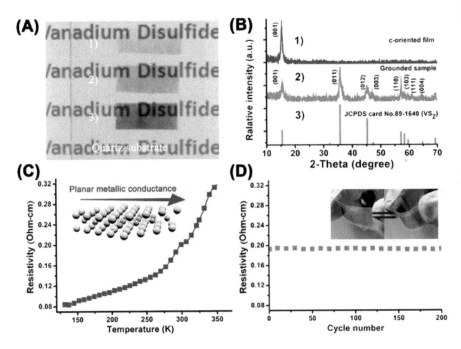

Fig. 7.56. (A) Photographs of the VS_2 thin films with different thicknesses transferred onto a quartz substrate, with translucent feature. (B) XRD pattern of the as-assembled VS_2 thin film, illustrating a well-defined c orientation. (C) Temperature dependence of planar resistivity of VS_2 thin film. (D) Conductance stability of a ribbon-shaped electrode based on the VS_2 thin film after repeated bending/extending deformation. Reproduced with permission from [75], Copyright © 2011, American Chemical Society.

(SDS) were put into a 40 mL Teflon-lined autoclave and then added with 30 mL deionized water, followed by sonication for 30 minutes. The mixture was subject to hydrothermal reaction at 220°C for 24 hours. After the reaction was finished, a dark powder was formed, which was collected and thoroughly washed with deionized water and ethanol. The product was dispersed in water, centrifuged and collected by using vacuum filtration through a 0.45 μm PVDF membrane. The final solid was dried in vacuum at 90°C overnight to obtain h-MoS$_2$ nanosheets. For liquid exfoliation of MoS2, molybdenum disulfide powder was dissolved in DMF, with the aid of sonication for 10 hours, followed by centrifugation and collection.

Figure 7.57 shows schematic diagram demonstrating fabrication process of the 2D MoS$_2$ film based microsupercapacitors, by using the laser patterning method, with the two groups of free-standing MoS$_2$ nanosheets, i.e., hydrothermal process and liquid exfoliation approach, denoted as h-MoS$_2$ and e-MoS$_2$, respectively [76]. Both the MoS$_2$ nanosheet powders were dispersed in a mixed solution of ethanol and water, with an ethanol/water volume ratio of 4:1, in order to form painting inks. The stable suspensions were sprayed onto Si\SiO$_2$ substrates. MoS$_2$ films with uniform thickness obtained, after the solvent was completely evaporated, i.e., steps 1 and 2, in Fig. 7.57(A). Finger-electrodes were patterned with laser writing, step 3 in Fig. 7.57(A), which led to microsupercapacitor device as shown in Fig. 7.57(C), with dimensions as stated above.

Figure 7.58(A) shows SEM image of an individual finger electrode, with the MoS$_2$ nanosheets to be well confined after laser patterning. Due to their atomic layered structures, the h-MoS$_2$ nanosheets had a crumpled morphology, with a visible layered structure, as seen in Fig. 7.58(B). In comparison, the exfoliated MoS$_2$ nanosheets were much flatter and rigid. The h-MoS$_2$ and e-MoS$_2$ nanosheets possessed thicknesses of 2–7 and 1–3 atomic layers, respectively, with an interlayer spacing of about 0.65 nm, as demonstrated in Fig. 7.58(C). HRTEM images indicated that both the h-MoS$_2$ and e-MoS$_2$ nanosheets exhibited well-defined hexagonal crystalline lattice. High purity characteristic of the two sulfide nanosheets was confirmed by XPS results, while their crystalline structure was evidenced by Raman spectra and XRD patterns. As observed in Fig. 7.58(D), the diffraction peaks could be indexed to the (100), (103), (110) planes of MoS$_2$.

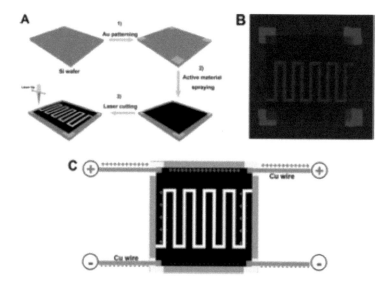

Fig. 7.57. (A) Fabrication process of the MoS$_2$ film microsupercapacitors in with three steps: (1) deposition of gold electrical conductive pads on silicon substrates used as current collectors, (2) masking and spray painting MoS$_2$ ink on the chips and (3) patterning interdigitated finger electrodes (five electrodes per polarity) on the area of active material by using laser writing. (B) Photograph of the MoS$_2$ film microsupercapacitor. (C) Top view of the electrochemical testing device. Cu wire was used for electrical contacts. The black area is the MoS$_2$ active material including patterned interdigitated finger electrodes and pads, while the white area between them is the separator which is filled with electrolyte for electrochemical testing. Reproduced with permission from [76], Copyright © 2013, WILEY-VCH Verlag GmbH & Co. KGaA, Weinheim.

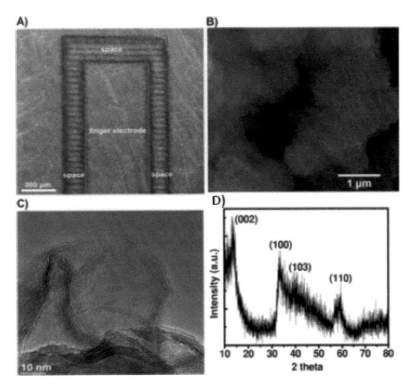

Fig. 7.58. Characterization of the h-MoS$_2$ finger electrodes and h-MoS$_2$ nanosheets. (A) Representative SEM image of a finger-like h-MoS$_2$ electrode (solid line area), with space to be produced by using laser writing (dot line area). (B) SEM image of the h-MoS$_2$ nanosheets used to form the h-MoS$_2$ film microsupercapacitor. (C) HRTEM image of the h-MoS$_2$ nanosheets with 2–7 atomic layers. (D) XRD pattern of the h-MoS$_2$ nanosheets. Reproduced with permission from [76], Copyright © 2013, WILEY-VCH Verlag GmbH & Co. KGaA, Weinheim.

The weak reflection of (002) implied that the nanosheets consisted of few layers stacked through van der Waals interactions, with a spacing of 0.65 nm, as evidenced by the TEM images.

3D architectures

3D architectures have been accepted as new types of electronic devices that have various functionalities for a variety of practical applications [77–79]. Recently, more and more microsupercapacitors have been reported to be assembled with 3D architectures. For example, a 3D framework has been fabricated by using the silicon fabrication technology, which was employed to assemble microsupercapacitors [80]. The 3D structures were developed by using the Deep Reactive Ion Etching (DRIE) method on Si wafer. SiO$_2$ layer and deposition Ti layer were deposited as the current collectors. A thin film of PPy was coated on the 3D structure by using electrochemical deposition to serve as the active material. Symmetric microsupercapacitors fabricated in this way exhibited a specific capacitance of 56 mF·cm^{-2} and a specific power of 0.56 mW·cm^{-2} at a scan rate of 20 mV·s^{-1}.

Figure 7.59 shows schematic diagrams of the 3D redox symmetric microsupercapacitor, which had a 3D 'through-structure', consisting two disconnected periodic beams coated with PPy films to serve as an anode and cathode of the capacitor device. The silicon substrate had a thickness of 525 μm and an area of 1 cm^2, on which interdigitated beams with a width of 100 μm where placed. The processing flow to fabricate the 3D microsupercapacitor is schematically demonstrated in Fig. 7.60.

To etch the silicon substrate, patterned Cr layer was used as the DRIE protection mask instead of SiO$_2$. Al layer on back side of the Si wafer was selected as the DRIE stop layer. The DRIE step was

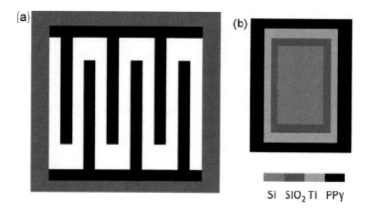

Fig. 7.59. (a) Schematic drawing of the microsupercapacitor and (b) cross-sectional view. Reproduced with permission from [80], Copyright © 2010, Elsevier.

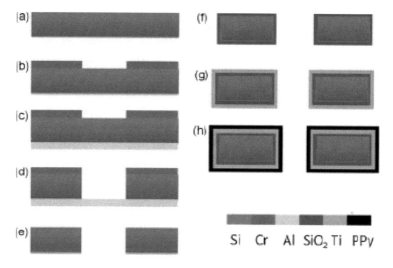

Fig. 7.60. Processing flow to fabricate the 3D MEMS microsupercapacitor. (a) Silicon wafer, (b) Cr mask deposition and patterning, (c) Al back-side deposition, (d) DRIE through the whole thickness of Si wafer, (e) metal layers removal, (f) thermal oxidation, (g) sputtering deposition of Ti current collectors and (h) electropolymerization of the PPy electrode. Reproduced with permission from [80], Copyright © 2010, Elsevier.

conducted with alternating SF_6 and C_4F_8 as the etching gas and passivation gas, respectively. An etching rate of about 7 μm·min^{-1} was achieved within the first 30 minutes. After that, 'black silicon' with grass-like structure was formed at bottom of the etched channel, which was attributed to the difference in etching rate in different opening area. In an area with a big width, as the heat generated could not be dissipated effectively, the substrate would experience uneven temperature distribution. In the area with narrower width, the organic passivation residue could be eliminated completely. Although the 'black silicon' could be avoided by adjusting the composition of the etching gas, the etching rate was slowed down. For instance, a total DRIE time of 210 minutes would be required. There is always a lag effect as the etched areas have different widths [81]. In this case, while the area with a large width was etched through, a layer of 68 μm silicon still remained in the region with a smaller width. To etch away the remaining silicon, an isotropic wet etching solution $HF/HNO_3/HAc$ (acetic acid) was used after removing the metal layers of Cr and Al.

After the DRIE step, a layer of thermal oxidized SiO_2 with a thickness of 1.5 μm was deposited to cover all the effective areas of the 3D structure as an insulator. After that, a continuous layer of Ti was deposited to cover all the top, bottom and side walls of the substrate, by using two-step sputtering (front and back sides). The substrate was rotated at 20 rotations per minute (rpm) during the sputtering process. Then, anodic and cathodic current collectors were formed, by selectively etching the Ti layer. Electroactive PPy films were deposited on the current collectors as the electrodes of the microsupercapacitor, by using electrochemical polymerization.

A series of 3D microsupercapacitors with carbon microrod arrays have been reported in the open literature, which were fabricated by using carbon microelectromechanical systems (C-MEMS) technology [39, 41, 82, 83]. In this approach, two interdigital electrodes and microrod arrays are deposited and patterned by using the typical photolithography processing technique. After that, a carbonization process was conducted, so as to develop the current collectors and the conductive electrodes from the photoresist. The carbon derived in this way usually has an amorphous structure and smooth surface but relatively low specific surface area. The surface area can be enlarged by using various strategies, such as post-synthesis activation [82] and varying synthetic conditions [83], which will be discussed as follows.

Interdigitated carbon micro-electrode arrays for microsupercapacitors have been fabricated by employing the C-MEMS technique, followed by carbonization of the patterned photoresist [82]. To improve performances of the electrodes, electrochemical activation was used to activate the carbon micro-electrode arrays. It was found that, with the application of the electrochemical activation, capacitance of the micro-electrode arrays was significantly increased by three orders of magnitude. The microsupercapacitors could achieve specific geometric capacitance of up to 75 mF·cm^{-2} at a scan rate of 5 mV·s^{-1}, as the electrochemical activation was applied for only 30 minutes. The initial capacitance could be retained by > 87% after 1000 charge-discharge cycles.

The C-MEMS micro-electrodes were fabricated on SiO_2 (2000 Å)/Si substrates, through a two-step photolithography process, followed by a post pyrolysis. Two negative tone photoresists, NANO™ SU-8 25 and SU-8 100, were selected for the lithography process. Firstly, development was conducted with a NANO™ SU-8 developer. Figure 7.61 shows a schematic diagram, describing detailed steps to develop the C-MEMS micro-electrode arrays. 2D interdigitated finger pattern was produced by using the photolithography of SU-8 25 photoresist, which was spin-coated on the substrate at an initial speed of 500 rpm and then accelerated to 3000 rpm, for 30 seconds. The spin-coated photoresist was baked at 65°C for 3 minutes and 95°C for 7 minutes, followed by patterning with a UV exposure dose of 300 mJ·cm^{-2}. Then, post-exposure bake was conducted at 65°C for 1 minute and 95°C for 5 minutes.

Secondly, cylindrical posts were produced on the patterned fingers, by applying another round of photolithography process, with the SU-8 100 photoresist, in a similar way [82]. After that, the SU-8 structures were pyrolyzed at 1000°C for 1 hour in a forming gas atmosphere, consisting of 95% N_2 and 5% H_2. The residual carbon between the fingers, after pyrolysis were removed by using oxygen plasma treatment at 400 mTorr with a power of 150 W for 20 seconds. It was found that electrical resistance between the interdigitated electrodes was at the scale of mega Ohms. The samples had 28 interdigitated C-MEMS micro-electrode arrays, as schematically illustrated in Fig. 7.62.

To electrochemically activate the electrodes of the C-MEMS microsupercapacitors, the two electrodes were connected by linking the contact pads, as seen in Fig. 7.62, with a piece of silver wire. The contact pads and the silver wire were fully covered with epoxy resin to avoid the attack by the electrolyte. After that, a three-electrode system was set up, with the devices acting as the working electrode, whereas reference and counter electrodes were Ag/AgCl and a Pt wire, respectively. Electrochemical activation was conducted in 0.5 M H_2SO_4 solution. A voltage of 1.9 V was applied to the electrodes for either 10 or 30 minutes, with a multichannel potentiostat/galvanostat. Then, the electrodes were negatively polarized at −0.3 V for 10 minutes. The electrochemically activated electrodes were thoroughly washed with DI water.

Figure 7.63 shows representative SEM images of the C-MEMS micro-electrodes [82]. Carbon posts were ideally aligned on the carbon fingers, while the capacitor had two 3D interdigitated electrodes. Every microsupercapacitor had a total footprint area of 9 mm × 9 mm, with 50 interdigitated fingers (25 fingers for each electrode) and a finger width of 100 μm. The post diameters were in the range of 53–68 μm after

502 *Nanomaterials for Supercapacitors*

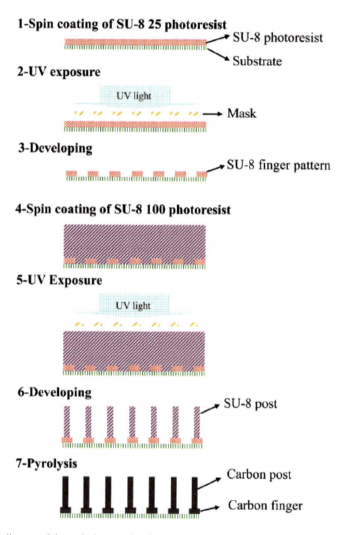

Fig. 7.61. Schematic diagram of the typical processing flow to fabricate the C-MEMS electrodes of microsupercapacitors. Reproduced with permission from [82], Copyright © 2011, Elsevier.

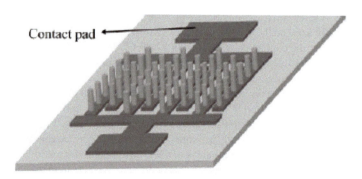

Fig. 7.62. Schematic diagram of 3D view of the microsupercapacitors after carbonization. Reproduced with permission from [82], Copyright © 2011, Elsevier.

Fig. 7.63. SEM images of (a) SU-8 structure and (b) carbonized structure, with the insets showing higher magnification images. Reproduced with permission from [82], Copyright © 2011, Elsevier.

the carbonization treatment. The carbon posts possessed heights varying in the range of 115–140 μm, with an average value of 130 μm.

Symmetric microsupercapacitors with 3D interdigital electrode configurations have been developed by using C-MEMS technology, with polypyrrole (PPy) films as the active materials and the 3D C-MEMS structure as current collector, as schematically shown in Fig. 7.64(a) [39]. Electrochemical performance of the microsupercapacitor was found to be influenced by structure of the 3D electrodes. Single PPy/C-MEMS electrodes had a specific capacitance of 162 mF·cm^{-2} and a specific power of 1.62 mW·cm^{-2} at the scan rate of 20 mV·s^{-1}. The symmetric microsupercapacitors demonstrated a specific capacitance of 78.35 mF·cm^{-2} and a specific power of 0.63 mW·cm^{-2} at the scan rate 20 mV·s^{-1}.

A similar three-electrode configuration was used to deposit PPy film on the C-MEMS current collectors [82]. To electrochemically deposit the PPy film on the C-MEMS, an electrolyte solution of 0.1 M pyrrole monomer and 0.5 M LiClO$_4$ supporting salt were used, with the deposition conducted at a constant current density of 1 mA·cm^{-2} for time durations in the range of 5–15 minutes. After the PPy film was deposited, the electrodes were thoroughly washed with DI water, followed by drying in air. Figure 7.64(b) shows a schematic diagram of the PPy/C-MEMS microsupercapacitor. Meanwhile, 2D film electrodes were also prepared by electrodepositing PPy on pyrolyzed photoresist carbon films, as illustrated in Fig. 7.64(a).

Figure 7.65 shows representative SEM images of the PPy/C-MEMS microsupercapacitors, clearly indicating the conformal coating of the PPy film on the interdigital C-MEMS electrodes. The microelectrodes had a finger width of about 80 μm, as well as average post diameter and height of 50 and 140

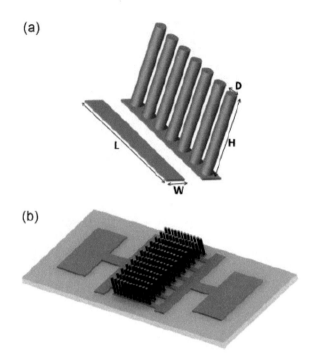

Fig. 7.64. (a) 2D and 3D configurations of the interdigital electrodes. (b) Schematic 3D view of the PPy/C-MEMS microsupercapacitor. Reproduced with permission from [39], Copyright © 2011, Elsevier.

μm, respectively. Figure 7.65(c) shows a high magnification image taken from the wall of the carbon post, revealing that the PPy film had a nanostructured granular raspberry-like morphology.

A micromolding strategy has been explored to fabricate 3D structured on-chip microsupercapacitors [84]. The structures were developed by injecting a mixture of activated carbon and a polymeric binder in pre-fabricated channels of a silicon chip which was used as the mold. The channels were formed by etching the silicon wafer with Inductively Coupled Plasma (ICP). After the active materials were injected into the channels, the silicon walls in between the electrodes were removed by using a second step etching with ICP. The 3D interdigital electrode had thicknesses in the range of 50–70 μm. The on-chip microsupercapacitors exhibited a specific capacitance of 90.7 mF·cm^{-2} and a power density of 51.5 mW·cm^{-2}.

Figure 7.66 shows a schematic diagram describing detailed fabrication steps for the 3D structured on-chip microsupercapacitors. Firstly, high-aspect-ratio interdigital channels were fabricated on the substrate, with walls in between the fingers. Secondly, an insulation layer and a metal layer were deposited, with the insulation layer to prevent current leakage and the metal layer to act as current collector. Thirdly, self-supporting electrode material was filled into the channels for the 3D structure, in which the walls in between the channels effectively isolated the two electrodes. Lastly, the walls were removed, so that the space left was used to host the electrolyte. This unique structure design, with the presence of the silicon walls, allowed the self-supporting material to be sufficiently thick and ideally separated.

To form interdigital channels, a layer of Al was coated on the Si substrate by using a sputtering method, which was used as a masking layer for the Inductively Coupled Plasma (ICP) processing. Each finger of the channels was 105 μm in width and 90 μm in depth, while the Si wall between the fingers was 15 μm in thickness. A 300 nm thick SiO$_2$ was deposited by using the Plasma Enhanced Chemical Vapor Deposition (PECVD) method, as the insulation layer, followed by the deposition of 10 nm/100 nm Ti/Au as the current collector layer by using evaporation deposition. The layers on top of the Si walls were partially etched out. Suspension of polymer binder, i.e., PVDF dissolved in N-methyl-2-pyrrolidone (NMP), was injected into the channels, followed by drying at 80–120°C. The inject-dry process was repeated several times until the channels were completely filled.

Microsupercapacitors 505

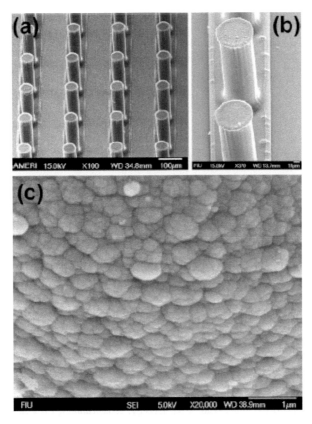

Fig. 7.65. SEM images of the microsupercapacitors: (a) as-pyrolyzed C-MEMS electrodes, (b) C-MEMS electrodes with PPy film and (c) high magnification image from the wall of the carbon post demonstrating nanostructure of PPy film. Reproduced with permission from [39], Copyright © 2011, Elsevier.

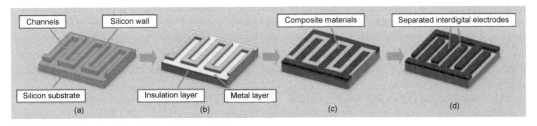

Fig. 7.66. Flow chart to show steps to form the 3D electrodes: (a) etching the substrate to form deep channels, (b) deposition of an insulation layer and then Ti/Au layer as the current collector, followed by removal of the unnecessary layers on the top of the walls through lithography and etching, (c) filling the channels with electrode materials and (d) etching the walls in between two electrodes. Reproduced with permission from [84], Copyright © 2011, Elsevier.

A second ICP was applied to etch the Si wall in between the electrodes, so that both the electrodes and current collectors would not be influenced anymore. Figure 7.67 shows SEM images of the separated electrodes. Due to the fluidity of the suspension and the heterogeneous heat distribution during the drying process, the surface of the electrodes became slightly rough. The final electrodes possessed thicknesses in the range of 50–70 μm, which were slightly thinner than initial depth of the channels.

506 *Nanomaterials for Supercapacitors*

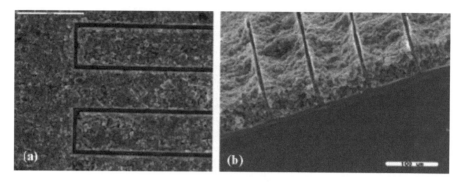

Fig. 7.67. SEM images of the cell after the removal of the Si walls in between the: (a) top view of the interdigital electrodes and (b) the cross-sectional view. Reproduced with permission from [84], Copyright © 2011, Elsevier.

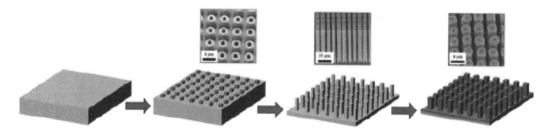

Fig. 7.68. Schematic illustration describing fabrication process of the 3D MnO_2 thin film electrode on Si square pillars. The process for the electrodes based on Si circular tubes is illustrated by the SEM pictures on the top panel. Reproduced with permission from [85], Copyright © 2015, WILEY-VCH Verlag GmbH & Co. KGaA, Weinheim.

A simple, binder-free fabrication method was developed to prepare electrodes of microsupercapacitors with high performances, which were based on MnO_2 thin films [85]. The 3D structured electrodes demonstrated an areal capacitance of 670 mF·cm^{-2} in aqueous electrolyte. In addition, the 3D MnO_2 based electrodes possessed a high cycling stability after more than 15000 cycles. Figure 7.68 shows a schematic diagram, demonstrating concept of the 3D microsupercapacitors. Firstly, 3D Si microstructures were fabricated by using a DRIE Bosch process. Secondly, the microstructures were covered with conformal layers of Al_2O_3 and Pt, by using atomic layer deposition, which served as the insulator and current collector, respectively. Then, MnO_2 thin films were deposited on the resulting 3D conducting architecture by using an electrodeposition method.

Vertically aligned micropillars or microtubes on Si wafers, with high aspect ratios, were fabricated by using the micromachining process, with a precise control over the size and dimension of the microstructures [86]. Firstly, patterns of pillars or tubes were defined on the Si wafer, by using the conventional photolithography method. Then, selective etching of the Si wafer was conducted, with a photoresist etching mask during the DRIE process. A Bosh process, consisting of cycles of alternating etching and passivation steps was selected to generate the microstructures. Two groups of samples were fabricated, (i) 55-μm length tubes (Si μtubes) and (ii) 75-μm length pillars (Si μpillars). The etching technique allowed an etching rate that was 50% lower inside than outside the tubes, so that the inside of the tubes was etched only to half their external length.

MnO_2 thin films were deposited by using anodic deposition with an aqueous plating solution consisting of 0.1 M $MnSO_4$ dissolved in 0.1 M H_2SO_4. Before deposition, the current collector was electrochemically cleaned with 20 voltammetric cycles in 0.1 M H_2SO_4 solution from –0.3 to 1.5 V vs. an Ag/AgCl reference electrode. After that, linear sweep voltammetry was conducted from Open Circuit Potential (OCP) to 1.5 V vs. Ag/AgCl, so that the suitable potential for electrodeposition was determined. Figure 7.69 shows SEM

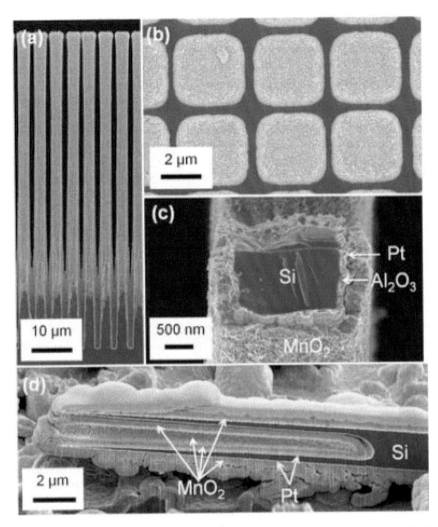

Fig. 7.69. (a) Cross-sectional and (b) top-view of the micropillars after the depositions of Al_2O_3, Pt and MnO_2, (c) tilted view of a broken pillar and (d) FIB cross-sectional view of a Pt and MnO_2 coated microtube. Reproduced with permission from [85], Copyright © 2015, WILEY-VCH Verlag GmbH & Co. KGaA, Weinheim.

images of the 3D microstructures deposited with the MnO_2 thin films. It was found that thickness of the MnO_2 thin films could be controlled in the range of 190–300 nm, with an average value of 275 nm for the Si micropillars, corresponding to 2700 electrodeposition pulses, while it was from 240 to 350 nm, with an average value of 300 nm for the Si microtubes, with 3075 electrodeposition pulses.

An Inductively Coupled Plasma (ICP) etching technique was used to fabricate microsupercapacitors, with 3D architectures [87]. Meanwhile, RuO_2–CNTs composite with a ripple-like morphology was prepared, by using the cathodic deposition method, with silica-based 3D microstructures as the template. The network structure of CNTs promoted the penetration of electrolyte and exchange/diffusion of protons. The 3D microelectrodes exhibited a specific capacitance of 272 mF·cm^{-2}, at a scan rate of 5 mV·s^{-1}, in a neutral Na_2SO_4 solution. During the accelerated cycle life testing at 80 mV·s^{-1}, a satisfactory cyclability was found. When incorporated on chips, the symmetric supercapacitor could reach specific capacitances of up to 37.23 mF·cm^{-2}, corresponding to specific power densities of up to 19.04 mW·cm^{-2}, at a current density of 50 mA·cm^{-2}.

Figure 7.70 shows a schematic diagram of structure of the on-chip microsupercapacitors based 3D arrays. A typical microsupercapacitor consisted of a 3D Si structure, a layer of SiO$_2$, a conformal Au layer as current collector and a continuous layer of electro-active composite materials. The 3D electrode typically had a dimension of 2 × 2 mm, consisting of 832 pillars with a diameter of 50 μm. The center distance between every adjacent pillars was 75 μm, while height of the pillars was 80 μm. Such 3D configurations created high effective surface area, as compared with 2D interdigital electrodes. The fabrication process included: (i) etching the Si substrate to form integrated 3D structures by using ICP, (ii) coating surface of the Si substrate with Au by using conformal radio-frequency (RF) sputtering, (iii) etching a deep electrical isolation trench with FIB and (iv) electrodepositing RuO$_2$–CNTs composite on the surface of the metallic layer.

To prepare the RuO$_2$–CNTs composite, 1 g purified CNTs, 0.05 g RuCl$_3$ and 0.85 g NaNO$_3$ were dispersed in 50 mL deionized water, with aid of sonication for 30 minutes, so that the Ru^{+3} ions were uniformly adsorbed by the CNTs. Electrodeposition was employed to coat the RuO$_2$/CNT composite onto the 3D MEMS Si/Au structures. The two electrodes in each device were connected, so that the device was used as the working electrode for electrochemical deposition. Pt wires were used as both the reference and counter electrodes. The deposition was conducted at a current density of 500 mA·cm^{-2} for time durations of 500–2000 seconds. After deposition, the samples were thoroughly cleaned and dried at 100°C.

Figure 7.71 shows representative SEM images of the 3D Si structures. After electrodeposition of RuO$_2$, a tubular topography was present [88]. At the bases of the pillars, which were 'grassy prominences', dense CNTs were grown, which suggested that the 'grassy prominences' played a critical role in 'blocking' the growth of the CNT composite. After deposition for 1000 seconds, density of the composite at the bases of the arrays was significantly increased. Deposition of 2000 seconds resulted in a situation that the entire 3D microstructure was uniformly covered with a thick layer of the CNT composite, with a thickness of > 5 μm. At the bottom ends of the pillars, the film was as thick as 10 μm. Importantly, no cracks or excoriations were observed, whereas the interface between active films and substrates was more coherent than that of the ruthenium oxide film. The composite film possessed a flower-shaped 3D microstructure, with ripple like surface profile. The RuO$_2$/CNT composite contained RuO$_2$ nanoparticles with uniform size and high surface coverage.

Figure 7.72 shows the desired schematic structure of the ripple-like composite (b) on a 3D microstructure (a). In this case, the presence of 'grassy prominences' was a very important criteria for the electrochemical codeposition. Generally, thin film electrode materials containing CNTs encounter the problem of poor bonding, which easily results in detachment from the substrates. However, 'grassy prominences' captured the CNTs, so that the RuO$_2$ nanoparticles could be tightly combined with the CNTs to form the micro-prominences. Due to the support of 'grassy prominence', the conductive CNT network was adhered strongly to the surfaces of the 3D arrays. As a result, the active materials could have relatively large thickness without compromising electrochemical performances of the microsupercapacitors.

Fig. 7.70. Proposed schematic structure of the 3D microsupercapacitors. Reproduced with permission from [87], Copyright © 2015, Elsevier.

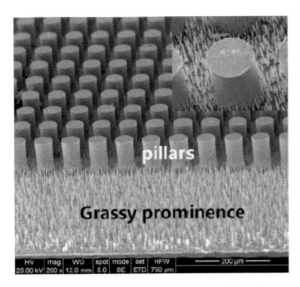

Fig. 7.71. Surface topography of the 3D microstructures with a large 'black silicon phenomenon' area. Reproduced with permission from [87], Copyright © 2015, Elsevier.

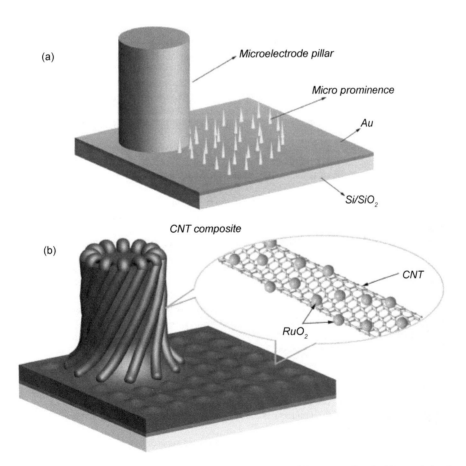

Fig. 7.72. Desired schematic structure of the CNT composite grown on the pillars. Reproduced with permission from [87], Copyright © 2015, Elsevier.

Concluding Remarks

With the development of miniaturized electronic devices, microsupercapacitors will become more and more important, because of the increasing demand of high power energy storage capability. Various architectures have been explored for the fabrication of microsupercapacitors. Electrochemical performances of microsupercapacitors are also closely related to properties and characteristics of the electrode materials, as well as the strategies used to fabricate them. Currently, the materials used in the conventional supercapacitors have all been studied for microsupercapacitors, which include nanocarbon materials (i.e., EDLC), such as AC, CNTs, OLC, graphene and graphene-derivatives, and pseudo-capacitive electrode materials, such as RuO_2, MnO_2, PPy and PANI. Comparatively, it is desired to have electrode materials that are highly conductive, binder-free and mechanically strong. Although, significant achievements have been made in design and fabrication of microsupercapacitors, more efforts should be put up to develop new fabrication methods and even new electrode materials.

Keywords: Microsupercapacitors, Micropower, Nanogenerator, Nanobatteries, Microdevice, Areal capacitance, On-chip microsupercapacitors, In-plane, 3D architectures, Interdigital electrodes, Onion-Like Carbon (OLC), Carbide Derived Carbon (CDC), Highly ordered pyrolytic graphite (HOPD), Sputtering, Chemical Vapor Deposition (CVD), Spray deposition, Ink-jet printing, Electrosynthesis, Layer-by-Layer (LbL)

References

[1] Beidaghi M, Gogotsi Y. Capacitive energy storage in micro-scale devices: recent advances in design and fabrication of micro-supercapacitors. Energy & Environmental Science. 2014; 7: 867–84.

[2] Kyeremateng NA, Brousse T, Pech D. Microsupercapacitors as miniaturized energy-storage components for on-chip electronics. Nature Nanotechnology. 2017; 12: 7–15.

[3] Wang ZL. Toward self-powered sensor networks. Nano Today. 2010; 5: 512–4.

[4] Wang ZL, Wu WZ. Nanotechnology-enabled energy harvesting for self-powered micro-/nanosystems. Angewandte Chemie-International Edition. 2012; 51: 11700–21.

[5] Wang ZL. Towards self-powered nanosystems: from nanogenerators to nanopiezotronics. Advanced Functional Materials. 2008; 18: 3553–67.

[6] Romani A, Filippi M, Tartagni M. Micropower design of a fully autonomous energy harvesting circuit for arrays of piezoelectric transducers. IEEE Transactions on Power Electronics. 2014; 29: 729–39.

[7] Vullers RJM, van Schaijk R, Doms I, Van Hoof C, Mertens R. Micropower energy harvesting. Solid-State Electronics. 2009; 53: 684–93.

[8] Wang ZL. Self-powered nanosensors and nanosystems. Advanced Materials. 2012; 24: 280–5.

[9] Simon P, Gogotsi Y. Capacitive energy storage in nanostructured carbon-electrolyte systems. Accounts of Chemical Research. 2013; 46: 1094–103.

[10] Yoo JJ, Balakrishnan K, Huang JS, Meunier V, Sumpter BG, Srivastava A et al. Ultrathin planar graphene supercapacitors. Nano Letters. 2011; 11: 1423–7.

[11] Honda Y, Haramoto T, Takeshige M, Shiozaki H, Kitamura T, Ishikawa M. Aligned MWCNT sheet electrodes prepared by transfer methodology providing high-power capacitor performance. Electrochemical and Solid State Letters. 2007; 10: A106–A10.

[12] Lee SW, Kim BS, Chen S, Shao-Horn Y, Hammond PT. Layer-by-layer assembly of all carbon nanotube ultrathin films for electrochemical applications. Journal of the American Chemical Society. 2009; 131: 671–9.

[13] Byon HR, Lee SW, Chen S, Hammond PT, Yang SH. Thin films of carbon nanotubes and chemically reduced graphenes for electrochemical micro-capacitors. Carbon. 2011; 49: 457–67.

[14] Du CS, Pan N. Supercapacitors using carbon nanotubes films by electrophoretic deposition. Journal of Power Sources. 2006; 160: 1487–94.

[15] Le LT, Ervin MH, Qiu HW, Fuchs BE, Lee WY. Graphene supercapacitor electrodes fabricated by inkjet printing and thermal reduction of graphene oxide. Electrochemistry Communications. 2011; 13: 355–8.

[16] Kaempgen M, Chan CK, Ma J, Cui Y, Gruner G. Printable thin film supercapacitors using single-walled carbon nanotubes. Nano Letters. 2009; 9: 1872–6.

[17] Ervin MH, Le LT, Lee WY. Inkjet-printed flexible graphene-based supercapacitor. Electrochimica Acta. 2014; 147: 610–6.

[18] Chen QL, Xue KH, Shen W, Tao FF, Yin SY, Xu W. Fabrication and electrochemical properties of carbon nanotube array electrode for supercapacitors. Electrochimica Acta. 2004; 49: 4157–61.

[19] Beidaghi M, Wang ZF, Gu L, Wang CL. Electrostatic spray deposition of graphene nanoplatelets for high-power thin-film supercapacitor electrodes. Journal of Solid State Electrochemistry. 2012; 16: 3341–8.

[20] Kim HK, Seong TY, Lim JH, Cho WI, Yoon YS. Electrochemical and structural properties of radio frequency sputtered cobalt oxide electrodes for thin-film supercapacitors. Journal of Power Sources. 2001; 102: 167–71.

[21] Sakurai Y, Okada S, Yamaki J, Okada T. Electrochemical-behavior of amorphous V_2O_5 ($-P_2O_5$) cathodes for lithium secondary batteries. Journal of Power Sources. 1987; 20: 173–7.

[22] Broughton JN, Brett MJ. Investigation of thin sputtered Mn films for electrochemical capacitors. Electrochimica Acta. 2004; 49: 4439–46.

[23] Li Y, Xie HQ, Li J, Wang JF. Magnetron sputtering deposited $MnO_{1.9}$ thin film for supercapacitor. Materials Letters. 2013; 102: 30–2.

[24] Jiang DE, Sumpter BG, Dai S. Unique chemical reactivity of a graphene nanoribbon's zigzag edge. Journal of Chemical Physics. 2007; 126: 134701.

[25] Jia XT, Hofmann M, Meunier V, Sumpter BG, Campos-Delgado J, Romo-Herrera JM et al. Controlled formation of sharp zigzag and armchair edges in graphitic nanoribbons. Science. 2009; 323: 1701–5.

[26] Srivastava A, Galande C, Ci L, Song L, Rai C, Jariwala D et al. Novel liquid precursor-based facile synthesis of large-area continuous, single, and few-layer graphene films. Chemistry of Materials. 2010; 22: 3457–61.

[27] Miller JR, Outlaw RA, Holloway BC. Graphene electric double layer capacitor with ultra-high-power performance. Electrochimica Acta. 2011; 56: 10443–9.

[28] Zhu MY, Wang JJ, Holloway BC, Outlaw RA, Zhao X, Hou K et al. A mechanism for carbon nanosheet formation. Carbon. 2007; 45: 2229–34.

[29] Sarker AK, Hong JD. Layer-by-layer self-assembled multilayer films composed of graphene/polyaniline bilayers: High-energy electrode materials for supercapacitors. Langmuir. 2012; 28: 12637–46.

[30] Dong XY, Wang L, Wang D, Li C, Jin J. Layer-by-layer engineered Co-Al hydroxide nanosheets/graphene multilayer films as flexible electrode for supercapacitor. Langmuir. 2012; 28: 293–8.

[31] Sk MM, Yue CY. Layer-by-layer (LbL) assembly of graphene with p-phenylenediamine (PPD) spacer for high performance supercapacitor applications. RSC Advances. 2014; 4: 19908–15.

[32] Chiang JC, Macdiarmid AG. Polyaniline—Protonic acid doping of the emeraldine from to the metallic regime. Synthetic Metals. 1986; 13: 193–205.

[33] Liu ZP, Ma RZ, Osada M, Iyi N, Ebina Y, Takada K et al. Synthesis, anion exchange, and delamination of Co-Al layered double hydroxide: Assembly of the exfoliated nanosheet/polyanion composite films and magneto-optical studies. Journal of the American Chemical Society. 2006; 128: 4872–80.

[34] Pech D, Brunet M, Durou H, Huang PH, Mochalin V, Gogotsi Y et al. Ultrahigh-power micrometre-sized supercapacitors based on onion-like carbon. Nature Nanotechnology. 2010; 5: 651–4.

[35] Du CS, Pan N. High power density supercapacitor electrodes of carbon nanotube films by electrophoretic deposition. Nanotechnology. 2006; 17: 5314–8.

[36] An SJ, Zhu YW, Lee SH, Stoller MD, Emilsson T, Park S et al. Thin film fabrication and simultaneous anodic reduction of deposited graphene oxide platelets by electrophoretic deposition. Journal of Physical Chemistry Letters. 2010; 1: 1259–63.

[37] Yan WB, Kim JY, Xing WD, Donavan KC, Ayvazian T, Penner RM. Lithographically patterned gold/manganese dioxide core/shell nanowires for high capacity, high rate, and high cyclability hybrid electrical energy storage. Chemistry of Materials. 2012; 24: 2382–90.

[38] Duay J, Gillette E, Liu R, Lee SB. Highly flexible pseudocapacitor based on freestanding heterogeneous MnO_2/conductive polymer nanowire arrays. Physical Chemistry Chemical Physics. 2012; 14: 3329–37.

[39] Beidaghi M, Wang CL. Micro-supercapacitors based on three dimensional interdigital polypyrrole/C-MEMS electrodes. Electrochimica Acta. 2011; 56: 9508–14.

[40] Portet C, Yushin G, Gogotsi Y. Electrochemical performance of carbon onions, nanodiamonds, carbon black and multiwalled nanotubes in electrical double layer capacitors. Carbon. 2007; 45: 2511–8.

[41] Chen W, Beidaghi M, Penmatsa V, Bechtold K, Kumari L, Li WZ et al. Integration of carbon nanotubes to C-MEMS for On-chip supercapacitors. IEEE Transactions on Nanotechnology. 2010; 9: 734–40.

[42] Wang GX, Shen XP, Yao J, Park J. Graphene nanosheets for enhanced lithium storage in lithium ion batteries. Carbon. 2009; 47: 2049–53.

[43] Singh M, Haverinen HM, Dhagat P, Jabbour GE. Inkjet printing-process and its applications. Advanced Materials. 2010; 22: 673–85.

[44] Li D, Mueller MB, Gilje S, Kaner RB, Wallace GG. Processable aqueous dispersions of graphene nanosheets. Nature Nanotechnology. 2008; 3: 101–5.

[45] Ersoy DA, McNallan MJ, Gogotsi Y. Carbon coatings produced by high temperature chlorination of silicon carbide ceramics. Materials Research Innovations. 2001; 5: 55–62.

[46] Chmiola J, Largeot C, Taberna PL, Simon P, Gogotsi Y. Monolithic carbide-derived carbon films for micro-supercapacitors. Science. 2010; 328: 480–3.

512 *Nanomaterials for Supercapacitors*

[47] Heon M, Lofland S, Applegate J, Nolte R, Cortes E, Hettinger JD et al. Continuous carbide-derived carbon films with high volumetric capacitance. Energy & Environmental Science. 2011; 4: 135–8.

[48] Wei L, Nitta N, Yushin G. Lithographically patterned thin activated carbon films as a new technology platform for on-chip devices. ACS Nano. 2013; 7: 6498–506.

[49] Yushin GN, Aleksov A, Wolter SD, Okuzumi F, Prater JT, Sitar Z. Wafer bonding of highly oriented diamond to silicon. Diamond and Related Materials. 2004; 13: 1816–21.

[50] Sung JH, Kim SJ, Lee KH. Fabrication of microcapacitors using conducting polymer microelectrodes. Journal of Power Sources. 2003; 124: 343–50.

[51] Sung JH, Kim S, Lee KH. Fabrication of all-solid-state electrochemical microcapacitors. Journal of Power Sources. 2004; 133: 312–9.

[52] Huang PH, Heon M, Pech D, Brunet M, Taberna PL, Gogotsi Y et al. Micro-supercapacitors from carbide derived carbon (CDC) films on silicon chips. Journal of Power Sources. 2013; 225: 240–4.

[53] Pech D, Brunet M, Taberna PL, Simon P, Fabre N, Mesnilgrente F et al. Elaboration of a microstructured inkjet-printed carbon electrochemical capacitor. Journal of Power Sources. 2010; 195: 1266–9.

[54] Lin J, Zhang CG, Yan Z, Zhu Y, Peng ZW, Hauge RH et al. 3-Dimensional graphene carbon nanotube carpet-based microsupercapacitors with high electrochemical performance. Nano Letters. 2013; 13: 72–8.

[55] Sun ZZ, Yan Z, Yao J, Beitler E, Zhu Y, Tour JM. Growth of graphene from solid carbon sources. Nature. 2010; 468: 549–52.

[56] Zhu Y, Li L, Zhang CG, Casillas G, Sun ZZ, Yan Z et al. A seamless three-dimensional carbon nanotube graphene hybrid material. Nature Communications. 2012; 3: 1225.

[57] Beidaghi M, Wang CL. Micro-supercapacitors based on interdigital electrodes of reduced graphene oxide and carbon nanotube composites with ultrahigh power handling performance. Advanced Functional Materials. 2012; 22: 4501–10.

[58] Liu WW, Feng YQ, Yan XB, Chen JT, Xue QJ. Superior micro-supercapacitors based on graphene quantum dots. Advanced Functional Materials. 2013; 23: 4111–22.

[59] Zhu SJ, Zhang JH, Tang SJ, Qiao CY, Wang L, Wang HY et al. Surface chemistry routes to modulate the photoluminescence of graphene quantum dots: From fluorescence mechanism to up-conversion bioimaging applications. Advanced Functional Materials. 2012; 22: 4732–40.

[60] Gao W, Singh N, Song L, Liu Z, Reddy ALM, Ci LJ et al. Direct laser writing of micro-supercapacitors on hydrated graphite oxide films. Nature Nanotechnology. 2011; 6: 496–500.

[61] Wei ZQ, Wang DB, Kim S, Kim SY, Hu YK, Yakes MK et al. Nanoscale tunable reduction of graphene oxide for graphene electronics. Science. 2010; 328: 1373–6.

[62] Punckt C, Pope MA, Liu J, Lin YH, Aksay IA. Electrochemical performance of graphene as effected by electrode porosity and graphene functionalization. Electroanalysis. 2010; 22: 2834–41.

[63] El-Kady MF, Strong V, Dubin S, Kaner RB. Laser scribing of high-performance and flexible graphene-based electrochemical capacitors. Science. 2012; 335: 1326–30.

[64] El-Kady MF, Kaner RB. Scalable fabrication of high-power graphene micro-supercapacitors for flexible and on-chip energy storage. Nature Communications. 2013; 4: 1475.

[65] Gilje S, Dubin S, Badakhshan A, Farrar J, Danczyk SA, Kaner RB. Photothermal deoxygenation of graphene oxide for patterning and distributed ignition applications. Advanced Materials. 2010; 22: 419–+.

[66] Cote LJ, Cruz-Silva R, Huang JX. Flash reduction and patterning of graphite oxide and its polymer composite. Journal of the American Chemical Society. 2009; 131: 11027–32.

[67] Strong V, Dubin S, El-Kady MF, Lech A, Wang Y, Weiller BH et al. Patterning and electronic tuning of laser scribed graphene for flexible all-carbon devices. ACS Nano. 2012; 6: 1395–403.

[68] Xue MQ, Xie Z, Zhang LS, Ma XL, Wu XL, Guo YG et al. Microfluidic etching for fabrication of flexible and all-solid-state micro supercapacitor based on MnO_2 nanoparticles. Nanoscale. 2011; 3: 2703–8.

[69] Lu HW, Zeng W, Li YS, Fu ZW. Fabrication and electrochemical properties of three-dimensional net architectures of anatase TiO_2 and spinel $Li_4Ti_5O_{12}$ nanofibers. Journal of Power Sources. 2007; 164: 874–9.

[70] Thavasi V, Singh G, Ramakrishna S. Electrospun nanofibers in energy and environmental applications. Energy & Environmental Science. 2008; 1: 205–21.

[71] Li D, Ouyang G, McCann JT, Xia YN. Collecting electrospun nanofibers with patterned electrodes. Nano Letters. 2005; 5: 913–6.

[72] Wang X, Myers BD, Yan J, Shekhawat G, Dravid V, Lee PS. Manganese oxide micro-supercapacitors with ultra-high areal capacitance. Nanoscale. 2013; 5: 4119–22.

[73] Liu CC, Tsai DS, Chung WH, Li KW, Lee KY, Huang YS. Electrochemical micro-capacitors of patterned electrodes loaded with manganese oxide and carbon nanotubes. Journal of Power Sources. 2011; 196: 5761–8.

[74] Eustache E, Frappier R, Porto RL, Bouhtiyya S, Pierson JF, Brousse T. Asymmetric electrochemical capacitor microdevice designed with vanadium nitride and nickel oxide thin film electrodes. Electrochemistry Communications. 2013; 28: 104–6.

[75] Feng J, Sun X, Wu CZ, Peng LL, Lin CW, Hu SL et al. Metallic few-layered VS_2 ultrathin nanosheets: High two-dimensional conductivity for in-plane supercapacitors. Journal of the American Chemical Society. 2011; 133: 17832–8.

[76] Cao LJ, Yang SB, Gao W, Liu Z, Gong YJ, Ma LL et al. Direct laser-patterned micro-supercapacitors from paintable MoS_2 films. Small. 2013; 9: 2905–10.

[77] Long JW, Rolison DR. Architectural design, interior decoration, and three-dimensional plumbing en route to multifunctional nanoarchitectures. Accounts of Chemical Research. 2007; 40: 854–62.

[78] Rolison DR, Long JW, Lytle JC, Fischer AE, Rhodes CP, McEvoy TM et al. Multifunctional 3D nanoarchitectures for energy storage and conversion. Chemical Society Reviews. 2009; 38: 226–52.

[79] Oudenhoven JFM, Baggetto L, Notten PHL. All-solid-state lithium-ion microbatteries: A review of various three-dimensional concepts. Advanced Energy Materials. 2011; 1: 10–33.

[80] Sun W, Zheng RL, Chen XY. Symmetric redox supercapacitor based on micro-fabrication with three-dimensional polypyrrole electrodes. Journal of Power Sources. 2010; 195: 7120–5.

[81] Tan YY, Zhou RC, Zhang HX, Lu GZ, Li ZH. Modeling and simulation of the lag effect in a deep reactive ion etching process. Journal of Micromechanics and Microengineering. 2006; 16: 2570–5.

[82] Beidaghi M, Chen W, Wang CL. Electrochemically activated carbon micro-electrode arrays for electrochemical micro-capacitors. Journal of Power Sources. 2011; 196: 2403–9.

[83] Hsia B, Kim MS, Vincent M, Carraro C, Maboudian R. Photoresist-derived porous carbon for on-chip micro-supercapacitors. Carbon. 2013; 57: 395–400.

[84] Shen CW, Wang XH, Zhang WF, Kang FY. A high-performance three-dimensional micro supercapacitor based on self-supporting composite materials. Journal of Power Sources. 2011; 196: 10465–71.

[85] Eustache E, Douard C, Retoux R, Lethien C, Brousse T. MnO_2 thin films on 3D scaffold: Microsupercapacitor electrodes competing with "bulk" carbon electrodes. Advanced Energy Materials. 2015; 5: 1500680.

[86] Eustache E, Tilmant P, Morgenroth L, Roussel P, Patriarche G, Troadec D et al. Silicon-microtube scaffold decorated with anatase TiO_2 as a negative electrode for a 3D litium-ion microbattery. Advanced Energy Materials. 2014; 4: 1301612.

[87] Wang XF, Yin YJ, Hao CL, You Z. A high-performance three-dimensional micro supercapacitor based on ripple-like ruthenium oxide-carbon nanotube composite films. Carbon. 2015; 82: 436–45.

[88] Wang XF, Yin YJ, Li XY, You Z. Fabrication of a symmetric micro supercapacitor based on tubular ruthenium oxide on silicon 3D microstructures. Journal of Power Sources. 2014; 252: 64–72.

Index

3D architectures 449, 499, 507, 510

A

α-MnO_2 185, 268
Activation 29, 33–43, 45–47, 49–51, 54, 57, 59, 61, 62, 68, 69, 71–77, 82, 86, 89, 96, 130, 148
Active carbon 75, 148
Alternating Current (AC) circuits 2, 3
Alternative energy 1–3
Areal capacitance 452, 453, 467, 487, 506, 510

B

Batteries 1–3

C

Capacitance 7, 8, 11, 14–17, 21, 22, 24–27
Capacitor 4–12, 17, 27
Carbide Derived Carbon (CDC) 475, 476, 481, 510
Carbon black 28, 97, 98, 148
Carbon cloth 440, 442
Carbon Fiber Papers (CFPs) 425, 442
Carbon fibers 424–427, 442
Carbon nanotubes (CNTs) 28, 97, 98, 102, 103, 106–109, 111–114, 116–120, 124–126, 145–148
Cellulose 436–438, 441, 442
Charge 5–7, 9–15, 18, 19, 21, 24, 26, 27, 66, 72–74, 76, 79–82, 86–88, 90, 91, 93, 94, 96–99, 101–104, 106–109, 112, 113, 119, 120, 127, 131, 132, 134, 139, 144, 145, 148
Chemical coprecipitation 230, 238
Chemical precipitation 228, 268
Chemical Solution Deposition (CSD) 311, 411
Chemical Vapor Deposition (CVD) 449, 454, 455, 481, 482, 504, 505, 510
Clean energy 1, 3
CNT fibers 148
CNT papers 117–119, 148
CNT-CoO 148
CNT-Fe_2O_3 148
CNT-MnO_2 148
CNT-NiO 148
CNTs 424, 425, 427–431, 436, 440, 442
Co_3O_4 284–286, 291, 294, 310, 317, 318, 326, 411
Conducting polymer 422, 424, 426, 427, 437, 439–442
Conventional energy 1, 3
CuO 277, 411
Cyclibility 268

Cyclic voltammetry 167, 208, 211, 216, 217, 226, 228, 268
Cyclic Voltammograms (CV) 170, 238

D

Dielectric constant 5–8, 13, 14, 16, 27
Dielectric loss 17, 27
Dielectric materials 5, 8, 27
Dielectric polarization 8, 27
Differential capacitance 14–16, 27
Direct Current (DC) circuits 2, 3
Discharge 11, 12, 21, 23, 27, 72–74, 79, 86–88, 90, 91, 94, 96–98, 102–104, 106, 107, 112, 113, 119, 120, 127, 131, 134, 145, 148
Double-layer potential 13, 27

E

Electric field 5, 6, 9, 11, 27
Electric potential 6, 9, 11, 27
Electrochemical double-layer (ECDL) 12, 17, 27
Electrochemical Impedance Spectroscopy (EIS) 167, 258, 268
Electrode 4, 6, 7, 9, 12–17, 21, 22, 24–27
Electrodeposition 277, 314–316, 372, 373, 375, 393, 406–411
Electrolyte 9, 12–17, 22, 24, 27
Electrosynthesis 479, 510
Energy density 2, 3, 21, 22, 23, 27
Energy efficiency 1, 3
Energy storage 1–3, 11, 27
Equivalent circuit 17, 27
Equivalent Series Resistance (ESR) 17, 27

F

Fe_2O_3 411
Fe_3O_4 411
Flexibility 425, 426, 429, 431–440, 442
Flexible electronics 442
Flexible supercapacitor 422–442

G

γ-MnO_2 185, 268
Galvanostatic charge 167, 168, 170, 191, 268
G-Fe_3O_4 148
G-MnO_2 148
Gouy-Chapman-Stern (GCS) model 14, 27
Graphene 28, 29, 44, 45, 93, 127, 129–132, 134–145, 147, 148, 424, 425, 429, 431–438, 442

516 *Nanomaterials for Supercapacitors*

Graphene fibers 148
Graphene oxide 28, 127, 130, 132, 134, 136, 139, 142, 145, 148
Graphene papers 148

H

Highly ordered pyrolytic graphite (HOPD) 510
Hybrids 411
Hydrothermal reaction 173, 174, 176–186, 189, 190, 246, 260, 268
Hydrothermal/solvothermal processing 323, 411

I

Ink-jet printing 449, 470, 471, 481, 510
In-plane 449, 454, 455, 478, 482–487, 495, 496, 510
Interdigital electrodes 466, 467, 478, 481, 482, 485–488, 501, 504–506, 508, 510

L

λ-MnO_2 268
Layer-by-Layer (LbL) 449, 454, 455, 458–461, 463, 465, 510

M

Manganese oxide 160–164, 170, 172, 188, 192, 193, 194, 196–199, 201, 202, 204, 205, 208–211, 216–223, 226–230, 232, 235, 237, 239, 240, 243, 247, 248, 254, 255, 257–260, 262, 268
Mesopore 29, 33, 36, 39, 40, 43, 47, 49, 50, 53, 57, 62–66, 68, 69, 75, 77, 82–84, 109, 148
Microdevice 466, 467, 476, 481, 482, 488, 489, 496, 510
Micropore 29–36, 39–47, 49–51, 57, 60–62, 64, 66–70, 72, 75, 77, 80–85, 148
Micropower 448, 510
Microsupercapacitors 448–510
$(Mn+Co)O_x \cdot nH_2O$ 231–233, 268
Mn_2O_3 185, 197, 268
Mn_3O_4 197, 256, 268
Mn-Co mixed oxides 231, 238
Mn-Fe mixed oxides 230, 268
$MnFe_2O_4$ 230, 231, 268
Mn-Ni mixed oxides 235, 268
MnO_2 166, 168, 169, 246, 251, 261, 268
MnO_2-C 251, 268
MnO_2-carbon nanotubes 250, 251, 268
MnO_2-CNTs 251, 268
MnO_2-MWCNTs 268
MnO_2-nanocarbon 250, 251, 268
MnO_2-poly(o-phenylenediamine) 248, 268
MnO_2-polyaniline 240, 268
MnO_2-polypyrrole 243, 268
MnO_2-polythiophene 247, 268
MnO_x 218, 237, 268
MnO_x-RuO_2 237, 268
MoO_3 277, 411
Multi-walled carbon nanotubes (MWCNTs) 97, 102, 106, 107, 109, 113, 114, 119, 123, 125, 126, 145, 146, 148
MWCNTs 442

N

Nanobatteries 510
Nanocellulose fiber (NCFs) 442
Nanocomposites 277, 334, 411
Nanoflakes 182, 199–201, 268
Nanoflower 194, 257, 259, 268
Nanogenerator 448, 510
Nanoneedle 201, 268
Nanopore 120, 138, 147, 148
Nanorod 168–170, 174, 176–179, 181–184, 190–192, 201, 241, 251, 253, 259, 260, 268
Nanorods 168, 170, 174, 176–178, 181–184, 190–192, 201, 241, 251, 253, 259, 260, 268
Nanosheet 163, 208, 230, 247, 258–260, 268
Nanostructures 278–282, 293, 295, 298, 299, 301, 303, 324, 330, 334, 335, 337, 338, 341, 342, 344–346, 348, 351, 352, 355, 357, 359–361, 363–365, 369, 371, 374, 376–379, 381, 384–389, 393, 394, 402, 409–411
Nanowhisker 176–179, 183, 220, 221, 268
New energy 1, 3
$NiCo_2O_4$ and ferrites 277, 411
NiO 277, 278, 330, 334–380, 397, 411
NiO_x 365, 411

O

On-chip microsupercapacitors 478, 504, 508, 510
Onion-Like Carbon (OLC) 449, 466, 467, 481, 510

P

Parallel-plate capacitor 5, 27
Plain weave 442
Polydimethylsiloxane (PDMS) 429, 442
Pore size 29–36, 39, 41, 43, 44, 46, 47, 49, 50, 52, 57, 59, 63–65, 67–69, 71, 75, 77, 81–83, 85, 98, 118, 129, 130, 138, 147, 148
Pore size distribution 29, 33–35, 39, 41, 43, 44, 46, 47, 49, 50, 52, 57, 59, 63, 67, 75, 81–83, 147, 148
Potential difference 6, 7, 9, 11, 13, 27
Power density 3, 22–24, 27
Power density 66, 84, 85, 91, 96, 101, 103, 109, 110, 116, 120, 122, 126, 127, 134, 138, 140, 142, 147, 149
Pseudocapacitive effect 277, 411
Pseudocapacitive reaction 162, 184, 216, 268
Pseudocapacitor 268, 278, 285, 372, 411

R

Ragone plot 23, 24, 27
Rate stability 75, 148
Reduced graphene oxide 127, 130, 145, 148
Reduction–oxidations (redox) 2, 3
Renewable energy 1, 3
$Ru_nMn_{1-n}O_x$ 268
RuO_2 277, 411

S

Single-walled carbon nanotubes (SWCNTs) 97, 98, 100, 102, 103, 106, 109, 112, 120, 122, 148
Sol-gel 170, 172, 196–199, 228, 229, 262, 268

Index 517

Sol-gel deposition 361, 411
Solvo reaction 268
Specific capacitance 62, 67–70, 72–76, 79, 80, 83–88, 90,
    92, 93, 96, 98–104, 106, 107, 109, 110, 112–114, 116,
    117, 120, 122, 124–127, 129–134, 136–139, 141, 142,
    145, 148
Specific surface area 29, 32, 33, 38, 44–46, 64, 67–70, 75,
    79, 83, 86, 88–90, 92, 96–98, 127, 129, 134, 147, 148
Spray deposition 321, 411, 449, 467–469, 482, 510
Sputtering 449–451, 475, 476, 495, 500, 501, 504, 505,
    508, 510
Strain weave 442
Stretchability 429, 442
Supercapacitor 1–4, 12, 16–24, 26–147, 162–268, 277–411
Sustainable energy 1, 3
SWCNTs 427–429, 431, 442

T

Thin film 291, 310, 313, 314, 321–323, 327, 363, 372,
    374, 411
Twill weave 425, 442

V

V_2O_5 411
Voltammetric charge 168, 268

W

Wearable electronics 424, 429, 436, 442
WO_3 277, 411

Z

ZnO 277, 411